MODELING AND OPTIMIZATION OF LCD OPTICAL PERFORMANCE

Wiley-SID Series in Display Technology

Series Editors:
Anthony C. Lowe and Ian Sage

Display Systems: Design and Applications
Lindsay W. MacDonald and Anthony C. Lowe (Eds.)

Electronic Display Measurement: Concepts, Techniques, and Instrumentation
Peter A. Keller

Reflective Liquid Crystal Displays
Shin-Tson Wu and Deng-Ke Yang

Colour Engineering: Achieving Device Independent Colour
Phil Green and Lindsay MacDonald (Eds.)

Display Interfaces: Fundamentals and Standards
Robert L. Myers

Digital Image Display: Algorithms and Implementation
Gheorghe Berbecel

Flexible Flat Panel Displays
Gregory Crawford (Ed.)

Polarization Engineering for LCD Projection
Michael G. Robinson, Jianmin Chen, and Gary D. Sharp

Introduction to Microdisplays
David Armitage, Ian Underwood, and Shin-Tson Wu

Mobile Displays: Technology and Applications
Achintya K. Bhowmik, Zili Li, and Philip Bos (Eds.)

Photoalignment of Liquid Crystalline Materials: Physics and Applications
Vladimir G. Chigrinov, Vladimir M. Kozenkov, and Hoi-Sing Kwok

Projection Displays, Second Edition
Matthew S. Brennesholtz and Edward H. Stupp

Introduction to Flat Panel Displays
Jiun-Haw Lee, David N. Liu, and Shin-Tson Wu

LCD Backlights
Shunsuke Kobayashi, Shigeo Mikoshiba, and Sungkyoo Lim (Eds.)

Liquid Crystal Displays: Addressing Schemes and Electro-Optical Effects, Second Edition
Ernst Lueder

Transflective Liquid Crystal Displays
Zhibing Ge and Shin-Tson Wu

Liquid Crystal Displays: Fundamental Physics and Technology
Robert H. Chen

3D Displays
Ernst Lueder

OLED Display Fundamentals and Applications
Takatoshi Tsujimura

Illumination, Color and Imaging: Evaluation and Optimization of Visual Displays
Peter Bodrogi and Tran Quoc Khanh

Interactive Displays: Natural Human-Interface Technologies
Achintya K. Bhowmik (Ed.)

Addressing Techniques of Liquid Crystal Displays
Temkar N. Ruckmongathan

Fundamentals of Liquid Crystal Devices, Second Edition
Deng-Ke Yang and Shin-Tson Wu

Modeling and Optimization of LCD Optical Performance
Dmitry A. Yakovlev, Vladimir G. Chigrinov, and Hoi-Sing Kwok

MODELING AND OPTIMIZATION OF LCD OPTICAL PERFORMANCE

Dmitry A. Yakovlev
Saratov State University, Russia

Vladimir G. Chigrinov
Hong Kong University of Science & Technology, Hong Kong

Hoi-Sing Kwok
Hong Kong University of Science & Technology, Hong Kong

WILEY

This edition first published 2015

Registered office
John Wiley & Sons Ltd, The Atrium, Southern Gate, Chichester, West Sussex, PO19 8SQ, United Kingdom

For details of our global editorial offices, for customer services and for information about how to apply for permission to reuse the copyright material in this book please see our website at www.wiley.com.

Library of Congress Cataloging-in-Publication Data

Yakovlev, Dmitry A.
Modeling and optimization of LCD optical performance / Dmitry A. Yakovlev, Vladimir G. Chigrinov, Hoi-Sing Kwok.
pages cm
Includes bibliographical references and index.
ISBN 978-0-470-68914-1 (hardback)
1. Liquid crystal displays. I. Chigrinov, V. G. (Vladimir G.) II. Kwok, Hoi-Sing. III. Title.
TK7872.L56Y35 2014
621.3815′422–dc23

2014001761

A catalogue record for this book is available from the British Library.

ISBN: 9780470689141

Set in 9/11pt Times by Aptara Inc., New Delhi, India

Printed and bound in Singapore by Markono Print Media Pte Ltd

1 2015

To our beloved wives: Larisa, Larisa, and Ying-Hung

Contents

Series Editor's Foreword

Liquid crystal displays are the bedrock of the flat panel display industry. Their success and their continued improvement in all aspects of performance are due in substantial part to improvements in the fundamental understanding of how liquid crystal structures interact with forces applied by external electrical fields and by the intrinsic potential differences which exist at boundaries between dissimilar materials.

Several computer modelling systems are commercially available. They enable users to predict the properties of displays, avoiding the necessity to test every new idea by experiment. They are essentially black boxes into which are inputted the properties of materials, cell dimensions, applied voltage and other data, and which output the optical properties of a display as functions of time, applied voltage, wavelength and viewing angle. Their use requires no fundamental understanding of the thermodynamics or mechanics of liquid crystal (LC) interactions and therein lies a potential problem. For reasons of efficiency and minimising computer time, most, if not all modelling routines operate on simplified and approximated formulæ. Under some circumstances these approximations can lead to unforeseen errors and this is a topic which is addressed in unprecedented detail in this volume. But first it contains an exposition of the fundamentals from a description of polarized light through the calculation of its interaction with LC layers by the Jones calculus to predict the properties of cell structures. Next are presented worked examples of different transmissive and reflective nematic and ferroelectric modes using modelling software developed by the authors. The second part of the book provides a more detailed analysis of mathematical methods, starting from the basic mathematics and matrix algebra specific to LC modelling. It then progresses from describing relatively simple models to a description of rigorous electromagnetic methods to describe the optics of 1D inhomogeneous media and their use for numerical modelling of LC optics. The impact of approximations on computational accuracy is discussed throughout. The final chapter of the book touches on layers which are anisotropic in two dimensions, an important topic for LCDs which increasingly use multi-domain pixel structures.

The detailed contents of each chapter are described by the authors in their introduction, but my purpose in presenting this briefer description here is to show what a comprehensive book this is. It goes even further because a companion website http://www.wiley.com/go/yakovlev/modelinglcd contains the well commented source code of the program library LMOPTICS, which is a collection of routines for calculating the optical characteristics of multilayer systems, based on the methods described in this book. It also contains a set of sample programs which exemplify the application of this library and the methods described in this book to modelling LCDs.

This book and its companion website provide a comprehensive operational base for scientists and engineers who wish to make reliable modelling experiments. It provides a wealth of information for academic researchers and students engaged in condensed matter physics which is of relevance not just to displays but to LC-based photonic devices in general.

Anthony Lowe
Braishfield, UK, 2014

Preface

Liquid crystal displays (LCDs) are ubiquitous nowadays. They are used in almost all electronic devices and information systems. This is the result of many years of research and development by dedicated scientists and technologists. Despite the relative maturity of LCD technologies, many improvements are still needed and research is being performed. For example, issues such as energy efficiency, simpler methods of achieving large viewing angles, lower manufacturing cost, and LC alignment techniques still have a lot of room for improvement. In this regard, computer modeling of LCDs is a very useful tool for designing new display modes and improving their performance.

Many monographs and textbooks have been written about LCDs. Some are at the pedagogical level, while others are at the more advanced engineering level. Some involve more physics, while others concentrate on the engineering aspects. It is our desire to add to this collection with a book devoted to computer modeling and optimization of the optical performance of LCDs. It is believed that there is a need for a book that is devoted to an in-depth treatment of this subject. Many useful methods and techniques as well as fine points not covered in previous books are considered here.

For three decades, the authors of this book have been dealing with the problems related to the practical application of liquid crystals, in particular, developing software for numerical modeling and optimization of LCDs. Wishing to make our software sufficiently versatile (applicable to most kinds of LCDs) and efficient (providing a high accuracy of modeling, fast, and provided with convenient optimization tools) and dealing with specific optimization problems, we have examined a great number of approaches, methods, and techniques. In this book we have tried to present a unified approach to the optical modeling of LCDs, which unites the most theoretically rigorous and efficient methods and determines how these methods should be used in different situations. We describe efficient algorithms for solving typical problems of LCD optics and give recommendations as to how to build a basic theoretical model and choose the mathematical tools to solve the problem at hand, considering the problem geometry, factors to be accounted for, and required accuracy. Much attention is given to analytical approaches to solving optimization and inverse problems.

Chapter 1 provides the basic knowledge necessary to proceed to optics of LCDs. Basic notions and concepts of polarization optics and optics of anisotropic media are presented. Particular attention is given to the classical Jones calculus, a method with the aid of which a great number of optical problems for LCDs were solved. The classical Jones calculus has many advantages and disadvantages. The main disadvantage is its conflict with electromagnetic theory in many respects. The main advantages are its simplicity, reliability in many important cases, and rich mathematical apparatus allowing one to analyze polarization-optical systems and solve many problems semi-analytically or analytically. This method is eminently suitable for demonstrating the benefits of using matrices and matrix analysis in polarization optics to the newcomer to this field. In Chapters 2 and 3 the Jones calculus is used for the analysis of the optical operation of LC layers and LCDs in terms of the simplest models. Applications of a parameter space approach and an optical equivalence theorem in LCD optics are demonstrated; these techniques provide a comprehensive picture of LC modes suitable for LCDs and LC photonic devices.

In Chapter 4 we consider various electro-optical effects used in LC displays as well as different kinds of LCDs, the features of their numerical modeling, and typical optimization problems. We give many examples of solving particular optimization problems with the help of computer modeling.

Chapter 5 begins with a brief review of notions and relations of matrix algebra as a foundation to understanding much of the theoretical material of this book. We purposely postponed the regular presentation of this mathematical material to this chapter, preferring to demonstrate first its usefulness, which we do in previous chapters. We included this material to make the book self-contained for the reader. Moreover, in this mathematical review we consider a specific kind of matrices, which is rarely considered in mathematical books but is important in our consideration of LCD optics. This is followed by definitions of some radiometric quantities, a summary of the optical conventions adopted in this book, and a section introducing several important notions concerning the characterization of wave fields by Stokes and Jones vectors.

In Chapter 6 we present a set of relatively simple approaches and representations useful in solving optimization and inverse problems for LCDs when normal incidence of light is considered. In typical situations, the approaches presented in this chapter have no contradictions with electromagnetic theory and can be used in conjunction with rigorous methods. The discussion is illustrated by experimental examples, which give a clear idea of the actual effect of various factors that are taken into account or neglected in different kinds of optical models of LCDs.

Chapters 7 through 10 are devoted to rigorous electromagnetic (EM) methods of optics of 1D-inhomogeneous media and their use for numerical modeling of the optical properties of LCDs.

In Chapter 7 we discuss different physical models used in modeling the LCD optics, models which determine the choice of EM methods and ways of their use. This chapter also presents two general algorithms for calculating transmission and reflection characteristics of layered structures with allowance for multiple reflections, namely, transfer matrix technique and adding technique. These techniques are employed in some EM methods considered in subsequent chapters. In the last section of Chapter 7, optical models of some basic elements of LCDs are considered.

In Chapters 8, 9, and 10 rigorous EM methods of optics of stratified media are discussed in detail. Along with the discussion of the EM methods, these chapters contain a description of the authors' program library LMOPTICS (Fortran 90), a collection of routines for calculating optical characteristics of stratified media based on these methods. This library, available on the companion website, greatly simplifies the development of program modules for accurate evaluation of the optical characteristics of LCDs, and we hope it will be useful to the reader.

One of the EM methods presented in Chapter 8 is a method referred to in this book as the eigenwave (EW) Jones matrix method. This is a rigorous method using transmission and reflection operators, represented by 2×2 matrices, to describe the optical effect of constituents of the layered system under consideration. One of the advantages of this method over the extended Jones matrix method variants described in earlier books on LCDs is better accuracy, especially in the case of oblique incidence. The EW Jones matrix method supplemented with a set of numerical techniques and approximate representations, which are considered in Chapters 11 and 12, is a convenient tool for solving optimization problems for LCDs and inverse problems for inhomogeneous LC layers. We show that in most practically interesting cases, this method provides nearly the same level of mathematical simplicity and the same possibilities to analyze as the classical Jones calculus does. Chapter 11 considers various ways of calculating transmission operators for inhomogeneous liquid crystal layers used in different variants of the Jones matrix method. Application of the EW Jones matrix method to inhomogeneous LC layers is discussed in detail. In Chapter 12 we consider some useful approximations and give examples of application of the EW Jones matrix method in solving optimization and inverse problems.

In Chapter 13 we discuss the potential and limitations of the EM methods of optics of inhomogeneous media in modeling LC displays with fine intra-pixel structure and demonstrate some capabilities of more general EM methods.

Appendix A provides examples of LCD modeling performed over the years by students at Hong Kong University of Science and Technology. Appendix B contains supplementary theoretical material.

Chapter 1 was written by D.A. Yakovlev (D.A.Y.) and H.S. Kwok (H.S.K.). Chapters 2 and 3 were written by H.S.K. Chapter 4 and Appendix A were written by V.G. Chigrinov. Chapters 5 through 13 and Appendix B were written by D.A.Y.

This book is mainly intended for engineers and researchers dealing with the development and application of LC devices. University researchers and students who are specialized in condensed matter physics and engaged in fundamental and applied research of liquid crystals may also find much useful information here.

It is our hope that this book will be helpful to developers of new generations of LC displays.

Vladimir G. Chigrinov
Hoi-Sing Kwok
Dmitry A. Yakovlev

Acknowledgments

We would like to thank Alex King, Genna Manaog, Baljinder Kaur, Thomas Tang, Anatoli A. Murauski, Evgeny Pozhidaev, Sergiy Valyukh, Xu Peizhi, Valery V. Tuchin, Alexander B. Pravdin, Dmitry D. Yakovlev, and Svetlana M. Moiseeva for their invaluable help in preparing this book.

Vladimir G. Chigrinov
Hoi-Sing Kwok
Dmitry A. Yakovlev

List of Abbreviations

AMM	approximating multilayer method
AR	antireflective
CJMM	classical Jones matrix method
DM	discretization method
DRA	direct-ray approximation
EAS	electrode–alignment layer system
ECB	electrically controlled birefringence
EJMM	extended Jones matrix method
EW	eigenwave
EWB	eigenwave basis
FLC	ferroelectric LC
FLCD	ferroelectric LCD
FP	Fabry–Perot
FPI	Fabry–Perot interference
GM	grating method
GOA	geometrical optics approximation
IPS	in-plane switching
JC	Jones calculus
LC	liquid crystal
LCD	liquid crystal display
MEF	modulation efficiency factor
MPW	monochromatic plane wave
NB	normally black
NBR	negligible bulk reflection (approximation)
NBRA	negligible-bulk-reflection approximation
NW	normally white
PBS	polarizing beam splitter
PCS	polarization-converting system
PDLC	polymer dispersed liquid crystal
PSM	power series method
QAA	quasiadiabatic approximation
QMPW	quasimonochromatic plane wave
QWP	quarter-wave plate
RVC	reflection–voltage curve, reflectance–voltage curve
SBA	small-birefringence approximation
SOP	state of polarization
STN	supertwisted nematic

TIR	total internal reflection
TN	twisted nematic
TVC	transmission–voltage curve, transmittance–voltage curve
VA	vertical alignment
WP	wave plate

About the Companion Website

This book is accompanied by a companion website:

www.wiley.com/go/yakovlev/modelinglcd

The website includes:

- Program library LMOPTICS (source code) – a collection of routines for calculating optical characteristics of multilayer systems, such as LCDs, based on methods considered in the book;
- A set of sample programs (source codes), which exemplify the application of this library and the methods described in the book; and
- Files with a short description of this program package and general comments.

1

Polarization of Monochromatic Waves. Background of the Jones Matrix Methods. The Jones Calculus

1.1 Homogeneous Waves in Isotropic Media

1.1.1 Plane Waves

Light is an electromagnetic radiation with frequencies ν lying in the range from $\sim 4 \times 10^{14}$ to $\sim 8 \times 10^{14}$ Hz. An elementary model of light is a plane monochromatic wave. The electric field of a plane monochromatic wave can be represented, in complex form, as

$$\mathbf{E}(\mathbf{r}, t) = \mathbf{E}_0 \mathrm{e}^{i(\mathbf{kr}-\omega t)}, \tag{1.1}$$

where $\omega = 2\pi\nu$ is the circular frequency and $\mathbf{k}$ is the wave vector of the wave, $\mathbf{r}$ is a position vector, and t is time. If the wave propagates in an isotropic nonabsorbing medium with refractive index n and is homogeneous (see Section 8.1.2), the vector $\mathbf{k}$ can be expressed as

$$\mathbf{k} = \frac{\omega}{c} n\mathbf{l}, \tag{1.2}$$

where $\mathbf{l}$ is the wave normal, a unit vector perpendicular to the wavefronts of the wave and indicating its propagation direction; c is the velocity of light in vacuum (free space). In this case, the wave is strictly transverse, satisfying the condition

$$\mathbf{l} \cdot \mathbf{E}_0 = 0. \tag{1.3}$$

The phase velocity of the wave is

$$c_n = \frac{\omega}{|\mathbf{k}|} = \frac{c}{n}. \tag{1.4}$$

Modeling and Optimization of LCD Optical Performance, First Edition.
Dmitry A. Yakovlev, Vladimir G. Chigrinov and Hoi-Sing Kwok.

Website Companion: www.wiley.com/go/yakovlev/modelinglcd

The true wavelength (λ_{true}) of the wave in the medium is defined as

$$\lambda_{\text{true}} \equiv c_n \tau,$$

where

$$\tau = \frac{1}{\nu} = \frac{2\pi}{\omega}$$

is the temporal period of the wave. Along with the true wavelength, one can associate with this wave the so-called *wavelength in free space*, defined as follows:

$$\lambda \equiv c\tau = \frac{c}{\nu} = \frac{2\pi c}{\omega}. \tag{1.5}$$

Throughout this book, speaking on monochromatic fields or monochromatic components of polychromatic fields, we will use the term "wavelength" only in the latter sense (often omitting "in free space"). Also, we will use the parameter

$$k_0 \equiv \frac{\omega}{c} = \frac{2\pi}{\lambda} \tag{1.6}$$

called the *wave number in free space*. In terms of λ and k_0, equation (1.1) can be rewritten as follows:

$$\mathbf{E}(\mathbf{r}, t) = \mathbf{E}_0 \mathrm{e}^{i(k_0 n \mathbf{l}\mathbf{r} - \omega t)} = \mathbf{E}_0 \mathrm{e}^{i\left(\frac{2\pi}{\lambda} n \mathbf{l}\mathbf{r} - \omega t\right)}. \tag{1.7}$$

The field (1.1) must satisfy the following wave equation [1]:

$$\nabla \times (\nabla \times \mathbf{E}) - k_0^2 \varepsilon \mathbf{E} = \widehat{\mathbf{0}}, \tag{1.8}$$

where ε is the electric permittivity of the medium, ∇ is the nabla operator, and $\widehat{\mathbf{0}}$ is the null vector. Throughout this book, we use the Gaussian system of units and consider only media that are nonmagnetic (i.e., having their magnetic permeability μ equal to 1) at optical frequencies. Substituting (1.1) into (1.8) gives the equation

$$\mathbf{k} \times (\mathbf{k} \times \mathbf{E}) + k_0^2 \varepsilon \mathbf{E} = \widehat{\mathbf{0}}, \tag{1.9a}$$

which can be rewritten as

$$\mathbf{k} \cdot (\mathbf{k} \cdot \mathbf{E}) - \mathbf{k}^2 \mathbf{E} + k_0^2 \varepsilon \mathbf{E} = \widehat{\mathbf{0}}, \tag{1.9b}$$

where $\mathbf{k}^2 \equiv \mathbf{k} \cdot \mathbf{k}$. Scalarly multiplying any of these equations by $\mathbf{k}$, we see that these equations include the condition

$$\mathbf{k} \cdot \mathbf{E} = 0; \tag{1.10}$$

this condition may also be derived from the Maxwell equation $\nabla (\varepsilon \mathbf{E}) = 0$. We should note that condition (1.10) is valid for inhomogeneous waves of the form (1.1) as well (see Sections 8.1.2 and 9.2). In the case of a homogeneous wave, condition (1.10) is tantamount to (1.3). In view of (1.10), equation (1.9b) can be reduced to the following one:

$$\left(k_0^2\varepsilon - \mathbf{k}^2\right)\mathbf{E} = \widehat{\mathbf{0}}. \tag{1.11}$$

This equation requires that

$$\sqrt{\mathbf{k}^2} = k_0\sqrt{\varepsilon}. \tag{1.12}$$

In the case of a homogeneous wave, equation (1.12) leads to (1.2) with

$$n = \sqrt{\varepsilon}. \tag{1.13}$$

With complex n and ε, equations (1.1)–(1.3) and (1.13) can be used to describe homogeneous waves propagating in absorbing media (see Section 8.1.2).

1.1.2 Polarization. Jones Vectors

Polarization Parameters

Let us consider a plane wave satisfying (1.3). We introduce a rectangular right-handed Cartesian system (x, y, z) with the z-axis codirectional with the wave normal $\mathbf{l}$. Denote the unit vectors indicating the positive directions of the axes x, y, and z by $\boldsymbol{x}$, $\boldsymbol{y}$, and $\boldsymbol{z}$. Using this coordinate system, we can represent the electric field of the wave as follows:

$$\mathbf{E}(\mathbf{r},t) = \mathbf{E}(z,t) = \left(\boldsymbol{x}\tilde{E}_x(z) + \boldsymbol{y}\tilde{E}_y(z)\right)\mathrm{e}^{-i\omega t} \tag{1.14a}$$

or

$$\mathbf{E}(\mathbf{r},t) = \left(\boldsymbol{x}\left|\tilde{E}_x(z)\right|\mathrm{e}^{i\delta_x} + \boldsymbol{y}\left|\tilde{E}_y(z)\right|\mathrm{e}^{i\delta_y}\right)\mathrm{e}^{-i\omega t}, \tag{1.14b}$$

where $\tilde{E}_x$ and $\tilde{E}_y$ are the scalar complex amplitudes, and δ_x and δ_y are the phases of the x-component and the y-component of the field. The quantity

$$\chi = \frac{\tilde{E}_y}{\tilde{E}_x} = \frac{|\tilde{E}_y|}{|\tilde{E}_x|}\mathrm{e}^{i\delta}, \tag{1.15}$$

where $\delta = \delta_y - \delta_x$, fully describes the state of polarization (SOP) of the wave. For completely polarized waves, which we consider here, the SOP is essentially the shape, orientation, and sense of the trajectory that is described with time by the end of the true electric vector [Re(**E**)] associated with a given point in space (**r**). It is well known that in general such a trajectory is an ellipse. With the help of Figure 1.1, we present basic parameters used for description of the SOP of completely polarized waves [1–3]:

1. The *azimuth* (*orientation angle*) γ_e of a polarization ellipse is defined as the angle between the positive direction of the x-axis and the major axis of the ellipse (Figure 1.1).
2. The *ellipticity* e_e is defined as

$$e_\mathrm{e} = \pm\frac{b}{a}, \tag{1.16}$$

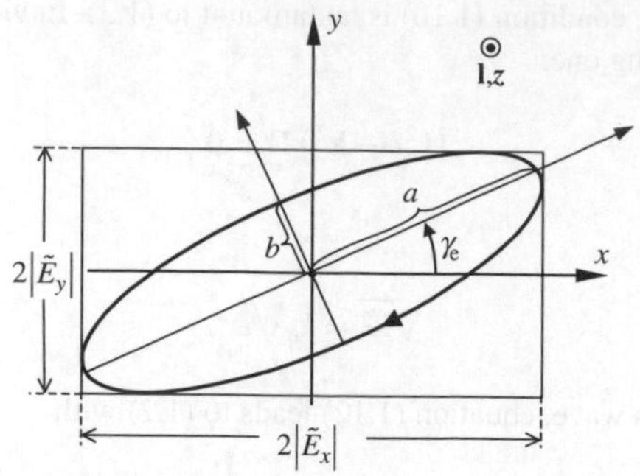

Figure 1.1 A polarization ellipse

where a and b are the lengths of the semimajor axis and semiminor axis of the ellipse, respectively. The ellipticity is taken positive if the polarization is right-handed and negative if the polarization is left-handed. The *handedness* of the polarization ellipse determines the sense in which the ellipse is described. In the literature, different conventions on the handedness of polarization are used. In this book, we use the convention adopted in the books [1, 2, 4]: the polarization is called *right-handed* if the polarization ellipse is described in the *clockwise* sense when looking against the direction of propagation of the light [this is the case in Figure 1.1 where the z-axis and the wave normal **l** are directed out of the page, toward the viewer] and *left-handed* otherwise. For a linearly polarized wave, $e_e = 0$. For right- and left-circularly polarized waves, e_e equals 1 and −1, respectively.

3. The *ellipticity angle* υ_e is defined by

$$e_e = \tan \upsilon_e. \tag{1.17}$$

The values of υ_e lie between $-\pi/4$ (left circular polarization) and $\pi/4$ (right circular polarization).

The azimuth γ_e and ellipticity angle υ_e are related to the complex polarization parameter χ as follows:

$$\cos 2\gamma_e = \frac{1 - |\chi|^2}{\sqrt{(1 - |\chi|^2)^2 + (2\,\mathrm{Re}\chi)^2}}, \quad \sin 2\gamma_e = \frac{2\,\mathrm{Re}\chi}{\sqrt{(1 - |\chi|^2)^2 + (2\,\mathrm{Re}\chi)^2}}, \tag{1.18}$$

$$\sin 2\upsilon_e = -\frac{2\,\mathrm{Im}(\chi)}{1 + |\chi|^2}. \tag{1.19}$$

Thus, given χ, the parameters γ_e, υ_e, and e_e can be calculated by formulas (1.18), (1.19), and (1.17). Note that for linearly polarized waves χ is purely real, while for circular polarizations it is purely imaginary ($\chi = -i$ for the right circular polarization and $\chi = i$ for the left circular polarization). We stress that relations (1.18) and (1.19) and all other relations for polarization parameters presented in this book correspond to the above choice of the convention on handedness and of the time factor in complex representation ($e^{-i\omega t}$).

The spatial evolution of the amplitudes $\tilde{E}_x$ and $\tilde{E}_y$ in (1.14) can be described by the following equations:

$$\tilde{E}_x(z) = \tilde{E}_x(z')e^{ik_0 n(z-z')}, \quad \tilde{E}_y(z) = \tilde{E}_y(z')e^{ik_0 n(z-z')}, \tag{1.20}$$

where z' is any given value of z. Even if the wave propagates in an absorbing medium (with complex n) and, consequently, is damped, its parameter χ is independent of z. This means that χ and the other polarization parameters listed above are spatially invariant and characterize the wave as a whole, that is, they are global characteristics of the wave.

Jones Vectors

The column

$$\tilde{J}(z) = \begin{pmatrix} \tilde{E}_x(z) \\ \tilde{E}_y(z) \end{pmatrix} \tag{1.21}$$

represents a Jones vector of the wave (1.14). Different kinds of Jones vectors are used in practice. Some of them are considered in Section 5.4 and Chapter 8. Definition (1.21) corresponds to one of those kinds. The Jones vector defined by (1.21) is a local characteristic of the wave, being dependent on z. According to (1.20), its values for two arbitrary values of z, z' and z'' ($z'' > z'$), are related by

$$\tilde{J}(z'') = e^{ik_0 n(z''-z')}\tilde{J}(z'). \tag{1.22}$$

This relation can be rewritten as

$$\tilde{J}(z'') = \boldsymbol{t}_{is,n}(z', z'')\tilde{J}(z'), \tag{1.23}$$

where

$$\boldsymbol{t}_{is,n}(z', z'') = \begin{pmatrix} e^{ik_0 n(z''-z')} & 0 \\ 0 & e^{ik_0 n(z''-z')} \end{pmatrix}. \tag{1.24}$$

The 2×2 matrix appearing here is a simple example of the Jones matrix.

If the medium where the wave propagates is nonabsorbing, the Jones vector $\tilde{J}(z)$ can be represented as

$$\tilde{J}(z) = a_\delta(z) a_{\mathrm{I}}\, \boldsymbol{J}, \tag{1.25}$$

where

$$\boldsymbol{J} = \begin{pmatrix} J_x \\ J_y \end{pmatrix} \tag{1.26}$$

is a spatially invariant Jones vector of the wave (see Section 5.4.3), a_δ is a scalar complex phase coefficient of unit magnitude ($a_\delta a_\delta^* = 1$), and a_{I} is a real coefficient that makes the following relation valid:

$$I = \boldsymbol{J}^\dagger \boldsymbol{J}, \tag{1.27}$$

where I represents a quantity (usually called *intensity*) that is regarded as a measure of irradiance (see Section 5.2) for waves in a particular problem or a method; the symbol † denotes the Hermitian conjugation operation (see Section 5.1.1). It is clear that, given $\boldsymbol{J}$, the complex polarization parameter χ of the wave can be calculated by the formula

$$\chi = \frac{J_y}{J_x}. \tag{1.28}$$

The use of such "global" and "fitted-to-intensity" [see (1.27)] Jones vectors for waves propagating in isotropic nonabsorbing media is a feature of the classical Jones calculus (JC) [5] (see Section 1.4). In JC, the quantity conventionally introduced to characterize irradiance is called *intensity*. Equation (1.27) is a standard expression for the intensity of a wave in terms of its Jones vector in this method. For many problems, the "global" Jones vector $\boldsymbol{J}$ of a wave contains all the information about the wave that is required for solving the problem, while the factors a_δ and a_I can be eliminated from the calculations. These factors are absent in standard algorithms based on JC. One should remember the differences between the vectors $\tilde{\boldsymbol{J}}$ and $\boldsymbol{J}$ when trying to use JC in combination with rigorous techniques derived from electromagnetic theory. Moreover, dealing with Jones vectors like $\tilde{\boldsymbol{J}}$, one should recognize that in many cases the use of the quantity

$$\tilde{I} = \tilde{\boldsymbol{J}}^\dagger \tilde{\boldsymbol{J}} = |\tilde{E}_x|^2 + |\tilde{E}_y|^2 \tag{1.29}$$

as a measure of irradiance is not justified. We will consider this issue in detail in Section 5.4. Here we restrict ourselves to the following example. Suppose that we use as *intensity* I FEFD irradiance (see Section 5.2), which is allowed by electromagnetic theory. In this case, the intensity I of the wave is expressed in terms of $\tilde{I}$ as follows:

$$I = \frac{cn}{8\pi}\tilde{I}. \tag{1.30}$$

As seen from (1.30), waves of equal $\tilde{I}$, propagating in media with different refractive indices, will have different "true" intensities I. Note that the coefficient a_I [see (1.25)] in this case is given by

$$a_\mathrm{I} = \sqrt{\frac{8\pi}{cn}}. \tag{1.31}$$

Polarization Jones Vector

Both the "global" and "fitted-to-intensity" Jones vector $\boldsymbol{J}$ and the local Jones vector $\tilde{\boldsymbol{J}}(z)$ can be represented as the product of a scalar factor and a unit vector

$$\boldsymbol{j} = \begin{pmatrix} j_x \\ j_y \end{pmatrix}, \tag{1.32}$$

unit in the sense that

$$\boldsymbol{j}^\dagger \boldsymbol{j} = 1. \tag{1.33}$$

The vector $\boldsymbol{j}$ carries information only on the polarization state of the wave ($\chi = j_y/j_x$) and may be called the *polarization Jones vector* (see Section 5.4.3). In solving practical problems, the polarization Jones vectors are often used to specify the polarization state of light incident on an optical system. Table 1.1 shows typical choices of the polarization vectors for different polarization states. The simplest choice of the vector $\boldsymbol{J}$ for incident light is

$$\boldsymbol{J} = \sqrt{I}\,\boldsymbol{j}. \tag{1.34}$$

A vector $\boldsymbol{J}'$ and the vector $\boldsymbol{J}'' = a\boldsymbol{J}'$, where a is a complex number of unit magnitude, can be regarded as equivalent apart from their phases. As a rule, when calculations for an optical system are performed in terms of "global" Jones vectors, the phases of these vectors are unimportant and can be assigned and transformed arbitrarily, owing to which there is a certain degree of freedom in choice of the vectors $\boldsymbol{j}$ and $\boldsymbol{J}$ for incident light and the Jones matrices describing the interaction of light with optical elements. In particular, this allows using reduced forms of Jones matrices for some kinds of elements (see, e.g., Sections 1.3.5 and 1.3.6), which simplifies the calculations.

Table 1.1 Variants of polarization Jones vectors for various polarization states

Polarization	Polarization Jones vector $\boldsymbol{j}$
Arbitrary elliptical	$\boldsymbol{j}_E(\gamma_e, \upsilon_e) \equiv \begin{pmatrix} \cos\gamma_e \cos\upsilon_e + i \sin\gamma_e \sin\upsilon_e \\ \sin\gamma_e \cos\upsilon_e - i \cos\gamma_e \sin\upsilon_e \end{pmatrix}$
Linear	$\boldsymbol{j}_P(\gamma_e) \equiv \begin{pmatrix} \cos\gamma_e \\ \sin\gamma_e \end{pmatrix}$
Right circular	$\boldsymbol{j}_R \equiv \begin{pmatrix} \frac{1}{\sqrt{2}} \\ -\frac{i}{\sqrt{2}} \end{pmatrix}$
Left circular	$\boldsymbol{j}_L \equiv \begin{pmatrix} \frac{1}{\sqrt{2}} \\ \frac{i}{\sqrt{2}} \end{pmatrix}$

Stokes Parameters

In many cases, it is convenient to use Stoke vectors as state characteristics of light. Stokes vector is a 4×1 column composed of the so-called Stokes parameters, four real quantities characterizing the intensity and polarization state of light. In this subsection we present some useful expressions for Stokes parameters of monochromatic plane waves in terms of the polarization parameters considered above. Definitions for different kinds of Stokes vectors are given in Section 5.3. In particular, in Section 5.3 we define two types of Stokes vectors for plane waves. The Stokes vectors of these types for a wave are simply related. In view of this, we consider here Stokes vectors of only one of these types, namely, intensity-based Stokes vectors.

Using the x-axis as the polarization reference axis (see Section 5.3), after substitution of (1.14) into (5.80) it is easy to obtain the following expression for the intensity-based Stokes vector of the wave (1.14):

$$S_{(I)} \equiv \begin{pmatrix} S_0 \\ S_1 \\ S_2 \\ S_3 \end{pmatrix} = \frac{cn}{8\pi} \begin{pmatrix} |\tilde{E}_x|^2 + |\tilde{E}_y|^2 \\ |\tilde{E}_x|^2 - |\tilde{E}_y|^2 \\ 2\,\mathrm{Re}\left(\tilde{E}_x \tilde{E}_y^*\right) \\ 2\,\mathrm{Im}\left(\tilde{E}_x \tilde{E}_y^*\right) \end{pmatrix}. \tag{1.35}$$

Since $\tilde{E}_x \tilde{E}_y^* = |\tilde{E}_x||\tilde{E}_y|\mathrm{e}^{-i\delta}$, we may rewrite this expression as follows:

$$S_{(I)} = \frac{cn}{8\pi} \begin{pmatrix} |\tilde{E}_x|^2 + |\tilde{E}_y|^2 \\ |\tilde{E}_x|^2 - |\tilde{E}_y|^2 \\ 2|\tilde{E}_x||\tilde{E}_y| \cos\delta \\ -2|\tilde{E}_x||\tilde{E}_y| \sin\delta \end{pmatrix}. \tag{1.36}$$

Another useful expression for $S_{(I)}$ can be obtained by using the following representation of the vector $\tilde{\boldsymbol{J}}(z)$:

$$\tilde{\boldsymbol{J}}(z) \equiv \begin{pmatrix} \tilde{E}_x(z) \\ \tilde{E}_y(z) \end{pmatrix} = a(z)\sqrt{\tilde{I}}\,\boldsymbol{j}_E(\gamma_e, \upsilon_e), \tag{1.37}$$

where a is a complex phase factor of unit magnitude and $\boldsymbol{j}_E(\gamma_e, \upsilon_e)$ is the polarization Jones vector given in Table 1.1. Substitution from (1.37) into (1.35) gives

$$S_{(I)} \equiv \begin{pmatrix} S_0 \\ S_1 \\ S_2 \\ S_3 \end{pmatrix} = \frac{cn}{8\pi} \begin{pmatrix} \tilde{I} \\ \tilde{I}\cos 2\gamma_e \cos 2\upsilon_e \\ \tilde{I}\sin 2\gamma_e \cos 2\upsilon_e \\ \tilde{I}\sin 2\upsilon_e \end{pmatrix} = \begin{pmatrix} I \\ I\cos 2\gamma_e \cos 2\upsilon_e \\ I\sin 2\gamma_e \cos 2\upsilon_e \\ I\sin 2\upsilon_e \end{pmatrix}, \tag{1.38}$$

where I is the intensity defined as the FEFD irradiance of the wave. This expression is convenient when there is a need to construct the Stokes vector for given γ_e and υ_e or, vice versa, to find γ_e and υ_e from calculated or measured Stokes parameters. Note that in the case of a quasimonochromatic partially polarized wave, its Stokes vector can be represented as

$$S_{(I)} \equiv \begin{pmatrix} S_0 \\ S_1 \\ S_2 \\ S_3 \end{pmatrix} = \begin{pmatrix} I \\ I_p\cos 2\gamma_e \cos 2\upsilon_e \\ I_p\sin 2\gamma_e \cos 2\upsilon_e \\ I_p\sin 2\upsilon_e \end{pmatrix}, \tag{1.39}$$

where I is the total intensity of the wave and I_p is the intensity of the completely polarized component of the wave. The intensity I_p is expressed in terms of the Stokes parameters as follows:

$$I_p = \sqrt{S_1^2 + S_2^2 + S_3^2}, \tag{1.40}$$

which allows one to easily find γ_e and υ_e from a given Stokes vector in this case as well.

If the Jones vector $\boldsymbol{J}$ is defined by (1.25) with a_I given by (1.31), the vector $S_{(I)}$ is expressed in terms of the $\boldsymbol{J}$ components as follows:

$$S_{(I)} \equiv \begin{pmatrix} S_0 \\ S_1 \\ S_2 \\ S_3 \end{pmatrix} = \begin{pmatrix} |J_x|^2 + |J_y|^2 \\ |J_x|^2 - |J_y|^2 \\ 2\,\mathrm{Re}\left(J_x J_y^*\right) \\ 2\,\mathrm{Im}\left(J_x J_y^*\right) \end{pmatrix} = \begin{pmatrix} |J_x|^2 + |J_y|^2 \\ |J_x|^2 - |J_y|^2 \\ 2|J_x||J_y|\cos\delta \\ -2|J_x||J_y|\sin\delta \end{pmatrix}. \tag{1.41}$$

Poincaré Sphere

Let us introduce the normalized Stokes parameters

$$s_1 = \frac{S_1}{S_0}, \quad s_2 = \frac{S_2}{S_0}, \quad s_3 = \frac{S_3}{S_0}. \tag{1.42}$$

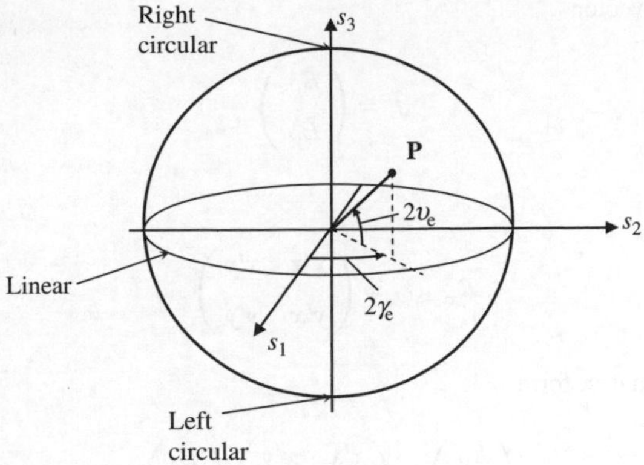

Figure 1.2 Representation of polarization states by points on the Poincaré sphere

According to (1.38), in the case of a completely polarized wave, these parameters can be expressed as follows:

$$s_1 = \cos 2\gamma_e \cos 2\upsilon_e, \quad s_2 = \sin 2\gamma_e \cos 2\upsilon_e, \quad s_3 = \sin 2\upsilon_e. \tag{1.43}$$

With γ_e and υ_e considered as free variables, equations (1.43) describe a unit sphere in a rectangular Cartesian coordinate system (s_1, s_2, s_3) (see Figure 1.2). This sphere is called the Poincaré sphere. The points of this sphere represent all possible SOPs of completely polarized light. The north and south poles on the Poincaré sphere represent the right and left circular polarizations, respectively. The equator represents linear polarization states and all the other points on the sphere represent elliptical polarization states. All left-handed polarization states are on the southern hemisphere, and the northern hemisphere corresponds to right-handed polarizations.

1.1.3 Coordinate Transformation Rules for Jones Vectors. Orthogonal Polarizations. Decomposition of a Wave into Two Orthogonally Polarized Waves

Coordinate Transformation Rules for Cartesian Jones Vectors

Let $\boldsymbol{x}'$ and $\boldsymbol{y}'$ be unit vectors directed along mutually orthogonal axes x' and y' perpendicular to the axis z. Using the reference frame (x', y', z) instead of (x, y, z), we can represent the wave (1.14) as

$$\mathbf{E}(\mathbf{r}, t) = \left(\boldsymbol{x}'\tilde{E}_{x'}(z) + \boldsymbol{y}'\tilde{E}_{y'}(z)\right) \mathrm{e}^{-i\omega t}. \tag{1.44}$$

According to (1.44) and (1.14a),

$$\boldsymbol{x}\tilde{E}_x + \boldsymbol{y}\tilde{E}_y = \boldsymbol{x}'\tilde{E}_{x'} + \boldsymbol{y}'\tilde{E}_{y'}. \tag{1.45}$$

Scalarly multiplying (1.45) by $\boldsymbol{x}'$ and $\boldsymbol{y}'$, we obtain the following equations:

$$\begin{aligned} \tilde{E}_{x'} &= (\boldsymbol{x}'\boldsymbol{x})\tilde{E}_x + (\boldsymbol{x}'\boldsymbol{y})\tilde{E}_y, \\ \tilde{E}_{y'} &= (\boldsymbol{y}'\boldsymbol{x})\tilde{E}_x + (\boldsymbol{y}'\boldsymbol{y})\tilde{E}_y. \end{aligned} \tag{1.46}$$

Introducing the column vector

$$\tilde{J}' = \begin{pmatrix} \tilde{E}_{x'} \\ \tilde{E}_{y'} \end{pmatrix} \tag{1.47}$$

and the matrix

$$\boldsymbol{R}_{xy\to x'y'} = \begin{pmatrix} \boldsymbol{x'x} & \boldsymbol{x'y} \\ \boldsymbol{y'x} & \boldsymbol{y'y} \end{pmatrix}, \tag{1.48}$$

we may write (1.46) in matrix form

$$\begin{pmatrix} \tilde{E}_{x'} \\ \tilde{E}_{y'} \end{pmatrix} = \begin{pmatrix} \boldsymbol{x'x} & \boldsymbol{x'y} \\ \boldsymbol{y'x} & \boldsymbol{y'y} \end{pmatrix} \begin{pmatrix} \tilde{E}_x \\ \tilde{E}_y \end{pmatrix} \tag{1.49}$$

or

$$\tilde{J}' = \boldsymbol{R}_{xy\to x'y'}\tilde{J}. \tag{1.50}$$

Considering the space of Jones vectors as a space of states of a wave where each Jones vector represents a unique state, we may say that the columns $\tilde{J}$ and $\tilde{J}'$ represent the same Jones vector (as they describe the same state) referred to different bases. Relation (1.49) represents the law of transformation of the elements of this Jones vector under the change of basis $(x, y) \to (x', y')$. In view of this, it would be more correct to rewrite relation (1.50) as follows:

$$\boldsymbol{J}_{x'y'} = \boldsymbol{R}_{xy\to x'y'}\boldsymbol{J}_{xy} \tag{1.51}$$

with obvious notation.

If the system (x', y', z), like the system (x, y, z), is right-handed (as in Figure 1.3), the matrix $\boldsymbol{R}_{xy\to x'y'}$ can be expressed as

$$\boldsymbol{R}_{xy\to x'y'} = \widehat{R}_{\mathrm{C}}(\phi), \tag{1.52}$$

where ϕ is the angle between the axes x and x' (Figure 1.3), and $\widehat{R}_{\mathrm{C}}$ is the rotation matrix defined as

$$\widehat{R}_{\mathrm{C}}(\alpha) \equiv \begin{pmatrix} \cos\alpha & \sin\alpha \\ -\sin\alpha & \cos\alpha \end{pmatrix} \tag{1.53}$$

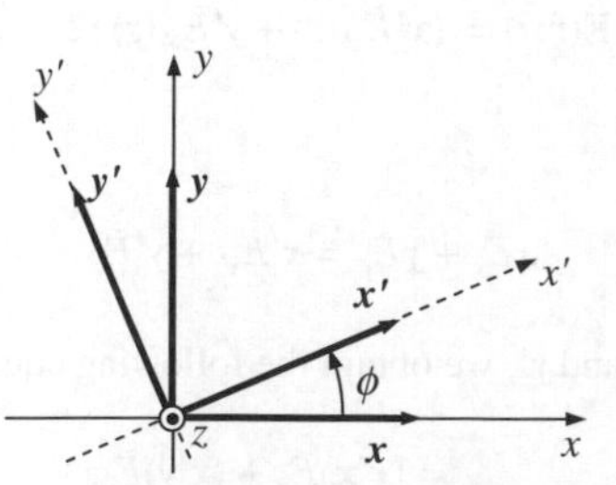

Figure 1.3 Reference frames (x, y, z) and (x', y', z)

for any α. Thus, in this case, the law of coordinate transformation can be expressed by the relation

$$\tilde{J}_{x'y'} = \widehat{R}_C(\phi)\tilde{J}_{xy}. \tag{1.54}$$

For the inverse change $(x', y') \to (x, y)$,

$$\tilde{J}_{xy} = \widehat{R}_C(\phi)^{-1}\tilde{J}_{x'y'} = \widehat{R}_C(-\phi)\tilde{J}_{x'y'}. \tag{1.55}$$

Expression (1.48) for the coordinate transformation matrix $\boldsymbol{R}_{xy\to x'y'}$ is valid irrespective of the handedness of the systems (x, y, z) and (x', y', z). For example, if the system (x, y, z) is, as before, right-handed, choosing the axes x' and y' so that $\boldsymbol{x}' = \boldsymbol{x}$ and $\boldsymbol{y}' = -\boldsymbol{y}$, we will obtain a left-handed system (x', y', z). In this case, equation (1.48) gives

$$\boldsymbol{R}_{xy\to x'y'} = \begin{pmatrix} 1 & 0 \\ 0 & -1 \end{pmatrix}. \tag{1.56}$$

We should note that many formulas presented in this book, in particular in the previous section, are valid for right-handed coordinate systems only. In this book, we deal with left-handed systems very rarely, and it is always stated; if the handedness of a coordinate system is not specified, this system is assumed to be right-handed.

Orthogonal Polarizations

Two waves propagating in the same direction are said to be *orthogonally polarized* if their ellipses of polarization have the same shape but mutually orthogonal major axes and are traced in opposite senses (Figure 1.4). The right circular polarization is orthogonal with respect to the left circular polarization. For a wave with $\gamma_e = \gamma'_e$, $\upsilon_e = \upsilon'_e$, and $\chi = \chi'$, where γ'_e, υ'_e, and χ' are arbitrary, a wave with the corresponding orthogonal polarization will have $\gamma_e = \gamma'_e \pm \pi/2$, $\upsilon_e = -\upsilon'_e$, and $\chi = -1/\chi'^*$ [2]. By checking that

$$\boldsymbol{j}_E(\gamma'_e \pm \pi/2, -\upsilon'_e)^\dagger \boldsymbol{j}_E(\gamma'_e, \upsilon'_e) = 0,$$

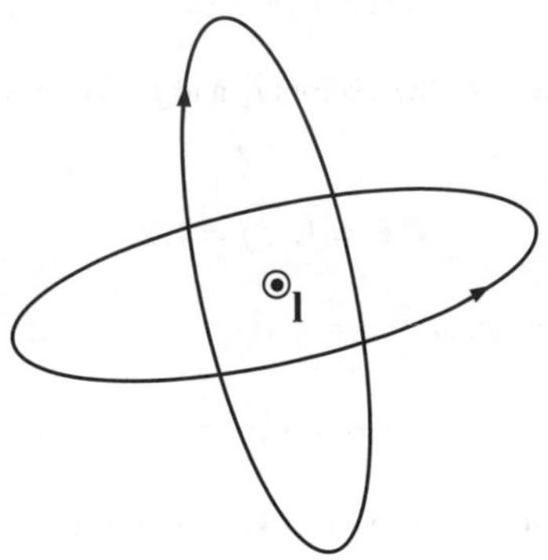

Figure 1.4 Polarization ellipses of mutually orthogonal polarizations

where j_E is the polarization vector defined in Table 1.1, it is easy to verify that the polarization Jones vectors of two orthogonally polarized waves, these vectors being denoted by j and j_{ort}, are orthogonal in the sense that

$$j_{ort}^{\dagger} j = j^{\dagger} j_{ort} = 0. \tag{1.57}$$

It is clear that the Jones vectors of the other above-mentioned kinds (J and $\tilde{J}$) for these waves will also be orthogonal in the same sense ($\tilde{J}_{ort}^{\dagger} \tilde{J} = 0, J_{ort}^{\dagger} J = 0$).

Decomposition of a Wave into Two Orthogonally Polarized Waves

The equation for the electric field of the wave (1.14) can be rewritten in the form

$$\mathbf{E}(\mathbf{r}, t) = \mathbf{E}_{(x)}(\mathbf{r}, t) + \mathbf{E}_{(y)}(\mathbf{r}, t), \tag{1.58}$$

where

$$\mathbf{E}_{(x)}(\mathbf{r}, t) = \boldsymbol{x}\,(\boldsymbol{x}\mathbf{E}(\mathbf{r}, t)) = \boldsymbol{x}\tilde{E}_x(z)\mathrm{e}^{-i\omega t},$$
$$\mathbf{E}_{(y)}(\mathbf{r}, t) = \boldsymbol{y}\,(\boldsymbol{y}\mathbf{E}(\mathbf{r}, t)) = \boldsymbol{y}\tilde{E}_y(z)\mathrm{e}^{-i\omega t}.$$

$\mathbf{E}_{(x)}(\mathbf{r}, t)$ and $\mathbf{E}_{(y)}(\mathbf{r}, t)$ represent linearly polarized plane waves, each satisfying the wave equation (1.8). These waves have mutually orthogonal polarizations: the field $\mathbf{E}_{(x)}(\mathbf{r}, t)$ vibrates along a line parallel to $\boldsymbol{x}$, while the field $\mathbf{E}_{(y)}(\mathbf{r}, t)$ oscillates along a line parallel to $\boldsymbol{y}$. Thus, we can regard the representation (1.58) as a decomposition of the wave $\mathbf{E}(\mathbf{r}, t)$ into two waves with given mutually orthogonal polarizations. A similar decomposition can be performed with the use of any other pair of orthogonal polarizations.

Let

$$j_1 = \begin{pmatrix} j_{1x} \\ j_{1y} \end{pmatrix} \quad \text{and} \quad j_2 = \begin{pmatrix} j_{2x} \\ j_{2y} \end{pmatrix}$$

be a pair of mutually orthogonal polarization Jones vectors ($j_1{}^{\dagger} j_2 = 0$). Introduce the vectors

$$\widehat{\mathbf{e}}_1 = j_{1x}\boldsymbol{x} + j_{1y}\boldsymbol{y},$$
$$\widehat{\mathbf{e}}_2 = j_{2x}\boldsymbol{x} + j_{2y}\boldsymbol{y},$$

which are three-dimensional analogs of the vectors j_1 and j_2. The vectors $\widehat{\mathbf{e}}_1$ and $\widehat{\mathbf{e}}_2$ are unit vectors in the sense that

$$\widehat{\mathbf{e}}_j^* \widehat{\mathbf{e}}_j = 1, \quad j = 1, 2, \tag{1.59}$$

and mutually orthogonal in the sense that

$$\widehat{\mathbf{e}}_1^* \widehat{\mathbf{e}}_2 = \widehat{\mathbf{e}}_2^* \widehat{\mathbf{e}}_1 = 0. \tag{1.60}$$

Using these vectors, we can represent the wave (1.14) as follows:

$$\mathbf{E}(\mathbf{r}, t) = \mathbf{E}_1(\mathbf{r}, t) + \mathbf{E}_2(\mathbf{r}, t), \tag{1.61}$$

where

$$\mathbf{E}_1(\mathbf{r},t) = \widetilde{\mathbf{e}}_1\left(\widetilde{\mathbf{e}}_1^*\mathbf{E}(\mathbf{r},t)\right) = \widetilde{\mathbf{e}}_1\tilde{A}_1(z)\mathrm{e}^{-i\omega t},$$
$$\mathbf{E}_2(\mathbf{r},t) = \widetilde{\mathbf{e}}_2\left(\widetilde{\mathbf{e}}_2^*\mathbf{E}(\mathbf{r},t)\right) = \widetilde{\mathbf{e}}_2\tilde{A}_2(z)\mathrm{e}^{-i\omega t}, \tag{1.62}$$
$$\tilde{A}_j(z) = \tilde{A}_j(z')\mathrm{e}^{ik_0 n(z-z')}, \quad j = 1,2.$$

$\mathbf{E}_1(\mathbf{r},t)$ and $\mathbf{E}_2(\mathbf{r},t)$ represent waves with polarizations $\boldsymbol{j}_1$ and $\boldsymbol{j}_2$, respectively. The column

$$\tilde{\boldsymbol{J}}_{j_1 j_2} = \begin{pmatrix} \tilde{A}_1 \\ \tilde{A}_2 \end{pmatrix} \tag{1.63}$$

is yet another representation of the Jones vector of the wave. From the relation

$$\boldsymbol{x}\tilde{E}_x + \boldsymbol{y}\tilde{E}_y = \widetilde{\mathbf{e}}_1\tilde{A}_1 + \widetilde{\mathbf{e}}_2\tilde{A}_2 = \left(j_{1x}\boldsymbol{x} + j_{1y}\boldsymbol{y}\right)\tilde{A}_1 + \left(j_{2x}\boldsymbol{x} + j_{2y}\boldsymbol{y}\right)\tilde{A}_2$$

it follows that

$$\begin{pmatrix} \tilde{E}_x \\ \tilde{E}_y \end{pmatrix} = \begin{pmatrix} \boldsymbol{x}\widetilde{\mathbf{e}}_1 & \boldsymbol{x}\widetilde{\mathbf{e}}_2 \\ \boldsymbol{y}\widetilde{\mathbf{e}}_1 & \boldsymbol{y}\widetilde{\mathbf{e}}_2 \end{pmatrix}\begin{pmatrix} \tilde{A}_1 \\ \tilde{A}_2 \end{pmatrix} = \begin{pmatrix} j_{1x} & j_{2x} \\ j_{1y} & j_{2y} \end{pmatrix}\begin{pmatrix} \tilde{A}_1 \\ \tilde{A}_2 \end{pmatrix} = \begin{pmatrix} \boldsymbol{j}_1 & \boldsymbol{j}_2 \end{pmatrix}\begin{pmatrix} \tilde{A}_1 \\ \tilde{A}_2 \end{pmatrix} = \boldsymbol{j}_1\tilde{A}_1 + \boldsymbol{j}_2\tilde{A}_2. \tag{1.64}$$

The column $\tilde{\boldsymbol{J}}_{j_1 j_2}$ can be expressed in terms of the column $\tilde{\boldsymbol{J}}_{xy}$ as follows:

$$\begin{pmatrix} \tilde{A}_1 \\ \tilde{A}_2 \end{pmatrix} = \begin{pmatrix} \boldsymbol{j}_1 & \boldsymbol{j}_2 \end{pmatrix}^{-1}\begin{pmatrix} \tilde{E}_x \\ \tilde{E}_y \end{pmatrix} = \begin{pmatrix} \boldsymbol{j}_1^\dagger \\ \boldsymbol{j}_2^\dagger \end{pmatrix}\begin{pmatrix} \tilde{E}_x \\ \tilde{E}_y \end{pmatrix}. \tag{1.65}$$

It is clear that the Cartesian Jones vectors $\tilde{\boldsymbol{J}}_{xy}$ and $\tilde{\boldsymbol{J}}_{x'y'}$ can also be defined in the same way as the vector $\tilde{\boldsymbol{J}}_{j_1 j_2}$: the vector $\tilde{\boldsymbol{J}}_{xy}$ corresponds to the choice

$$\boldsymbol{j}_1 = \begin{pmatrix} 1 \\ 0 \end{pmatrix}, \quad \boldsymbol{j}_2 = \begin{pmatrix} 0 \\ 1 \end{pmatrix}$$

($\widetilde{\mathbf{e}}_1 = \boldsymbol{x}$, $\widetilde{\mathbf{e}}_2 = \boldsymbol{y}$), and the vector $\tilde{\boldsymbol{J}}_{x'y'}$ to

$$\boldsymbol{j}_1 = \begin{pmatrix} \cos\phi \\ \sin\phi \end{pmatrix}, \quad \boldsymbol{j}_2 = \begin{pmatrix} -\sin\phi \\ \cos\phi \end{pmatrix}$$

($\widetilde{\mathbf{e}}_1 = \boldsymbol{x}'$, $\widetilde{\mathbf{e}}_2 = \boldsymbol{y}'$) in the coordinate system (x, y, z,).

The representation of wave fields in terms of basis wave modes (basis eigenwaves) is widely used in rigorous methods of polarization optics and optics of stratified media (see Chapter 8). State vectors introduced in the same manner as $\tilde{\boldsymbol{J}}_{j_1 j_2}$ [see (1.61)–(1.63)] are natural elements of these methods, where they are employed for description of homogeneous waves propagating in isotropic media as well as homogeneous waves propagating along the optic axis in uniaxial media. Choosing the basis polarization vectors in such a way that the Jones vector can be treated as a Cartesian Jones vector referred to a right-handed coordinate system makes it possible to use the formulas relating the components of Cartesian Jones vectors and the polarization ellipse parameters of Section 1.1.2 in such calculations.

General Coordinate Transformation Rules for Jones Vectors

The column $\tilde{\boldsymbol{J}}_{j_1 j_2}$ [see (1.63)] is a particular representation of the Jones vector of the wave; to introduce this column we used the polarization basis $(\boldsymbol{j}_1, \boldsymbol{j}_2)$ [or, what is the same, $(\widehat{\mathbf{e}}_1, \widehat{\mathbf{e}}_2)$]. Let $(\boldsymbol{j}'_1, \boldsymbol{j}'_2)$ [$(\widehat{\mathbf{e}}'_1, \widehat{\mathbf{e}}'_2)$] be another polarization basis [with $\boldsymbol{j}'^{\dagger}_1 \boldsymbol{j}'_2 = 0$ ($\widehat{\mathbf{e}}'^{*}_1 \widehat{\mathbf{e}}'_2 = 0$)], and let the column $\tilde{\boldsymbol{J}}_{j'_1 j'_2}$ represent the same Jones vector in this new basis. One can show that

$$\tilde{\boldsymbol{J}}_{j'_1 j'_2} = \begin{pmatrix} \boldsymbol{j}'^{\dagger}_1 \boldsymbol{j}_1 & \boldsymbol{j}'^{\dagger}_1 \boldsymbol{j}_2 \\ \boldsymbol{j}'^{\dagger}_2 \boldsymbol{j}_1 & \boldsymbol{j}'^{\dagger}_2 \boldsymbol{j}_2 \end{pmatrix} \tilde{\boldsymbol{J}}_{j_1 j_2} \tag{1.66}$$

or, equivalently,

$$\tilde{\boldsymbol{J}}_{j'_1 j'_2} = \begin{pmatrix} \widehat{\mathbf{e}}'^{*}_1 \widehat{\mathbf{e}}_1 & \widehat{\mathbf{e}}'^{*}_1 \widehat{\mathbf{e}}_2 \\ \widehat{\mathbf{e}}'^{*}_2 \widehat{\mathbf{e}}_1 & \widehat{\mathbf{e}}'^{*}_2 \widehat{\mathbf{e}}_2 \end{pmatrix} \tilde{\boldsymbol{J}}_{j_1 j_2}. \tag{1.67}$$

Relation (1.66) can readily be derived by using (1.64) and (1.65).

1.2 Interface Optics for Isotropic Media

Many problems of LCD optics involve considering the optical effect of interfaces. In this book, we will deal with interfaces of different kinds—from interfaces between isotropic media to those between arbitrary anisotropic media. The simplest problem, the problem on reflection and transmission of a plane monochromatic wave incident on a plane interface between isotropic media, is considered in detail in many textbooks (e.g., [1, 4]). In Section 1.2.1, we present, without derivation, the basic laws and formulas relating to this problem. In Section 1.2.2, we use this problem to show some options of modern variants of the Jones matrix method.

1.2.1 Fresnel's Formulas. Snell's Law

Let a homogeneous plane monochromatic wave propagating in an isotropic homogeneous nonabsorbing medium with refractive index n_1 be obliquely incident at angle β_{inc} on a plane surface of another isotropic homogeneous nonabsorbing medium with refractive index n_2. First we consider the case when $n_1 < n_2$, which is illustrated by Figure 1.5. In this case, at any β_{inc}, the reflected and transmitted fields will be homogeneous plane waves. Considering amplitude relations between the incident, reflected, and transmitted waves, it is convenient to decompose each of these waves into two linearly polarized constituents: the wave with its electric field vector parallel to the plane of incidence, it is the so-called *p-polarized* component, and the wave with electric field vector perpendicular to the plane of incidence, it is the so-called *s-polarized* component (*the plane of incidence* is the plane containing the incident light wave vector and a normal to the interface). One can use the following variant of decomposition of the electric fields of the incident, reflected, and transmitted wave fields:

$$\begin{aligned} &\text{Incident wave:} \quad \mathbf{E}_{\text{inc}}(\mathbf{r}, t) = \left[\mathbf{e}_{\text{p}}^{(\text{inc})} A_{\text{p}}^{(\text{inc})}(\mathbf{r}) + \mathbf{e}_{\text{s}}^{(\text{inc})} A_{\text{s}}^{(\text{inc})}(\mathbf{r})\right] \mathrm{e}^{-i\omega t}, \\ &\text{Reflected wave:} \quad \mathbf{E}_{\text{ref}}(\mathbf{r}, t) = \left[\mathbf{e}_{\text{p}}^{(\text{ref})} A_{\text{p}}^{(\text{ref})}(\mathbf{r}) + \mathbf{e}_{\text{s}}^{(\text{ref})} A_{\text{s}}^{(\text{ref})}(\mathbf{r})\right] \mathrm{e}^{-i\omega t}, \\ &\text{Transmitted wave:} \quad \mathbf{E}_{\text{tr}}(\mathbf{r}, t) = \left[\mathbf{e}_{\text{p}}^{(\text{tr})} A_{\text{p}}^{(\text{tr})}(\mathbf{r}) + \mathbf{e}_{\text{s}}^{(\text{tr})} A_{\text{s}}^{(\text{tr})}(\mathbf{r})\right] \mathrm{e}^{-i\omega t}, \end{aligned} \tag{1.68}$$

where $\mathbf{e}_{\text{p}}^{(\text{inc})}$, $\mathbf{e}_{\text{s}}^{(\text{inc})}$, $\mathbf{e}_{\text{p}}^{(\text{ref})}$, $\mathbf{e}_{\text{s}}^{(\text{ref})}$, $\mathbf{e}_{\text{p}}^{(\text{tr})}$, and $\mathbf{e}_{\text{s}}^{(\text{tr})}$ are unit real vectors which specify vibration directions of the electric fields of the p- and s-components of the waves and are oriented as indicated in Figure 1.5, and

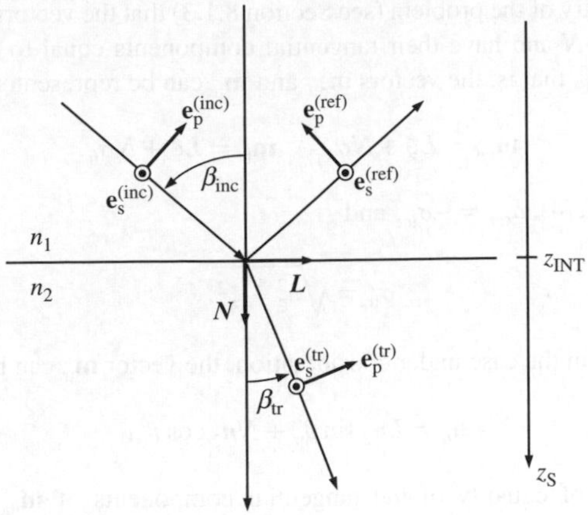

Figure 1.5 Transmission and reflection at a plane interface between isotropic media. Geometry of the problem

$A_p^{(inc)}, A_s^{(inc)}, A_p^{(ref)}, A_s^{(ref)}, A_p^{(tr)}$, and $A_s^{(tr)}$ are the scalar complex amplitudes of these components. The spatial evolution of the scalar amplitudes in the regions where the corresponding waves exist can be described by the equations

$$A_j^{(inc)}(\mathbf{r}) = A_j^{(inc)}(\mathbf{r}')\mathrm{e}^{ik_0\mathbf{m}_{inc}(\mathbf{r}'-\mathbf{r})}, \quad A_j^{(ref)}(\mathbf{r}) = A_j^{(ref)}(\mathbf{r}'')\mathrm{e}^{ik_0\mathbf{m}_{ref}(\mathbf{r}''-\mathbf{r})},$$
$$A_j^{(tr)}(\mathbf{r}) = A_j^{(tr)}(\mathbf{r}''')\mathrm{e}^{ik_0\mathbf{m}_{tr}(\mathbf{r}'''-\mathbf{r})}, \qquad (1.69)$$
$$j = \mathrm{s}, \mathrm{p},$$

where $\mathbf{m}_{inc}$, $\mathbf{m}_{ref}$, and $\mathbf{m}_{tr}$ are the refraction vectors (see Section 8.1.2) of the incident, reflected, and transmitted waves, respectively. The refraction vectors are related to the corresponding wave vectors by the equations

$$\mathbf{m}_{inc} = k_0^{-1}\mathbf{k}_{inc}, \quad \mathbf{m}_{ref} = k_0^{-1}\mathbf{k}_{ref}, \quad \mathbf{m}_{tr} = k_0^{-1}\mathbf{k}_{tr}. \qquad (1.70)$$

Using the quantities

$$\zeta \equiv \mathbf{m}_{inc}\boldsymbol{L} = n_1 \sin\beta_{inc}, \quad \mathbf{b} = \boldsymbol{L}\zeta, \quad \sigma_{inc} \equiv \mathbf{m}_{inc}\boldsymbol{N} = n_1\cos\beta_{inc}, \qquad (1.71)$$

where $\boldsymbol{N}$ and $\boldsymbol{L}$ are unit vectors oriented as shown in Figure 1.5 ($\boldsymbol{N}$ is normal to the interface surface; $\boldsymbol{L}$ is tangent to this surface), one may represent the vector $\mathbf{m}_{inc}$ as follows:

$$\mathbf{m}_{inc} = \boldsymbol{L}n_1\sin\beta_{inc} + \boldsymbol{N}n_1\cos\beta_{inc} = \boldsymbol{L}\zeta + \boldsymbol{N}\sigma_{inc} = \mathbf{b} + \boldsymbol{N}\sigma_{inc}. \qquad (1.72)$$

According to (1.12),

$$\mathbf{m}_{inc}\mathbf{m}_{inc} = n_1^2, \quad \mathbf{m}_{ref}\mathbf{m}_{ref} = n_1^2, \quad \mathbf{m}_{tr}\mathbf{m}_{tr} = n_2^2. \qquad (1.73)$$

It follows from the symmetry of the problem (see Section 8.1.3) that the vectors $\mathbf{m}_{\text{ref}}$ and $\mathbf{m}_{\text{tr}}$ are coplanar with the vectors $\mathbf{m}_{\text{inc}}$ and $\boldsymbol{N}$ and have their tangential components equal to the tangential component ($\mathbf{b} = \boldsymbol{L}\zeta$) of the vector $\mathbf{m}_{\text{inc}}$, that is, the vectors $\mathbf{m}_{\text{ref}}$ and $\mathbf{m}_{\text{tr}}$ can be represented as follows:

$$\mathbf{m}_{\text{ref}} = \boldsymbol{L}\zeta + \boldsymbol{N}\sigma_{\text{ref}}, \quad \mathbf{m}_{\text{tr}} = \boldsymbol{L}\zeta + \boldsymbol{N}\sigma_{\text{tr}}. \tag{1.74}$$

According to (1.73) and (1.74), $\sigma_{\text{ref}} = -\sigma_{\text{inc}}$ and

$$\sigma_{\text{tr}} = \sqrt{n_2^2 - \zeta^2}. \tag{1.75}$$

If n_2 is real and $\zeta < n_2$, as in the case under consideration, the vector $\mathbf{m}_{\text{tr}}$ can be represented as

$$\mathbf{m}_{\text{tr}} = \boldsymbol{L}n_2 \sin\beta_{\text{tr}} + \boldsymbol{N}n_2 \cos\beta_{\text{tr}}. \tag{1.76}$$

Then from the condition of equality of the tangential components of $\mathbf{m}_{\text{inc}}$ and $\mathbf{m}_{\text{tr}}$ it follows that

$$n_2 \sin\beta_{\text{tr}} = n_1 \sin\beta_{\text{inc}}, \tag{1.77}$$

which is the well-known *Snell's law*.

Let the plane of the interface coincide with the plane $z_S = z_{\text{INT}}$ in a rectangular Cartesian coordinate system (x_S, y_S, z_S) with the z_S-axis directed as shown in Figure 1.5. From the requirement of continuity of the tangential components of the electric and magnetic fields across the interface surface (see Section 8.1.1), one can find that amplitudes of the p-polarized components of the transmitted and reflected waves depend only on the amplitude of the p-polarized component of the incident wave and the same is true for the s-polarized components and that the ratios

$$\begin{aligned} t_{\text{pp}} &\equiv \frac{A_{\text{p}}^{(\text{tr})}(x_S, y_S, z_{\text{INT}}+0)}{A_{\text{p}}^{(\text{inc})}(x_S, y_S, z_{\text{INT}}-0)}, \quad t_{\text{ss}} \equiv \frac{A_{\text{s}}^{(\text{tr})}(x_S, y_S, z_{\text{INT}}+0)}{A_{\text{s}}^{(\text{inc})}(x_S, y_S, z_{\text{INT}}-0)}, \\ r_{\text{pp}} &\equiv \frac{A_{\text{p}}^{(\text{ref})}(x_S, y_S, z_{\text{INT}}-0)}{A_{\text{p}}^{(\text{inc})}(x_S, y_S, z_{\text{INT}}-0)}, \quad r_{\text{ss}} \equiv \frac{A_{\text{s}}^{(\text{ref})}(x_S, y_S, z_{\text{INT}}-0)}{A_{\text{s}}^{(\text{inc})}(x_S, y_S, z_{\text{INT}}-0)}, \end{aligned} \tag{1.78}$$

where $z_S = z_{\text{INT}} - 0$ and $z_S = z_{\text{INT}} + 0$ stand for the sides of the plane $z_S = z_{\text{INT}}$ facing the half-spaces $z_S < z_{\text{INT}}$ and $z_S > z_{\text{INT}}$ respectively (or for corresponding planes infinitely close to the plane $z_S = z_{\text{INT}}$), are independent of x_S and y_S and can be expressed as follows:

$$t_{\text{pp}} = \frac{2n_1 n_2 \cos\beta_{\text{inc}}}{n_1\sqrt{n_2^2 - n_1^2 \sin^2\beta_{\text{inc}}} + n_2^2 \cos\beta_{\text{inc}}}, \tag{1.79}$$

$$t_{\text{ss}} = \frac{2n_1 \cos\beta_{\text{inc}}}{n_1 \cos\beta_{\text{inc}} + \sqrt{n_2^2 - n_1^2 \sin^2\beta_{\text{inc}}}}, \tag{1.80}$$

$$r_{\text{pp}} = -\frac{n_1\sqrt{n_2^2 - n_1^2 \sin^2\beta_{\text{inc}}} - n_2^2 \cos\beta_{\text{inc}}}{n_1\sqrt{n_2^2 - n_1^2 \sin^2\beta_{\text{inc}}} + n_2^2 \cos\beta_{\text{inc}}}, \tag{1.81}$$

$$r_{\text{ss}} = \frac{n_1 \cos\beta_{\text{inc}} - \sqrt{n_2^2 - n_1^2 \sin^2\beta_{\text{inc}}}}{n_1 \cos\beta_{\text{inc}} + \sqrt{n_2^2 - n_1^2 \sin^2\beta_{\text{inc}}}}. \tag{1.82}$$

The quantities t_{pp}, t_{ss}, r_{pp}, and r_{ss} are called *the amplitude transmission* and *reflection coefficients*. Expressions (1.79)–(1.82) are *the Fresnel formulas* written in a special form.

In the case under consideration (nonabsorbing media, $n_1 < n_2$), the coefficients t_{pp}, t_{ss}, r_{pp}, and r_{ss} have real values at any β_{inc}. At $\beta_{inc} \neq 0$, the amount of the reflected light and that of the transmitted light depend on the polarization state of the incident light.

Transmissivity and Reflectivity of the Interface

Let $E^{(inc)}(z_{INT} - 0)$ be the irradiance produced by the incident wave on the plane $z_S = z_{INT} - 0$, $E^{(ref)}(z_{INT} - 0)$ the irradiance produced by the reflected wave on the same plane, and $E^{(tr)}(z_{INT} + 0)$ the irradiance produced by the transmitted wave on the plane $z_S = z_{INT} + 0$ (note that we deal here with another kind of irradiance than FEFD irradiance used in Section 1.1.2; see Sections 5.2, 5.4.2, and 8.5). The quantities

$$T_I \equiv \frac{E^{(tr)}(z_{INT} + 0)}{E^{(inc)}(z_{INT} - 0)} \text{ and } R_I \equiv \frac{E^{(ref)}(z_{INT} - 0)}{E^{(inc)}(z_{INT} - 0)} \tag{1.83}$$

are called respectively the *transmissivity* and *reflectivity* of the interface. In the case under consideration, the irradiances entering into (1.83) can be expressed as follows:

$$E^{(inc)}(z_{INT} - 0) = \frac{cn_1 \cos \beta_{inc}}{8\pi} \left(\left| A_p^{(inc)}(x_S, y_S, z_{INT} - 0) \right|^2 + \left| A_s^{(inc)}(x_S, y_S, z_{INT} - 0) \right|^2 \right), \tag{1.84a}$$

$$E^{(ref)}(z_{INT} - 0) = \frac{cn_1 \cos \beta_{inc}}{8\pi} \left(\left| A_p^{(ref)}(x_S, y_S, z_{INT} - 0) \right|^2 + \left| A_s^{(ref)}(x_S, y_S, z_{INT} - 0) \right|^2 \right), \tag{1.84b}$$

$$E^{(tr)}(z_{INT} + 0) = \frac{cn_2 \cos \beta_{tr}}{8\pi} \left(\left| A_p^{(tr)}(x_S, y_S, z_{INT} + 0) \right|^2 + \left| A_s^{(tr)}(x_S, y_S, z_{INT} + 0) \right|^2 \right) \tag{1.84c}$$

at arbitrary x_S and y_S. Using the above formulas, it is easy to find that if the incident wave is p-polarized,

$$T_I = T_{pp} \equiv \frac{n_2 \cos \beta_{tr}}{n_1 \cos \beta_{inc}} |t_{pp}|^2, \tag{1.85a}$$

$$R_I = R_{pp} \equiv |r_{pp}|^2 \tag{1.85b}$$

and, if the incident wave is s-polarized,

$$T_I = T_{ss} \equiv \frac{n_2 \cos \beta_{tr}}{n_1 \cos \beta_{inc}} |t_{ss}|^2, \tag{1.86a}$$

$$R_I = R_{ss} \equiv |r_{ss}|^2. \tag{1.86b}$$

Here we have denoted the transmissivities and reflectivities of the interface for a p-polarized incident wave by T_{pp} and R_{pp} and those for an s-polarized incident wave by T_{ss} and R_{ss}. As an illustration, Figure 1.6 shows the dependences of these transmissivities and reflectivities on the angle of incidence β_{inc} at $n_1 = 1$ (vacuum or air) and $n_2 = 1.5$ (e.g., glass).

At any polarization of the incident wave and at any β_{inc},

$$T_I + R_I = 1. \tag{1.87}$$

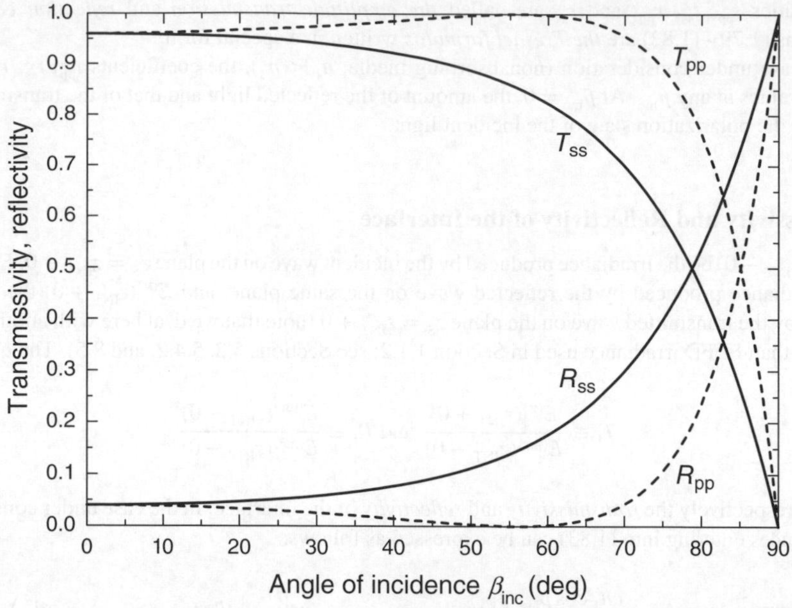

Figure 1.6 Transmissivities T_{pp} and T_{ss} and reflectivities R_{pp} and R_{ss} versus the angle of incidence β_{inc} at $n_1 = 1$ and $n_2 = 1.5$

The Brewster Angle

The angle

$$\beta_B = \arctan \frac{n_2}{n_1} \tag{1.88}$$

is called the *polarizing* or *Brewster angle*. As can be seen from (1.81), at $\beta_{inc} = \beta_B$ the coefficient r_{pp} is equal to zero, as is the reflectivity R_{pp} [see (1.85b)]. If $\beta_{inc} = \beta_B$, whatever the polarization of the incident wave, the reflected wave will be s-polarized. In the example illustrated by Figure 1.6 ($n_1 = 1$ and $n_2 = 1.5$), $\beta_B \approx 56.3°$.

The Case $n_1 > n_2$. Critical Angle

So far it has been assumed that $n_1 < n_2$. All the formulas presented above for the case $n_1 < n_2$ are also valid in the case $n_1 > n_2$ for $\beta_{inc} < \beta_c$, where

$$\beta_c = \arcsin\left(\frac{n_2}{n_1}\right) \tag{1.89}$$

is *the critical angle of total internal reflection*. At $\beta_{inc} > \beta_c$, in contrast to the case $\beta_{inc} < \beta_c$, the vector $\mathbf{m}_{tr}$ will be complex and have nonparallel real and imaginary parts [from (1.74) and (1.75) it is easy to see that $\text{Re}(\mathbf{m}_{tr})$ and $\text{Im}(\mathbf{m}_{tr})$ will be parallel to $\boldsymbol{L}$ and $\boldsymbol{N}$, respectively], that is, the transmitted wave will be *inhomogeneous* (see Section 8.1.2). In this case, decomposing the field $\mathbf{E}_{tr}$ [see (1.68)], we can use the same real vector $\mathbf{e}_{\mathrm{s}}^{(\mathrm{tr})}$ as in the above cases but cannot use a real vector $\mathbf{e}_{\mathrm{p}}^{(\mathrm{tr})}$ since with a real $\mathbf{e}_{\mathrm{p}}^{(\mathrm{tr})}$ $\mathbf{E}_{\mathrm{tr}}$ will not meet (1.10). To satisfy (1.10), one can take the following vector $\mathbf{e}_{\mathrm{p}}^{(\mathrm{tr})}$:

$$\mathbf{e}_{\mathrm{p}}^{(\mathrm{tr})} = \frac{1}{\sqrt{\mathbf{m}_{\mathrm{tr}}\mathbf{m}_{\mathrm{tr}}^{*}}}\mathbf{e}_{\mathrm{s}}^{(\mathrm{tr})} \times \mathbf{m}_{\mathrm{tr}} \tag{1.90}$$

with $\mathbf{e}_{\mathrm{s}}^{(\mathrm{tr})}$ being chosen the same as in the previous cases (i.e., real, unit, and oriented as shown in Figure 1.5). The vector $\mathbf{e}_{\mathrm{p}}^{(\mathrm{tr})}$ given by (1.90) is such that $\mathbf{m}_{\mathrm{tr}}\mathbf{e}_{\mathrm{p}}^{(\mathrm{tr})} = 0$, which is necessary for (1.10) to be satisfied, and unit in the sense that $\sqrt{\mathbf{e}_{\mathrm{p}}^{(\mathrm{tr})}\mathbf{e}_{\mathrm{p}}^{(\mathrm{tr})*}} = 1$. With the choice of $\mathbf{e}_{\mathrm{p}}^{(\mathrm{inc})}$, $\mathbf{e}_{\mathrm{s}}^{(\mathrm{inc})}$, $\mathbf{e}_{\mathrm{p}}^{(\mathrm{ref})}$, $\mathbf{e}_{\mathrm{s}}^{(\mathrm{ref})}$, and $\mathbf{e}_{\mathrm{s}}^{(\mathrm{tr})}$ as in Figure 1.5 and $\mathbf{e}_{\mathrm{p}}^{(\mathrm{tr})}$ as in (1.90) [note that the vector $\mathbf{e}_{\mathrm{p}}^{(\mathrm{tr})}$ used above in the case of real $\mathbf{m}_{\mathrm{tr}}$ satisfies (1.90)], expressions (1.80)–(1.82) for the coefficients t_{ss}, r_{pp}, and r_{ss} remain valid in the case $\beta_{\mathrm{inc}} > \beta_{\mathrm{c}}$ (but these coefficients become complex), while the expression for t_{pp} takes a more general form, namely,

$$t_{\mathrm{pp}} = C_{n2}\frac{2n_1n_2\cos\beta_{\mathrm{inc}}}{n_1\sqrt{n_2^2 - n_1^2\sin^2\beta_{\mathrm{inc}}} + n_2^2\cos\beta_{\mathrm{inc}}}, \tag{1.91}$$

where

$$C_{n2} = \sqrt{C_{\beta n2}^{*}C_{\beta n2} + S_{\beta n2}^{*}S_{\beta n2}} \tag{1.92}$$

with

$$C_{\beta n2} = \frac{1}{n_2}\sqrt{n_2^2 - n_1^2\sin^2\beta_{\mathrm{inc}}}, \quad S_{\beta n2} = \left(\frac{n_1}{n_2}\right)\sin\beta_{\mathrm{inc}}.$$

As seen from these formulas, at $\beta_{\mathrm{inc}} < \beta_{\mathrm{c}}$, $C_{n2} = 1$ and expression (1.91) becomes identical to (1.79).

Total Internal Reflection (TIR)

In the case $\beta_{\mathrm{inc}} > \beta_{\mathrm{c}}$, it is convenient to rewrite expressions (1.81) and (1.82) as follows:

$$r_{\mathrm{pp}} = -\frac{in_1\sqrt{n_1^2\sin^2\beta_{\mathrm{inc}} - n_2^2} - n_2^2\cos\beta_{\mathrm{inc}}}{in_1\sqrt{n_1^2\sin^2\beta_{\mathrm{inc}} - n_2^2} + n_2^2\cos\beta_{\mathrm{inc}}}, \tag{1.93}$$

$$r_{\mathrm{ss}} = \frac{n_1\cos\beta_{\mathrm{inc}} - i\sqrt{n_1^2\sin^2\beta_{\mathrm{inc}} - n_2^2}}{n_1\cos\beta_{\mathrm{inc}} + i\sqrt{n_1^2\sin^2\beta_{\mathrm{inc}} - n_2^2}}. \tag{1.94}$$

It is easy to see from (1.93) and (1.94) that $|r_{\mathrm{pp}}|=|r_{\mathrm{ss}}|=1$. Since, as before, the incident and reflected waves are assumed to be homogeneous and the medium where they propagate to be nonabsorbing, expressions (1.84a) and (1.84b) and hence (1.85b) and (1.86b) remain applicable. According to (1.85b) and (1.86b), when $|r_{\mathrm{pp}}|=|r_{\mathrm{ss}}|=1$, $R_{\mathrm{pp}} = R_{\mathrm{ss}} = 1$, that is, *total reflection* takes place. Expression (1.84c) is not applicable when $\beta_{\mathrm{inc}} > \beta_{\mathrm{c}}$ because in this case the transmitted wave is inhomogeneous. One can show that at $\beta_{\mathrm{inc}} > \beta_{\mathrm{c}}$, $E^{(\mathrm{tr})} = 0$ and consequently $T_{\mathrm{pp}} = T_{\mathrm{ss}} = 0$ (although t_{pp} and t_{ss} are different from zero). Even at small deviations β_{inc} from β_{c} and n_2 from n_1, the transmitted wave, having an imaginary $\sigma_{\mathrm{tr}} = i\sqrt{n_1^2\sin^2\beta_{\mathrm{inc}} - n_2^2}$, has an appreciable amplitude only near the interface. Such waves are called *surface* or *evanescent waves*.

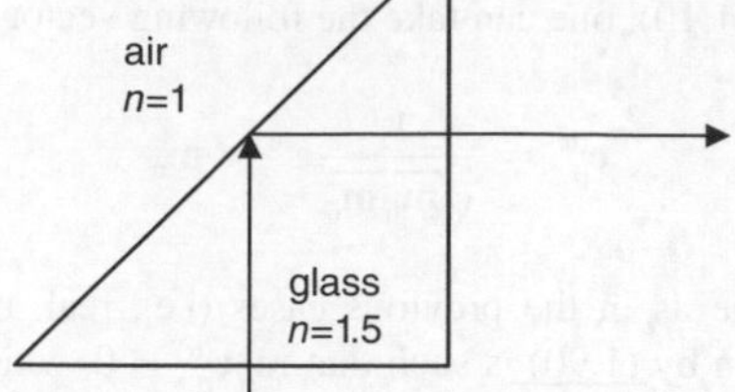

Figure 1.7 A prism reflector using the TIR phenomenon

At $\beta_{\rm inc} > \beta_{\rm c}$, $r_{\rm pp}$ and $r_{\rm ss}$, being complex, are different in phase and the difference of the phases of $r_{\rm pp}$ and $r_{\rm ss}$ gradually changes with $\beta_{\rm inc}$. This means that the phase shifts introduced into the p- and s-components of the reflected wave at reflection are different and that the difference of these phase shifts (and hence the shape of the polarization ellipse of the reflected light) can be controlled by choosing $\beta_{\rm inc}$. The latter is used in polarization-transforming devices such as the Fresnel rhomb.

For a glass–air interface with $n_1 = 1.5$ and $n_2 = 1$, $\beta_{\rm c} \approx 41.8°$. Therefore a right-angle glass prism can be used as a high-efficiency reflector as shown in Figure 1.7. Such a reflector may be almost lossless provided that the entrance and exit surfaces have antireflection coatings. The TIR phenomenon is used in many kinds of optical elements and devices. It is the principle of waveguides and optical fibers. In liquid crystal display applications, TIR is exploited in elements of backlight units, in projection systems, in beam steering, and so on. In Section 4.3, we will deal with an application of the TIR phenomenon in the intensity-modulating unit of an LCD.

Incidence of a Homogeneous Wave from a Nonabsorbing Medium on an Absorbing One

Formulas (1.80)–(1.82) and (1.91) can also be used for calculating the amplitude transmission and reflection coefficients in the case when the second medium is absorbing; in this case, n_2 is assumed to be complex. These formulas correspond to the choice of the vectors $\mathbf{e}_{\rm p}^{\rm (inc)}$, $\mathbf{e}_{\rm s}^{\rm (inc)}$, $\mathbf{e}_{\rm p}^{\rm (ref)}$, $\mathbf{e}_{\rm s}^{\rm (ref)}$, $\mathbf{e}_{\rm p}^{\rm (tr)}$, and $\mathbf{e}_{\rm s}^{\rm (tr)}$ in accordance with the same rules that were just used in the case of TIR. The transmitted wave in the absorbing medium will be inhomogeneous at any nonzero $\beta_{\rm inc}$ and has nonzero Re $\sigma_{\rm tr}$ and Im $\sigma_{\rm tr}$ at any $\beta_{\rm inc}$.

1.2.2 Reflection and Transmission Jones Matrices for a Plane Interface between Isotropic Media

In all the above cases, the interaction of the incident light with the interface can be described by the relations

$$\tilde{\boldsymbol{J}}^{\rm (tr)}(x_{\rm S}, y_{\rm S}, z_{\rm INT} + 0) = \tilde{\boldsymbol{t}}_{\rm I}\tilde{\boldsymbol{J}}^{\rm (inc)}(x_{\rm S}, y_{\rm S}, z_{\rm INT} - 0), \tag{1.95}$$

$$\tilde{\boldsymbol{J}}^{\rm (ref)}(x_{\rm S}, y_{\rm S}, z_{\rm INT} - 0) = \tilde{\boldsymbol{r}}_{\rm I}\tilde{\boldsymbol{J}}^{\rm (inc)}(x_{\rm S}, y_{\rm S}, z_{\rm INT} - 0), \tag{1.96}$$

where

$$\tilde{\boldsymbol{J}}^{\rm (inc)} = \begin{pmatrix} A_{\rm p}^{\rm (inc)} \\ A_{\rm s}^{\rm (inc)} \end{pmatrix}, \quad \tilde{\boldsymbol{J}}^{\rm (tr)} = \begin{pmatrix} A_{\rm p}^{\rm (tr)} \\ A_{\rm s}^{\rm (tr)} \end{pmatrix}, \quad \tilde{\boldsymbol{J}}^{\rm (ref)} = \begin{pmatrix} A_{\rm p}^{\rm (ref)} \\ A_{\rm s}^{\rm (ref)} \end{pmatrix} \tag{1.97}$$

are Jones vectors of the incident, transmitted, and reflected waves, and

$$\tilde{t}_{\mathrm{I}} = \begin{pmatrix} t_{\mathrm{pp}} & 0 \\ 0 & t_{\mathrm{ss}} \end{pmatrix}, \quad \tilde{r}_{\mathrm{I}} = \begin{pmatrix} r_{\mathrm{pp}} & 0 \\ 0 & r_{\mathrm{ss}} \end{pmatrix} \tag{1.98}$$

are the transmission and reflection Jones matrices of the interface corresponding to the representation (1.97) of the Jones vectors. The vectors $\tilde{\boldsymbol{J}}^{(\mathrm{inc})}$ and $\tilde{\boldsymbol{J}}^{(\mathrm{ref})}$ in all the considered cases as well as the vector $\tilde{\boldsymbol{J}}^{(\mathrm{tr})}$ when it characterizes a homogeneous wave are Jones vectors of the same kind as the vector $\tilde{\boldsymbol{J}}$ considered in Section 1.1.2. It is clear that the presented variant of transmission and reflection Jones matrices for the interface is not unique. Other kinds and representations of Jones matrices for interfaces may be more suitable in solving particular problems. For example, when considering transmission and reflection at an interface between nonabsorbing media in a situation where the waves in both media are homogeneous, it may be convenient to deal with the transmission and reflection matrices corresponding to the following Jones vectors:

$$\tilde{\boldsymbol{J}}_F^{(\mathrm{inc})} = \sqrt{2n_1 \cos\beta_{\mathrm{inc}}}\,\tilde{\boldsymbol{J}}^{(\mathrm{inc})}, \quad \tilde{\boldsymbol{J}}_F^{(\mathrm{ref})} = \sqrt{2n_1 \cos\beta_{\mathrm{inc}}}\,\tilde{\boldsymbol{J}}^{(\mathrm{ref})}, \quad \tilde{\boldsymbol{J}}_F^{(\mathrm{tr})} = \sqrt{2n_2 \cos\beta_{\mathrm{tr}}}\,\tilde{\boldsymbol{J}}^{(\mathrm{tr})}. \tag{1.99}$$

We denote these Jones matrices by $\tilde{t}_{\mathrm{I}(F)}$ and $\tilde{r}_{\mathrm{I}(F)}$. From (1.95), (1.96) and the relations

$$\tilde{\boldsymbol{J}}_F^{(\mathrm{tr})}(x_{\mathrm{S}}, y_{\mathrm{S}}, z_{\mathrm{INT}} + 0) = \tilde{t}_{\mathrm{I}(F)}\,\tilde{\boldsymbol{J}}_F^{(\mathrm{inc})}(x_{\mathrm{S}}, y_{\mathrm{S}}, z_{\mathrm{INT}} - 0), \tag{1.100}$$

$$\tilde{\boldsymbol{J}}_F^{(\mathrm{ref})}(x_{\mathrm{S}}, y_{\mathrm{S}}, z_{\mathrm{INT}} - 0) = \tilde{r}_{\mathrm{I}(F)}\,\tilde{\boldsymbol{J}}_F^{(\mathrm{inc})}(x_{\mathrm{S}}, y_{\mathrm{S}}, z_{\mathrm{INT}} - 0), \tag{1.101}$$

it follows that

$$\tilde{t}_{\mathrm{I}(F)} = \frac{\sqrt{n_2 \cos\beta_{\mathrm{tr}}}}{\sqrt{n_1 \cos\beta_{\mathrm{inc}}}}\tilde{t}_{\mathrm{I}}, \quad \tilde{r}_{\mathrm{I}(F)} = \tilde{r}_{\mathrm{I}}. \tag{1.102}$$

According to (1.84), (1.97), and (1.99), the irradiances $E^{(\mathrm{inc})}$, $E^{(\mathrm{ref})}$, and $E^{(\mathrm{tr})}$can be expressed as follows:

$$\begin{aligned}
E^{(\mathrm{inc})} &= \frac{cn_1 \cos\beta_{\mathrm{inc}}}{8\pi}\tilde{\boldsymbol{J}}^{(\mathrm{inc})\dagger}\tilde{\boldsymbol{J}}^{(\mathrm{inc})} = \frac{c}{16\pi}\tilde{\boldsymbol{J}}_F^{(\mathrm{inc})\dagger}\tilde{\boldsymbol{J}}_F^{(\mathrm{inc})}, \\
E^{(\mathrm{ref})} &= \frac{cn_1 \cos\beta_{\mathrm{inc}}}{8\pi}\tilde{\boldsymbol{J}}^{(\mathrm{ref})\dagger}\tilde{\boldsymbol{J}}^{(\mathrm{ref})} = \frac{c}{16\pi}\tilde{\boldsymbol{J}}_F^{(\mathrm{ref})\dagger}\tilde{\boldsymbol{J}}_F^{(\mathrm{ref})}, \\
E^{(\mathrm{tr})} &= \frac{cn_2 \cos\beta_{\mathrm{tr}}}{8\pi}\tilde{\boldsymbol{J}}^{(\mathrm{tr})\dagger}\tilde{\boldsymbol{J}}^{(\mathrm{tr})} = \frac{c}{16\pi}\tilde{\boldsymbol{J}}_F^{(\mathrm{tr})\dagger}\tilde{\boldsymbol{J}}_F^{(\mathrm{tr})}.
\end{aligned} \tag{1.103}$$

Substitution of these expressions into (1.83) gives the following expressions for the transmissivity T_{I} and reflectivity R_{I} of the interface in terms of the Jones vectors:

$$T_{\mathrm{I}} = \left(\frac{n_2 \cos\beta_{\mathrm{tr}}}{n_1 \cos\beta_{\mathrm{inc}}}\right)\frac{\tilde{\boldsymbol{J}}^{(\mathrm{tr})}(\mathbf{r}_{\mathrm{INT}}^{+})^{\dagger}\tilde{\boldsymbol{J}}^{(\mathrm{tr})}(\mathbf{r}_{\mathrm{INT}}^{+})}{\tilde{\boldsymbol{J}}^{(\mathrm{inc})}(\mathbf{r}_{\mathrm{INT}}^{-})^{\dagger}\tilde{\boldsymbol{J}}^{(\mathrm{inc})}(\mathbf{r}_{\mathrm{INT}}^{-})} = \frac{\tilde{\boldsymbol{J}}_F^{(\mathrm{tr})}(\mathbf{r}_{\mathrm{INT}}^{+})^{\dagger}\tilde{\boldsymbol{J}}_F^{(\mathrm{tr})}(\mathbf{r}_{\mathrm{INT}}^{+})}{\tilde{\boldsymbol{J}}_F^{(\mathrm{inc})}(\mathbf{r}_{\mathrm{INT}}^{-})^{\dagger}\tilde{\boldsymbol{J}}_F^{(\mathrm{inc})}(\mathbf{r}_{\mathrm{INT}}^{-})}, \tag{1.104}$$

$$R_{\mathrm{I}} = \frac{\tilde{\boldsymbol{J}}^{(\mathrm{ref})}(\mathbf{r}_{\mathrm{INT}}^{-})^{\dagger}\tilde{\boldsymbol{J}}^{(\mathrm{ref})}(\mathbf{r}_{\mathrm{INT}}^{-})}{\tilde{\boldsymbol{J}}^{(\mathrm{inc})}(\mathbf{r}_{\mathrm{INT}}^{-})^{\dagger}\tilde{\boldsymbol{J}}^{(\mathrm{inc})}(\mathbf{r}_{\mathrm{INT}}^{-})} = \frac{\tilde{\boldsymbol{J}}_F^{(\mathrm{ref})}(\mathbf{r}_{\mathrm{INT}}^{-})^{\dagger}\tilde{\boldsymbol{J}}_F^{(\mathrm{ref})}(\mathbf{r}_{\mathrm{INT}}^{-})}{\tilde{\boldsymbol{J}}_F^{(\mathrm{inc})}(\mathbf{r}_{\mathrm{INT}}^{-})^{\dagger}\tilde{\boldsymbol{J}}_F^{(\mathrm{inc})}(\mathbf{r}_{\mathrm{INT}}^{-})}, \tag{1.105}$$

where $\mathbf{r}_{\rm INT}^{-} = (x_S, y_S, z_{\rm INT} - 0)$ and $\mathbf{r}_{\rm INT}^{+} = (x_S, y_S, z_{\rm INT} + 0)$. Defining the length $|\tilde{J}|$ of a Jones vector $\tilde{J}$ as

$$|\tilde{J}| \equiv \sqrt{\tilde{J}^{\dagger}\tilde{J}}, \tag{1.106}$$

we can rewrite expressions (1.104) and (1.105) in the following form:

$$T_{\rm I} = \left(\frac{n_2 \cos\beta_{\rm tr}}{n_1 \cos\beta_{\rm inc}}\right) \frac{\left|\tilde{J}^{(\rm tr)}(\mathbf{r}_{\rm INT}^{+})\right|^2}{\left|\tilde{J}^{(\rm inc)}(\mathbf{r}_{\rm INT}^{-})\right|^2} = \frac{\left|\tilde{J}_F^{(\rm tr)}(\mathbf{r}_{\rm INT}^{+})\right|^2}{\left|\tilde{J}_F^{(\rm inc)}(\mathbf{r}_{\rm INT}^{-})\right|^2}, \tag{1.107}$$

$$R_{\rm I} = \frac{\left|\tilde{J}^{(\rm ref)}(\mathbf{r}_{\rm INT}^{-})\right|^2}{\left|\tilde{J}^{(\rm inc)}(\mathbf{r}_{\rm INT}^{-})\right|^2} = \frac{\left|\tilde{J}_F^{(\rm ref)}(\mathbf{r}_{\rm INT}^{-})\right|^2}{\left|\tilde{J}_F^{(\rm inc)}(\mathbf{r}_{\rm INT}^{-})\right|^2}. \tag{1.108}$$

Denote a polarization Jones vector of the incident wave in the basis $(\mathbf{e}_{\rm p}^{(\rm inc)}, \mathbf{e}_{\rm s}^{(\rm inc)})$ by $\boldsymbol{j}^{(\rm inc)}$. By definition, the vectors $\tilde{J}^{(\rm inc)}(\mathbf{r}_{\rm INT}^{-})$ and $\tilde{J}_F^{(\rm inc)}(\mathbf{r}_{\rm INT}^{-})$ are related to $\boldsymbol{j}^{(\rm inc)}$ as follows:

$$\tilde{J}^{(\rm inc)}(\mathbf{r}_{\rm INT}^{-}) = a(\mathbf{r}_{\rm INT}^{-})\boldsymbol{j}^{(\rm inc)}, \quad \tilde{J}_F^{(\rm inc)}(\mathbf{r}_{\rm INT}^{-}) = a_F(\mathbf{r}_{\rm INT}^{-})\boldsymbol{j}^{(\rm inc)}, \tag{1.109}$$

where $a(\mathbf{r}_{\rm INT}^{-})$ and $a_F(\mathbf{r}_{\rm INT}^{-})$ are scalar factors. Substitution from (1.109) into (1.95), (1.96), (1.100), and (1.101) gives expressions for the Jones vectors of the transmitted and reflected waves in terms of $\boldsymbol{j}^{(\rm inc)}$. Substituting these expressions into (1.107) and (1.108) and using the fact that $|\tilde{J}^{(\rm inc)}(\mathbf{r}_{\rm INT}^{-})|^2 = |a(\mathbf{r}_{\rm INT}^{-})|^2$ and $|\tilde{J}_F^{(\rm inc)}(\mathbf{r}_{\rm INT}^{-})|^2 = |a_F(\mathbf{r}_{\rm INT}^{-})|^2$, we obtain the following expressions for the transmissivity and reflectivity: in terms of $\tilde{t}_{\rm I}$ and $\tilde{r}_{\rm I}$,

$$T_{\rm I} = \left(\frac{n_2 \cos\beta_{\rm tr}}{n_1 \cos\beta_{\rm inc}}\right) \left|\tilde{t}_{\rm I}\boldsymbol{j}^{(\rm inc)}\right|^2, \tag{1.110}$$

$$R_{\rm I} = \left|\tilde{r}_{\rm I}\boldsymbol{j}^{(\rm inc)}\right|^2 \tag{1.111}$$

and, in terms of $\tilde{t}_{{\rm I}(F)}$ and $\tilde{r}_{{\rm I}(F)}$,

$$T_{\rm I} = \left|\tilde{t}_{{\rm I}(F)}\boldsymbol{j}^{(\rm inc)}\right|^2, \tag{1.112}$$

$$R_{\rm I} = \left|\tilde{r}_{{\rm I}(F)}\boldsymbol{j}^{(\rm inc)}\right|^2. \tag{1.113}$$

Employing the Jones vectors and matrices labeled by the subscript $_F$, we include all the information required for finding $T_{\rm I}$, apart from that contained in $\boldsymbol{j}^{(\rm inc)}$, in the Jones matrix and can use the unified and algebraically simplest expressions for calculating the transmissivity and reflectivity from the corresponding Jones matrices. Note that we could introduce the vectors $\tilde{J}_F^{(\rm inc)}$, $\tilde{J}_F^{(\rm tr)}$, and $\tilde{J}_F^{(\rm ref)}$ as

$$\tilde{J}_F^{(\rm inc)} = \begin{pmatrix} A_{\rm p}^{(\rm inc)} \\ A_{\rm s}^{(\rm inc)} \end{pmatrix}, \quad \tilde{J}_F^{(\rm tr)} = \begin{pmatrix} A_{\rm p}^{(\rm tr)} \\ A_{\rm s}^{(\rm tr)} \end{pmatrix}, \quad \tilde{J}_F^{(\rm ref)} = \begin{pmatrix} A_{\rm p}^{(\rm ref)} \\ A_{\rm s}^{(\rm ref)} \end{pmatrix} \tag{1.114}$$

[see (1.68)] by adopting the following normalization conditions for the basis vibration vectors:

$$\mathbf{e}_\mathrm{p}^{(\mathrm{inc})*}\mathbf{e}_\mathrm{p}^{(\mathrm{inc})} = \mathbf{e}_\mathrm{s}^{(\mathrm{inc})*}\mathbf{e}_\mathrm{s}^{(\mathrm{inc})} = \mathbf{e}_\mathrm{p}^{(\mathrm{ref})*}\mathbf{e}_\mathrm{p}^{(\mathrm{ref})} = \mathbf{e}_\mathrm{s}^{(\mathrm{ref})*}\mathbf{e}_\mathrm{s}^{(\mathrm{ref})} = \frac{1}{2n_1\cos\beta_\mathrm{inc}}, \tag{1.115}$$

$$\mathbf{e}_\mathrm{p}^{(\mathrm{tr})*}\mathbf{e}_\mathrm{p}^{(\mathrm{tr})} = \mathbf{e}_\mathrm{s}^{(\mathrm{tr})*}\mathbf{e}_\mathrm{s}^{(\mathrm{tr})} = \frac{1}{2n_2\cos\beta_\mathrm{tr}}. \tag{1.116}$$

Special normalizations of the basis vibration vectors, like this one, able to simplify a problem are considered in Chapters 8–12.

1.3 Wave Propagation in Anisotropic Media

Needless to say, the propagation of electromagnetic waves in optically anisotropic (birefringent) media and transmission characteristics of anisotropic layers are extremely important subjects to LCD optics. These subjects are considered in detail in Chapters 8 and 9, where we discuss rigorous methods of optics of stratified media applicable to both isotropic and anisotropic media. In the present section, we want to give an overview of basic features of light propagation in anisotropic media and shortly discuss transmission properties of anisotropic layers at normal incidence of light. The latter is directly concerned with the classical Jones matrix method (CJMM). In this section and almost everywhere in this book, we restrict our attention to anisotropic media that are nonmagnetic and nongyrotropic in the optical region.

1.3.1 Wave Equations

The basic difference of anisotropic media from isotropic ones from the standpoint of the Maxwell electromagnetic theory lies in relation between the electric field strength vector **E** and the electric displacement vector **D** (see Section 8.1.1). In the case of an arbitrary nongyrotropic medium, the vector **D** can be expressed in terms of the vector **E** as follows:

$$\mathbf{D} = \boldsymbol{\varepsilon}\mathbf{E}, \tag{1.117}$$

where $\boldsymbol{\varepsilon}$ is the permittivity tensor, $\boldsymbol{\varepsilon}$ being symmetric ($\boldsymbol{\varepsilon} = \boldsymbol{\varepsilon}^\mathrm{T}$, where $^\mathrm{T}$ denotes the matrix transposition). If the medium is isotropic, the tensor $\boldsymbol{\varepsilon}$ can be represented as $\boldsymbol{\varepsilon} = \varepsilon\mathbf{U}$, where ε is a scalar (the permittivity coefficient) and **U** is the unit matrix. This, in particular, means that **D** is parallel to **E** and that the ratio $|\mathbf{D}|/|\mathbf{E}|$ is independent of the direction of **E**. In the case of an anisotropic medium, the representation $\boldsymbol{\varepsilon} = \varepsilon\mathbf{U}$ is not applicable, **D** and **E** may be unparallel, and the ratio $|\mathbf{D}|/|\mathbf{E}|$ depends on the **E** direction.

An analogue of equation (1.8) for the case of a homogeneous anisotropic medium is

$$\nabla\times(\nabla\times\mathbf{E}) - k_0^2\boldsymbol{\varepsilon}\mathbf{E} = \widehat{\mathbf{0}}. \tag{1.118}$$

The wave vectors and vibration modes of the electric field of plane waves that can exist inside the anisotropic medium—such waves are called *natural waves*, *eigenwaves*, or *proper waves*—can be found from the equation

$$\mathbf{k}\times(\mathbf{k}\times\mathbf{E}) + k_0^2\boldsymbol{\varepsilon}\mathbf{E} = \widehat{\mathbf{0}} \tag{1.119}$$

which can be obtained by substituting (1.1) into (1.118). It is convenient to rewrite this equation in terms of the refraction vector $\mathbf{m} = \mathbf{k}/k_0$ and electric vibration vector $\mathbf{e}$ [$\mathbf{E}(\mathbf{r},t) = \mathbf{e}A(\mathbf{r},t)$, see (8.38) and definitions in Section 8.1.2]:

$$\mathbf{m} \times (\mathbf{m} \times \mathbf{e}) + \boldsymbol{\varepsilon}\mathbf{e} = \widehat{\mathbf{0}}. \tag{1.120}$$

This equation can be written in the following form:

$$\mathbf{Q_E e} = \widehat{\mathbf{0}}. \tag{1.121}$$

The matrix $\mathbf{Q_E}$ is expressed in terms of the elements of

$$\mathbf{m} \equiv \begin{pmatrix} m_1 \\ m_2 \\ m_3 \end{pmatrix} \text{ and } \boldsymbol{\varepsilon} \equiv \begin{pmatrix} \varepsilon_{11} & \varepsilon_{12} & \varepsilon_{13} \\ \varepsilon_{12} & \varepsilon_{22} & \varepsilon_{23} \\ \varepsilon_{13} & \varepsilon_{23} & \varepsilon_{33} \end{pmatrix}$$

as follows:

$$\mathbf{Q_E} = \mathbf{Q_m} + \boldsymbol{\varepsilon} = \begin{pmatrix} \varepsilon_{11} - m_2^2 - m_3^2 & \varepsilon_{12} + m_1 m_2 & \varepsilon_{13} + m_1 m_3 \\ \varepsilon_{12} + m_1 m_2 & \varepsilon_{22} - m_1^2 - m_3^2 & \varepsilon_{23} + m_2 m_3 \\ \varepsilon_{13} + m_1 m_3 & \varepsilon_{23} + m_2 m_3 & \varepsilon_{33} - m_1^2 - m_2^2 \end{pmatrix}, \tag{1.122}$$

where

$$\mathbf{Q_m} = \begin{pmatrix} -m_2^2 - m_3^2 & m_1 m_2 & m_1 m_3 \\ m_1 m_2 & -m_1^2 - m_3^2 & m_2 m_3 \\ m_1 m_3 & m_2 m_3 & -m_1^2 - m_2^2 \end{pmatrix}. \tag{1.123}$$

In some cases, it is simpler to use the following form of equation (1.120):

$$\mathbf{Q_D d} = \widehat{\mathbf{0}}, \tag{1.124}$$

where $\mathbf{d} = \boldsymbol{\varepsilon}\mathbf{e}$ is the displacement vibration vector [$\mathbf{D}(\mathbf{r},t) = \mathbf{d}A(\mathbf{r},t)$, see (8.38)], and

$$\mathbf{Q_D} = \mathbf{Q_E}\boldsymbol{\varepsilon}^{-1} = \mathbf{Q_m}\boldsymbol{\varepsilon}^{-1} + \mathbf{U}. \tag{1.125}$$

The vector $\mathbf{d}$ (as well as $\mathbf{D}$) of a plane wave is always orthogonal to its refraction vector $\mathbf{m}$ in the sense that

$$\mathbf{m} \cdot \mathbf{d} = 0, \tag{1.126}$$

as it follows from the Maxwell equation $\nabla\mathbf{D} = 0$. According to (1.120),

$$\mathbf{d} = -\mathbf{m} \times (\mathbf{m} \times \mathbf{e}) = \mathbf{e}(\mathbf{m} \cdot \mathbf{m}) - \mathbf{m}(\mathbf{m} \cdot \mathbf{e}),$$

that is, the vector $\mathbf{d}$ is a linear combination of the vectors $\mathbf{e}$ and $\mathbf{m}$. If the wave is homogeneous and linearly polarized, this means simply that the vectors $\mathbf{d}$, $\mathbf{e}$, and $\mathbf{m}$ are coplanar.

Equations (1.120) and (1.121) have a nontrivial solution only if

$$\det \mathbf{Q_E} = 0. \tag{1.127}$$

This condition can also be written as

$$\det \mathbf{Q_D} = 0. \tag{1.128}$$

From (1.127) or (1.128), the refraction vectors of natural waves are found.

In the next two sections we will consider some situations when the above equations are readily solved.

1.3.2 Waves in a Uniaxial Layer

In the case of a uniaxial medium with optic axis parallel to a unit vector **c**, the tensor ε can be represented as

$$\varepsilon \equiv \begin{pmatrix} \varepsilon_{11} & \varepsilon_{12} & \varepsilon_{13} \\ \varepsilon_{12} & \varepsilon_{22} & \varepsilon_{23} \\ \varepsilon_{13} & \varepsilon_{23} & \varepsilon_{33} \end{pmatrix} = \begin{pmatrix} \varepsilon_\perp + \Delta\varepsilon c_1^2 & \Delta\varepsilon c_1 c_2 & \Delta\varepsilon c_1 c_3 \\ \Delta\varepsilon c_1 c_2 & \varepsilon_\perp + \Delta\varepsilon c_2^2 & \Delta\varepsilon c_2 c_3 \\ \Delta\varepsilon c_1 c_3 & \Delta\varepsilon c_2 c_3 & \varepsilon_\perp + \Delta\varepsilon c_3^2 \end{pmatrix}, \quad \Delta\varepsilon = \varepsilon_\| - \varepsilon_\perp, \tag{1.129}$$

where $\varepsilon_\|$ and $\varepsilon_\perp$ are the principal permittivities of the medium ($\mathbf{D} = \varepsilon_\| \mathbf{E}$ if $\mathbf{E} \| \mathbf{c}$, and $\mathbf{D} = \varepsilon_\perp \mathbf{E}$ if $\mathbf{E} \perp \mathbf{c}$), and c_j ($j = 1,2,3$) are the elements of the vector $\mathbf{c} \equiv \begin{pmatrix} c_1 \\ c_2 \\ c_3 \end{pmatrix}$. The principal permittivities are related to the principal refractive indices of the medium, $n_\|$ and $n_\perp$, by

$$\varepsilon_\| = n_\|^2, \quad \varepsilon_\perp = n_\perp^2. \tag{1.130}$$

Ordinary and Extraordinary Waves

Natural waves in uniaxial media are divided into two classes: *ordinary waves* and *extraordinary waves*. The refraction vectors of the *ordinary* waves are independent of the optic axis orientation and satisfy the equation $\mathbf{m}\cdot\mathbf{m} = \varepsilon_\perp$. The refraction vectors of the *extraordinary* waves depend on the optic axis orientation and meet the equation $\mathbf{m}\cdot(\varepsilon\mathbf{m}) = \varepsilon_\perp\varepsilon_\|$ (see Section 9.3). Let $\mathbf{m}_o$ and $\mathbf{e}_o$ be the refraction vector and an electric vibration vector of an *ordinary wave*, and let $\mathbf{m}_e$ and $\mathbf{e}_e$ be those of an extraordinary wave. In general, the vector $\mathbf{e}_o$ satisfies the conditions $\mathbf{c}\cdot\mathbf{e}_o = 0$ and $\mathbf{m}_o\cdot\mathbf{e}_o = 0$, while the vector $\mathbf{e}_e$ can be represented as a linear combination of the vectors $\mathbf{m}_e$ and **c**. If the medium is nonabsorbing and the waves are homogeneous (not evanescent), the vectors $\mathbf{m}_o$ and $\mathbf{m}_e$ are real (see Section 9.3). In this case, the electric field of the ordinary wave performs oscillations along a straight line perpendicular to **c** and $\mathbf{m}_o$, while the electric field of the extraordinary wave vibrates along a straight line parallel to the plane spanned by the vectors $\mathbf{m}_e$ and **c** (Figure 1.8). For homogeneous waves, the plane containing the wave normal and **c** is referred to as *the principal plane* [1].

If a natural wave is homogeneous, one can associate with it a refractive index (see Section 8.1.2). For a homogeneous wave, the vector **m** can be represented as $\mathbf{m} = n_w\mathbf{l}$, where **l** is the wave normal and n_w is the refractive index for the wave. We will denote refractive indices for ordinary and extraordinary waves

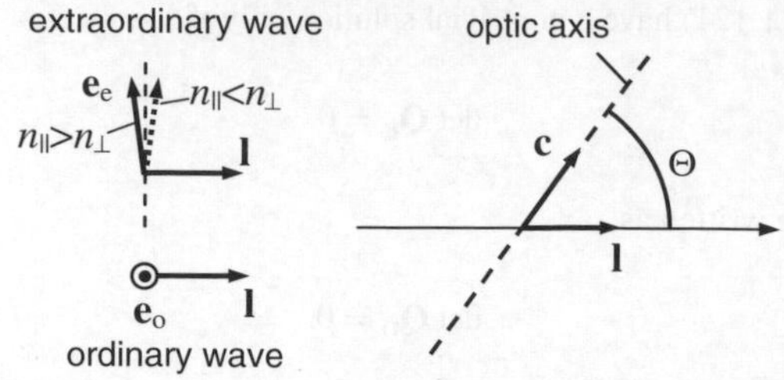

Figure 1.8 Homogeneous natural waves in a nonabsorbing uniaxial medium. **l** is the wave normal

by n_o and n_e, respectively. The refractive indices of a homogeneous ordinary wave and a homogeneous extraordinary wave can be expressed as

$$n_o = \sqrt{\varepsilon_\perp} = n_\perp, \tag{1.131}$$

$$n_e = \frac{\sqrt{\varepsilon_\perp \varepsilon_\parallel}}{\sqrt{\varepsilon_\perp + \Delta\varepsilon \cos^2 \Theta}} = \frac{n_\parallel n_\perp}{\sqrt{n_\parallel^2 \cos^2 \Theta + n_\perp^2 \sin^2 \Theta}}, \tag{1.132}$$

where Θ is the angle between the wave normal of the extraordinary wave and the optic axis (Figure 1.8). At $\Theta = 90°$, $n_e = n_\parallel$. At $\Theta = 0$, $n_e = n_\perp$, and the extraordinary wave turns into an ordinary one. Waves propagating along the optic axis (**m**||**c**) can have different polarizations (linear, elliptical, circular) as if the medium were isotropic.

Geometry of the Problem for a Layer

Let us consider a homogeneous uniaxial layer whose boundaries coincide with the planes $z_c = z_1$ and $z_c = z_2$ $(z_2 > z_1)$ in a coordinate system (x_c, y_c, z_c) and whose optic axis is parallel to the x_c–z_c plane (Figure 1.9a). In this case, the vector **c** can be represented as

$$\mathbf{c} = \begin{pmatrix} \cos\theta \\ 0 \\ \sin\theta \end{pmatrix}, \tag{1.133}$$

where θ is the angle between the x_c–y_c plane and the vector **c**, and, according to (1.129),

$$\boldsymbol{\varepsilon} = \begin{pmatrix} \varepsilon_\perp + \Delta\varepsilon\cos^2\theta & 0 & \Delta\varepsilon\cos\theta\sin\theta \\ 0 & \varepsilon_\perp & 0 \\ \Delta\varepsilon\cos\theta\sin\theta & 0 & \varepsilon_\perp + \Delta\varepsilon\sin^2\theta \end{pmatrix} \tag{1.134}$$

in the system (x_c, y_c, z_c). Let this layer be surrounded by a homogeneous nonabsorbing isotropic medium with refractive index n_1, and let a plane homogeneous wave with refraction vector $\mathbf{m}_{inc}$ fall on this layer from the half-space $z_c < z_1$. As in Section 1.2.1, we represent the vector $\mathbf{m}_{inc}$ as

$$\mathbf{m}_{inc} = \boldsymbol{L} n_1 \sin\beta_{inc} + \boldsymbol{N} n_1 \cos\beta_{inc} = \boldsymbol{L}\zeta + \boldsymbol{N}\sigma_{inc} \tag{1.135}$$

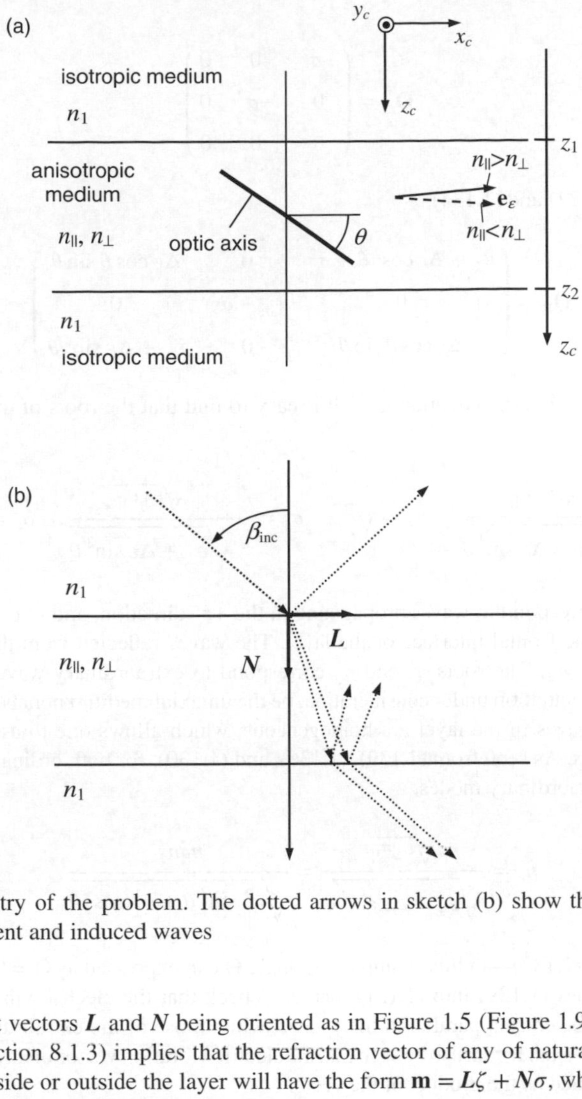

Figure 1.9 Geometry of the problem. The dotted arrows in sketch (b) show the directions of wave normals of the incident and induced waves

[see (1.72)], the unit vectors $\boldsymbol{L}$ and $\boldsymbol{N}$ being oriented as in Figure 1.5 (Figure 1.9b). The symmetry of the problem (see Section 8.1.3) implies that the refraction vector of any of natural waves produced by the incident wave inside or outside the layer will have the form $\mathbf{m} = \boldsymbol{L}\zeta + \boldsymbol{N}\sigma$, where $\zeta = n_1 \sin\beta_{\text{inc}}$. In particular, this means that all emergent waves in the half-space $z_c > z_2$, the components of the transmitted field, will have the same refraction vector, which allows considering any combination of these waves as a single plane wave. The same can be said about emergent waves propagating in the half-space $z_c < z_1$.

Normal Incidence

In the case of normal incidence ($\beta_{\text{inc}} = 0$), the refraction vectors of the waves propagating inside the layer, being represented in the system (x_c, y_c, z_c), will have the form

$$\mathbf{m} = \begin{pmatrix} 0 \\ 0 \\ \sigma \end{pmatrix}. \tag{1.136}$$

With such **m**,

$$\mathbf{Q_m} = \begin{pmatrix} -\sigma^2 & 0 & 0 \\ 0 & -\sigma^2 & 0 \\ 0 & 0 & 0 \end{pmatrix} \tag{1.137}$$

and, according to (1.122) and (1.134),

$$\mathbf{Q_E} = \begin{pmatrix} \varepsilon_\perp + \Delta\varepsilon\cos^2\theta - \sigma^2 & 0 & \Delta\varepsilon\cos\theta\sin\theta \\ 0 & \varepsilon_\perp - \sigma^2 & 0 \\ \Delta\varepsilon\cos\theta\sin\theta & 0 & \varepsilon_\perp + \Delta\varepsilon\sin^2\theta \end{pmatrix}. \tag{1.138}$$

Equation (1.127) is a quartic equation in σ. It is easy to find that the roots of this equation with $\mathbf{Q_E}$ given by (1.138) are

$$\sigma_1 = \frac{\sqrt{\varepsilon_\perp\varepsilon_{||}}}{\sqrt{\varepsilon_\perp + \Delta\varepsilon\sin^2\theta}}, \quad \sigma_2 = \sqrt{\varepsilon_\perp}, \quad \sigma_3 = -\frac{\sqrt{\varepsilon_\perp\varepsilon_{||}}}{\sqrt{\varepsilon_\perp + \Delta\varepsilon\sin^2\theta}}, \quad \sigma_4 = -\sqrt{\varepsilon_\perp}. \tag{1.139}$$

The first two roots correspond to waves propagating in the $+z_c$-direction, and in particular to the waves transmitted through the frontal interface of the layer. The waves reflected from the rear interface will have $\sigma = \sigma_3$ and $\sigma = \sigma_4$. The roots σ_1 and σ_3 correspond to extraordinary waves, and σ_2 and σ_4 to ordinary waves. In the situation under consideration, be the uniaxial medium nonabsorbing or absorbing, the induced natural waves in the layer are homogeneous, which allows one to associate with each of them a refractive index. As seen from (1.139), (1.136), and (1.130), for both ordinary modes, as it must, $n_o = n_\perp$. For both extraordinary modes,

$$n_e = \frac{\sqrt{\varepsilon_\perp\varepsilon_{||}}}{\sqrt{\varepsilon_\perp + \Delta\varepsilon\sin^2\theta}} = \frac{n_{||}n_\perp}{\sqrt{n_{||}^2\sin^2\theta + n_\perp^2\cos^2\theta}}, \tag{1.140}$$

which conforms with (1.132)—in this example, the angle Θ can expressed as $\Theta = 90°-\theta$.

Substituting solutions (1.139) into (1.121), one can check that the electric vibration vectors of the ordinary waves must be chosen parallel to the y_c-axis, while those of the extraordinary waves must be perpendicular to the y_c-axis. It can also be seen that the electric vibration vector of any of the extraordinary waves can be represented as the product of the vector

$$\mathbf{e}_\varepsilon = \begin{pmatrix} \varepsilon_\perp + \Delta\varepsilon\sin^2\theta \\ 0 \\ -\Delta\varepsilon\cos\theta\sin\theta \end{pmatrix} \tag{1.141}$$

[in the system (x_c, y_c, z_c)] and a scalar. Note that the vector $\mathbf{e}_\varepsilon$ is parallel to the x_c-axis only when either $\cos\theta$ or $\sin\theta$ is equal to zero. In any other case, this vector is not perpendicular to the refraction vectors. If the medium is nonabsorbing, the vector $\mathbf{e}_\varepsilon$ is real and the electric fields of the extraordinary modes vibrate along a line parallel to $\mathbf{e}_\varepsilon$. The fact that the vibration direction of the electric field of such a wave is not perpendicular to its refraction vector, in particular, suggests that the direction of energy transfer by the wave—this direction is perpendicular to the electric field vector [see (8.16)]—is different from the direction of the refraction vector. For any homogeneous ordinary wave, the direction of energy transfer coincides with the direction of its refraction vector. A difference of the directions of energy transfer for an ordinary wave and an extraordinary wave having codirectional refraction vectors is a manifestation of the phenomenon of *double refraction* or *birefringence*. A difference in refractive indices for these waves is another manifestation of this phenomenon.

If the uniaxial medium is absorbing and the vector $\mathbf{e}_\varepsilon$ is not parallel to the x_c-axis, the vector $\mathbf{e}_\varepsilon$ is complex and $\mathrm{Re}\,\mathbf{e}_\varepsilon$ is in general not parallel to $\mathrm{Im}\mathbf{e}_\varepsilon$. It implies that the end of the true (real) electric field vector of the wave describes with time an ellipse in the plane parallel to the x_c-z_c plane. In contrast to the elliptically polarized waves considered in Section 1.1, for which the plane of the vibration ellipse is perpendicular to the refraction vector, in this case the vibration ellipse plane is parallel to the refraction vector. Really, we have dealt with waves having a similar polarization in some examples of Section 1.2.1. These are the "p-polarized" waves in the second medium in the cases where these waves are inhomogeneous (TIR mode, absorbing medium at oblique incidence). Such waves cannot be called linearly polarized. At the same time, the term "plane-polarized wave" as applied to them seems acceptable. The linearly polarized waves are also often called plane-polarized. Where convenient, we will also do so.

Thus, if the optic axis is not perpendicular to the layer boundaries, be the layer nonabsorbing or absorbing, all natural waves induced inside it by a normally incident plane wave are plane-polarized. The plane of polarization of the extraordinary waves is the x_c-z_c plane (the principal plane), and that of the ordinary waves is the y_c-z_c plane.

Oblique Incidence

When a plane wave falls obliquely from an isotropic medium on a plane interface with an anisotropic medium, it produces in general two transmitted waves with nonparallel wave normals in the anisotropic medium. This is one more manifestation of double refraction. As an illustration, returning to the uniaxial layer, we consider the simple situation when the plane of incidence is parallel to the x_c-z_c plane, that is, the optic axis is parallel to the plane of incidence. Let the vector $\boldsymbol{L}$ be codirectional with the positive x_c-axis (Figure 1.10). In this case, the refraction vectors of the natural waves induced in the layer, being represented in the system (x_c, y_c, z_c), have the form

$$\mathbf{m} = \begin{pmatrix} \zeta \\ 0 \\ \sigma \end{pmatrix}, \tag{1.142}$$

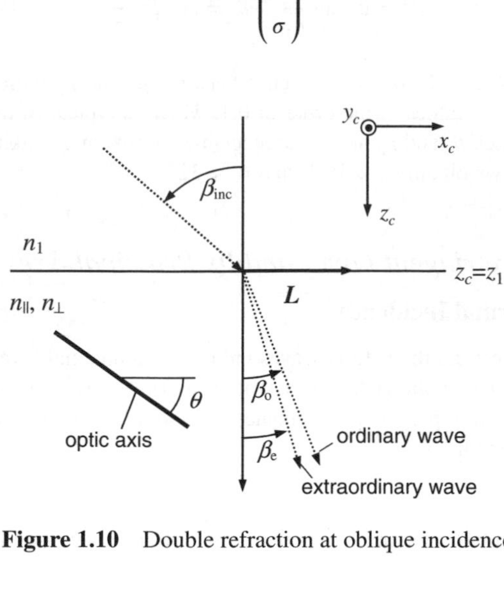

Figure 1.10 Double refraction at oblique incidence

where $\zeta = n_1 \sin\beta_{inc}$. As seen from (1.123), (1.122), and (1.134), with such **m**,

$$\mathbf{Q_E} = \begin{pmatrix} \varepsilon_\perp + \Delta\varepsilon\cos^2\theta - \sigma^2 & 0 & \Delta\varepsilon\cos\theta\sin\theta + \zeta\sigma \\ 0 & \varepsilon_\perp - \zeta^2 - \sigma^2 & 0 \\ \Delta\varepsilon\cos\theta\sin\theta + \zeta\sigma & 0 & \varepsilon_\perp + \Delta\varepsilon\sin^2\theta - \zeta^2 \end{pmatrix}.$$

The solutions of (1.127) with this $\mathbf{Q_E}$ that correspond to waves propagating away from the frontal boundary of the layer are

$$\sigma_1 = \frac{-\zeta\Delta\varepsilon\cos\theta\sin\theta + \sqrt{(\zeta\Delta\varepsilon\cos\theta\sin\theta)^2 + \left(\varepsilon_\perp + \Delta\varepsilon\sin^2\theta\right)\left[\varepsilon_\perp\varepsilon_\parallel - \zeta^2\left(\varepsilon_\perp + \Delta\varepsilon\cos^2\theta\right)\right]}}{\varepsilon_\perp + \Delta\varepsilon\sin^2\theta} \tag{1.143}$$

for extraordinary waves and

$$\sigma_2 = \sqrt{\varepsilon_\perp - \zeta^2} = \sqrt{n_\perp^2 - \zeta^2} \tag{1.144}$$

for ordinary waves. If the optic axis is parallel to the layer boundaries ($\theta = 0$), σ_1 can be expressed as follows:

$$\sigma_1 = \sqrt{\frac{\varepsilon_\parallel\left(\varepsilon_\perp - \zeta^2\right)}{\varepsilon_\perp}} = \frac{n_\parallel}{n_\perp}\sqrt{n_\perp^2 - \zeta^2}. \tag{1.145}$$

If the uniaxial medium is nonabsorbing and ζ is such that the radicands in the above expressions for σ_1 and σ_2 are positive (e.g., this is the case at any β_{inc} when $n_\parallel$ and $n_\perp$ is greater than n_1), the corresponding waves are homogeneous. In this case, the angle of refraction for the transmitted extraordinary wave, β_e, and that for the transmitted ordinary wave, β_o (see Figure 1.10), can be calculated by the formulas

$$\beta_e = \arctan\frac{\zeta}{\sigma_1}, \quad \beta_o = \arctan\frac{\zeta}{\sigma_2}. \tag{1.146}$$

As clearly seen from (1.144)–(1.146), the difference between β_e and β_o increases with increasing the ratio $\delta_n = |n_\parallel - n_\perp|/n_\perp$. At δ_n values of the order of 0.1, which is typical of the liquid crystals used in LCDs, the difference between β_e and β_o may be appreciable. For example, taking $n_1 = 1$, $n_\parallel = 1.7$, $n_\perp = 1.5$, $\theta = 0$, and $\beta_{inc} = 60°$, we obtain $\beta_o \approx 35.3°$ and $\beta_e \approx 32°$.

1.3.3 A Simple Birefringent Layer and Its Principal Axes

A Biaxial Layer at Normal Incidence

As noted in the previous section, the natural waves induced in a uniaxial layer by a normally incident plane wave are in general plane-polarized, and each of these waves has its polarization plane coincident with one of two fixed, mutually perpendicular, planes. The same can be said about natural waves in a biaxial layer if the biaxial medium

(i) is nonabsorbing or
(ii) being absorbing has a plane of symmetry perpendicular to the layer boundaries.

To show this we refer to (1.124). If any of the two conditions is satisfied, there exists a coordinate system (x_c, y_c, z_c), with the z_c-axis perpendicular to the layer boundaries, such that the components $\bar{\varepsilon}_{12}$ and $\bar{\varepsilon}_{21}$ of the tensor $\varepsilon^{-1} \equiv [\bar{\varepsilon}_{jk}]$ in this system are zero[1]. In this coordinate system, the matrix $\mathbf{Q_m}$ [see (1.123)] has the form (1.137), and, according to (1.125), the matrix $\mathbf{Q_D}$ can be written as follows:

$$\mathbf{Q_D} = \begin{pmatrix} 1-\sigma^2\bar{\varepsilon}_{11} & 0 & -\sigma^2\bar{\varepsilon}_{13} \\ 0 & 1-\sigma^2\bar{\varepsilon}_{22} & -\sigma^2\bar{\varepsilon}_{23} \\ 0 & 0 & 1 \end{pmatrix}. \tag{1.147}$$

The roots of equation (1.128) in this case are

$$\sigma_1 = \frac{1}{\sqrt{\bar{\varepsilon}_{11}}}, \quad \sigma_2 = \frac{1}{\sqrt{\bar{\varepsilon}_{22}}}, \quad \sigma_3 = -\frac{1}{\sqrt{\bar{\varepsilon}_{11}}}, \quad \sigma_4 = -\frac{1}{\sqrt{\bar{\varepsilon}_{22}}}. \tag{1.148}$$

It is easily seen from (1.147) that at $\bar{\varepsilon}_{22} \neq \bar{\varepsilon}_{11}$, the displacement vibration vectors $\mathbf{d}$ for the waves with σ equal to σ_1 and σ_3 are parallel to the x_c-axis, and, consequently, the electric vibration vectors ($\mathbf{e} = \varepsilon^{-1}\mathbf{d}$) of these waves lie in the x_c–z_c plane [recall that $\bar{\varepsilon}_{12} = \bar{\varepsilon}_{21} = 0$ in the system (x_c, y_c, z_c)], while the waves with σ equal to σ_2 and σ_4 have displacement vibration vectors parallel to the y_c-axis, and, consequently, their electric vibration vectors lie in the y_c–z_c plane, which is what we set out to prove.

A Simple Birefringent Layer and Its Principal Axes. Fast and Slow Axes

Thus, there is a broad class of homogeneous anisotropic layers such that any natural wave induced in the layer by a normally incident plane monochromatic wave is plane-polarized and has its polarization plane parallel to one of two fixed mutually perpendicular planes. Such layers will be called *simple birefringent layers*. The two fixed planes showing the possible orientations of the polarization planes of natural waves will be called the *basic planes of the layer*. Two mutually orthogonal axes each of which is parallel to the layer boundaries and one of the basic planes of the layer are called *the principal axes* of the layer. The principal axes of a layer should not be confused with the principal axes of the medium in the layer, although in many cases a principal axis of a layer is parallel to a principal axis of the medium. For example, the principal axis of a uniaxial medium is its optic axis. For the uniaxial layer considered in the previous section (see Figure 1.9a), one of the principal axes of the layer is parallel to the x_c-axis, and the other to the y_c-axis (as well as in the above example for a biaxial layer). If the optic axis is parallel to the layer boundaries, the former principal axis of the layer is parallel to its optic axis. A principal axis of a simple birefringent layer is called the *fast axis* or the *slow axis* according to whether the phase velocity of the natural waves with polarization plane parallel to this axis is greater or smaller than that of the natural waves whose polarization plane is perpendicular to this axis. For a layer of a nonabsorbing positive uniaxial medium ($n_\parallel > n_\perp$), the fast axis is perpendicular to the optic axis. For a layer of a nonabsorbing medium with negative birefringence ($n_\parallel < n_\perp$), the slow axis is perpendicular to the optic axis (in both cases, we assume that the optic axis is not perpendicular to the layer boundaries).

If the wave normally incident on a simple birefringent layer is linearly polarized along one of the principal axes of the layer, this wave induces in the layer only waves with polarization plane coincident with the polarization plane of the incident wave, and the wave transmitted by the layer has the same polarization state as the incident wave. The truth of this assertion can be proved by using the requirement of continuity of the tangential components of the electric and magnetic fields across interfaces (see Sections 8.1.1 and 12.2). This property of the simple birefringent layers is one of the cornerstones of the classical JC [5] where it is used in the mathematical description of the optical action of anisotropic

[1] In the presence of the plane of symmetry, the x_c-axis of such a system is parallel or perpendicular to this plane.

homogeneous layers (plates, films, etc.) functioning as linear retarders and linear polarizers in optical systems.

1.3.4 Transmission Jones Matrices of a Simple Birefringent Layer at Normal Incidence

Consider a simple birefringent layer sandwiched between nonabsorbing isotropic media. As in the above examples, we assume that the boundaries of the layer coincide with the planes $z_c = z_1$ and $z_c = z_2$ ($z_2 > z_1$) in a coordinate system (x_c, y_c, z_c) whose x_c-axis and y_c-axis are parallel to the principal axes of the layer. Let a plane monochromatic wave fall in the normal direction on the boundary $z_c = z_1$ of the layer from the medium of refractive index n_1. The refractive index of the medium beyond the layer will be denoted by n_2. Let reference frames (x, y, z) and (x', y', z) be introduced as in Section 1.1 (Figure 1.3) to represent the Jones vectors of the incident and transmitted waves. In the case under consideration, the z-axis is codirectional with the z_c-axis. Let the axes of the frame (x', y') be parallel to the principal axes of the layer (the x'-axis may be parallel to the x_c-axis or y_c-axis). We denote the Jones vectors—of the kind (1.21), referred to the system (x', y')—of the incident and transmitted waves by

$$\tilde{\boldsymbol{J}}'^{(\mathrm{inc})} \equiv \begin{pmatrix} \tilde{J}_{x'}^{(\mathrm{inc})} \\ \tilde{J}_{y'}^{(\mathrm{inc})} \end{pmatrix} \text{ and } \tilde{\boldsymbol{J}}'^{(\mathrm{tr})} \equiv \begin{pmatrix} \tilde{J}_{x'}^{(\mathrm{tr})} \\ \tilde{J}_{y'}^{(\mathrm{tr})} \end{pmatrix}, \tag{1.149}$$

respectively. Since the axes x' and y' are parallel to the principal axes, the components of the vector $\tilde{\boldsymbol{J}}'^{(\mathrm{tr})}$ are related to those of $\tilde{\boldsymbol{J}}'^{(\mathrm{inc})}$ by

$$\tilde{J}_{x'}^{(\mathrm{tr})}(z_2+0) = \tilde{t}_{\mathrm{L}x'}\tilde{J}_{x'}^{(\mathrm{inc})}(z_1-0), \quad \tilde{J}_{y'}^{(\mathrm{tr})}(z_2+0) = \tilde{t}_{\mathrm{L}y'}\tilde{J}_{y'}^{(\mathrm{inc})}(z_1-0), \tag{1.150}$$

where $\tilde{t}_{\mathrm{L}x'}$ and $\tilde{t}_{\mathrm{L}y'}$ are transmission coefficients depending on parameters of the layer and the wavelength λ. Here the Jones vectors are considered as functions of z_c. According to (1.150),

$$\tilde{\boldsymbol{J}}'^{(\mathrm{tr})}(z_2+0) = \tilde{\boldsymbol{t}}'_{\mathrm{L}}\tilde{\boldsymbol{J}}'^{(\mathrm{inc})}(z_1-0), \tag{1.151}$$

where

$$\tilde{\boldsymbol{t}}'_{\mathrm{L}} = \begin{pmatrix} \tilde{t}_{\mathrm{L}x'} & 0 \\ 0 & \tilde{t}_{\mathrm{L}y'} \end{pmatrix}. \tag{1.152}$$

The matrix $\tilde{\boldsymbol{t}}'_{\mathrm{L}}$ is the transmission Jones matrix of the layer, corresponding to the chosen kind and representation of the Jones vectors. Let us find the equivalent Jones matrix relating the input and output Jones vectors referred to the frame (x, y). Denote the Jones vectors of the incident and transmitted waves referred to the (x, y) frame by $\tilde{\boldsymbol{J}}^{(\mathrm{inc})}$ and $\tilde{\boldsymbol{J}}^{(\mathrm{tr})}$, respectively. According to (1.54),

$$\tilde{\boldsymbol{J}}'^{(\mathrm{inc})} = \widehat{R}_{\mathrm{C}}(\phi)\tilde{\boldsymbol{J}}^{(\mathrm{inc})}, \quad \tilde{\boldsymbol{J}}'^{(\mathrm{tr})} = \widehat{R}_{\mathrm{C}}(\phi)\tilde{\boldsymbol{J}}^{(\mathrm{tr})}, \tag{1.153}$$

where ϕ is the angle between the axes x and x' (see Figure 1.3). On substituting (1.153) into (1.151) and premultiplying the obtained equation by $\widehat{R}_{\mathrm{C}}(-\phi)$ [recall that $\widehat{R}_{\mathrm{C}}(-\phi) = \widehat{R}_{\mathrm{C}}(\phi)^{-1}$], we have

$$\tilde{\boldsymbol{J}}^{(\mathrm{tr})}(z_2+0) = \widehat{R}_{\mathrm{C}}(-\phi)\tilde{\boldsymbol{t}}'_{\mathrm{L}}\widehat{R}_{\mathrm{C}}(\phi)\tilde{\boldsymbol{J}}^{(\mathrm{inc})}(z_1-0). \tag{1.154}$$

The transmission Jones matrix $\tilde{t}_{\mathrm{L}}$ of the layer for the input and output Jones vectors represented in the system (x, y) is defined by the relation

$$\tilde{J}^{(\mathrm{tr})}(z_2+0)=\tilde{t}_{\mathrm{L}}\tilde{J}^{(\mathrm{inc})}(z_1-0). \tag{1.155}$$

From (1.154) it is seen that the matrix $\tilde{t}_{\mathrm{L}}$ can be expressed in terms of the matrix $\tilde{t}'_{\mathrm{L}}$ as follows:

$$\tilde{t}_{\mathrm{L}}=R_{\mathrm{C}}(-\phi)\tilde{t}'_{\mathrm{L}}R_{\mathrm{C}}(\phi). \tag{1.156}$$

In principle, in modeling of a polarization system, the matrix $\tilde{t}'_{\mathrm{L}}$ of an optical element can be defined in such a way as to take account of the whole variety of the optical effects involved in the process of light propagation through the layer, including multiple reflections from the boundaries of the layer. But usually, when employing the Jones matrix method, the multiple reflections are neglected and the transmitted light is considered as a result of the following sequence of operations: transmission of the frontal boundary of the layer → transmission of the bulk of the layer → transmission of the rear boundary of the layer. In this case, the amplitude transmission coefficients $\tilde{t}_{\mathrm{L}x'}$ and $\tilde{t}_{\mathrm{L}y'}$ [see (1.150)] of the layer can be expressed as follows:

$$\tilde{t}_{\mathrm{L}x'}=\tilde{t}_{\mathrm{LB}x'}\exp\left(ik_0n_{\mathrm{w}x'}d\right),\quad \tilde{t}_{\mathrm{L}y'}=\tilde{t}_{\mathrm{LB}y'}\exp\left(ik_0n_{\mathrm{w}y'}d\right), \tag{1.157}$$

where the factors $\tilde{t}_{\mathrm{LB}x'}$ and $\tilde{t}_{\mathrm{LB}y'}$ describe the transmission of the boundaries, $n_{\mathrm{w}x'}$ is the refractive index for the natural waves of the layer that have polarization planes parallel to the x'-axis, $n_{\mathrm{w}y'}$ is that for the natural waves whose polarization planes are parallel to the y'-axis, and $d=z_1-z_2$ is the thickness of the layer. The z_c-dependences of the electric fields of the natural waves traveling inside the layer from the plane $z_c=z_1$ toward the plane $z_c=z_2$ are given by

$$\mathbf{E}_{(x')}(x_c,y_c,z_c,t)=\mathbf{E}_{(x')}(x_c,y_c,z_1+0,t)\exp\left[ik_0n_{\mathrm{w}x'}\left(z_c-z_1\right)\right] \tag{1.158}$$

for a wave polarized in the plane parallel to the x'-axis and

$$\mathbf{E}_{(y')}(x_c,y_c,z_c,t)=\mathbf{E}_{(y')}(x_c,y_c,z_1+0,t)\exp\left[ik_0n_{\mathrm{w}y'}\left(z_c-z_1\right)\right] \tag{1.159}$$

for a wave polarized in the plane parallel to the y'-axis, which explains the presence and the form of the exponential factors in (1.157). If in the above examples for the uniaxial and biaxial layers we direct the x'-axis along the x_c-axis, the refractive indices $n_{\mathrm{w}x'}$ and $n_{\mathrm{w}y'}$ can be expressed as

$$n_{\mathrm{w}x'}=n_{\mathrm{e}}=\frac{n_{||}n_{\perp}}{\sqrt{n_{||}^2\sin^2\theta+n_{\perp}^2\cos^2\theta}},\quad n_{\mathrm{w}y'}=n_{\mathrm{o}}=n_{\perp} \tag{1.160}$$

for the uniaxial layer and

$$n_{\mathrm{w}x'}=\sigma_1,\quad n_{\mathrm{w}y'}=\sigma_2 \tag{1.161}$$

with σ_1 and σ_2 given by (1.148) for the biaxial layer.

The transmittance[2] (or the transmissivity) of the layer can be expressed as

$$T_{\mathrm{L}} = \frac{n_2}{n_1}\left|\tilde{t}_{\mathrm{L}}\boldsymbol{j}^{(\mathrm{inc})}\right|^2 = \frac{n_2}{n_1}\left|\tilde{t}'_{\mathrm{L}}\boldsymbol{j}'^{(\mathrm{inc})}\right|^2 \tag{1.162}$$

[cf. (1.110)], where $\boldsymbol{j}^{(\mathrm{inc})}$ is the polarization Jones vector of the incident wave referred to the frame (x, y) and $\boldsymbol{j}'^{(\mathrm{inc})}$ is the same vector but referred to the frame (x', y').

Let us consider analogous relations for other kinds of Jones vectors, namely, for the local "fitted-to-irradiance" (see Section 5.4.2) Jones vectors of the incident and transmitted waves defined by analogy with (1.99) as

$$\tilde{\boldsymbol{J}}_F^{(\mathrm{inc})} = \sqrt{2n_1}\tilde{\boldsymbol{J}}^{(\mathrm{inc})}, \quad \tilde{\boldsymbol{J}}_F^{(\mathrm{tr})} = \sqrt{2n_2}\tilde{\boldsymbol{J}}^{(\mathrm{tr})} \tag{1.163}$$

and the "global" Jones vectors of these waves, $\boldsymbol{J}^{(\mathrm{inc})}$ and $\boldsymbol{J}^{(\mathrm{tr})}$, defined in the same way as the vector $\boldsymbol{J}$ in Section 1.1.2. Neglecting the multiple reflections, the matrix $\tilde{t}_{\mathrm{L}(F)}$ such that

$$\tilde{\boldsymbol{J}}_F^{(\mathrm{tr})}(z_2+0) = \tilde{t}_{\mathrm{L}(F)}\,\tilde{\boldsymbol{J}}_F^{(\mathrm{inc})}(z_1-0) \tag{1.164}$$

can be represented as follows:

$$\tilde{t}_{\mathrm{L}(F)} = \widehat{R}_{\mathrm{C}}(-\phi)\tilde{t}'_{\mathrm{L}(F)}\widehat{R}_{\mathrm{C}}(\phi), \tag{1.165}$$

where

$$\tilde{t}'_{\mathrm{L}(F)} = \begin{pmatrix} \tilde{t}_{\mathrm{L}x'(F)} & 0 \\ 0 & \tilde{t}_{\mathrm{L}y'(F)} \end{pmatrix} \tag{1.166}$$

with

$$\tilde{t}_{\mathrm{L}x'(F)} = \tilde{t}_{\mathrm{LB}x'(F)}\exp\left(ik_0 n_{\mathrm{w}x'}d\right), \quad \tilde{t}_{\mathrm{L}y'(F)} = \tilde{t}_{\mathrm{LB}y'(F)}\exp\left(ik_0 n_{\mathrm{w}y'}d\right). \tag{1.167}$$

The transmittances of the layer for waves linearly polarized along its principal axes will be referred to as the *principal transmittances* of the layer. In the case under consideration, the principal transmittances can be expressed as $T_{\mathrm{L}x'} = \tilde{t}^*_{\mathrm{L}x'(F)}\tilde{t}_{\mathrm{L}x'(F)}$ and $T_{\mathrm{L}y'} = \tilde{t}^*_{\mathrm{L}y'(F)}\tilde{t}_{\mathrm{L}y'(F)}$. For any given polarization of the incident wave, the transmittance of the layer can be calculated by the formula

$$T_{\mathrm{L}} = \left|\tilde{t}_{\mathrm{L}(F)}\boldsymbol{j}^{(\mathrm{inc})}\right|^2 = \left|\tilde{t}'_{\mathrm{L}(F)}\boldsymbol{j}'^{(\mathrm{inc})}\right|^2. \tag{1.168}$$

One of the principal transmittances is equal to the maximum value of T_{L} over all possible polarization states of the incident wave, and the other to the minimum one. The quantities $T_{\mathrm{LB}x'} \equiv \tilde{t}^*_{\mathrm{LB}x'(F)}\tilde{t}_{\mathrm{LB}x'(F)}$ and $T_{\mathrm{LB}y'} \equiv \tilde{t}^*_{\mathrm{LB}y'(F)}\tilde{t}_{\mathrm{LB}y'(F)}$ are equal to the products of the transmittances of the frontal and rear boundaries of the layer for the corresponding polarizations of the incident wave. If the layer is nonabsorbing, $\tilde{t}_{\mathrm{LB}x'(F)}$ and $\tilde{t}_{\mathrm{LB}y'(F)}$ are real. For absorbing layers, these coefficients are in general complex but most often have

[2] The term "transmittance" which is commonly used in CJMM corresponds to the treatment of the incident light as a beam of finite diameter (see Section 7.1). At the same time, CJMM uses a plane-wave approximation which involves the possibility to evaluate a transmittance as the corresponding transmissivity (see Section 7.1). The notion of transmittance is closer to practice than transmissivity and we will use it where convenient, even when this implies an approximation (as in this case).

very small imaginary parts and, to a good approximation, can be considered real (see examples in Section 12.2). Therefore, almost always, the matrix $\tilde{t}'_{L(F)}$ can be represented as

$$\tilde{t}'_{L(F)} = \begin{pmatrix} \sqrt{T_{LBx'}} \exp\left(ik_0 n_{wx'} d\right) & 0 \\ 0 & \sqrt{T_{LBy'}} \exp\left(ik_0 n_{wy'} d\right) \end{pmatrix}. \tag{1.169}$$

The representations (1.166) and (1.169) take into account polarization-dependent losses (*diattenuation*) at the interfaces. However, in most cases of practical interest the coefficients $\tilde{t}_{LBx'(F)}$ and $\tilde{t}_{LBy'(F)}$ are of the order of 1 and very close to each other (see Section 12.2), which allows one to neglect the diattenuation at the interfaces and to use the following approximation:

$$t_{LBx'} = t_{LBy'} = t_{LB}, \tag{1.170}$$

where t_{LB} is the average over the actual values of $\sqrt{T_{LBx'}}$ and $\sqrt{T_{LBy'}}$. With this approximation, the matrix $\tilde{t}'_{L(F)}$ can be written as

$$\tilde{t}'_{L(F)} = t_{LB} \begin{pmatrix} \exp\left(ik_0 n_{wx'} d\right) & 0 \\ 0 & \exp\left(ik_0 n_{wy'} d\right) \end{pmatrix}. \tag{1.171}$$

On omitting the factor t_{LB}, we arrive at the form of $\tilde{t}'_{L(F)}$ usual for the classical JC, namely,

$$\tilde{t}'_{L(F)} = \tilde{t}'_{LU}, \tag{1.172}$$

where

$$\tilde{t}'_{LU} \equiv \begin{pmatrix} \exp\left(ik_0 n_{wx'} d\right) & 0 \\ 0 & \exp\left(ik_0 n_{wy'} d\right) \end{pmatrix}. \tag{1.173}$$

Omission of the factor t_{LB} is often quite a reasonable step, but one should remember that, almost always, this step is far out of the rigorous theory. In the case of a nonabsorbing layer, one can avoid serious contradictions with the rigorous theory by using the matrix $\tilde{t}'_{LU}$ as the operator relating the *polarization* Jones vectors of the incident and transmitted waves:

$$j'^{(tr)} = \tilde{t}'_{LU} j'^{(inc)}, \tag{1.174}$$

where both vectors are referred to the frame (x', y'). In terms of the polarization Jones vectors of the incident and transmitted waves referred to the frame (x, y), respectively $j^{(inc)}$ and $j^{(tr)}$, the same relation can be written as

$$j^{(tr)} = \tilde{t}_{LU} j^{(inc)}, \tag{1.175}$$

where

$$\tilde{t}_{LU} = \widehat{R}_C(-\phi)\tilde{t}'_{LU}\widehat{R}_C(\phi). \tag{1.176}$$

Jones matrices relating polarization Jones vectors will be called *polarization Jones matrices*. Note that polarization Jones matrices are always unitary because polarization Jones vectors are unit in the sense (1.33) (see Section 5.1.3).

The Jones matrix $\boldsymbol{t}_{\mathrm{L}}$ intended for linking the "global" Jones vectors of the incident and transmitted waves,

$$\boldsymbol{J}^{(\mathrm{tr})} = \boldsymbol{t}_{\mathrm{L}}\,\boldsymbol{J}^{(\mathrm{inc})}, \tag{1.177}$$

can be taken equal to $\tilde{\boldsymbol{t}}_{\mathrm{L}(F)}$. Since the phase of "global" Jones vectors is unimportant, any other matrix $\boldsymbol{t}_{\mathrm{L}}$ representable as $\boldsymbol{t}_{\mathrm{L}} = a_{\mathrm{P}}\tilde{\boldsymbol{t}}_{\mathrm{L}(F)}$, where a_{P} is a complex number of unit magnitude ($|a_{\mathrm{P}}| = 1$), can also be used for this purpose. The same can be said about matrices relating polarization Jones vectors.

The above representations are used in constructing transmission Jones matrices of various polarization elements, in particular linear retarders and linear absorptive polarizers.

In closing, we note the following relations. In general, the coefficients $\tilde{t}_{\mathrm{LB}x'}$ and $\tilde{t}_{\mathrm{LB}y'}$ entering into (1.157) are related to the coefficients $\tilde{t}_{\mathrm{LB}x'(F)}$ and $\tilde{t}_{\mathrm{LB}y'(F)}$ as follows:

$$\tilde{t}_{\mathrm{LB}x'} = \sqrt{\frac{n_1}{n_2}}\tilde{t}_{\mathrm{LB}x'(F)}, \quad \tilde{t}_{\mathrm{LB}y'} = \sqrt{\frac{n_1}{n_2}}\tilde{t}_{\mathrm{LB}y'(F)}. \tag{1.178}$$

When the refractive indices n_1 and n_2 differ greatly from each other, the coefficients $\tilde{t}_{\mathrm{LB}x'}$ and $\tilde{t}_{\mathrm{LB}y'}$, even when the reflection losses are small and $\tilde{t}_{\mathrm{LB}x'(F)}$ and $\tilde{t}_{\mathrm{LB}y'(F)}$ are close to unity, may differ greatly from unity. For example, if $n_1 = 1$, $n_2 = 1.5$, and the principal refractive indices of the layer are real and about 1.5, the coefficients $\tilde{t}_{\mathrm{LB}x'(F)}$ and $\tilde{t}_{\mathrm{LB}y'(F)}$ will be about 0.98, while the coefficients $\tilde{t}_{\mathrm{LB}x'}$ and $\tilde{t}_{\mathrm{LB}y'}$ will be close to 0.8. If, with the same layer, $n_1 = 1.5$ and $n_2 = 1$, $\tilde{t}_{\mathrm{LB}x'}$ and $\tilde{t}_{\mathrm{LB}y'}$ will be about 1.2, while the coefficients $\tilde{t}_{\mathrm{LB}x'(F)}$ and $\tilde{t}_{\mathrm{LB}y'(F)}$ will be the same as in the previous case. The matrices $\tilde{\boldsymbol{t}}_{\mathrm{L}}$ and $\tilde{\boldsymbol{t}}_{\mathrm{L}(F)}$ are related by

$$\tilde{\boldsymbol{t}}_{\mathrm{L}} = \sqrt{\frac{n_1}{n_2}}\tilde{\boldsymbol{t}}_{\mathrm{L}(F)} \tag{1.179}$$

and are equal to each other at $n_1 = n_2$. Let (x'', y'') be a reference frame with the x''-axis parallel to the x'-axis and the y''-axis parallel to the y'-axis. Whatever the values of n_1 and n_2, the transmission Jones matrix of the layer for the local "fitted-to-irradiance" Jones vectors referred to the system (x'', y'') for the reverse propagation direction (that is, for the case when the incident wave normally falls on the layer from the half-space $z_c > z_2$) is equal to the matrix $\tilde{\boldsymbol{t}}'_{\mathrm{L}(F)}$. For the Jones matrices associated with the Jones vectors of the kind (1.21), such a relation will take place only at $n_1 = n_2$.

1.3.5 Linear Retarders

Linear retarders—retardation films and retardation plates—are common optical elements used to convert the polarization state of passing light. Retardation films are used in LCDs for color dispersion compensation and to improve the viewing angle characteristics. Detailed discussion of the standard applications of retarders in polarization optics and terminology connected with retarders can be found in the books [2, 6] and many others. Here we briefly discuss the action of linear retarders at normal incidence.

A simple linear retarder is a nonabsorbing birefringent layer. When light enters such a layer, in general it splits into two plane-polarized natural waves propagating through the layer with different phase velocities. These waves experience different phase retardation as they propagate through the layer and, upon exiting the layer, recombine into a new wave with a new polarization state. This is the operating principle of linear retarders. The most important characteristic of a retarder is the relative phase retardation

$$\Gamma = \frac{2\pi(n_{\mathrm{s}} - n_{\mathrm{f}})d}{\lambda}, \tag{1.180}$$

where n_s and n_f are the refractive indices for the natural waves with polarization planes parallel to the slow axis and to the fast axis, respectively; d is the thickness of the birefringent layer. If, at a given λ, $(n_s - n_f)d = \lambda/4$ and, consequently, $\Gamma = \pi/2$, the retarder is called *a quarter-wave plate (film)* for the given λ. The retarders with $(n_s - n_f)d = \lambda/2$ ($\Gamma = \pi$) are called *half-wave plates (films)*.

Let the x'-axis of the frame (x', y') attached to the principal axes of a nonabsorbing simple birefringent layer be oriented along its slow axis. In this case, the polarization Jones matrix of the layer for the input and output Jones vectors referred to the frame (x', y') [see (1.174)] may be written as

$$\tilde{t}'_{LU} = \begin{pmatrix} e^{i\frac{2\pi n_s d}{\lambda}} & 0 \\ 0 & e^{i\frac{2\pi n_f d}{\lambda}} \end{pmatrix} \tag{1.181}$$

or

$$\tilde{t}'_{LU} = e^{i\frac{\pi(n_s+n_f)d}{\lambda}} \begin{pmatrix} e^{i\delta} & 0 \\ 0 & e^{-i\delta} \end{pmatrix}, \tag{1.182}$$

where

$$\delta \equiv \frac{\Gamma}{2} = \frac{\pi(n_s - n_f)d}{\lambda}. \tag{1.183}$$

Since the phase of a polarization Jones vector is inessential, we can omit the common exponential factor in expression (1.182) to deal with the mathematically simplest expression for $\tilde{t}'_{LU}$:

$$\tilde{t}'_{LU} = \begin{pmatrix} e^{i\delta} & 0 \\ 0 & e^{-i\delta} \end{pmatrix}. \tag{1.184}$$

On substituting (1.184) into (1.176), we obtain

$$\tilde{t}_{LU} = \widehat{R}_C(-\phi)\tilde{t}'_{LU}\widehat{R}_C(\phi) = \begin{pmatrix} \cos\delta + i\sin\delta\cos 2\phi & i\sin\delta\sin 2\phi \\ i\sin\delta\sin 2\phi & \cos\delta - i\sin\delta\cos 2\phi \end{pmatrix}; \tag{1.185}$$

here ϕ can be treated as the angle between the x-axis of the frame (x, y), to which the input and output polarization Jones vectors are referred [see (1.175)], and the slow axis of the layer. This is a general expression for the polarization Jones matrix of the linear retarder in a reference frame arbitrarily oriented with respect to its principal axes.

Let us illustrate the ability of retarders to convert polarization by some examples, using the Jones matrix method.

Half-Wave Plate

In this case, $\delta = \pi/2$ and the matrix $\tilde{t}'_{LU}$ can be written as

$$\tilde{t}'_{LU} = i\begin{pmatrix} 1 & 0 \\ 0 & -1 \end{pmatrix}. \tag{1.186}$$

Suppose that the frame (x, y) coincides with the frame (x', y'), so that $\tilde{t}_{\mathrm{LU}} = \tilde{t}'_{\mathrm{LU}}$. Assume that the incident wave has an arbitrary elliptical polarization. Taking the polarization vector $j^{(\mathrm{inc})}$ in the form

$$j^{(\mathrm{inc})} = j_E(\gamma_{\mathrm{inc}}, \upsilon_{\mathrm{inc}}) \tag{1.187}$$

(see Table 1.1), where γ_{inc} and υ_{inc} are the values of the azimuth γ_{e} and ellipticity angle υ_{e} (see Section 1.1.2) of the incident wave, it is easy to find that

$$j^{(\mathrm{tr})} = \tilde{t}_{\mathrm{LU}} j^{(\mathrm{inc})} = i \begin{pmatrix} 1 & 0 \\ 0 & -1 \end{pmatrix} j_E(\gamma_{\mathrm{inc}}, \upsilon_{\mathrm{inc}}) = i j_E(-\gamma_{\mathrm{inc}}, -\upsilon_{\mathrm{inc}}). \tag{1.188}$$

Since the vector $j_E(-\gamma_{\mathrm{inc}}, -\upsilon_{\mathrm{inc}})$ represents just the same polarization state as the vector $j^{(\mathrm{tr})} = i j_E(-\gamma_{\mathrm{inc}}, -\upsilon_{\mathrm{inc}})$, we may conclude that the transmitted wave will have an azimuth $\gamma_{\mathrm{e}} = -\gamma_{\mathrm{inc}}$ and an ellipticity angle $\upsilon_{\mathrm{e}} = -\upsilon_{\mathrm{inc}}$. If the incident wave is linearly polarized, the transmitted wave will also be linearly polarized, the polarization plane of the transmitted wave being the mirror image of that of the incident wave with respect to the x'–z plane. If the incident wave has the left circular polarization, the transmitted wave will have the right circular polarization and vice versa.

Quarter-Wave Plate

The main application of quarter-wave plates is in transforming linearly polarized light into circularly polarized one and vice versa. To illustrate these options, we again, for simplicity, assume that the frames (x, y) and (x', y') are coincident. For a quarter-wave plate, $\delta = \pi/4$ and the matrix $\tilde{t}_{\mathrm{LU}}$ can be represented as

$$\tilde{t}_{\mathrm{LU}} = \mathrm{e}^{i\frac{\pi}{4}} \begin{pmatrix} 1 & 0 \\ 0 & -i \end{pmatrix}. \tag{1.189}$$

It is easy to verify that the polarization vectors from Table 1.1 satisfy the following relations:

$$\begin{gathered} \begin{pmatrix} 1 & 0 \\ 0 & -i \end{pmatrix} j_P\left(\frac{\pi}{4}\right) = j_R, \quad \begin{pmatrix} 1 & 0 \\ 0 & -i \end{pmatrix} j_P\left(-\frac{\pi}{4}\right) = j_L, \\ \begin{pmatrix} 1 & 0 \\ 0 & -i \end{pmatrix} j_R = j_P\left(-\frac{\pi}{4}\right), \quad \begin{pmatrix} 1 & 0 \\ 0 & -i \end{pmatrix} j_L = j_P\left(\frac{\pi}{4}\right). \end{gathered} \tag{1.190}$$

As is seen from these relations, a quarter-wave plate can perform the following conversions:

$$P_{\pi/4} \to R, \quad P_{-\pi/4} \to L, \quad R \to P_{-\pi/4}, \quad L \to P_{\pi/4}, \tag{1.191}$$

where the symbols $P_{\pi/4}$, $P_{-\pi/4}$, R, and L denote respectively the linear polarization with $\gamma_{\mathrm{e}} = \pi/4$, the linear polarization with $\gamma_{\mathrm{e}} = -\pi/4$, the right circular polarization, and the left circular polarization.

1.3.6 Jones Matrices of Absorptive Polarizers. Ideal Polarizer

Absorptive polarizers are used in most kinds of liquid crystal displays. The main element of the usual absorptive polarizer is an absorbing anisotropic film exhibiting high diattenuation due to absorption anisotropy. In the spectral region where this film acts effectively as polarizer, one of the two principal transmittances of the film is close to zero, while the other is sufficiently high (ideally, equal to 1). The principal axis of the film corresponding to the higher principal transmittance is called *the transmission axis* of the polarizer [6].

A standard optical model of the polarizing film or the polarizer as a whole is a uniaxial layer whose optic axis is parallel to the layer boundaries (see Section 7.3). In the rigorous methods which are considered in Chapters 8–10, the specification of such a model includes the specification of the principal complex refractive indices of the layer. In calculations performed for the case of normal incidence using the classical Jones calculus, as a rule, simpler variants of specification of polarizers are used. Here we consider some of them.

Let the x'-axis of the reference frame (x', y') be parallel to the transmission axis of the layer being a model of the polarizer. We denote the principal transmittances of the layer by $t_{\|}$ and $t_{\perp}$, where $t_{\|}$ corresponds to the polarization along the transmission axis. Assuming that $\mathrm{Re}(n_{wx'}) = \mathrm{Re}(n_{wy'})$, in accordance with (1.171) we may write the matrix $\tilde{t}'_{L(F)}$ of the layer as follows:

$$\tilde{t}'_{L(F)} = \exp\left[ik_0 \,\mathrm{Re}\left(n_{wx'}\right) d\right] \begin{pmatrix} t_{LB} \exp\left[-k_0 \,\mathrm{Im}\left(n_{wx'}\right) d\right] & 0 \\ 0 & t_{LB} \exp\left[-k_0 \,\mathrm{Im}\left(n_{wy'}\right) d\right] \end{pmatrix}. \tag{1.192}$$

In this case, the principal transmittances of the layer can be expressed as

$$t_{\|} = t_{LB}^2 \exp\left[-2k_0 \,\mathrm{Im}\left(n_{wx'}\right) d\right], \quad t_{\perp} = t_{LB}^2 \exp\left[-2k_0 \,\mathrm{Im}\left(n_{wy'}\right) d\right], \tag{1.193}$$

and consequently the matrix $\tilde{t}'_{L(F)}$ can be represented as follows:

$$\tilde{t}'_{L(F)} = \exp\left[ik_0 \,\mathrm{Re}\left(n_{wx'}\right) d\right] \begin{pmatrix} \sqrt{t_{\|}} & 0 \\ 0 & \sqrt{t_{\perp}} \end{pmatrix}. \tag{1.194}$$

According to (1.194), the simplest variant of the Jones matrix of the polarizer for the "global" Jones vectors referred to the system (x', y') is

$$t'_L = \begin{pmatrix} \sqrt{t_{\|}} & 0 \\ 0 & \sqrt{t_{\perp}} \end{pmatrix} \tag{1.195}$$

[see the remark after (1.177)]. The corresponding Jones matrix for the "global" Jones vectors referred to the system (x, y) rotated with respect to the system (x', y') can be calculated by the formula

$$t_L = \widehat{R}_C(-\phi) t'_L \widehat{R}_C(\phi), \tag{1.196}$$

where ϕ is the angle between the x-axis and the x'-axis (the transmission axis of the polarizer). Thus, in this case, to specify the polarizer we need only the principal transmittances and orientation angle ϕ. It is sometimes convenient to represent the principal transmittances $t_{\|}$ and $t_{\perp}$ as follows:

$$t_{\|} = C_p t_{\|p}, \quad t_{\perp} = C_p t_{\perp p}, \tag{1.197}$$

where $t_{\|p}$ and $t_{\perp p}$ are the principal bulk transmittances of the layer,

$$t_{\|p} = \exp\left[-2k_0 \,\mathrm{Im}\left(n_{wx'}\right) d\right], \quad t_{\perp p} = \exp\left[-2k_0 \,\mathrm{Im}\left(n_{wy'}\right) d\right], \tag{1.198}$$

and $C_p = t_{LB}^2$ is a factor taking account of the reflection losses at the boundaries.

As a rule, the real parts of the refractive indices $n_{wx'}$ and $n_{wy'}$ of a real polarizing film are different. To take this circumstance into account one can use the following form of the matrix $\boldsymbol{t}'_{\mathrm{L}}$:

$$\boldsymbol{t}'_{\mathrm{L}} = \begin{pmatrix} \sqrt{t_{\parallel}} \exp\left(i\delta_{\mathrm{w}}\right) & 0 \\ 0 & \sqrt{t_{\perp}} \exp\left(-i\delta_{\mathrm{w}}\right) \end{pmatrix}, \tag{1.199}$$

where $\delta_{\mathrm{w}} = \pi \left[\mathrm{Re}\left(n_{\mathrm{w}x'}\right) - \mathrm{Re}\left(n_{\mathrm{w}y'}\right)\right] d/\lambda$. Although the situation when $\mathrm{Re}n_{\mathrm{w}x'} \neq \mathrm{Re}n_{\mathrm{w}y'}$ is common, in solving typical problems for LCDs, as a rule, there is no need to use the representation (1.199) instead of (1.195) because the phase factors in (1.199) contribute nothing to the quantities to be estimated, such as the transmittance of the LCD panel, or their influence on the LCD characteristics is negligible.

The above matrices $\boldsymbol{t}'_{\mathrm{L}}$ at $t_{\perp} \neq 0$ describe partial polarizers. All real absorptive polarizers are partial ones. However, for many practical polarizers, $t_{\perp}$ is so small that it can be taken as zero in calculations. In such a case, the matrix $\boldsymbol{t}'_{\mathrm{L}}$ can be written as

$$\boldsymbol{t}'_{\mathrm{L}} = \begin{pmatrix} \sqrt{t_{\parallel}} & 0 \\ 0 & 0 \end{pmatrix} = \sqrt{t_{\parallel}} \begin{pmatrix} 1 & 0 \\ 0 & 0 \end{pmatrix}. \tag{1.200}$$

Often a still further idealized model of a linear polarizer is used. This model is *an ideal linear polarizer* whose matrix $\boldsymbol{t}'_{\mathrm{L}}$ is as follows:

$$\boldsymbol{t}'_{\mathrm{L}} = \begin{pmatrix} 1 & 0 \\ 0 & 0 \end{pmatrix}. \tag{1.201}$$

The matrix $\boldsymbol{t}_{\mathrm{L}}$ (1.196) in this case can be written as

$$\boldsymbol{t}_{\mathrm{L}} = \begin{pmatrix} \cos^2\phi & \cos\phi\sin\phi \\ \cos\phi\sin\phi & \sin^2\phi \end{pmatrix}. \tag{1.202}$$

The concept of *an ideal polarizer* as an ideal device that transmits the light of a given polarization only, without losses, is applied to polarizers extracting an elliptical or a circular polarization as well [2].

With a given matrix $\boldsymbol{t}_{\mathrm{L}}$ of a polarizer, the transmittance of the polarizer for an incident wave with a given polarization Jones vectors $\boldsymbol{j}^{(\mathrm{inc})}$ can be calculated by the following general formula:

$$T_{\mathrm{L}} = \left|\boldsymbol{t}_{\mathrm{L}} \boldsymbol{j}^{(\mathrm{inc})}\right|^2. \tag{1.203}$$

In the case of *an ideal polarizer*, a simpler expression for the transmittance can be used:

$$T_{\mathrm{L}} = \left|\boldsymbol{j}_{\mathrm{tp}}^{\dagger} \boldsymbol{j}^{(\mathrm{inc})}\right|^2, \tag{1.204}$$

where $\boldsymbol{j}_{\mathrm{tp}}$ is the polarization Jones vector of waves that are transmitted by the polarizer. For example, the vector $\boldsymbol{j}_{\mathrm{tp}}$ for the ideal linear polarizer with matrix $\boldsymbol{t}_{\mathrm{L}}$ given by (1.202) can be expressed as $\boldsymbol{j}_{\mathrm{tp}} = \boldsymbol{j}_P(\phi)$ (see Table 1.1) in the system (x, y). Assuming that the light incident on this polarizer is linearly polarized and taking $\boldsymbol{j}^{(\mathrm{inc})} = \boldsymbol{j}_P(\gamma)$, we readily obtain from (1.204)

$$T_{\mathrm{L}} = \cos^2\gamma_{\phi}, \tag{1.205}$$

where $\gamma_\phi = \gamma - \phi$ is the angle between the transmission axis of the polarizer and the polarization direction of the incident light. Equation (1.205) expresses the familiar *law of Malus*. For a partial polarizer whose matrix t'_{L} is expressed by (1.195) or (1.199) the dependence of the polarizer transmittance on γ_ϕ is as follows:

$$T_{\mathrm{L}} = t_{||} \cos^2 \gamma_\phi + t_\perp \sin^2 \gamma_\phi. \tag{1.206}$$

This expression can easily be derived by using the following representation of T_{L}:

$$T_{\mathrm{L}} = \left| t'_{\mathrm{L}}\, j_P(\gamma_\phi) \right|^2 = j_P(\gamma_\phi)^\dagger \left(t'^\dagger_{\mathrm{L}} t'_{\mathrm{L}} \right) j_P(\gamma_\phi). \tag{1.207}$$

1.4 Jones Calculus

The classical Jones matrix method (CJMM) includes two fundamental methods. The first method is a calculus for treatment of optical systems containing plane-parallel layers of anisotropic materials, homogeneous or with continuously varying parameters [5, 7, 8]. The second is a general method of description of the interaction of polarized light with nondepolarizing linear optical systems [9]: an action of the optical system is described by a 2 × 2 matrix (**t**) relating the Jones vector of a wave incident on the system ($\mathbf{J}_{\mathrm{inc}}$) and the Jones vector of the wave emerging from the system that is considered as the result of this action with respect to the incident wave ($\mathbf{J}_{\mathrm{out}}$) as follows:

$$\mathbf{J}_{\mathrm{out}} = \mathbf{t}\mathbf{J}_{\mathrm{inc}}. \tag{1.208}$$

Jones matrices are adequate characteristics in any situation where waves incident on a system and emerging from it can be adequately represented by Jones vectors. For instance, in optics of stratified media, Jones matrices are commonly used to characterize transmission and reflection of such media, as transmission and reflection operators, including the case of oblique light incidence. If the incident and emergent waves are homogeneous and propagate in isotropic nonabsorbing media, they can be described by classical Cartesian Jones vectors. The description in terms of Jones vectors and Jones matrices is entirely consistent with electromagnetic theory. The rigorous methods discussed in Chapter 8 enable calculation of transmission and reflection Jones matrices of layered systems in strict accordance with this theory. The transmission and reflection Jones matrices for the interface between isotropic media in Section 1.2.2 are examples of exact Jones matrices.

In contrast to the matrix description [9], the Jones calculus (JC) is a semiempirical method and is limited to the case of normal incidence. This method was developed for calculating transmission characteristics of optical systems consisting of retarders and polarizers and other systems for which the transformation of the polarization state of the passing light by their elements is of paramount importance. In JC, the effect of an optical element of an optical system on a light beam is considered as a transformation of a plane wave incident on the element into a plane wave emerging from it and is characterized by a Jones matrix that relates Jones vectors of these waves. The action of an optical system consisting of two or more elements is considered as a chain of such transformations. JC is not strongly tied to electromagnetic theory and takes into account only basic functions of the elements and basic optical effects connected with performing these functions by the elements. In contrast to the electromagnetic methods where elements of an optical system are specified by their material parameters, in JC the elements are specified through description of their transfer characteristics which are specified using material parameters where it is convenient. When JC is used in solving optimization problems for finding optimal values of key parameters of polarization elements (such as orientation angles for polarizers, orientation angles and retardances for retarders, the twist angle and thickness for an LC layer), as a rule, the model system for direct analysis is composed of ideal elements such as an ideal polarizer, an ideal retarder, an ideal lossless LC cell, and so on.

JC has played and is playing a very important role in LCD optics. Many fundamental formulas in optics of liquid crystals and LCD optics were derived and many optimization problems were solved using this method. A lot of extremely useful and beautiful mathematics have appeared in polarization optics thanks to JC. That is why much attention in this book is given to this method and its applications in modeling and optimization of LCDs.

There are many points in JC that seem to be or really are inconsistent with the rigorous theory. On the other hand, based on the rigorous theory, one can prove that JC gives accurate results for many practical optical systems including LCDs. Starting from Maxwell's equations, using approximations that are fully justified in the context of electromagnetic theory, one may arrive at a technique which is mathematically (but not in every respect physically) equivalent to JC. This will be shown in Chapters 8, 11, and 12. The formal equivalence of JC and the more rigorous technique allows one to use the mathematical apparatus of JC, very rich and convenient, in the latter technique, or, what is practically the same, to use JC as it is but taking into account the amendments and refinements concerning the physical interpretation of some quantities and procedures involved in this method. Note that many helpful mathematical elements of JC are successfully used within the more rigorous method in considering both normal and oblique light incidence (see Chapter 11 and Section 12.4).

In this section, we consider some basic concepts of JC as well as some mathematical tricks useful when JC is applied to LCDs.

1.4.1 Basic Principles of the Jones Calculus

As has been said, in JC the action of an optical system is considered as a series of transformations to which the light is subjected as it passes through the system. Each of these elementary transformations is characterized by a Jones matrix. The Jones matrices are chosen in such a way that the output Jones vector for the Jones matrix describing the first or any intermediate transformation is the input Jones vector for the Jones matrix of the next transformation, which allows one to relate the Jones vector of the light incident on the system ($\mathbf{J}_{\rm inc}$) and that of the light emerging from the system ($\mathbf{J}_{\rm out}$) by the following chain of equations:

$$\mathbf{J}_1 = \mathbf{t}_1\mathbf{J}_{\rm inc},\ \mathbf{J}_2 = \mathbf{t}_2\mathbf{J}_1, \ldots,\ \mathbf{J}_{M-1} = \mathbf{t}_{M-1}\mathbf{J}_{M-2},\ \mathbf{J}_{\rm out} = \mathbf{t}_M\mathbf{J}_{M-1}, \tag{1.209}$$

where M is the number of the elementary transformations and $\mathbf{t}_j$ is the Jones matrix of the jth transformation ($j = 1,2,\ldots, M$). The substitutions of the expression for $\mathbf{J}_1$ in (1.209) (the first equation) into the second equation, of the obtained expression for $\mathbf{J}_2$ in terms of $\mathbf{J}_{\rm inc}$ into the third equation, and so on lead to the following relation:

$$\mathbf{J}_{\rm out} = \mathbf{t}_M\mathbf{t}_{M-1}\cdots\mathbf{t}_2\mathbf{t}_1\mathbf{J}_{\rm inc}. \tag{1.210}$$

Due to the associativity of the matrix product, this relation can be rewritten as

$$\mathbf{J}_{\rm out} = (\mathbf{t}_M\mathbf{t}_{M-1}\ldots\mathbf{t}_2\mathbf{t}_1)\mathbf{J}_{\rm inc} \tag{1.211}$$

or

$$\mathbf{J}_{\rm out} = \mathbf{t}_{\rm sys}\mathbf{J}_{\rm inc}, \tag{1.212}$$

where

$$\mathbf{t}_{\rm sys} = \mathbf{t}_M\mathbf{t}_{M-1}\ldots\mathbf{t}_2\mathbf{t}_1 \tag{1.213}$$

is the matrix that is regarded in JC as the Jones matrix of the system. Thus, the validity of (1.209) allows one to calculate the Jones matrix of the system by multiplying the Jones matrices of the elementary transformations in accordance with (1.213).

In Sections 1.3.4–1.3.6, we gave many expressions for Jones matrices of different optical elements, which can be used in such calculations. In all the cases considered in those sections, we assumed that the light incident on an element and the light emerging from the element propagate in isotropic media, so that we could legitimately use usual Cartesian Jones vectors to describe the waves regarded as the operand and the result of the transformation performed by the element. A peculiarity of the classical JC is that in any case the Jones matrix describing the transformation performed by an element is calculated as if the input and output media for this transformation (i.e., the medium from which the light falls on the element and the medium into which the transformed light passes leaving the element) were isotropic. Thus, for example, the transmission Jones matrix of a system consisting of two contiguous anisotropic layers is calculated as if there were an isotropic layer between the anisotropic layers but ignoring the effect of this intermediate isotropic layer on the passing light. It is clear that this approach is somewhat artificial. Some arguments for this approach from the standpoint of electromagnetic theory can be found in Chapter 12.

In principle, in considerations using the above algorithm, different kinds of Jones vectors (see Section 1.1.1) can be used. In the classical JC, the ordinary Jones vectors are assumed to be "fitted-to-intensity," the following relation between the Jones vector **J** and intensity I of a wave being adopted:

$$I = |\mathbf{J}|^2 \equiv \mathbf{J}^\dagger \mathbf{J}. \tag{1.214}$$

In the further consideration of JC and its applications, we will adhere to this convention and other prescriptions and principles of the classical variant of this method.

Standard Definition and Usual Representations of Transmittance in the Jones Calculus. Average Transmittance. "Unpolarized" Transmittance

The transmittance t of a device (a system or an element) is defined as

$$t \equiv I_{\text{out}}/I_{\text{inc}}, \tag{1.215}$$

where I_{inc} and I_{out} are the intensities of the light incident on the device and the light transmitted by the device, respectively. According to (1.214) and (1.215), the transmittance t can be expressed as

$$t = |\mathbf{J}_{\text{out}}|^2/|\mathbf{J}_{\text{inc}}|^2, \tag{1.216}$$

where $\mathbf{J}_{\text{inc}}$ and $\mathbf{J}_{\text{out}}$ are the Jones vectors of the incident light and transmitted light, respectively.

The substitution of the expression

$$\mathbf{J}_{\text{out}} = \mathbf{t}\mathbf{J}_{\text{inc}}, \tag{1.217}$$

where $\mathbf{t}$ is the Jones matrix of the device, into (1.216) gives the following expression for t:

$$t = |\mathbf{t}\mathbf{J}_{\text{inc}}|^2/|\mathbf{J}_{\text{inc}}|^2. \tag{1.218}$$

Yet another standard expression for the transmittance is

$$t = |\mathbf{t}\boldsymbol{j}_{\text{inc}}|^2, \tag{1.219}$$

where j_{inc} is the polarization Jones vector of the incident light ($|j_{inc}| = 1$). We have dealt with expressions of this kind in the previous sections. The product $\mathbf{t}\,j_{inc}$ is a normalized Jones vector whose squared norm is equal to the transmittance t.

Let $t_1 = |\mathbf{t}\,j_1|^2$ and $t_2 = |\mathbf{t}\,j_2|^2$ be the values of the transmittance of the device for two arbitrary mutually orthogonal polarizations of the incident light, described by polarization Jones vectors j_1 and j_2 ($j_2^\dagger j_1 = 0$). It is easy to verify that the magnitude of the *average transmittance* of the device defined as $t_{avr} = (t_1 + t_2)/2$ is independent of the choice of the pair of incident orthogonal polarizations and

$$t_{avr} = \frac{1}{2}\left(t_{11}^* t_{11} + t_{12}^* t_{12} + t_{21}^* t_{21} + t_{22}^* t_{22}\right) = \frac{1}{2}\,\|\mathbf{t}\|_E^2\,, \tag{1.220}$$

where t_{jk} are elements of the matrix $\mathbf{t}$ and $\|\mathbf{t}\|_E$ is the Euclidean norm of $\mathbf{t}$ (see Section 5.1.4). The *transmittance* of the device *for quasimonochromatic unpolarized incident light*, t_{unp}, according to prescriptions of JC, is calculated as t_{avr} in (1.220), that is, by the formula

$$t_{unp} = \frac{1}{2}\left(t_{11}^* t_{11} + t_{12}^* t_{12} + t_{21}^* t_{21} + t_{22}^* t_{22}\right). \tag{1.221}$$

The unpolarized quasimonochromatic incident wave can be represented as a superposition of two mutually incoherent quasimonochromatic orthogonally polarized waves of equal intensity, with polarization Jones vectors j_1 and j_2. Denoting the transmittances of the device for these polarized constituents as t_1 and t_2, we may express t_{unp} as $t_{unp} = (t_1 + t_2)/2$. Then the assumption that the transmittances t_1 and t_2 can be calculated as $t_1 = |\mathbf{t}\,j_1|^2$ and $t_2 = |\mathbf{t}\,j_2|^2$, that is, just as in the case of monochromatic waves, leads us to (1.221).

Lossless Transformations and Transformations Without Diattenuation

Solving many problems is significantly simplified by using specific mathematical properties of Jones matrices describing certain kinds of transformations. Here we consider two important classes of transformations. One of them is the class of transformations for which the output light intensity is equal to the input light intensity whatever the SOP of the incident light. Such transformations are called *lossless*. Definition (1.214) of intensity determines that the Jones matrix describing such a transformation is a unitary matrix (see Section 5.1.3). Actually, let $\mathbf{t}$ be the Jones matrix of an operation, and let $\mathbf{J}_{inc}$ and $\mathbf{J}_{out} = \mathbf{t}\mathbf{J}_{inc}$ be the Jones vectors of the incident and output waves for this operation. Then, according to (1.214), the condition of equality of intensities of the incident and output waves can be written as

$$\mathbf{J}_{out}^\dagger \mathbf{J}_{out} = \mathbf{J}_{inc}^\dagger \mathbf{J}_{inc} \tag{1.222}$$

or

$$(\mathbf{t}\mathbf{J}_{inc})^\dagger \mathbf{t}\mathbf{J}_{inc} = \mathbf{J}_{inc}^\dagger \mathbf{J}_{inc}. \tag{1.223}$$

Using the identity $(\mathbf{t}\mathbf{J}_{inc})^\dagger = \mathbf{J}_{inc}{}^\dagger \mathbf{t}^\dagger$ [see (5.15)], we can rewrite (1.223) as follows:

$$\mathbf{J}_{inc}^\dagger (\mathbf{t}^\dagger \mathbf{t}) \mathbf{J}_{inc} = \mathbf{J}_{inc}^\dagger \mathbf{J}_{inc}. \tag{1.224}$$

This relation holds at any $\mathbf{J}_{inc}$ only if

$$\mathbf{t}^\dagger \mathbf{t} = \mathbf{U}, \tag{1.225}$$

where $\mathbf{U}$ is the unit matrix. A square matrix $\mathbf{A}$ satisfying the condition $\mathbf{A}^{\dagger}\mathbf{A} = \mathbf{U}$ is called *unitary*. A summary of properties of unitary matrices is given in Section 5.1.3. Devices that are assumed to perform lossless transformations are often called *lossless* or *unitary*.

Lossless transformations belong to the class of *transformations without diattenuation* (i.e., without polarization-dependent losses). A transformation can be called a transformation without diattenuation if the ratio of the output light intensity to the input light intensity is independent of the SOP of the incident light. This determining condition implies that at any $\mathbf{J}_{\text{inc}}$,

$$\mathbf{J}_{\text{inc}}^{\dagger}(\mathbf{t}^{\dagger}\mathbf{t})\mathbf{J}_{\text{inc}} = t_l \mathbf{J}_{\text{inc}}^{\dagger}\mathbf{J}_{\text{inc}}, \tag{1.226}$$

where $\mathbf{t}$ is the Jones matrix of the transformation, t_l is a real constant independent of $\mathbf{J}_{\text{inc}}$. In the presence of losses, $t_l < 1$. The transmittance t [see (1.216)] associated with this transformation in any case is equal to t_l. Relation (1.226) will hold at any $\mathbf{J}_{\text{inc}}$ only if

$$\mathbf{t}^{\dagger}\mathbf{t} = t_l \mathbf{U}. \tag{1.227}$$

Any matrix satisfying (1.227) can be represented as $\mathbf{t} = \varsigma \mathbf{t}_{\text{U}}$, where $\mathbf{t}_{\text{U}}$ is a unitary matrix and ς is a scalar factor such that $\varsigma\varsigma^* = t_l$. In this book, such matrices are referred to as STU matrices (see Section 5.1.3).

A chain of lossless transformations is a lossless transformation. The product of unitary matrices is always a unitary matrix. A chain of transformations without diattenuation is a transformation without diattenuation. The product of STU matrices is always an STU matrix.

An interesting feature of transformations without diattenuation is that under such transformations orthogonally polarized waves are converted into orthogonally polarized ones: if $\mathbf{t}$ is an STU matrix and $\mathbf{J}_{\text{inc1}}$ and $\mathbf{J}_{\text{inc2}}$ are arbitrary mutually orthogonal Jones vectors ($\mathbf{J}_{\text{inc1}}{}^{\dagger}\mathbf{J}_{\text{inc2}} = 0$), the vectors $\mathbf{J}_{\text{out1}} = \mathbf{t}\mathbf{J}_{\text{inc1}}$ and $\mathbf{J}_{\text{out2}} = \mathbf{t}\mathbf{J}_{\text{inc2}}$ will be also mutually orthogonal ($\mathbf{J}_{\text{out1}}{}^{\dagger}\mathbf{J}_{\text{out2}} = 0$) (see Section 5.1.3). This feature explains the following well-known property of transmissive devices (layers or layered systems) without diattenuation. If such a device is placed between linear polarizers (ideal or with zero transmittance for the unwanted polarization), the transmittance of the polarizer–device–polarizer system is invariant under rotations of the device about the axis of light propagation by 90°. Actually, due to the mentioned feature of transformations without diattenuation, such a rotation changes only the handedness of the polarization ellipse of the light emerging from the device. The transmittance of a linear polarizer is independent of the handedness of the polarization of light incident on it. Therefore, the intensity of the light transmitted by the second polarizer will remain unchanged after the rotation of the device.

Many practical optical elements and systems whose purpose is to convert the SOP of light with minimal losses (wave plates, polarization rotators, LC layers in most kinds of LCDs, compensation systems in LCDs, etc.) can be considered to a good approximation as devices that transmit light, at normal incidence, without diattenuation.

Idealized Systems in the Jones Calculus. Unitary Systems

As a rule, the object for JC is an idealized system whose transmittance multiplied by a certain attenuation factor is considered to be equal to the transmittance of a real (realistic) lossy system of interest. The attenuation factor may take account of absorption losses in isotropic layers of the lossy system, reflection losses, and some other kinds of losses. Almost always, the losses on the polarization-converting elements that are considered to perform transformations without diattenuation are taken into account in the attenuation factor, so that these elements are represented in the idealized system by lossless elements. An idealized system consisting of only lossless elements is clearly lossless. Such systems are called *unitary systems*. Representing the Jones matrix of a realistic lossy system with negligible diattenuation in the form $\varsigma\mathbf{t}_{\text{U}}$, where $\mathbf{t}_{\text{U}}$ is a unitary matrix and ς is a scalar, we can use the matrix $\mathbf{t}_{\text{U}}$ as operator relating polarization Jones vectors of the incident and emergent waves (we have used this in Section 1.3.5). If

$\mathbf{t}_U$ is the Jones matrix of a unitary system associated with the lossy system, we may regard this unitary system as a model system that transforms polarization in the same manner as the realistic lossy system. The concept of a unitary system is widely used in LCD optics (see Chapters 2, 3, 6, and 12).

The typical idealized optical system for JC is a sequence of elements each of which is able to convert the polarization state of light. The effect of spaces between the elements is usually disregarded, because, as a rule, there is no need to trace the changes in the absolute phase of the passing light.

1.4.2 Three Useful Theorems for Transmissive Systems

The usual model of an inhomogeneous LC layer is a pile of homogeneous birefringent layers (see Sections 2.1 and 11.1.1). The standard idealized model of a transmissive LCD to treat by means of JC is also a pile of homogeneous anisotropic layers. In this section, we present three theorems showing how the transmission Jones matrix of such a system changes under certain transformations of the system. Applied to inhomogeneous LC layers, these theorems are useful when there is a need to compare the optical properties of similar layers whose structures (LC director fields) are mapped into each other by a rotation, a reflection, or the inversion (see, e.g., [10]). For systems invariant under any of the transformations considered here, by using these theorems, it is easy to find restrictions imposed by this invariance on the Jones matrices of these systems. Knowledge of such restrictions simplifies solving some optimization problems for LCDs (see Chapter 6).

Consider a system S consisting of N simple birefringent layers (Figure 1.11) (say, a system of linear polarizers and linear retarders) whose boundaries are perpendicular to an axis z. The effect of spaces between the layers will be ignored here. Let the elements of the system (birefringent layers) be numbered as shown in Figure 1.11, and let a light wave $\vec{\mathbf{X}}_i$ propagating in the positive z direction be incident on the system (Figure 1.11a). We can calculate the Jones matrix of the system,

$$\vec{t}_S \equiv \begin{pmatrix} \vec{t}_{S11} & \vec{t}_{S12} \\ \vec{t}_{S21} & \vec{t}_{S22} \end{pmatrix}, \tag{1.228}$$

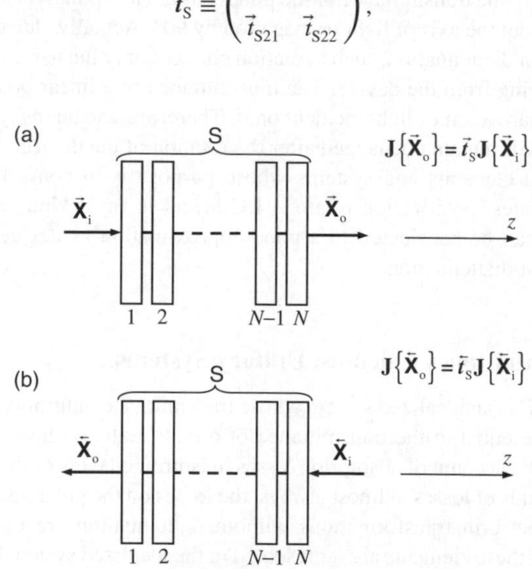

Figure 1.11 A transmissive system of birefringent layers. $\mathbf{J}\{\mathbf{X}\}$ stands for the Jones vector of a wave **X**. Parts (a) and (b) show the two cases compared in Jones's reversibility theorem

as

$$\vec{t}_{\mathrm{S}} = \vec{t}_N \vec{t}_{N-1} \ldots \vec{t}_2 \vec{t}_1, \tag{1.229}$$

where $\vec{t}_j$ is a Jones matrix of layer number j, if the input reference frame[3] of the matrix $\vec{t}_j$ (at j=2, … , N) is the same as the output reference frame of the matrix $\vec{t}_{j-1}$. This condition will be satisfied if we use a fixed frame as the input and output one for all the matrices $\vec{t}_j$. Take the frame (x_{I}, y_{I}) of a rectangular right-handed Cartesian system (x_{I}, y_{I}, z_{I}) with the z_{I}-axis codirectional with the z-axis as such a fixed frame. Let (x'_j, y'_j) be a frame whose axes are parallel to the principal axes of the jth layer, and let

$$t'_j = \begin{pmatrix} t_{xj} & 0 \\ 0 & t_{yj} \end{pmatrix} \tag{1.230}$$

be the transmission Jones matrix of the jth layer for Jones vectors referred to the frame (x'_j, y'_j). Then the matrices $\vec{t}_j$ can be represented as

$$\vec{t}_j = \widehat{R}_{\mathrm{C}}(-\phi_j) t'_j \widehat{R}_{\mathrm{C}}(\phi_j), \tag{1.231}$$

where ϕ_j is the angle between the axes x_{I} and x'_j.

Note that at any ϕ_j, the matrix $\vec{t}_j$ is symmetric, that is,

$$\vec{t}_j = \vec{t}_j^{\,\mathrm{T}}. \tag{1.232}$$

Actually, according to (1.231),

$$\vec{t}_j^{\,\mathrm{T}} = \left(\widehat{R}_{\mathrm{C}}(-\phi_j) t'_j \widehat{R}_{\mathrm{C}}(\phi_j)\right)^{\mathrm{T}}. \tag{1.233}$$

Using matrix identity (5.14), the relation $\widehat{R}_{\mathrm{C}}(\phi)^{\mathrm{T}} = \widehat{R}_{\mathrm{C}}(-\phi)$, and the fact that $t_j'^{\mathrm{T}} = t'_j$, we can rewrite this expression as

$$\vec{t}_j^{\,\mathrm{T}} = \widehat{R}_{\mathrm{C}}(\phi_j)^{\mathrm{T}} t_j'^{\mathrm{T}} \widehat{R}_{\mathrm{C}}(-\phi_j)^{\mathrm{T}} = \widehat{R}_{\mathrm{C}}(-\phi_j) t'_j \widehat{R}_{\mathrm{C}}(\phi_j). \tag{1.234}$$

Comparing (1.234) and (1.231), we see that $\vec{t}_j^{\,\mathrm{T}} = \vec{t}_j$.

Theorem 1.1 The Jones matrix $\vec{t}_{\mathrm{S}'}$ of a system S′ that can be obtained from the system S by the permutation of the elements that provides the inverse order of the elements and, possibly, by rotating some elements by 180° about the z-axis is related to the Jones matrix of the system S as follows:

$$\vec{t}_{\mathrm{S}'} = \vec{t}_{\mathrm{S}}^{\,\mathrm{T}}. \tag{1.235}$$

[3] Considering a Jones matrix, we will call the reference frames to which the input and output Jones vectors for this matrix are referred respectively the *input frame* and *output frame* of this Jones matrix. A frame that is used as both the input one and the output one for a Jones matrix will be called *the input and output frame*.

Proof. The rotation of any element about the z-axis by 180° does not change the Jones matrix of this element. Therefore, in any case, the matrix $\vec{t}_{S'}$ can be expressed in terms of the Jones matrices of the elements of the system S as

$$\vec{t}_{S'} = \vec{t}_1\vec{t}_2 \ldots \vec{t}_{N-1}\vec{t}_N. \tag{1.236}$$

Using the fact that all the matrices $\vec{t}_j$ are symmetric and identity (5.14), we can transform this expression as follows:

$$\vec{t}_{S'} = \vec{t}_1^{\,\mathrm{T}}\vec{t}_2^{\,\mathrm{T}} \ldots \vec{t}_{N-1}^{\,\mathrm{T}}\vec{t}_N^{\,\mathrm{T}} = \left(\vec{t}_N\vec{t}_{N-1} \ldots \vec{t}_2\vec{t}_1\right)^{\mathrm{T}}. \tag{1.237}$$

As is seen from (1.237) and (1.229), the matrix $\vec{t}_{S'}$ is really equal to $\vec{t}_S^{\,\mathrm{T}}$.

If the system S is such that $\vec{t}_N = \vec{t}_1$, $\vec{t}_{N-1} = \vec{t}_2$, and so on, the inversion of the order of its elements will give a system whose Jones matrix is equal to $\vec{t}_S$. It follows from Theorem 1.1 that the matrix $\vec{t}_S$ in this case satisfies the condition $\vec{t}_S = \vec{t}_S^{\,\mathrm{T}}$, that is, it is symmetric.

Going to the next theorem, denote the values of the azimuthal angles ϕ_j and matrices $\vec{t}_j$ (j=1,2, … , N) for the system S by $\phi_j^{(S)}$ and $\vec{t}_j^{(S)}$ respectively. With this notation, the matrix $\vec{t}_S$ is expressed as

$$\vec{t}_S = \vec{t}_N^{(S)}\vec{t}_{N-1}^{(S)} \ldots \vec{t}_2^{(S)}\vec{t}_1^{(S)}, \tag{1.238}$$

where

$$\vec{t}_j^{(S)} = \hat{R}_C\left(-\phi_j^{(S)}\right)t'_j\hat{R}_C\left(\phi_j^{(S)}\right). \tag{1.239}$$

Theorem 1.2 Suppose that a system S′ consists of the same layers as the system S and their order is the same as in S, but the layers are rotated about the z-axis so that for the jth layer (j = 1,2, … , N) the angle ϕ_j is equal to $-\phi_j^{(S)}$ or $-\phi_j^{(S)} + 180°$. Then the Jones matrices of the systems S′ and S are related by

$$\vec{t}_{S'} = \mathbf{I}_1\vec{t}_S\mathbf{I}_1, \tag{1.240}$$

where

$$\mathbf{I}_1 = \begin{pmatrix} 1 & 0 \\ 0 & -1 \end{pmatrix}. \tag{1.241}$$

Note that $\mathbf{I}_1\mathbf{I}_1 = \mathbf{U}$, where, as before, $\mathbf{U}$ is the unit matrix, that is, $\mathbf{I}_1^{-1} = \mathbf{I}_1$. According to (1.240),

$$\vec{t}_{S'} = \begin{pmatrix} \vec{t}_{S11} & -\vec{t}_{S12} \\ -\vec{t}_{S21} & \vec{t}_{S22} \end{pmatrix}.$$

Proof. The Jones matrix $\vec{t}_j^{(S')}$ of the jth layer of the system S′ for Jones vectors referred to the frame (x_I, y_I), whether ϕ_j for this layer be equal to $-\phi_j^{(S)}$ or $-\phi_j^{(S)} + 180°$, can be expressed as follows:

$$\vec{t}_j^{(S')} = \hat{R}_C\left(\phi_j^{(S)}\right)t'_j\hat{R}_C\left(-\phi_j^{(S)}\right). \tag{1.242}$$

It is easy to check that, at any ϕ, $\widehat{R}_{\rm C}(\phi) = \mathbf{I}_1 \widehat{R}_{\rm C}(-\phi)\mathbf{I}_1$. Using this relation, we can rewrite (1.242) as

$$\vec{t}_j^{(\mathrm{S}')} = \mathbf{I}_1 \widehat{R}_{\rm C}\left(-\phi_j^{(\mathrm{S})}\right)\mathbf{I}_1 t'_j \mathbf{I}_1 \widehat{R}_{\rm C}\left(\phi_j^{(\mathrm{S})}\right)\mathbf{I}_1. \tag{1.243}$$

Since t'_j is a diagonal matrix, $\mathbf{I}_1 t'_j \mathbf{I}_1 = t'_j$. Consequently, from (1.243) we have

$$\vec{t}_j^{(\mathrm{S}')} = \mathbf{I}_1 \widehat{R}_{\rm C}\left(-\phi_j^{(\mathrm{S})}\right) t'_j \widehat{R}_{\rm C}\left(\phi_j^{(\mathrm{S})}\right)\mathbf{I}_1. \tag{1.244}$$

From (1.244) and (1.239), we see that

$$\vec{t}_j^{(\mathrm{S}')} = \mathbf{I}_1 \vec{t}_j^{(\mathrm{S})} \mathbf{I}_1. \tag{1.245}$$

In the case under consideration, the matrix $\vec{t}_{\mathrm{S}'}$ is expressed in terms of the matrices $\vec{t}_j^{(\mathrm{S}')}$ as follows:

$$\vec{t}_{\mathrm{S}'} = \vec{t}_N^{(\mathrm{S}')} \vec{t}_{N-1}^{(\mathrm{S}')} \ldots \vec{t}_2^{(\mathrm{S}')} \vec{t}_1^{(\mathrm{S}')}. \tag{1.246}$$

On substituting from (1.245) into (1.246), we obtain

$$\vec{t}_{\mathrm{S}'} = \mathbf{I}_1 \vec{t}_N^{(\mathrm{S})} \mathbf{I}_1 \mathbf{I}_1 \vec{t}_{N-1}^{(\mathrm{S})} \mathbf{I}_1 \ldots \mathbf{I}_1 \vec{t}_2^{(\mathrm{S})} \mathbf{I}_1 \mathbf{I}_1 \vec{t}_1^{(\mathrm{S})} \mathbf{I}_1 = \mathbf{I}_1 \left(\vec{t}_N^{(\mathrm{S})} \vec{t}_{N-1}^{(\mathrm{S})} \ldots \vec{t}_2^{(\mathrm{S})} \vec{t}_1^{(\mathrm{S})}\right) \mathbf{I}_1, \tag{1.247}$$

where we have made use of the property $\mathbf{I}_1 \mathbf{I}_1 = \mathbf{U}$.

Theorem 1.3 Suppose that a system S″ differs from a system S′ that satisfies the conditions of the previous theorem only in that it has the inverse order of elements, and, consequently, the Jones matrix of the system S″, $\vec{t}_{\mathrm{S}''}$, can be expressed in terms of the matrices $\vec{t}_j^{(\mathrm{S}')}$ as follows:

$$\vec{t}_{\mathrm{S}''} = \vec{t}_1^{(\mathrm{S}')} \vec{t}_2^{(\mathrm{S}')} \ldots \vec{t}_{N-1}^{(\mathrm{S}')} \vec{t}_N^{(\mathrm{S}')}. \tag{1.248}$$

Then the matrix $\vec{t}_{\mathrm{S}''}$ is related to the Jones matrix $\vec{t}_{\mathrm{S}}$ of the system S by

$$\vec{t}_{\mathrm{S}''} = \mathbf{I}_1 \vec{t}_{\mathrm{S}}^{\mathrm{T}} \mathbf{I}_1. \tag{1.249}$$

According to (1.249),

$$\vec{t}_{\mathrm{S}''} = \begin{pmatrix} \vec{t}_{\mathrm{S}11} & -\vec{t}_{\mathrm{S}21} \\ -\vec{t}_{\mathrm{S}12} & \vec{t}_{\mathrm{S}22} \end{pmatrix}. \tag{1.250}$$

Proof. By Theorem 1.1, $\vec{t}_{\mathrm{S}''} = \vec{t}_{\mathrm{S}'}^{\mathrm{T}}$. According to Theorem 1.2, $\vec{t}_{\mathrm{S}'} = \mathbf{I}_1 \vec{t}_{\mathrm{S}} \mathbf{I}_1$. Therefore,

$$\vec{t}_{\mathrm{S}''} = \left(\mathbf{I}_1 \vec{t}_{\mathrm{S}} \mathbf{I}_1\right)^{\mathrm{T}} = \mathbf{I}_1^{\mathrm{T}} \vec{t}_{\mathrm{S}}^{\mathrm{T}} \mathbf{I}_1^{\mathrm{T}} = \mathbf{I}_1 \vec{t}_{\mathrm{S}}^{\mathrm{T}} \mathbf{I}_1.$$

Note that a system S″ satisfying the conditions of Theorem 1.3 can be obtained by the rotation of the system S by 180° about an axis parallel to the $x_{\rm I}$-axis. Thus, Theorem 1.3 makes clear how the Jones matrix of a system of birefringent layers is transformed under such a rotation. Starting from Theorem 1.3, by means of standard basis transformations, it is easy to find the rule of transformation of the Jones matrix of such a system under the 180° rotation of this system about a given axis perpendicular to the light propagation direction for the case of an arbitrary orientation of this axis with respect to the axes of the reference frame for the Jones matrix.

If the rotation of the system S by 180° about an axis parallel to the x_I-axis maps the system S into itself, that is, yields a system that is equivalent to S in its initial state, then, according to Theorem 1.3,

$$\vec{t}_S = \mathbf{I}_1 \vec{t}_S^{\,T} \mathbf{I}_1, \tag{1.251}$$

which implies the following form of the matrix $\vec{t}_S$:

$$\vec{t}_S = \begin{pmatrix} \vec{t}_{S11} & \vec{t}_{S12} \\ -\vec{t}_{S12} & \vec{t}_{S22} \end{pmatrix}. \tag{1.252}$$

Applying this conclusion to the standard model of an inhomogeneous LC layer as a pile of homogeneous uniaxial layers with a varying, from layer to layer, orientation of the optic axis (see Section 11.1.1), one can readily show that the transmission Jones matrix of an LC layer that is invariant with respect to the 180° rotation about an axis parallel to the layer boundaries (this kind of symmetry is typical of LC layers of practical LCDs, see Figure 6.7 and Section 6.2.3) has the form (1.252) if the axis x_I of a reference frame (x_I, y_I) which is used as the input and output one for this Jones matrix is parallel to the symmetry axis (axis C_2 in Figure 6.7).

Certainly, the matrix $\vec{t}_S$ has the form (1.252) not only when the system S is symmetrical in the mentioned sense. For any variant of S for which $\mathbf{t}'_{N-j+1} = \mathbf{t}'_j$ and $\phi^{(S)}_{N-j+1}$ is equal to $-\phi_j^{(S)}$ or $-\phi_j^{(S)} \pm 180°$ ($j = 1,2, \ldots, N$), the matrix $\vec{t}_S$ will be of the form (1.252).

For completeness, we must also mention here the following obvious relation. If a system S′ is composed of the same elements as the system S, arranged in the same order, but these elements are rotated about the z-axis so that for them the angles ϕ_j are equal to $\phi_j^{(S)} + \alpha_R$ or $\phi_j^{(S)} + \alpha_R + 180°$, where α_R is a fixed angle, the matrices $\vec{t}_{S'}$ and $\vec{t}_S$ are related by

$$\vec{t}_{S'} = \hat{R}_C(-\alpha_R)\vec{t}_S \hat{R}_C(\alpha_R). \tag{1.253}$$

1.4.3 Reciprocity Relations. Jones's Reversibility Theorem

In the previous section, we supposed that light is incident on the system S in the positive direction of the z-axis. Denote the transmission Jones matrix of this system for light incident on this system from the other side in the opposite direction (Figure 1.11b) by $\overleftarrow{t}_S$. Using Theorem 1.3 of the previous section, we can easily determine the relation between the matrices $\vec{t}_S$ and $\overleftarrow{t}_S$. Let a system S″ be identical to the system S rotated by 180° about an axis parallel to the x_I-axis and let a coordinate system (x_R, y_R, z_R) whose frame (x_R, y_R) is used as the input and output basis of the matrix $\overleftarrow{t}_S$ be identical to the system (x_I, y_I, z_I) rotated by 180° about the x_I-axis (Figure 1.12a). With this choice of the frame (x_R, y_R) the matrix $\overleftarrow{t}_S$ is obviously equal to the matrix $\vec{t}_{S''}$ which is referred to the frame (x_I, y_I). Since the system S″ satisfies the conditions of Theorem 1.3, $\vec{t}_{S''} = \mathbf{I}_1 \vec{t}_S^{\,T} \mathbf{I}_1$ and, consequently,

$$\overleftarrow{t}_S = \mathbf{I}_1 \vec{t}_S^{\,T} \mathbf{I}_1.$$

Considering three choices of the basis (x_R, y_R, z_R) that are shown in Figure 1.12 and named C1, C2, and C3, the relationship between the matrices $\vec{t}_S$ and $\overleftarrow{t}_S$ can be expressed as follows:

$$\overleftarrow{t}_S = \mathbf{U}_r \vec{t}_S^{\,T} \mathbf{U}_r, \tag{1.254}$$

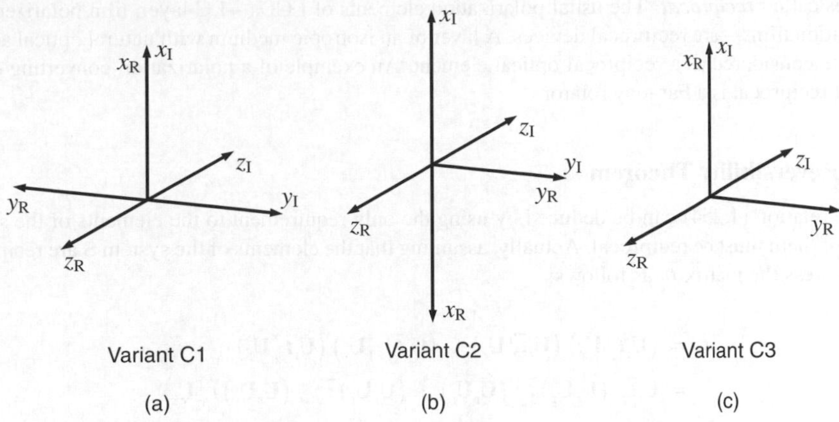

Figure 1.12 Three choices of reference frames for the Jones vectors of waves propagating in opposite directions. The axes z_I and z_R indicate the propagation directions of the waves. The system (x_I, y_I, z_I) is right-handed. The system (x_R, y_R, z_R) is right-handed in the cases C1 and C2 and left-handed in the case C3

where

$$\mathbf{U}_r = \begin{cases} \begin{pmatrix} 1 & 0 \\ 0 & -1 \end{pmatrix} & \text{in the case C1} \\ \begin{pmatrix} -1 & 0 \\ 0 & 1 \end{pmatrix} & \text{in the case C2} \\ \begin{pmatrix} 1 & 0 \\ 0 & 1 \end{pmatrix} & \text{in the case C3.} \end{cases} \quad (1.255)$$

In the cases C1 and C2, the matrix $\overleftarrow{t}_S$ is expressed in terms of the elements of $\vec{t}_S$ as

$$\overleftarrow{t}_S = \begin{pmatrix} \vec{t}_{S11} & -\vec{t}_{S21} \\ -\vec{t}_{S12} & \vec{t}_{S22} \end{pmatrix}. \quad (1.256)$$

In the case C3,

$$\overleftarrow{t}_S = \begin{pmatrix} \vec{t}_{S11} & \vec{t}_{S21} \\ \vec{t}_{S12} & \vec{t}_{S22} \end{pmatrix}. \quad (1.257)$$

Equations that relate a transfer characteristic of an optical system to the characteristic of the same kind but for the reverse passage of light through the system, such as (1.254), are usually called *reciprocity relations*.

In the literature, reciprocity relations for Jones matrices of polarization devices are often written in the form $\overleftarrow{t} = \vec{t}^{\,T}$ (see, e.g., [11]) and correspond to the situation when the input reference frame for the matrix $\vec{t}$ is the same (geometrically) as the output reference frame for the matrix $\overleftarrow{t}$ and vice versa (as is the case in the above example for the variant C3). Devices for which such a reciprocity relation holds are sometimes called *reciprocal*. The usual polarization elements of LCDs—LC layer, film polarizers, and compensation films—are reciprocal devices. A layer of an isotropic medium with natural optical activity can also be considered as a reciprocal optical element. An example of a polarization-converting device that is not reciprocal is a Faraday rotator.

Jones's Reversibility Theorem

Certainly, relation (1.254) can be deduced by using the only requirement to the elements of the system S—each of them must be reciprocal. Actually, assuming that the elements of the system S are reciprocal, we can express the matrix $\overleftarrow{t}_S$ as follows:

$$\begin{aligned}\overleftarrow{t}_S &= (\mathbf{U}_r\vec{t}_1^{\,T}\mathbf{U}_r)(\mathbf{U}_r\vec{t}_2^{\,T}\mathbf{U}_r)\ldots(\mathbf{U}_r\vec{t}_{N-1}^{\,T}\mathbf{U}_r)(\mathbf{U}_r\vec{t}_N^{\,T}\mathbf{U}_r)\\ &= \mathbf{U}_r\vec{t}_1^{\,T}(\mathbf{U}_r\mathbf{U}_r)\vec{t}_2^{\,T}(\mathbf{U}_r\mathbf{U}_r)\ldots(\mathbf{U}_r\mathbf{U}_r)\vec{t}_{N-1}^{\,T}(\mathbf{U}_r\mathbf{U}_r)\vec{t}_N^{\,T}\mathbf{U}_r,\end{aligned}$$

where the product $\mathbf{U}_r\vec{t}_j^{\,T}\mathbf{U}_r$ represents the Jones matrix of the jth element for the reverse direction of light propagation. For all the three variants of $\mathbf{U}_r$, $\mathbf{U}_r\mathbf{U}_r = \mathbf{U}$. Consequently,

$$\overleftarrow{t}_S = \mathbf{U}_r\vec{t}_1^{\,T}\vec{t}_2^{\,T}\ldots\vec{t}_{N-1}^{\,T}\vec{t}_N^{\,T}\mathbf{U}_r = \mathbf{U}_r\left(\vec{t}_N\vec{t}_{N-1}\ldots\vec{t}_2\vec{t}_1\right)^T\mathbf{U}_r = \mathbf{U}_r\vec{t}_S^{\,T}\mathbf{U}_r,$$

which shows that the system S is reciprocal. The statement that a system composed of reciprocal elements is reciprocal expresses the essence of *Jones's reversibility theorem* [5, 11].

In Section 8.6.2, we consider analogous reciprocity relations of the rigorous electromagnetic theory of light propagation in stratified media.

The reciprocity relations for Jones matrices are used, for example, in calculations for reflective devices, and in particular RLCDs (see, e.g., [12]).

Application to Reflective Devices

Consider a reflective device consisting of a transmissive system S and a specular reflector (mirror) R which reflects the light transmitted by the system S back to S (Figure 1.13). Denote the transmission Jones matrices of the system S for the propagation directions toward the reflector and from it by $\vec{t}_S$ and $\overleftarrow{t}_S$, respectively. The Jones matrix describing reflection from the mirror will be denoted by r_R. Let a frame (x_I, y_I), chosen as in the above consideration, be used as the input and output reference frame for the matrix $\vec{t}_S$ and the input reference frame for the matrix r_R, and let a reference frame (x_R, y_R) be used as the input and output one for the matrix $\overleftarrow{t}_S$ and the output one for the matrix r_R. Then we can express

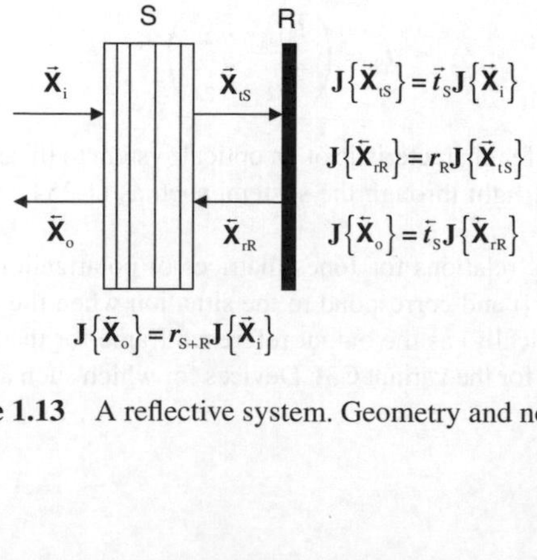

Figure 1.13 A reflective system. Geometry and notation

the Jones matrix $\boldsymbol{r}_{S+R}$ that relates the Jones vector of the wave $\vec{\mathbf{X}}_i$ incident on the reflective system and that of the wave $\overleftarrow{\mathbf{X}}_o$ emerging from this system (see Figure 1.13) as follows:

$$\boldsymbol{r}_{S+R} = \overleftarrow{\boldsymbol{t}}_S \boldsymbol{r}_R \vec{\boldsymbol{t}}_S. \tag{1.258}$$

For the three variants of the frame (x_R, y_R) shown in Figure 1.12, the Jones matrix of the reflector can be represented as

$$\boldsymbol{r}_R = r_R \mathbf{U}_r, \tag{1.259}$$

where $r_R = \sqrt{R_R}$ with R_R being the reflectivity of the reflector, and $\mathbf{U}_r$, as before, is the matrix defined by (1.255). In the case of an ideal lossless reflector, one can take

$$\boldsymbol{r}_R = \mathbf{U}_r. \tag{1.260}$$

Using the reciprocity relation $\overleftarrow{\boldsymbol{t}}_S = \mathbf{U}_r \vec{\boldsymbol{t}}_S^{\mathrm{T}} \mathbf{U}_r$ and (1.259), we can modify expression (1.258) as follows:

$$\boldsymbol{r}_{S+R} = \overleftarrow{\boldsymbol{t}}_S \boldsymbol{r}_R \vec{\boldsymbol{t}}_S = \left(\mathbf{U}_r \vec{\boldsymbol{t}}_S^{\mathrm{T}} \mathbf{U}_r\right)\left(r_R \mathbf{U}_r\right) \vec{\boldsymbol{t}}_S = r_R \mathbf{U}_r \left(\vec{\boldsymbol{t}}_S^{\mathrm{T}} \vec{\boldsymbol{t}}_S\right). \tag{1.261}$$

Thus, one can compute the matrix $\boldsymbol{r}_{S+R}$ without dealing with the matrix $\overleftarrow{\boldsymbol{t}}_S$. Note that the matrix $\vec{\boldsymbol{t}}_S^{\mathrm{T}} \vec{\boldsymbol{t}}_S$ is symmetric, as is the matrix $\boldsymbol{r}_{S+R}$ in the case C3. In the cases C1 and C2, the off-diagonal elements of $\boldsymbol{r}_{S+R}$ are equal but opposite in sign.

The following theorem is also useful in considering RLCDs.

1.4.4 Theorem of Polarization Reversibility for Systems Without Diattenuation

Let $\mathbf{X}_d$ and $\mathbf{X}_r$ be plane monochromatic waves of the same frequency propagating in an isotropic medium in opposite directions. We will say that the polarization of the wave $\mathbf{X}_r$ is *reverse* with respect to the polarization of the wave $\mathbf{X}_d$, or that the waves $\mathbf{X}_d$ and $\mathbf{X}_r$ are *reversely polarized,* if the shape and orientation of the polarization ellipses of these waves are identical, but these ellipses are described in opposite senses (Figure 1.14). Note that the handedness of the polarization ellipses of waves with mutually reverse polarizations is the same (recall that oppositely propagating waves are compared here). For example, the waves $\mathbf{X}_d$ and $\mathbf{X}_r$ can be called reversely polarized if they both have the right circular polarization or left circular polarization. If the waves $\mathbf{X}_d$ and $\mathbf{X}_r$ are linearly polarized and have the same polarization plane, they can also be called reversely polarized. If waves $\mathbf{X}_d$ and $\mathbf{X}_r$ have mutually

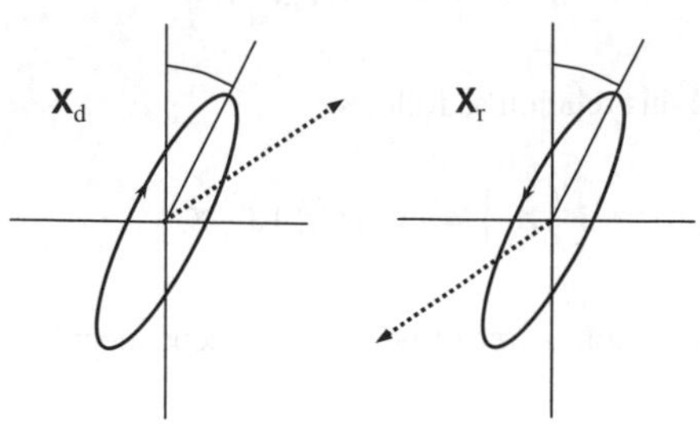

Figure 1.14 Reversely polarized waves. The dotted arrows show the propagation directions of the waves

reverse polarizations and reference frames to which the Jones vectors of these waves, $\mathbf{J}\{\mathbf{X}_d\}$ and $\mathbf{J}\{\mathbf{X}_r\}$, are referred, are chosen as in Figure 1.12, the relationship between these vectors can be expressed as follows:

$$\mathbf{J}\{\mathbf{X}_d\} = k\mathbf{U}_r\mathbf{J}\{\mathbf{X}_r\}^*, \tag{1.262}$$

where k is a scalar factor depending on the intensities and phases of the waves; the matrix $\mathbf{U}_r$ is defined in (1.255).

If a reciprocal system is free of polarization-dependent losses, for this system a theorem, which we will call *the theorem of polarization reversibility*, is valid [11, 12]. With the notation of Figure 1.11 for the incident ($\overrightarrow{\mathbf{X}}_i$, $\overleftarrow{\mathbf{X}}_i$) and transmitted ($\overrightarrow{\mathbf{X}}_o$, $\overleftarrow{\mathbf{X}}_o$) waves, this theorem can be formulated as follows: whatever the polarization of $\overrightarrow{\mathbf{X}}_i$, if the polarization of $\overleftarrow{\mathbf{X}}_i$ is reverse with respect to the polarization of $\overrightarrow{\mathbf{X}}_o$, the polarization of $\overleftarrow{\mathbf{X}}_o$ will be reverse with respect to that of $\overrightarrow{\mathbf{X}}_i$. This theorem can be proved in the following way.

Suppose that the polarization of the wave $\overleftarrow{\mathbf{X}}_i$ is reverse to that of the wave $\overrightarrow{\mathbf{X}}_o$. By making use of (1.262), we can express the Jones vector of $\overleftarrow{\mathbf{X}}_i$ as follows:

$$\mathbf{J}\left\{\overleftarrow{\mathbf{X}}_i\right\} = k\mathbf{U}_r\mathbf{J}\left\{\overrightarrow{\mathbf{X}}_o\right\}^*. \tag{1.263}$$

By definition,

$$\mathbf{J}\left\{\overrightarrow{\mathbf{X}}_o\right\} = \overrightarrow{t}_S\mathbf{J}\left\{\overrightarrow{\mathbf{X}}_i\right\}, \tag{1.264}$$

$$\mathbf{J}\left\{\overleftarrow{\mathbf{X}}_o\right\} = \overleftarrow{t}_S\mathbf{J}\left\{\overleftarrow{\mathbf{X}}_i\right\}. \tag{1.265}$$

On substituting (1.263) into (1.265), we have

$$\mathbf{J}\left\{\overleftarrow{\mathbf{X}}_o\right\} = k\overleftarrow{t}_S\mathbf{U}_r\mathbf{J}\left\{\overrightarrow{\mathbf{X}}_o\right\}^*. \tag{1.266}$$

According to (1.264) and identity (5.13),

$$\mathbf{J}\left\{\overrightarrow{\mathbf{X}}_o\right\}^* = \overrightarrow{t}_S^*\mathbf{J}\left\{\overrightarrow{\mathbf{X}}_i\right\}^*.$$

Substitution of this expression into (1.266) leads to the following relation:

$$\mathbf{J}\left\{\overleftarrow{\mathbf{X}}_o\right\} = k\overleftarrow{t}_S\mathbf{U}_r\overrightarrow{t}_S^*\mathbf{J}\left\{\overrightarrow{\mathbf{X}}_i\right\}^*. \tag{1.267}$$

Using (1.254), we can rewrite this relation as follows:

$$\mathbf{J}\left\{\overleftarrow{\mathbf{X}}_o\right\} = k\mathbf{U}_r\left(\overrightarrow{t}_S^{\mathrm{T}}\overrightarrow{t}_S^*\right)\mathbf{J}\left\{\overrightarrow{\mathbf{X}}_i\right\}^*. \tag{1.268}$$

Since the system under consideration is free of polarization-dependent losses, the matrix $\overrightarrow{t}_S$ satisfies the relations

$$\overrightarrow{t}_S^{\dagger}\overrightarrow{t}_S = (\overrightarrow{t}_S^{\dagger}\overrightarrow{t}_S)^* = \overrightarrow{t}_S^{\mathrm{T}}\overrightarrow{t}_S^* = t_S\mathbf{U}, \tag{1.269}$$

where t_S is the transmittance of the system. From (1.268) and (1.269), we obtain

$$\mathbf{J}\left\{\overleftarrow{\mathbf{X}}_o\right\} = k'\mathbf{U}_r\mathbf{J}\left\{\overrightarrow{\mathbf{X}}_i\right\}^*, \tag{1.270}$$

where $k' = kt_S$. Relation (1.270) [cf. (1.262)] shows that with the chosen polarization of $\overleftarrow{\mathbf{X}}_i$, the polarization of $\overleftarrow{\mathbf{X}}_o$ is really reverse to that of $\overrightarrow{\mathbf{X}}_i$.

This theorem explains the following properties of reflective systems without diattenuation, which are very important in considering single-polarizer reflective LCDs and transflective LCDs.

Two Important Properties of Reflective Systems Without Polarization-Dependent Losses

To present these properties, we return to the problem illustrated by Figure 1.13 and assume that the system S is free of diattenuation.

Property 1 Suppose that the wave $\overrightarrow{\mathbf{X}}_i$ incident on the reflective system is linearly polarized and the transmitted wave $\overrightarrow{\mathbf{X}}_{tS}$ is also linearly polarized. Then the reflected wave $\overleftarrow{\mathbf{X}}_{rR}$ is linearly polarized and has the same polarization plane as $\overrightarrow{\mathbf{X}}_{tS}$, that is, the waves $\overleftarrow{\mathbf{X}}_{rR}$ and $\overrightarrow{\mathbf{X}}_{tS}$ are reversely polarized. According to the theorem of polarization reversibility, the wave $\overleftarrow{\mathbf{X}}_o$ in this case is linearly polarized and has the same polarization plane as the incident wave $\overrightarrow{\mathbf{X}}_i$.

Property 2 Let the wave $\overrightarrow{\mathbf{X}}_i$ incident on the system be linearly polarized and let the transmitted wave $\overrightarrow{\mathbf{X}}_{tS}$ be circularly polarized. In this case, the output wave $\overleftarrow{\mathbf{X}}_o$ will be linearly polarized and have a polarization plane perpendicular to that of $\overrightarrow{\mathbf{X}}_i$. To elucidate this situation, we assume, for definiteness, that the wave $\overrightarrow{\mathbf{X}}_{tS}$ has the right circular polarization. In this case, the reflected wave $\overleftarrow{\mathbf{X}}_{rR}$ will have the left circular polarization. It follows from the theorem of polarization reversibility that if the wave $\overleftarrow{\mathbf{X}}_{rR}$ had the right circular polarization, the output wave $\overleftarrow{\mathbf{X}}_o$ would have the linear polarization and the same polarization plane as $\overrightarrow{\mathbf{X}}_i$. However, $\overleftarrow{\mathbf{X}}_{rR}$ has polarization orthogonal to the right circular one, and the wave $\overleftarrow{\mathbf{X}}_o$, being linearly polarized, will have its polarization plane orthogonal to the polarization plane of $\overrightarrow{\mathbf{X}}_i$, which is clear in view of the fact that transformations without diattenuation convert orthogonally polarized waves into orthogonally polarized ones (see item *Lossless transformations and transformations without diattenuation* in Section 1.4.1).

1.4.5 *Particular Variants of Application of the Jones Calculus. Cartesian Jones Vectors for Wave Fields in Anisotropic Media*

Reduced Transmittance of a System

When dealing with optical devices in which the input and output elements are polarizers (e.g., double-polarizer LCDs, single-polarizer reflective LCDs, reflective LCDs with polarizing beam splitters), the following approach is often used.

A scheme of light passage through an idealized system used in considering such a device can be written as follows: input polarizer → polarization-converting system → output polarizer. As for LCDs, typical elements of polarization-converting systems (PCSs), along with LC layer, are compensation films (retarders) and reflector in the case of reflective LCDs. On the assumption that the polarizers are ideal, the transmittance T defined as

$$T \equiv I_{out}/I_{incPCS}, \tag{1.271}$$

where I_{incPCS} is the intensity of the light incident on the PCS and I_{out} is the intensity of the light emerging from the output polarizer, is considered as a key characteristic of the system. This kind of transmittance will be referred to as *reduced transmittance*. The reduced transmittance can be expressed in terms of the Jones matrix of the PCS, $\boldsymbol{t}_{PCS}$, as

$$T = \left| \boldsymbol{j}_{tp2}^{\dagger} \boldsymbol{t}_{PCS} \boldsymbol{j}_{tp1} \right|^2, \tag{1.272}$$

where $\boldsymbol{j}_{tp1}$ and $\boldsymbol{j}_{tp2}$ are the polarization Jones vectors of waves that are transmitted by the input polarizer and the output polarizer, respectively [cf. (1.204)]. Sometimes, equation (1.272) is directly used for computation of T. When using this expression, it should be remembered that the vectors $\boldsymbol{j}_{tp1}$ and $\boldsymbol{j}_{tp2}$ must be referred, respectively, to the input and output reference frames of the matrix $\boldsymbol{t}_{PCS}$. In Chapter 6, we give convenient explicit expressions for the reduced transmittance in terms of orientation angles of the polarizers for different kinds of LCDs and present optimization methods using these expressions.

Unimodular Representation of Unitary Jones Matrices

In Chapters 2 and 3 and some other places of this book, PCSs of LCDs are considered as systems of lossless optical elements, that is, as unitary systems. The absence of losses allows one to calculate the Jones matrices of PCSs dealing with only unitary Jones matrices. Such calculations as well as further analysis and calculations are simplified when all elements of the PCS are represented by unimodular unitary (UU) Jones matrices, because such matrices are simple in form and their product is a matrix of a simple form (see Section 5.1.3). Any optical element that can be represented by a unitary Jones matrix can be represented in such calculations by a UU Jones matrix that describes the same transformation of polarization. In the most compact and convenient variants of representation of Jones matrices for lossless elements, these matrices are unimodular (see, e.g., expressions (1.184) and (1.185) for wave plates). The product of UU matrices is a UU matrix. Therefore, the Jones matrix of a unitary system that is calculated as the product of UU Jones matrices is also a UU matrix. By definition, the determinant of any unimodular matrix is equal to 1 or –1. All UU 2×2 matrices of determinant 1 have the form (5.31). This is the case, for example, for rotation matrices $\widehat{R}_C$ [see (1.53)] and Jones matrices for wave plates given by (1.184) and (1.185). The product of such UU matrices is always a matrix of determinant 1, that is, a matrix of the form (5.31). All UU 2×2 matrices of determinant -1 have the form (5.33). This is the case, for example, for the reflection Jones matrix of a lossless reflector given by (1.260) in the cases C1 and C2. In calculations involving such Jones matrices, the sign of the determinant of the resultant matrix of the system and, consequently, the form of this matrix can be predicted by using property (5.17) of determinants. In the context of the optical equivalence theorem that is presented in Section 3.1, it is important that any unitary system can be represented by a UU Jones matrix with determinant 1 (the multiplication of a UU 2×2 matrix by the imaginary unit gives a UU matrix with the opposite sign of the determinant) and that only three real parameters are in general required to fully specify such a matrix [(see (5.32)].

Cartesian Jones Vector for a Wave Field Propagating in an Anisotropic Medium

So far we dealt only with Cartesian Jones vectors that describe waves propagating in isotropic media, for example, in an isotropic medium surrounding an optical system or in isotropic spaces between optical elements. In the classical JC, Cartesian Jones vectors are used to characterize wave fields propagating inside anisotropic regions as well. In particular, this variant of description underlies the differential JC [8] (see Sections 2.1 and 11.1.1) which is used for treatment of inhomogeneous layers whose local optical parameters are continuous functions of spatial coordinates, such as inhomogeneous LC layers. The use of Cartesian Jones vectors for describing wave fields propagating in anisotropic media raises some questions. To explain, we return to the example illustrated by Figure 1.9a.

Suppose that the light falls on the uniaxial layer in the normal direction and is polarized so that both ordinary and extraordinary waves are induced. In accordance with the classical JC, the state of the wave field consisting of the forward propagating ordinary and extraordinary waves at an arbitrary point inside the layer can be described by a "fitted-to-intensity" Cartesian Jones vector $\boldsymbol{J} = (J_{x''}\ J_{y''})^{\mathrm{T}}$ referred to an arbitrary rectangular coordinate system (x'', y'', z'') with the z''-axis directed along the wave normal of the incident wave. A Cartesian Jones vector of any kind is a column composed of two Cartesian components of a vector collinear to the complex electric field strength vector of the wave field to be characterized. In our case, the wave field to be characterized is a superposition of two waves and its electric field strength vector, we denote it by $\mathbf{E}_{\mathrm{e+o}}$, is equal to $\mathbf{E}_{\mathrm{e}} + \mathbf{E}_{\mathrm{o}}$, where $\mathbf{E}_{\mathrm{e}}$ and $\mathbf{E}_{\mathrm{o}}$ are the electric fields strength vectors of the extraordinary wave and ordinary wave, respectively. By definition, we have

$$J_{x''} = b(\mathbf{x}''\mathbf{E}_{\mathrm{e+o}}), \quad J_{y''} = b(\mathbf{y}''\mathbf{E}_{\mathrm{e+o}}), \tag{1.273}$$

where $\mathbf{x}''$ and $\mathbf{y}''$ are unit vectors along the axes x'' and y'', and b is a complex coefficient. If the optic axis of the layer is parallel to its boundaries and, consequently, the vector $\mathbf{E}_{\mathrm{e+o}}$ is perpendicular to the z''-axis, the Jones vector $\boldsymbol{J}$ characterizes the wave field to the same extent as the Jones vector, of the same kind, characterizing a wave propagating in an isotropic medium. However, there is a serious difference. The contributions of the extraordinary and ordinary components into the intensity, with any reasonable choice of the physical quantity considered as intensity (see Sections 5.2 and 5.4), depend on their phase velocities which are different. Therefore the ratio of $|\boldsymbol{J}|^2$ to the intensity is dependent on $\boldsymbol{J}$. This means that the vector $\boldsymbol{J}$ cannot be "fitted-to-intensity" in principle. This vector can be considered to be approximately "fitted-to-intensity" only when the principal refractive indices of the anisotropic medium are very close to each other or, more precisely, when $|n_{\|} - n_{\perp}| \ll n_{\|}, n_{\perp}$. Thus, defining a Cartesian Jones vector as in (1.273) and postulating that this vector is "fitted-to-intensity," we thereby restrict the consideration to the case of a weakly anisotropic medium. The assumption that the medium is weakly anisotropic also allows us to disregard the fact that at $\theta \neq 0, 90°$ the field $\mathbf{E}_{\mathrm{e}}$ has a nonzero z''-component [see (1.141)], since at $|n_{\|} - n_{\perp}| \ll n_{\|}, n_{\perp}$ this component is very small compared with the transverse constituent of $\mathbf{E}_{\mathrm{e}}$. Note that liquid crystals in most display applications cannot be considered as a weakly anisotropic medium.

It is possible to remove the mentioned restriction by using another, somewhat artificial, definition of Cartesian Jones vector for anisotropic media. To illustrate this, we proceed with the above example.

To define the Cartesian vector $\boldsymbol{J}(\xi)$ at points of a plane $z_c = \xi$ inside the uniaxial layer, we may imagine that we replaced the rest of the layer beyond this plane by an isotropic medium and let the light pass the boundary $z_c = \xi$ without losses. Then we may take as $\boldsymbol{J}(\xi)$ the Jones vector of the emergent wave just beyond the plane $z_c = \xi$. It is clear that this kind of definition of Jones vectors is applicable in considering inhomogeneous layers as well. We should note that this definition, where the Jones vector characterizes the wave field inside the anisotropic medium indirectly, is to the greatest extent consistent with the standard apparatus of JC developed for considering continuously inhomogeneous media, which is used in LCD optics for calculating Jones matrices for inhomogeneous LC layers (see Sections 2.1 and 11.1.1).

References

[1] M. Born and E. Wolf, *Principles of Optics*, 7th ed. (Pergamon Press, New York, 1999).

[2] R. M. A. Azzam and N. M. Bashara, *Ellipsometry and Polarized Light* (North-Holland, Amsterdam, 1977).

[3] A. Yariv and P. Yeh, *Optical Waves in Crystals* (Wiley, New York, 2003).

[4] E. Hecht, *Optics*, 4th ed. (Addison Wesley, San Francisco, 2002).

[5] R. C. Jones, "A new calculus for the treatment of optical systems. I. Description and discussion of the calculus," *J. Opt. Soc. Am.* **31**, 488 (1941).

[6] W. A. Shurcliff, *Polarized Light: Production and Use* (Harvard University Press, Cambridge, 1962).

[7] R. C. Jones, "A new calculus for the treatment of optical systems. IV," *J. Opt. Soc. Am.* **32**, 486 (1942).

[8] R. C. Jones, "A new calculus for the treatment of optical systems. VII. Properties of the N-matrices," *J. Opt. Soc. Am.* **38**, 671 (1948).
[9] R. C Jones, "A new calculus for the treatment of optical systems. V. A more general formulation, and description of another calculus," *J. Opt. Soc. Am.* **38**, 671 (1948).
[10] N. Hiji, Y. Ouchi, H. Takezoe, and A. Fukuda, "Structures of twisted states in ferroelectric liquid crystals studied by microspectrophotometry and numerical calculations," *Jpn. J. Appl. Phys.* **27**, 8 (1988).
[11] N. Vansteenkiste, P. Vignolo, and A. Aspect, "Optical reversibility theorems for polarization: application to remote control of polarization," *J. Opt. Soc. Am. A.* **10**, 2240 (1993).
[12] C. R. Fernández-Pousa, I. Moreno, N. Bennis, and C. Gómez-Reino, "Generalized formulation and symmetry properties of reciprocal nonabsorbing polarization devices: application to liquid-crystal displays," *J. Opt. Soc. Am. A* **17**, 2074 (2000).

2

The Jones Calculus: Solutions for Ideal Twisted Structures and Their Applications in LCD Optics

In most practical types of LC displays, LC layers have a twisted structure. These are twisted nematic (TN) LCDs, supertwisted nematic (STN) LCDs, and so on. As a rule, in the field-off state, the structure of the LC layer in such a display is close to an ideal twisted one. Here we use the term "an ideal twisted LC structure" for structures that have a uniform tilt of the LC director throughout the LC layer and a uniform twist, a uniform rotation of the LC director through the layer thickness. An ideal twisted structure is one of a few kinds of inhomogeneous LC structures for which analytical solutions of optical problems can be obtained. In this chapter, we consider analytical solutions that were obtained for LC layers with ideal twisted structure by using the Jones calculus. These solutions enable one to gain an insight into a variety of existing modes of displays with twisted LC layers and help in finding new modes. Furthermore, they underlie many measurement methods for determining the parameters of twisted LC layers, including those used for process control in the production of LCD panels.

2.1 Jones Matrix and Eigenmodes of a Liquid Crystal Layer with an Ideal Twisted Structure

Consider a nonabsorbing nematic layer with an ideal twisted structure. Let (x, y, z) be a right-handed Cartesian coordinate system with the planes $z = 0$ and $z = d$, where d is the thickness of the layer, coincident with the boundaries of the layer, and the x–z plane parallel to the LC director at the boundary $z = 0$. Then the orientation of the LC director (and local optic axis) in the layer can be represented as follows:

$$\varphi(z) = \left(\frac{z}{d}\right)\Phi, \qquad \theta(z) = \theta_0, \tag{2.1}$$

where θ is the tilt angle of the LC director, φ is the azimuthal angle of the LC director measured from the x–z plane, Φ is the twist angle of the LC structure, and θ_0 is the pretilt angle.

Modeling and Optimization of LCD Optical Performance, First Edition.
Dmitry A. Yakovlev, Vladimir G. Chigrinov and Hoi-Sing Kwok.

Website Companion: www.wiley.com/go/yakovlev/modelinglcd

The Jones calculus gives the following expression for a unimodular transmission Jones matrix of this layer, referred to the frame (x, y), for light propagating in the positive z direction:

$$t_{\text{LC}} = \begin{pmatrix} a+ib & -c+id \\ c+id & a-ib \end{pmatrix}, \tag{2.2}$$

where

$$\begin{aligned} a &= \cos\chi\cos\phi + \frac{\Phi}{\chi}\sin\chi\sin\Phi, \\ b &= \frac{\delta}{\chi}\sin\chi\cos\Phi, \\ c &= \cos\chi\sin\Phi - \frac{\Phi}{\chi}\sin\chi\cos\Phi, \\ d &= \frac{\delta}{\chi}\sin\chi\sin\Phi, \end{aligned} \tag{2.3}$$

with

$$\chi = \sqrt{\Phi^2 + \delta^2} \tag{2.4}$$

and δ being the half phase retardation of the LC layer, defined as

$$\delta = \frac{\pi d \Delta n}{\lambda}, \tag{2.5}$$

$$\Delta n = n_{\text{e}} - n_{\perp},\ n_{\text{e}} = \frac{n_{||}n_{\perp}}{\sqrt{n_{\perp}^2\cos^2\theta_0 + n_{||}^2\sin^2\theta_0}}, \tag{2.6}$$

where $n_{||}$ and $n_{\perp}$ are the principal refractive indices of the liquid crystal. The matrix t_{LC} can also be represented as

$$t_{\text{LC}} = \widehat{R}_{\text{C}}(-\Phi)\,t'_{\text{LC}}, \tag{2.7}$$

where

$$t'_{\text{LC}} = \begin{pmatrix} \cos\chi + i\dfrac{\delta}{\chi}\sin\chi & \dfrac{\Phi}{\chi}\sin\chi \\ -\dfrac{\Phi}{\chi}\sin\chi & \cos\chi - i\dfrac{\delta}{\chi}\sin\chi \end{pmatrix} \tag{2.8}$$

and $\widehat{R}_{\text{C}}$ is the rotation matrix defined in (1.53). t'_{LC} is the Jones matrix of the layer whose input reference frame is (x, y) and output one is a frame (x_{Φ}, y_{Φ}) that is obtained by rotation of the frame (x, y) by the angle Φ.

There are several ways to derive the above expressions [1–6]. For example, following an approach used in References 2–5, the LC layer can be considered as a stack of N ($N \to \infty$) identical homogeneous birefringent layers each of which produces the phase retardation equal to $2\delta/N$; the azimuthal angle of the optic axis of the j-th layer ($j = 1, 2, \ldots, N$), measured from the x–z plane, ϕ_j, is assumed to be equal to $j\Phi/N$ (Figure 2.1).

In this case, the matrix t_{LC} can be expressed as

$$t_{\text{LC}} = \widehat{R}_{\text{C}}(-\phi_N)t'\widehat{R}_{\text{C}}(\phi_N)\widehat{R}_{\text{C}}(-\phi_{N-1})t'\widehat{R}_{\text{C}}(\phi_{N-1})\cdot\ \ldots\ \cdot\widehat{R}_{\text{C}}(-\phi_2)t'\widehat{R}_{\text{C}}(\phi_2)\widehat{R}_{\text{C}}(-\phi_1)t'\widehat{R}_{\text{C}}(\phi_1), \tag{2.9}$$

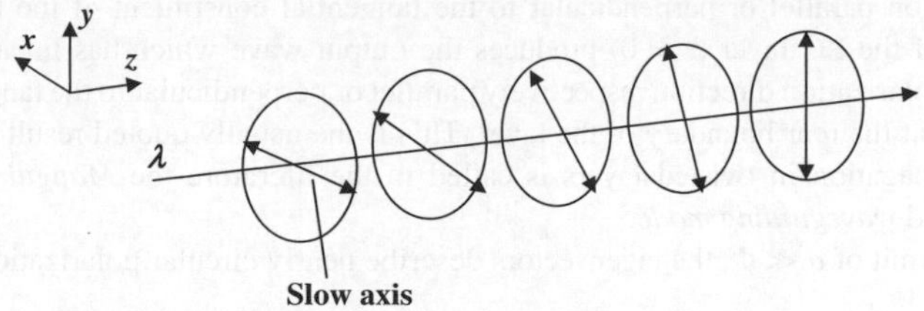

Figure 2.1 Stack of plate approximation for the LC cell

where

$$t' = \begin{pmatrix} e^{i\delta/N} & 0 \\ 0 & e^{-i\delta/N} \end{pmatrix}. \tag{2.10}$$

Since $\hat{R}_C(\phi_j)\hat{R}_C(-\phi_{j-1}) = \hat{R}_C(\Phi/N)$, $\hat{R}_C(\phi_1) = \hat{R}_C(\Phi/N)$, and $\hat{R}_C(\phi_N) = \hat{R}_C(\Phi)$, Equation (2.9) can be rewritten in the form (2.7) with

$$t'_{LC} = \left[t'\hat{R}_C(\Phi/N)\right]^N. \tag{2.11}$$

The use of Sylvester's matrix theorem leads to relatively simple explicit formulas for the elements of a matrix being a power of a 2 × 2 matrix **A** in terms of the elements of **A**, which are especially simple when the determinant of **A** is equal to 1, as in the case under consideration. Using such formulas to express the matrix $[t'\hat{R}_C(\Phi/N)]^N$ and proceeding to the limit as $N \to \infty$, one can arrive at (2.8) [2–5].

The fact that the optical constants of the liquid crystal, thickness of the layer, and light wavelength enter into (2.8) only in the combination $d\Delta n/\lambda$ [see (2.5)] much simplifies understanding the effect of these parameters and is often used in analysis of optical properties of twisted LC layers.

It is useful to examine the eigenvectors of t'_{LC} (see Section 5.1.1), which can provide some interesting insights on the optical properties of twisted LC layers. The eigenvalue problem is given by

$$t'_{LC}\boldsymbol{J}_\kappa = \kappa \boldsymbol{J}_\kappa, \tag{2.12}$$

where $\boldsymbol{J}_\kappa$ is an eigenvector and κ is the eigenvalue corresponding to $\boldsymbol{J}_\kappa$. The solutions of this equation are

$$\kappa_1 = e^{i\chi},\ \kappa_2 = e^{-i\chi}, \tag{2.13}$$

$$\boldsymbol{J}_{\kappa_1} = \frac{1}{\sqrt{(\delta+\chi)^2+\Phi^2}}\begin{pmatrix} \delta+\chi \\ i\Phi \end{pmatrix},\quad \boldsymbol{J}_{\kappa_2} = \frac{1}{\sqrt{(\delta+\chi)^2+\Phi^2}}\begin{pmatrix} i\Phi \\ \delta+\chi \end{pmatrix}. \tag{2.14}$$

Here the eigenvectors are normalized so that $\boldsymbol{J}^\dagger_{\kappa_j}\boldsymbol{J}_{\kappa_j} = 1$ $(j = 1,2)$. It is easy to see that $\boldsymbol{J}^\dagger_{\kappa_1}\boldsymbol{J}_{\kappa_2} = 0$ (eigenvectors corresponding to different eigenvalues of a unitary matrix are always orthogonal). Thus the eigenvectors $\boldsymbol{J}_{\kappa_1}$ and $\boldsymbol{J}_{\kappa_2}$ of t'_{LC} describe orthogonal elliptical polarizations. The output reference frame of the matrix is rotated with respect to the input one by Φ. Therefore, it follows from (2.12) that if the incident light has polarization described by $\boldsymbol{J}_{\kappa_1}$ or $\boldsymbol{J}_{\kappa_2}$, the transmitted light will have the same polarization ellipse but rotated by Φ.

There are two limits of the above solution that are noteworthy. It can be checked that in the limit of $\delta \gg \Phi$, the eigenvectors describe almost linear polarizations: $\boldsymbol{J}_{\kappa_1}$ along x at entry and along x_Φ at exit, and $\boldsymbol{J}_{\kappa_2}$ along y at entry and along y_Φ at exit. In this case, a linearly polarized incident wave with polarization direction parallel or perpendicular to the tangential constituent of the LC director at the frontal boundary of the LC layer ($z = 0$) produces the output wave which has linear or almost linear polarization with polarization direction respectively parallel or perpendicular to the tangential constituent of the LC director at the rear boundary of the layer. This is the usually quoted result for TN cells. This mode of light propagation in twisted layers is called in the literature the *Mauguin mode*, *adiabatic following mode*, and *waveguiding mode*.

In the opposite limit of $\delta \ll \Phi$, the eigenvectors describe nearly circular polarizations.

Derivation by using the differential Jones Calculus

The same results can easily be obtained by using the differential Jones calculus [1, 6] (see also Section 11.1.1). Consider a very thin slice of the LC layer, occupying the space between an arbitrary plane $z = \tau$ and the plane $z = \tau + \Delta z$ ($\Delta z \to 0$). The optical effect of this slice may be represented by the relation

$$\boldsymbol{J}(\tau + \Delta z) = \boldsymbol{t}_{\tau,\tau+\Delta z}\boldsymbol{J}(\tau), \tag{2.15}$$

where $\boldsymbol{J}(\tau)$ and $\boldsymbol{J}(\tau + \Delta z)$ are the Jones vectors of the passing light, referred to the frame (x, y), at $z = \tau$ and $z = \tau + \Delta z$, $\boldsymbol{t}_{\tau,\tau+\Delta z}$ is the corresponding Jones matrix of the slice. Then, neglecting the variation of the angle φ [see (2.1)] within the slice, we can express the matrix $\boldsymbol{t}_{\tau,\tau+\Delta z}$ as

$$\boldsymbol{t}_{\tau,\tau+\Delta z} = \widehat{R}_{\mathrm{C}}(-q\tau)\begin{pmatrix} e^{ik\Delta z} & 0 \\ 0 & e^{-ik\Delta z} \end{pmatrix}\widehat{R}_{\mathrm{C}}(q\tau) \tag{2.16}$$

where $k = \pi\Delta n/\lambda$ and $q = \Phi/d$. To the first order in Δz,

$$\boldsymbol{t}_{\tau,\tau+\Delta z} = \mathbf{U} + ik\widehat{R}_{\mathrm{C}}(-q\tau)\begin{pmatrix} 1 & 0 \\ 0 & -1 \end{pmatrix}\widehat{R}_{\mathrm{C}}(q\tau)\,\Delta z, \tag{2.17}$$

where $\mathbf{U}$ is the unit matrix. Substitution of (2.17) into (2.15) gives

$$\boldsymbol{J}(z + \Delta z) = \boldsymbol{J}(z) + ik\widehat{R}_{\mathrm{C}}(-q\tau)\begin{pmatrix} 1 & 0 \\ 0 & -1 \end{pmatrix}\widehat{R}_{\mathrm{C}}(q\tau)\,\Delta z\boldsymbol{J}(z). \tag{2.18}$$

Rearrangement of (2.18) and proceeding to the limit as $\Delta z \to 0$ lead to the following equation:

$$\left.\frac{\mathrm{d}\boldsymbol{J}}{\mathrm{d}z}\right|_{z=\tau} = \lim_{\Delta z\to 0}\frac{\boldsymbol{J}(\tau + \Delta z) - \boldsymbol{J}(\tau)}{\Delta z} = ik\widehat{R}_{\mathrm{C}}(-q\tau)\begin{pmatrix} 1 & 0 \\ 0 & -1 \end{pmatrix}\widehat{R}_{\mathrm{C}}(q\tau)\boldsymbol{J}(\tau). \tag{2.19}$$

Since τ is arbitrary, this equation can be rewritten as

$$\frac{\mathrm{d}\boldsymbol{J}(z)}{\mathrm{d}z} = \boldsymbol{H}(z)\boldsymbol{J}(z), \tag{2.20}$$

where

$$\boldsymbol{H}(z) = ik\widehat{R}_{\mathrm{C}}(-qz)\begin{pmatrix} 1 & 0 \\ 0 & -1 \end{pmatrix}\widehat{R}_{\mathrm{C}}(qz). \tag{2.21}$$

A transformation of variable from $\boldsymbol{J}(z)$ to

$$\boldsymbol{J}_{\mathrm{R}}(z) = \widehat{R}_{\mathrm{C}}(qz)\boldsymbol{J}(z) \tag{2.22}$$

can be made (see Section 11.1.1). The propagation equation (2.20) is thus converted to a simpler looking equation:

$$\frac{\mathrm{d}\boldsymbol{J}_{\mathrm{R}}(z)}{\mathrm{d}z} = \begin{pmatrix} ik & q \\ -q & -ik \end{pmatrix} \boldsymbol{J}_{\mathrm{R}}(z) = \boldsymbol{H}' \boldsymbol{J}_{\mathrm{R}}(z), \tag{2.23}$$

where

$$\boldsymbol{H}' = \begin{pmatrix} ik & q \\ -q & -ik \end{pmatrix}. \tag{2.24}$$

Since the matrix $\boldsymbol{H}'$ is independent of z, Equation (2.23) can be easily solved. At any τ, the Jones vector $\boldsymbol{J}_{\mathrm{R}}(\tau)$ represents the same state of the light as the vector $\boldsymbol{J}(\tau)$ but, in contrast to $\boldsymbol{J}(\tau)$, is referred to a local reference frame $(x'(\tau), y'(\tau))$ that can be obtained by rotation of the frame (x, y) by the angle $\varphi(\tau) = q\tau$.

A fundamental system of solutions for (2.23) can be composed of the following linearly independent particular solutions of this equation:

$$\boldsymbol{J}_{\mathrm{R}}^{+}(z) = \boldsymbol{J}_{\mathrm{R}}^{+}(0)\, e^{i\xi z},\ \boldsymbol{J}_{\mathrm{R}}^{-}(z) = \boldsymbol{J}_{\mathrm{R}}^{-}(0)\, e^{-i\xi z}, \tag{2.25}$$

where

$$\xi = \sqrt{k^2 + q^2}, \tag{2.26}$$

$$\boldsymbol{J}_{\mathrm{R}}^{+}(0) = \frac{1}{\sqrt{(k+\xi)^2 + q^2}} \begin{pmatrix} k+\xi \\ iq \end{pmatrix}, \quad \boldsymbol{J}_{\mathrm{R}}^{-}(0) = \frac{1}{\sqrt{(k+\xi)^2 + q^2}} \begin{pmatrix} iq \\ k+\xi \end{pmatrix}. \tag{2.27}$$

The vectors $\boldsymbol{J}_{\mathrm{R}}^{+}$ and $\boldsymbol{J}_{\mathrm{R}}^{-}$ are eigenvectors of $\boldsymbol{H}'$. The corresponding eigenvalues of $\boldsymbol{H}'$ are equal to $i\xi$ and $-i\xi$ respectively. Note that $\boldsymbol{J}_{\mathrm{R}}^{+}(0) = \boldsymbol{J}_{\kappa_1}$ and $\boldsymbol{J}_{\mathrm{R}}^{-}(0) = \boldsymbol{J}_{\kappa_2}$. Any particular solution of (2.23) can be written as

$$\boldsymbol{J}_{\mathrm{R}}(z) = c_1 \boldsymbol{J}_{\mathrm{R}}^{+}(z) + c_2 \boldsymbol{J}_{\mathrm{R}}^{-}(z), \tag{2.28}$$

where c_1 and c_2 are scalar factors.

By definition,

$$\boldsymbol{J}_{\mathrm{R}}(d) = \boldsymbol{t}'_{\mathrm{LC}} \boldsymbol{J}_{\mathrm{R}}(0). \tag{2.29}$$

By making use of (2.25), it is easy to find the elements of $\boldsymbol{t}'_{\mathrm{LC}} \equiv [t'_{\mathrm{LC}jl}]$. This can be done in the following way. If we take $\boldsymbol{J}_{\mathrm{R}}(0) = \begin{pmatrix} 1 \\ 0 \end{pmatrix}$, the output Jones vector $\boldsymbol{J}_{\mathrm{R}}(d)$, according to (2.28), will be equal to $\begin{pmatrix} t'_{\mathrm{LC11}} \\ t'_{\mathrm{LC21}} \end{pmatrix}$. The values of the coefficients c_1 and c_2 for the particular solution with $\boldsymbol{J}_{\mathrm{R}}(0) = \begin{pmatrix} 1 \\ 0 \end{pmatrix}$ can be found from equation

$$c_1 \boldsymbol{J}_{\mathrm{R}}^{+}(0) + c_2 \boldsymbol{J}_{\mathrm{R}}^{-}(0) = \begin{pmatrix} 1 \\ 0 \end{pmatrix}. \tag{2.30}$$

The solution of (2.30) is

$$c_1 = \frac{k+\xi}{\sqrt{(k+\xi)^2+q^2}}, \quad c_2 = \frac{-iq}{\sqrt{(k+\xi)^2+q^2}}. \tag{2.31}$$

From (2.28) and (2.25), we have

$$\boldsymbol{J}_{\mathrm{R}}(d) = c_1 \boldsymbol{J}_{\mathrm{R}}^{+}(0)\, e^{i\xi d} + c_2 \boldsymbol{J}_{\mathrm{R}}^{-}(0)\, e^{-i\xi d}. \tag{2.32}$$

Upon substituting (2.31) in (2.32), we obtain the following expression:

$$\begin{pmatrix} t'_{\mathrm{LC11}} \\ t'_{\mathrm{LC21}} \end{pmatrix} = \frac{k+\xi}{(k+\xi)^2+q^2}\begin{pmatrix} k+\xi \\ iq \end{pmatrix} e^{i\xi d} - \frac{iq}{(k+\xi)^2+q^2}\begin{pmatrix} iq \\ k+\xi \end{pmatrix} e^{-i\xi d}, \tag{2.33}$$

from which we find that

$$t'_{\mathrm{LC11}} = \cos(\xi d) + i\frac{k}{\xi}\sin(\xi d), \tag{2.34}$$

$$t'_{\mathrm{LC21}} = -\frac{q}{\xi}\sin(\xi d). \tag{2.35}$$

The elements t'_{LC12} and t'_{LC22} in this case can be expressed as follows:

$$t'_{\mathrm{LC12}} = \frac{q}{\xi}\sin(\xi d), \tag{2.36}$$

$$t'_{\mathrm{LC22}} = \cos(\xi d) - i\frac{k}{\xi}\sin(\xi d). \tag{2.37}$$

The expressions (2.34)–(2.37) are identical to those given by (2.8) because

$$\begin{aligned} \xi d &= \sqrt{\delta^2+\Phi^2} = \chi, \\ \frac{q}{\xi} &= \frac{\Phi}{\sqrt{\delta^2+\Phi^2}} = \frac{\Phi}{\chi}, \\ \frac{k}{\xi} &= \frac{\delta}{\sqrt{\delta^2+\Phi^2}} = \frac{\delta}{\chi}. \end{aligned} \tag{2.38}$$

The presented expressions for the matrix $\boldsymbol{t}_{\mathrm{LC}}$ or other ones equivalent to them underlie much of the analytical analysis of twisted LC layers in LCDs that does not rely on full numerical calculations.

2.2 LCD Optics and the Gooch–Tarry Formulas

The simplest liquid crystal display consists of an input polarizer, the LC cell, and an output polarizer. The definition of the various angles are shown in Figure 2.2, where the input director azimuthal direction is defined as the x-axis, as in the previous section.

The transmittance of this system is given by

$$T = T(\alpha,\gamma,\phi,\delta) = \left| (\cos\gamma \quad \sin\gamma)\cdot \boldsymbol{t}_{\mathrm{LC}} \cdot \begin{pmatrix} \cos\alpha \\ \sin\alpha \end{pmatrix} \right|^2, \tag{2.39}$$

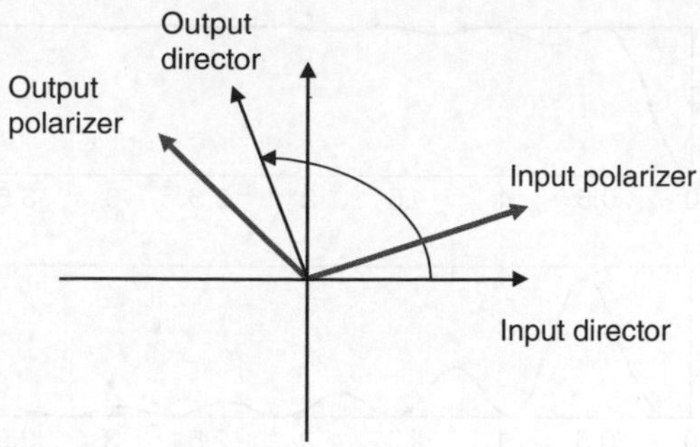

Figure 2.2 Definition of various directions in an LCD

where α and γ are the input and output polarizer directions [see (1.272)]. In all considerations in this chapter and Chapter 3, transmittances and reflectances of LCDs are defined as the reduced transmittance in (1.271).

Gooch and Tarry [7, 8] analyzed the optics of twisted nematic cells and derived certain important results related to waveguiding modes. In the original Gooch–Tarry configuration, the input polarizer for a twist angle Φ cell is along the input director and the output polarizer is perpendicular to the output director of the LC cell as shown in Figure 2.3.

Thus the transmittance is given by

$$T = \left| (\cos(\Phi + \pi/2) \quad \sin(\Phi + \pi/2)) \cdot t_{\mathrm{LC}} \cdot \begin{pmatrix} 1 \\ 0 \end{pmatrix} \right|^2. \tag{2.40}$$

The substitution of (2.2) in (2.40) gives the following expression for the transmittance:

$$T = \frac{\Phi^2}{\chi^2} \sin^2 \chi = \frac{1}{1+u^2} \sin^2 \left(\Phi \sqrt{1+u^2} \right), \tag{2.41}$$

where

$$u = \frac{\delta}{\Phi} = \frac{\pi d \Delta n}{\lambda \Phi}. \tag{2.42}$$

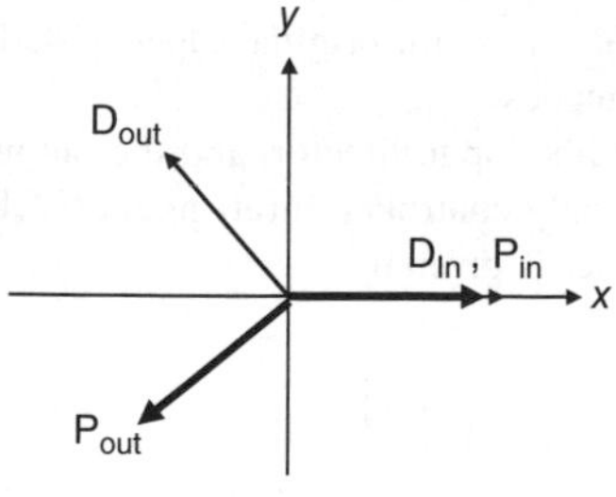

Figure 2.3 Gooch and Tarry conditions

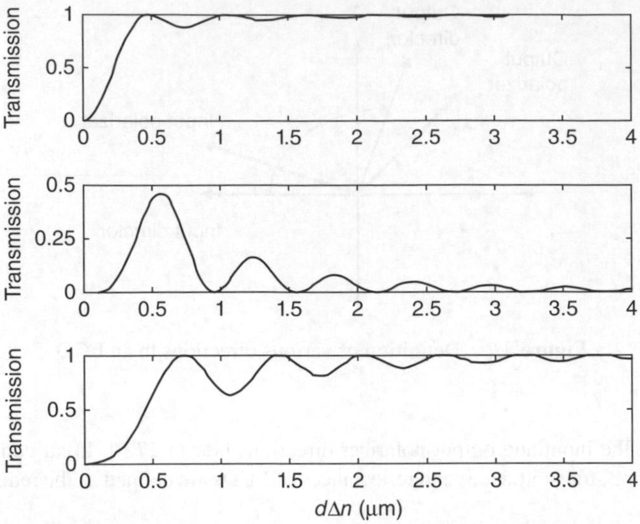

Figure 2.4 Transmittance of the LCD as a function of $d\Delta n$ for $\Phi = 90°$, 180°, and 270°

This is identical to the original Gooch–Tarry result, as it should. The minimum transmittance points of this equation ($T = 0$) are the Mauguin minima, and are given by

$$\frac{d\Delta n}{\lambda} = \frac{\Phi}{\pi}\sqrt{N^2\pi^2 - 1}, \tag{2.43}$$

for integer values of N. In particular for a 90° TN cell, the first two Mauguin minima at $\lambda = 550$ nm are given by $d\Delta n = 0.475$, 1.065 μm. They are known as the first and second minima for TN displays. In practice, the output polarizer is along the output director to achieve a normally white state. Thus the transmittance is the complementary value, that is,

$$T = 1 - \frac{1}{1+u^2}\sin^2\left(\Phi\sqrt{1+u^2}\right). \tag{2.44}$$

So in fact the Mauguin minima correspond to maximum transmission of the LCD. The name Mauguin minima is still used though by common practice. Figure 2.4 shows the transmittance of the LC cell as a function of $d\Delta n$ for three typical twist angles. It can be seen that the transmittance approaches unity always for large values of $d\Delta n$. These modes are therefore referred to as the waveguiding modes. Notice that the 90° twist case is normally called the TN LCD, the 180° case is exactly the OMI (optical mode interference) display [9], and the 270° case is the original supertwisted birefringence effect (SBE) display [10, 11]. They are all waveguiding modes.

If the input polarizer is at 45° to the input director, and the output polarizer is perpendicular to the input polarizer, we have the electrically controlled birefringent (ECB) mode display. These modes are interference modes. The transmittance is given by

$$\begin{aligned} T &= \frac{1}{4}\left|(1 \quad -1)\cdot t_{LC}\cdot\begin{pmatrix}1\\1\end{pmatrix}\right|^2 \\ &= \sin^2\chi\cos^2\Phi + \cos^2\chi\sin^2\Phi - \frac{2\Phi}{\chi}\sin\chi\cos\chi\sin\Phi\cos\Phi. \end{aligned} \tag{2.45}$$

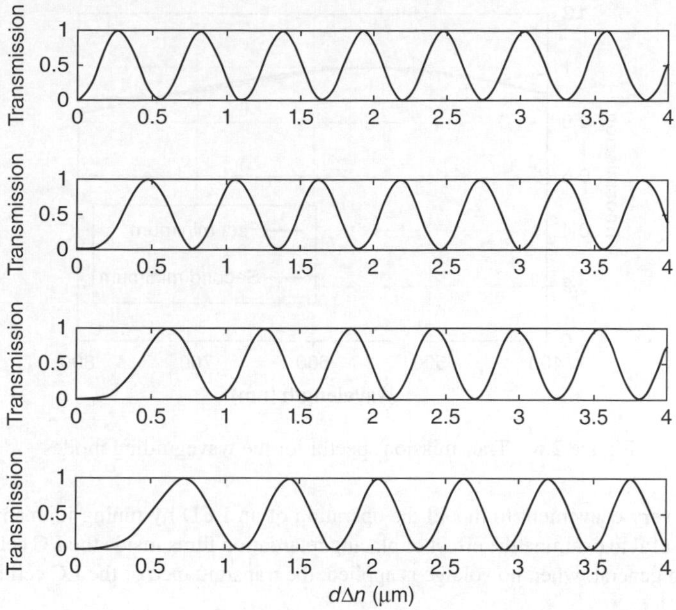

Figure 2.5 Transmission as a function of $d\Delta n$ for $\Phi = 0, 90°, 180°$, and $270°$

For twist angles that are even multiples of $\pi/2$, the transmittance reduces to

$$T = \sin^2 \sqrt{\Phi^2 + \delta^2}. \tag{2.46}$$

For twist angles that are odd multiples of $\pi/2$, the transmittance is given by

$$T = \cos^2 \sqrt{\Phi^2 + \delta^2}. \tag{2.47}$$

For these interference modes, unlike the waveguiding modes, the transmittance never reaches a constant value as $d\Delta n$ increases. This is shown in Figure 2.5 for twist angles of 0, 90°, 180°, and 270°. It should be noted that the ordinary STN LCD is really closer to an ECB mode than a TN mode. With an input polarizer at 45° and a twist angle near 240°, this STN LCD is rather dispersive.

The waveguiding and the interference modes are different not only in the behavior of the transmittance at large $d\Delta n$. They are also very different in the wavelength dependence of the output. The waveguiding modes have small wavelength dependence while the interference modes are quite wavelength dispersive. The transmission spectra of the waveguiding and interference modes are shown in Figures 2.6 and 2.7. It can be seen that for a 90° TN display, both the first minimum and the second minimum give rather flat spectra. However, for the ECB mode in Figure 2.7, the transmittance has strong wavelength dependence for both the bright and dark states. Notice that the optical mode interference (OMI) mode is actually a waveguiding mode and the word "interference" used in its name is a misnomer.

2.3 Interactive Simulation

The Jones matrix formulation of the liquid crystal cell is useful in modeling of the display under no voltage bias conditions. Because of its simplicity, simulation results can almost be obtained in real

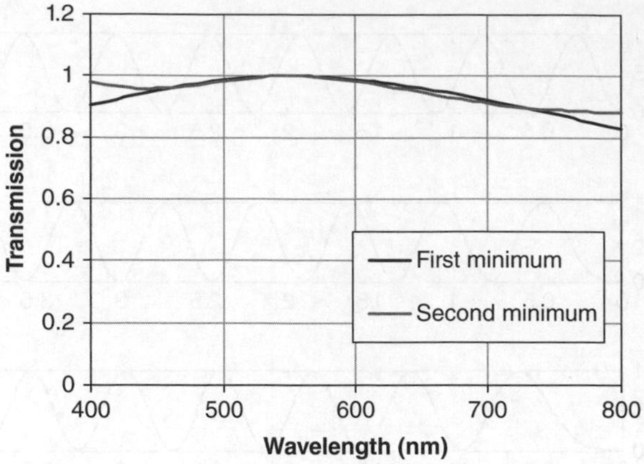

Figure 2.6 Transmission spectra for the waveguiding modes

time. Thus, it is very convenient to model the operation of an LCD by tuning its many parameters. In particular, it is useful to examine the effect of placing retardation films inside the LC cell for wavelength compensation. In general, when no voltage is applied, the transmittance of the LC cell is given by

$$T = \left| \begin{pmatrix} \cos\gamma & \sin\gamma \end{pmatrix} \cdot \ldots \cdot t_{\mathrm{FILM}j+1} \cdot t_{\mathrm{LC}} \cdot t_{\mathrm{FILM}j} \cdot \cdots \cdot \begin{pmatrix} \cos\alpha \\ \sin\alpha \end{pmatrix} \right|^2, \tag{2.48}$$

where a number of films can be placed inside the LC display between the input and output polarizers, in addition to the liquid crystal layer. Each Jones matrix in (2.48) represents an optical element. In this equation, there are four parameters α, γ, δ_{LC}, Φ, for the LC layer. For each retardation film, there are two additional parameters χ, δ, where χ is the angle of the film axis relative to the x-axis and δ is the retardation value. Since matrix multiplication can be performed very fast in a personal computer, it is possible to write a relatively simple software to calculate the transmittance spectrum as a function of all the parameters, α, γ, δ_{LC}, Φ, χ, δ. In such a program, the parameters can be varied as a slide bar as shown in Figure 2.8. The transmission spectrum can be shown instantly as the parameters are varied one at a time. Good user graphic interface can be constructed as well. The color

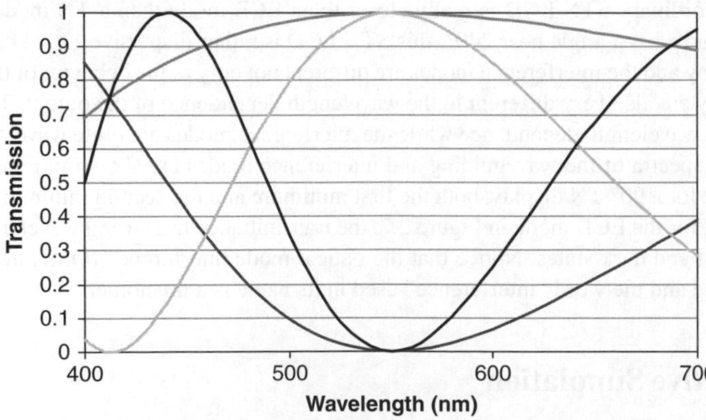

Figure 2.7 Transmission spectra for the interference modes

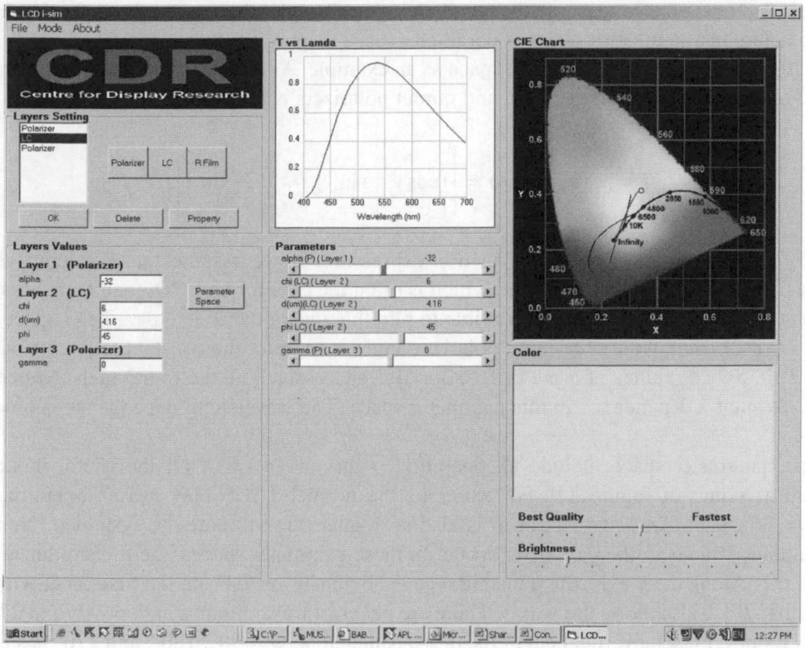

Figure 2.8 User graphics interface for a software that calculates (2.48) in real time

corresponding to the particular spectrum and the standard CIE coordinates can be given instantly. This software is useful for obtaining birefringent color displays or for compensating the dispersion of the transmittance.

The idea of "instantaneous simulator" is a powerful one. Due to the increasing processing and computational power of computers, solutions of differential equations can be obtained very rapidly. Thus it is possible to use the concept of a "sliding bar" as in Figure 2.8 to vary various parameters and examine their effects instantaneously. This is extremely useful both as a pedagogical tool and as a practical simulator for device optimization. One such possible simulator is the solution of the Euler–Lagrange equations for obtaining the elastic deformation results of an LC cell under an applied voltage. Using appropriate approximations to simplify the nonlinear coupled equations, it is possible to obtain the elastic deformation of the LC cell in less than 1-sec computation time using a desktop computer. Though approximate, this is a very useful tool to visualize the effects of changing the various parameters on the deformation of the LC alignment inside the LC cell. Traditional non-interactive method of setting the values of various parameters and performing a computer run in order to see the results are only needed when more accurate and exact solutions are desired.

2.4 Parameter Space

The transmittance or reflectance of an LC cell depends on many parameters. In Section 2.3, we describe one approach to examine the parametric dependence of the optics of an LC cell on the various variables α, γ, δ_{LC}, Φ, χ, δ using an instantaneous solution simulator method. The entire transmission or reflectance spectrum can be visualized instantaneously. However, if one needs to know only the reflectance or transmittance at a single wavelength, another approach can be used to obtain the parametric dependence, that is, the use of a parameter space. In the parameter space, we plot the T or R of an LC cell as a function of all parameters in a single or a series of parameter space diagrams.

Let us use the transmittance parameter space as an example. As indicated before, the transmittance of an LC cell sandwiched between the input and output polarizers is given by

$$T = T(\alpha, \gamma, \phi, \delta) = \left| (\cos\gamma \quad \sin\gamma)\, t_{\mathrm{LC}} \begin{pmatrix} \cos\alpha \\ \sin\alpha \end{pmatrix} \right|^2. \tag{2.49}$$

There are four independent parameters in this expression. If we assume that the input and output polarizers are perpendicular to each other, which is often the case for an LCD, then there are only three independent parameters left. Thus it is possible to plot the transmittance in a parameter space as shown in Figure 2.9 [12]. Here we fix the input polarizer angle α and plot the constant transmittance contour curves of $T(\Phi, \delta)$. The values of α are 0, 15°, 30°, 45° successively in the four panels. Notice also that there is no explicit λ dependence in this parameter space. The wavelength dependence is hidden in the value of δ.

Since the parameter space include all operating points of an LCD, all the useful modes can be represented as points in Figure 2.9. For example, the normal TN display has an operating point of $\alpha = 0$, $\Phi = 90°$, and $d\Delta n = 0.475$ μm. It is also very interesting to note the systematic trend of how the transmittance varies in the parameter space. In these parameter spaces, the transmittance increases in steps of 0.1. Along the x-axis, the transmittance is obviously zero. From the first panel with $\alpha = 0$, it is evident that $T = 0$ also along the y-axis. There are peaks of transmittance at $\Phi = 90°$ and 270°. These are the TN modes. In fact a vertical cut of the $\alpha = 0$ panel along $\Phi = 90°$, 180°, and 270° has been given in Figure 2.4.

Now as α increases, it can be seen that the peaks and valleys begin to break up and move sideways. The $T = 1$ point moves in a circle in the parameter space. At $\alpha = 45°$, the parameter space is totally different

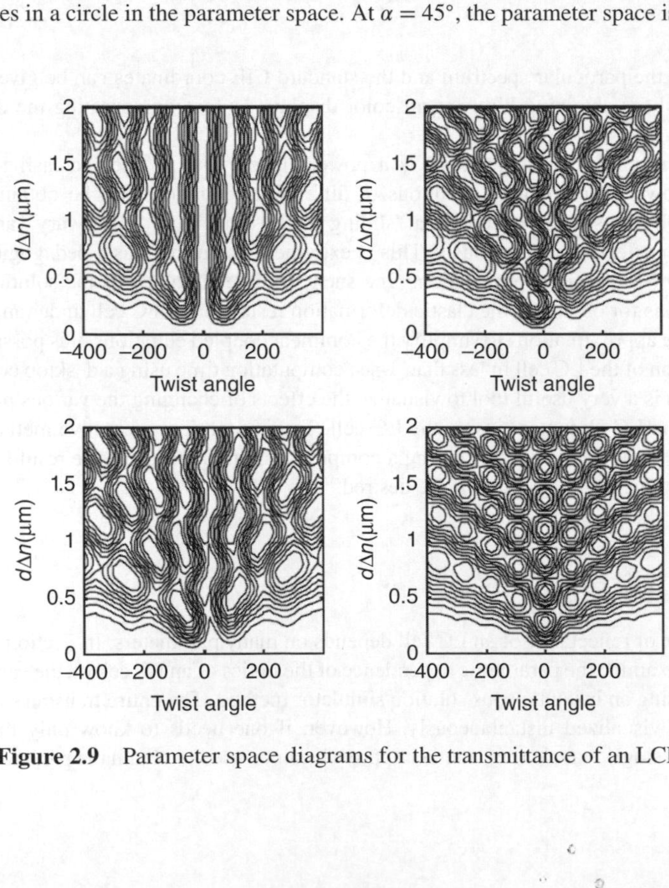

Figure 2.9 Parameter space diagrams for the transmittance of an LCD

in nature. In fact a cut along $\phi = 0$, 90°, 180°, and 360° again will be exactly as the transmittance curves given in Figure 2.5. These are the interference modes. The exact trajectory of the $T = 1$ point in the parameter space as α is varied in fact can be obtained easily. We simply calculate the $T = 0$ points for parallel polarizers. They are identical to the $T = 1$ points for perpendicular polarizers. So

$$T = \left| (\cos\alpha \quad \sin\alpha)\, t_{\mathrm{LC}} \begin{pmatrix} \cos\alpha \\ \sin\alpha \end{pmatrix} \right|^2 = 0. \tag{2.50}$$

Equating both the real and imaginary parts to zero, we obtain two equations

$$\cos\phi\cos\chi + \frac{\phi}{\chi}\sin\phi\sin\chi = 0, \tag{2.51}$$

$$\sin\chi\cos(\phi - 2\alpha) = 0. \tag{2.52}$$

As α is varied, these two equations describe a trajectory in the (Φ, δ) parameter space. In the next chapter, we shall see that these are actually the LP1 polarization conserving modes.

We can also examine the parameter space for reflective displays with one polarizer [13–17]. The single polarizer reflective display can either be used in conjunction with a polarizing beam splitter (PBS) or in direct view as shown in Figure 2.10.

For the former case of using a PBS, the reflectance is given by

$$R = \left| (\sin\alpha \quad -\cos\alpha)\, t_{\mathrm{LC}}^{\mathrm{T}} t_{\mathrm{LC}} \begin{pmatrix} \cos\alpha \\ \sin\alpha \end{pmatrix} \right|^2 \tag{2.53}$$

[see (1.272) and (1.261)]. For the latter, the reflectance is given by

$$R = \left| (\cos\alpha \quad \sin\alpha)\, t_{\mathrm{LC}}^{\mathrm{T}} t_{\mathrm{LC}} \begin{pmatrix} \cos\alpha \\ \sin\alpha \end{pmatrix} \right|^2. \tag{2.54}$$

They are actually complementary to each other. Obviously the PBS case is used for projectors and the direct view case is for direct view applications. In both cases, R depends on only three independent parameters (α, Φ, δ). The output polarizer is either perpendicular or parallel to the input polarizer (they are the same polarizer). The parameter space can be obtained readily as shown in Figure 2.11.

In Figure 2.11, the value of α also varies as 0°, 15°, 30°, and 45°. The constant reflectance contours increase in steps of 0.1. The reflectance of the x-axis is $R = 1$ for the case of (2.53) and $R = 0$ for the case of (2.54). There are peaks in R which correspond to $R = 1$ modes for projection applications. In this

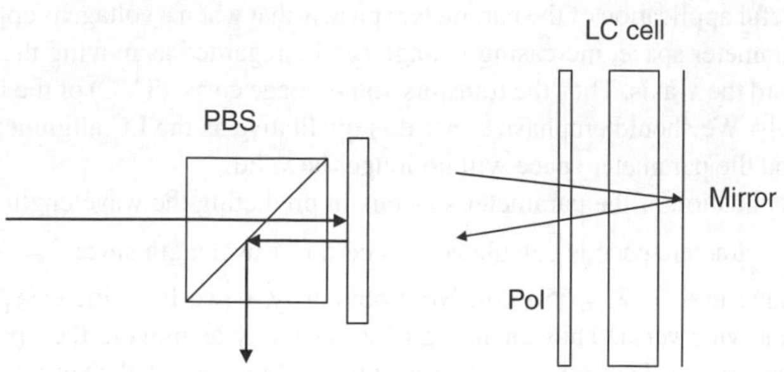

Figure 2.10 Different viewing options for the reflective display

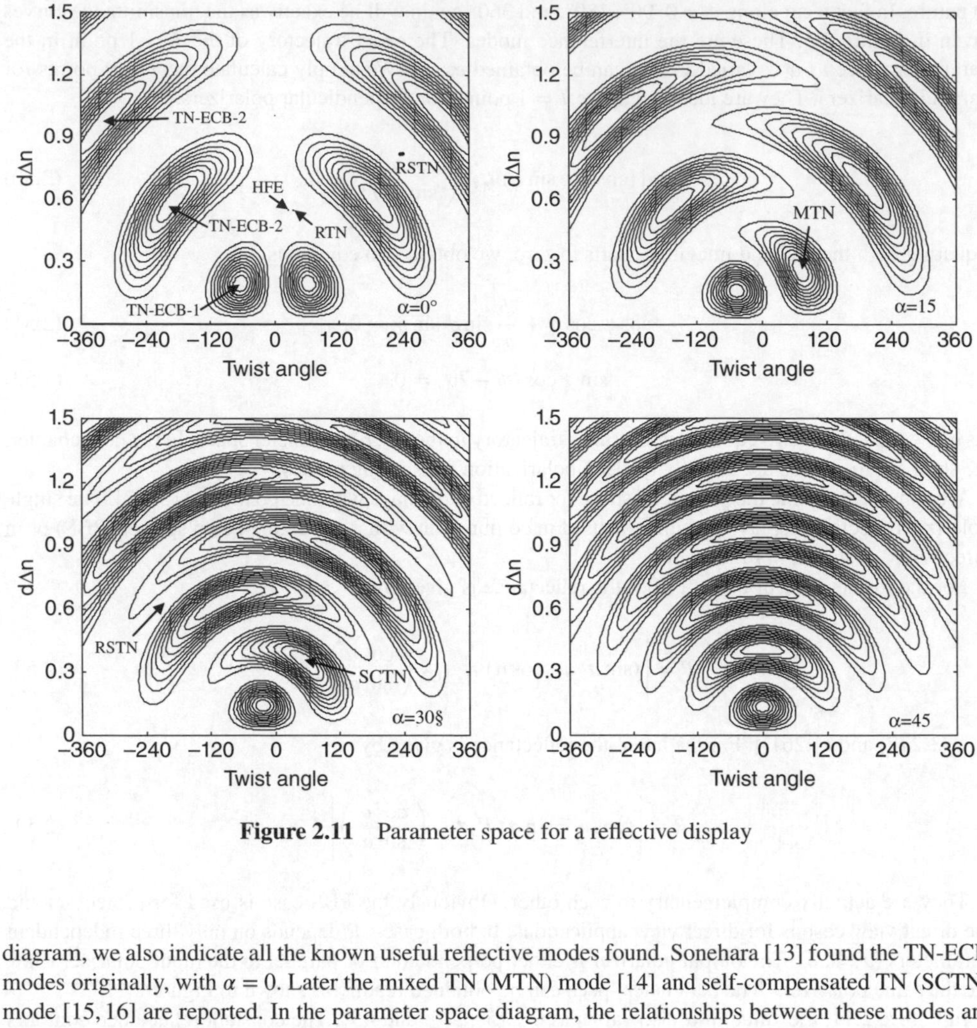

Figure 2.11 Parameter space for a reflective display

diagram, we also indicate all the known useful reflective modes found. Sonehara [13] found the TN-ECB modes originally, with $\alpha = 0$. Later the mixed TN (MTN) mode [14] and self-compensated TN (SCTN) mode [15, 16] are reported. In the parameter space diagram, the relationships between these modes are clearly seen. In fact, there are a lot of modes with $R = 1$ that can be used for projection application. They can be collectively called the mixed TN-birefringent (MTB) modes [17–21]. In the same figure, we also indicate the reflective TN (RTN), reflective STN (RSTN) and hybrid field effect (HFE) modes. These modes are off the peaks and are the operating modes for direct view applications.

One somewhat useful application of the parameter space is that when a voltage is applied, δ approaches zero. Thus in the parameter space, increasing voltage can be regarded as moving the point of operation vertically down toward the x-axis. Thus the transmission–voltage curve (TVC) of the LCD can be simply predicted qualitatively. We should emphasize that it is qualitative as the LC alignment will change as a voltage is applied and the parameter space will no longer be valid.

Another useful application of the parameter space is in predicting the wavelength dependence of the transmittance. The parameter space is calculated at a certain wavelength since $\delta = \dfrac{\pi d \Delta n}{\lambda}$. For Figures 2.9 and 2.11, the value used is $\lambda = 550$ nm. Now obviously, if we let λ increase, it is equivalent to making δ smaller, and vice versa. Thus changing of λ is exactly as moving the operating point in the parameter space up and down. Thus how rapidly the value of T changes in the vertical direction will give the wavelength dependence of the transmittance spectrum.

References

[1] R. C. Jones, "A new calculus for the treatment of optical systems. VII. Properties of the N-matrices," *J. Opt. Soc. Am.* **38**, 671 (1948).

[2] S. Chandrasekhar, *Liquid Crystals*, 2nd ed. (Cambridge University Press, 1992).

[3] A. Yariv and P. Yeh, *Optical Waves in Crystals-Propagation and Control of Laser Radiation* (John Wiley & Sons, Inc., 1984).

[4] P. Yeh and C. Gu, *Optics of Liquid Crystal Displays* (John Wiley & Sons, Inc., 1999).

[5] E. P. Raynes and R. J. A. Tough, "The guiding of plane polarized light by twisted liquid crystal layers," *Mol. Cryst. Liq. Cryst. Lett.* **2**(5), 139–145 (1985).

[6] P. Allia, C. Oldano and L Trossi, "Linear optics: matrix methods," in *Physics of Liquid Crystalline Materials*, Chapter 13, edited by I. C. Khoo and F. Simoni (Gordon and Breach, Philadelphia, 1992).

[7] C. H. Gooch and H. A. Tarry, "Optical characteristics of twisted nematic liquid-crystal films," *Electro. Lett.* **10**(1), 2–4 (1974).

[8] C. H. Gooch and H. A. Tarry, "The optical properties of twisted nematic liquid crystal structure with twist angles $\leq 90°$," *J. Phys. D: Appl. Phys.* **8**, 1575–1584 (1975).

[9] M. Schadt and F. Leenhouts, "Electrooptic performance of a new black/white highly multiplexable display," *Appl. Phys. Lett.* **50**, 236 (1987).

[10] F. Leenhouts, M. Schadt and H. J. Fromm, "Electrooptic characteristics of a new liquid crystal display with an improved grayscale capability," *Appl. Phys. Lett.* **50**, 1468 (1987).

[11] T. J. Scheffer and J. Nehring, "Investigation of the electro-optical properties of 270° chiral nematic layers in the birefringent mode," *J. Appl. Phys.* **58**, 3022 (1985).

[12] H. S. Kwok, "Parameter space representation of liquid crystal displays in transmission and reflection," *J Appl. Phys.* **80**, 3687–3693 (1996).

[13] T. Sonehara, "Photo-addressed liquid crystal SLM with twisted nematic ECB (TN-ECB) mode," *Jap. J. Appl. Phys.* **29**(7), L1231–L1234 (1990).

[14] S. T. Wu and C. S. Wu, "Mixed-mode twisted nematic liquid crystal cells for reflective displays," *Appl. Phys. Lett.* **68**(11), 1455–1457 (1996).

[15] K. H. Yang, "A self-compensated twisted nematic mode for reflective light valve," *Euro Display* 449–451 (1996).

[16] K. H. Yang and Minhua Lu, "Nematic liquid crystal modes for Si wafer-based reflective spatial light modulators," *Displays* **20**, 211–219 (1999).

[17] S. T. Tang, F. H. Yu, J. Chen, M. Wong, H. C. Huang and H. S. Kwok, "Reflective nematic liquid crystal displays I. Retardation compensation," *J. Appl. Phys.* **81**, 5924–5929 (1997).

[18] F. H. Yu, J. Chen, S. T. Tang, and H. S. Kwok, "Reflective liquid crystal displays II—elimination of retardation film and rear polarizer," *J. Appl. Phys.* **82**, 5287–5285 (1997).

[19] H. S. Kwok, J. Chen, F. H. Yu, and H. C. Huang, "Generalized mixed mode reflective LCDs with large cell gaps and high contrast ratios," *J. Soc. Inf. Display* **7**, 127–133 (1999).

[20] S. T. Tang, PhD Dissertation, *Polarization Optics of Liquid Crystal and Its Applications*, The Hong Kong University of Science and Technology, 2001.

[21] H. S. Kwok, "Reflective liquid crystal displays with mixed twisted nematic and birefringent modes," US Patent 6,341,001, January 22, 2002.

3

Optical Equivalence Theorem

3.1 General Optical Equivalence Theorem

As noted in Section 1.4.5, any lossless (unitary) polarization system may be represented by a unitary Jones matrix of the form

$$\boldsymbol{t} = \begin{pmatrix} A & -B^* \\ B & A^* \end{pmatrix}, \tag{3.1}$$

or equivalently,

$$\boldsymbol{t} = \begin{pmatrix} a+ib & -c+id \\ c+id & a-ib \end{pmatrix}, \tag{3.2}$$

where a, b, c, and d are real numbers such that $a^2 + b^2 + c^2 + d^2 = 1$ [1]. Suppose that a transmissive unitary system is characterized by a Jones matrix $\boldsymbol{t}$ of the form (3.1), referred to a Cartesian reference frame (x_I, y_I). The general optical equivalence theorem [2–8] states that the matrix $\boldsymbol{t}$ can be represented in the form

$$\boldsymbol{t} = \boldsymbol{R}(\Omega)\,\boldsymbol{WP}(\Gamma,\psi), \tag{3.3}$$

where

$$\boldsymbol{R}(\Omega) = \begin{pmatrix} \cos\Omega & -\sin\Omega \\ \sin\Omega & \cos\Omega \end{pmatrix}$$

is the Jones matrix of a polarization rotator rotating a polarization ellipse by angle Ω, and

$$\boldsymbol{WP}(\Gamma,\psi) = \begin{pmatrix} \cos\psi & -\sin\psi \\ \sin\psi & \cos\psi \end{pmatrix} \begin{pmatrix} \exp\left(-\dfrac{i\Gamma}{2}\right) & 0 \\ 0 & \exp\left(\dfrac{i\Gamma}{2}\right) \end{pmatrix} \begin{pmatrix} \cos\psi & \sin\psi \\ -\sin\psi & \cos\psi \end{pmatrix} \tag{3.4}$$

Modeling and Optimization of LCD Optical Performance, First Edition.
Dmitry A. Yakovlev, Vladimir G. Chigrinov and Hoi-Sing Kwok.

Website Companion: www.wiley.com/go/yakovlev/modelinglcd

is the Jones matrix of a wave plate (WP) with phase retardation Γ and with its fast axis at an angle ψ to the x_I-axis. In other words, the action of a unitary system on monochromatic light of a given wavelength is equivalent to the action of a wave plate in tandem with a polarization rotator.

The relations between parameters of a unitary optical system and its equivalent wave plate and rotator can be derived readily. According to (3.3),

$$t = \begin{pmatrix} \cos\Omega & -\sin\Omega \\ \sin\Omega & \cos\Omega \end{pmatrix} \begin{pmatrix} \cos\psi & -\sin\psi \\ \sin\psi & \cos\psi \end{pmatrix} \begin{pmatrix} \exp\left(-\frac{i\Gamma}{2}\right) & 0 \\ 0 & \exp\left(\frac{i\Gamma}{2}\right) \end{pmatrix} \begin{pmatrix} \cos\psi & \sin\psi \\ -\sin\psi & \cos\psi \end{pmatrix}. \tag{3.5}$$

Expanding (3.5), we obtain

$$A = \cos\frac{\Gamma}{2}\cos\Omega - i\sin\frac{\Gamma}{2}\cos(\Omega + 2\psi), \tag{3.6}$$

$$B = -\cos\frac{\Gamma}{2}\sin\Omega - i\sin\frac{\Gamma}{2}\sin(\Omega + 2\psi). \tag{3.7}$$

Consequently,

$$a = \cos\frac{\Gamma}{2}\cos\Omega, \tag{3.8}$$

$$b = -\sin\frac{\Gamma}{2}\cos(\Omega + 2\psi), \tag{3.9}$$

$$c = \cos\frac{\Gamma}{2}\sin\Omega, \tag{3.10}$$

$$d = -\sin\frac{\Gamma}{2}\sin(\Omega + 2\psi). \tag{3.11}$$

We can also invert (3.8) and (3.11) to express the equivalent WP and rotator parameters in terms of the a, b, c, d coefficients.

$$\cos^2\frac{\Gamma}{2} = a^2 + c^2, \tag{3.12}$$

$$\tan\Omega = \frac{c}{a}, \tag{3.13}$$

$$\tan 2\psi = \frac{ad - bc}{ab + cd}. \tag{3.14}$$

The existence of these solutions indicates that we can always determine the equivalent wave plate and polarization rotator parameters once we know the unitary matrix of an optical system. The exception is for the case of $a = c = 0$. Under this condition, Equations (3.8)–(3.11) are reduced to

$$b = -\sin\frac{\Gamma}{2}\cos(\Omega + 2\psi), \tag{3.15}$$

$$\cos\frac{\Gamma}{2} = 0, \tag{3.16}$$

$$b^2 + d^2 = 1, \tag{3.17}$$

which indicates that the equivalent wave plate is a half-wave plate with $\Gamma = (2N + 1)\pi$, and Ω and ψ cannot be uniquely determined. There is an infinite set of pairs of values of Ω and ψ satisfying (3.15).

Of the three parameters in the equivalent model, Ω and Γ are independent of the orientation of the reference frame (x_I, y_I). Ω is called *the characteristic angle* and Γ *the characteristic phase* of the unitary optical system [9].

This optical equivalence theorem is very useful to study some LCD modes as well as in developing LCD measurement methods.

3.2 Optical Equivalence for the Twisted Nematic Liquid Crystal Cell

In the previous chapter it is shown that the Jones matrix t_{LC} of a twisted LC layer referred to a reference frame (x, y) chosen as indicated in Section 2.3 (the x axis is parallel to the tangential constituent of the input LC director) is uniquely determined by the parameters Φ and δ of the LC layer. If we choose the reference frame (x_I, y_I) so that it is coincident with the system (x, y), we have

$$a = \cos\chi\cos\phi + \frac{\Phi}{\chi}\sin\chi\sin\Phi, \tag{3.18}$$

$$b = \frac{\delta}{\chi}\sin\chi\cos\Phi, \tag{3.19}$$

$$c = \cos\chi\sin\Phi - \frac{\Phi}{\chi}\sin\chi\cos\Phi, \tag{3.20}$$

$$d = \frac{\delta}{\chi}\sin\chi\sin\Phi \tag{3.21}$$

[see (2.3)], and, according to (3.12)–(3.14),

$$\cos^2\frac{\Gamma}{2} = \cos^2\chi + \frac{\Phi^2}{\chi^2}\sin^2\chi, \tag{3.22}$$

$$\tan\Omega = \frac{\tan\Phi - \dfrac{\Phi}{\chi}\tan\chi}{1 + \dfrac{\Phi}{\chi}\tan\Phi\tan\chi}, \tag{3.23}$$

$$\tan 2\psi = \frac{\Phi}{\chi}\tan\chi. \tag{3.24}$$

Thus, given the parameters (Φ, δ) of the LC cell, the equivalent WP and polarization rotator parameters can be determined. We emphasize that in this case, ψ is the angle between the fast axis of the equivalent wave plate and the x axis that is attached to the input LC director.

It is interesting to examine graphically the relationship between the parameters (Φ, δ) and (Ω, Γ). The results are shown in Figure 3.1. For any pair of values of (Φ, δ), the corresponding values of (Ω, Γ) can be obtained, and vice versa. In that figure, the circular contours are constant retardation curves, while the other set of contours is the constant Ω curves. Retardation values are given in units of $d\Delta n/\lambda$. For example, the first minimum TN mode will have a coordinate of (90°, 0.9), which corresponds to a Γ of 0 and a Ω of 90° in Figure 3.1. As a matter of fact, the $\Gamma = 0$ contour in Figure 3.1 gives all the Gooch–Tarry waveguiding modes for a general twisted nematic LC layer. This is a useful nomograph similar to a Smith chart in microwave design [10, 11]. It can be also applied to LCD measurements as will be shown later.

3.3 Polarization Conserving Modes

Applied to a twisted liquid crystal layer, the optical equivalence theorem can be used to obtain the conditions for achieving particular output polarization state. If a linearly polarized light is provided at

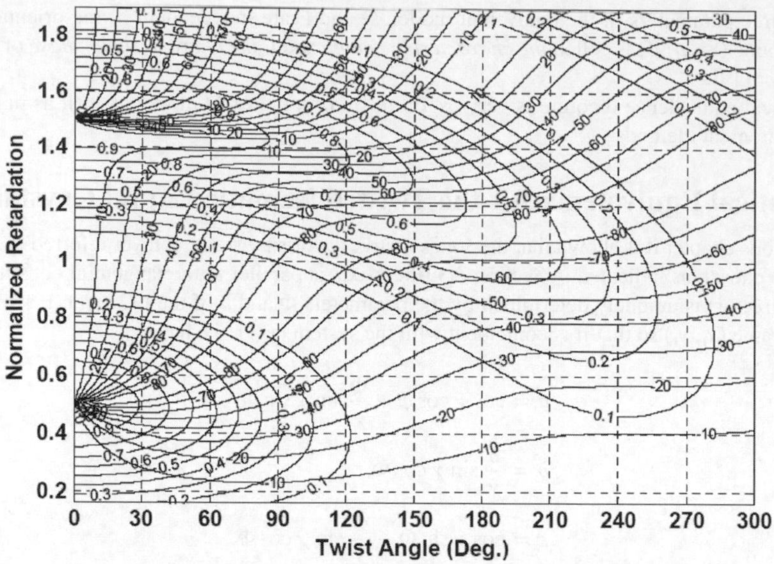

Figure 3.1 Nomograph of the equivalence parameters

the input, the output from the LC cell will in general be elliptically polarized. However, there are some conditions such that the output polarization state is either linearly polarized or circularly polarized. We shall call these, the *polarization preserving modes*. They are of interest in analyzing the operation of an LCD, as well as in measurement of LC cell parameters.

In the following consideration, we assume that the light incident on the layer is linearly polarized. The angle between the polarization plane of the incident light and the x axis (the input polarizer angle) is denoted by α.

3.3.1 LP1 Modes

If the equivalent wave plate has a retardation value that is a multiple of 2π, the LC layer acts as a pure polarization rotator regardless of the value of α. This is exactly the waveguiding, or Gooch–Tarry condition [12, 13]. Using (3.22), it can be seen that this even wave plate condition gives

$$\cos^2\chi + \frac{\Phi^2}{\chi^2}\sin^2\chi = 1 \tag{3.25}$$

or

$$\left(1 - \frac{\Phi^2}{\chi^2}\right)\sin^2\chi = 0. \tag{3.26}$$

This is possible only if $\chi = N\pi$, or

$$\frac{d\Delta n}{\lambda} = \sqrt{N^2 - \left(\frac{\Phi}{\pi}\right)^2} \tag{3.27}$$

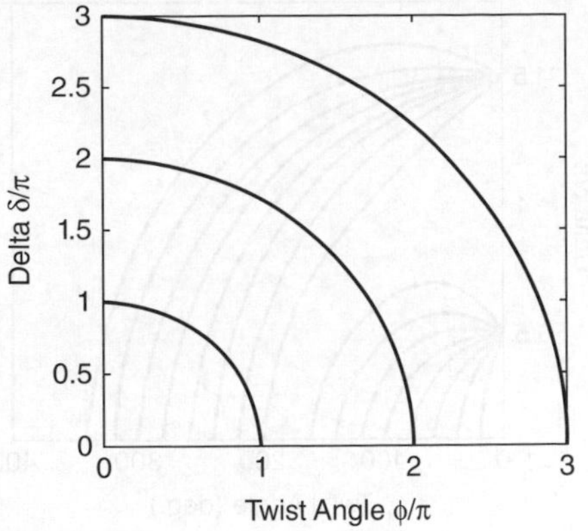

Figure 3.2 Solution space for LP1 modes

for integer values of N. This is exactly the classical Mauguin minima condition. Thus the LP1 modes are the generalization of the waveguiding modes. From (3.23),

$$\Omega = \Phi. \tag{3.28}$$

Thus the output polarization angle γ is given by

$$\gamma = \Phi + \Omega. \tag{3.29}$$

That is, the polarization plane of linearly polarized input light is rotated by an angle of Φ which is the twist angle of LC cell. This is again the same as the previously derived result for the waveguiding mode. The beauty of this optical equivalence approach is that one can find generalizations of the Gooch–Tarry waveguiding modes easily. Basically, for any input polarizer angle, there are a number of waveguiding modes that are possible. The solution space of LP1 modes is shown in Figure 3.2. From (3.27) the solution consists of circles in the $(\delta/\pi, \Phi/\pi)$ space. Figure 3.2 shows all the combinations of Φ and δ that can give waveguiding effects. Each point on the LP1 curve corresponds to a value of α that will give the waveguiding effect [14–17]. One can in fact imagine a solution space for the LP1 modes in three dimensions. The solutions are then spirals in the vertical direction with a period of 2π.

3.3.2 LP2 Modes

Another interesting condition is when the input light is linearly polarized and is parallel to the fast axis of the equivalent wave plate, that is, $\psi = \alpha$. In this case, the LC cell behaves as a pure polarization rotator and the output light will also be linearly polarized. From (3.24), we have

$$\tan 2\alpha = \frac{\Phi}{\chi} \tan \chi. \tag{3.30}$$

Using (3.23), it is easy to see that

$$\tan \Omega = \tan(\Phi - 2\alpha). \tag{3.31}$$

Thus the output polarization γ is given by

$$\gamma = \Omega + \alpha = \Phi - \alpha. \tag{3.32}$$

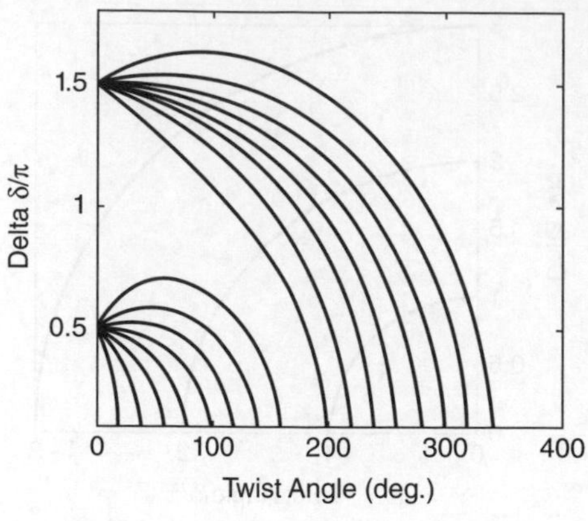

Figure 3.3 Solution space for LP2 modes

Notice that this case gives a different output polarization orientation from the LP1 solution. This fact can be applied to the design of bistable twisted nematic displays which will be discussed in a later section. For each value of α, Equation (3.30) can be solved to get a relationship between δ and Φ. Thus the parameter space for LP2 mode is an area. This is different from the LP1 mode where the solution space is described by a line. The solution space for LP2 modes is shown in Figure 3.3. Again, each value of α corresponds to a set of (Φ, δ) values that will provide the LP2 type linear polarization output. It should be noted in this case, there is no waveguiding effect.

From (3.32) it can be seen that if we require the input and output polarizations to be perpendicular to each other, as in most LCD, then α has to be 45°. Thus the LP2 modes correspond to the interference modes for LCD operation as discussed in Section 2.2. It should be mentioned that the STN LCD belongs to this class of LP2 solution. For such modes, Equation (3.30) gives

$$\chi = \left(N - \frac{1}{2}\right)\pi. \tag{3.33}$$

Thus the retardation values of such special LP2 modes are given by

$$\frac{d\Delta n}{\lambda} = \sqrt{\left(N - \frac{1}{2}\right)^2 - \left(\frac{\Phi}{\pi}\right)^2}. \tag{3.34}$$

The solution looks similar to (3.27) for the LP1 modes except for the odd integer inside the square root.

3.3.3 *LP3 Modes*

Yet another possibility for a linearly polarized output is when the equivalent wave plate is a half-wave retardation plate. The output will also be linearly polarized. From (3.22), this is possible only if

$$\cos^2\chi + \frac{\Phi^2}{\chi^2}\sin^2\chi = 0. \tag{3.35}$$

The only way for (3.35) to hold is when $\chi = (N - 1/2)\pi$ and $\Phi = 0$. Thus the equivalent wave plate is a half-wave plate only if the LC layer is a uniform cell and has a half-wave retardation. This is actually obvious. The LP3 solutions are therefore a trivial subset of the LP1 and LP2 solutions. But the interesting conclusion is that there are no other conditions that can lead to a half-wave retardation situation.

Another way of expressing this result is to find the condition for $\Omega = 0$, that is, when the system is a pure retardation plate without polarization rotation. From (3.22), it is seen that this condition is possible only if $\Phi = 0$. This result is the same as the conclusion reached in (3.35).

3.3.4 CP Modes

Finally, let us examine the conditions for obtaining a circularly polarized output, given a linearly polarized input light. In the equivalence model, the output will be circularly polarized if the equivalent wave plate has a quarter-wave retardation and the input light is polarized at 45° to its *c*-axis. From (3.22) and (3.23), we therefore obtain

$$\frac{\delta}{\chi}\sin\chi = \frac{1}{\sqrt{2}} \tag{3.36}$$

and

$$\frac{\Phi}{\chi}\tan\chi = \cot 2\alpha. \tag{3.37}$$

Therefore, given the polarization of the incoming light, conditions for (Φ, δ) can be obtained by solving (3.35) and (3.37). These are the same conditions for the MTB (mixed TN-birefringence) modes derived previously in Chapter 2 for single polarizer reflective liquid crystal displays [10]. The solution space for the CP modes is shown in Figure 3.4. They are actually the same as the MTB modes. Notice that when $\alpha = 0$, Equation (3.37) will imply that $\chi = (N - 1/2)\pi$. Substituting in (3.36) therefore gives $\Phi = \delta$. These solutions correspond exactly to the TN-ECB modes derived by Sonehara [18, 19]. This is indicated in Figure 3.4 as the tip of the solution contour touching the $\delta = \Phi$ line. Thus it can be seen again that the TN-ECB modes are a subset of the more general MTB modes. Such MTB modes are essential for the operation of reflective LCD such as those used in liquid crystal on silicon (LCOS) devices. Of course, the results discussed here are for a single wavelength. More numerical analysis is needed for the optimization of reflective modes used in LCOS, using the CP and MTB modes as guidance. This will be discussed in more detail in Section 3.5.

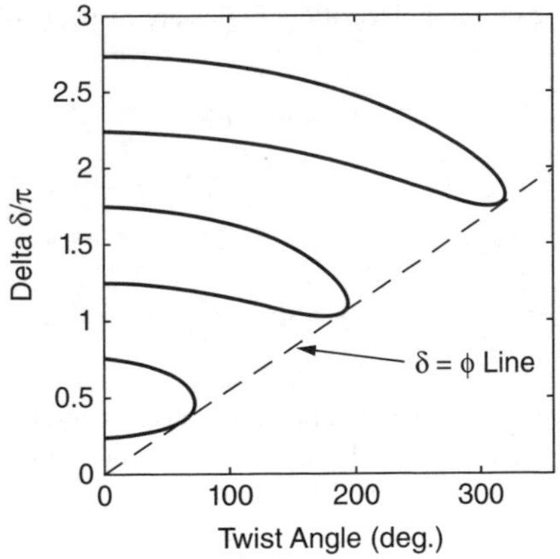

Figure 3.4 Solution space for CP modes

The LP1, LP2, and CP modes are most useful in the analysis of reflective LCD and in bistable displays. They are also valuable in developing techniques to measure the characteristic parameters of an unknown LC cell. We shall discuss them in the following sections.

3.4 Application to Nematic Bistable LCDs

Bistable LCDs are interesting because they can be used as e-paper [20–35]. There are many types of bistable LCDs such as cholesteric and ferroelectric LC. One particular class of bistable LCD is based on nematic LC cell which has two stable alignment states under zero bias voltage. There are quite a number of possibilities for this to happen [36, 37]. Essentially since the elastic deformation of the liquid crystal layer can either be splay (S), bend (B), or twist (T), it is possible to have bistable bend-splay (BBS), bistable splay-twist (BST), bistable bend-twist (BBT), and bistable twist-twist (BTT). BTT is also traditionally known as bistable TN (BTN) display. We shall keep the traditional name BTN even though the name BTT is more accurate.

For the optical optimization to be discussed here, we shall only be concerned with the BTN display. This can be possible if the *d*/*P* ratio of the LC cell is such that both LC alignment configurations have the same elastic energy. There are two possible BTN configurations. In the original discovery of bistability, it was found that the zero twist and 2π twist states can have the same deformation elastic energy and can be bistable [20]. They shall be called the 2π-BTN display. It was found that the zero twist and π-twist states can also be bistable. They shall be known as the π-BTN display. In this book, we shall not be concerned with the physics of the bistable configuration or their switching mechanisms. We shall just examine the optics of such displays using the polarization conserving modes discussed above to optimize the optical properties of such displays. Since the BTNs have no applied voltage, they are ideally suited for such optimization studies.

3.4.1 2π Bistable TN Displays

It was found that in addition to $(0, 2\pi)$ bistability, it is possible to have $(\Phi_0, \Phi_0 + 2\pi)$ bistability. As long as the two twist states differ by an angle of 2π, bistable conditions can be maintained. Here Φ_o can be any angle. Thus Φ_o can be used as an independent parameter to optimize the optical properties of the 2π-BTN.

From the above discussions on polarization conserving modes, it is obvious that if the input light is linearly polarized, then an optimal condition can be achieved if one of the bistable states is an LP1 mode and the other an LP2 mode. The output polarization angles are respectively

$$\gamma = \Phi + \alpha \tag{3.38}$$

and

$$\gamma = \Phi - \alpha \tag{3.39}$$

for the two conditions. So if we let the two output polarizations be perpendicular to each other, the optical property will be the best. One twist state will have $T = 1$ and other will have $T = 0$. Suppose we let the Φ_o twist state be LP2 and the $\Phi_o + 2\pi$ state be LP1, then

$$\alpha = (2N - 1)\frac{\pi}{4}. \tag{3.40}$$

Table 3.1 2π–BTN LCD construction conditions using LP solutions

Mode number	$d\Delta n$ (μm)	Φ_1 (°)	Φ_2 (°)	α (°)	γ (°)
1	0.3995	−124	236	45	−79
2	0.765	−101	259	45	34
3	1.0735	−79	281	45	−34
4	1.364	−57	303	45	78
5	1.647	−33	327	45	12
6	0.273	−11	349	45	−56
7	1.925	−11	349	45	−56
8	2.200	11	371	45	56
9	2.473	34	394	45	−11
10	2.745	56	416	45	−79
11	0.522	56	416	45	−79
12	0.733	124	484	45	79
13	0.932	191	551	45	56

That is, the input polarizer should be placed at odd multiples of $\pi/4$ to the input director. Without loss of generality, we can set $\alpha = \pi/4$. From (3.27) and (3.34), the retardation of the LC cell is given by

$$\delta^2 + \Phi_o^2 = \left(N - \frac{1}{2}\right)^2 \pi^2, \tag{3.41}$$

$$\delta^2 + \left(\Phi_o + 2\pi\right)^2 = M^2\pi^2, \tag{3.42}$$

where M, N are integers. So every possible combination of M and N, there will be a solution of the 2π-BTN where one twist state will have $T = 0$ and the other twist state has $T = 1$. Table 3.1 shows some possible solutions.

One can also reverse the situation and allow the Φ_o twist state be LP1 and the $\Phi_o + 2\pi$ state be LP2 solutions. Then another set of possible operating conditions can be obtained. Figure 3.5 shows the optical transmittance of the on- and off-states of several 2π-BTN modes.

3.4.2 π Bistable TN Displays

The 2π-BTN LCD is not truly bistable as both twist states will always decay to the middle π-twist state which is more stable [22,23]. The π-BTN LCD is therefore a more attractive alternative. Here the stable twist states are zero twist and π-twist. Again, we can generalize the stable twist states to be (Φ_o, $\Phi_o + \pi$). Again, solutions for Φ_o can be obtained readily if we assume one twist state to be an LP1 mode and the other to be an LP2 mode.

As in the case of 2π-BTN LCD, the input polarizer angle should be given by $\alpha = \pi/4$. The retardation value of the cell is also governed by the LP1 and LP2 solutions as

$$\delta^2 + \Phi_o^2 = \left(N - \frac{1}{2}\right)^2 \pi^2, \tag{3.43}$$

$$\delta^2 + \left(\Phi_o + \pi\right)^2 = M^2\pi^2, \tag{3.44}$$

for integer values of M and N. Table 3.2 shows some of the solutions and Figure 3.6 shows the transmission spectra of the first two of these solutions. The first solution of mode #1 is very good in giving an almost TN mode like low dispersion transmission spectra.

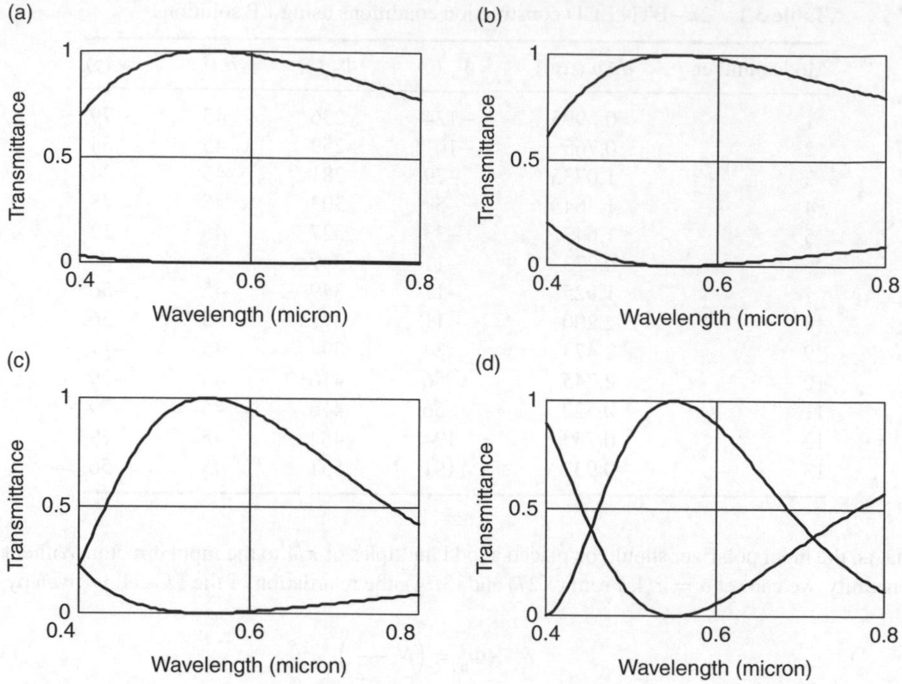

Figure 3.5 Optically optimized 2π-BTN LCD

One can also derive conditions for the single polarizer bistable TN display. In this case, one of the modes should be a CP mode. Thus the two modes should be LP1-CP or LP2-CP. This is left as an exercise for the reader.

3.5 Application to Reflective Displays

A reflective display is defined as one with only a front polarizer [38–52]. There is a reflector in the back but no rear polarizer. They are useful when either the substrate of the LCD is opaque or when the reflector is inside the LC cell. The former is the case for silicon substrate and the latter case will arise for transflective displays. Such reflective displays are applied in two possible optical arrangements as shown in Figure 2.10. The case of using a PBS is used in reflective microdisplay projectors. The direct view case is used in a transflective display where a single polarizer reflective sub-pixel may be needed.

Table 3.2 The optical optimized π-configuration parameters

Mode	Φ_1 (°)	Φ_2 (°)	$d\Delta n$ (μm)	α (°)	γ (°)
#1	−22.5	157.5	0.266	45	−67.5
#2	22.5	202.5	0.546	45	67.5
#3	67.5	247.5	0.799	45	22.5
#4	112.5	292.5	1.045	45	−22.5
#5	157.5	337.5	1.288	45	−67.5

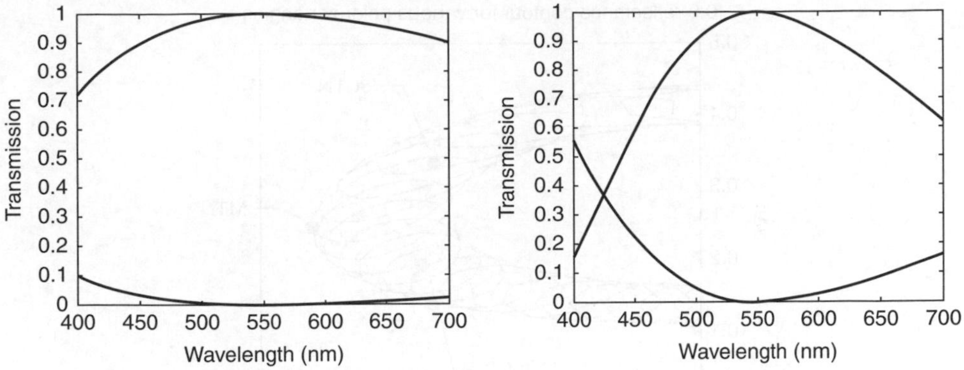

Figure 3.6 Transmission spectra of the first two π-BTN modes

For the reflective display, the LC cell has to function as a quarter-wave plate (QWP). Then linearly polarized input light will be turned into circular polarization after going through the LC cell. Upon reflection, the handedness of the wave changes. Therefore when it goes through the LC cell again, it become linearly polarized but is rotated by 90°. This situation is described in Section 1.4.4 (*property* 2).

Thus for a reflective LCD, the LC cell parameters (α, Φ, δ) should be that of a CP mode. The CP mode parameter space is shown in Figure 3.4. It shows the solution for $R = 0$ with a direct view configuration or $R = 1$ for a PBS projector configuration. The situation is shown in Table 3.3 which indicates the reflectance of the LCD for solutions given in Figure 3.4 which corresponds to the CP mode.

If we want to have a normally bright display for direct view, then a totally different mode needs to be used. That mode has to situate between the CP mode contours of Figure 3.4 and will not be discussed here. Now each point on the solution contour corresponds to a particular value of α. Any point along the solution contour can function as a reflective mode. Thus there are infinite combinations of (α, Φ, δ) that can be used. Obviously the parameter space for CP mode is only valid for the no bias voltage situation. For the design of a reflective display, the reflectance–voltage curve has to be taken into consideration. So one should take all possible solutions of (α, Φ, δ) and then perform the full voltage-on simulation using numerical software as shown in Chapter 4 to obtain the RVC. This is obviously a very tedious procedure.

Another consideration is that instead of requiring $R = 1$ for the no-bias voltage condition, it may be better to allow for some compromises in R. For example, it may be better to allow R to be 0.9 and achieve a better contrast. The complete regions of $R > 0.9$ for various (α, Φ, δ) are shown in the parameter space in Section 2.4. Here we examine the case of the reflective LCD near the first TN-ECB mode. The $R = 0.9$ curves are shown in Figure 3.7 for various values of α. All (α, Φ, δ) inside the small circles are possible solutions. Indeed, the TN-ECB mode of Sonehara [40], the MTN mode of Wu and Wu [45,46], and the SCTN mode of Yang [47,48] can all be found inside this parameter space. The practical reflective mode used in commercial products is also within this range of parameters.

It should also be mentioned that single polarizer reflective modes can also be useful in the analysis of transflective liquid crystal displays [53–56]. In such transflective displays, the reflective mode quite

Table 3.3 Reflectance of the single polarizer LCD

	Projector	Direct view
No voltage applied	$R = 1$	$R = 0$
Voltage applied	$R = 0$	$R = 1$

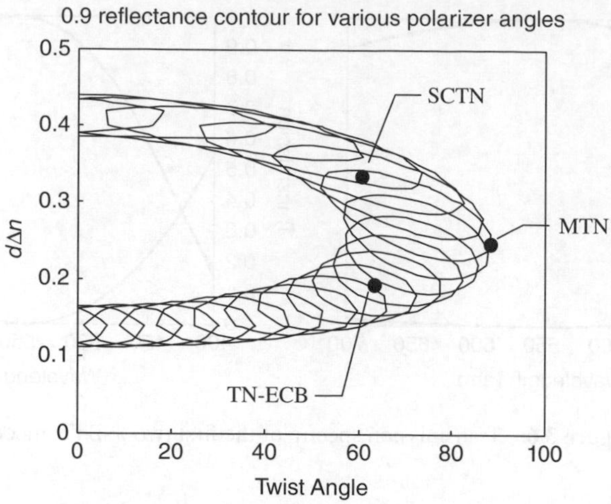

Figure 3.7 Parameter space for reflective LCD

often is of a single polarizer design as in the present case. There is only one front polarizer and the rear reflector is partially reflecting. Thus is is useful to invoke CP modes in the design of transflective displays as well.

3.6 Measurement of Characteristic Parameters of an LC Cell

The optical equivalence theorem can also be applied to measure the properties of an unknown LC cell with unknown twist angle and unknown retardation value. Using the optical equivalence theorem, one can understand better the operational principles of conventional techniques of LC cell measurement [57–65], and discover new methods for such measurements. Here we shall discuss the idea of measuring the characteristic angle Ω and the characteristic phase Γ of the equivalent WP and polarization rotator. Equations (3.11) and (3.12) can then be used to find Φ and δ. It is also possible to simply read the results of the nomograph in Figure 3.1. Here the retardation is normalized by the incident wavelength. It is important to note that the sign of the characteristic angle Ω is critical. It includes information on the twist sense of the LC layer. In Figure 3.1, the sense of the LC twist angles is used as the reference. Characteristic angle of the same sense are said to be positive and vice versa.

There have been some methods proposed for the determination of characteristic parameters of an optical gadget [3–5]. The simplest method to determine the characteristic angle Ω is the iterative method of Srinath and Keshavan [65]. A modified Senarmont method is suitable for determining the characteristic phase Γ. When determining the characteristic angle Ω, the direction of the fast axis of the equivalent wave plate, $\mathbf{F}_{WP}$, is also determined.

3.6.1 *Characteristic Angle Ω*

The characteristic angle is determined if we can align the input polarizer with the axis of the equivalent WP. The experimental set-up is shown in Figure 3.8. The unknown LC cell S and the analyzer A are mounted on rotary stages. The two rotary stages were rotated iteratively until a minimum transmission (should be zero theoretically) is obtained. Then the polarizer direction is parallel (or perpendicular) to $\mathbf{F}_{WP}$, and the angle between the analyzer and the polarizer is equal to $\Omega \pm 90°$. In other words, the LC cell and the input polarizer can be regarded as a single element operating in the LP2 condition.

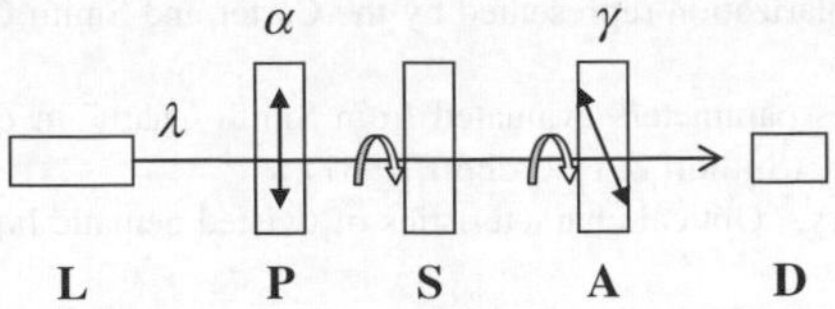

Figure 3.8 Measurement setup for the characteristic parameters

The trick for this measurement is that there are two rotation stages. They have to be varied systematically to find the minimum. By using the iterative procedure, the final analyzer direction can converge by a factor approximately equal to $\sin^2(\Gamma/2)$. Usually two to three iterations will be needed to arrive at the final analyzer position.

3.6.2 Characteristic Phase Γ

The conventional Sernamont method of finding the phase value of a waveplate is as follows. It makes use of a quarter wave plate (QWP) as an analyzer. The input polarizer is set parallel to the QWP and then the unknown wave plate is oriented at 45° to the input polarizer. The analyzer is then rotated to find the null or maximum transmission direction. The unknown wave plate phase angle Γ is equal to two times the angle between the maximum transmission analyzer direction and the input polarizer direction. To obtain the equivalent wave plate characteristic phase Γ, it is possible to employ a modified Sernamont method. The set-up is the same as in Figure 3.8. The equivalent wave plate axis is again set 45° to the input polarizer. Then the following procedure can be used:

1. Complete the procedure to obtain the characteristic angle Ω first.
2. Further rotate the unknown optical media by 45°.
3. Insert a QWP in between the optical media and the analyzer. The QWP axis is parallel to the analyzer. That is at Ω angle to the input polarizer.
4. Rotate the analyzer to get the minimum transmission (should be zero theoretically). Then rotate 90° to get the maximum transmission direction.
5. The angle between the QWP and the analyzer is equal to half the characteristic phase Γ.

References

[1] E. Collett, *Polarized Light-Fundamentals and Applications* (Marcel Dekker, Inc., 1992).

[2] R. C. Jones, "A new calculus for the treatment of optical systems. II Proof of Three General Equivalent Theorems," *J. Opt. Soc. Am.* **31**, 493 (1941).

[3] L. S. Srinath and A. V. S. S. S. R. Sarma, "Determination of the optically equivalent model in three dimensional photoelasticity," *Exp. Mec.* **14**, 118–122 (1974).

[4] L. S. Srinath, K. Ramesh, and V. Ramamurti, "Determination of characteristic parameters in three dimensional photoelasticity," *Opt. Eng.* **27**, 225–230 (1988).

[5] H. Aben, *Integrated Photoelasticity* (McGraw Hill Inc., 1979).

[6] H. Aben, "Optical phenomena in photoelastic models by the rotation of principal axes," *Exp. Mec.* **6**, 13–22 (1966).

[7] A. Kuske and G. Robertson, *Photoelastic Stress Analysis*, Chapter 13 (John Wiley & Sons, Inc., 1974).

[8] P. S. Theocaris and E. E. Gdoutos, *Matrix Theory of Photoelasticity* (Springer-Verlag, 1979).

[9] H. Aben, "Characteristic directions in optics of twisted birefringence media," *J. Opt. Soc. Am. A* **3**, 1414–1421 (1986).

[10] P. S. Theocaris, "Elliptic polarization represented by the Carter and Smith Charts," *Appl. Opt.* **18**(23), 4017–4024 (1979).

[11] P. S. Theocaris, "The stokes parameters evaluated from Smith Chart," in *Optical Methods in Mechanics of Solids*, edited by A. Lagarde, (Sijthoff & Noordhoff, 1981).

[12] C. H. Gooch and H. A. Tarry, "Optical characteristics of twisted nematic liquid-crystal films," *Electron. Lett.* **10**(1), 2–4 (1974).

[13] C. H. Gooch and H. A. Tarry, "The optical properties of twisted nematic liquid crystal structure with twist angles $\leq 90°$," *J. Phys. D Appl. Phys.* **8**, 1575–1584 (1975).

[14] S. T. Tang, PhD dissertation, *Polarization Optics of Liquid Crystal and Its Applications* The Hong Kong University of Science and Technology, 2001.

[15] S. T. Tang and H. S. Kwok, "3×3 Matrix for unitary optical systems," *J. Opt. Soc. Am. A* **18**, 2138–2145 (2001).

[16] S. T. Tang and H. S. Kwok, "A new method for measuring liquid crystal cell parameters," *Soc. Inf. Display Symp. Dig*. 552–555 (1998).

[17] S. T. Tang and H. S. Kwok, "A new method of designing LCDs with optimal optical properties," *Soc. Inf. Display Symp. Dig*. 195–197 (1999).

[18] T. Sonehara and O. Okumura, "A new twisted nematic ECB (TN-ECB) mode for the a reflective light valve," Japan Display'89, in Proceedings of the 9th International Display Research Conference, pp. 192–195 (1989).

[19] T. Sonehara, "Photo-addressed liquid crystal SLM with twisted nematic ECB (TN-ECB) mode," *Jpn. J. Appl. Phys*. **29**(7), L1231-L1234 (1990).

[20] D. W. Berremen and W. R. Heffer, "New bistable liquid-crystal twist cell," *J. Appl. Phys*. **52**, 3032 (1981).

[21] T. Z. Qian, Z. L. Xie, H. S. Kwok, and P. Sheng, "Dynamic flow and switching bistability in twisted nematic liquid crystal cells," *Appl. Phys. Lett*. **71**, 632 (1997).

[22] I. Dozov, M. Nobili, and G. Durand, "Fast bistable nematic display using monostable surface switching," *Appl. Phys. Lett.* **70**, 1179 (1997).

[23] R. Barbcri, M. Giocondo, J. Li, R. Bartolino, I. Dozov, and G. Durand, "Fast bistable nematic display with grey scale," *Appl. Phys. Lett.* **71**, 3459 (1997).

[24] C. D. Hoke, J. Li, J. R. Kelly, and P. J. Bos, "Matrix addressing of a 0-360 degree bistable twist cell," *SID Symp. Dig*. **28**, 29 (1997).

[25] Z. L. Xie and H. S. Kwok, "New bistable twisted nematic liquid crystal displays," *J. Appl. Phys*. **84**, 77 (1998).

[26] Z. L. Xie and H. S. Kwok, "Reflective bistable twisted nematic liquid crystal display," *Jpn. J. Appl. Phys*. **37**, 2572 (1998).

[27] S. W. Suh, Z. Zhuang, and J. Patel, "36.4: propagation and optimization of stokes parameters for arbitrary twisted nematic liquid crystal," *SID Dig*. 997 (1998).

[28] J. Chen, F. H. Fu, H. C. Huang, and H. S. Kwok, "Reflective supertwisted liquid crystal displays," *Jpn. J. Appl. Phys*. **37**, 217 (1998).

[29] G.-D. Lee, H.-S. Kim, T.-H. Yoon, J. C. Kim, and E.-S. Lee, "High-speed-addressing method of a bistable twisted-nematic LCD," *SID Symp. Dig*. **29**, 842 (1998).

[30] H. Bock, "Random domain formation in 0°–360° bistable nematic twist cells," *Appl. Phys. Lett*. **73**, 2905 (1998).

[31] T. Tanaka, T. Obikawa, Y. Sato, H. Nomura, and S. Iino, "An advanced driving method for bistable twisted nematic (BTN) LCD," *IDRC* (Asia Display) 295–298 (1998).

[32] Y. J. Yim, Z. Zhuang, and J. Patel, "Reflective single-polarizer bistable nematic liquid crystal display with optimum twist," *SID Dig*. 868 (1999).

[33] S. T. Tang, H. W. Chiu, and H. S. Kwok, "Optically optimized transmittive and reflective bistable twisted nematic liquid crystal displays," *J. Appl. Phys*. **87**, 632 (2000).

[34] J. X. Guo, Z. G. Meng, M. Wong, and H. S. Kwok, "Three-terminal bistable twisted nematic liquid crystal displays," *Appl. Phys. Lett*. **77**, 3716 (2000).

[35] T. Z. Qian, Z. L. Xie, H. S. Kwok, and P. Sheng, "Dynamic flow, broken surface anchoring and bistability in three-terminal twisted nematic liquid crystal displays," *J. Appl. Phys*. **90**, 3121 (2001).

[36] H. S. Kwok and S. T. Tang, "Bistable twisted nematic liquid crystal displays," US Patent No. 6,707,527, March 2004.

[37] H. S. Kwok and J. X. Guo, "Optically optimized permanently bistable twisted nematic liquid crystal displays," US Patent No. 6,784,955B2, August 2004.

[38] J. Grinberg, A. Jacobson, W. Bleha, L. Miller, L. Fraas, D. Boswell, and G. Myer, "A new real-time non-coherent to coherent light image converter the hybrid field effect liquid crystal light valve," *Opt. Eng*. **14**(3), 217–225 (1975).

[39] R. Kmetz, "A single-polarizer twisted nematic display," *Proc. SID* **21**, 63–65 (1980).
[40] T. Sonehara, "Photo-addressed liquid crystal SLM with twisted nematic ECB (TN-ECB) mode," *Jpn. J. Appl. Phys*. **29**(7), L1231–L1234 (1990).
[41] V. Konovalov, A. Muravksi, S. Yakovenko, A. Smirnov, and A. Usenok, "A single-polarizer color reflective AMLCD with improved contrast," *SID dig*. 615–615 (1994).
[42] H. C. Huang, P. W. Cheng, W. C. Yip, H. S. Kwok, C. S. Li, and Y. Liao, "AMLCD on a Silicon Substrate with Integrated Digital Data Driver," in Proceedings of The 15th International Display Research Conference (Asia Display '95), pp. 481–484 (1995).
[43] I. Fukuda, E. Sakai, Y. Kotani, M. Kitamura, and T. Uchida, "A new achromatic reflective STN-LCD with one polarizer and one retardation," *J. SID*. **3**(2), 83–87 (1995).
[44] K. Lu and B. E. A. Saleh, "A single polarizer reflective LCD with wide-viewing-angle range for gray scales," "SID '96, Digest of Applications Papers, 63–66 (1996).
[45] S. T. Wu and C. S. Wu, "Mixed-mode twisted nematic liquid crystal cells for reflective displays," *Appl. Phys. Lett*. **68**(11), 1455–1457 (1996).
[46] S. T. Wu, C. S. Wu, and C. L. Kuo, "Reflective direct view and projection displays using twisted nematic LCD," *Jpn. J. Appl. Phys*. **36**, 2721–2727 (1997).
[47] K. H. Yang, "A self-compensated twisted nematic mode for reflective light valve," *Euro. Display* 449–451 (1996).
[48] K. H. Yang and M. Lu, "Nematic liquid crystal modes for Si wafer-based reflective spatial light modulators," *Displays* **20**, 211–219 (1999).
[49] S. Stallinga, "Berreman 4 × 4 matrix method for reflective liquid crystal displays," *J. Appl. Phys*. **85**, 3023–3031, (1999).
[50] S. T. Tang, F. H. Yu, J. Chen, M. Wong, H. C. Huang, and H. S. Kwok, "Reflective nematic liquid crystal displays I. retardation compensation," *J. Appl. Phys*. **81**, 5924–5929 (1997).
[51] F. H. Yu, J. Chen, S. T. Tang, and H. S. Kwok, "Reflective liquid crystal displays II—elimination of retardation film and rear polarizer," *J. Appl. Phys.* **82**, 5287 (1997).
[52] H. S. Kwok, J. Chen, F. H. Yu, and H. C. Huang, "Generalized mixed mode reflective LCDs with large cell gaps and high contrast ratios," *J. Soc. Inf. Disp.* **7**, 127–133, (1999).
[53] J. Kim, D. W. Kim, C. J. Yu, and S. D. Lee, "New configuration of a transflective liquid crystal display having a single cell gap and a single liquid crystal mode," *Jpn. J. Appl. Phys.* **43**, L1369 (2004).
[54] H. Y. Mak, P. Xu, X. Li, V. Chigrinov, and H. S. Kwok, "Single cell gap transflective liquid crystal display without sub pixel separation," *Jpn. J. Appl. Phys*. **46**, 7798 (2007).
[55] T. Du, H. Y. Mak, P. Xu, V. Chigrinov, and H. S. Kwok, "Single twist nematic mode single cell gap transflective liquid crystal display," *Jpn. J. Appl. Phys*. **48**, 010209 (2009).
[56] T. Du, H. Y. Mak, P. Xu, V. Chigrinov, and H. S. Kwok, "Transflective liquid crystal display using low twist nematic and electrically controlled birefringence modes by photoalignment technology," *Jpn. J. Appl. Phys*. **48**, 010210 (2009).
[57] T. Inoue, "Method of measuring thickness of liquid crystal cells," US Patent 5,239,365, 1993.
[58] A. Lien and H. Takano, "Cell gap measurement of filled twisted nematic liquid crystal displays by a phase compensation method," *J. Appl. Phys*. **69**, 1304–1309 (1991).
[59] A. Lien, "Simultaneous measurement of twist angle and cell gap of a twisted nematic cell by an optical method," *IDRC* 192–194 (1991), 191.
[60] X. Shao, T. Yu, Z. Wang, J. Yuan, J. Guo, and X. Huang, "A new method to measure thickness of twist nematic liquid crystal cells," *EuroDisplay*, 309–312 (1996).
[61] H. T. Jessop, "On the Tardy and Senarmont methods of measuring fractional relative retardations," *Br. J. Appl. Phys*. **45**, 138–141 (1953).
[62] N. Vansteenkiste, P. Vignoolo, and A. Aspect, "Optical reversibility theorems for polarization: application to remote control of polarization," *J. Opt. Soc. Am. A* **10**(10), 2240–2245 (1993).
[63] S. T. Tang and H. S. Kwok, "Reflective method of LCD measurement," *Int'l Disp Works*, 109–111 (2000).
[64] S. T. Tang and H. S. Kwok, "Transmissive liquid crystal cell parameters measurement by spectroscopic ellipsometry," *J. Appl. Phys*. **89**, 80–85 (2001).
[65] L. S. Srinath and S. Y. Keshavan, "A fast iterative procedure to determine photoelastic characteristic parameters," *Mech. Res. Comm*. **5**, 159–165 (1978).

4

Electro-optical Modes: Practical Examples of LCD Modeling and Optimization

In this chapter, we consider different LC electro-optical modes which are used in various LC devices. Many practical examples of modeling and optimization of the LCD performance for different kinds of LCD are presented.

4.1 Optimization of LCD Performance in Various Electro-optical Modes

4.1.1 Electrically Controlled Birefringence

Director Distribution

The most important geometries of electrically controlled birefringence (ECB) are shown in Figure 4.1.

A compromise between dielectric and elastic torques results in the reorientation of the director from the initial alignment $\theta(z)$ with the maximum deviation θ_m at the center of the layer (the Fréedericksz transition). The effect occurs when the electric field exceeds a certain threshold value:

$$U_F = \pi(4\pi K_{ii}/\Delta\varepsilon)^{1/2}, \tag{4.1}$$

where the LC elastic constant $K_{ii} = K_{11}$ or K_{33} for the splay (S) and bend (B) Fréedericksz transitions, respectively (Figures 4.1a and b), and $\Delta\varepsilon = \varepsilon_{\parallel} - \varepsilon_{\perp}$ is the LC dielectric anisotropy. The initial director alignment is homogeneous planar for the S-effect ($\mathbf{E} \perp \mathbf{L}_0$) and homeotropic for the B-effect ($\mathbf{E} \parallel \mathbf{L}_0$). The action of the electric field on the LC layer results in the deformation of the initial molecular (LC director) distribution and a corresponding variation in the LC cell optical properties [1–3]. Director **L** reorients in an electric field under the action of the dielectric torque, which is proportional to the dielectric anisotropy $\Delta\varepsilon$. The corresponding contribution g_ε to the density of the nematic free energy gives [1–3]

$$g_\varepsilon = -\mathbf{DE}/8\pi = -\varepsilon_\perp E^2/8\pi - \Delta\varepsilon\,(\mathbf{EL})^2/8\pi, \tag{4.2}$$

Modeling and Optimization of LCD Optical Performance, First Edition.
Dmitry A. Yakovlev, Vladimir G. Chigrinov and Hoi-Sing Kwok.

Website Companion: www.wiley.com/go/yakovlev/modelinglcd

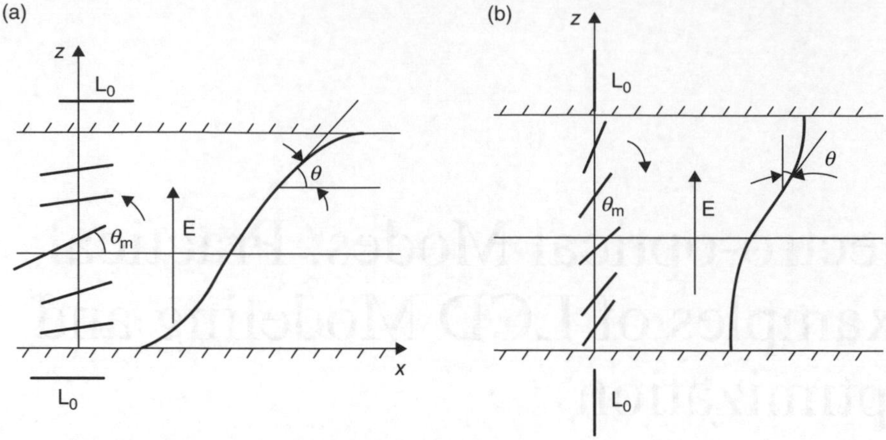

Figure 4.1 Electrically controlled birefringence (ECB) mode for various LC configurations: (a) S-effect (homogeneous planar alignment, electric field vertical or perpendicular to the substrates); (b) B-effect or vertically aligned nematic (VAN) effect (homeotropic alignment) [1–3]

that is, director **L** tends to align itself along the field ($\mathbf{L}\|\mathbf{E}$) if $\Delta\varepsilon > 0$ and is perpendicular to it ($\mathbf{L} \perp \mathbf{E}$), provided that $\Delta\varepsilon < 0$.

The elastic torque supports the initial director orientation, fixed by the boundary conditions on the surface (S), which in case of a strong surface anchoring is

$$\mathbf{L}|_S = \mathbf{L}_0. \tag{4.3}$$

As a result of this a compromised director profile appears, which satisfies to the condition of the minimum free energy:

$$F_V = \int\limits_V (g_k + g_\varepsilon)\mathrm{d}\tau, \tag{4.4}$$

where

$$g_k = {}^1\!/_2\{K_{11}(\nabla \cdot \mathbf{L})^2 + K_{22}[\mathbf{L} \cdot (\nabla \times \mathbf{L}) - q_0]^2 + K_{33}[\mathbf{L} \times (\nabla \times \mathbf{L})]^2\}$$

is the elastic energy density and K_{ii} are the elastic moduli; $q_0 = 2\pi/p_0$ characterizes the "natural chirality" and equals zero in pure nematics. p_0 is the natural helix pitch in the LC mixture induced by a chiral dopant; usually the value of the helix pitch is inversely proportional to the concentration of the dopant [1–3].

In the more general case of a finite director anchored at the boundaries, we can write the total energy F of LCs as follows:

$$F = F_v + F_s, \tag{4.5}$$

where F_s is the surface energy, usually characterized by different anchoring terms known as "polar" and "azimuthal" anchoring energies [1–3]. The corresponding contributions to the anchoring F_s on the surface S are often written as (Figure 4.2)

$$W_\theta = {}^1\!/_2\, W_{\theta 0} \sin^2(\theta - \theta_d), \quad W_\varphi = {}^1\!/_2\, W_{\varphi 0} \sin^2(\varphi - \varphi_d), \tag{4.6}$$

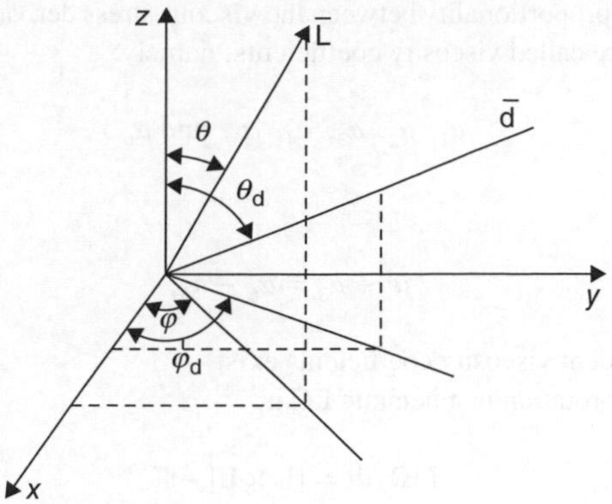

Figure 4.2 LC interaction with a solid surface [1–3]

where θ and φ are the corresponding polar and azimuthal angles of the LC director at the surface S; θ_d and φ_d are, respectively, the polar and azimuthal angles of the so-called easy direction (easy axis) [1–3] on S; and W_θ and W_φ are the corresponding anchoring energies.

The dynamics of nematic LCs is described by (i) the director field $\mathbf{L}(\mathbf{r},t)$ and (ii) velocities of the centers of the molecules $\mathbf{V}(\mathbf{r},t)$. These variables in general obey the following equations:

1. The equation of continuity in incompressible liquids:

$$\nabla \cdot \mathbf{V} = 0. \tag{4.7}$$

2. The Navier–Stokes equation in an anisotropic viscous liquid:

$$\rho(\partial v_i/\partial t + \partial v_i/\partial x_k) = f_i + \partial\sigma'_{ki}/\partial x_k, \tag{4.8}$$

where i, k, x_i, $x_k = x, y, z$, and

$$f_i = -\partial P/\partial x_i + QE_i \tag{4.9}$$

is the external force in the anisotropic liquid dielectric and σ'_{ki} is the viscous stress tensor:

$$\sigma'_{ki} = \alpha_1 L_k L_i A_{mn} L_m L_n + \alpha_2 L_k N_i + \alpha_3 L_i N_k + \alpha_4 A_{ki} + \alpha_5 L_k L_m A_{mi} + \alpha_6 L_i L_m A_{km}. \tag{4.10}$$

P is an external pressure

$$A_{ij} = \tfrac{1}{2}(\partial v_i/\partial x_j + \partial v_j/\partial x_i) \tag{4.11}$$

and is analogous to the viscous stress tensor for the anisotropic liquid, and

$$\mathbf{N} = \mathrm{d}\mathbf{L}/\mathrm{d}t - \tfrac{1}{2}[\mathbf{L} \times (\nabla \times \mathbf{v})] \tag{4.12}$$

is the rate of motion of the director $\mathbf{L}$, which vanishes when the entire fluid is under uniform rotation with an angular velocity $^1/_2\nabla \times \mathbf{v}$.

The coefficients of the proportionality between the viscous stress derivatives and the time derivative of velocity in (4.10) are called viscosity coefficients, namely:

$$\alpha_1,\ \alpha_2,\ \alpha_3,\ \alpha_4,\ \alpha_5,\ \text{and}\ \alpha_6.$$

It can be shown that

$$\alpha_2 + \alpha_3 = \alpha_6 - \alpha_5,$$

that is, only five independent viscosity coefficients exist.

3. The equation for director rotation in a nematic LC is

$$I\mathrm{d}\Omega/\mathrm{d}t = [\mathbf{L} \times \mathbf{h}] - \Gamma, \tag{4.13}$$

where $\Omega = [\mathbf{L} \times \mathrm{d}\mathbf{L}/\mathrm{d}t]$ is the angular velocity of the director rotation; I is the moment of inertia for the molecular reorientation, normalized to a unit volume;

$$\mathbf{h} = -\delta F/\delta \mathbf{L} \tag{4.14}$$

is the functional derivative of the LC volume free energy with respect to the director components $\mathbf{L}$ or the so-called molecular field; and

$$\Gamma = [\mathbf{L} \times (\gamma_1 \mathbf{N} + \gamma_2 \mathrm{A} \cdot \mathbf{L}] \tag{4.15}$$

is the frictional torque, which is analogous to the viscous term in the Navier–Stokes equation, while the matrix $\mathrm{A} = [A_{ij}]$ is defined by (4.11) and vector $\mathbf{N}$ by (4.12). Here $\gamma_1 = \alpha_3 - \alpha_2$, $\gamma_2 = \alpha_3 + \alpha_2$ are viscosity coefficients, and

$$\gamma_1 = \alpha_3 - \alpha_2 \tag{4.16}$$

is the so-called rotational viscosity, which characterizes the pure rotation of the nematic LC director without any movement of the centers of the molecules, that is, the so-called "back-flow" effect.

Typical plots of LC director distributions for the ECB effect are shown in Figure 4.3.

Effect of a Weak Anchoring at the Boundaries

In the case of planar and homeotropic initial orientations,[1] the threshold for the Fréedericksz transition remains for a weak director boundary anchoring, but has a lower value. The Fréedericksz transition thresholds for finite (W) and infinite (∞) polar anchoring energies, respectively $U_\mathrm{F}(W)$ and $U_\mathrm{F}(\infty)$, are related by

$$\mathrm{cotan}[\pi U_\mathrm{F}(W)/U_\mathrm{F}(\infty)] = \pi K_{ii} U_\mathrm{F}(W)/[W d U_\mathrm{F}(\infty)] \tag{4.17}$$

[1–3]; the effective elastic coefficient K_{ii} and $U_\mathrm{F}(\infty)$ are defined in (4.1). For large anchoring energies, expression (4.17) reduces to

$$U_\mathrm{F}(W) = U_\mathrm{F}(\infty)[1 - 2K_{\mathrm{ii}}/(Wd)], \quad Wd/K_{ii} \gg 1. \tag{4.18}$$

[1] In the case of a tilted orientation there is no threshold, irrespective of the surface anchoring [2].

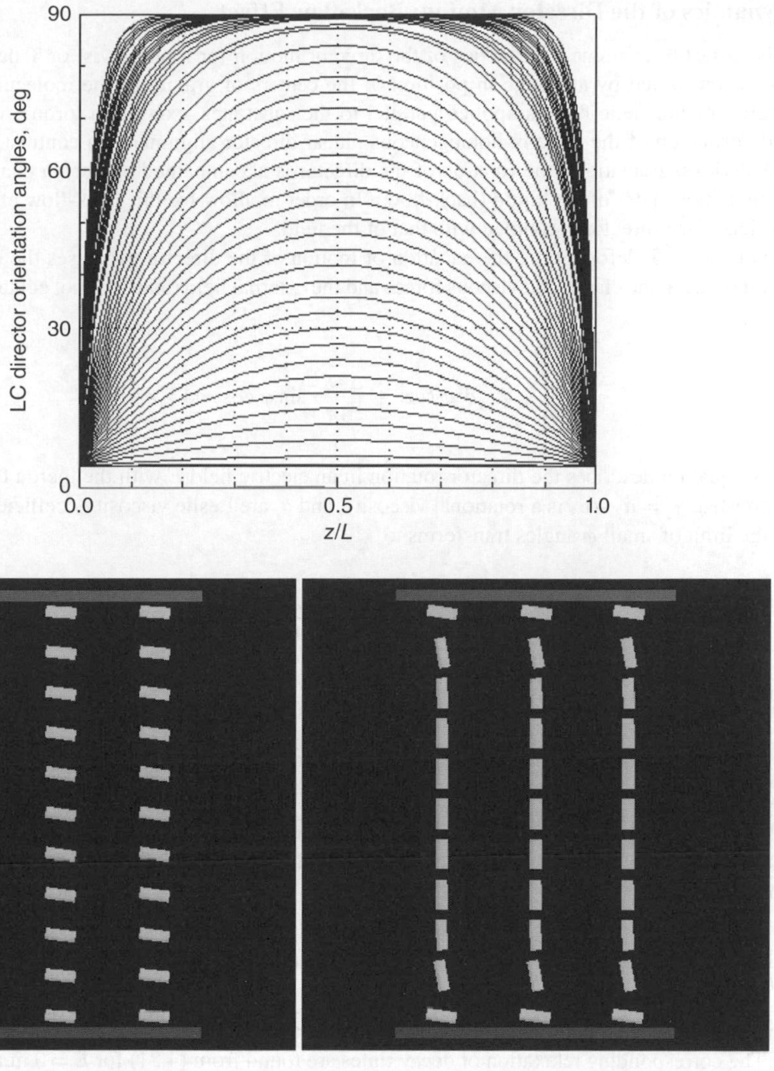

Figure 4.3 Top: LC director distributions for the relative thickness of the cell, $0 \leq z/d \leq 1$. Bottom: LC molecular arrangement inside LC cell

For sufficiently small values of anchoring energy, the electro-optical response of the nematic cell becomes infinitely steep, so that for a certain critical value of *W*, a hysteresis and first-order Fréedericksz transition becomes possible [1–3].

In the case of finite anchoring, there also exists a saturation voltage for the total reorientation of the director, parallel (Figure 4.1a) or perpendicular (Figure 4.1b) to the field, when the boundary regions disappear [1–3]. Development of the nematic cells with a good and reliable control of anchoring energy is of great importance for applications, as very steep transmission–voltage curves, memory states, and improved response times could be realized [1–3].

Dynamics of the Director Motion; Back-flow Effect

It is easiest to examine the kinetics of the director motion for a pure twist or T deformation, which is not accompanied by a change in position of the centers of gravity of the molecules. In this case, the electric or magnetic field is applied parallel to the substrates, exerting a torque, which causes a pure T deformation of the initially homogeneous planar director alignment. In contrast, for S- and B-effect Fréedericksz transitions, the rotation of the director is accompanied by such a change in position, that is, by a movement of the liquid (back-flow). In order to allow for this back-flow effect, the equation of motion of the director is coupled with that of the fluid.

For a pure T deformation, the equation of motion of the director expresses the balance between the torques due to the elastic and viscous forces and the external field (and does not contain the fluid velocity) [1–3]:

$$K_{22}\partial^2\varphi/\partial z^2 + \frac{\Delta\varepsilon E^2}{4\pi}\sin\varphi\cos\varphi = \gamma_1\partial\varphi/\partial t. \tag{4.19}$$

This equation describes the director rotation in an electric field E with the inertia term $I\partial^2\varphi/\partial t^2$, which is omitted, $\gamma_1 = \alpha_3 - \alpha_2$ is a rotational viscosity, and α_i are Leslie viscosity coefficients. Equation (4.19) in the limit of small φ angles transforms to

$$K_{22}\partial^2\varphi/\partial z^2 + \frac{\Delta\varepsilon E^2}{4\pi}\varphi = \gamma_1\partial\varphi/\partial t \tag{4.20}$$

with the solution

$$\varphi = \varphi_{\rm m}\exp(t/\tau_{\rm r})\sin(\pi z/d), \tag{4.21}$$

where $\tau_{\rm r} = \gamma_1/\left(\Delta\varepsilon E^2/4\pi - K_{22}\pi^2/d^2\right)$ is the reaction or switching-on time, and $\varphi_{\rm m}$ is the maximum twist angle at the center of the layer. Solution (4.21) satisfies the strong anchoring boundary conditions

$$\varphi(z = 0) = \varphi(z = d) = \varphi_{\rm m}$$

and assumes a maximum value at the center of the layer $\varphi(z = d/2) = \varphi_{\rm m}$.

The corresponding relaxation or decay times are found from (4.21) for $E = 0$ in a similar way:

$$\tau_{\rm d} = \gamma_1 d^2/(K_{22}\pi^2). \tag{4.22}$$

Unlike T deformation, the reorientation of the director in S- and B-effects is always accompanied by a macroscopic flow of a nematic LC (back-flow) with velocity $\mathbf{V} = (V(z), 0, 0)$, where the z-axis is perpendicular to the substrates and the deformations take place in the xz-plane. The velocity V includes only the x-component, because the z-component is zero according to the continuity equation ($\nabla \cdot \mathbf{V} = 0$), and the y-component vanishes due to the symmetry of the problem.

For small variations of the angle θ, the *characteristic* times of the S- and B-effects can be found from solutions of the coupled dynamic equations of the nematic director $\theta(z)$ and velocity $V(z)$ in the following form:

$$\tau_{\rm r} = \gamma_1^*/(\Delta\varepsilon E^2/4\pi - K_{ii}\pi^2/d^2), \quad \tau_{\rm d} = \gamma_1^*/(K_{ii}\pi^2/d^2), \tag{4.23}$$

where γ_1^* is the effective rotational viscosity. We have

$$\begin{aligned} \gamma_1^* &= \gamma_1 - 2\alpha_3^2/(\alpha_3 + \alpha_4 + \alpha_6), \quad K_{ii} = K_{11} \text{ for the S-effect}, \\ \gamma_1^* &= \gamma_1 - 2\alpha_2^2/(\alpha_4 + \alpha_5 - \alpha_2), \quad K_{ii} = K_{33} \text{ for the B-effect}. \end{aligned} \tag{4.24}$$

The relationships in (4.24) show that back-flow considerably alters the response times and should be taken into account.

The relative difference $(\gamma_1 - \gamma_1^*)/\gamma_1$ is close to zero for voltages slightly above the threshold in the S-effect and for very high voltages in the B-effect, but can hardly exceed 50% within the whole voltage interval. The results of computer simulation were confirmed by experiment [1–3].

Weak boundary anchoring decreases a rise time τ_r and increases the relaxation time τ_d. This can be easily understood by making the substitution $K_{ii}\,\pi^2/d^2 \Rightarrow 2W/d$ in (4.22)–(4.23), which proves to be correct for sufficiently low anchoring energies of the director [1–3].

Optical Response

Splay Mode (S-effect)

To understand the optical characteristics of a LC layer in the ECB effect, let us consider the geometry of Figure 4.1a with the initial homogeneous director orientation along the *x*-axis. If the applied voltage is below the threshold, the nematic LC layers manifest birefringence, $\Delta n = n_e - n_o = n_\| - n_\perp$. When the field exceeds its threshold value, the director deviates from its orientation along the *x*-axis while remaining perpendicular to the *y*-axis. The refractive index for the ordinary wave remains unchanged, $n_o = n_\perp$. At the same time, the refractive index for the extraordinary wave (n_e) decreases, tending toward $n_\perp$. The local extraordinary refractive index $n_e(z)$ can be expressed in terms of the angle of orientation of the director $\theta(z)$ as follows:

$$n_e(z) = n_\| n_\perp / \sqrt{n_\|^2 \sin^2\theta(z) + n_\perp^2 \cos^2\theta(z)} \tag{4.25}$$

[1–3] (see Section 1.3.2). The phase difference between the extraordinary and the ordinary waves for monochromatic light of wavelength λ is found by integrating over the layer depth:

$$\Delta\Phi = (2\pi/\lambda)\int_0^d (n_e(z) - n_o)\mathrm{d}z = 2\pi d\,\langle\Delta n(z)\rangle/\lambda, \tag{4.26}$$

where

$$\langle\Delta n(z)\rangle = \frac{1}{d}\int_0^d (n_e(z) - n_o)\mathrm{d}z$$

and d is the thickness of the LC layer. The intensity of the light passing through the cell and the output polarizer (analyzer) depends on the angle φ_0 between the polarization vector of the incident wave and the initial orientation of the director of the nematic LC,

$$I = I_0 \sin^2 2\varphi_0 \sin^2(\Delta\Phi/2), \tag{4.27}$$

where I_0 is the intensity of the plane polarized light incident on the cell. Hence, the external magnetic or electric field changes the orientation of the director and, consequently, the values of $\langle\Delta n\rangle$ and $\Delta\Phi$.

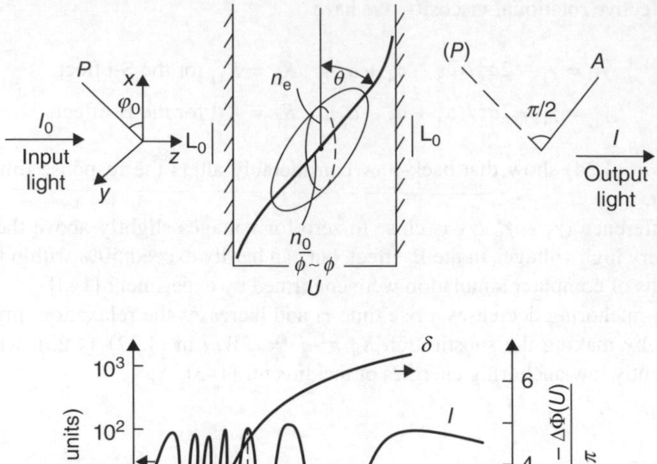

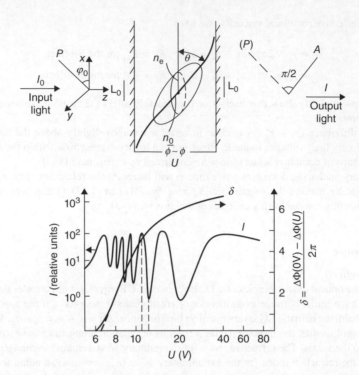

Figure 4.4 ECB in a homogeneous LC cell (S-effect). Polarizer (P) is crossed with analyzer (A) and set at the angle φ_0 with respect to the initial LC director. The intensity I of the transmitted light and phase difference δ are shown below versus the applied voltage. The phase π-switching regime is indicated, which results in the variation of the intensity from I_{max} to I_{min} [1–3]

A change in the phase difference $\Delta\Phi$, in turn, results in an oscillatory dependence of the optical signal at the exit of the analyzer. The maximum amplitude of these oscillations corresponds to an angle $\varphi_0 = 45°$ and the maximum possible number of oscillations (e.g., the number of maxima during a complete reorientation of the director) is approximately $(n_{\parallel} - n_{\perp})d/\lambda$.

The characteristic curves of transmitted intensity in the ECB effect are shown in Figure 4.4. The light emerging from the LC cell, in general, becomes elliptically polarized, so that its ellipticity e and the angle ψ, between the long ellipse axis and the polarizer, also depend on φ_0 and $\Delta\Phi$ [1–3]:

$$\begin{aligned} e &= \tan[(1/2)\arcsin(\sin 2\varphi_0 \sin \Delta\Phi)], \\ \tan 2\psi &= \tan 2\varphi_0 \cos \Delta\Phi. \end{aligned} \tag{4.28}$$

The experimental dependences of $e(U)$ and $\psi(U)$ on the applied voltage resemble those shown in Figure 4.4 for the intensity curve $I(U)$ [1–3].

The effect of phase modulation for an initial planar orientation of the director (along x) with positive dielectric anisotropy ($\Delta\varepsilon > 0$) and with the field applied along the z-axis (Figure 4.1a) is called the S-effect [1–3], since the initial deformation is a splay deformation, even though a bend deformation is also induced above the threshold.

The applied field causes a phase difference to arise between the ordinary and extraordinary waves and the intensity of the light oscillates in accordance with (4.27). As the bend deformation is now in the initial stages of its development (Figure 4.1b), the corresponding electro-optical effect is called the

B-effect, occurring for negative values of the dielectric anisotropy ($\Delta\varepsilon < 0$). However, here the final orientation of the director is not defined (degenerate), and the sample does not remain as monodomain and contains many specific defects [1–3]. In principle, the preferred direction of the final orientation of the director can be established, and these defects in the structure are eliminated by a special preparation of the surfaces with a slight pretilt.

The important point of the ECB effect is to provide full switching from I_{max} to I_{min} with minimal response times. According to (4.27), such a switching is stipulated when the phase difference $\Delta\Phi$ is changed by as much as π. If a nematic cell is subjected to a voltage U corresponding to a maximum intensity I_{max}, then to attain another state with I_{min}, we have to supply an additional voltage $U_\pi \geq \Delta U$, where ΔU is the minimum possible value of U_π (Figure 4.4).

The main disadvantage of the S-effect for display applications is the strong dependence of the transmitted intensity on the light wavelength, and the nonuniform transmission–voltage characteristics at oblique light incidence. It is possible, however, to avoid them by placing a compensating birefringent plate between the LC cell and one of the polarizers or by using two nematic S-cells in series which have perpendicular initial directors [1–3].

π-cells

Special attention should be paid to the so-called π-cells when the intensity is switched in the last fall of the oscillation curve (see Figures 4.4 and 4.5) [1–3]. In this case, the switching is attained due to the very slight variation of the director distribution within the narrow regions near the boundaries, thus resulting in a very fast response speed. The corresponding switching times can be estimated according to the formula [1–3]

$$\tau \approx (\Delta/\pi/2\pi)^2\ 1/(1 - \beta U_\pi/U_0)^2\ \gamma_1 \lambda^2/K_{11}\Delta n^2, \tag{4.29}$$

where $\Delta/\pi \approx 1$ is a relative phase difference for the last intensity fall, and $\beta \approx 1$ is the LC material constant. As we can see, the response time does not depend on the cell thickness. By combining π-cells with phase retardation plates, both with a positive and negative phase shifts, it is possible to optimize the contrast and color uniformity of a LC device [1–3].

Bend Mode (B-effect)

The B-effect in homeotropic or quasi-homeotropic (slightly tilted) nematic samples remains attractive for applications, including displays with high information content [1–3]. With good homeotropic orientation of a nematic LC, the B-effect is characterized by a steep growth in optical transmission with voltage, that is, the threshold is very sharp. This is due to the very weak light scattering of the homeotropically

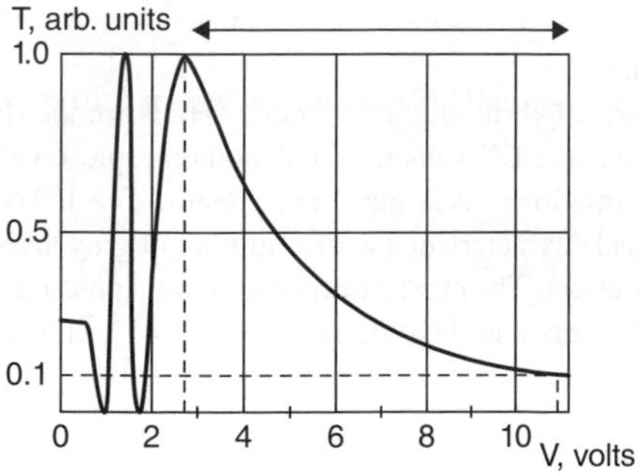

Figure 4.5 The regime of switching in a π-cell. For the last intensity fall the controlling voltage has to change from U_π to U_0 [1–3]

oriented layer and the complete absence of birefringence in the initial state [1–3]. Basically, the patterns observed experimentally are similar to the S-effect, including dynamic behavior. Let us consider the transmission–voltage curves for the B-effect at voltages slightly exceeding the threshold value, which are important for applications. To avoid degeneracy in the director reorientation a slight initial pretilt from the normal to the substrate is needed, $\theta_0 \approx 0.5°–1°$ (Figure 4.1b).[2] The electro-optical response is very sensitive to the θ_0 value: that is, small $\theta_0 < 0.5°$ do not allow us to avoid defects, while larger angles, $\theta_0 > 1°$, strongly reduce the contrast [1–3]. According to (4.19), for voltages slightly exceeding the threshold, the contrast is

$$C = I_{on}/I_{off} \approx \sin^2(A\langle\theta^2_{on}\rangle)/\sin^2(A\theta^2_{off}) \approx \theta^4_m/4\theta^2_0, \tag{4.30}$$

where $\langle\theta^2_{on}\rangle = d^{-1}\int_0^d \theta^2(z)dz \approx \theta^2_m/2$ for small θ and $\theta_{off} = \theta_0$ are the director angles in the switched-on and switched-off states; $A = \pi d n_\perp(1 - n_\perp^2/n_\parallel^2)/(2\lambda)$ is a phase factor. The contrast ratio C crucially depends on the pretilt angle θ_0. We can obtain the following relation for θ^2_m in the B-effect:

$$\theta^2_m = 4(U/U_B - 1)(K_{11}/K_{33} + \varepsilon_\parallel/\varepsilon_\perp - 1)^{-1}, \tag{4.31}$$

where $U_B = \pi(4\pi\ K_{33}/|\Delta\varepsilon|)^{1/2}$ is the B-effect threshold voltage. Using expression (4.27) for $\varphi_0 = \pi/4$, we can derive the following relationship for the optical transmission $T = I/I_0$ in the ECB effect in a homeotropic nematic (B-effect):

$$\begin{aligned} T &= \sin^2 \pi d\Delta n\lambda^{-1}\langle\theta_{on}\rangle^2 \approx 4A^2(U/U_B - 1)^2(K_{11}/K_{33} + \varepsilon_\perp/\varepsilon_\parallel - 1)^{-2} \\ &\approx [2\pi d\lambda^{-1}(n_\parallel - n_\perp)/(K_{11}/K_{33} + \varepsilon_\perp/\varepsilon_\parallel - 1)]^2(U/U_B - 1)^2. \end{aligned} \tag{4.32}$$

As can be seen from (4.32), a steep electro-optical response of the cell is attained for sufficiently large values of the optical path difference $d(n_\parallel - n_\perp)$ and the elasticity anisotropy K_{33}/K_{11}, as well as for small dielectric anisotropy $|\Delta\varepsilon|/\varepsilon_\parallel$. However, the values of d cannot be too large, since the latter results in an increase in response times (see (4.23)), while small values of the dielectric anisotropy lead to a growth in operating voltages. Thus, to develop a good ECB material, a compromise is needed. In order to obtain the steep electro-optical characteristics required for displays with high information content, together with a fast response and uniformity of transmission for the oblique light incidence, the following parameters of nematic cells for the B-effect can be proposed [1–3]: small thickness, $d < 5$ μm; small pretilt angle $\theta_0 \approx 0.5°–1°$; large optical retardation $d\Delta n \approx 1$ micron; large K_{33}/K_{11} ratio; and small dielectric anisotropy $|\Delta\varepsilon|/\varepsilon_\parallel < 0.5$.

Hybrid Aligned Nematic Mode

It is also interesting to consider a hybrid aligned nematic (HAN) mode (Figure 4.6). In this mode, the director alignment is planar on one LC substrate and homeotropic on the other (Figure 4.6), so the combination of S and B deformations takes place as a result. The HAN mode has no threshold and possesses smooth electro-optical characteristics with a number of gray levels (Figure 4.6). The geometry of this mode is similar to the S-effect. The maximum possible transmission is obtained when the structure is placed between crossed polarizers and the polarizer angle is 45° with respect to the plane of S and B deformation (Figure 4.6).

---

[2] Note that when considering the B-effect here we measure the angles θ, θ_0, and θ_m from the z-axis (Figure 4.1b) rather than from the xy-plane as usual (see Figure 4.1a). The changes in (4.25), connected with this redefinition, are obvious.

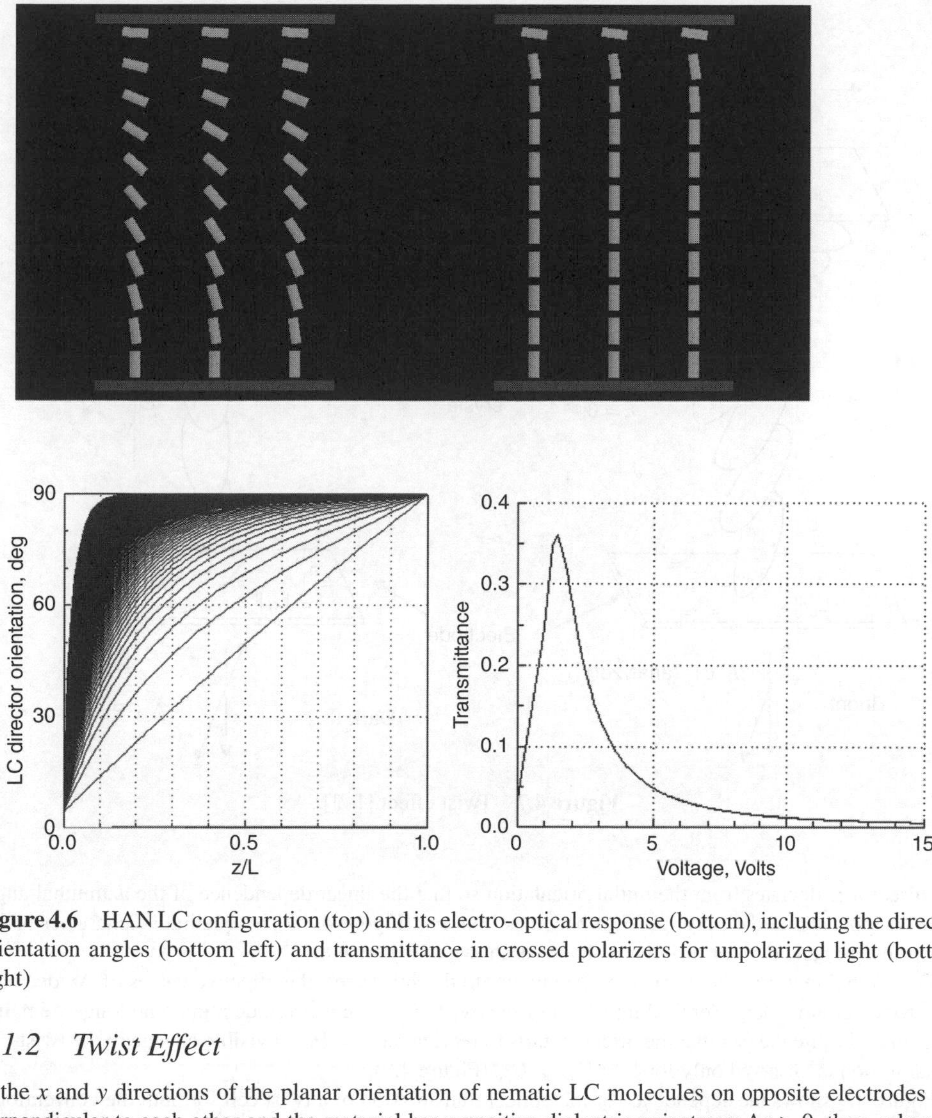

Figure 4.6 HAN LC configuration (top) and its electro-optical response (bottom), including the director orientation angles (bottom left) and transmittance in crossed polarizers for unpolarized light (bottom right)

4.1.2 Twist Effect

If the x and y directions of the planar orientation of nematic LC molecules on opposite electrodes are perpendicular to each other and the material has a positive dielectric anisotropy $\Delta\varepsilon > 0$, then, when an electric field is applied along the z-axis (Figure 4.7), a reorientation effect occurs that is a combination of the S, B, and T deformations [1–3].

In the absence of the field, the light polarization vector follows the director and, consequently, the structure rotates the polarization plane up to the angle characterizing the structure, $\varphi_m = \pi/2$ (Figure 4.7). This specific waveguide regime (the Mauguin regime) takes place when

$$\Delta nd/\lambda \gg 1. \tag{4.33}$$

When the applied voltage exceeds a certain threshold value,

$$U_{tw} = \pi[\pi(4K_{11} + K_{33} - 2K_{22})/\Delta\varepsilon]^{1/2}, \tag{4.34}$$

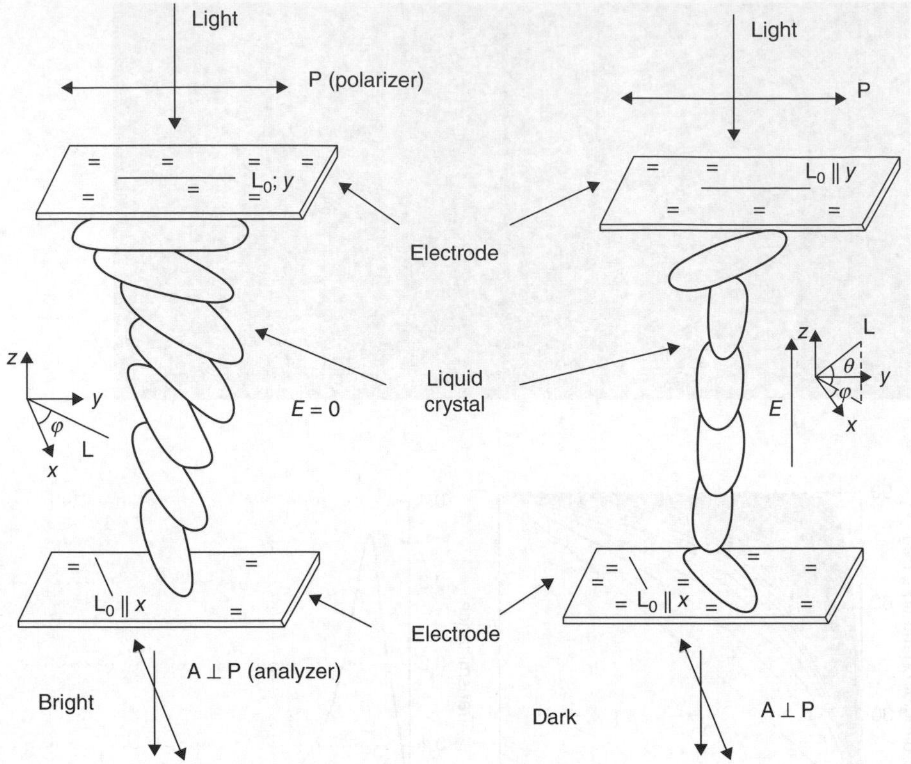

Figure 4.7 Twist effect [1–3]

the director **L** deviates from the initial orientation so that the linear dependence of the azimuthal angle $\varphi(z)$ disappears and the tilt angle $\theta(z)$ becomes nonzero (Figure 4.8). The qualitative character of the functions $\varphi(z)$ and $\theta(z)$ for different voltages is shown in Figure 4.8.

Since the director tends to orient perpendicular to the substrates, the effective values of Δn decrease and, for a certain voltage (optical threshold of the twist U_{opt}), the waveguide regime no longer remains. Note that, despite the fact that the director starts to reorient at $U = U_{tw}$, a visible change in the twist-cell transmission is observed only for $U = U_{opt} > U_{tw}$ (Figure 4.9).

Figure 4.9 shows the dependence of the optical transmission of a twist cell for both the conventional geometry $\mathbf{P} \parallel \mathbf{L}(0)$ and when the polarizer transmission axis **P** forms an angle of 45° with respect to the orientation of the director at $z = 0$ [1–3]. The deformation threshold, $U_{tw} = 6$ V, determined by extrapolating the linear section of the phase delay curve to $\delta(U) = 0$, coincides with that calculated from (4.34). The optical threshold for the twist effect increases on decreasing the wavelength ($U_{opt} = 8.9$ V and 10.2 V for $\lambda = 750$ and 450 nm, respectively), since the cut-off implied by the Mauguin condition occurs at higher voltages for shorter wavelengths (see (4.33)).

Twist-Cell Geometry for Zero Voltage; Mauguin Conditions

A twist cell is usually formed by placing orienting glasses on top of each other. Then twist directions at angles $\pi/2$ and $-\pi/2$ are equally probable. This degeneracy in the sign of the twist can be removed if small amounts of optically active material are added to the nematic LC. In this case, the walls disappear

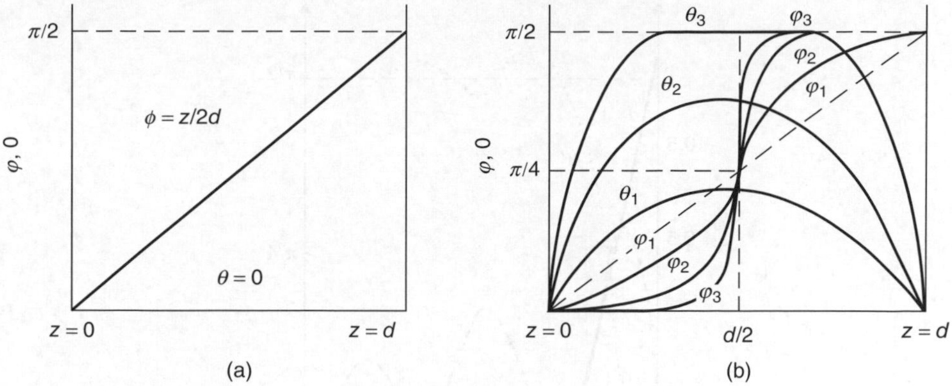

Figure 4.8 Distribution of the director angles $\varphi(z)$ and $\theta(z)$ in the twist effect for different voltages: (a) $U \leq U_{tw}$; (b) $U_{tw} < U_1 < U_2 < U_3$ [1–3]

and the twist cell has a uniform structure throughout its entire area. A nonzero tilt of the molecules to the cell surface θ_0 also results in a nonuniform twisting.

At zero field, the waveguide (Mauguin) regime is violated for small values of $\Delta nd/\lambda$. Figure 4.10 shows the corresponding dependence of the transmitted light intensity (I), ellipticity (e), and rotation angle (ψ) of the major ellipse axis with respect to the polarizer transmission axis (P) on the parameter $\Delta nd/\lambda$ [1–3]. (The director at the first substrate is parallel (or perpendicular) to the transmission axis of the polarizer (P) and analyzer (A), and the analyzer (A) is parallel to the polarizer (P)). As seen from Figure 4.10, the exact Mauguin mode conditions

$$I = 0, \quad \psi = \pi/2, \quad e = 0 \tag{4.35}$$

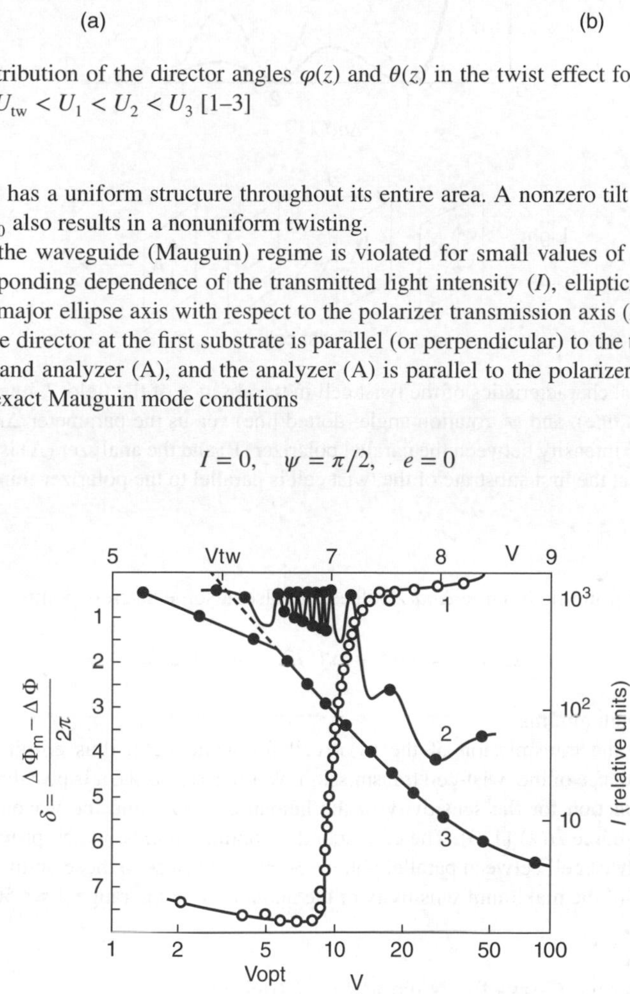

Figure 4.9 Optical response of the twist cell between parallel polarizers [1–3]: curve 1, polarizers are parallel to the director on the input surface of the cell (conventional orientation); curve 2, polarizers are at an angle of 45° to the director on the input surface (maximum birefringence intensity); curve 3, phase retardation in (4.26) calculated from curve 2 [1–3]

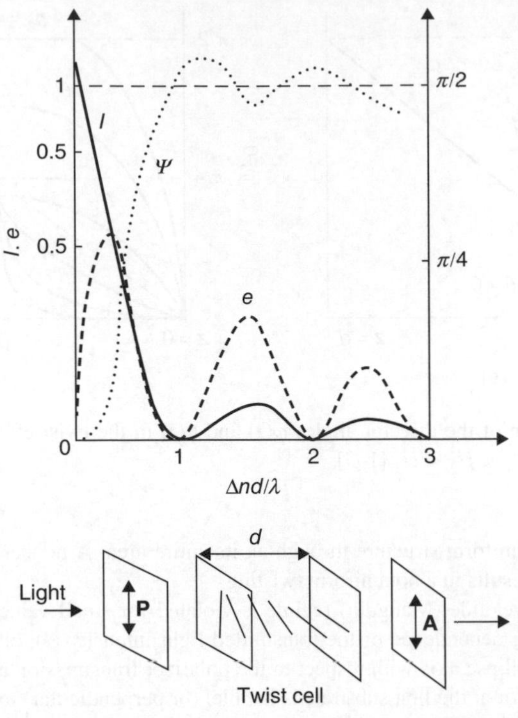

Figure 4.10 Optical characteristics of the twist cell in the absence of the field. *I*, intensity (solid line), *e*, ellipticity (dashed line), and ψ, rotation angle (dotted line) versus the parameter $\Delta nd/\lambda$. Without the twist cell ($d = 0$), the intensity between the parallel polarizer (P) and the analyzer (A) is taken to be equal to one. The director at the first substrate of the twist cell is parallel to the polarizer transmission axis (P) [1–3]

take place not only for infinitely large $\Delta nd/\lambda$ values, but also at some discrete points

$$\Delta nd/\lambda = (4m^2 - 1)^{1/2}/2, \quad m = 1, 2, 3, \tag{4.36}$$

usually called Mauguin minima.

We can consider the transmission of the twist cell for white light, thus eliminating λ from the characteristic dependence of the twist-cell transmission. Writing $T(\Delta nd/\lambda)$, it is possible to average $T(\lambda)$, together with the function for the sensitivity of the human eye $\bar{y}(\lambda)$, and the wavelength distribution of the illumination source $H(\lambda)$ [1–3]. The corresponding optimal points, which provide the minimum transmission of the twist cell between parallel polarizers, are very close to those defined by (4.36), if we take the wavelength of the maximum sensitivity of the human eye in the range $\lambda \approx 550$–580 nm.

Transmission–Voltage Curve for Normal Light Incidence

The typical transmission–voltage curve (TVC) of the twist effect for normal light incidence is shown in Figure 4.11 for a twist cell placed between parallel polarizers. As has been mentioned, a significant visible variation in the transmittance of the polarizer–twist-cell–analyzer system at normal incidence begins at a larger voltage (optical threshold) than U_{tw}. In view of the Mauguin requirement (see (4.33)),

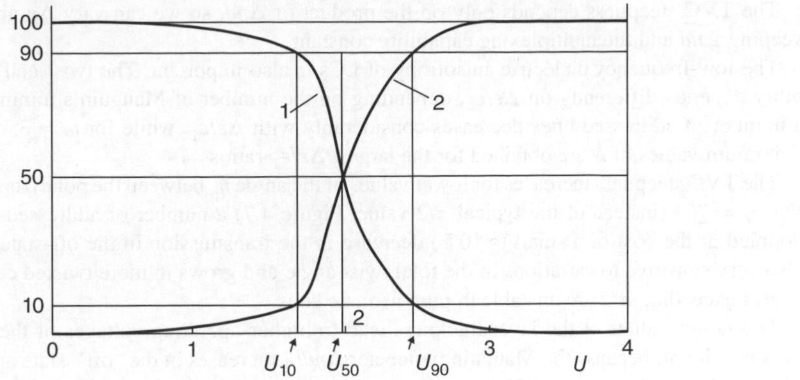

Figure 4.11 Optical transmission of the twist cell versus voltage for crossed (curve 1) and parallel (curve 2) polarizers [1–3]. The voltages U_{90}, U_{50}, and U_{10} correspond to the 90%, 50%, and 10% transmission levels of curve 1, respectively [1–3]

the optical threshold of the twist effect decreases for smaller values of the cell thickness and the optical anisotropy of the LC [1–3].

One of the most important parameters of a twist cell is the steepness of the TVC. Usually, the steepness parameter p is defined from the TVC of the twist cell, placed between crossed polarizers (Figure 4.11),

$$p_{50} = U_{50}/U_{90} - 1, \quad p_{10} = U_{10}/U_{90} - 1, \tag{4.37}$$

where U_{90}, U_{50}, and U_{10} correspond to the 90%, 50%, and 10% levels, respectively, of the optical transmission. As in (4.37), the steeper TVCs correspond to the smaller values of p_{50} and p_{10}.

The steepness of the TVC can be optimized for any specific case. There is no need to perform numerous experiments as a computer simulation can solve the problem to a high degree of accuracy. At present, many researchers are considering the problem of calculating the performance of twist nematic (TN) LCDs [1–3]. The common procedure for the simulation of the electro-optical behavior of TN-LCDs and most other kinds of LCDs involves two steps. First, the distributions of the LC director are found for given conditions. The second step is the calculation of the optical performance of the LCD with the obtained LC director distributions. Different optical methods are used to accomplish the latter step [4–6]. We will consider these methods and their applications to LCDs in detail in Chapters 7–11. At present, several groups [6–9] have provided a set of computer programs as commercially available products for the simulation of the electro-optical behavior of TN-LCDs and other kinds of LCDs. MOUSE-LCD (Modeling Universal System of LCD Electrooptics) [4], developed by the authors of this book, is one of these programs. The detailed analysis of TVC steepness p_{50} and p_{10}, based on computer simulation, was proposed in [1–3].

Note that a number of addressing lines N in the matrix LCDs, with a high information content or multiplexing capability sharply increases for steep TVCs, that is, low p values. The precise dependence $N(p)$ is defined by the type of driving scheme and will be discussed below.

The number of addressing lines N can be calculated from the relation [1–3]

$$N = [(1+p)^2 + 1]^2/[(1+p)^2 - 1]^2, \quad N \approx 1/p^2 \quad \text{for } p \ll 1, \tag{4.38}$$

which is the result of optimization of the driving conditions.

Our calculations show that decreasing K_{33}/K_{11} from 2 to 0.5 results in a considerable growth in the number of addressed lines of the passively addressed LCD.

The TVC steepness depends only on the product of Δnd, so we can vary Δn and d independently, keeping Δnd and the multiplexing capability constant.

The low-frequency dielectric anisotropy of LCs is also important. The twist-cell multiplexing capability depends differently on $\Delta\varepsilon/\varepsilon_\perp$, depending on the number of Mauguin's minimum, m. For $m = 1$ a number of addressed lines decreases considerably with $\Delta\varepsilon/\varepsilon_\perp$, while for $m = 3$, on the contrary, the maximum values of N are obtained for the largest $\Delta\varepsilon/\varepsilon_\perp$ ratios.

The TVC steepness increases for lower values of the angle η_p between the polarizer and analyzer [1–3]. For $\eta_p = 70°$ (instead of the typical $\pi/2$ value, Figure 4.7) a number of addressed lines are, however, doubled at the cost of a small ($\approx$10%) decrease in the transmission in the off-state. TVC steepness is also very sensitive to variations in the total twist angle and grows in more twisted cells. However, twist angles exceeding $\pi/2$ are unstable in pure nematic cells.

The larger values of the layer thickness lead to higher operating voltages of the twist effect [1–3]. This is evident, because the Mauguin parameter $\Delta nd/\lambda$ increases in the "off"-state and in order to break Mauguin's condition (see (4.33)) smaller Δn values in the "on"-state are needed, corresponding to the stronger director deformation in higher fields. As mentioned above, TVCs in parallel and perpendicular polarizers are complementary (Figure 4.11). This is not true, however, for contrast ratios as functions of applied voltages for a nonmonochromatic (white) light. The contrast ratio C is defined as the ratio of transmitted luminances in the on- and off-state [1–3]:

$$C = B_{\mathrm{on}}/B_{\mathrm{off}}, \tag{4.39}$$

where

$$B = \int_\lambda H(\lambda)\bar{y}(\lambda)t(\lambda)\mathrm{d}\lambda \Big/ \int_\lambda H(\lambda)\bar{y}(\lambda)\mathrm{d}\lambda.$$

The LCD transmittances $t(\lambda)$ in the on- and off-states are averaged with the function for the sensitivity of the human eye $\bar{y}(\lambda)$ and the energy distribution of illumination source $H(\lambda)$ over the visible spectrum (380–780 nm). The electro-optical effect in the twist cell placed between parallel and crossed polarizers is called the "normally black" and "normally white" mode [1–3], in accordance with the appearance of the twist cell in the off-state (dark or bright). Contrast ratios in the white mode are considerably higher than in the black mode, as the transmission in the on-state for a normally white mode can be very small, limited only by the quality of polarizers and LC orientation.

Viewing Angle Dependences of Twist LCD

Twist-cell transmission at oblique incidence depends on the values of the polar i_θ and azimuthal i_φ angles of light incidence (Figure 4.12). This can be interpreted in terms of the corresponding Mauguin parameter $\Delta nd/\lambda$, which becomes a function of the light direction. The Mauguin parameter in the direction **e** is estimated by averaging the value of $p\Delta n/\lambda$ along **e**, where $p = (\mathrm{d}\varphi/\mathrm{d}e)^{-1}$ is a local value of the pitch, that is, the distance of the total director azimuthal rotation by 2π (for **e** parallel to the twist axis $p = 4d$) and the effective optical anisotropy $\Delta n = (\sin^2\theta/n_\parallel^2 + \cos^2\theta/n_\perp^2)^{-1/2} - n_\perp$ with the polar θ and azimuthal φ angles of the director with respect to the **e**-axis.

The characteristics of transmission for oblique incidence $i_\theta \neq 0$ can be described in terms of the azimuthal dependence of the transmittance $T(i_\varphi)$ for a given polar angle of incidence i_θ and applied voltage U. For directors parallel to the boundaries the twist-cell transmission is symmetric with respect to the plane located at an angle of 45° to the director orientation on the boundaries [1–3]. However, for nonzero director pretilt angles, the symmetry is broken. In the TVCs obtained for oblique light incidence there appears to be a minimum of transmission, which goes to lower voltages for higher incidence angles (Figure 4.13). Indeed, for a certain voltage the direction of light propagation may coincide with the

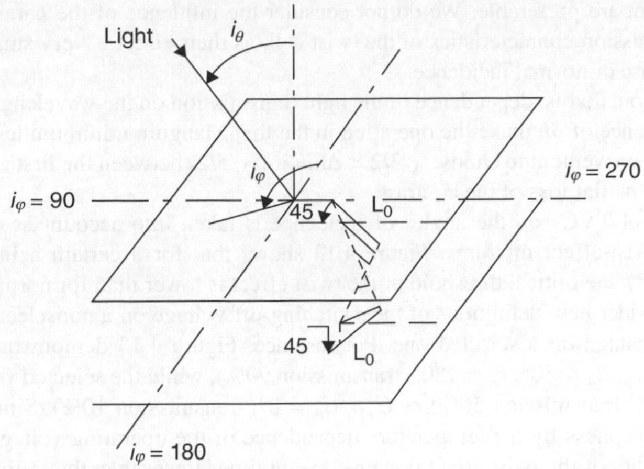

Figure 4.12 The geometry of the twist cell for oblique light incidence. The angle i_θ is the polar angle of light incidence and the angle i_φ is the corresponding azimuthal angle [1–3]

director at the center of the layer, thus providing very low values of the Mauguin parameter. As seen in Figure 4.13, a twist cell for oblique light incidence is most sensitive to an external voltage for azimuthal angles $i_\varphi = 180°$ (Figure 4.12), while $i_\varphi = 0°$ corresponds to quite the opposite case.

The most crucial parameter that affects the uniformity of transmission is the Mauguin number m (see (4.36)). For low $\Delta nd/\lambda$ values, the anisotropy of transmission for oblique incidence is weak [1–3]. According to this, LC mixtures with low Δn values and the first Mauguin minimum $\Delta nd/\lambda = \sqrt{3/2}$

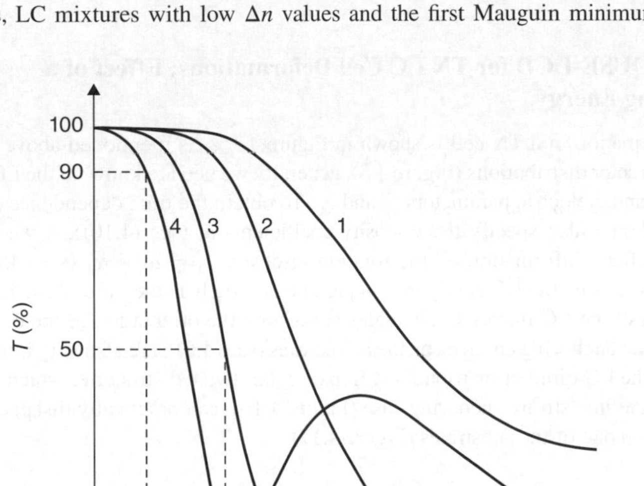

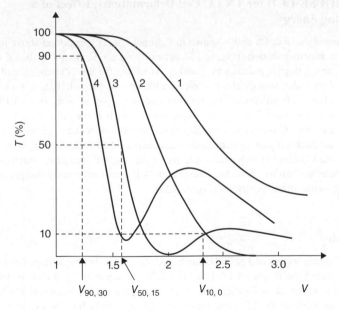

Figure 4.13 TVC of the twist cell for oblique light incidence [1–3]. Curve 1 corresponds to the light incidence angles $i_\theta = 15°$, $i_\varphi = 0°$; curve 2 to $i_\theta = 0°$, $i_\varphi = 0°$; curve 3 to $i_\theta = 15°$, $i_\varphi = 180°$; and curve 4 to $i_\theta = 30°$, $i_\varphi = 180°$ [1–3]

as an operating point are preferable. We do not consider the influence of the parameters K_{33}/K_{11} and $\Delta\varepsilon/\varepsilon_\perp$ on the transmission characteristics of the twist cell, as their effect is very small compared to the above-mentioned case of normal incidence.

Note that coloration, that is, dependence of the light transmission on the wavelength, and the stronger temperature dependence of δn make the operation in the first Mauguin minimum less attractive. Sometimes it seems more convenient to choose $\sqrt{3}/2 < \Delta nd/\lambda < \sqrt{5}/2$ (between the first and second minima) even at the cost of a partial loss of the contrast.

The dependence of TVCs on the angles of incidence is taken into account as one of the positive characteristics of twist-effect mixtures. Figure 4.13 shows that for a certain azimuth of an oblique incidence ($i_\varphi = 180°$), the optical threshold of the twist effect is lower than for normal incidence. Thus, it is possible to consider new definitions of the switching-off voltage on a nonselected display element and switching-on voltage on a selected one. For instance, Figure 4.13 demonstrates that the optical threshold can be $U_{90,30}$ ($i_\theta = 30°$, $i_\varphi = 180°$, transmission 90%), while the selected voltage can be $U_{50,15}$ ($i_\theta = 15°$, $i_\varphi = 180°$, transmission 50%) or $U_{10,0}$ ($i_\theta = 0°$, transmission 10%). Similar limitations are imposed on TVC steepness by the temperature dependence of the operating voltages. As a result, the multiplexing capability of the twist effect remains several times lower than that estimated from TVC at normal incidence and room temperature.

For applications it is convenient to evaluate the angular dependence of the transmission by isocontrast curves, which show the levels of equal contrast ratio for different angles of incidence. Examples of these curves for the normally white mode (twist cell between crossed polarizers) and normally black mode (twist cell between parallel polarizers) are given in Figure 4.14. The radial coordinate in the isocontrast diagram defines the value of the incidence angle i_θ, while the azimuthal one defines the azimuthal incidence angle i_φ. As seen in Figure 4.14, the normally black mode provides wider viewing and more uniform viewing angles than the normally black mode in twist LC cells.

Interface of MOUSE-LCD for TN LC Cell Deformations; Effect of a Weak Anchoring Energy

The director deformation in a TN cell is shown in Figure 4.15. As mentioned above in calculating the equilibrium LC director distributions (Figure 4.8), generally we need to know all the LC elastic constants K_{11}, K_{22}, and K_{33} and dielectric parameters $\varepsilon_\parallel$ and $\varepsilon_\perp$. To obtain the time dependence of the LC director distribution, we should also specify the viscosity coefficients α_i (see (4.10)). If we are not interested in the back-flow effect, information on the rotation viscosity $\gamma_1 = \alpha_3 - \alpha_2$ (see (4.16)) will be quite sufficient. Needless to say, the LC configuration parameters, such as the ratio of the LC thickness to the natural helix pitch of the LC material, the angles describing the orientation of the easy axes as well as azimuthal and planar anchoring energies on the boundaries (see (4.6)), are also very important parameters which determine the LC director distributions. In particular, twisted structure, which still remains even at very high voltages in a strong anchoring case (Figure 4.16), can practically disappear for a low polar anchoring energy on one of the substrates (Figure 4.17).

Stress TN Mode

A special "stress" configuration of TN-LCDs can both decrease the driving voltage (power consumption) and improve response time (Figure 4.18) [10, 11]. The twist sense of TN-LCDs is determined by the directions of pretilt angles on both alignment layers. In commercially available TN-LCDs, the chiral material is used to stabilize the LC twist sense. The LC materials for TN-LCDs possess a twisting property in the same direction as the one determined by the combination of pretilt angle directions. On the contrary, by adding a chiral reagent in which the twist direction is opposite to the one determined by the combination of pretilt angle directions, the splayed twist state is formed (Figure 4.18). The stability

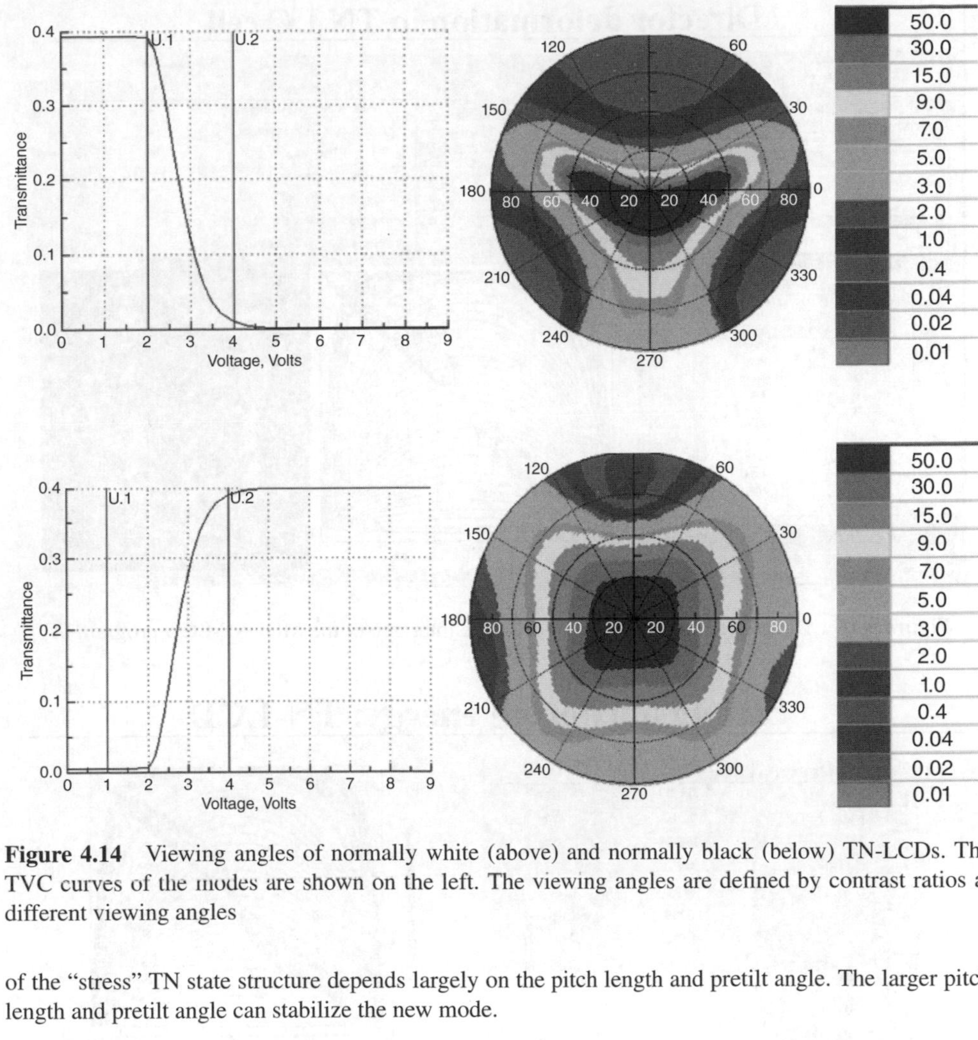

Figure 4.14 Viewing angles of normally white (above) and normally black (below) TN-LCDs. The TVC curves of the modes are shown on the left. The viewing angles are defined by contrast ratios at different viewing angles

of the "stress" TN state structure depends largely on the pitch length and pretilt angle. The larger pitch length and pretilt angle can stabilize the new mode.

4.1.3 Supertwist Effect

A picture of the supertwist LC configuration [1–3] is shown in Figure 4.19. A small voltage difference transforms a highly twisted (supertwist) LC state to almost a homeotropic configuration.

The TVCs in a supertwist LC cell become steeper, and the angle dependence sometimes becomes smoother, than in a TN LC cell (Figure 4.20).

The general scheme of realization of supertwist LC cells is shown in Figure 4.21. Here the general scheme of supertwist display geometry shows the input and output orientations of the molecules or their projections (L_{in}, L_{out}) as well as the input and ouput orientations of the polarizers (P_{in}, P_{out}). The angle φ_m is the twist angle, and β and η define the location of the polarizers: η is the angle between the polarizers and β the angle of the first polarizer with respect to the director on the front substrate. As seen in Figure 4.21, various supertwist geometries can be obtained by altering the supertwist angle φ_m and polarizer angles β, η. Furthermore, we can change such parameters of the supertwist mixtures as the optical path difference Δnd and the director pretilt at the boundaries θ_0.

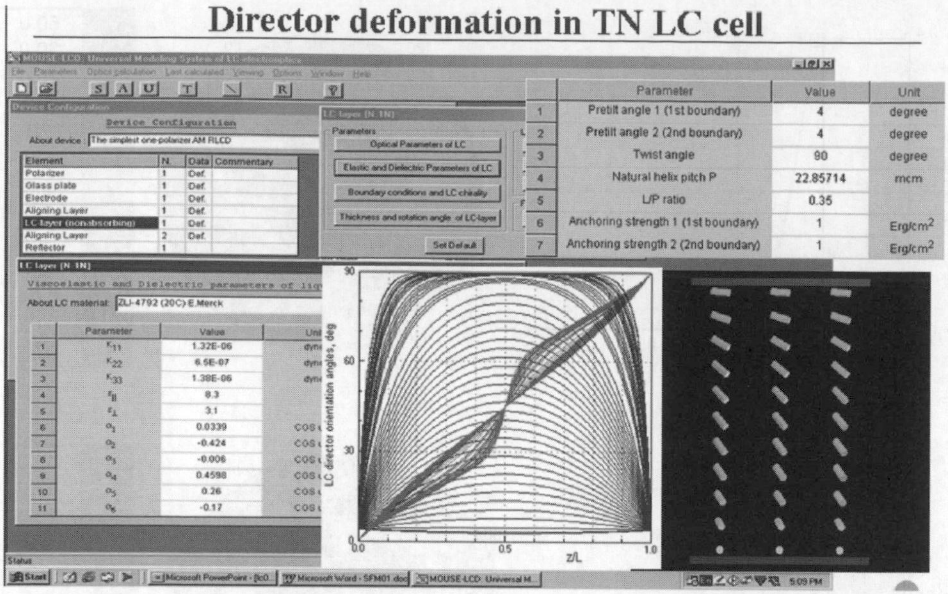

Figure 4.15 Director deformation in TN cell. The interface of the MOUSE-LCD program [4]

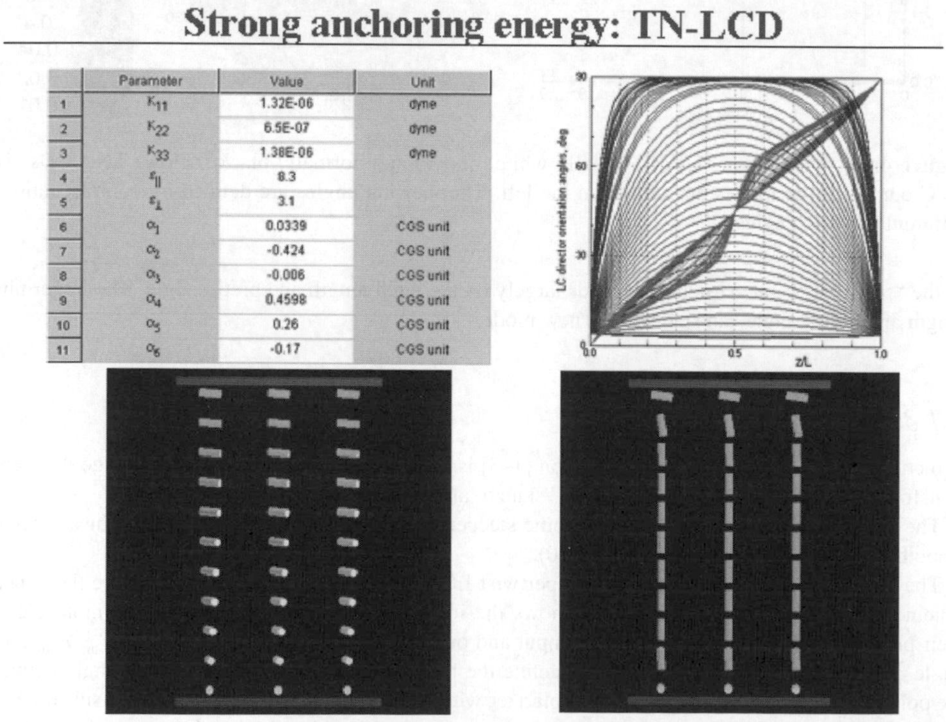

	Parameter	Value	Unit
1	K_{11}	1.32E-06	dyne
2	K_{22}	6.5E-07	dyne
3	K_{33}	1.38E-06	dyne
4	$\varepsilon_{\parallel}$	8.3	
5	$\varepsilon_{\perp}$	3.1	
6	α_1	0.0339	CGS unit
7	α_2	-0.424	CGS unit
8	α_3	-0.006	CGS unit
9	α_4	0.4598	CGS unit
10	α_5	0.26	CGS unit
11	α_6	-0.17	CGS unit

Figure 4.16 Evolution of the TN structure for a sufficiently high anchoring energy. The 90° angle between LC directors at the boundaries remains the same, even for a very high electric field

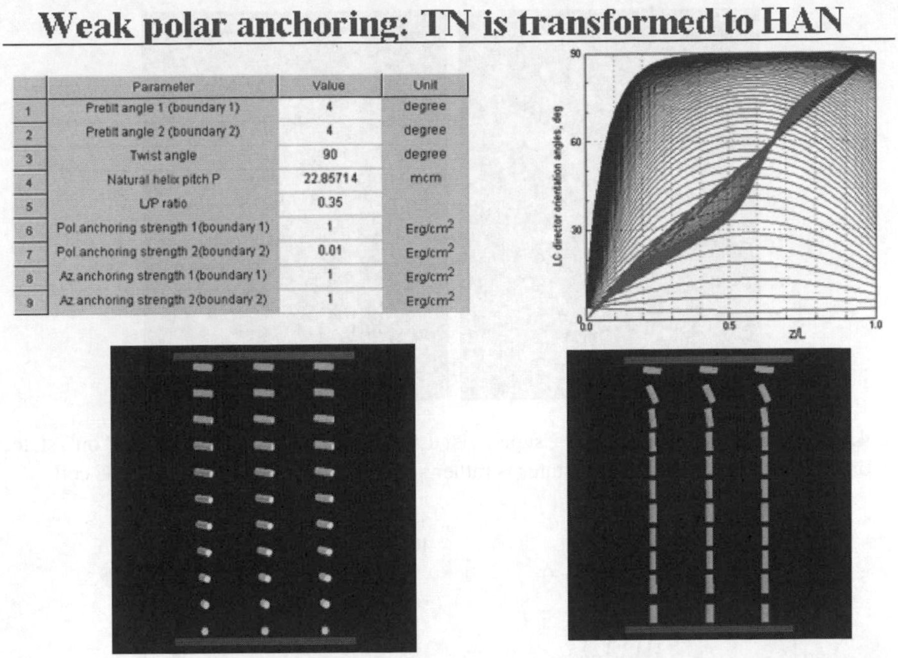

	Parameter	Value	Unit
1	Pretilt angle 1 (boundary 1)	4	degree
2	Pretilt angle 2 (boundary 2)	4	degree
3	Twist angle	90	degree
4	Natural helix pitch P	22.85714	mcm
5	L/P ratio	0.35	
6	Pol.anchoring strength 1(boundary 1)	1	Erg/cm^2
7	Pol.anchoring strength 2(boundary 2)	0.01	Erg/cm^2
8	Az.anchoring strength 1(boundary 1)	1	Erg/cm^2
9	Az.anchoring strength 2(boundary 2)	1	Erg/cm^2

Figure 4.17 Evolution of the TN structure in the case of a low polar anchoring energy on one of the substrates. The HAN structure occurs at very high voltages

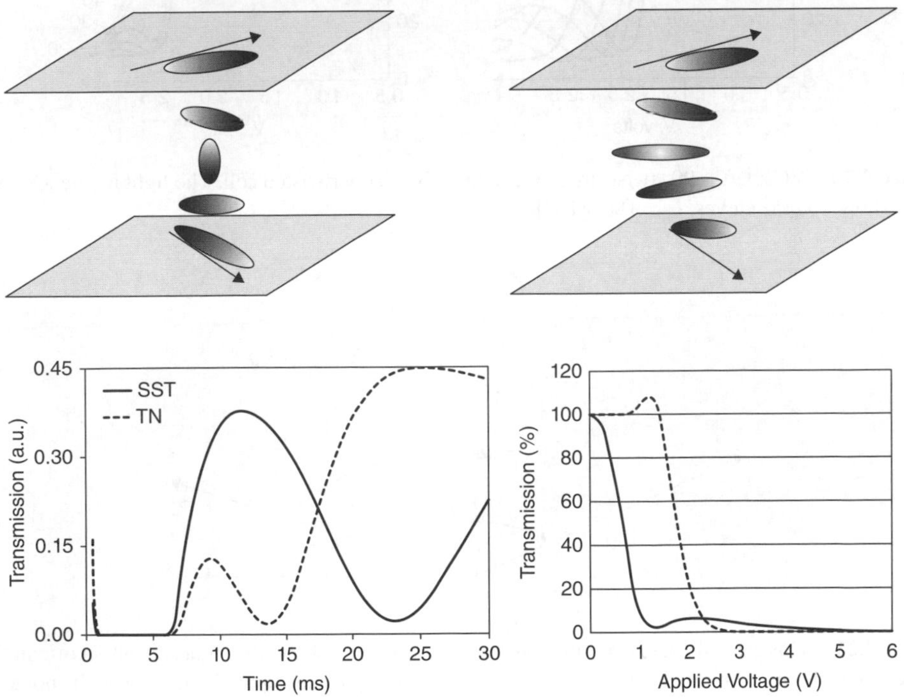

Figure 4.18 Stress splay twist (SST) [11] or Reverse TN [10] mode. Top: common TN (left), SST/Reverse TN mode (right). Bottom: comparison of response time (left) [11] and applied voltage (right) [10] between common TN and SST/Reverse TN mode

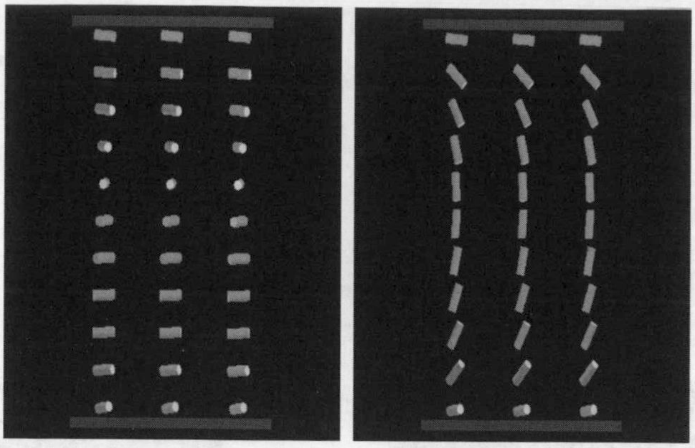

Figure 4.19 Director deformation in a supertwist LC cell. Left: "off"-state; right: "on"-state. The voltage difference between off- and on-states is rather small in comparison with a TN LC cell

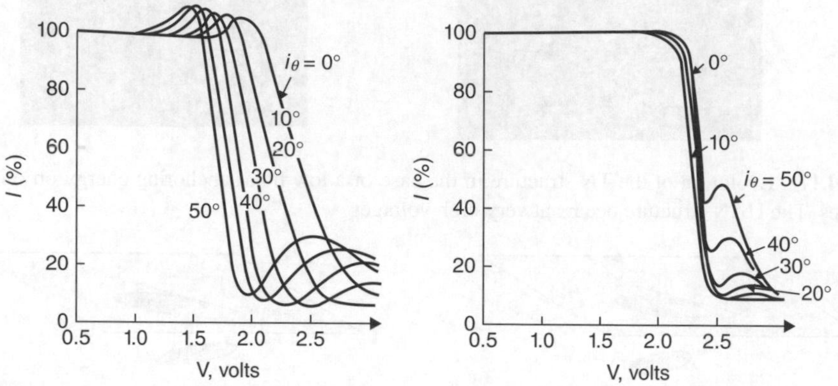

Figure 4.20 VCs: left, −90° twist structure; right, a 200° supertwisted cell. The light incidence angles i_θ are shown on the curves, $i_\varphi = 180°$ [1–3]

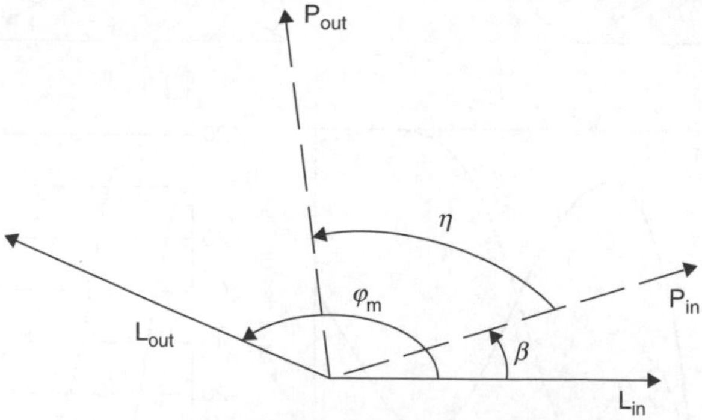

Figure 4.21 This general scheme of supertwist display geometry shows the input and output orientations of the molecules or their projections ($\mathbf{L}_{in}$, $\mathbf{L}_{out}$) as well as the input and ouput orientations of the polarizers ($\mathbf{P}_{in}$, $\mathbf{P}_{out}$). The angle φ_m is the twist angle; β and η define the location of the polarizers [1–3]

Table 4.1 Supertwist structures for passively addressed high-information-content LCDs [1–3]

Supertwist mode	Supertwist angle	Thickness-to-pitch ratio, d/P	Optical path length, Δnd (μm)	Polarizer angles β, η	Pretilt angles on the substrates, θ_0
Supertwist birefringent effect (SBE) [12]	270°	0.75	0.85	−32.5°, 65°	28°
Supertwist nematic effect (STN)	240°	0.66	0.82	45°, 60°	4°–6°
Optical mode interference (OMI) effect [13]	180°	−0.3	0.46	0°, 90°	2°–3°
Optical mode interference (OMI) effect	240°	0.43	0.54		4°–6°

Various realizations of supertwist structures for passively addressed high-information-content LCDs are presented in Table 4.1 [1–3].

New methods for the realization of the electro-optical effects in supertwist LC cells such as supertwist nematic (STN) or optical mode interference (OMI) avoid certain limitations and disadvantages which are observed in supertwist birefringence (SBE) LC cells. For example, the boundary tilt angles for preventing the appearance of domain structures are not as large as in the SBE case and the requirements of thickness nonuniformity become softer. Consequently, the manufacture of LCDs becomes easier. Both STN and OMI mixtures have the same or even better steepness as the SBE prototype, which in accordance with (4.38) means addressing the larger number of lines in LCDs with passive addressing [1–3].

The OMI effect provides a weak wavelength dependence of the transmission in the visible region [13]. The effect requires a low optical path difference, which leads to a strong interference of two polarization modes when propagating through an OMI cell. (Mauguin's waveguide regime (4.33) is not valid.) The transmission spectra of the OMI cell enable us to realize the black and white appearance of the two display states, which is impossible in STN or SBE cells. The electro-optical characteristics of the OMI cells are much more tolerant to cell gap nonuniformity than in the STN case and are less temperature dependent. However, one of the main disadvantages of the OMI display is low brightness in the off-state. For instance, if we take the brightness of two parallel polarizers equal to 100%, then the off-states of the 90° twist cell and the 240° STN cell would correspond to 95% and 64%, respectively, while the brightness of the 180° OMI cell does not exceed 40% [14]. However, the brightness of OMI displays can be greatly improved up to 77% by increasing the twist angles up to 270°, if we make an appropriate choice of Δnd and the angular position of the polarizers [15]. Further, OMI displays possess response times about 1.5 times lower than STN-LCDs for the same value of the contrast and multiplexing capability. Surprisingly, the viewing angles of OMI displays for high-information-content screens are not much wider than in the STN case, despite the lower values of Δnd. Thus OMI displays are competitive with STN displays for high-information-content passively addressed LCDs.

Supertwisted LCDs with Improved Characteristics

Parameter Space Approach

The parameter space approach was proposed in [16] to help optimize the characteristics of LCDs that use twisted (and supertwisted) LC structures (see Section 2.4). This approach often simplifies the localization

Table 4.2 Various optimal modes for LCD operation [17]

Acronym	Name	Twist angle	Polarizer angle	Retardation NW	Retardation NB
HFE	Hybrid field effect	45°	0°	0.362	0.968
RTN	Reflective twisted nematic	54°	0°	0.359	0.9515
TN-ECB	TN electrically controlled birefringence	63.7°	0°	0.354	0.935
MTN	Mixed mode twisted nematic	~80°	~20°	0.45	0.864
SCTN	Self-compensated twisted nematic	60°	30°	0.61	0.943

NW, Normally White; NB, Normally Black.

of optimal values of Δnd and twist angle for such devices. Various modes of transmissive and reflective LCDs are considered using the parameter space approach in Section 2.4. Some of these modes are presented in Table 4.2.

Film-Compensated STN

One of the main goals is to attain black and white switching of a supertwist cell with a high contrast ratio and an acceptable brightness. At the same time, we need to maintain a sufficiently high steepness of the TVC to enable a sufficient resolution of STN-LCD (see (4.38)). One possible solution of the problem is to use phase retardation plates in combination with a STN panel. Supertwisted displays with one and two phase retardation plates have been reported [1–3]. The orientation of the retardation plates and their optical path differences are optimized to provide both black and white switching and wide viewing angles. The double plates provide better achromatic appearance and contrast ratio than a single plate, especially when they are placed on both sides of the STN cell. Thin polymer films (polycarbonate, polyvinyl alcohol, etc.) are used today as phase retardation plates. Biaxial compensator films and optically negative polymeric films composed of discotic molecules have also been developed [18]. The contrast ratios of the STN-LCD with negative birefringence films achieve 100:1 for normally incident light. However, it is very difficult to realize such a film with a uniform phase retardation over a large surface area.

The typical performance of a high-resolution film-compensated STN (FSTN)-LCD with phase retardation plates and antireflective layers is shown in Figure 4.22. The calculations were done using MOUSE-LCD [4] which offers many useful optimization utilities (theoretical approaches and representations underlying some of them are presented in Chapters 6 and 12). The multiplexing duty ratio $N = 32{:}1$ (the number of addressing lines can be $2N = 64$ [1–3]), and the contrast ratio is more than 300:1 for normally incident light.

Double-STN-Cell Configuration

The main problem with the STN-LCD using the phase retardation plate is a limited temperature range over which a good compensation is possible due to the different temperature dependence of Δn in the LC and polymer layers [14]. The latter disadvantage is avoided when, instead of the phase compensator, another supertwist cell of the same thickness is used which has no electrode and is twisted in the opposite direction. Thus the second passive layer optically compensates the active STN layer in the off-state and the light passing through the two cells becomes linearly polarized perpendicular to the input polarizer direction. As a result, this light is absorbed by the analyzer, which is crossed with the polarizer (Figure 4.23) [1–3].

The device consisting of two supertwist cells is called a double-layer STN-LCD or double STN (DSTN)-LCD and in the off-states looks dark for all wavelengths in the visible region. DSTN-LCDs demonstrate both higher contrast ratios and wider viewing angles than STN-LCDs with phase retarders. Due to the double-cell construction, the requirement of gap nonuniformity in DSTN-LCDs is more strict than in STN-LCDs. Other drawbacks are increased display thickness and weight, which are not

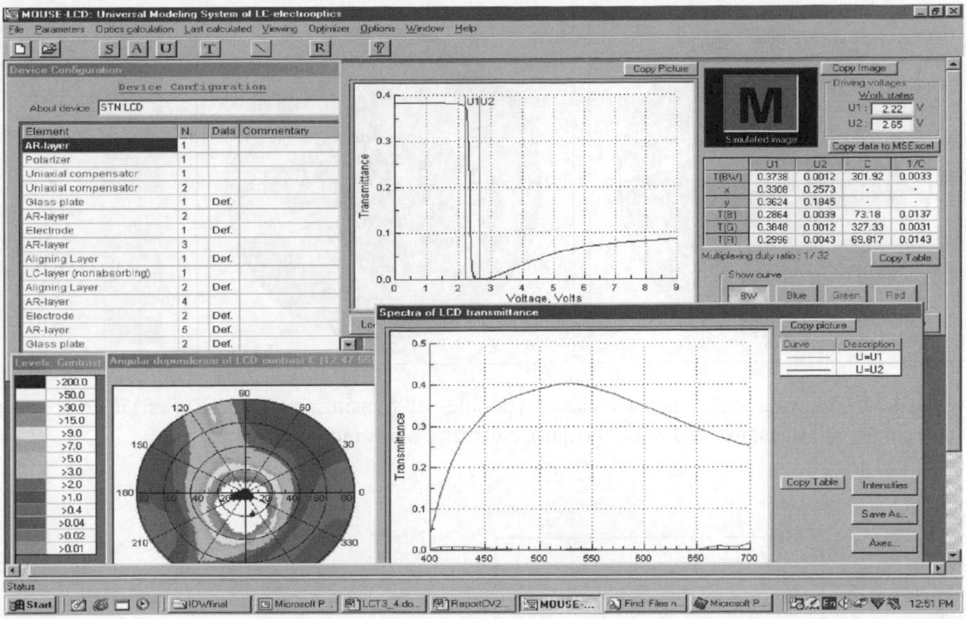

Figure 4.22 An optimized variant of a FSTN-LCD

convenient, especially in applications in portable computers. The TVC of a DSTN cell possesses the same steepness as that for a STN-LCD, but the viewing angles are considerably wider [14]. Of course supertwist LCDs in general have much higher contrast ratios and better viewing angles than TN-LCDs for passive matrix addressing with a high information content. However, TN-LCDs in a double-cell configuration, similar to that shown in Figure 4.23, also have much better viewing angle characteristics than a single TN cell (Figure 4.24) because the same phase compensation principle within the whole visible spectra works in this case too.

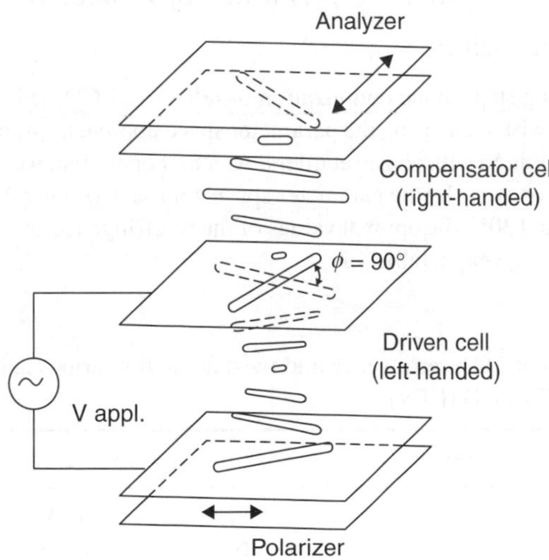

Figure 4.23 DSTN-LCD configuration [1–3]

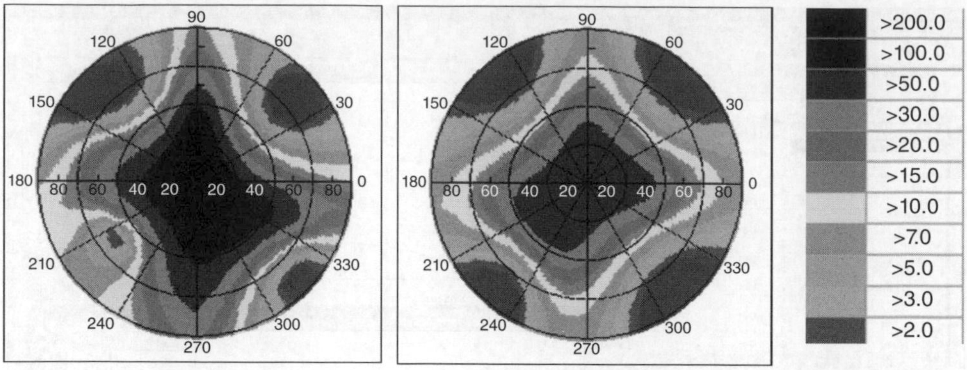

Figure 4.24 Viewing angle characteristics of a double-cell transmissive TN-LCD (left) in comparison with a single-cell transmissive TN-LCD (right). Normally black mode is used

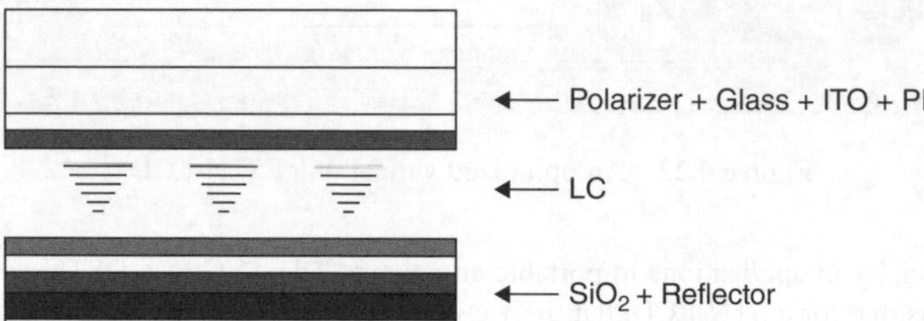

Figure 4.25 Reflective LCD [19]

4.1.4 Optimization of Optical Performance of Reflective LCDs

Typical Optimization Procedure

Let us consider the following steps in the optimization of reflective LCDs (Figure 4.25): (i) selection of optimal birefringence and twist angle using the parameter space approach; (ii) insertion of the antireflective layers; and (iii) compensation with phase retarders. As a first operation, we choose the optimal values of LC birefringence and twist angle by the parameter space approach (Figure 2.11) for various angles of the polarizer, $\alpha = 0°$, 15°, and 30°. The optimal values of the birefringence and twist angle calculated for $\lambda = 550$ nm (green light) are given in Table 4.3.

Table 4.3 Optimal values of LC birefringence and twist angle for various angles of the polarizer calculated for a reflective TN-LCD (RTN)

	$\alpha = 0°$		$\alpha = 15°$		$\alpha = 30°$	
Twist angle (deg)	63	190	73	190	55	−90
$d\Delta n$ (μm)	0.2	0.585	0.26	0.685	0.38	0.65

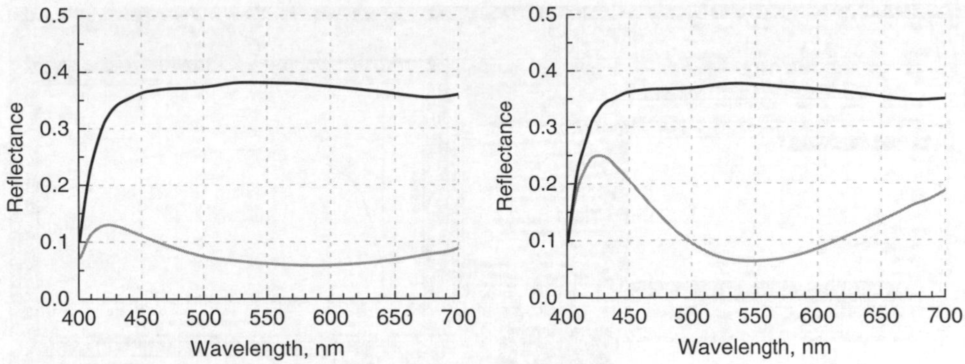

Figure 4.26 Spectral dependence of the RTN-LCD for $\alpha = 0°$ and two optimal LC twist angles $\varphi = 63°$ (left) and 190° (right). The contrast ratio for the green light $\lambda = 550$ nm is the same

The corresponding spectral dependence of the RTN-LCD for $\alpha = 0°$ and two optimal LC twist angles $\varphi = 63°$ and 190° are shown in Figure 4.26. It is clear that the contrast for the green light $\lambda = 550$ nm is the same, while for other wavelengths of the visible spectrum it is considerably different. It looks like LC twist angles $\varphi = 63°$ provide more uniform and higher contrast in the whole visible spectrum (Figure 4.26).

The second step is the optical matching of various films of the RTN-LCD with antireflective (AR) layers (Figure 4.27). The idea of such matching is well known [20]. To eliminate the reflection on the boundary with two layers having the refractive indexes $n = n_1$ and $n = n_2$ at the wavelength λ, we need to insert between them a quarter-wave plate with the refractive index $n_m = \sqrt{n_1 n_2}$ a thickness $\lambda/4n_m$. It is most important in this case to avoid parasitic reflection on the first boundary (air–glass), where usually about 4% of light is reflected. Figure 4.28 shows a considerable improvement in the spectral dependence and the contrast ratio of the RTN-LCD after insertion of the AR layers shown in Figure 4.27.

The third step of the optimization operation is the application of phase retardation plates, which considerably improves the contrast ratio of the RTN-LCD and makes the dark state almost independent of the wavelength in the whole visible spectrum (Figure 4.29). Both the angles of the compensators, their

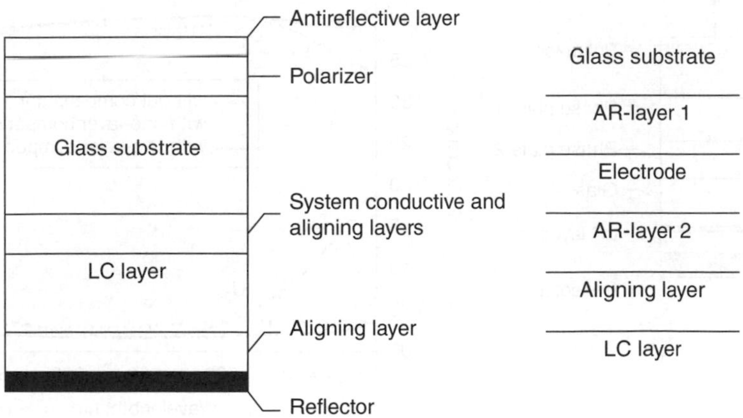

Figure 4.27 RTN-LCD with antireflective (AR) layers [19]

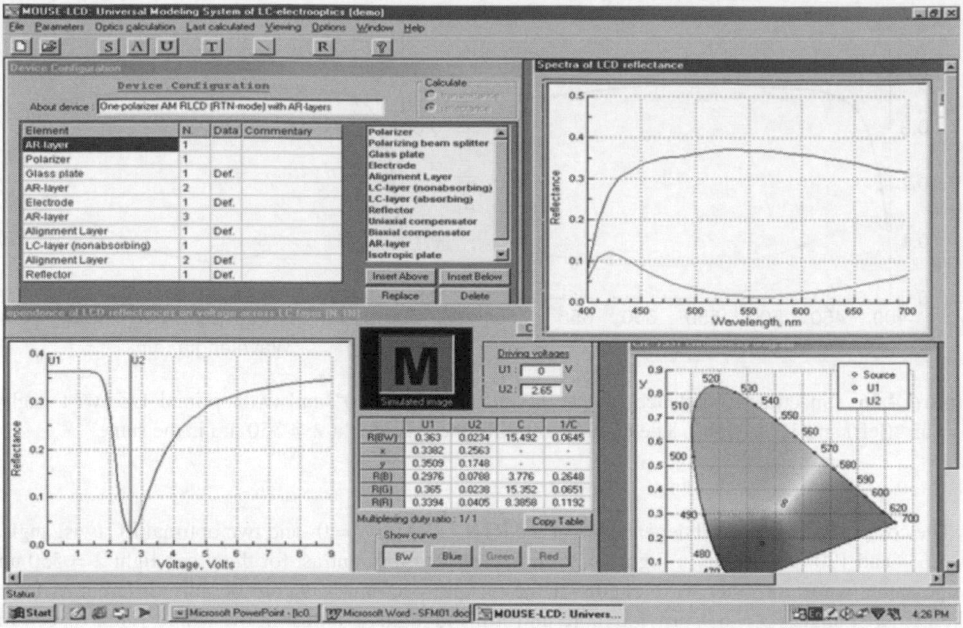

Figure 4.28 RTN-LCD after insertion of the AR layers. The calculations were done using MOUSE-LCD [19]

thickness, and the polarizer angle can be optimized. The best parameters of the phase compensation are obtained when we apply the two uniaxial compensators.

The three-step optimization operations performed for the RTN-LCD can in principle be applied for any LCD electro-optical modes described above, namely, ECB, HAN, VAN, TN, and STN, which are used in both transmissive and reflective regimes (see Table 4.4).

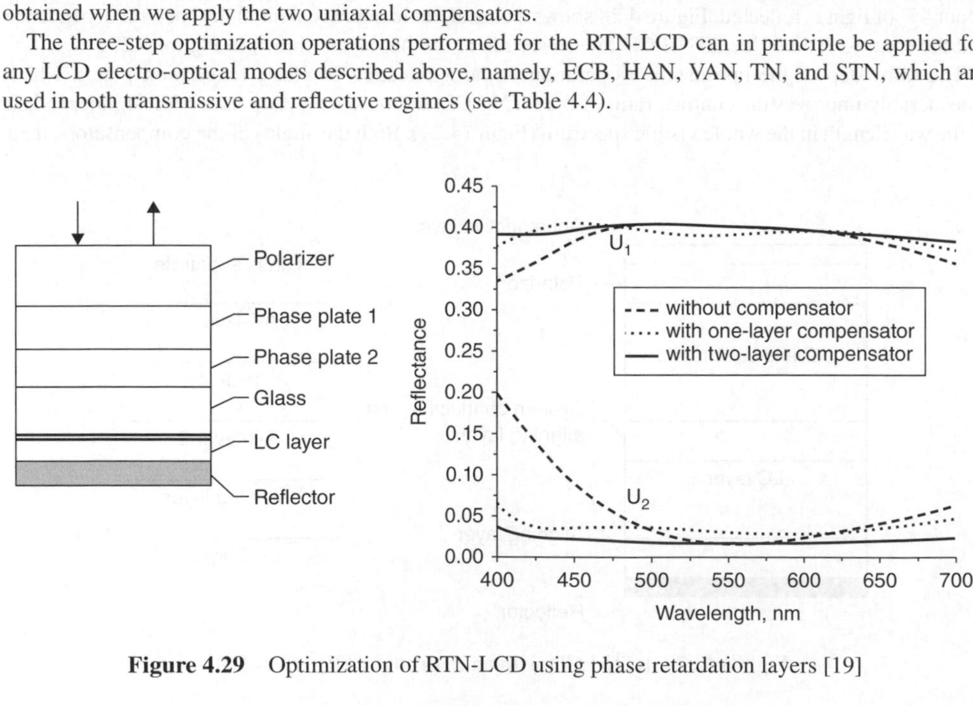

Figure 4.29 Optimization of RTN-LCD using phase retardation layers [19]

Table 4.4 Contrast ratio of various LC electro-optical modes after three-step optimization

Electro-optical mode	Transmissive LCD	Reflective LCD
TN	433	79
STN	302	34
VAN	408	38
HAN	434	50
ECB	397	52

4.2 Transflective LCDs

Transflective LCDs combine the characteristics of transmissive and reflective LCDs. The pixels of the transflective LCD are divided into transmissive and reflective sub-pixels. The transmissive sub-pixels in a transflective display transmit the backlight illumination and the reflective sub-pixels reflect light from the environment under ambient illumination (Figure 4.30).

Conventional transflective LCDs are commonly fabricated using a double-cell-gap approach to maintain the same optical characteristics on increasing the voltage. In the double-cell-gap approach, the cell gap for the transmissive mode is double that for the reflective mode. Thus, both reflective and transmissive modes have the same optical path difference. On the other hand, fabrication would be much more complicated. In a single-cell-gap approach matching the TVC with the reflectance–voltage curve (RVC), different LC modes in the transmissive and reflective sub-pixels have been suggested recently and their optical parameters optimized [21].

In the next two subsections, we consider two transflective configurations based on the dual-mode single-cell-gap approach as well as another two transflective configurations based on the single-mode single-cell-gap approach. The simulations show that all the configurations work very well, as was confirmed by experiment [22].

4.2.1 Dual-Mode Single-Cell-Gap Approach

In dual-mode single-cell-gap transflective LCD configurations, different LC modes can be applied to the transmissive and reflective sub-pixels of the display. Ideally all conventional LC modes can be applied and adjusted in this approach, including ECB, TN, optically compensated birefringence (OCB), VAN, and in-plane switching (IPS). However, for easy fabrication purposes, ECB and TN modes with different twist angles are studied first. Two dual-mode configurations are introduced.

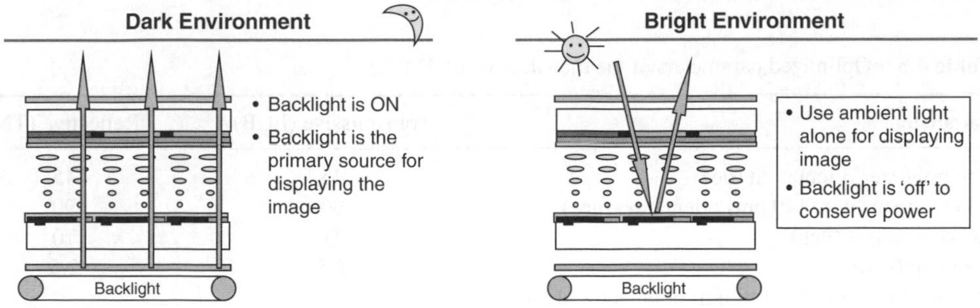

Figure 4.30 Transflective LCD [21]

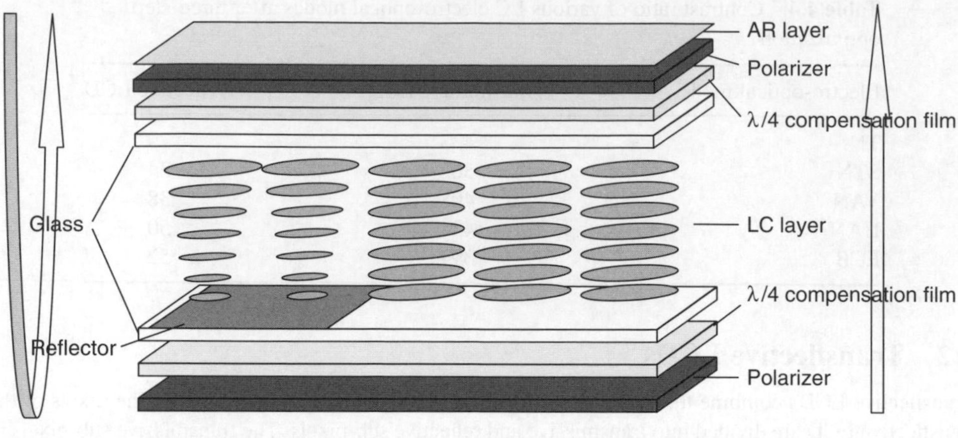

Figure 4.31 TN-ECB single-cell-gap transflective LCD configuration

TN-ECB Configuration

The first one is called the TN-ECB configuration, which applies ECB mode in the transmissive sub-pixel and TN mode in the reflective sub-pixel. The LC material in all the simulation processes is ZLI-4792 from E. Merck. The principal refractive indexes of ZLI-4792 are $n_\perp = 1.4939$ and $n_\parallel = 1.5987$, $n_\perp = 1.4819$ and $n_\parallel = 1.5809$, and $n_\perp = 1.4774$ and $n_\parallel = 1.5734$ at wavelengths of 436, 546, and 633 nm, respectively. The low-frequency dielectric anisotropy and the elastic constants are $\Delta\varepsilon = 5.2$, $K_{11} = 1.32 \times 10^{-6}$, $K_{22} = 6.5 \times 10^{-7}$, and $K_{33} = 1.38 \times 10^{-6}$ dynes, respectively. The cell gap for both transmissive and reflective parts is 2.5 μm, and their surface pretilt angle is 2°. The spectrum of the light source used in the simulation (in all examples considered in this and the next subsection) is that of the standard illuminant D65 for both the transmissive and reflective parts. Typical absorption characteristics of polarizers and typical dispersion characteristics of the compensation films were used in the calculations (in all the examples).

Figure 4.31 shows the configuration of the TN-ECB single-cell-gap transflective LCD. The twist angle in the reflective part is equal to 70°, while there is no twist in the transmissive part. The top polarizer is coated with an AR layer to lower the surface reflection. One $\lambda/4$ compensation film is added in between the front glass and front polarizer. Another $\lambda/4$ compensation film is added in between the rear glass and rear polarizer, which works only for the transmissive part since the light in the reflective part is blocked by the reflector. The optimized parameters are given in Table 4.5.

Figure 4.32 shows the simulated performance of the TN-ECB transflective LCD. The experimental results confirmed that the TVC and RVC match very well.

Table 4.5 Optimized parameters of the transflective LCD [22]

Parameter	Transmissive (ECB)	Reflective (TN)
Top polarizer orientation (deg)	45	45
Top compensator (140 nm) orientation (deg)	90	90
LC twist angle (deg)	0	70
Cell gap (μm)	2.5	2.5
Bottom compensator (140 nm) orientation (deg)	0	—
Bottom polarizer orientation (deg)	−45	—

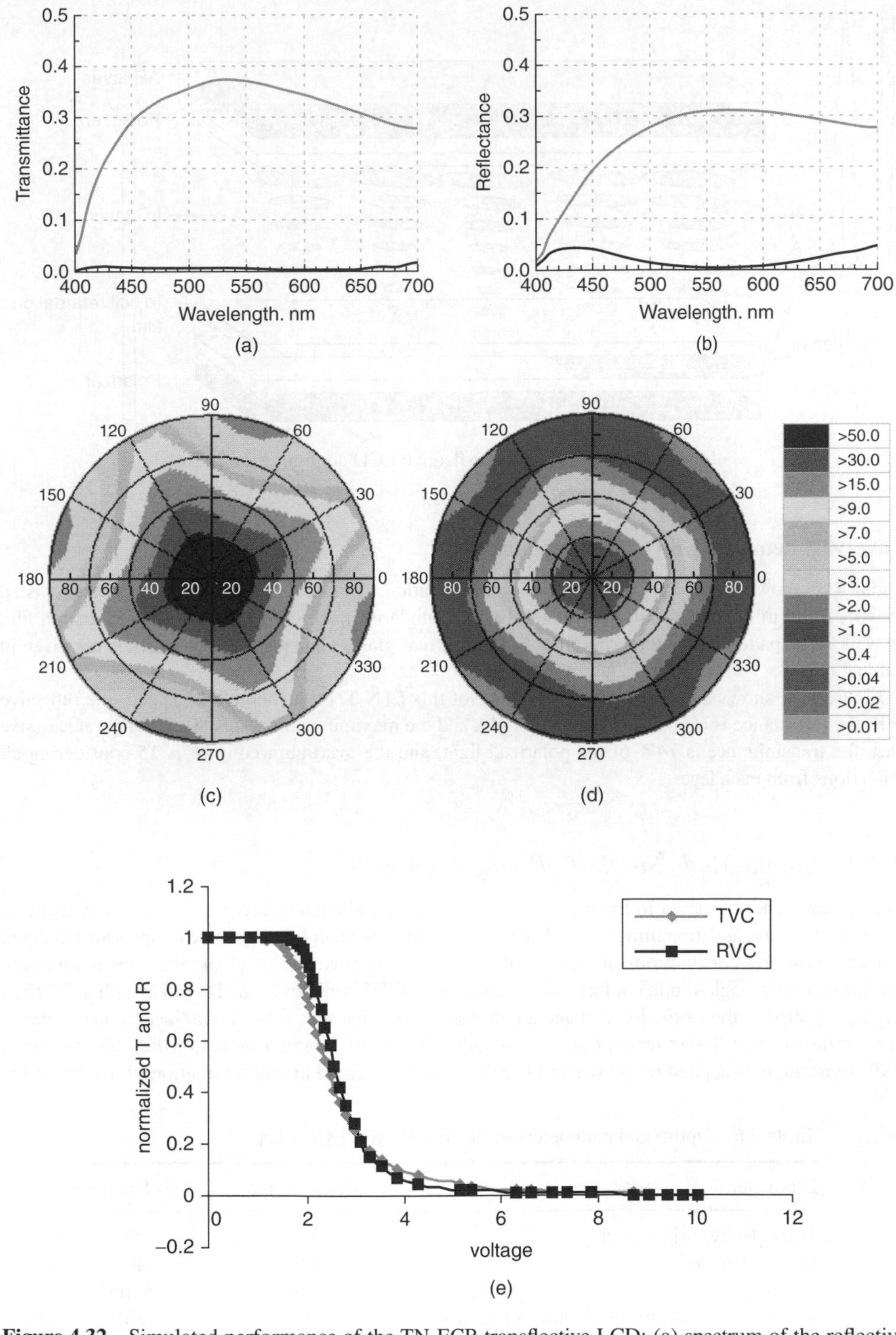

Figure 4.32 Simulated performance of the TN-ECB transflective LCD: (a) spectrum of the reflective part; (b) spectrum of the transmissive part; (c) contrast ratio distribution of the reflective part; (d) contrast ratio distribution of the transmissive part; (e) TVC and RVC

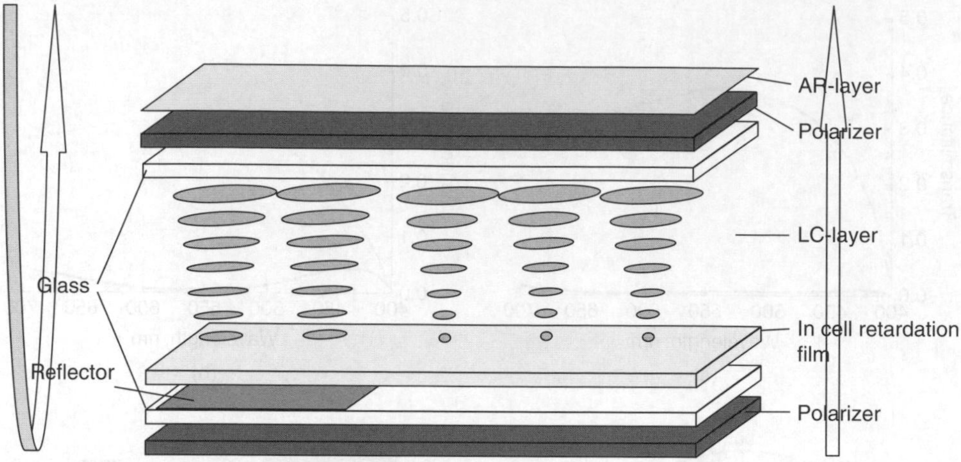

Figure 4.33 LTN-TN transflective LCD configuration [22]

Low twist nematic (LTN)-TN Configuration

Figure 4.33 shows the scheme for the LTN-TN configuration. The LC in the reflective sub-pixel is twisted by 40°, while in the transmissive sub-pixel the twist angle is equal to 90°. The in-cell retardation film as well as the patterned reflector are produced on the rear glass. The optimized parameters are given in Table 4.6.

Figure 4.34 shows the simulated performance of this LTN-TN transflective LCD. For the reflective part, the reflectance is 60% of the polarized light, and the maximum contrast is 80. For the transmissive part, the transmittance is 74% of the polarized light, and the maximum contrast is 15 considering all reflections from each layer.

4.2.2 Single-Mode Single-Cell-Gap Approach

In this approach, in contrast to the previous case, a patterned alignment layer is not necessary. Instead, in-cell patterned retardation film is used. The in-cell retarders in the following two examples are patterned films with different orientations of the optical axis in the reflective sub-pixel and the transmissive sub-pixel. In the first single-mode configuration, called the TN 75° configuration, TN mode with a 75° twist angle is applied to the whole LC cell and a patterned retardation film is used to adjust the performance of the reflective part. The other configuration is called the TN 90° configuration, in which TN mode with a 90° twist angle is applied to the whole LC cell and two patterned in-cell retardation films are used to

Table 4.6 Optimized parameters of the transflective LCD [22]

Parameter	Transmissive	Reflective
Top polarizer orientation	90°	90°
LC twist angle	90°	40°
Cell gap	5 μm	5 μm
In-cell retardation film (132 nm) orientation	88°	88°
Bottom polarizer orientation	90°	—

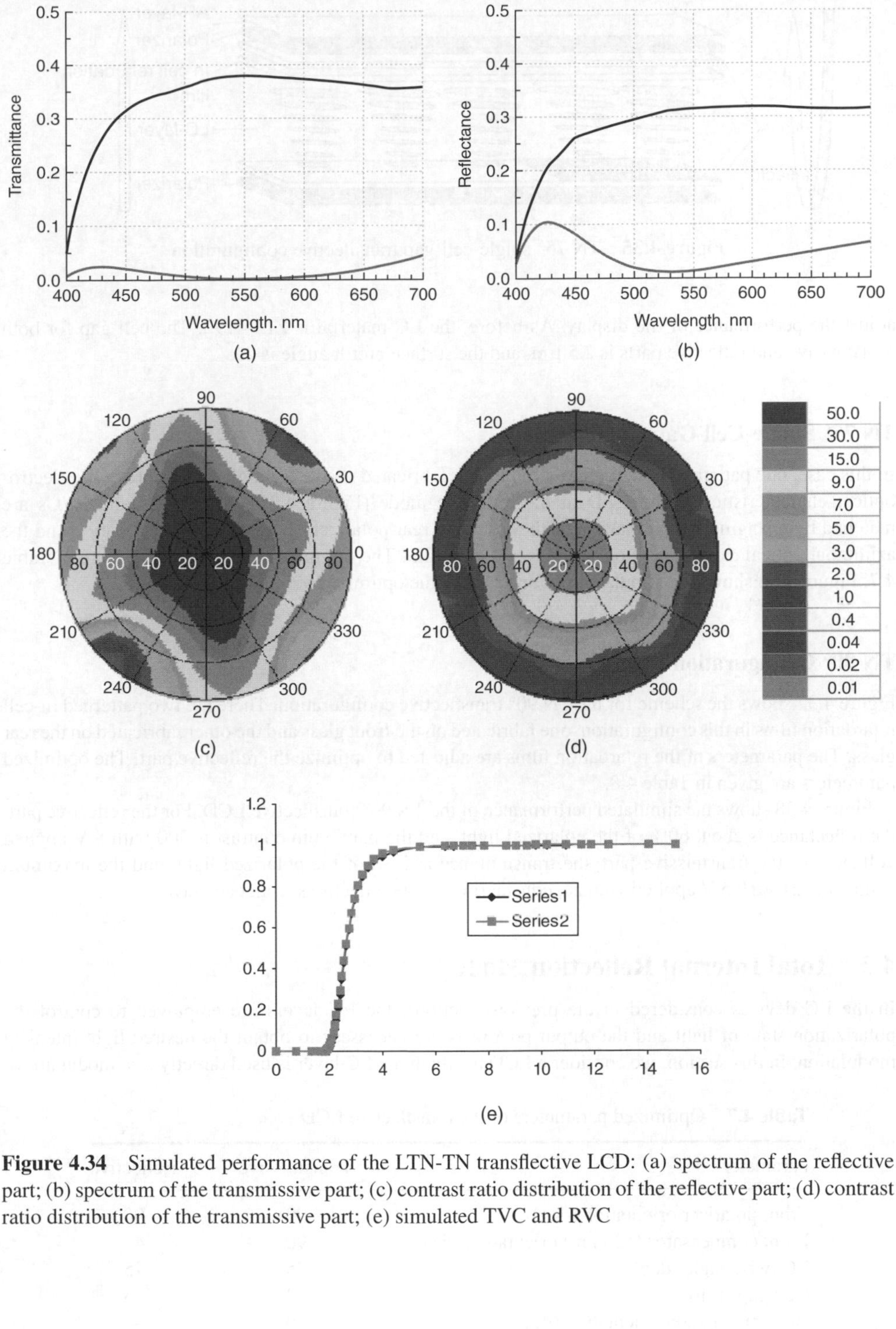

Figure 4.34 Simulated performance of the LTN-TN transflective LCD: (a) spectrum of the reflective part; (b) spectrum of the transmissive part; (c) contrast ratio distribution of the reflective part; (d) contrast ratio distribution of the transmissive part; (e) simulated TVC and RVC

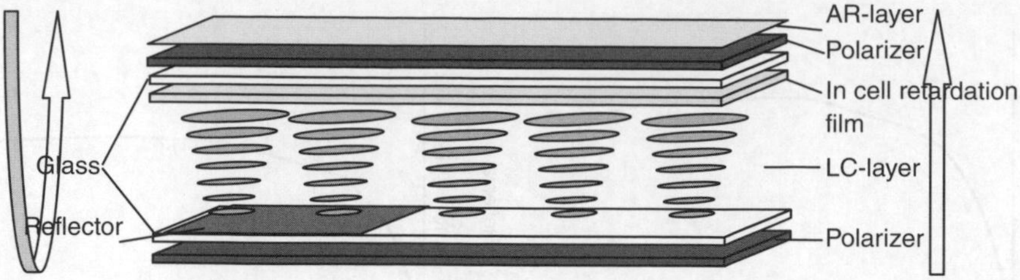

Figure 4.35 TN 75° single-cell-gap transflective configuration

adjust the performance of the display. As before, the LC material is ZLI-4792. The cell gap for both transmissive and reflective parts is 2.5 μm, and the surface pretilt angle is 2°.

TN 75° Single-Cell-Gap Configuration

In this case, one patterned in-cell retardation film is fabricated on the front glass to improve the electro-optical characteristics of the LCD in the reflective mode (Figure 4.35). The TVCs and RVCs are matched by changing the orientation of the front and rear polarizers, as well as the retardation and the azimuthal optical axis orientation of the retardation film. The optimized parameters are given in Table 4.7. Figure 4.36 shows the simulated performance of the optimized variant of the LCD.

TN 90° Configuration

Figure 4.37 shows the scheme for the TN 90° transflective configuration. There are two patterned in-cell retardation films in this configuration, one fabricated on the front glass and the other fabricated on the rear glass. The parameters of the retardation films are adjusted to optimize the reflective part. The optimized parameters are given in Table 4.8.

Figure 4.38 shows the simulated performance of the TN 90° transflective LCD. For the reflective part, the reflectance is about 60% of the polarized light, and the maximum contrast is 300 with 5 V applied voltage. For the transmissive part, the transmittance is 74% of the polarized light, and the maximum contrast is 21 with 5 V applied voltage, considering all the reflections from each layer.

4.3 Total Internal Reflection Mode

In the LC devices considered in the previous sections, the LC layers are employed to control the polarization state of light and the output polarizers are necessary to obtain the desired light intensity modulation. In this section, we consider a LCD in which the LC layer is used directly as a modulator of

Table 4.7 Optimized parameters of the transflective LCD

Parameter	Transmissive	Reflective
Front polarizer orientation (deg)	90	90
Front compensator (125 nm) orientation (deg)	90	45
LC twist angle (deg)	75	75
Cell gap (μm)	3	3
Bottom polarizer orientation (deg)	0	—

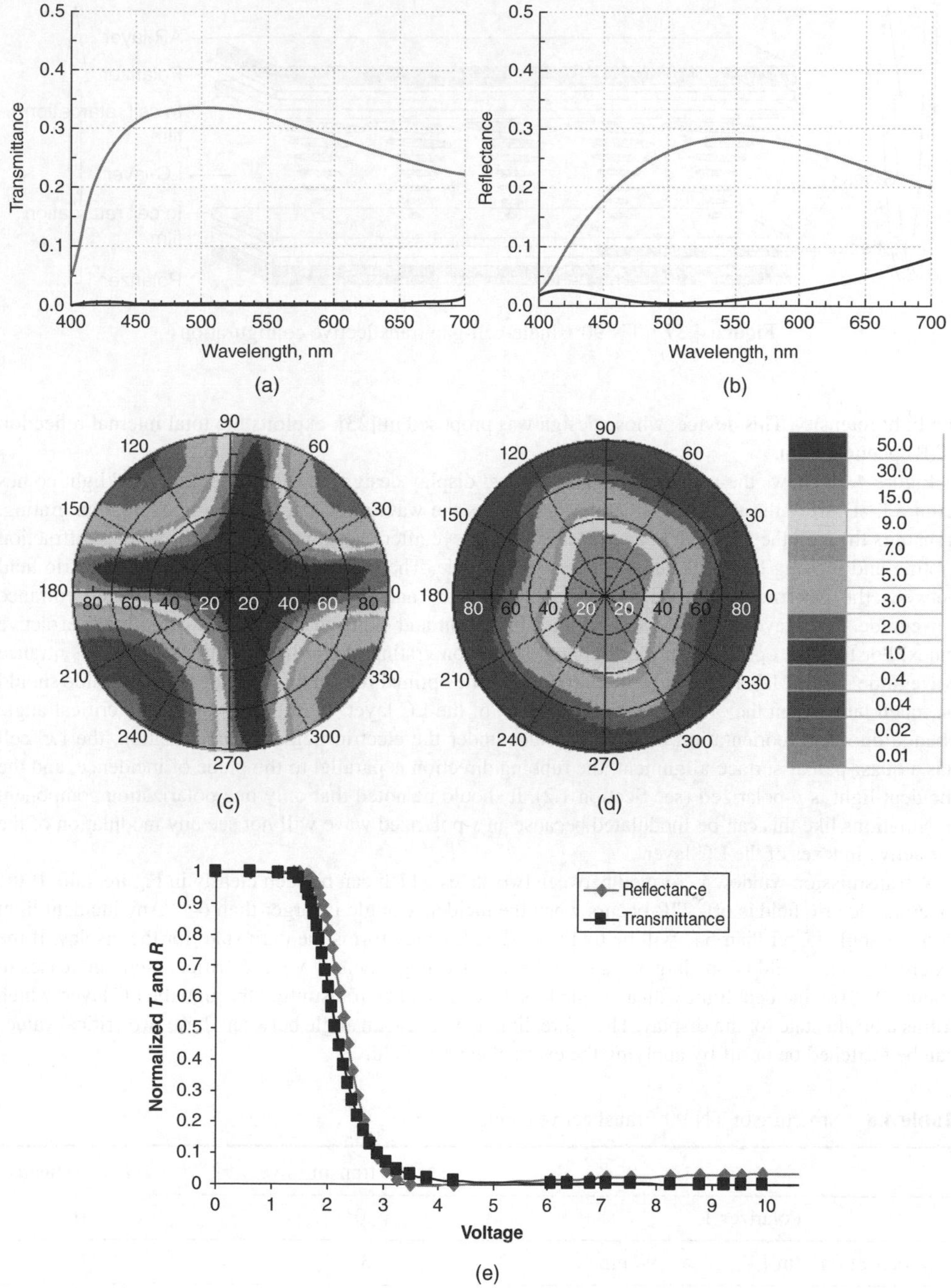

Figure 4.36 Simulated performance of the TN 75° transflective LCD: (a) spectrum of the reflective part; (b) spectrum of the transmissive part; (c) contrast ratio distribution of the reflective part; (d) contrast ratio distribution of the transmissive part; (e) simulated TVC and RVC

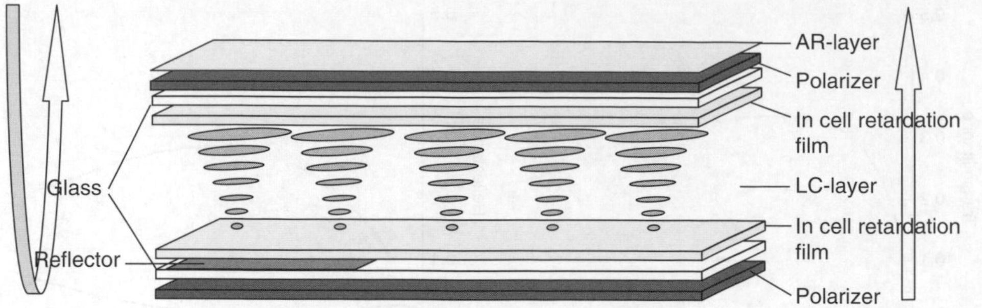

Figure 4.37 TN 90° single-cell-gap transflective configuration

the light intensity. This device, whose design was proposed in [23], exploits the total internal reflection (TIR) phenomenon.

Figure 4.39 shows the scheme of our TIR-based display design. In this design, the backlight comes from a LED array through a collimating lens, enters the waveguide via a coupling diffraction grating, transmits through the LC switch, and reaches human eye after passing through a decoupling diffraction grating and moving lens attached to the exit waveguide. The LC switch works under the electric field between the patterned electrode on the exit waveguide and the common electrode on the entrance waveguide. The waveguide used in both the simulation and experiment is glass with a high refractive index. For feasible operation of the coupling diffraction grating, the critical angle between the entrance waveguide and LC layer cannot be too large. Thus, the optimal refractive index of the waveguide should be much larger than the ordinary refractive index of the LC layer. Figure 4.40 shows the critical angle change due to a reorientation of LC molecules under the electric field. In this situation, the LC cell has a quasi-planar surface alignment, the rubbing direction is parallel to the plane of incidence, and the incident light is p-polarized (see Section 1.2). It should be noted that only one polarization component in situations like this can be modulated because an s-polarized wave will not see any modulation of the refractive indexes of the LC layer.

A transmission window is formed between two states which can be seen clearly in Figure 4.40. If the external electric field is off, TIR occurs when the incidence angle is larger than 64°. Any incident light with an angle larger than 64° will be totally reflected, which forms the dark state for the display. If the external electric field is on (larger than the threshold voltage, here 4 V), the critical angle increases to about 80°. Any incident light with an angle less than 80° will be transmitted through the LC layer, which forms a bright state for the display. Therefore, light incident at an angle between these two critical values can be switched on or off by applying the external electric field.

Table 4.8 Structure of TN 90° transflective configuration

		TN 90° transmissive	TN 90° reflective
Polarizer 1		0°	0°
Compensation film 1	90 nm	34°	0°
LC cell	Cell gap Twist angle	5 μm 90°	5 μm 90°
Compensation film 2	60 nm	48°	90°
Polarizer 2		90°	

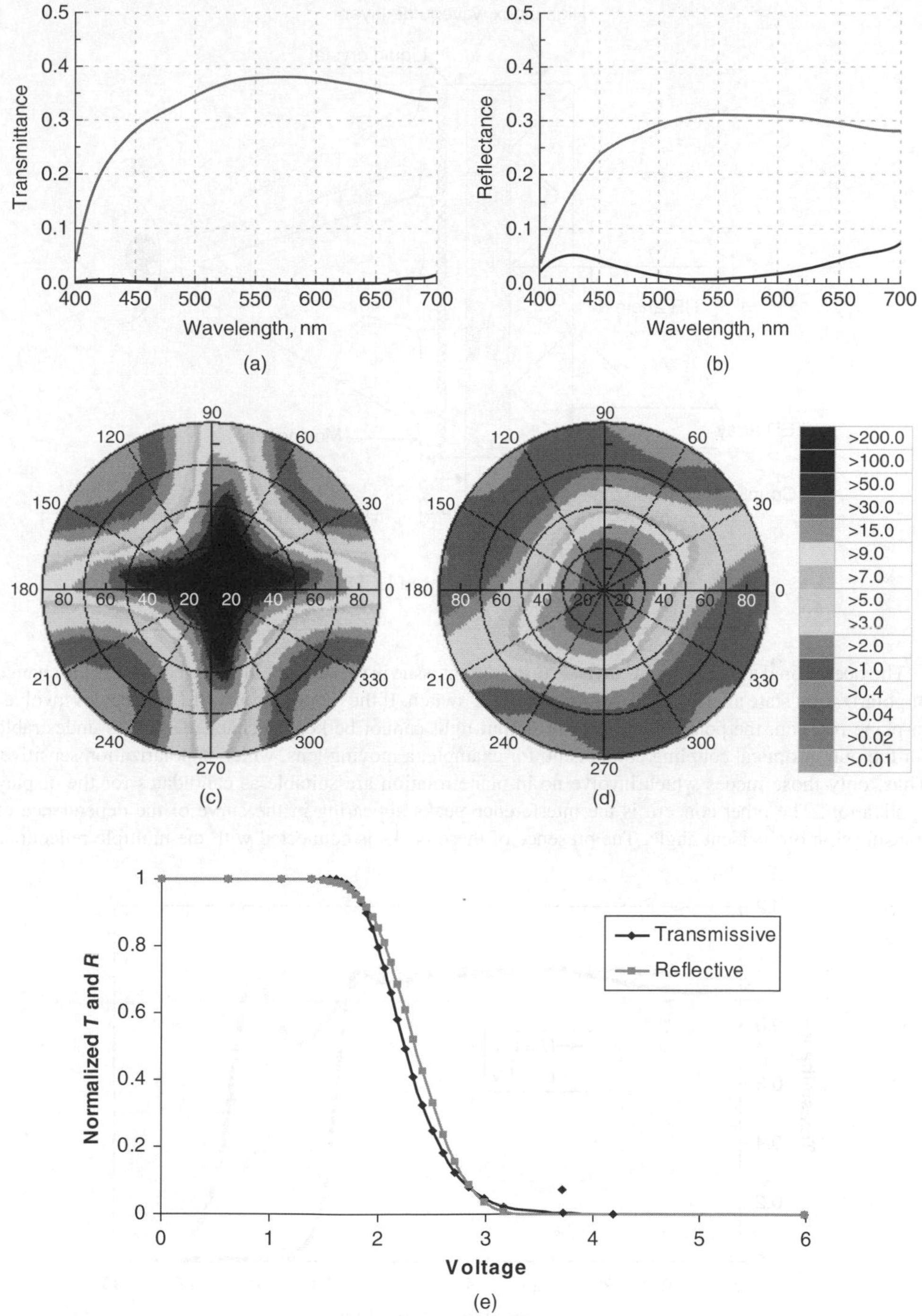

Figure 4.38 Simulated performance of the TN 90° transflective LCD: (a) spectrum of the reflective part; (b) spectrum of the transmissive part; (c) contrast ratio distribution of the reflective part; (d) contrast ratio distribution of the transmissive part; (e) simulated TVC and RVC

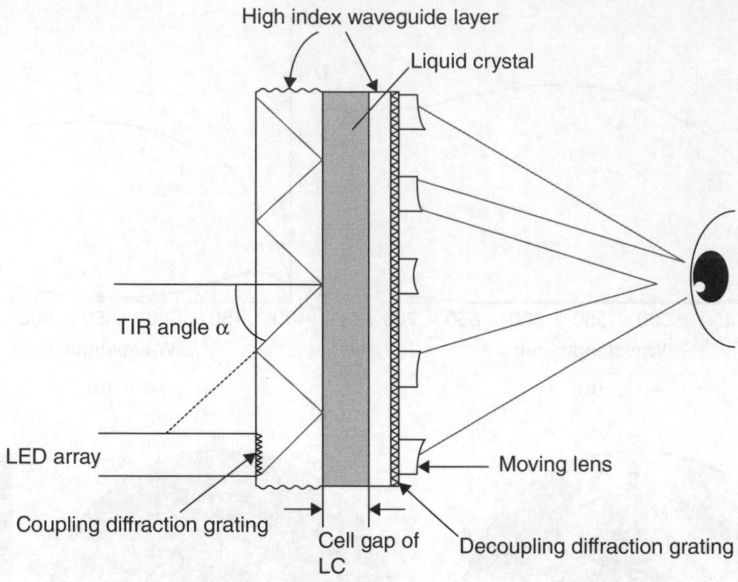

Figure 4.39 TIR-based LCD [23]

The operational mode of the LC layer is an important issue in the design [24]. One concern is the change of polarization state after light goes through the LC switch. If the reorientation of LC molecules involves in-plane rotation, the polarization state of output light cannot be kept unchanged. This is undesirable for the output optical coupling component, for example, a moving lens, which is polarization sensitive. Thus, only those modes which involve no in-plane rotation are suitable as candidates for the display application. The other concern is the interference peaks appearing in the curve of the dependence of transmission on incident angle. The presence of these peaks is connected with the multiple reflections

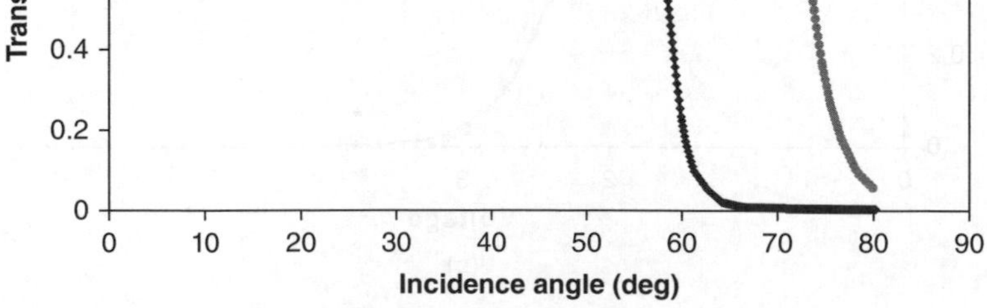

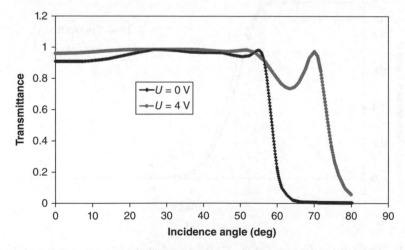

Figure 4.40 LC cell TIR transmission for two applied voltages as a function of the incidence angle [23]

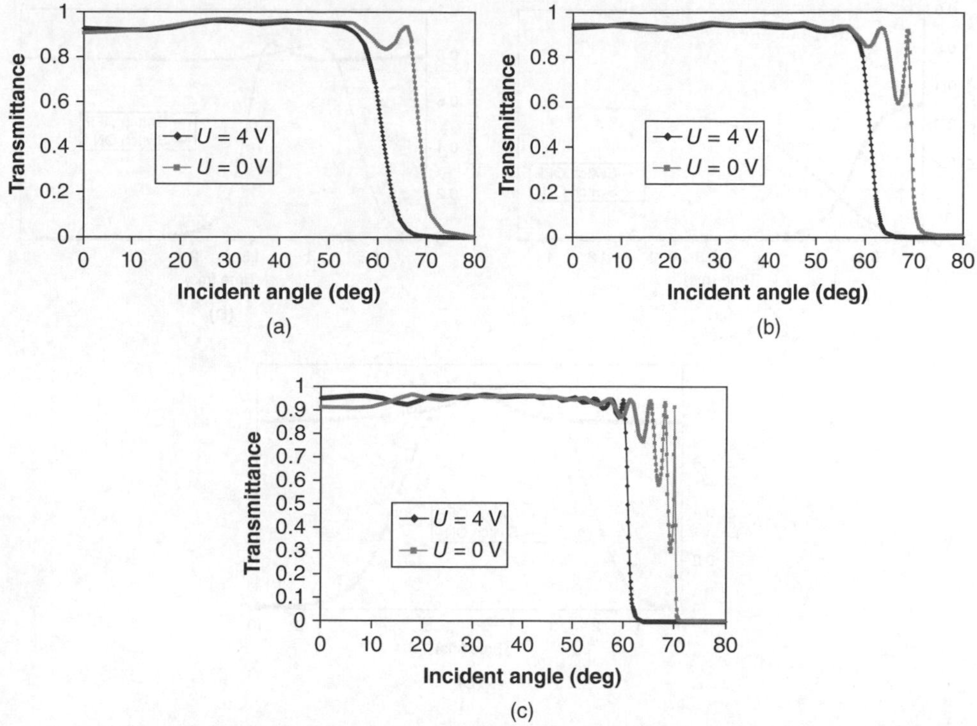

Figure 4.41 Dependence of TIR transmittance of LC cell on incident angle for different cell gaps: (a) $d = 0.5$ μm; (b) $d = 1$ μm; (c) $d = 2$ μm [23]

in the layered structure sandwiched between the glass plates. The interference pattern may significantly deteriorate the bright state transmission. The negative effect of this interference can be minimized by using quasi-homeotropic surface alignment, that is, the VAN mode. The next example deals with a VAN-mode TIR-based display.

To obtain a wide transmission window for the display, large optical anisotropy Δn is preferable for LC material such as MLC-7029 from Merck KGaA. The properties of MLC-7029 are as follows: optical anisotropy $\Delta n = 0.1265$ and extraordinary refractive index $n_{\parallel} = 1.6157$ at $\lambda = 589.3$ nm; low-frequency dielectric anisotropy $\Delta\varepsilon = -3.6$, low-frequency dielectric constant $\varepsilon_{\parallel} = 3.6$; and rotational viscosity $\gamma_1 = 175$ mPa s. Preparation of the LC cell includes no twist, and the pretilt angle is 88° for both top and bottom substrates. To reduce the interference pattern and minimize response time for high-resolution driving, the cell gap was chosen to be from 0.5 to 2.0 micron [23]. The glass used in the calculation is S-TIH1, whose refractive index $n = 1.7118$ at 650 nm. The incident light in the simulation is a p-polarized monochromatic wave with $\lambda = 633$ nm. The voltage applied for the on-state is $V_{on} = 0$ V, and $V_{off} = 4$ V for the off-state.

Figures 4.41a–c show the dependence of transmittance on incident angle for different cell gaps. It can be seen from Figure 4.41a that when the cell gap is 0.5 μm, no very obvious interference pattern can be observed near the critical angle. This is because the cell gap of the LC layer is comparable to the wavelength of the incident light. Further, a skew transmission window can be observed in the figure. The non-steep transmission curve is caused by the increasing reflectivity near the critical angle and light leakage of the evanescent wave due to a wavelength-order cell gap in the LC layer. The skew transmission curves narrow down the effective transmission window. To make the display with this cell gap work, the

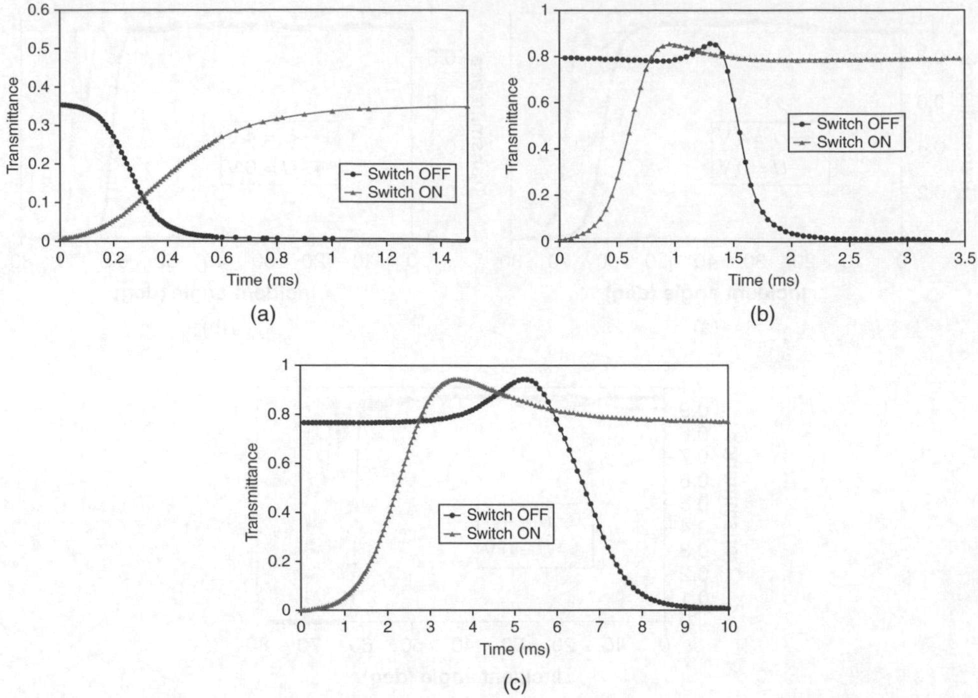

Figure 4.42 Response time of TIR LC switch for different cell gaps: (a) $d = 0.5$ μm; (b) $d = 1$ μm; (c) $d = 2$ μm [23]

chosen incident angle has to be larger than the critical angle $\theta_c = \sin^{-1}(n_o/n_g) = \sin^{-1}(1.4892/1.7118) = 60.45°$. Also the contrast ratio deteriorates due to the dimmer bright state at this incident angle. When the cell gap increases to 1 μm, the interference pattern consequently becomes more obvious and the skewness of the transmission window improves as shown in Figure 4.41b. Strong interference results in a significant vibration of the transmission curve in the vicinity of the critical angle, which impairs the width of the transmission window. Similarly, more interference peaks and steeper transmission curves can be observed in Figure 4.41c for a larger cell gap of LC layer where $d = 2$ μm. From Figure 4.41, it can be deduced that too small a cell gap, less than 1 μm, is not appropriate for the display application since the effective transmission window is narrow due to the skewness. The difficulty of preparing a LC cell using a spacer of less than 1 μm is also an issue in fabrication.

Figures 4.42a–c show the response time of the LC switch for different cell gaps. It should be noted that the dynamic response behavior of the LC switch is related to the chosen incident angle. For instance, in Figure 4.42a, since the transmission window is skew, an incident angle larger than critical angle $\theta_c = \sin^{-1}(n_o/n_g) = 60.45°$ has to be chosen to achieve a good dark state. In the calculation, the switch is sufficiently dark when the incident angle is 69°. However, at this incident angle, the transmission for a bright state is less than half of the maximum value. Due to the advantage of a small cell gap, a total response time about 1.8 ms can be achieved for the 0.5 μm LC switch, which is very fast in application a of nematic LCs. Although the total response times of the 1 and 2 μm LC switch are much slower compared to that of the 0.5 μm one, due to the quadratic relationship between response time and thickness of LC cell, a 10–90% gray-level response is still fast seen from, as can be seen in Table 4.9. To obtain a compromise between a fast response time and wide transmission window, a cell gap range from 1 to 2 μm is a good choice for fabricating TIR LCDs.

Table 4.9 Total and 10–90% response time of TIR LC switch for different cell gaps [23]

	0.5 μm cell (ms)	1 μm cell (ms)	2 μm cell (ms)
Dark to full white[a]	1	1	2.95
Full white to dark[a]	0.8	2.95	9.8
10–90%	0.59	0.45	1.7
90–10%	0.24	0.4	3

[a]0.1% of transmission is defined as full dark.

Features of Optical Calculations for TIR-Based LC Devices

Adequate simulation of TIR-based LC devices like that described above can only be performed with the aid of rigorous electromagnetic methods which accurately describe transmission and reflection at interfaces and multiple reflections. It is significant that some methods satisfying the mentioned requirements, namely, methods belonging to the class of transfer matrix methods (see Sections 7.2.1, 8.2.2, and 8.3), such as the Berreman method [25] and the 4×4 matrix method proposed by Yeh [26], are inapplicable to such devices because of the numerical instability of these methods in the TIR mode [27] (see Section 7.2.1). Appropriate methods for solving such problems are the scattering matrix method developed by Ko and Sambles [27] and a method using the adding (S-matrix) technique see also [28], which is described in Section 8.4.3. The application of the method in [27] to the above TIR-based LC device was described in detail in Xu *et al.* 23. In the MOUSE-LCD software [4], the method presented in Section 8.4.3 is used for calculations of this kind. For the above LC device and, as far as we know, in any other case, these methods give the same results. Useful information concerning the calculation of the optical characteristics of layered systems in which the TIR mode is realized can also be found in Section 8.1.3.

4.4 Ferroelectric LCDs

4.4.1 Basic Physical Properties

Symmetry

The symmetry of the ferroelectric smectic C* phase corresponds to the polar symmetry group C_2, Figure 4.43, so that when going along the z-coordinate parallel to a helix axis and perpendicular to the smectic layers, the director **L** and the polarization vector **P**, directed along the C_2-axis, rotate, as follows:

$$\mathbf{L}(z + R) = \mathbf{L}(z), \quad \mathbf{P}(z + R) = \mathbf{P}(z), \tag{4.40}$$

that is, the helix pitch R is equal to a spatial period of the FLC structure. In the absence of external fields, the FLC equilibrium helix pitch is R_0 and the average polarization of the FLC volume is equal to zero, Figure 4.43.

Main Physical Parameters

The main physical parameters which define FLC electro-optical behavior are:

(i) tilt angle, θ;
(ii) spontaneous polarization, P_s;
(iii) helix pitch, R_0;
(iv) rotational viscosity, γ_φ;

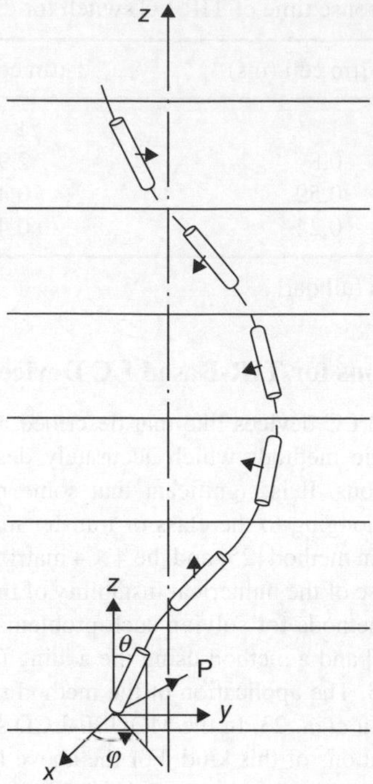

Figure 4.43 Ferroelectric liquid crystal (FLC) structure [1–3]

(v) dielectric anisotropy, $\Delta\varepsilon$;
(vi) optical anisotropy, Δn;
(vii) elastic moduli; and
(viii) anchoring energy of the director with a solid substrate.

Let us briefly characterize each of these parameters.

Tilt Angle

The value of the tilt angle can vary from several degrees to $\theta \approx 45°$ in some FLCs. Usually, in electro-optical FLC materials, $\theta \approx 22.5°$ is an operating temperature range. However, for some electro-optical applications, it is desirable to have the value of θ as high as possible. The temperature (T) dependence near the phase transition point T_c to the more symmetric LC phase (smectic A, nematic, or chiral nematic)

$$\theta \propto (T_c - T)^{1/2} \tag{4.41}$$

is typical of second-order phase transitions.

Spontaneous Polarization

The value of the spontaneous polarization depends on the molecular characteristics of the FLC substance itself and the achiral dopant introduced into the matrix, and can vary from 1 to more than 200 nC/cm^2. The value of the spontaneous polarization is one of the main FLC characteristics which define the FLC electro-optical response.

Rotational Viscosities

The switching times of the electro-optical effects in FLCs is defined by the rotational viscosity γ_φ which characterizes the energy dissipation in the director reorientation process. According to FLC symmetry, two viscosity coefficients should be taken into account, γ_θ and γ_φ, which determine the corresponding response rates with respect to the director angle φ (Figure 4.43). The relevant dynamic equations take the form [1–3]

$$\gamma_\varphi \, \mathrm{d}\varphi/\mathrm{d}t - P_\mathrm{s} E \sin \varphi = 0, \quad \tau_\varphi = \gamma_\varphi/(P_\mathrm{s} E), \tag{4.42}$$

where τ_φ are the characteristic response times for the FLC director. The viscosity coefficient γ_φ can be rewritten as

$$\gamma_\varphi = \gamma'_\varphi \sin^2 \theta, \tag{4.43}$$

where γ'_φ is independent of the angle θ. According to (4.43), $\gamma_\varphi \Rightarrow 0$ for $\theta \Rightarrow 0$, that is, γ_φ is very low for small tilt angles θ.

Near T_c one can change the angle θ, for example, by applying an electric field E; this effect is known as the electroclinic effect [29]. However, we will not consider the electroclinic effect in this book.

The rotational viscosity γ_φ can be estimated from the experimental dependence of the electro-optical response as follows [30]:

$$\begin{aligned} \gamma_\varphi &= P_\mathrm{s} E \tau_\varphi, \\ \tau_\varphi &= (t_{90} - t_{50})/\ln \sqrt{5}, \end{aligned} \tag{4.44}$$

where t_{90} and t_{50} are the corresponding times for 90% and 50% transmission from the maximum level (Figure 4.44).

One possible way to increase the switching rate in FLCs includes minimizing the viscosity with a simultaneous rise in polarization. However, this way is not very promising, because (i) the increase in polarization is hardly compatible with the lower values of rotational viscosity [30], and (ii) the repolarization component of the alternating current grows rapidly with the value of polarization, that is,

$$i_\mathrm{P} \sim P_\mathrm{s}/\tau_\varphi \sim P_\mathrm{s}^2/\gamma_\varphi, \tag{4.45}$$

which is always undesirable for applications.

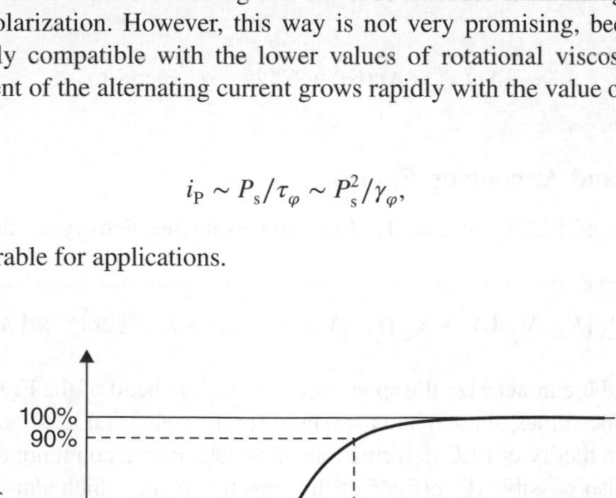

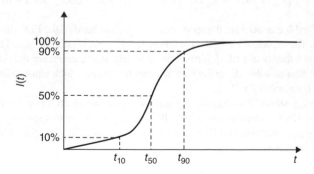

Figure 4.44 Dynamics of FLC cell transmission [1–3]

Helix Pitch

The helix pitch value R_0 is easily controlled by varying the concentration of a chiral dopant in a smectic C matrix. To vary the helix pitch from 0.1 more than to 100 μm, we require a chiral dopant with a high twisting power and good solubility. At the same time, the chiral dopant should not depress the smectic C temperature range [7]. Moreover, the growth of polarization should be more pronounced than that of the rotational viscosity γ_φ with an increasing concentration of the chiral dopant. The inverse pitch of the FLC helix, for reasonably large concentrations of the helix of the chiral dopant c_D, is proportional to the concentration:

$$R_0^{-1} \propto c_D. \tag{4.46}$$

Dielectric and Optical Properties

It is important for applications to have negative dielectric anisotropy, $\Delta\varepsilon = \varepsilon_\parallel - \varepsilon_\perp < 0$, which stabilizes the relevant director structures [1–3]. Usually values of the negative dielectric anisotropy $\Delta\varepsilon$ in FLC mixtures are between −0.5 and −2 in the kilohertz region, decreasing at higher frequencies [1–3].

The effect of nonzero dielectric anisotropy can be taken into account, by inserting an additional term into (4.42) for the azimuthal director motion in the electric field E [1–3]:

$$\gamma_\varphi \mathrm{d}\varphi/\mathrm{d}t = P_s E \sin\varphi + (\Delta\varepsilon/4\pi)E^2 \sin^2\theta \sin\varphi \cos\varphi. \tag{4.47}$$

The dielectric tensor of a FLC at *optical frequencies* can be regarded as uniaxial [1–3] (see also Section 7.3). Then the FLC possesses only two refractive indexes, $n_\parallel$ along the director and $n_\perp$ perpendicular to it. The electro-optical behavior of a FLC is mainly defined by the optical anisotropy $\Delta n = n_\parallel - n_\perp$. The birefringence value Δn can be obtained from electro-optical measurements, where the dependence of the transmitted intensity on the phase factor $\Delta nd/\lambda$ (d is the cell thickness and λ is the light wavelength) is used [1–3]. The dispersion law for the optical birefringence can be represented by the two-coefficient Cauchy formula [20]

$$\Delta n(\lambda) = \Delta n(\infty) + C/\lambda^2, \quad C = \text{const.} \tag{4.48}$$

Elastic Properties and Anchoring Energy

The elastic properties of FLCs are usually discussed using the density of the elastic energy g_{el} as follows [1–3]:

$$g_{el} = \tfrac{1}{2}[K_{11}(\nabla \cdot \mathbf{L})^2 + K_{22}(\mathbf{L} \cdot (\nabla \times \mathbf{L}) - t)^2 + K_{33}(L \times (\nabla \times \mathbf{L}) - \mathbf{b})^2], \tag{4.49}$$

where parameters t and $\mathbf{b}$ characterize the spontaneous twist and bend of the FLC director $\mathbf{L}$, and K_{ii} are FLC elastic moduli. The values of $t > 0$ and $t < 0$ hold for the right- and left-handed FLCs, respectively. The general continuum theory of FLC deformations must take into account not only deformations of the FLC director $\mathbf{L}$, but also possible distortions of the smectic layers, which can be described in terms of the variation of the layer normal $\mathbf{v}$ [1–3].

The FLC free energy should also include the surface terms, which are the polar $w_p = -W_p(\mathbf{Pv})$ and dispersion $w_d = -W_d(\mathbf{Pv})^2$ contributions, where W_p and W_d are the corresponding anchoring strength coefficients, $\mathbf{v}$ is the layer normal, and $\mathbf{P}$ is the FLC polarization [31]. Thus the total free energy F_d of the FLC director deformations is

$$F_d = \int_v g_{el}\,\mathrm{d}\tau + \int_s (w_p + w_d)\mathrm{d}\sigma. \tag{4.50}$$

In the presence of the external field, the total free energy includes the energy of the director deformations F_d and the energy F_E of the interaction of the ferroelectric phase with the field **E**:

$$F = F_d + F_E = F_d + \int_v [-(\mathbf{PE}) - \mathbf{DE}/8\pi]\, d\tau. \tag{4.51}$$

Various textures of FLC are observed [1–3]: a helical state in sufficiently large thickness $d \gg R_0$, where R_0 is the helix pitch and d is the FLC layer thickness; uniform states with $R_0 > d$ due to the parallel orientation order effect on the substrates ("up" and "down" states differ by the direction of the FLC polarization **P**); and a twisted state for large values of $W_p/W_d \gg 1$, which favors antiparallel surface aligning of the FLC director.

The anchoring energy of FLC can be determined either by measuring the width of the coercivity loop ΔV in the $P(E)$ dependence in a static field according to the relation $\Delta V = 8W_d/P_s$, or by measuring the free relaxation times τ_r of the FLC director in the bistable states $\tau_r = \gamma_\varphi d/4W_d$, where γ_φ is the FLC rotational viscosity (Figure 4.44) [32]. The results of these two methods coincide with each other to an accuracy of 30%.

4.4.2 Electro-optical Effects in FLC Cells

Clark–Lagerwall Effect

Let us consider the main electro-optical phenomena in FLC. The best known is the Clark–Lagerwall effect [33], which results in the reorientation of the director from one bistable state to the other when an external electric field changes its sign (Figure 4.45). In this case, the FLC layers are perpendicular to the substrates and the director moves along the surface of a cone whose axis is normal to the layers and parallel to the cell substrates. In each final position of its deviation, the director remains parallel to the substrates, thus transforming the FLC cell into a uniaxial phase plate. The origin of electro-optical switching in the FLC cell is the interaction of the polarization **P** perpendicular to the director with the electric field **E**. The maximum variation of the transmitted intensity is achieved when the FLC cell is placed between crossed polarizers, so that an axis of the input polarizer coincides with one of the final director positions. The total angle of switching equals the double tilt angle θ (Figure 4.45). The Clark–Lagerwall effect is observed in so-called surface-stabilized FLC (SSFLC) structures. In SSFLC cells $d \ll R_0$ and the situation arises where the existence of the helix is unfavorable, that is, the helix is unwound by the walls.

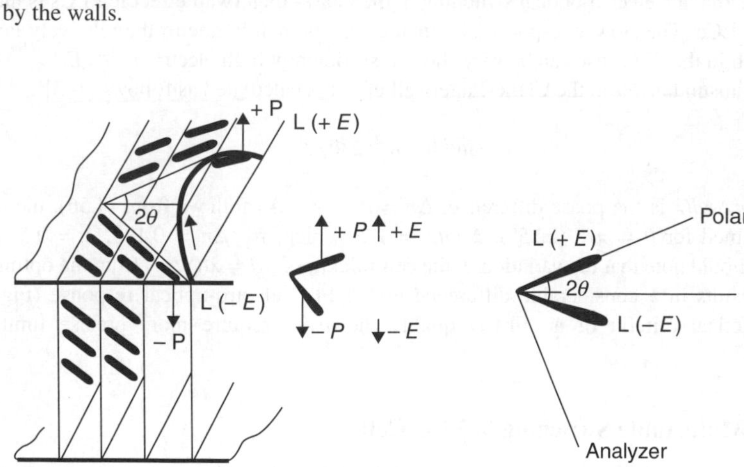

Figure 4.45 Clark–Lagerwall effect in FLC cell [1–3]

The variation of the azimuthal director angle φ in the Clark–Lagerwall effect (Figure 4.45) is described by the equation of the torque equilibrium (see (4.47)), which comes from the condition of the minimum of the FLC free energy (see (4.51)). The total model of FLC reorientation in a "bookshelf" geometry (Figure 4.45) with appropriate boundary conditions is

$$\begin{aligned} \gamma_\varphi \partial\varphi/\partial t &= K\partial^2\varphi/\partial x^2 + P_s E \sin\varphi + [\Delta\varepsilon/(4\pi)]E^2 \sin^2\theta \sin\varphi \cos\varphi, \\ K\partial\varphi/\partial x &+ W_p \sin\varphi \pm W_d \sin 2\varphi \,|_{z=0,d} = 0. \end{aligned} \tag{4.52}$$

For typical values of polarizations, $P_s \approx 20$ nC/cm^2, the driving fields $E \approx 10$ V/μm, and the dielectric anisotropy $\Delta\varepsilon \approx 1$, we have

$$|\Delta\varepsilon E/(4\pi)| < P_s, \tag{4.53}$$

and, consequently, the dielectric term in (4.52) can be omitted. If the inequality (4.53) is invalid, which occurs for sufficiently high fields, then for $|\Delta\varepsilon\, E/(4\pi)| \cong P_s$ the response times of the Clark–Lagerwall effect sharply increase for positive $\Delta\varepsilon$ values [1–3]. Experiments show that for negative values of $\Delta\varepsilon$ the slope of the FLC dynamic response increases, that is, the corresponding switching times become shorter. This is especially important for practical applications, because it promotes an increase in the information capacity of FLCDs. If the driving field increases, the FLC response time passes through a minimum, then grows, passes through a maximum at $|\Delta\varepsilon\, E/(4\pi)| \cong P_s$, and then decreases again.

For $|\Delta\varepsilon\, E/(4\pi)| \gg P_s$ the FLC switching times τ are approximately governed by the field squared, $\tau \cong 4\pi\gamma_\varphi/(\Delta\varepsilon\, E^2)$, as in the ECB effect in nematic LC (see (4.23)). FLC mixtures with negative dielectric anisotropy $\Delta\varepsilon < 0$ are also used for the dielectric stabilization of the initial orientation in FLCDs [34].

As shown in (4.42), response times in the Clark–Lagerwall effect are determined by $\tau_\varphi = \gamma_\varphi/(P_s E)$. The values of $t_{90} - t_{10}$ and $t_{90} - t_{50}$, measured in experiments (Figure 4.44), are fairly close to τ_φ. Comparing the response times in nematics (N)

$$\tau_r^{(N)} \cong 4\pi\gamma_1/(\Delta\varepsilon\, E^2), \quad \tau_d^{(N)} \cong d^2\gamma_1/(K\pi^2) \tag{4.54}$$

and FLCs

$$\tau_r^{(FLC)} \cong \tau_d^{(FLC)} \cong \tau_\varphi \cong \gamma_\varphi/(P_s E), \tag{4.55}$$

we can conclude that the electro-optical switching in the Clark–Lagerwall effect in FLCs is much faster than in nematic LCa. The slower response of nematic LCs is mainly due to the relatively large decay times $\tau_d^{(N)}$, which in the FLC case can be very short in sufficiently high electric fields E.

The optical transmittance I in the Clark–Lagerwall effect is calculated as follows [1–3]:

$$I = \sin^2 4\theta\, \sin^2(\Delta\Phi/2), \tag{4.56}$$

where $\Delta\Phi = 2\pi\Delta n d/\lambda$ is the phase difference, $\Delta n = n_\parallel - n_\perp$. As follows from (4.56), the maximum contrast is obtained for $\theta = \pi/8$ (22.5°), $\Delta nd/\lambda = 1/2$, which, for $\Delta n = 0.125$, $\lambda = 0.5$ μm, gives $d = 2$ μm. We should note that the variation in the cell thickness $\Delta d = \lambda/(8\Delta n)$ from the optimum value $d = \lambda/(2\Delta n)$ results in a considerable difference in the FLC electro-optical response (Figure 4.46) [1–3]. The practical criteria of an FLCD quality, however, require more precise limitations of $d = 2 \pm 0.2$ μm.

Bistable and Multistable Switching in FLC Cells

Bistable switching in the SSFLC geometry (Figure 4.45) takes place above a certain threshold field $E_{th} \propto W_d/K^{1/2}$, where K is an average elastic constant and W_d is a dispersion anchoring energy with the polar

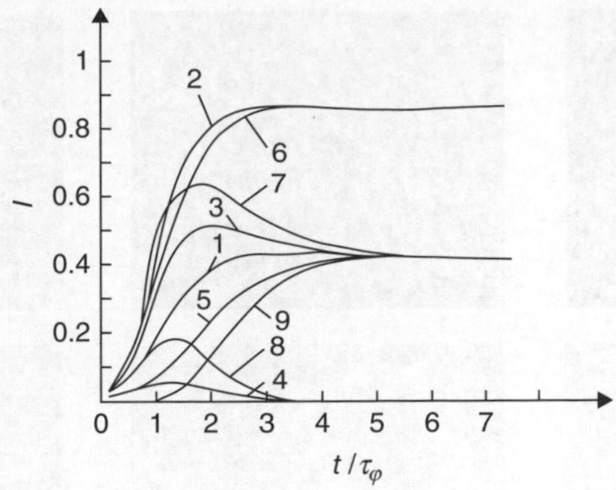

Figure 4.46 Electro-optical response of a FLC cell in the Clark–Lagerwall effect for different phase factors $d\Delta n/\lambda = k/4$, $k = 1, \ldots, 9$ [1–3]

anchoring energy taken equal to zero, $W_p = 0$ [1–3]. Thus with increasing anchoring, we have to increase the switching amplitude of the electric field. As the energy of switching electric torque is proportional to the product of P_sE, the bistability threshold is inversely proportional to the value of the FLC spontaneous polarization P_s.

The problems we meet in using the Clark–Lagerwall effect include not only severe restrictions on the optimum layer thickness and requirements for defect-free samples, but also difficulties in the realization of a perfect bistability or optical memory switched by the electric field and also providing the gray scale. The latter problem is a most crucial one, because it is very inconvenient to provide the gray scale either using complicated driving circuits or increasing the number of working elements (pixels) in FLCD [34]. This problem arises in the Clark–Lagerwall effect because the level of transmission is not defined by the amplitude of the driving voltage pulse U, but by the product $U\tau$, where τ is the electric pulse duration.

A SSFLC structure with bistable switching cannot provide an intrinsic continuous gray scale, unless a time- or space-averaging process is applied [35]. The inherent physical gray scale of passively addressed FLC cells can be obtained if the FLC possesses multistable electro-optical switching with a sequence of ferroelectric domains, which appear if the spontaneous polarization P_s is high enough [36]. Ferroelectric domains in a helix-free FLC form a quasi-periodic structure with a variable optical density as it appears between crossed polarizers [36] (Figure 4.47). The bookshelf configuration (Figure 4.45) of smectic layers is preferable for the observation of these domains. If the duration of the electric pulse applied to a helix-free SSFLC layer containing ferroelectric domains is shorter than the total FLC switching time, the textures shown in Figure 4.47 are memorized after switching off this pulse and short-circuiting the FLC cell electrodes. The domains appear as a quasi-regular structure of bright and dark stripes parallel to the smectic layer planes. The bright stripes indicate spatial regions with a complete switching of the FLC director, while the dark stripes indicate regions that remain in the initial state. The sharp boundaries between the black and white domain stripes seem to illustrate the fact that only two stable director orientations exist. The variation in the occupied area between bright and dark stripes depends on the energy of the applied driving pulses. The total light transmission of the structure is the result of a spatial averaging over the aperture of the light passing through the FLC cell and is always much larger than the period of the ferroelectric domains. Both the amplitude and the duration of the driving pulses can be varied to change the switching energy, which defines the memorized level of FLC cell transmission in a multistable electro-optical response. Therefore, any level of the FLC cell transmission, intermediate between the maximum and minimum transmissions, can be memorized after switching off the voltage pulses and short-circuiting of cell electrodes (Figure 4.47).

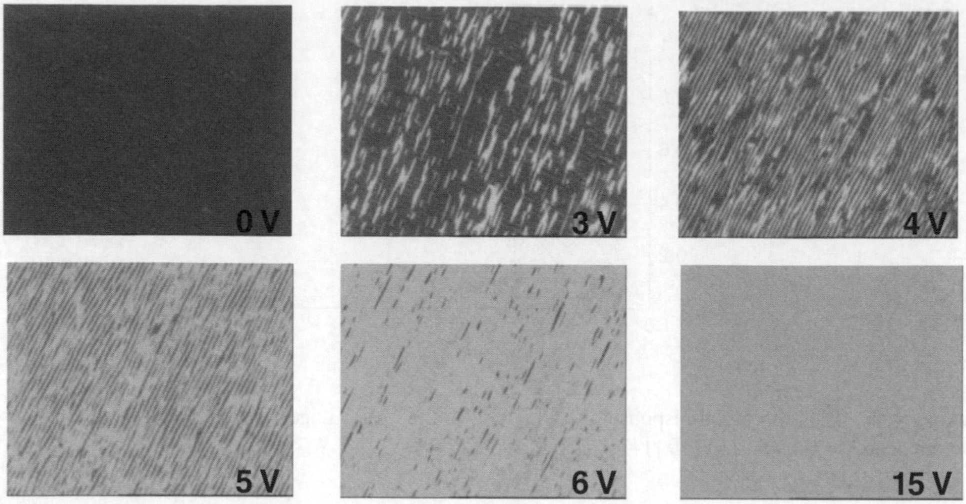

Figure 4.47 Continuous variation of the width of ferroelectric domains with a change in the applied voltage of the FLC layer between crossed polarizers [36]

Optical Transmission of Multistable FLC Cell

The necessary conditions of multistable switching modes are (i) sufficiently high FLC spontaneous polarization $P_s > 50$ nC/cm^2 and (ii) a relatively low energy of the boundaries between the two FLC states existing in FLC domains (Figure 4.47), which is usually typical for the antiferroelectric phase [1–3]. The multistability is responsible for three new electro-optical modes with different shapes of the gray-scale curve that can be either S-shaped (double or single depending upon the applied voltage pulse sequence and boundary conditions) or V-shaped depending upon the boundary conditions and FLC cell parameters (Figure 4.48).

The simple approach describes well the experimental curves of average intensity of multistable FLC states (S- and V-shapes) for various positions of the FLC texture between crossed polarizers. Consider a homogeneous smectic C* layer as shown in Figure 4.49. The analyzer is crossed with the polarizer. The normal of the smectic C* layer (z-axis) is placed at an angle β to the polarizer.

The director **L** of the LC rotates around the z-axis with cone angle θ_0, which can be described by (4.42), where elastic terms have been neglected since the applied field E and FLC spontaneous polarization P_s are sufficiently large in our case.

Equation (4.42) can be solved as

$$\varphi(t) = 2\arctan\left(A \cdot \exp\left(\frac{E\tau_p P_s}{\gamma_\varphi}\right)\right), \tag{4.57}$$

where $A = \tan(\varphi_0/2)$ with φ_0 ($\varphi(t = 0)$) is the initial phase angle, and τ_p is the pulse duration of the driving rectangular voltage. The relative light transmission $I = J/J_0$ (where J and J_0 are the FLC cell input and output intensity) of the smectic C* layer placed between crossed polarizers (Figure 4.49) can be written as [35]

$$I = \sin^2 2(\beta - \theta_0 \cos\varphi) \sin^2\left(\frac{\pi \Delta n d_{FLC}}{\lambda}\right), \tag{4.58}$$

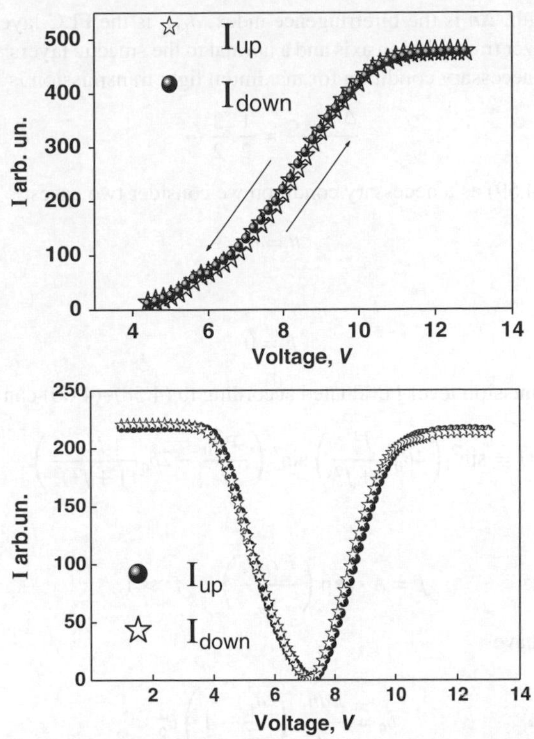

Figure 4.48 S-shaped (above) and V-shaped (below) FLC multistable switching [35]

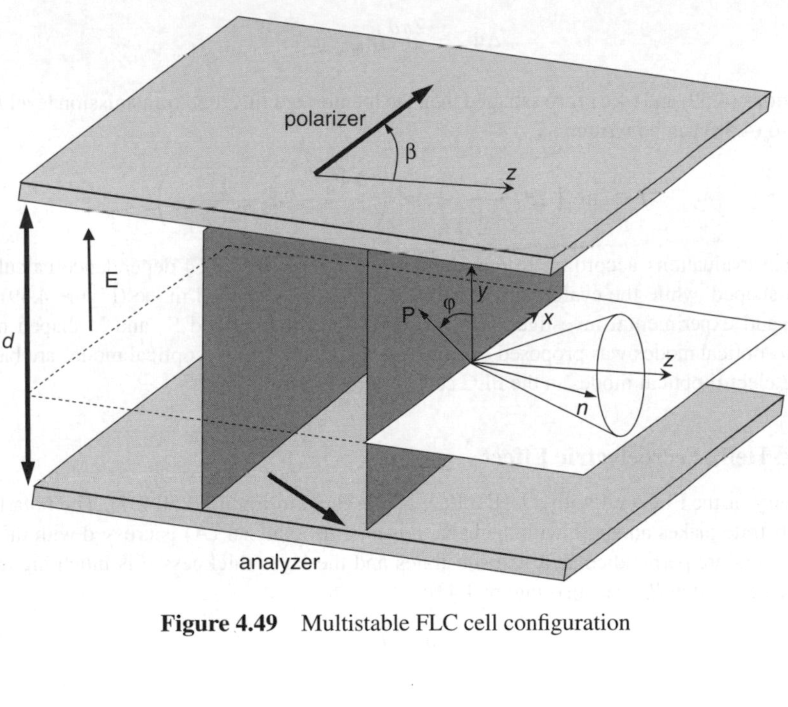

Figure 4.49 Multistable FLC cell configuration

where λ is the wavelength, Δn is the birefringence index, d_{FLC} is the FLC layer thickness, and β is an angle between the polarizer transmission axis and a normal to the smectic layers as shown in Figure 4.49. According to (4.58) the necessary condition for maximum light transmission is

$$\frac{\Delta n d_{FLC}}{\lambda} = \frac{1}{2}, \frac{3}{2} \cdots \tag{4.59}$$

Taking into account (4.59) as a necessary condition we consider two cases:

$$\beta = \theta_0 \tag{4.60}$$

and

$$\beta = 0 \tag{4.61}$$

The memorized transmission level I evaluated according to (4.58)–(4.60) can be written as

$$I = \sin^2\left(4\theta_0 \frac{f^2}{1+f^2}\right) \sin^2\left(\frac{\Delta\Phi_0}{2} - 2\delta_0 \frac{f^2}{(1+f^2)^2}\right), \tag{4.62}$$

where

$$f = A \cdot \exp\left(\frac{E\tau_p P_s}{\gamma_\varphi}\right), \quad f \gg 1. \tag{4.63}$$

Further, in (4.62) we have

$$\delta_0 = \frac{\pi d n_{||}}{\lambda}\left(\frac{n_{||}^2}{n_\perp^2} - 1\right)\theta_0^2, \tag{4.64}$$

where $n_{||}$ and $n_\perp$ are the principal refractive indexes, parallel and perpendicular to the long molecular axis correspondingly, and

$$\Delta\Phi_0 = \frac{2\pi d}{\lambda}(n_{||} - n_\perp). \tag{4.65}$$

If conditions (4.59) and (4.61) are satisfied then the memorized FLC cell transmission level I evaluated according to (4.58) can be written as

$$I = \sin^2\left(2\theta_0 \frac{f^2-1}{f^2+1}\right) \sin^2\left(\frac{\Delta\Phi_0}{2} - 2\delta_0 \frac{f^2}{(1+f^2)^2}\right). \tag{4.66}$$

Numerical evaluations according to our simple model show that $I(E)$ dependence calculated from (4.62) is S-shaped, while the evaluations from (4.66) give the V-shaped mode (Figure 4.49). Thus the theoretical and experimental investigation of reversible and memorized S- and V-shaped multistable FLC electro-optical modes was proposed on this basis [35]. New electro-optical modes are based on the multistable electro-optical modes in the FLC cell (Figure 4.47).

Deformed Helix Ferroelectric Effect

The geometry of the FLC cell with a DHF effect is presented in Figure 4.50 [37]. The polarizer (P) on the first substrate makes an angle with the helix axis and the analyzer (A) is crossed with the polarizer. The FLC layers are perpendicular to the substrates and the layer thickness d is much higher than the value of the helix pitch R_0 (see also Figure 4.43):

$$d \gg R_0. \tag{4.67}$$

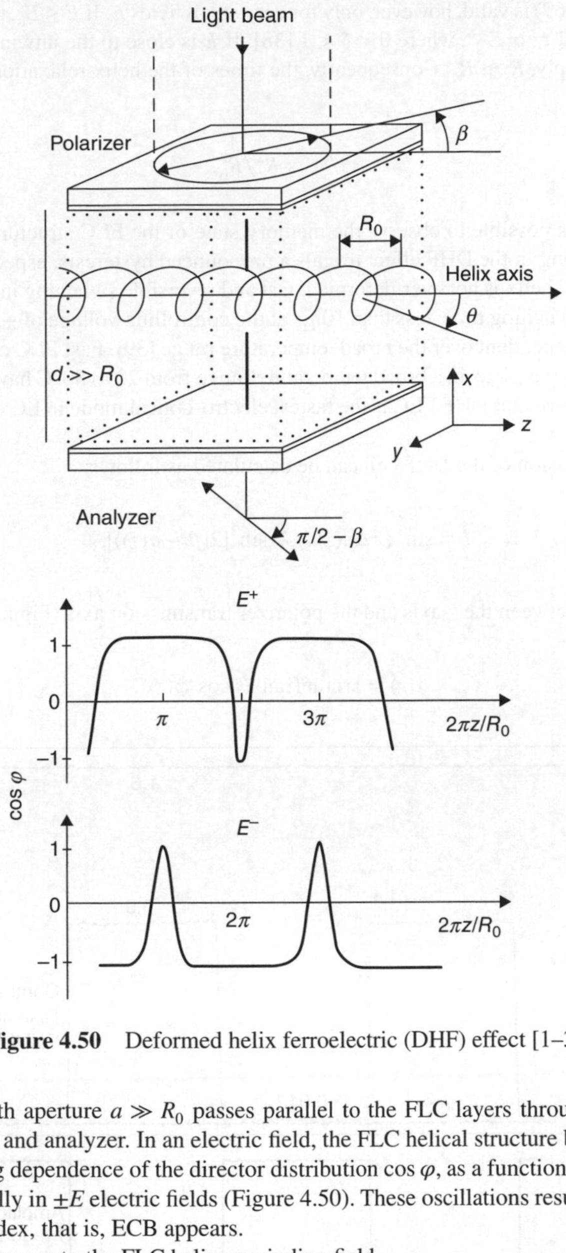

Figure 4.50 Deformed helix ferroelectric (DHF) effect [1–3]

The light beam with aperture $a \gg R_0$ passes parallel to the FLC layers through a FLC cell placed between the polarizer and analyzer. In an electric field, the FLC helical structure becomes deformed, so that the corresponding dependence of the director distribution cos φ, as a function of coordinate $2\pi z/R_0$, oscillates symmetrically in $\pm E$ electric fields (Figure 4.50). These oscillations result in a variation of the effective refractive index, that is, ECB appears.

The effect takes place up to the FLC helix unwinding field

$$E_u = (\pi^2/16)\, K_{22} q_o^2 / P_s, \tag{4.68}$$

where K_{22} is the FLC twist elastic constant, and $q_o = 2\pi/R_0$ is the helix wave vector.

The characteristic response times τ_c of the effect in small fields $E/E_u \ll 1$ are independent of the FLC polarization P_s and the field E, and defined only by the rotational viscosity γ_φ and the helix pitch R_0:

$$\tau_c = \gamma_\varphi / K_{22} q_o^2. \tag{4.69}$$

The dependence (4.69) is valid, however, only for very small fields E. If $E \leq E_u$, the FLC helix becomes strongly deformed and $\tau_c \propto E^{-\delta}$, where $0 < \delta < 1$ [38]. If E is close to the unwinding field E_u the helix pitch R increases sharply, $R \gg R_o$. Consequently, the times of the helix relaxation τ_d to the initial state also rise,

$$\tau_d/\tau_c \propto R^2/R_o^2, \tag{4.70}$$

that is, for $E \approx E_u$ it is possible to observe the memory state of the FLC structure [37]. In this regime, electro-optical switching in the DHF effect reveals a pronounced hysteresis, especially for $E \Rightarrow E_u$.

However, if the FLC helix is not over-deformed, fast and reversible switching in the DHF mode can be obtained [1–3]. The switching time, less than 10 μs at the controlling voltage of ±20 V, can be provided, and is temperature independent over the broad temperature range [39]. Fast FLC cells (DHF effect) with a response time of less than 1 μs in a broad temperature range from 20 to 80°C have been developed [40] (Figure 4.51). We believe that DHF FLC is the fastest electro-optical mode in LC cells for photonics and display applications.

The optical transmission of the DHF cell can be calculated as follows:

$$I = \sin^2(\pi \Delta n(z) d/\lambda) \sin^2[2(\beta - \alpha(z))], \tag{4.71}$$

where β is the angle between the z-axis and the polarizer transmission axis (Figure 4.50),

$$\alpha(z) = \arctan(\tan\,\theta\,\cos\,\varphi(z)) \tag{4.72}$$

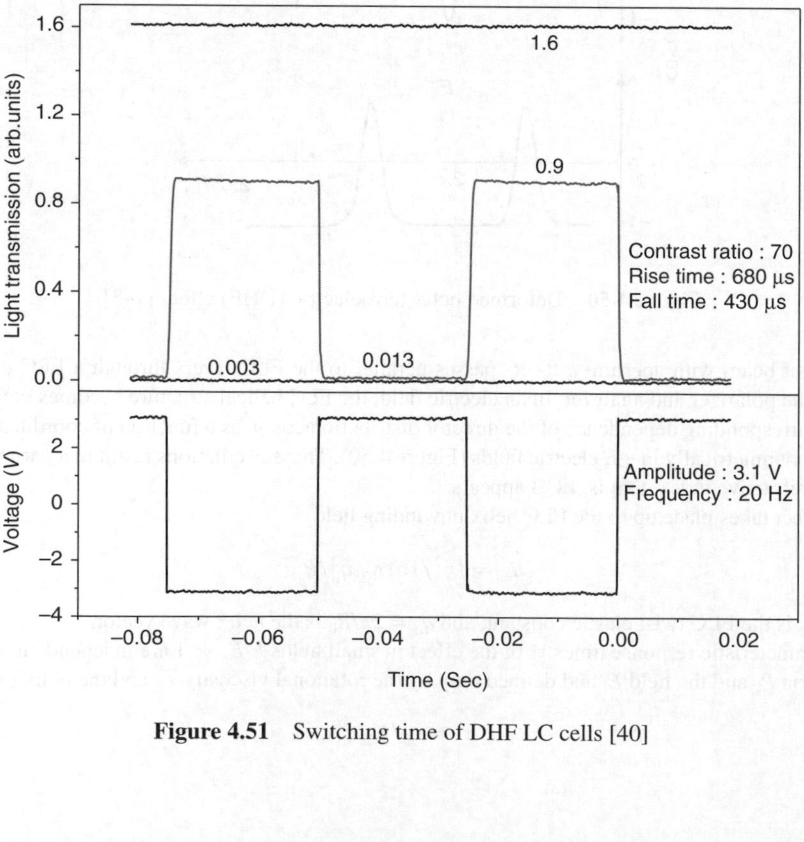

Figure 4.51 Switching time of DHF LC cells [40]

is the angle between the projection of the optical axis on the *yz*-plane and the *z*-axis, and $\Delta n(z) = n_{\text{eff}}(z) - n_{\perp}$ is the effective birefringence:

$$n_{\text{eff}} = n_{\|}n_{\perp}/[n_{\perp}^2 + (n_{\|}^2 - n_{\perp}^2)\sin^2\theta\sin^2\varphi]^{1/2}. \tag{4.73}$$

In the case of small angles $|\theta| \ll 1$, the transmission in (4.71) can be expanded as a series in θ,

$$I \propto (\sin^2 2\beta - 2\theta\ \sin 4\beta\cos\varphi + 4\theta^2\ \cos 4\beta\ \cos^2\varphi)\sin^2(\pi\Delta nd/\lambda) \tag{4.74}$$

As shown in [1–3], for small values of the applied field $\cos\varphi \propto E/E_u$ and it changes its sign for the field reversal $E \Rightarrow -E$ (Figure 4.50). Thus according to (4.74), for $\sin 4\beta = 0$ we have a quadratic gray scale, that is,

$$\Delta I \propto \theta^2\cos^2\varphi \propto \theta^2 E^2/E_u^2 \tag{4.75}$$

and for other values of θ the gray scale is linear. For $\cos 4\beta = 0$ the quadratic component in the modulated intensity I is absent, that is,

$$\Delta I \propto \theta\cos\varphi \propto \theta E/E_u. \tag{4.76}$$

If $E(t) = E_o\cos(wt)$, then in the case of (4.75) we come to the modulation regime, which doubles the frequency of the applied field. Relationships (4.75) and (4.76) were confirmed by experiment [37]. Using a "natural" gray scale of the DHF mode, many gray levels have been obtained with fast switching between them [41]. New ferroelectric mixtures with a helix pitch $R_0 < 0.3$ μm and tilt angle $\theta > 30°$ have recently been developed for the DHF effect [41, 42]. The helix unwinding voltage was about 2–3 V. Short-pitch FLC mixtures can also be used to obtain pseudo-bistable switching in FLC samples. Using these new FLCs, electrically controlled V-shaped switching in the DHF mode can be applied for new active-matrix LCDs with field sequential colors (FSCs).

A geometry with $\beta = 0$ was selected in all experiments to provide a non-sensitive electro-optical response to the driving voltage polarity. Maximum light transmission under this condition, as follows from (4.71), occurs if $\alpha(z) = 45°$ and $\Delta n(z)d = \lambda/2$. It is easy to show from (4.72) that the tilt angle θ of the FLC should be close to 45° to provide maximum light transmission at $\beta = 0$. Typical V-shaped symmetrical (voltage-sign-independent) DHF switching is shown in Figure 4.52 [41].

Let us point out certain advantages of the DHF electro-optical effect for applications as compared to the Clark–Lagerwall mode:

1. High operating speed is achieved for low driving voltages. This takes place because a slight distortion of the helix near the equilibrium state results in a considerable change in the transmission. Consequently, an instantaneous response of the FLC cell is provided without the so-called delay time inherent in the Clark–Lagerwall effect.
2. The DHF effect is also less sensitive to surface treatment and more tolerant of cell gap inhomogeneity. As follows from experiment and qualitative estimations [1–3], the effective birefringence value Δn_{eff} is approximately twice as low as $\Delta n = n_{\|} - n_{\perp}$ in the Clark–Lagerwall effect.
3. The DHF effect allows the implementation of a "natural" gray scale (i.e., dependent on voltage amplitude) that is both linear and quadratic in voltage. Moreover, at $E \approx E_u$ long-term optical memory states are possible.

Transflective FLCD

A new optical configuration of transflective bistable display using FLC cells with a single cell gap has been developed [43, 44]. This configuration provides high brightness and high contrast ratio for both

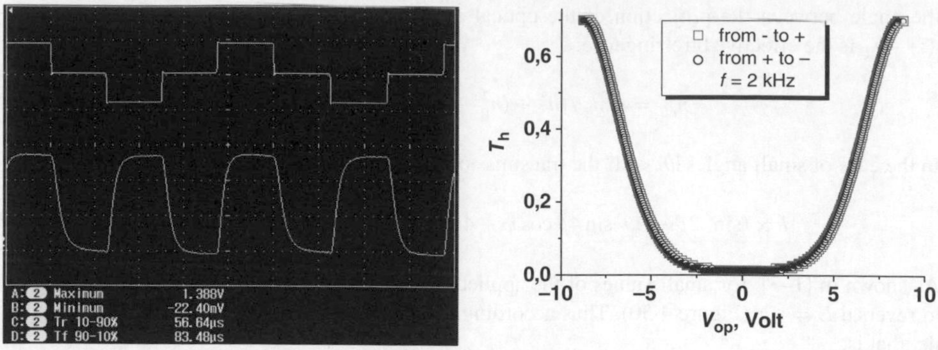

Figure 4.52 Symmetric (voltage-sign-independent) electro-optical response of DHF-FLC [41]. Left: top, the driving voltage waveform applied to the cell; bottom, the electro-optical response of the cell. Right: V-shaped mode in the envelope curve of light transmission saturation states measured at electro-optical response frequency 2 kHz. Light transmission T_h evaluated in comparison with transmission of empty cell placed between parallel polarizers; this transmission is defined as $T_h = 1$

reflective and transmissive modes. Since there are no double-cell-gap structure and patterning polarizers or retarders, this configuration is very easy to fabricate (Figure 4.53). The optimized parameters of a transflective FLC configuration are given in Table 4.10.

The structure of a transflective FLCD is shown in Figure 4.53. It is composed of two polarizers, a retardation film, a transflective film, and a FLC cell. The transflective film is used as a reflector in the sunlight or a bright place and as a transmitter at night or in a dark place. An antireflection layer is inserted at the top of the configuration in order to reduce surface reflectance. The optical anisotropy for the FLC material is as follows: $\Delta n = 0.0875 + (12\,251\ \text{nm}^2)/\lambda^2$. This FLC has a spontaneous polarization of $P_s \sim 100\ \text{nC/cm}^2$ and a tilt angle of $\theta = 26°$ at $T = 23°\text{C}$. MOUSE-LCD was used for the simulations [43, 44]. The optimal values of the parameters are given in Table 4.10. The single-cell-gap FLC cell serves as a quarter-wave plate. Figure 4.54 shows the simulated spectrum for bright and dark states of the transflective FLCDs.

The viewing angle for the reflective and transmissive modes respectively in a transflective FLCD are shown in Figure 4.55. The transmissive part has a wider viewing angle than the reflective one. For the normal light incidence, the contrast can be as large as 28:1 for the reflective and 200:1 for the transmissive part of the transflective FLCD [43, 44].

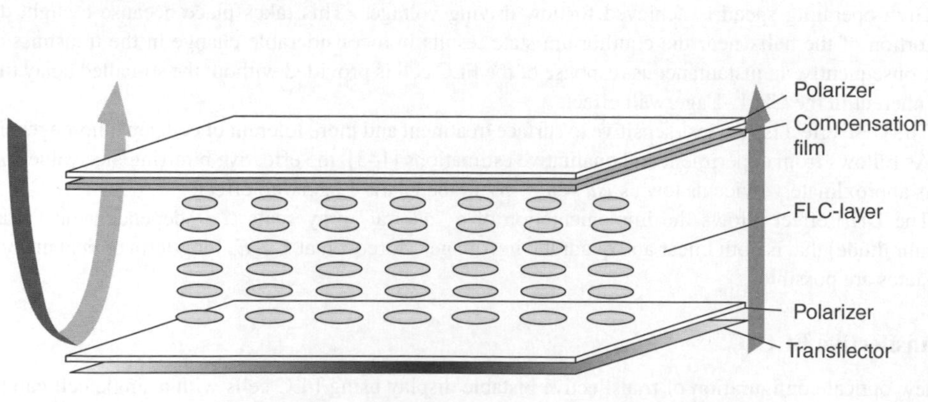

Figure 4.53 The structure of a transflective FLCD

Table 4.10 Optimized parameters of the configuration of a transflective FLCD

Angle of first polarizer (transmission axis)	25°
Angle of compensation film (slow axis)	90°
Retardation of compensation film	140 nm
Thickness of FLC514 layer	1 μm
Angle of FLC514 layer	0°
Angle of second polarizer (transmission axis)	−65°

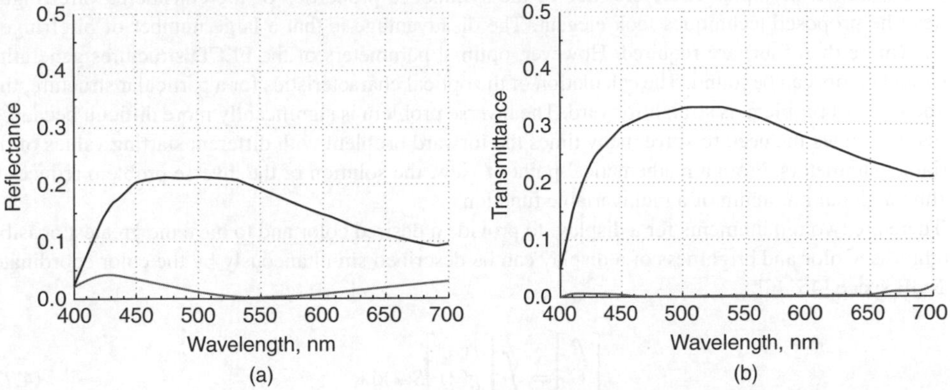

Figure 4.54 The spectrum of bright and dark states for transflective FLCDs: (a) reflective part and (b) transmissive part

4.5 Birefringent Color Generation in Dichromatic Reflective FLCDs

The potential of simple constructions of dichromatic bistable FLCDs using just phase retardation plates and one or two polarizers has been studied [45]. The results obtained will allow us to judge the possibilities of a FLCD to achieve any two desired colors and will clear up the question of how many retardation plates are enough to generate these colors.

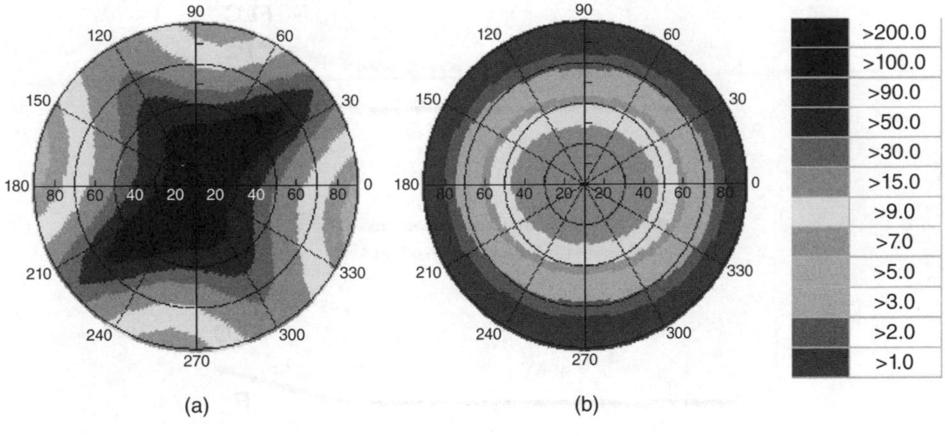

Figure 4.55 Simulated angular dependence of the contrast ratio for (a) the transmissive and (b) the reflective part of our transflective FLCD [44]

The structures of the FLCD are shown in Figure 4.56. The double-polarizer FLCD (Figure 4.56, top) consists of a top polarizer (P1), several retardation plates (Ret1, … ,RetN), a layer of FLC, a rear polarizer (P2), and a reflector (R). The single-polarizer FLCD has a similar structure (Figure 4.56, bottom) except for the rear polarizer. Using only one polarizer allows us to increase reflectivity of the bright state and avoid parallax effects [1–3].

The spectrum of the reflected light depends on retardations and orientations of the FLC and birefringent plates as well as orientations of the polarizers, and the spectrum of an illuminator. Earlier, several techniques were proposed for the synthesis of birefringent optical filters [1–3, 20]. As a rule, the filter consists of a set of identical retardation plates having different orientations or the retardation values have a common multiple. Therefore, due to the symmetric properties of the considered birefringent plates, the proposed techniques look elegant. The disadvantage is that a large number of birefringent plates (more than four) are required. However, optimal parameters of the FLCD structures generating desirable colors can be found. The calculation of the optical characteristics for a particular structure, that is, the forward problem, is straightforward. The inverse problem is significantly more difficult because it already includes the need to solve many times the forward problem with different starting values of the variable parameters. From a mathematical point of view, the solution of the inverse problem reduces to finding a global minimum of a multivariable function.

There are two requirements for a display: to provide a desired color and to have the greatest possible brightness. Color and brightness of a display can be described simultaneously by the color coordinates in RGB space [45,46]:

$$\begin{bmatrix} R \\ G \\ B \end{bmatrix} = \int \begin{bmatrix} r(\lambda) \\ g(\lambda) \\ b(\lambda) \end{bmatrix} S(\lambda)\mathrm{d}\lambda, \tag{4.77}$$

where $r(\lambda)$, $g(\lambda)$, $b(\lambda)$ are the spectral tristimulus values, and $S(\lambda)$ is the spectral distribution of the intensity of the light reflected from the display. The relative values of the coordinates correspond to a display color, whereas the absolute values are proportional to the brightness.

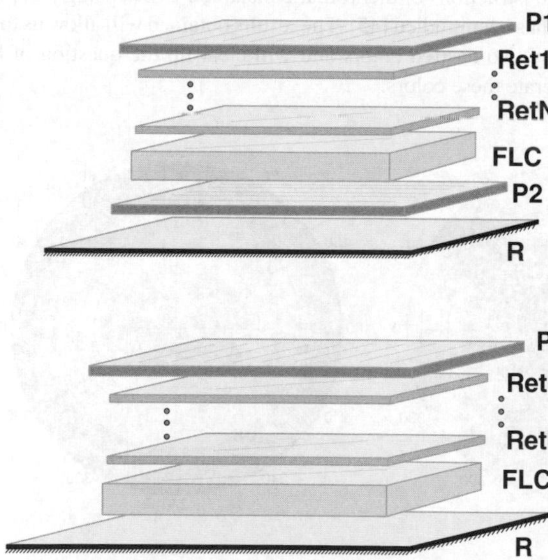

Figure 4.56 Structures of the considered reflective FLCD. Top: double-polarizer FLCD. Bottom: single-polarizer FLCD [45]

Let the desirable color of the first stable state of the FLCD be characterized by the coordinates

$$\begin{bmatrix} R_1^{\mathrm{d}} \\ G_1^{\mathrm{d}} \\ B_1^{\mathrm{d}} \end{bmatrix},$$

whereas the desirable color of the second stable state is characterized by the coordinates

$$\begin{bmatrix} R_2^{\mathrm{d}} \\ G_2^{\mathrm{d}} \\ B_2^{\mathrm{d}} \end{bmatrix}.$$

By varying the parameters of the FLCD, we achieve the limits:

$$\begin{bmatrix} R_1 \\ G_1 \\ B_1 \end{bmatrix} \rightarrow \begin{bmatrix} R_1^{\mathrm{d}} \\ G_1^{\mathrm{d}} \\ B_1^{\mathrm{d}} \end{bmatrix} \text{ and } \begin{bmatrix} R_2 \\ G_2 \\ B_2 \end{bmatrix} \rightarrow \begin{bmatrix} R_2^{\mathrm{d}} \\ G_2 \\ B_2^{\mathrm{d}} \end{bmatrix}, \tag{4.78}$$

where

$$\begin{bmatrix} R_1 \\ G_1 \\ B_1 \end{bmatrix} \text{ and } \begin{bmatrix} R_2 \\ G_2 \\ B_2 \end{bmatrix}$$

and are coordinates of the FLCD in the first and the second states, respectively.

Let $R(\alpha, \beta, d\Delta n(\lambda), \varphi, \Gamma_1, \gamma_1, \Gamma_2, \gamma_2, \ldots, \Gamma_N, \gamma_N, \lambda)$ be the reflectivity of a FLCD that depends on the polarizer orientations described by the angles α, β, the retardation of the FLC $d\Delta n(\lambda)$, the orientation of the optical axis of the FLC described by the angle φ, and the orientations and retardations of the birefringent plates $\gamma_1, \gamma_2, \ldots, \gamma_N$ and $\Gamma_1, \Gamma_2, \ldots, \Gamma_N$, respectively. The spectral intensity $S(\lambda)$ of the reflected light in this case is expressed as

$$S(\lambda) = L^r(\lambda) R(\alpha, \beta, d\Delta n(\lambda), \varphi, \Gamma_1, \gamma_1, \Gamma_2, \gamma_2, \ldots, \Gamma_N, \gamma_N \lambda), \tag{4.79}$$

where $L^r(\lambda)$ is the spectral intensity of the illuminant light. Reflectivity $R(\alpha, \beta,\ d\Delta n(\lambda), \varphi, \Gamma_1, \gamma_1, \Gamma_2, \gamma_2, \ldots, \Gamma_N, \gamma_N \lambda)$ consists mainly of two components: the Fresnel reflectivity of the top surface of the FLCD $r_{\mathrm{ar}}(\lambda)$ and the reflectivity $\rho(\alpha, \beta, d\Delta n(\lambda), \varphi, \Gamma_1, \gamma_1, \Gamma_2, \gamma_2, \ldots, \Gamma_N, \gamma_N \lambda)$ caused by the reflector of the FLCD:

$$R(\alpha, \beta, d\Delta n(\lambda), \varphi, \Gamma_1, \gamma_1, \ldots, \Gamma_N, \gamma_N \lambda) = r_{\mathrm{ar}}(\lambda) + \rho(\alpha, \beta, \Delta nd, \varphi, \Gamma_1, \gamma_1, \ldots, \Gamma_N, \gamma_N, \lambda).$$

The reflected components caused by the reflection from the interfaces between the layers can be neglected. Applying the formalism of the Jones matrices, the reflectivity $\rho(\alpha, \beta, d\Delta n(\lambda), \varphi, \Gamma_1, \gamma_1, \Gamma_2, \gamma_2, \ldots, \Gamma_N, \gamma_N \lambda)$ can be expressed as

$$\rho(\alpha, \beta, d\Delta n(\lambda), \varphi, \Gamma_1, \gamma_1, \Gamma_2, \gamma_2, \ldots, \Gamma_N, \gamma_N \lambda) = \frac{1}{2} \left\| \hat{M} \right\|_E^2, \tag{4.80}$$

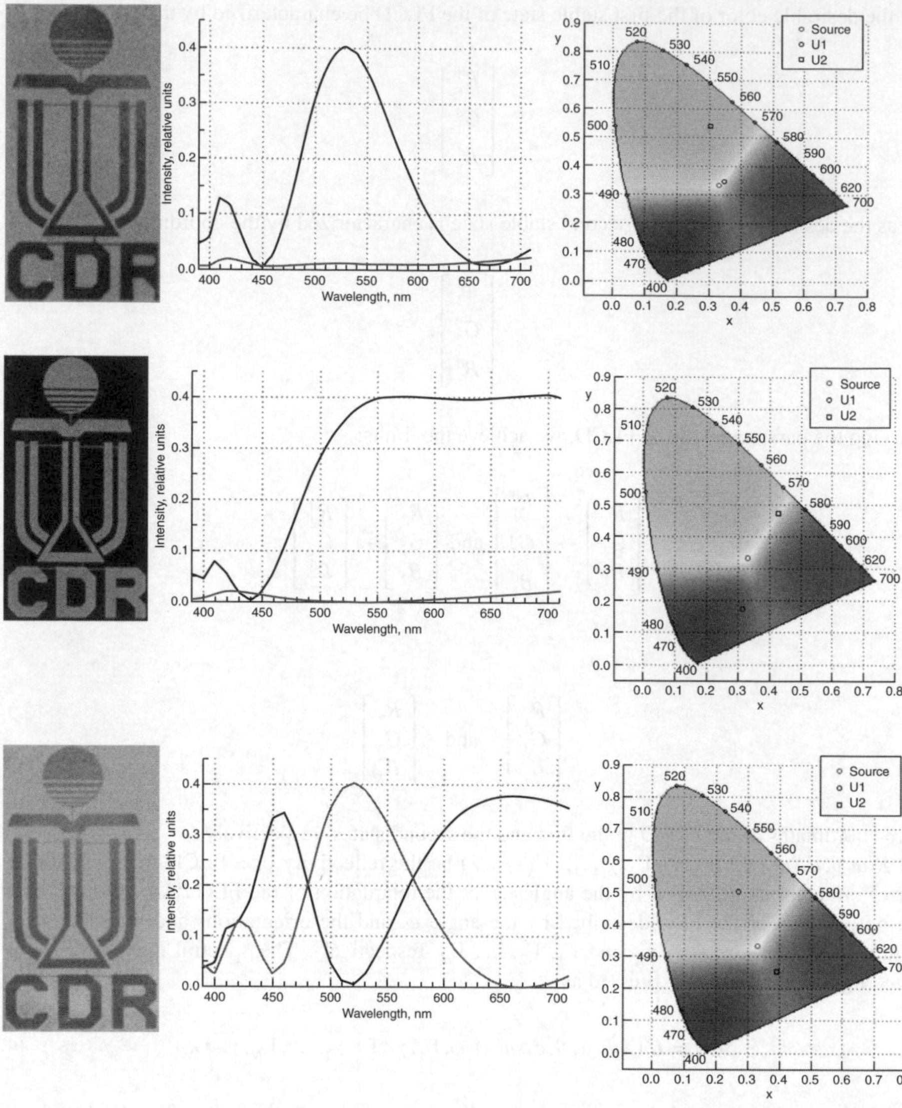

Figure 4.57 Dichromatic FLCD: from top to bottom, green/black, yellow/black, and green/pink. Left: experimental samples and spectral characteristics of the two bistable states. Right: color coordinates

where $\|\hat{M}\|_E$ is the Euclidean norm (see Section 5.1.4 and Equation 5.44 there) of the Jones matrix $\hat{M}$ of the FLCD [47] that is calculated as [47]

$$\hat{M} = r_R \hat{M}^T_{LCP} \hat{M}_{LCP}, \tag{4.81}$$

where r_R is the reflectivity of the reflector, ^{T} denotes transposition, and $\hat{M}_{LCP}$ is the transmission Jones matrix of the LCD panel (except the reflector). In the case of the double-polarizer FLCD (Figure 4.56, top),

$$\hat{M}_{LCP} = \hat{M}_{p2} \hat{M}_{FLC} \hat{M}_{\Gamma_1} \hat{M}_{\Gamma_2} \ldots, \hat{M}_{\Gamma_N} \hat{M}_{p1},$$

and in the case of the single-polarizer FLCD (Figure 4.56, bottom), and

$$\hat{M}_{\rm LCP} = \hat{M}_{\rm FLC}\hat{M}_{\Gamma_1}\hat{M}_{\Gamma_2}\ldots,\hat{M}_{\Gamma_N}\hat{M}_{\rm p1}.$$

where $\hat{M}_{\rm FLC}$ and $\hat{M}_{\Gamma_j}(j = 1, \ldots, N)$ are the Jones matrices of the FLC layer and the jth retardation plate, respectively; $\hat{M}_{\rm p1}$ and $\hat{M}_{\rm p2}$ are the Jones matrices of the top and rear polarizers, respectively.

There are two vector functions or six coordinates (scalar functions) in the optimization criterion (Equation 4.78). In order to use one variable as a criterion for the minimization task (Equation 4.78), let us introduce the scalar function

$$f = \sum_{i=1,2}\sum_{J=R,G,B}\left(J_i - J_i^d\right)^2, \tag{4.82}$$

where J_i and J_i^d are the calculated and desired color coordinates, respectively.

It can be seen that the function f reaches its global minimum (zero in the ideal case) when condition (4.78) is satisfied. Thus, Equation (4.78) will be equivalent to the requirement

$$f(\alpha, \beta, d\Delta n(\lambda), \varphi, \Gamma_1, \gamma_1, \Gamma_2, \gamma_2, \ldots, \Gamma_N, \gamma_N, \lambda) \rightarrow 0. \tag{4.83}$$

Thus, the further task under consideration reduces to finding the global minimum of the multivariable function $f(\alpha, \beta, d\Delta n(\lambda), \varphi, \Gamma_1, \gamma_1, \Gamma_2, \gamma_2, \ldots, \Gamma_N, \gamma_N, \lambda)$.

According to the procedure described above, any dichromatic reflective LCD can be modeled and reproduced experimentally. Figure 4.57 shows several examples: green/black, yellow/black, and green/pink. It was demonstrated that good color characteristics and high brightness can be achieved simultaneously using the structures including one or two retardation plates only.

References

[1] V. G. Chigrinov, *Liquid Crystal Devices: Physics and Applications* (Artech House, 1999).

[2] L. M. Blinov and V. G. Chigrinov, *Electrooptical Effects in Liquid Crystalline Materials* (Springer-Verlag, 1994).

[3] S. V. Pasechnik, V. G. Chigrinov, and D. V. Shmeliova, *Liquid Crystals: Viscous and Elastic Properties* (Wiley-VCH Verlag GmbH, 2009).

[4] V. Chigrinov, D. Yakovlev, and H. S. Kwok, "Optimization and modeling of liquid crystal displays," *In. Disp.* **20**, 26 (2004).

[5] P. Yeh and C. Gu, *Optics of Liquid Crystal Displays* (John Wiley & Sons, Inc., 1999).

[6] H. Wohler and M. E. Becker, "Numerical modeling of LCD electrooptical performance," *Opto-Electron. Rev.* **10**, 23 (2002).

[7] www.autronic-melchers.com/ (accessed February 15, 2014).

[8] www.shintech.jp/(accessed February 15, 2014).

[9] www.sanayisystem.com(accessed February 15, 2014).

[10] K. Takatoh, M. Akimoto, H. Kaneko, K. Kawashima, and S. Kobayashi, "The low driving voltage, the molecular arrangement and the stability of reverse TN mode," *IDW'08 Digest*, 1567 (2008).

[11] Y. W. Li, L. Tan, and H. S. Kwok, "Field-sequential-color LCDs based on transient modes," *SID'08 Digest*, 32 (2008).

[12] T. J. Scheffer and J. Nehring, "A new, highly multiplexable liquid crystal display," *Appl. Phys. Lett.* **45**, 1021 (1984).

[13] M. Schadt and F. Leenhouts, "Electro-optical performance of a new, black-white and highly multiplexable liquid crystal display," *Appl. Phys. Lett.* **50**, 236 (1987).

[14] T. J. Scheffer and J. Nehring, "Twisted nematic and supertwisted nematic mode LCDs," in *Liquid Crystals: Applications and Uses*, ed. B. Bahadur (World Scientific, 1991).

[15] J. Li, J. Kelly, Ch. Hoke, and P. J. Bos, "Optimization of luminance for optical-mode-interference (OMI) displays," *J. Soc. Inf. Disp.* **4/2**, 95 (1996).
[16] S. T. Tang and H. S. Kwok, "New method of designing LCD with optimal optical properties," *SID'99 Digest* 195–197 (1999).
[17] http://www.sidmembers.org/proc/MD2000/27Three-FiveSystemsLCModesfor.pdf (accessed February 15, 2014).
[18] H. Mori, M. Nagai, H. Nakayama, Y. Itoh, K. Kamada, K. Arakawa, and K. Kawata, "Novel optical compensation method based upon a discotic optical compensation film for wide-viewing-angle LCDs," *SID'03 Digest*, 1058 (2003).
[19] D. A. Yakovlev, V. G. Chigrinov, and H. S. Kwok, "Optimization of one-polarizer reflective LCDs with phase compensator," *Mol. Cryst. Liq. Cryst.* **410**, 351 (2004).
[20] E. Hecht, *Optics* (Addison-Wesley, 2002).
[21] D.-K. Yang and S.-T. Wu, *Fundamentals of Liquid Crystal Devices* (John Wiley & Sons, Ltd., 2006).
[22] T. Du, L. Yao, V. G. Chigrinov, and H.-S. Kwok, "Single-cell-gap transflective liquid-crystal display and the use of photoalignment technology," *J. Soc. Inf. Disp.* **18/6**, 421 (2010).
[23] P. Xu, V. Chigrinov, and H. S. Kwok, "Optical analysis of a liquid-crystal switch system based on total internal reflection," *J. Opt. Soc. Am.* **25**, 866 (2008).
[24] V. G. Chigrinov, T. V. Korkishko, and B. A. Umanskii, "Total internal reflection during a Freedericksz transition in a nematic liquid crystal," *Opt. Spectrosc.* **57**, 283 (1984).
[25] D. W. Berreman, "Optics in stratified and anisotropic media: 4 × 4-matrix formulation," *J. Opt. Soc. Am.* **62**, 502 (1972).
[26] E. Merzbacher, *Quantum Mechanics* (John Wiley & Sons, Inc., 1970).
[27] D. Y. K. Ko and J. R. Sambles, "Scattering matrix method for propagation of radiation in stratified media: attenuated total reflection studies of liquid crystals," *J. Opt. Soc. Am. A* **5**, 1863 (1988).
[28] K. L. Woon, M. O'Neill, G. J. Richards, M. P. Aldred, and S. M. Kelly, "Stokes parameter analysis of the polarization of light transmitted through a chiral nematic liquid-crystal cell," *J. Opt. Soc. Am. A* **22**, 760 (2005).
[29] S. Garoff and B. Meyer, "Electroclinic effect at the A-C phase change in a chiral smectic liquid crystal," *Phys. Rev. A.* **19**, 338 (1979).
[30] E. P. Pozhidaev, M. A. Osipov, V. G. Chigrinov, V. A. Baikalov, L. M. Blinov, and L. A. Beresnev, "Rotational viscosity of smectic C* phase of ferroelectric liquid crystals," *Sov. Phys.–JETP* **67**, 283 (1988).
[31] G. S. Chilaya and V. G. Chigrinov, "Optics and electrooptics of chiral smectic C liquid crystals," *Phys.–Usp.* **36**, 909 (1993).
[32] V. Chigrinov, Yu. Panarin, V. Vorflusev, and E. Pozhidaev, "Aligning properties and anchoring strength of ferroelectric liquid crystals," *Ferroelectrics* **178**, 145 (1996).
[33] N. A. Clark and S. T. Lagerwall, "Submicrosecond bistable electrooptic switching in liquid crystals," *Appl. Phys. Lett.* **36**, 899 (1980).
[34] S. T. Lagerwall, "Ferroelectric and antiferroelectric liquid crystals," *Ferroelectrics* **301**, 15 (2004).
[35] E. Pozhidaev, V. Chigrinov, G. Hegde, and P. Xu, "Multistable electro-optical modes in ferroelectric liquid crystals," *J. Soc. Inf. Disp.* **17/1**, 35 (2009).
[36] E. P. Pozhidaev and V. G. Chigrinov, "Bistable and multistable states in ferroelectric liquid crystals," *Cryst. Rep.* **51**, 1030 (2006).
[37] L. A. Beresnev, V. G. Chigrinov, D. I. Dergachev, E. P. Pozhidaev, J. Funfschilling, and M. Schadt, "Deformed helix ferroelectric liquid crystal display: a new electrooptic mode in ferroelectric chiral smectic C liquid crystal," *Liq. Cryst.* **5**, 1171 (1989).
[38] Yu. Panarin, E. Pozhidaev, and V. Chigrinov, "Dynamics of controlled birefringence in an electric field deformed helical structure of a ferroelectric liquid crystal," *Ferroelectrics* **114**, 181 (1991).
[39] E. P. Pozhidaev, "Electrooptical properties of deformed-helix ferroelectric liquid crystal display cells," *SPIE Dig.* **4511**, 92 (2001).
[40] V. Presnyakov, Z. Liu, and V. G. Chigrinov, "Fast optical retarder using deformed-helical ferroelectric liquid crystals," *Proc. SPIE* **5970**, 426 (2005).
[41] E. Pozhidaev and V. Chigrinov, "Fast photo-aligned V-shape ferroelectric LCD based on DHF mode," *SID'10 Digest*, 387 (2010).
[42] G. Hegde, P. Xu, E. Pozhidaev, V. Chigrinov, and H. S. Kwok, "Electrically controlled birefringence colours in deformed helix ferroelectric liquid crystals," *Liq. Cryst.* **35**, 1137 (2008).
[43] S. Valyukh, J. Osterman, K. Skarp, and V. Chigrinov, "Optical performance of bistable reflective and transflective ferroelectric LCDs," *SID'05 Digest*, 702 (2005).

[44] X. Li, P. Xu, A. Murauki, and V. G. Chigrinov, "A new optical configuration for transflective ferroelectric liquid crystal display with single cell gap," *IDW'07 Digest*, 1563 (2007).
[45] S. Valyukh, I. Valyukh, P. Xu, and V. Chigrinov, "Study on birefringent color generation for a reflective ferroelectric liquid crystal display," *Jpn. J. Appl. Phys.* **45**, 7819 (2006).
[46] P. A. Keller, *Electronic Display Measurements* (John Wiley & Sons, Inc., 1997).
[47] S.-T. Wu and D.-K. Yang, *Reflective Liquid Crystal Displays* (John Wiley & Sons, Inc., 2001).

5

Necessary Mathematics. Radiometric Terms. Conventions. Various Stokes and Jones Vectors

It is needless to stress the importance of matrix algebra in LCD optics. For the convenience of the reader, in Section 5.1 we give a brief review of the notions and relations of matrix algebra used in this book. This section can be considered as a mathematical appendix to the preceding part of the book and a mathematical preliminary to what follows.

In Section 5.2, we give definitions of some radiometric quantities and a summary of the optical conventions adopted in this book. In Sections 5.3 and 5.4, some notions concerning the characterization of wave fields by Stokes and Jones vectors are introduced which are important in our consideration of LCD optics.

5.1 Some Definitions and Relations from Matrix Algebra

5.1.1 General Definitions

An $m \times n$ matrix $\mathbf{A}$ is an array of numbers:

$$\mathbf{A} = [a_{jk}] = \begin{pmatrix} a_{11} & a_{12} & \cdots & a_{1n} \\ a_{21} & a_{22} & \dots & a_{2n} \\ \vdots & \vdots & \ddots & \vdots \\ a_{m1} & a_{m2} & \cdots & a_{mn} \end{pmatrix}.$$

Its entries a_{jk} ($j = 1, 2, \dots, m; k = 1, 2, \dots, n$) are called matrix elements; the element a_{jk} is situated in the jth row and in the kth column of the matrix $\mathbf{A}$. Sometimes we use the notation $[\mathbf{A}]_{jk} \equiv a_{jk}$.

$n \times 1$ matrices are called column matrices [column (n-) vectors].

$1 \times n$ matrices are row matrices [row (n-) vectors].

A matrix having the same number of rows and columns is called a *square* matrix.

A *scalar* may be formally considered as a 1×1 matrix.

Modeling and Optimization of LCD Optical Performance, First Edition.
Dmitry A. Yakovlev, Vladimir G. Chigrinov and Hoi-Sing Kwok.

Website Companion: www.wiley.com/go/yakovlev/modelinglcd

A square matrix $\mathbf{A} = [a_{jk}]$ is said to be *diagonal* if $a_{jk} = 0$ for all j and k at $k \neq j$, that is, if

$$\mathbf{A} = \begin{pmatrix} a_{11} & 0 & \cdots & 0 \\ 0 & a_{22} & \cdots & 0 \\ \vdots & \vdots & \ddots & \vdots \\ 0 & 0 & \cdots & a_{nn} \end{pmatrix}.$$

A diagonal matrix $\mathbf{U} = [u_{jk}]$ with all diagonal elements equal to 1, that is,

$$\mathbf{U} = \begin{pmatrix} 1 & 0 & \cdots & 0 \\ 0 & 1 & \cdots & 0 \\ \vdots & \vdots & \ddots & \vdots \\ 0 & 0 & \cdots & 1 \end{pmatrix},$$

is called a unit matrix or an identity matrix.

Matrices all of whose elements are equal to 0 are called *zero* (or *null*) *matrices*.

The *transpose* of an $m \times n$ matrix $\mathbf{A} = [a_{jk}]$ is the $n \times m$ matrix $\mathbf{A}^{\mathrm{T}} \equiv [a_{kj}]$, that is, the rows of $\mathbf{A}$ are the columns of $\mathbf{A}^{\mathrm{T}}$.

A matrix $\mathbf{A} = [a_{jk}]$ is said to be *symmetric* if $\mathbf{A} = \mathbf{A}^{\mathrm{T}}$, that is, if $a_{jk} = a_{kj}$.

The *complex conjugate* of an $m \times n$ matrix $\mathbf{A} = [a_{jk}]$ is the $m \times n$ matrix $\mathbf{A}^* \equiv [a_{jk}^*]$.

The *Hermitian conjugate* of a matrix $\mathbf{A} = [a_{jk}]$ is defined as the complex conjugate of the transpose of the matrix $\mathbf{A}$, that is,

$$\mathbf{A}^{\dagger} = (\mathbf{A}^{\mathrm{T}})^* = [a_{kj}^*].$$

Addition and Subtraction of Matrices

Let $\mathbf{A} = [a_{jk}]$ and $\mathbf{B} = [b_{jk}]$ be $m \times n$ matrices. The sum $\mathbf{A} + \mathbf{B}$ and difference $\mathbf{A} - \mathbf{B}$ of these matrices are the $m \times n$ matrices defined as

$$\mathbf{A} + \mathbf{B} \equiv [a_{jk} + b_{jk}], \quad \mathbf{A} - \mathbf{B} \equiv [a_{jk} - b_{jk}].$$

Multiplication by a Scalar

The product of an $m \times n$ matrix $\mathbf{A} = [a_{jk}]$ by a scalar α is the $m \times n$ matrix

$$\alpha\mathbf{A} \equiv \mathbf{A}\alpha \equiv [\alpha a_{jk}].$$

Matrix Product

The matrix product of an $m \times n$ matrix $\mathbf{A} = [a_{jk}]$ and an $n \times r$ matrix $\mathbf{B} = [b_{kl}]$ is the $m \times r$ matrix

$$\mathbf{AB} = \left[c_{jl} \equiv \sum_{k=1}^{n} a_{jk} b_{kl} \right].$$

Example 1 Let $\mathbf{A}$ and $\mathbf{B}$ be 2×2 matrices:

$$\mathbf{A} = \begin{pmatrix} a_{11} & a_{12} \\ a_{21} & a_{22} \end{pmatrix}, \quad \mathbf{B} = \begin{pmatrix} b_{11} & b_{12} \\ b_{21} & b_{22} \end{pmatrix}.$$

For these matrices,

$$\mathbf{AB} = \begin{pmatrix} a_{11}b_{11} + a_{12}b_{21} & a_{11}b_{12} + a_{12}b_{22} \\ a_{21}b_{11} + a_{22}b_{21} & a_{21}b_{12} + a_{22}b_{22} \end{pmatrix}$$

and

$$\mathbf{BA} = \begin{pmatrix} b_{11}a_{11} + b_{12}a_{21} & b_{11}a_{12} + b_{12}a_{22} \\ b_{21}a_{11} + b_{22}a_{21} & b_{21}a_{12} + b_{22}a_{22} \end{pmatrix}.$$

It is seen that in general $\mathbf{BA} \neq \mathbf{AB}$. But, taking the product $\mathbf{A}^{\mathrm{T}}\mathbf{B}^{\mathrm{T}}$,

$$\mathbf{A}^{\mathrm{T}}\mathbf{B}^{\mathrm{T}} = \begin{pmatrix} a_{11} & a_{21} \\ a_{12} & a_{22} \end{pmatrix}\begin{pmatrix} b_{11} & b_{21} \\ b_{12} & b_{22} \end{pmatrix} = \begin{pmatrix} a_{11}b_{11} + a_{21}b_{12} & a_{11}b_{21} + a_{21}b_{22} \\ a_{12}b_{11} + a_{22}b_{12} & a_{12}b_{21} + a_{22}b_{22} \end{pmatrix},$$

one may notice that for any **A** and **B**

$$\mathbf{BA} = (\mathbf{A}^{\mathrm{T}}\mathbf{B}^{\mathrm{T}})^{\mathrm{T}}.$$

Example 2 Let **A** be a 2×2 matrices, and **B** a 2×1 column vector:

$$\mathbf{A} = \begin{pmatrix} a_{11} & a_{12} \\ a_{21} & a_{22} \end{pmatrix}, \quad \mathbf{B} \equiv \begin{pmatrix} b_{11} \\ b_{21} \end{pmatrix} \equiv \begin{pmatrix} b_1 \\ b_2 \end{pmatrix}.$$

The matrix product **AB** in this case is the column vector

$$\mathbf{AB} = \begin{pmatrix} a_{11}b_1 + a_{12}b_2 \\ a_{21}b_1 + a_{22}b_2 \end{pmatrix}.$$

The matrix product of the row vector $\mathbf{B}^{\mathrm{T}}$ and the matrix **A** is the row vector

$$\mathbf{B}^{\mathrm{T}}\mathbf{A} = (\, a_{11}b_1 + a_{21}b_2 \quad a_{12}b_1 + a_{22}b_2 \,).$$

Example 3 Let **A** be a 1×2 row vector, and **B** a 2×1 column vector:

$$\mathbf{A} \equiv (\, a_{11} \quad a_{12} \,) \equiv (\, a_1 \quad a_2 \,), \quad \mathbf{B} \equiv \begin{pmatrix} b_{11} \\ b_{21} \end{pmatrix} \equiv \begin{pmatrix} b_1 \\ b_2 \end{pmatrix}.$$

The matrix product **AB** of these vectors is the scalar

$$\mathbf{AB} = a_1b_1 + a_2b_2.$$

Example 4 *Matrix representation of operations with Euclidean vectors.* By definition, the *scalar product* of Euclidean vectors $\mathbf{a} = (a_1, a_2, a_3)$ and $\mathbf{b} = (b_1, b_2, b_3)$ is the scalar

$$\mathbf{a} \cdot \mathbf{b} \equiv \mathbf{ab} \equiv a_1b_1 + a_2b_2 + a_3b_3.$$

Representing the vectors **a** and **b** by the column vectors

$$\mathbf{A} = \begin{pmatrix} a_1 \\ a_2 \\ a_3 \end{pmatrix} \quad \text{and} \quad \mathbf{B} = \begin{pmatrix} b_1 \\ b_2 \\ b_3 \end{pmatrix},$$

we can represent the scalar product $\mathbf{a} \cdot \mathbf{b}$ as a matrix product:

$$\mathbf{a} \cdot \mathbf{b} = \mathbf{A}^{\mathrm{T}}\mathbf{B} = \mathbf{B}^{\mathrm{T}}\mathbf{A}.$$

The *vector* product of the vectors **a** and **b** is the vector

$$\mathbf{a} \times \mathbf{b} = \left(a_2b_3 - a_3b_2, \quad a_3b_1 - a_1b_3, \quad a_1b_2 - a_2b_1\right).$$

Writing the vector $\mathbf{a} \times \mathbf{b}$ as the column

$$\mathbf{C} = \begin{pmatrix} a_2b_3 - a_3b_2 \\ a_3b_1 - a_1b_3 \\ a_1b_2 - a_2b_1 \end{pmatrix},$$

we can represent the vector product as follows:

$$\mathbf{C} = \begin{pmatrix} 0 & -a_3 & a_2 \\ a_3 & 0 & -a_1 \\ -a_2 & a_1 & 0 \end{pmatrix} \begin{pmatrix} b_1 \\ b_2 \\ b_3 \end{pmatrix} = \mathbf{A}^{\times}\mathbf{B},$$

where

$$\mathbf{A}^{\times} = \begin{pmatrix} 0 & -a_3 & a_2 \\ a_3 & 0 & -a_1 \\ -a_2 & a_1 & 0 \end{pmatrix}.$$

The standard operation $\boldsymbol{\alpha}\mathbf{b}$, where $\boldsymbol{\alpha}$ is a tensor, is usually treated as the matrix multiplication of a matrix $\boldsymbol{\Lambda}$ representing the tensor $\boldsymbol{\alpha}$ and a column vector **B** representing the vector **b**, $\boldsymbol{\Lambda}\mathbf{B}$. Following this line, the operation $\mathbf{b}\boldsymbol{\alpha}$, which appears in some equations of Chapters 8 and 9, can be treated as the multiplication of the row vector $\mathbf{B}^{\mathrm{T}}$ and the matrix $\boldsymbol{\Lambda}$, $\mathbf{B}^{\mathrm{T}}\boldsymbol{\Lambda}$.

For any $n \times n$ matrix **A** and the unit $n \times n$ matrix **U**, $\mathbf{AU} = \mathbf{UA} = \mathbf{A}$.

Trace

The trace (spur) of an $n \times n$ matrix $\mathbf{A} = [a_{jk}]$ is

$$\mathrm{Tr}\mathbf{A} = \sum_{j=1}^{n} a_{jj}.$$

Determinant

The determinant of an $n \times n$ matrix $\mathbf{A} = [a_{jk}]$ is the sum of the $n!$ terms $(-1)^r a_{1k_1} a_{2k_2} \ldots a_{nk_n}$ each corresponding to one of the $n!$ different ordered sets $\{k_1, k_2, \ldots, k_n\}$ obtained by r interchanges of elements from the set $\{1, 2,\ldots, n\}$. The number n is called *the order of the determinant*. The nth order determinant can be represented in terms of the $(n-1)$th order determinants. For instance, the determinant of the $n \times n$ matrix $\mathbf{A}$ may be written as

$$\det \mathbf{A} = \sum_{k=1}^{n} (-1)^{l+k} a_{lk} \det \mathbf{A}_{lk} = \sum_{j=1}^{n} (-1)^{j+l} a_{jl} \det \mathbf{A}_{jl} \tag{5.1}$$

at any $l \in \{1, 2,\ldots, n\}$ (simple Laplace development), where $\mathbf{A}_{jk}$ is the $(n-1) \times (n-1)$ matrix obtained from the matrix $\mathbf{A}$ by erasing the jth row and the kth column.

Examples

$$\det(a_{11}) = a_{11},$$

$$\det \begin{pmatrix} a_{11} & a_{12} \\ a_{21} & a_{22} \end{pmatrix} = a_{11}a_{22} - a_{12}a_{21},$$

$$\det \begin{pmatrix} a_{11} & a_{12} & a_{13} \\ a_{21} & a_{22} & a_{23} \\ a_{31} & a_{32} & a_{33} \end{pmatrix} = a_{11}a_{22}a_{33} + a_{12}a_{31}a_{23} + a_{13}a_{21}a_{32} - a_{11}a_{23}a_{32} - a_{12}a_{21}a_{33} - a_{13}a_{22}a_{31},$$

$$\det \begin{pmatrix} a_{11} & a_{12} & a_{13} & a_{14} \\ a_{21} & a_{22} & a_{23} & a_{24} \\ a_{31} & a_{32} & a_{33} & a_{34} \\ a_{41} & a_{42} & a_{43} & a_{44} \end{pmatrix} = a_{11} \det \begin{pmatrix} a_{22} & a_{23} & a_{24} \\ a_{32} & a_{33} & a_{34} \\ a_{42} & a_{43} & a_{44} \end{pmatrix} - a_{12} \det \begin{pmatrix} a_{21} & a_{23} & a_{24} \\ a_{31} & a_{33} & a_{34} \\ a_{41} & a_{43} & a_{44} \end{pmatrix}$$

$$+ a_{13} \det \begin{pmatrix} a_{21} & a_{22} & a_{24} \\ a_{31} & a_{32} & a_{34} \\ a_{41} & a_{42} & a_{44} \end{pmatrix} - a_{14} \det \begin{pmatrix} a_{21} & a_{22} & a_{23} \\ a_{31} & a_{32} & a_{33} \\ a_{41} & a_{42} & a_{43} \end{pmatrix}.$$

We note the following properties of determinants:

$$\det(\mathbf{A}^{\mathrm{T}}) = \det\mathbf{A}, \tag{5.2}$$

$$\det(\mathbf{A}^{\dagger}) = (\det\mathbf{A})^{*}, \tag{5.3}$$

$$\det(\mathbf{A}^{-1}) = (\det\mathbf{A})^{-1}. \tag{5.4}$$

If $\mathbf{A}$ is an $n \times n$ matrix,

$$\det(\alpha\mathbf{A}) = \alpha^{n}\det\mathbf{A}, \tag{5.5}$$

where α is an arbitrary number.

The determinant of any diagonal matrix is equal to the product of the diagonal elements of this matrix. The determinant of a unit matrix is equal to 1.

Adjugate

The adjugate of an $n \times n$ matrix $\mathbf{A} = [a_{jk}]$ is the $n \times n$ matrix

$$\mathrm{adj}\mathbf{A} \equiv [c_{jk} \equiv (-1)^{j+k} \det\mathbf{A}_{kj}],$$

where the $(n-1) \times (n-1)$ matrices $\mathbf{A}_{jk}$ are defined as in (5.1).

Examples

$$\mathrm{adj}\begin{pmatrix} a_{11} & a_{12} \\ a_{21} & a_{22} \end{pmatrix} = \begin{pmatrix} a_{22} & -a_{12} \\ -a_{21} & a_{11} \end{pmatrix},$$

$$\mathrm{adj}\begin{pmatrix} a_{11} & a_{12} & a_{13} \\ a_{21} & a_{22} & a_{23} \\ a_{31} & a_{32} & a_{33} \end{pmatrix} = \begin{pmatrix} a_{22}a_{33} - a_{23}a_{32} & a_{32}a_{13} - a_{12}a_{33} & a_{12}a_{23} - a_{13}a_{22} \\ a_{23}a_{31} - a_{21}a_{33} & a_{11}a_{33} - a_{13}a_{31} & a_{13}a_{21} - a_{11}a_{23} \\ a_{21}a_{32} - a_{31}a_{22} & a_{12}a_{31} - a_{11}a_{32} & a_{11}a_{22} - a_{21}a_{12} \end{pmatrix}.$$

The main feature of the matrix adj**A** is that this matrix satisfies the relations

$$\mathbf{A}\mathrm{adj}\mathbf{A} = \mathrm{adj}(\mathbf{A})\mathbf{A} = \det(\mathbf{A})\mathbf{U}. \tag{5.6}$$

When dealing with the equation

$$\mathbf{AX} = \mathbf{0}, \tag{5.7}$$

where $\mathbf{A} = [a_{jk}]$ is a given $n \times n$ matrix with det**A** = 0, $\mathbf{X} = [x_j]$ is the $n \times 1$ column vector of unknowns x_j ($j = 1,2,\ldots, n$), and **0** is the null $n \times 1$ column vector,

$$\mathbf{X} = \begin{pmatrix} x_1 \\ x_2 \\ \vdots \\ x_n \end{pmatrix}, \quad \mathbf{0} = \begin{pmatrix} 0 \\ 0 \\ \vdots \\ 0 \end{pmatrix},$$

we can choose a solution of (5.7) among the columns of the matrix adj**A** because, according to (5.6), any column of adj**A** taken as **X** satisfies (5.7).

Inverse

The inverse of an $n \times n$ matrix $\mathbf{A} = [a_{jk}]$ is the matrix $\mathbf{A}^{-1}$ such that

$$\mathbf{A}^{-1}\mathbf{A} = \mathbf{A}\mathbf{A}^{-1} = \mathbf{U},$$

where **U** is the unit $n \times n$ matrix. The matrix **A** has the inverse if det**A** $\neq 0$. According to (5.6), the matrix $\mathbf{A}^{-1}$ may be expressed as follows:

$$\mathbf{A}^{-1} = \frac{1}{\det \mathbf{A}}\mathrm{adj}\mathbf{A}.$$

Example

$$\begin{pmatrix} a_{11} & a_{12} \\ a_{21} & a_{22} \end{pmatrix}^{-1} = \frac{1}{a_{11}a_{22} - a_{12}a_{21}}\begin{pmatrix} a_{22} & -a_{12} \\ -a_{21} & a_{11} \end{pmatrix}.$$

Eigenvalues

The *characteristic* (*secular*) *equation* of an $n \times n$ matrix $\mathbf{A} = [a_{jk}]$ is

$$\det(\mathbf{A} - \lambda\mathbf{U}) \equiv \det \begin{pmatrix} a_{11} - \lambda & a_{12} & \cdots & a_{1n} \\ a_{21} & a_{22} - \lambda & \cdots & a_{2n} \\ \vdots & \vdots & \ddots & \vdots \\ a_{n1} & a_{n2} & \cdots & a_{nn} - \lambda \end{pmatrix} = 0, \quad (5.8)$$

where $\mathbf{U}$ is the unit $n \times n$ matrix. The roots of this equation are called *eigenvalues* of the matrix $\mathbf{A}$.

Example The characteristic equation of a 2×2 matrix $\mathbf{A} = [a_{jk}]$ is

$$\lambda^2 - \lambda \mathrm{Tr}\mathbf{A} + \det \mathbf{A} = 0. \quad (5.9)$$

It may also be written as

$$\lambda^2 - \lambda(a_{11} + a_{22}) + (a_{11}a_{22} - a_{12}a_{21}) = 0. \quad (5.10)$$

The roots of (5.9), λ_1 and λ_2, the eigenvalues of $\mathbf{A}$, are

$$\lambda_{1,2} = \frac{\mathrm{Tr}\mathbf{A} \pm \sqrt{(\mathrm{Tr}\mathbf{A})^2 - 4\det \mathbf{A}}}{2} = \frac{a_{11} + a_{22} \pm \sqrt{(a_{11} + a_{22})^2 - 4(a_{11}a_{22} - a_{12}a_{21})}}{2}. \quad (5.11)$$

Eigenvectors

For an $n \times n$ matrix $\mathbf{A} = [a_{jk}]$ with an eigenvalue λ, a nonzero $n \times 1$ vector $\mathbf{X}$ satisfying the equation

$$\mathbf{AX} = \lambda\mathbf{X}$$

or, equivalently,

$$(\mathbf{A} - \lambda\mathbf{U})\mathbf{X} = \mathbf{0},$$

where $\mathbf{0}$ is the zero $n \times 1$ vector [see (5.7)], is called an *eigenvector* of $\mathbf{A}$ corresponding to or associated with the eigenvalue λ.

Singular Values

The singular values $\sigma_j(\mathbf{A})$ ($j = 1, 2, \ldots$) of a square matrix $\mathbf{A}$ are the nonnegative square roots of the eigenvalues of the matrix $\mathbf{A}^\dagger\mathbf{A}$.

Example The singular values of a 2×2 matrix $\mathbf{A}$, $\sigma_1(\mathbf{A})$ and $\sigma_2(\mathbf{A})$, may be expressed as follows:

$$\sigma_{1,2}(\mathbf{A}) = \sqrt{\frac{\|\mathbf{A}\|_{\mathrm{E}}^2 \pm \sqrt{\|\mathbf{A}\|_{\mathrm{E}}^4 - 4\,|\det \mathbf{A}|^2}}{2}}, \quad (5.12)$$

where $\|\mathbf{A}\|_{\mathrm{E}}$ is the Euclidean norm of $\mathbf{A}$ [see Section 5.1.4 and (5.44)].

5.1.2 Some Important Properties of Matrix Products

For any matrices **A**, **B**, and **C** of appropriate dimensions

$$\mathbf{A}(\mathbf{BC}) = (\mathbf{AB})\mathbf{C},$$

therefore the products **A**(**BC**) and (**AB**)**C** may be written simply as **ABC**.

As already noted, in general $\mathbf{AB} \neq \mathbf{BA}$.

Let a matrix **T** be the product of (not necessarily square) matrices $\mathbf{T}_1$, $\mathbf{T}_2$,..., and $\mathbf{T}_N$:

$$\mathbf{T} = \mathbf{T}_1\mathbf{T}_2\mathbf{T}_3 \ldots \mathbf{T}_{N-1}\mathbf{T}_N.$$

The following relations are valid:

$$\mathbf{T}^* \equiv (\mathbf{T}_1\mathbf{T}_2\mathbf{T}_3 \ldots \mathbf{T}_{N-1}\mathbf{T}_N)^* = \mathbf{T}_1^*\mathbf{T}_2^*\mathbf{T}_3^* \ldots \mathbf{T}_{N-1}^*\mathbf{T}_N^*, \tag{5.13}$$

$$\mathbf{T}^\mathrm{T} \equiv (\mathbf{T}_1\mathbf{T}_2\mathbf{T}_3 \ldots \mathbf{T}_{N-1}\mathbf{T}_N)^\mathrm{T} = \mathbf{T}_N^\mathrm{T}\mathbf{T}_{N-1}^\mathrm{T} \ldots \mathbf{T}_3^\mathrm{T}\mathbf{T}_2^\mathrm{T}\mathbf{T}_1^\mathrm{T}, \tag{5.14}$$

$$\mathbf{T}^\dagger \equiv (\mathbf{T}_1\mathbf{T}_2\mathbf{T}_3 \ldots \mathbf{T}_{N-1}\mathbf{T}_N)^\dagger = \mathbf{T}_N^\dagger\mathbf{T}_{N-1}^\dagger \ldots \mathbf{T}_3^\dagger\mathbf{T}_2^\dagger\mathbf{T}_1^\dagger, \tag{5.15}$$

$$\mathbf{T}^{-1} \equiv (\mathbf{T}_1\mathbf{T}_2\mathbf{T}_3 \ldots \mathbf{T}_{N-1}\mathbf{T}_N)^{-1} = \mathbf{T}_N^{-1}\mathbf{T}_{N-1}^{-1} \ldots \mathbf{T}_3^{-1}\mathbf{T}_2^{-1}\mathbf{T}_1^{-1}, \tag{5.16}$$

$$\det(\mathbf{T}_1\mathbf{T}_2\mathbf{T}_3 \ldots \mathbf{T}_{N-1}\mathbf{T}_N) = \det\mathbf{T}_1 \cdot \det\mathbf{T}_2 \cdot \det\mathbf{T}_3 \ldots \det\mathbf{T}_{N-1} \cdot \det\mathbf{T}_N. \tag{5.17}$$

5.1.3 Unitary Matrices. Unimodular Unitary 2 × 2 Matrices. STU Matrices

Unitary Matrices

A square matrix **A** is said to be *unitary* if for this matrix

$$\mathbf{A}^{-1} = \mathbf{A}^\dagger, \tag{5.18}$$

that is,

$$\mathbf{A}^\dagger\mathbf{A} = \mathbf{A}\mathbf{A}^\dagger = \mathbf{U}, \tag{5.19}$$

where **U** is the unit matrix.

A key property of a unitary $n \times n$ matrix **A** is that for any $n \times 1$ vector **X** and the vector $\mathbf{Y} = \mathbf{AX}$ the following relation holds:

$$\mathbf{Y}^\dagger\mathbf{Y} = \mathbf{X}^\dagger\mathbf{X}. \tag{5.20}$$

Using the notion of the Euclidean norm of a vector (see Section 5.1.4), this property can be expressed by the relation

$$\|\mathbf{AX}\|_\mathrm{E} = \|\mathbf{X}\|_\mathrm{E}\,. \tag{5.21}$$

Another important property of the unitary matrix **A** is the following. For any $n \times 1$ vectors $\mathbf{X}_1$ and $\mathbf{X}_2$

$$\mathbf{Y}_2^\dagger\mathbf{Y}_1 = \mathbf{X}_2^\dagger\mathbf{X}_1, \tag{5.22}$$

where $\mathbf{Y}_1 = \mathbf{AX}_1$ and $\mathbf{Y}_2 = \mathbf{AX}_2$. Relation (5.22) can easily be derived from the defining relation (5.19). Using property (5.15) of matrix products, we may represent the Hermitian conjugate of the vector $\mathbf{Y}_2$ as follows:

$$\mathbf{Y}_2^\dagger = (\mathbf{AX}_2)^\dagger = \mathbf{X}_2^\dagger\mathbf{A}^\dagger. \tag{5.23}$$

Using (5.23) and (5.19), we obtain

$$\mathbf{Y}_2^\dagger\mathbf{Y}_1 = \mathbf{X}_2^\dagger\mathbf{A}^\dagger\mathbf{AX}_1 = \mathbf{X}_2^\dagger(\mathbf{A}^\dagger\mathbf{A})\mathbf{X}_1 = \mathbf{X}_2^\dagger\mathbf{UX}_1 = \mathbf{X}_2^\dagger\mathbf{X}_1. \tag{5.24}$$

Since relation (5.22) is equivalent to (5.20) at $\mathbf{X}_1 = \mathbf{X}_2 = \mathbf{X}$, derivation (5.24) also proves (5.20).

The product of any number of unitary matrices is a unitary matrix. To prove this it suffices to show that the product of two unitary matrices is a unitary matrix. Let $\mathbf{C} = \mathbf{AB}$, where $\mathbf{A}$ and $\mathbf{B}$ are arbitrary unitary matrices. Again using (5.15), we represent $\mathbf{C}^\dagger$ as $\mathbf{B}^\dagger\mathbf{A}^\dagger$. Then, using relations $\mathbf{A}^\dagger\mathbf{A} = \mathbf{U}$ and $\mathbf{B}^\dagger\mathbf{B} = \mathbf{U}$, we obtain

$$\mathbf{C}^\dagger\mathbf{C} = (\mathbf{B}^\dagger\mathbf{A}^\dagger)(\mathbf{AB}) = \mathbf{B}^\dagger(\mathbf{A}^\dagger\mathbf{A})\mathbf{B} = \mathbf{B}^\dagger\mathbf{B} = \mathbf{U},$$

that is, the matrix $\mathbf{C}$ satisfies the condition $\mathbf{C}^\dagger\mathbf{C} = \mathbf{U}$ and hence is a unitary matrix.

According to (5.17) and (5.3),

$$\det(\mathbf{A}^\dagger\mathbf{A}) = \det(\mathbf{A}^\dagger)\det\mathbf{A} = (\det\mathbf{A})^*\det\mathbf{A}. \tag{5.25}$$

From (5.19) and (5.25), we see that $|\det\mathbf{A}| = 1$ for any unitary matrix $\mathbf{A}$.

Unimodular Matrices

Matrices whose determinant equals 1 or –1 are sometimes called *unimodular*. We, throughout this book, also use the term *unimodular matrix* in this sense.

As is seen from (5.17), the product of any number of unimodular matrices is a unimodular matrix. The determinant of the product of unimodular matrices $\mathbf{T}_1$, $\mathbf{T}_2$,..., and $\mathbf{T}_N$ may be expressed as

$$\det(\mathbf{T}_1\mathbf{T}_2\ldots\mathbf{T}_N) = (-1)^{N'},$$

where N' is the number of matrices with negative determinant in the set $\{\mathbf{T}_j; j = 1,2,\ldots, N\}$.

Any unitary matrix $\mathbf{A}$ can be represented as

$$\mathbf{A} = a_{\mathrm{U}}\mathbf{A}_{\mathrm{UM}}, \tag{5.26}$$

where a_{U} is a number with $|a_{\mathrm{U}}| = 1$ and $\mathbf{A}_{\mathrm{UM}}$ is a unimodular unitary matrix. For instance, if $\mathbf{A}$ is an $n \times n$ matrix, we may take

$$\mathbf{A}_{\mathrm{UM}} = (\det\mathbf{A})^{-1/n}\,\mathbf{A}, \tag{5.27}$$

$$a_{\mathrm{U}} = (\det\mathbf{A})^{1/n}, \tag{5.28}$$

because, according to (5.5),

$$\det\left((\det\mathbf{A})^{-1/n}\,\mathbf{A}\right) = \left[(\det\mathbf{A})^{-1/n}\right]^n \det\mathbf{A} = (\det\mathbf{A})^{-1}\det\mathbf{A} = 1,$$

and $\left|(\det \mathbf{A})^{1/n}\right| = 1$ as $|\det\mathbf{A}| = 1$. In this case, $\det\mathbf{A}_{\text{UM}} = 1$. An alternative choice of the factors in (5.26), with $\det\mathbf{A}_{\text{UM}} = -1$, is as follows:

$$\mathbf{A}_{\text{UM}} = (-\det \mathbf{A})^{-1/n}\,\mathbf{A}, \tag{5.29}$$

$$a_{\text{U}} = (-\det \mathbf{A})^{1/n}. \tag{5.30}$$

Unitary Unimodular 2 × 2 Matrices

Any unitary 2 × 2 matrix whose determinant is equal to 1 has the form

$$\begin{pmatrix} a & b \\ -b^* & a^* \end{pmatrix}, \tag{5.31}$$

where a and b are complex numbers such that $a^*a + b^*b = 1$, and can be represented as

$$\begin{pmatrix} e^{i\varphi}\cos\rho & e^{i\psi}\sin\rho \\ -e^{-i\psi}\sin\rho & e^{-i\varphi}\cos\rho \end{pmatrix}, \tag{5.32}$$

where ρ, φ, and ψ are real numbers. Any unitary 2 × 2 matrix with determinant –1 has the form

$$\begin{pmatrix} a & b \\ b^* & -a^* \end{pmatrix}, \tag{5.33}$$

where also $a^*a + b^*b = 1$, and can be represented as

$$\begin{pmatrix} e^{i\varphi}\cos\rho & e^{i\psi}\sin\rho \\ e^{-i\psi}\sin\rho & -e^{-i\varphi}\cos\rho \end{pmatrix}. \tag{5.34}$$

STU Matrices

The class of matrices that we call *STU matrices* is not a standard class of matrices in matrix algebra. Nevertheless the notion of an STU matrix is very important in our consideration of LCD optics. We say that a matrix is an *STU matrix* (*a Scalar Times a Unitary* matrix) if this matrix can be represented as the product of a unitary matrix and a scalar. If a matrix $\mathbf{A}$ is represented as $\mathbf{A} = a\mathbf{A}_{\text{U}}$, where $\mathbf{A}_{\text{U}}$ is a unitary matrix and a is a (possibly complex) scalar, the scalar multiplier a and matrix $\mathbf{A}_{\text{U}}$ will be called the *loss factor* and *base matrix* of the STU matrix $\mathbf{A}$, respectively. Peculiar properties of STU matrices are clearly connected with properties of unitary matrices.

Most important for us are the following properties of STU matrices.

An arbitrary $n \times n$ STU matrix $\mathbf{A}$ with a loss factor a and a base matrix $\mathbf{A}_{\text{U}}$ satisfies the relations

$$\mathbf{A}^\dagger\mathbf{A} = \mathbf{A}\mathbf{A}^\dagger = |a|^2\mathbf{U}. \tag{5.35}$$

Actually,

$$\mathbf{A}^\dagger\mathbf{A} = (a\mathbf{A}_{\text{U}})^\dagger a\mathbf{A}_{\text{U}} = a^*\mathbf{A}_{\text{U}}^\dagger a\mathbf{A}_{\text{U}} = a^*a\mathbf{A}_{\text{U}}^\dagger\mathbf{A}_{\text{U}} = |a|^2\mathbf{U}.$$

For any $n \times 1$ vector $\mathbf{X}$ and the vector $\mathbf{Y} = \mathbf{A}\mathbf{X}$,

$$\mathbf{Y}^\dagger\mathbf{Y} = |a|^2\mathbf{X}^\dagger\mathbf{X} \tag{5.36}$$

[cf. (5.20)]. For any $n \times 1$ vectors $\mathbf{X}_1$ and $\mathbf{X}_2$ and the vectors $\mathbf{Y}_1 = \mathbf{A}\mathbf{X}_1$ and $\mathbf{Y}_2 = \mathbf{A}\mathbf{X}_2$,

$$\mathbf{Y}_2^\dagger \mathbf{Y}_1 = |a|^2 \mathbf{X}_2^\dagger \mathbf{X}_1 \tag{5.37}$$

[cf. (5.22)]. One may notice that, as in the case of unitary matrices,

$$\mathbf{Y}_2^\dagger \mathbf{Y}_1 = 0 \text{ if } \mathbf{X}_2^\dagger \mathbf{X}_1 = 0. \tag{5.38}$$

The product of any number of STU matrices is an STU matrix. If $n \times n$ STU matrices $\mathbf{T}_1, \mathbf{T}_2, \ldots,$ and $\mathbf{T}_N$ are given by their loss factors t_j and base matrices $\mathbf{T}_{Uj}$, the loss factor t and base matrix $\mathbf{T}_U$ of the matrix $\mathbf{T} = \mathbf{T}_1 \mathbf{T}_2 \ldots \mathbf{T}_N$ can evidently be calculated as

$$t = t_1 t_2 \ldots t_N, \quad \mathbf{T}_U = \mathbf{T}_{U1} \mathbf{T}_{U2} \ldots \mathbf{T}_{UN}. \tag{5.39}$$

The loss factor and base matrix representing an STU matrix can always be chosen so that the base matrix is unimodular and has a desired determinant (1 or –1).

5.1.4 *Norms of Vectors and Matrices*

Norms are real-valued nonnegative-valued quantities introduced to characterize the "magnitude" of vectors and matrices. There are many types of norms for vectors and matrices. In this book we use one type of vector norms (Euclidean norm) and two types of matrix norms (Euclidean norm and spectral norm).

Euclidean Norm and Length of a Vector

The Euclidean norm of an n-vector $\mathbf{A} = (a_1, a_2, \ldots, a_n)$ is defined as

$$\|\mathbf{A}\|_E = \sqrt{\sum_{j=1}^{n} |a_j|^2} = \sqrt{\sum_{j=1}^{n} a_j^* a_j}. \tag{5.40}$$

If $\mathbf{A}$ is a column vector, its Euclidean norm may be expressed as

$$\|\mathbf{A}\|_E = \sqrt{\mathbf{A}^\dagger \mathbf{A}}. \tag{5.41}$$

If $\mathbf{A}$ is a Euclidean vector, $\|\mathbf{A}\|_E$ may be represented as follows:

$$\|\mathbf{A}\|_E = \sqrt{\mathbf{A}^* \cdot \mathbf{A}} = \sqrt{\mathbf{A} \cdot \mathbf{A}^*}, \tag{5.42}$$

where the dot "·" denotes the scalar (dot) multiplication. The *length* of a vector is usually defined as the Euclidean norm of this vector.

Euclidean Norm of a Matrix

The Euclidean norm of an $m \times n$ matrix $\mathbf{A} = [a_{jk}]$ is calculated as

$$\|\mathbf{A}\|_E = \sqrt{\sum_{j=1}^{m} \sum_{k=1}^{n} \left| a_{jk} \right|^2} = \sqrt{\sum_{j=1}^{m} \sum_{k=1}^{n} a_{jk}^* a_{jk}}. \tag{5.43}$$

Example If $\mathbf{A}$ is a 2×2 matrix,

$$\|\mathbf{A}\|_E = \sqrt{a_{11}^* a_{11} + a_{12}^* a_{12} + a_{21}^* a_{21} + a_{22}^* a_{22}}. \tag{5.44}$$

The Euclidean norm of a square matrix $\mathbf{A}$ may be represented as

$$\|\mathbf{A}\|_E = \sqrt{\mathrm{Tr}(\mathbf{A}^\dagger \mathbf{A})}. \tag{5.45}$$

From (5.45), the following useful expression can easily be obtained:

$$\|\mathbf{C} - \mathbf{B}\|_E^2 = \|\mathbf{C}\|_E^2 + \|\mathbf{B}\|_E^2 - \mathrm{Tr}\left(\mathbf{C}^\dagger \mathbf{B} + \mathbf{B}^\dagger \mathbf{C}\right), \tag{5.46}$$

where $\mathbf{C}$ and $\mathbf{B}$ are arbitrary square matrices.

For a unitary $n \times n$ matrix $\mathbf{A}_U$, by definition, $\mathbf{A}_U^\dagger \mathbf{A}_U = \mathbf{U}$ and, consequently, $\mathrm{Tr}(\mathbf{A}_U^\dagger \mathbf{A}_U) = n$. Therefore, according to (5.45), $\|\mathbf{A}_U\|_E = \sqrt{n}$.

Spectral Norm of a Matrix

The spectral norm of a square matrix $\mathbf{A}$ may be defined as

$$\|\mathbf{A}\|_S \equiv \sqrt{\lambda_{\max}[\mathbf{A}^\dagger \mathbf{A}]} \equiv \sigma_{\max}(\mathbf{A}), \tag{5.47}$$

where $\lambda_{\max}[\mathbf{A}^\dagger \mathbf{A}]$ is the largest eigenvalue of the matrix $\mathbf{A}^\dagger \mathbf{A}$, and $\sigma_{\max}(\mathbf{A})$ is the largest singular value of $\mathbf{A}$.

Example 1 For a 2×2 matrix $\mathbf{A}$,

$$\|\mathbf{A}\|_S = \sqrt{\frac{\|\mathbf{A}\|_E^2 + \sqrt{\|\mathbf{A}\|_E^4 - 4\,|\det \mathbf{A}|^2}}{2}}. \tag{5.48}$$

Example 2 For a unitary matrix $\mathbf{A}_U$, ${\mathbf{A}_U}^\dagger \mathbf{A}_U = \mathbf{U}$. Hence all singular values of $\mathbf{A}_U$ are equal to 1, and, consequently, $\|\mathbf{A}_U\|_S = 1$.

A very useful property of spectral norm is

$$\|\mathbf{AX}\|_E \leq \|\mathbf{A}\|_S \, \|\mathbf{X}\|_E, \tag{5.49}$$

where $\mathbf{A}$ is an arbitrary $n \times n$ matrix and $\mathbf{X}$ is an arbitrary $n \times 1$ vector.

For any 2 × 2 matrix **A**,

$$\frac{1}{\sqrt{2}}\,\|\mathbf{A}\|_{\mathrm{E}} \le \|\mathbf{A}\|_{\mathrm{S}} \le \|\mathbf{A}\|_{\mathrm{E}}. \tag{5.50}$$

Some Properties of Norms

For any norm,

$$\|\mathbf{A}\| > 0 \text{ if } \mathbf{A} \ne \mathbf{0}, \quad \|\mathbf{0}\| = 0, \tag{5.51}$$

$$\|\alpha\mathbf{A}\| = |\alpha| \cdot \|\mathbf{A}\|, \tag{5.52}$$

$$\|\mathbf{A}+\mathbf{B}\| \le \|\mathbf{A}\| + \|\mathbf{B}\|, \tag{5.53}$$

$$\|\mathbf{A}\mathbf{C}\| \le \|\mathbf{A}\| \cdot \|\mathbf{C}\|, \tag{5.54}$$

where α is an arbitrary number, **A**, **B**, and **C** are arbitrary matrices of appropriate dimensions, and **0** is a zero matrix. These are defining properties of matrix norms.

According to (5.53), for any $m \times n$ matrices **C** and **D**,

$$\|\mathbf{C}\| = \|(\mathbf{C}+\mathbf{D}) - \mathbf{D}\| \le \|\mathbf{C}+\mathbf{D}\| + \|\mathbf{D}\|, \tag{5.55}$$

$$\|\mathbf{D}\| = \|(\mathbf{C}+\mathbf{D}) - \mathbf{C}\| \le \|\mathbf{C}+\mathbf{D}\| + \|\mathbf{C}\|. \tag{5.56}$$

From (5.55) we have

$$\|\mathbf{C}\| - \|\mathbf{D}\| \le \|\mathbf{C}+\mathbf{D}\|.$$

On the other hand, according to (5.56),

$$\|\mathbf{D}\| - \|\mathbf{C}\| \le \|\mathbf{C}+\mathbf{D}\|.$$

Therefore, in general,

$$\big|\|\mathbf{C}\| - \|\mathbf{D}\|\big| \le \|\mathbf{C}+\mathbf{D}\|. \tag{5.57}$$

Combining (5.53) and (5.57), we may write the following general relation for norms of the sum of two matrices:

$$\big|\|\mathbf{A}\| - \|\mathbf{B}\|\big| \le \|\mathbf{A}+\mathbf{B}\| \le \|\mathbf{A}\| + \|\mathbf{B}\|. \tag{5.58}$$

This relation allows one to approximately estimate $\|\mathbf{A}+\mathbf{B}\|$ when $\|\mathbf{A}\|$ and $\|\mathbf{B}\|$ are known.

In this book, we also use the following properties of Euclidean and spectral matrix norms:

$$\|\mathbf{A}\mathbf{B}\|_{\mathrm{E}} \le \|\mathbf{A}\|_{\mathrm{S}}\,\|\mathbf{B}\|_{\mathrm{E}}, \quad \|\mathbf{A}\mathbf{B}\|_{\mathrm{E}} \le \|\mathbf{A}\|_{\mathrm{E}}\,\|\mathbf{B}\|_{\mathrm{S}}, \tag{5.59}$$

$$\|\mathbf{C}_{\mathrm{U}}\mathbf{A}\mathbf{D}_{\mathrm{U}}\|_{\mathrm{E}} = \|\mathbf{A}\|_{\mathrm{E}}, \tag{5.60}$$

$$\|\mathbf{C}_{\mathrm{U}}\mathbf{A}\mathbf{D}_{\mathrm{U}}\|_{\mathrm{S}} = \|\mathbf{A}\|_{\mathrm{S}}, \tag{5.61}$$

where **A** and **B** are arbitrary square matrices and $\mathbf{C}_{\mathrm{U}}$ and $\mathbf{D}_{\mathrm{U}}$ are arbitrary unitary matrices of the same size.

5.1.5 Kronecker Product of Matrices

In this book, along with the usual matrix product considered above, we sometimes deal with another type of product of matrices, namely, with the *Kronecker* (or *direct*) *product*.

For any $n \times m$ matrix $\mathbf{A}$ and $p \times q$ matrix $\mathbf{B}$,

$$\mathbf{A} = \begin{pmatrix} a_{11} & \cdots & a_{1m} \\ \vdots & \ddots & \vdots \\ a_{n1} & \cdots & a_{nm} \end{pmatrix}, \quad \mathbf{B} = \begin{pmatrix} b_{11} & \cdots & b_{1q} \\ \vdots & \ddots & \vdots \\ b_{p1} & \cdots & b_{pq} \end{pmatrix} \tag{5.62}$$

their Kronecker product, denoted as $\mathbf{A} \otimes \mathbf{B}$, is the following $np \times mq$ matrix (in block form):

$$\mathbf{A} \otimes \mathbf{B} = \begin{pmatrix} a_{11}\mathbf{B} & \cdots & a_{1m}\mathbf{B} \\ \vdots & \ddots & \vdots \\ a_{n1}\mathbf{B} & \cdots & a_{nm}\mathbf{B} \end{pmatrix}. \tag{5.63}$$

Example 1 For

$$\mathbf{A} = \begin{pmatrix} a_{11} & a_{12} \\ a_{21} & a_{22} \end{pmatrix} \quad \text{and} \quad \mathbf{B} = \begin{pmatrix} b_{11} & b_{12} \\ b_{21} & b_{22} \end{pmatrix},$$
$$\mathbf{A} \otimes \mathbf{B} = \begin{pmatrix} a_{11}b_{11} & a_{11}b_{12} & a_{12}b_{11} & a_{12}b_{12} \\ a_{11}b_{21} & a_{11}b_{22} & a_{12}b_{21} & a_{12}b_{22} \\ a_{21}b_{11} & a_{21}b_{12} & a_{22}b_{11} & a_{22}b_{12} \\ a_{21}b_{21} & a_{21}b_{22} & a_{22}b_{21} & a_{22}b_{22} \end{pmatrix}. \tag{5.64}$$

Example 2 For

$$\mathbf{A} = \begin{pmatrix} a_1 \\ a_2 \end{pmatrix} \quad \text{and} \quad \mathbf{B} = \begin{pmatrix} b_1 \\ b_2 \end{pmatrix},$$
$$\mathbf{A} \otimes \mathbf{B} = \begin{pmatrix} a_1 b_1 \\ a_1 b_2 \\ a_2 b_1 \\ a_2 b_2 \end{pmatrix}. \tag{5.65}$$

Example 3 The *dyadic product* [see (8.26)] can be regarded as a special case of the Kronecker product. For

$$\mathbf{A} = \begin{pmatrix} a_1 \\ a_2 \\ a_3 \end{pmatrix} \quad \text{and} \quad \mathbf{B} = \begin{pmatrix} b_1 & b_2 & b_3 \end{pmatrix},$$
$$\mathbf{A} \otimes \mathbf{B} = \begin{pmatrix} a_1 b_1 & a_1 b_2 & a_1 b_3 \\ a_2 b_1 & a_2 b_2 & a_2 b_3 \\ a_3 b_1 & a_3 b_2 & a_3 b_3 \end{pmatrix}. \tag{5.66}$$

Note the following properties of the Kronecker product. For any matrices $\mathbf{A}$ and $\mathbf{B}$,

$$(\mathbf{A} \otimes \mathbf{B})^{\mathrm{T}} = \mathbf{A}^{\mathrm{T}} \otimes \mathbf{B}^{\mathrm{T}}, \tag{5.67}$$
$$(\mathbf{A} \otimes \mathbf{B})^{\dagger} = \mathbf{A}^{\dagger} \otimes \mathbf{B}^{\dagger}. \tag{5.68}$$

If matrices **A** and **B** are square and nonsingular (i.e., det**A** $\neq 0$ and det**B** $\neq 0$),

$$(\mathbf{A} \otimes \mathbf{B})^{-1} = \mathbf{A}^{-1} \otimes \mathbf{B}^{-1}. \tag{5.69}$$

For any matrices **A**, **B**, **C**, and **D** of appropriate dimensions,

$$(\mathbf{AB}) \otimes (\mathbf{CD}) = (\mathbf{A} \otimes \mathbf{C})(\mathbf{B} \otimes \mathbf{D}). \tag{5.70}$$

5.1.6 *Approximations*

For a square matrix **H** with $\|\mathbf{H}\|_E < 1$, the following relation is valid:

$$(\mathbf{U} + \mathbf{H})^{-1} = \mathbf{U} + \sum_{k=1}^{\infty} (-\mathbf{H})^k, \tag{5.71}$$

where **U** is the unit matrix. Suppose that **A** is a nonsingular matrix for which the inverse $\mathbf{A}^{-1}$ is known and we need to estimate the inverse of the matrix **A** + **B**, where **B** is a matrix with $\|\mathbf{B}\|_E \ll \|\mathbf{A}\|_E$ (i.e., **B** may be considered as a small perturbation for **A**). We may represent the matrix **A** + **B** as

$$\mathbf{A} + \mathbf{B} = \mathbf{A}(\mathbf{U} + \mathbf{H}), \tag{5.72}$$

where $\mathbf{H} = \mathbf{A}^{-1}\mathbf{B}$, and the inverse of **A** + **B** as

$$(\mathbf{A} + \mathbf{B})^{-1} = (\mathbf{U} + \mathbf{H})^{-1}\mathbf{A}^{-1}. \tag{5.73}$$

Then, provided that $\|\mathbf{H}\|_E < 1$, we may use (5.71) to get

$$(\mathbf{A} + \mathbf{B})^{-1} = (\mathbf{U} + \mathbf{H})^{-1}\,\mathbf{A}^{-1} = \left(\mathbf{U} + \sum_{k=1}^{\infty} (-\mathbf{H})^k\right)\mathbf{A}^{-1} = \mathbf{A}^{-1} + \left(\sum_{k=1}^{\infty} (-\mathbf{H})^k\right)\mathbf{A}^{-1}. \tag{5.74}$$

If $\|\mathbf{H}\|_E \ll 1$, we may neglect the terms with $k > 1$ in (5.74), which gives the following approximate expression:

$$(\mathbf{A} + \mathbf{B})^{-1} \approx \mathbf{A}^{-1} - \mathbf{A}^{-1}\mathbf{B}\mathbf{A}^{-1}. \tag{5.75}$$

Sources for Section 5.1: References [1–3]

5.2 Some Radiometric Quantities. Conventions

Radiant energy, Q, is the amount of energy propagating onto, through, or emerging from a specified surface of given area in a given period of time [4].

Spectral radiant energy, Q_λ, is the spectral density of radiant energy:

$$Q_\lambda \equiv \frac{\mathrm{d}Q}{\mathrm{d}\lambda}.$$

Radiant flux (it is also called *radiant power*), Φ, is the time rate of the flow of radiant energy:

$$\Phi \equiv \frac{dQ}{dt}.$$

Radiant flux characterizes the amount of energy transferred through a surface or region of space per unit time.

Spectral radiant flux (*power*), Φ_λ, is the spectral density of radiant flux:

$$\Phi_\lambda \equiv \frac{d\Phi}{d\lambda} = \frac{dQ_\lambda}{dt}.$$

Irradiance, E, at a point on a surface S (physical or geometrical) is the ratio of the radiant flux incident on an element dS of the surface S containing this point to the area ds of the element dS:

$$E \equiv \frac{d\Phi}{ds}.$$

Irradiance depends not only on properties of the radiation but also on the orientation of the surface element dS. One can specify this orientation by indicating the direction of the normal to dS. Therefore, specifying a particular irradiance, one may use the term *irradiance along a direction*, understanding the direction of the normal to dS. In textbooks on physical optics, the term "irradiance" is often used only in a narrow sense, in the sense of "irradiance along the direction of energy flow" (this quantity is often called *intensity*). In this book, we use the term "irradiance" in the standard sense that is prescribed by the definition given at the beginning of this paragraph. The mentioned irradiance along the direction of energy flow is called, throughout this book, *fitted-to-energy-flow-direction* (*FEFD*) *irradiance* or sometimes, in what follows, simply *intensity*. The applicability of FEFD irradiance is limited to simple wave fields that have a unique direction of energy flow in each point. In this book, we often deal with the net irradiance produced by a set of waves carrying energy in different directions. Considering light propagation in layered media, we will deal mainly with irradiances along the stratification direction. Such irradiances (irradiances along the stratification direction) will be denoted simply by symbol E (we hope that the reader will not confuse irradiance with electric field strength which is denoted by similar symbols). Stokes vectors defined in terms of such irradiances will be labeled by subscript (E) (see, e.g., Sections 5.3, 7.1, and 10.1).

Conventions

In Chapters 1–3 and in what follows:

- For time periodic fields, in complex representation, the time-dependent factor $\exp(-i\omega t)$ is assumed.
- The polarization of light is called right-handed (left-handed) if to an observer looking against the direction of propagation of the light, the end-point of the electric vector would appear to describe the ellipse in the clockwise (anticlockwise) sense (this convention is adopted, e.g., in [5–7]).
- The term *wavelength* and the symbol λ are used only for wavelengths in free space, that is, wavelengths defined as $\lambda = c/v$, where c is the velocity of light in vacuum and v is the frequency.

5.3 Stokes Vectors of Plane Waves and Collimated Beams Propagating in Isotropic Nonabsorbing Media

In this book, we use several different characteristics called "Stokes vector." Really, in modern optics, Stokes vector is treated rather as a form of representation of properties of wave fields than as a particular characteristic. In different cases, irradiances, radiances, radiant fluxes, and so on are represented by Stokes vectors. There are a few defining forms of Stokes vectors. In this section, we will deal with two of them, most important for us.

Stokes Vectors of a Plane Wave

Consider an undamped monochromatic or quasimonochromatic plane wave $\mathbf{X}$ with wave normal $\mathbf{l}$ propagating in an isotropic nonabsorbing homogeneous medium with refractive index n. Let $\tilde{\mathbf{x}}$ be the unit vector directed along a reference axis *PRA* (*Polarization Reference Axis*) perpendicular to $\mathbf{l}$. The wave $\mathbf{X}$ allows the following decompositions:

$$\mathbf{X} = \mathbf{X}_{\|} + \mathbf{X}_{\perp} = \mathbf{X}_{+\pi/4} + \mathbf{X}_{-\pi/4} = \mathbf{X}_{\mathrm{r}} + \mathbf{X}_{\mathrm{l}}, \tag{5.76}$$

where $\mathbf{X}_{\|}$, $\mathbf{X}_{\perp}$, $\mathbf{X}_{+\pi/4}$, $\mathbf{X}_{-\pi/4}$, $\mathbf{X}_{\mathrm{r}}$, and $\mathbf{X}_{\mathrm{l}}$ are fully polarized components of $\mathbf{X}$:

$\mathbf{X}_{\|}$ is linearly polarized along $\tilde{\mathbf{x}}$,

$\mathbf{X}_{\perp}$ is linearly polarized along $\tilde{\mathbf{y}}$, where $\tilde{\mathbf{y}} = \mathbf{l} \times \tilde{\mathbf{x}}$,

$\mathbf{X}_{+\pi/4}$ is linearly polarized along $\tilde{\mathbf{x}} + \tilde{\mathbf{y}}$,

$\mathbf{X}_{-\pi/4}$ is linearly polarized along $\tilde{\mathbf{x}} - \tilde{\mathbf{y}}$,

$\mathbf{X}_{\mathrm{r}}$ has the right circular polarization, and

$\mathbf{X}_{\mathrm{l}}$ has the left circular polarization.

The usual, *intensity-based*, Stokes vector of the wave is defined as

$$S_{(I)}\{\mathbf{X}\} = \begin{pmatrix} I\{\mathbf{X}\} \\ I\{\mathbf{X}_{\|}\} - I\{\mathbf{X}_{\perp}\} \\ I\{\mathbf{X}_{+\pi/4}\} - I\{\mathbf{X}_{-\pi/4}\} \\ I\{\mathbf{X}_{\mathrm{r}}\} - I\{\mathbf{X}_{\mathrm{l}}\} \end{pmatrix}, \tag{5.77}$$

where $I\{\ldots\}$ is the intensity (FEFD irradiance) (see Section 5.2) of the wave field indicated in the curly brackets. In subsequent chapters, when considering layered media, we deal mainly with Stokes vectors defined in terms of irradiances along the stratification direction ($\mathbf{z}$) of the layered medium under consideration (see Section 5.2). The Stokes vector of this kind for the wave $\mathbf{X}$ is defined as follows:

$$S_{(E)}\{\mathbf{X}\} = \begin{pmatrix} E\{\mathbf{X}\} \\ E\{\mathbf{X}_{\|}\} - E\{\mathbf{X}_{\perp}\} \\ E\{\mathbf{X}_{+\pi/4}\} - E\{\mathbf{X}_{-\pi/4}\} \\ E\{\mathbf{X}_{\mathrm{r}}\} - E\{\mathbf{X}_{\mathrm{l}}\} \end{pmatrix}, \tag{5.78}$$

where E is irradiance along $\mathbf{z}$. Such Stokes vectors will be called *irradiance-based*. The vectors $S_{(E)}\{\mathbf{X}\}$ and $S_{(I)}\{\mathbf{X}\}$ are related by

$$S_{(E)}\{\mathbf{X}\} = |\mathbf{lz}| S_{(I)}\{\mathbf{X}\}. \tag{5.79}$$

The vectors $S_{(E)}\{\mathbf{X}\}$ and $S_{(I)}\{\mathbf{X}\}$ may be expressed in terms of the electric field of the wave:

$$S_{(I)}\{\mathbf{X}\} = \frac{cn}{8\pi}\begin{pmatrix} \langle \mathcal{E}_{\tilde{x}}(\mathbf{r},t)\mathcal{E}_{\tilde{x}}(\mathbf{r},t)^*\rangle + \left\langle \mathcal{E}_{\tilde{y}}(\mathbf{r},t)\mathcal{E}_{\tilde{y}}(\mathbf{r},t)^*\right\rangle \\ \langle \mathcal{E}_{\tilde{x}}(\mathbf{r},t)\mathcal{E}_{\tilde{x}}(\mathbf{r},t)^*\rangle - \left\langle \mathcal{E}_{\tilde{y}}(\mathbf{r},t)\mathcal{E}_{\tilde{y}}(\mathbf{r},t)^*\right\rangle \\ 2\,\mathrm{Re}\left\langle \mathcal{E}_{\tilde{x}}(\mathbf{r},t)\mathcal{E}_{\tilde{y}}(\mathbf{r},t)^*\right\rangle \\ 2\,\mathrm{Im}\left\langle \mathcal{E}_{\tilde{x}}(\mathbf{r},t)\mathcal{E}_{\tilde{y}}(\mathbf{r},t)^*\right\rangle \end{pmatrix}, \tag{5.80}$$

$$S_{(E)}\{\mathbf{X}\} = \frac{cn\,|\mathbf{lz}|}{8\pi}\begin{pmatrix} \langle \mathcal{E}_{\tilde{x}}(\mathbf{r},t)\mathcal{E}_{\tilde{x}}(\mathbf{r},t)^*\rangle + \left\langle \mathcal{E}_{\tilde{y}}(\mathbf{r},t)\mathcal{E}_{\tilde{y}}(\mathbf{r},t)^*\right\rangle \\ \langle \mathcal{E}_{\tilde{x}}(\mathbf{r},t)\mathcal{E}_{\tilde{x}}(\mathbf{r},t)^*\rangle - \left\langle \mathcal{E}_{\tilde{y}}(\mathbf{r},t)\mathcal{E}_{\tilde{y}}(\mathbf{r},t)^*\right\rangle \\ 2\,\mathrm{Re}\left\langle \mathcal{E}_{\tilde{x}}(\mathbf{r},t)\mathcal{E}_{\tilde{y}}(\mathbf{r},t)^*\right\rangle \\ 2\,\mathrm{Im}\left\langle \mathcal{E}_{\tilde{x}}(\mathbf{r},t)\mathcal{E}_{\tilde{y}}(\mathbf{r},t)^*\right\rangle \end{pmatrix}, \tag{5.81}$$

$$\mathcal{E}_{\tilde{x}}(\mathbf{r},t) \equiv \tilde{\mathbf{x}}\mathbf{E}(\mathbf{r},t), \quad \mathcal{E}_{\tilde{y}}(\mathbf{r},t) \equiv \tilde{\mathbf{y}}\mathbf{E}(\mathbf{r},t),$$

where $\mathbf{E}(\mathbf{r},t)$ is the electric field strength vector of the wave $\mathbf{X}$ in complex representation and the brackets $\langle\,\rangle$ denote time averaging. In the case under consideration (an undamped plane wave), $S_{(I)}\{\mathbf{X}\}$ and $S_{(E)}\{\mathbf{X}\}$ are independent of $\mathbf{r}$.

The Stokes vectors $S_{(I)}\{\mathbf{X}\}$ and $S_{(E)}\{\mathbf{X}\}$ as defined by (5.77) and (5.78) correspond to the following defining form of Stokes vectors:

$$\begin{pmatrix} X\{\mathbf{X}\} \\ X\{\mathbf{X}_{\parallel}\} - X\{\mathbf{X}_{\perp}\} \\ X\{\mathbf{X}_{+\pi/4}\} - X\{\mathbf{X}_{-\pi/4}\} \\ X\{\mathbf{X}_{\mathrm{r}}\} - X\{\mathbf{X}_{\mathrm{l}}\} \end{pmatrix}, \tag{5.82}$$

where X is a scalar characteristic. The template (5.82) is usually used to introduce Stokes-vector-form analogs of various radiometric characteristics (irradiance, radiant flux, radiance, etc.).

Expressions (5.80) and (5.81) show another defining form of Stokes vectors, namely,

$$k\begin{pmatrix} \langle a_1(t)a_1(t)^*\rangle + \langle a_2(t)a_2(t)^*\rangle \\ \langle a_1(t)a_1(t)^*\rangle - \langle a_2(t)a_2(t)^*\rangle \\ 2\,\mathrm{Re}\,\langle a_1(t)a_2(t)^*\rangle \\ 2\,\mathrm{Im}\,\langle a_1(t)a_2(t)^*\rangle \end{pmatrix}, \tag{5.83}$$

where $a_1(t)$ and $a_2(t)$ are scalar complex functions and k is a real constant. This form is widely used in statistical optics. We use it in Section 10.1 to introduce *eigenwave (EW) Stokes vectors.*

Flux-Based Stokes Vector of a Beam

The form (5.82) originates from historically the first, phenomenological, definition of the Stokes parameters for a beam [8]. The Stokes vectors of beams corresponding to that definition of the Stokes parameters will be called *flux-based Stokes vectors*. Let $\mathbf{X}$ be a well-collimated (having a very narrow angular spectrum) monochromatic or quasimonochromatic beam with nominal propagation direction $\mathbf{l}$ propagating in an isotropic nonabsorbing medium. On the assumption that the decompositions (5.76) of the beam **X** are possible, the flux-based Stokes vector of this beam is defined as

$$S_{(\Phi)}\{\mathbf{X}\} = \begin{pmatrix} \Phi\{\mathbf{X}\} \\ \Phi\{\mathbf{X}_{\parallel}\} - \Phi\{\mathbf{X}_{\perp}\} \\ \Phi\{\mathbf{X}_{+\pi/4}\} - \Phi\{\mathbf{X}_{-\pi/4}\} \\ \Phi\{\mathbf{X}_{\mathrm{r}}\} - \Phi\{\mathbf{X}_{\mathrm{l}}\} \end{pmatrix}, \tag{5.84}$$

where Φ is radiant flux (power) across a plane perpendicular to **l** or across any tilted geometrical plane crossed by **X** (the incidence of **X** on which is not glancing).

5.4 Jones Vectors

In this book, we consider several variants of the Jones matrix method and different variants of their application, which use different types of Jones vectors. In this section, we discuss some important general aspects of application of Jones vectors for describing wave fields and introduce some terminology. Discussing here Jones vectors of wave fields propagating in anisotropic media, we restrict ourselves to considering fields induced in an anisotropic layer by a normally incident wave. The application of Jones vectors in the case of oblique incidence is considered in Chapters 8, 11, and 12. It should be noted that in this section we deal only with Jones vectors that are precisely defined from the standpoint of electromagnetic theory and examine them in the context of this theory. We have discussed the application of Jones vectors for describing fields in anisotropic media within the framework of the classical Jones calculus in Section 1.4.5 and have noted its drawbacks. We will not return to that discussion here, because the current discussion is oriented to methods that are more consistent with electromagnetic theory than the classical Jones calculus.

5.4.1 Fitted-to-Electric-Field Jones Vectors and Fitted-to-Transverse-Component-of-Electric-Field Jones Vectors

Isotropic medium. A homogeneous (see Section 8.1.2) plane monochromatic wave propagating in an isotropic nonabsorbing medium is strictly electrically transverse, that is, the electric field strength vector **E** of the wave is strictly perpendicular to its wave normal, and therefore the electric field of the wave can be represented as

$$\mathbf{E}(\mathbf{r},t) = \left(\widehat{\mathbf{x}}A_{0\widehat{x}} + \widehat{\mathbf{y}}A_{0\widehat{y}}\right)\mathrm{e}^{i(k_0 n\mathbf{l}\mathbf{r}-\omega t)} = \left(\widehat{\mathbf{x}}A_{\widehat{x}}(\mathbf{r}) + \widehat{\mathbf{y}}A_{\widehat{y}}(\mathbf{r})\right)\mathrm{e}^{-i\omega t}, \tag{5.85}$$

where **l** is the wave normal, $\widehat{\mathbf{x}}$ and $\widehat{\mathbf{y}}$ are mutually orthogonal unit vectors both perpendicular to **l**, **r** is a position vector, and n is the refractive index of the medium. The column vector composed of the scalar complex amplitudes $A_{\widehat{x}}$ and $A_{\widehat{y}}$

$$\mathbf{J}_{\widehat{x}\widehat{y}} = \begin{pmatrix} A_{\widehat{x}} \\ A_{\widehat{y}} \end{pmatrix} \tag{5.86}$$

is one of the common variants of Jones vector (see Section 1.1).

Uniaxial medium. Suppose that the wave (5.85) is incident normally on a layer of a uniaxial nonabsorbing medium occupying the space between the planes $z = z_1$ and $z = z_2$ in a Cartesian coordinate system (x, y, z). To be specific, we assume that the z-axis is codirectional with the wave normal **l** and $z_1 < z_2$. If the optic axis of the layer—we specify the direction of this axis by a unit vector **c**—is not parallel to $\mathbf{l}$, the wave field propagating in the positive z direction inside the layer will in general be a superposition of two plane waves having identical wave normals but different propagation speeds, namely, an extraordinary wave with electric field

$$\mathbf{E}_e(\mathbf{r}, t) = \widehat{\mathbf{e}}_e A_{0e} e^{i(k_0 n_e \mathbf{l}\mathbf{r} - \omega t)} = \widehat{\mathbf{e}}_e A_e(\mathbf{r}) e^{-i\omega t} \tag{5.87}$$

and an ordinary wave with electric field

$$\mathbf{E}_o(\mathbf{r}, t) = \widehat{\mathbf{e}}_o A_{0o} e^{i(k_0 n_o \mathbf{l}\mathbf{r} - \omega t)} = \widehat{\mathbf{e}}_o A_o(\mathbf{r}) e^{-i\omega t}, \tag{5.88}$$

where $\widehat{\mathbf{e}}_e$ and $\widehat{\mathbf{e}}_o$ are unit vectors specifying the vibration directions of the electric fields of these waves, and n_e and n_o are the corresponding refractive indices. The vector $\widehat{\mathbf{e}}_o$, whatever the orientation of the optic axis $\mathbf{c}$, is perpendicular to $\mathbf{l}$ and $\mathbf{c}$. The vector $\widehat{\mathbf{e}}_e$ is coplanar with $\mathbf{l}$ and $\mathbf{c}$ and, except for the case $\mathbf{l} \perp \mathbf{c}$, is not perpendicular to $\mathbf{l}$, that is, the extraordinary wave is in general not strictly electrically transverse (see Section 1.3.2). Here we note two variants of Jones vectors, which can be used to describe this pair of waves. One is

$$\mathbf{J}_{eo} = \begin{pmatrix} A_e \\ A_o \end{pmatrix}. \tag{5.89}$$

The other is

$$\mathbf{J}_{eo\text{-}t} = \begin{pmatrix} A_{et} \\ A_o \end{pmatrix}, \tag{5.90}$$

where

$$A_{et} = A_e \cos \gamma_{et},$$

with γ_{et} being the angle between $\widehat{\mathbf{e}}_e$ and the x–y plane. The amplitude A_e directly characterizes the magnitude of the field $\mathbf{E}_e$ ($|A_e|$ is equal to the length $|\mathbf{E}_e|$ of the vector $\mathbf{E}_e$; $|\mathbf{E}_e| \equiv (\mathbf{E}_e^* \cdot \mathbf{E}_e)^{1/2}$, where "$\cdot$" denotes the dot multiplication), while the amplitude A_{et} directly characterizes the magnitude of the transverse component of this field $\mathbf{E}_{et} = \mathbf{E}_e - \mathbf{l}(\mathbf{l} \cdot \mathbf{E}_e)$ ($|A_{et}| = |\mathbf{E}_{et}| \equiv (\mathbf{E}_{et}^* \cdot \mathbf{E}_{et})^{1/2}$). Since the ordinary wave is strictly transverse, the amplitude A_o directly characterizes both the magnitude of the field $\mathbf{E}_o$ and that of its transverse component. In what follows, the Jones vectors whose components directly characterize the magnitudes of electric fields will be referred to as *fitted-to-electric-field (FEF) Jones vectors*. The Jones vectors whose components directly characterize the magnitudes of the transverse components of electric fields will be called *fitted-to-transverse-component-of-electric-field (FTCEF) Jones vectors*. The column $\mathbf{J}_{eo}$ [see (5.89)] is an example of FEF Jones vector. The column $\mathbf{J}_{eo\text{-}t}$ [see (5.90)] is an FTCEF Jones vector. Since the wave in the isotropic medium [see (5.85)] in the above example is strictly transverse, the Jones vector $\mathbf{J}_{\widehat{x}\widehat{y}}$ [see (5.86)] is both FEF and FTCEF.

5.4.2 Fitted-to-Irradiance Jones Vectors

One of the important aspects of light propagation is energy transfer. In considering the interaction of a plane wave and a stratified medium, the basic quantity characterizing the amount of energy carried by a wave field is the irradiance along the stratification direction, E (see Sections 5.2 and 5.3). This irradiance may be expressed in terms of the time-averaged Poynting vector of the field (see Section 8.1)

$$\langle \mathbf{S} \rangle = \frac{c}{8\pi} \mathrm{Re}\,(\mathbf{E} \times \mathbf{H}^*), \tag{5.91}$$

where **E** and **H** are respectively the electric and magnetic field strength vectors (the field is assumed to be time-harmonic), and c is the velocity of light in free space, as follows:

$$E = |\mathbf{z}\,\langle \mathbf{S}\rangle|\,,$$

where **z** is the unit vector along the stratification direction.

It is usual to employ for characterization of plane waves propagating in isotropic media the irradiance along the wave normal, which is expressed as

$$E_{\mathbf{l}} = |\mathbf{l}\,\langle \mathbf{S}\rangle|\,,$$

where **l** is the wave normal. Irradiance of this kind may also be used to characterize a pair of *equinormal* (having identical wave normals) natural waves propagating in an anisotropic layer, such as (5.87) and (5.88). In the case of normal incidence, **l**||**z** and consequently

$$E_{\mathbf{l}} = E.$$

The magnetic field of a homogeneous plane wave propagating in an isotropic nonabsorbing medium with refractive index n can be expressed as

$$\mathbf{H} = n(\mathbf{l} \times \mathbf{E}), \tag{5.92}$$

where, as before, **l** is the wave normal. Substitution of (5.92) into (5.91) gives

$$\langle \mathbf{S}\rangle = \frac{cn}{8\pi}\mathrm{Re}\,(\mathbf{E} \times (\mathbf{l} \times \mathbf{E}^*)) = \frac{cn}{8\pi}\mathrm{Re}\,(\mathbf{l}\,(\mathbf{E}\cdot\mathbf{E}^*) - \mathbf{E}^*\,(\mathbf{l}\cdot\mathbf{E}))\,. \tag{5.93}$$

Since the wave is strictly transverse, $\mathbf{l} \cdot \mathbf{E} = 0$ and consequently

$$\langle \mathbf{S}\rangle = \frac{c}{8\pi} n\mathbf{l}\,(\mathbf{E}\cdot\mathbf{E}^*)\,. \tag{5.94}$$

For the wave (5.85), $\langle \mathbf{S}\rangle$ can be expressed in terms of $A_{\bar{x}}$ and $A_{\bar{y}}$:

$$\langle \mathbf{S}\rangle = \frac{cn}{8\pi}\mathbf{l}\left(A_{\bar{x}}A^*_{\bar{x}} + A_{\bar{y}}A^*_{\bar{y}}\right). \tag{5.95}$$

The quantity $A_{\bar{x}}A^*_{\bar{x}} + A_{\bar{y}}A^*_{\bar{y}}$ may be written as

$$A_{\bar{x}}A^*_{\bar{x}} + A_{\bar{y}}A^*_{\bar{y}} = \mathbf{J}^\dagger_{\bar{x}\bar{y}}\mathbf{J}_{\bar{x}\bar{y}} \equiv \left|\mathbf{J}_{\bar{x}\bar{y}}\right|^2, \tag{5.96}$$

where $\left|\mathbf{J}_{\bar{x}\bar{y}}\right| \equiv \left\|\mathbf{J}_{\bar{x}\bar{y}}\right\|_{\mathrm{E}}$ is the length (the Euclidean norm) of the vector $\mathbf{J}_{\bar{x}\bar{y}}$ (see Section 5.1.4). Therefore, the time-averaged Poynting vector of the wave (5.85) and irradiance $E_{\mathbf{l}}$ can be expressed in terms of its FEF Jones vector as follows:

$$\langle \mathbf{S}\rangle = \frac{cn}{8\pi}\mathbf{l}\left(\mathbf{J}^\dagger_{\bar{x}\bar{y}}\mathbf{J}_{\bar{x}\bar{y}}\right), \tag{5.97}$$

$$E_{\mathbf{l}} = \frac{cn}{8\pi}\mathbf{J}^\dagger_{\bar{x}\bar{y}}\mathbf{J}_{\bar{x}\bar{y}}. \tag{5.98}$$

In the case of normal incidence, the irradiance E can also be expressed as

$$E = \frac{cn}{8\pi}\mathbf{J}^\dagger_{\bar{x}\bar{y}}\mathbf{J}_{\bar{x}\bar{y}}. \tag{5.99}$$

One may notice that the factor required to calculate irradiances from the FEF Jones vector depends on the refractive index of the medium where the wave propagates. Two waves with the same irradiance E_l propagating in media with different refractive indices will be characterized by FEF Jones vectors of different length [see (5.95) and (5.96)]. For example, a wave propagating in glass with $n = 1.5$ will have a $\sqrt{1.5}$ times smaller, in length, FEF Jones vector than a wave of the same irradiance E_l propagating in air ($n = 1$). Therefore, we cannot compare the irradiances of waves propagating in different media comparing their FEF Jones vectors only.

In the case of an anisotropic medium, the relationship between FEF Jones vectors and irradiances is still more complicated. For example, in the simplest situation when $\mathbf{c}\perp\mathbf{l}$, the irradiance E of the wave field consisting of the waves (5.87) and (5.88) ($\mathbf{l}\|\mathbf{z}$) is given by

$$E = \frac{c}{8\pi}\left(n_{\|}A_e A_e^* + n_{\perp}A_o A_o^*\right), \tag{5.100}$$

where $n_{\|}$ and $n_{\perp}$ are the principal refractive indices of the uniaxial medium. As is seen from (5.100), in this case, if $|A_e|$ is equal to $|A_o|$, the contributions of the ordinary wave and extraordinary wave to the irradiance E will be different. According to (5.100), the irradiance may be expressed in terms of the FEF Jones vector $\mathbf{J}_{eo}$ [see (5.89)] as follows:

$$E = \frac{c}{16\pi}\mathbf{J}_{eo}^{\dagger}\mathbf{n}_0\mathbf{J}_{eo}, \tag{5.101}$$

where

$$\mathbf{n}_0 = \begin{pmatrix} 2n_{\|} & 0 \\ 0 & 2n_{\perp} \end{pmatrix}. \tag{5.102}$$

The factor 2 is introduced in $\mathbf{n}_0$ to provide the consistency of these formulas with analogous formulas in next sections. Matrices specifying the relationship between amplitude characteristics and power characteristics of wave fields, such as $\mathbf{n}_0$, will be called *metric matrices*. Relation (5.99) may be rewritten using the metric matrix

$$\mathbf{n}_0 = \begin{pmatrix} 2n & 0 \\ 0 & 2n \end{pmatrix}$$

as

$$E = \frac{c}{16\pi}\mathbf{J}_{\bar{x}\bar{y}}^{\dagger}\mathbf{n}_0\mathbf{J}_{\bar{x}\bar{y}}. \tag{5.103}$$

In the above cases and many other cases, it is possible to define Jones vectors so that the metric matrix is equal to the unit matrix. For instance, in the above cases we may define the Jones vector as follows:

$$\mathbf{J} = \begin{pmatrix} A_1 \\ A_2 \end{pmatrix}, \tag{5.104}$$

where

$$A_1 = \sqrt{2n}A_{\bar{x}}, \quad A_2 = \sqrt{2n}A_{\bar{y}} \tag{5.105}$$

for the field in the isotropic medium, and

$$A_1 = \sqrt{2n_{||}}A_e, \quad A_2 = \sqrt{2n_{\perp}}A_o \tag{5.106}$$

for the field in the uniaxial medium. In both cases, the irradiance E can be expressed as

$$E = \frac{c}{16\pi}\mathbf{J}^{\dagger}\mathbf{J}. \tag{5.107}$$

Jones vectors defined so that relation (5.107) is applicable to them will be referred to as *fitted-to-irradiance* (*FI*) Jones vectors. An FI Jones vector contains all the information required to estimate the corresponding irradiance. Wave fields of equal irradiance are characterized by FI Jones vectors of equal length. If a wave field normally incident on a layered system and the wave field transmitted by the system are represented by their FI Jones vectors, $\mathbf{J}_{inc}$ and $\mathbf{J}_{tr}$ respectively, the transmittance t of this system for this incident field may be calculated by the formula

$$t(\mathbf{J}_{inc}) = \frac{\mathbf{J}_{tr}^{\dagger}\mathbf{J}_{tr}}{\mathbf{J}_{inc}^{\dagger}\mathbf{J}_{inc}} \tag{5.108}$$

even when the media in which the incident and transmitted fields travel are different. Another profit of the use of FI Jones vectors is a simple mathematical form of corresponding Jones matrices when they characterize lossless operations or operations without diattenuation (just as in the classical Jones calculus, see Section 1.4). In the former case, these matrices are unitary. In the latter case, they are STU matrices (see Sections 5.1.3). In general this is not the case for Jones matrices operating with FEF Jones vectors. The fact that the operations performed by certain elements of the optical system being considered are described by unitary or STU Jones matrices often greatly simplifies analysis and calculations. The reader can find many examples of this in previous and next chapters.

5.4.3 Conventional Jones Vectors

It is often convenient to use, as state characteristics, Jones vectors (J) related to corresponding true Jones vectors ($\mathbf{J}$) as follows:

$$\mathbf{J} = t_J J, \tag{5.109}$$

where t_J is a scalar quantity which is usually complex-valued and is a function of position. The use of such vectors significantly simplifies solving many problems and is customary for the classical Jones matrix method (see Chapter 1). We will call such vectors *conventional Jones vectors*. Let us consider some types of conventional Jones vectors.

Polarization Jones Vectors

Conventional Jones vectors j of unit length, i.e., such that

$$j^{\dagger}j = 1, \tag{5.110}$$

are usually called *polarization Jones vectors*. This name fully corresponds to the meaning of such vectors when they are used for description of waves propagating in isotropic media, because in this case such a vector really characterizes the state of polarization of a wave (see Section 1.1.2). In any case, we will use the following general definition: the *polarization Jones vector* of a wave field is a unit vector collinear to the true FI Jones vector of the field. This definition removes an ambiguity appearing when wave fields in anisotropic media are considered. In this book, the polarization Jones vectors of plane waves propagating in isotropic homogeneous media are always considered as spatially invariant.

True-Phase and Prescribed-Phase Jones Vectors

The Jones vectors (5.86), (5.89), and (5.90) contain information about the magnitude and relative phase of the field components as well as about the absolute phase of the corresponding fields. Such vectors may be called *true-phase Jones vectors*. In many cases, information about the absolute phase of wave fields is unnecessary. This permits one to perform calculations without tracing the absolute phase and to operate with Jones vectors having an arbitrary, no matter what or some convenient, phase. Such Jones vectors will be called *prescribed-phase*.

Spatially Invariant Jones Vectors

The Jones vectors (5.86), (5.89), and (5.90) are local characteristics of the wave fields, being functions of position. When a homogeneous plane wave propagates in an isotropic nonabsorbing medium, its true-phase Jones vectors at different points differ from each other only in absolute phase, and it is possible to define a conventional, prescribed-phase, Jones vector of this wave so that this vector is spatially invariant. The use of such Jones vectors allows one to avoid the necessity of tracing the spatial evolution of Jones vectors inside isotropic nonabsorbing media. The polarization Jones vector of a plane wave propagating in a homogeneous isotropic medium can be defined as spatially invariant even if the medium is absorbing, thanks to normalization (5.110).

References

[1] R. A. Horn and C. R. Johnson, *Matrix Analysis* (Cambridge University Press, Cambridge, 1986).
[2] G. A. Korn and T. M. Korn, *Mathematical Handbook for Scientists and Engineers*, 2nd ed. (Dover, New York, 2000).
[3] H. Lütkepohl, *Handbook of Matrices* (Wiley, Chichester, 1996).
[4] W. R. McCluney, *Introduction to Radiometry and Photometry* (Artech House, Boston, 1994).
[5] M. Born and E. Wolf, *Principles of Optics*, 7th ed. (Pergamon Press, New York, 1999).
[6] R. M. A. Azzam and N. M. Bashara, *Ellipsometry and Polarized Light* (North-Holland, Amsterdam, 1977).
[7] E. Hecht, *Optics*, 4th ed. (Addison Wesley, San Francisco, 2002).
[8] W. A. Shurcliff, *Polarized Light: Production and Use* (Harvard University Press, Cambridge, 1962).

6

Simple Models and Representations for Solving Optimization and Inverse Optical Problems. Real Optics of LC Cells and Useful Approximations

In many kinds of LCDs (TN LCDs, STN LCDs, etc.), the liquid crystal layer in its working state has a distorted chiral structure and, because of this, exhibits rather intricate polarization-optical properties. Optimization of the optical performance of such LCDs involves finding optimal parameters of the polarizers and elements of the compensation system (compensation films) and is a complicated multiparametric problem. The main aim of this chapter is to present a set of concepts, models, and representations helpful in solving such optimization problems. Some of the representations considered here are also useful in solving inverse problems for LC layers with twisted structure (experimental determination of the twist angle, surface orientation of the LC director, thickness of LC layers, etc.). In this chapter, we consider only the case of normal incidence of light. Possible applications of some concepts presented here in the case of oblique incidence are discussed in succeeding chapters.

In Chapters 1 and 2, some simple optical models for LCDs were considered. In this chapter, we will arrive at the same and allied models and approximations underlying these models starting from real optical properties of LC cells. The discussion of the real optics of LC cells is illustrated by experimental examples. These experimental examples are interesting not only in the context of this chapter. They allow one to estimate the effects of some secondary factors that are taken into account or neglected in different optical models of LCDs.

Modeling and Optimization of LCD Optical Performance, First Edition.
Dmitry A. Yakovlev, Vladimir G. Chigrinov and Hoi-Sing Kwok.

Website Companion: www.wiley.com/go/yakovlev/modelinglcd

6.1 Polarization Transfer Factor of an Optical System

Let us consider a layered optical system M described by a Mueller matrix $M = [m_{ij}]$ that relates the Stokes vector $\mathbf{S}_\mathrm{I}$ of a quasimonochromatic light beam normally incident on this system and the Stokes vector $\mathbf{S}_\mathrm{O}$ of a light beam emerging from the system:

$$\mathbf{S}_\mathrm{O} = M\mathbf{S}_\mathrm{I}. \tag{6.1}$$

The media where the incident and output beams propagate are assumed to be isotropic and nonabsorbing. The Stokes vectors $\mathbf{S}_\mathrm{I}$ and $\mathbf{S}_\mathrm{O}$ are assumed to be flux-based ones (see Section 5.3). Let the polarization reference axis (Section 5.3) for the vector $\mathbf{S}_\mathrm{I}$ be codirectional with the axis x_I of a right-handed coordinate system (x_I, y_I, z_I) with the axis z_I directed along the nominal propagation direction of the incident beam and that for the vector $\mathbf{S}_\mathrm{O}$ be codirectional with the axis x_O of a right-handed coordinate system $(x_\mathrm{O}, y_\mathrm{O}, z_\mathrm{O})$ whose axis z_O is directed along the nominal propagation direction of the output beam. Suppose that the incident beam is linearly polarized with polarization direction at angle ϑ from the axis x_I (the positive direction for ϑ is taken to be from the positive x_I-axis toward the positive y_I-axis; we will use this rule in all similar cases in what follows). The vector $\mathbf{S}_\mathrm{I}$ is then can be represented as

$$\mathbf{S}_\mathrm{I} = \Phi_\mathrm{I}\begin{pmatrix} 1 \\ \cos 2\vartheta \\ \sin 2\vartheta \\ 0 \end{pmatrix}, \tag{6.2}$$

where Φ_I is the radiant flux carried by the incident beam along z_I. Let us assume that the output light is passed through an ideal linear polarizer (analyzer). We denote the angle between the axis x_O and the transmission axis of the analyzer by ϑ'. According to the Stokes vector–Mueller matrix formalism [1], the radiant flux Φ_OA (along z_O) transmitted by the analyzer can be expressed as follows:

$$\Phi_\mathrm{OA} = \frac{1}{2}(1 \quad \cos 2\vartheta' \quad \sin 2\vartheta' \quad 0)\mathbf{S}_\mathrm{O}. \tag{6.3}$$

The quantity

$$t_\mathrm{PA}(\vartheta,\ \vartheta') = \Phi_\mathrm{OA}/\Phi_\mathrm{I} \tag{6.4}$$

will be called *the polarization transfer factor of the system*. From (6.1)–(6.4), one can derive the following expression for the polarization transfer factor in terms of elements of the matrix M and the angles ϑ and ϑ':

$$\begin{aligned} t_\mathrm{PA}(\vartheta, \vartheta') = {} & B_0 + B_1 \cos\eta^- + B_2 \cos\eta^+ + B_3 \sin\eta^- + B_4 \sin\eta^+ \\ & + B_5 \cos 2\vartheta + B_6 \sin 2\vartheta + B_7 \cos 2\vartheta' + B_8 \sin 2\vartheta', \end{aligned} \tag{6.5}$$

$$\eta^- = 2(\vartheta - \vartheta'), \quad \eta^+ = 2(\vartheta + \vartheta'), \tag{6.6}$$

where

$$\begin{aligned} B_0 &= m_{11}/2, \\ B_1 &= (m_{22} + m_{33})/4, \\ B_2 &= (m_{22} - m_{33})/4, \\ B_3 &= (m_{23} - m_{32})/4, \\ B_4 &= (m_{23} + m_{32})/4, \\ B_5 &= m_{12}/2,\ B_6 = m_{13}/2, \\ B_7 &= m_{21}/2,\ B_8 = m_{31}/2. \end{aligned} \tag{6.7}$$

From (6.7), we see an unambiguous correspondence between the parameters B_j ($j = 0,1,\ldots, 8$) and nine elements m_{jk} ($j, k = 1,2,3$) of the Mueller matrix. Expression (6.5) is very general. It is applicable to both nondepolarizing and depolarizing optical systems, including systems with polarization-dependent losses. The coefficients B_j will be referred to as *polarization transport coefficients* of the system.

Assuming that the incident linearly polarized light is obtained by using an ideal polarizer, we will call the angle ϑ *the polarizer orientation angle*. The angle ϑ' will be called *the analyzer orientation angle*.

Polarization-Dependent Losses (Diattenuation)

Suppose that the light incident on the system M is unpolarized. Then the transmittance of the combined system consisting of the system M and the ideal polarizer (analyzer) can be expressed as follows:

$$t_{\text{UP-A}}(\vartheta') = \frac{1}{2}\left(t_{\text{PA}}(\upsilon, \vartheta') + t_{\text{PA}}(\upsilon + 90°, \vartheta')\right) \tag{6.8}$$

at any υ (υ is an arbitrary value of the variable ϑ). Substituting (6.5) into (6.8), we obtain

$$t_{\text{UP-A}}(\vartheta') = B_0 + B_7 \cos 2\vartheta' + B_8 \sin 2\vartheta'. \tag{6.9}$$

It is clear that $t_{\text{UP-A}}$ can vary with ϑ' only if the light transmitted by M is polarized, at least, partially, that is, when the system M is able to polarize the light, i.e., exhibits polarization-dependent losses (diattenuation). On the other hand, if the light incident on M is linearly polarized, the transmittance of M can be expressed as follows:

$$t_{\text{P}}(\vartheta) = t_{\text{PA}}(\vartheta, \upsilon') + t_{\text{PA}}(\vartheta, \upsilon' + 90°) \tag{6.10}$$

at any υ'. From (6.10) and (6.5), we find that

$$t_{\text{P}}(\vartheta) = 2(B_0 + B_5 \cos 2\vartheta + B_6 \sin 2\vartheta). \tag{6.11}$$

The dependence of this transmittance on ϑ also implies the presence of polarization-dependent losses. It is thus obvious that the coefficients B_5, B_6, B_7, and B_8 are nonzero only for systems exhibiting polarization-dependent losses and characterize their amount and effect. See also [2].

In the absence of diattenuation, the function $t_{\text{PA}}(\vartheta, \vartheta')$ in general has the following form:

$$t_{\text{PA}}(\vartheta, \vartheta') = B_0 + B_1 \cos 2(\vartheta - \vartheta') + B_2 \cos 2(\vartheta + \vartheta') + B_3 \sin 2(\vartheta - \vartheta') + B_4 \sin 2(\vartheta + \vartheta'). \tag{6.12}$$

Analyzing this expression, one can see that the minimum ($t_{\text{PA min}}$) and maximum ($t_{\text{PA max}}$) values of the function $t_{\text{PA}}(\vartheta, \vartheta')$ are:

$$t_{\text{PA min}} = B_0 - B_{13} - B_{24}, \quad t_{\text{PA max}} = B_0 + B_{13} + B_{24}, \tag{6.13}$$

where

$$B_{13} = \sqrt{B_1^2 + B_3^2}, \quad B_{24} = \sqrt{B_2^2 + B_4^2}. \tag{6.14}$$

Hence, the range of the function $t_{\text{PA}}(\vartheta, \vartheta')$ has a width equal to $2\left(\sqrt{B_1^2+B_3^2}+\sqrt{B_2^2+B_4^2}\right)$. Moreover, it can be noticed that, according to (6.12), the function $t_{\text{PA}}(\vartheta, \vartheta')$ in this case satisfies the following equation:

$$t_{\text{PA}}(\upsilon, \upsilon') = t_{\text{PA}}(\upsilon \pm 90°, \upsilon' \pm 90°), \qquad (6.15)$$

where υ and υ' are arbitrary values of ϑ and ϑ'. This is the well-known general property of optical systems without diattenuation, which has been mentioned in Section 1.4.1.

Coefficients B_j and Jones Matrix

Let us assume that the matrix M is a Mueller–Jones matrix, that is, a matrix representable in the form

$$M = \mathbf{L}(\mathbf{t} \otimes \mathbf{t}^*)\mathbf{L}^{-1}, \qquad (6.16)$$

where $\mathbf{t}$ is a Jones matrix, $\mathbf{L}$ is the 4×4 matrix given by (10.6) (Section 10.1), $\otimes$ denotes the Kronecker matrix multiplication (see Section 5.1.5). Conditions under which this representation is adequate are discussed in detail in subsequent chapters. Here we only note that this representation is often used for Mueller matrices of LC cells when the analysis is performed with the use of the classical Jones calculus (JC) [3] (Section 1.4). It is clear that the Jones matrix in (6.16) is assumed to have the input and output bases identical to those of the matrix M.

According to (6.7) and (6.16), the coefficients B_j can be expressed in terms of the elements of the matrix $\mathbf{t} = [t_{jk}]$ as follows:

$$\begin{aligned}
B_0 &= \frac{1}{4}\left(|t_{11}|^2+|t_{12}|^2+|t_{21}|^2+|t_{22}|^2\right),\\
B_1 &= \frac{1}{8}\left(|t_{11}+t_{22}|^2-|t_{12}-t_{21}|^2\right),\\
B_2 &= \frac{1}{8}\left(|t_{11}-t_{22}|^2-|t_{21}+t_{12}|^2\right),\\
B_3 &= \frac{1}{4}\text{Re}\left(\left(t_{12}-t_{21}\right)\left(t_{11}^*+t_{22}^*\right)\right),\\
B_4 &= \frac{1}{4}\text{Re}\left(\left(t_{12}+t_{21}\right)\left(t_{11}^*-t_{22}^*\right)\right),\\
B_5 &= \frac{1}{4}\left(|t_{11}|^2-|t_{12}|^2+|t_{21}|^2-|t_{22}|^2\right), \quad B_6 = \frac{1}{2}\text{Re}\left(t_{11}^*t_{12}+t_{21}^*t_{22}\right),\\
B_7 &= \frac{1}{4}\left(|t_{11}|^2+|t_{12}|^2-|t_{21}|^2-|t_{22}|^2\right), \quad B_8 = \frac{1}{2}\text{Re}\left(t_{11}t_{21}^*+t_{12}t_{22}^*\right).
\end{aligned} \qquad (6.17)$$

This representation implies that the system does not manifest any depolarizing action, that is, the output light is always completely polarized when the incident light is completely polarized.

Unitary Approximation. Unitary Systems

Suppose that the matrix $\mathbf{t}$ in (6.16) can be represented as

$$\mathbf{t} = \varsigma \mathbf{t}_{\text{U}}, \qquad (6.18)$$

where $\mathbf{t}_{\mathrm{U}}$ is a unitary matrix, and ς is a scalar, possibly complex, coefficient. With this representation, the operation described by the matrices M and $\mathbf{t}$ is without depolarization and diattenuation. Possible polarization-independent losses in the system are taken into account via the coefficient ς. Rewrite expression (6.18) in the form

$$\mathbf{t} = \varsigma_{\mathrm{UM}} \mathbf{t}_{\mathrm{UM}}, \tag{6.19}$$

where

$$\varsigma_{\mathrm{UM}} = \varsigma \sqrt{\det \mathbf{t}_{\mathrm{U}}}, \quad \mathbf{t}_{\mathrm{UM}} = \frac{1}{\sqrt{\det \mathbf{t}_{\mathrm{U}}}} \mathbf{t}_{\mathrm{U}}. \tag{6.20}$$

The matrix $\mathbf{t}_{\mathrm{UM}}$ is unitary and has determinant 1, and, consequently [see (5.31)], can be written as

$$\mathbf{t}_{\mathrm{UM}} = \begin{pmatrix} a' + ia'' & b' + ib'' \\ -b' + ib'' & a' - ia'' \end{pmatrix}, \tag{6.21}$$

where a', a'', b', and b'' are real scalars satisfying the equation

$$a'^2 + a''^2 + b'^2 + b''^2 = 1. \tag{6.22}$$

Since $|\det \mathbf{t}_{\mathrm{U}}| = 1$ (see Section 5.3),

$$|\varsigma_{\mathrm{UM}}| = |\varsigma|. \tag{6.23}$$

By making use of (6.19)–(6.23), we can obtain the following expressions for the coefficients B_j for this case:

$$\begin{gathered} B_0 = \frac{K}{2}, \quad B_1 = \frac{K}{2}(a'^2 - b'^2), \quad B_2 = \frac{K}{2}(a''^2 - b''^2), \\ B_3 = Ka'b', \quad B_4 = Ka''b'', \\ B_5 = B_6 = B_7 = B_8 = 0, \end{gathered} \tag{6.24}$$

where $K = \varsigma\varsigma^*$. The coefficient K is equal to the ratio of the radiant flux of the beam emerging from the system M to that of the incident beam. As can be seen from (6.24) and (6.22), coefficients B_j in this case satisfy the following relation:

$$B_0 = \sqrt{B_1^2 + B_3^2} + \sqrt{B_2^2 + B_4^2}. \tag{6.25}$$

According to (6.23), (6.13), and (6.14), for the case under consideration,

$$t_{\mathrm{PA\,min}} = 0, \tag{6.26}$$

$$t_{\mathrm{PA\,max}} = 2B_0 = K. \tag{6.27}$$

The zero t_{PA} is reached when ϑ [at $B_{13} \neq 0$, see (6.14)] and ϑ' (at $B_{24} \neq 0$) are such that

$$\begin{aligned} \cos 2(\vartheta - \vartheta') &= -\frac{B_1}{B_{13}}, \quad \sin 2(\vartheta - \vartheta') = -\frac{B_3}{B_{13}}, \\ \cos 2(\vartheta + \vartheta') &= -\frac{B_2}{B_{24}}, \quad \sin 2(\vartheta + \vartheta') = -\frac{B_4}{B_{24}}. \end{aligned} \tag{6.28}$$

The maximum t_{PA} is reached at ϑ and ϑ' satisfying the relations

$$\cos 2(\vartheta - \vartheta') = \frac{B_1}{B_{13}}, \quad \sin 2(\vartheta - \vartheta') = \frac{B_3}{B_{13}},$$
$$\cos 2(\vartheta + \vartheta') = \frac{B_2}{B_{24}}, \quad \sin 2(\vartheta + \vartheta') = \frac{B_4}{B_{24}}. \tag{6.29}$$

The assumption that a Jones matrix can be represented in the form (6.18) will be called the *unitary approximation*. Recall that in Section 5.1.3 matrices representable in the form (6.18) have been called STU matrices. The unitary approximation naturally leads to the concepts of a *unitary system* (see Section 1.4), a system which changes only the state of polarization of the passing light. An imaginary optical system whose action is described by the unitary matrix $\mathbf{t}_U$ in (6.18), by definition, is a *unitary system*. Unitary Jones matrices, such as $\mathbf{t}_U$ and $\mathbf{t}_{UM}$, can be regarded as *polarization Jones matrices*. Recall that polarization Jones matrices were defined in Section 1.3.4 as operators that relate the polarization Jones vectors (see Sections 1.1.2 and 5.4.3) of the input and output light.

In the case of the unitary approximation, the factor t_{PA} can be expressed as follows:

$$t_{PA}(\vartheta, \vartheta') = K[B_{U0} + B_{U1}\cos 2(\vartheta - \vartheta') + B_{U2}\cos 2(\vartheta + \vartheta') + B_{U3}\sin 2(\vartheta - \vartheta') + B_{U4}\sin 2(\vartheta + \vartheta')], \tag{6.30}$$

where

$$B_{U0} = \frac{1}{2}, \quad B_{U1} = \frac{a'^2 - b'^2}{2}, \quad B_{U2} = \frac{a''^2 - b''^2}{2}, \quad B_{U3} = a'b', \quad B_{U4} = a''b''. \tag{6.31}$$

The coefficients B_{Uj} here meet the condition

$$\sqrt{B_{U1}^2 + B_{U3}^2} + \sqrt{B_{U2}^2 + B_{U4}^2} = 0.5 \tag{6.32}$$

and characterize the imaginary unitary system associated with the system in question. These coefficients will be called *unitary polarization transport coefficients*. Polarization transfer factors of the associated unitary system ($t_{PAU} = t_{PA}/K$) will be referred to as *unitary polarization transfer factors*.

Now we will give some experimental examples in which representation (6.5) is used to describe transmission properties of LC cells. These examples will be used by us as illustrations in discussions of different models, approximations, and methods. In passing, we will note some important properties of the coefficients B_j for transmissive systems.

6.2 Optics of LC Cells in Terms of Polarization Transport Coefficients

In the experiments the results of which are presented in this section, we used the standard measurement scheme shown in Figure 6.1. The light beam from a broadband light source passed through the polarizer and normally fell on the LC cell. The light transmitted by the LC cell passed through the second polarizer (analyzer), and, after that, its spectrum was registered by a spectrometer. The devices employed in the measurements are described in [4]. The reference systems (x_I, y_I, z_I) and (x_O, y_O, z_O) are taken to be

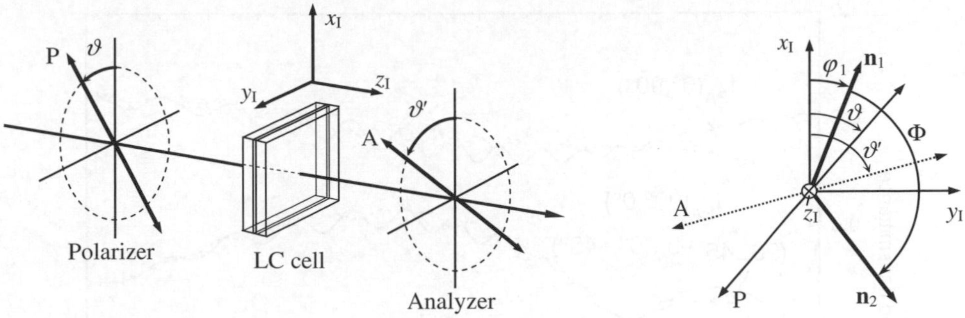

Figure 6.1 Geometry of the experiment. P and A are the transmission axes of the polarizer and analyzer, respectively

equivalent. The polarization transfer factors measured using this scheme will be called the *polarized transmittances*. The polarized transmittance associated with particular values of the angles ϑ and ϑ' (see Figure 6.1), say $\vartheta = \upsilon$ and $\vartheta' = \upsilon'$, was calculated as

$$t_{PA}(\upsilon, \upsilon') = I_{t\upsilon\upsilon'}/I_{i\upsilon}, \tag{6.33}$$

where $I_{t\upsilon\upsilon'}$ is the registered intensity of the light transmitted by the polarizer–LC cell–analyzer system at $\vartheta = \upsilon$ and $\vartheta' = \upsilon'$, and $I_{i\upsilon}$ is the registered intensity of the light transmitted by the polarizer–analyzer system at $\vartheta' = \vartheta = \upsilon$ for the same incident light. The spectral resolution of the spectrometer in the measurements was about 4 nm. Therefore the measured value of t_{PA} for a wavelength $\lambda = \tilde{\lambda}$ on the spectrometer scale can be regarded as an estimate of t_{PA} for a quasimonochromatic incident light with mean wavelength equal to $\tilde{\lambda}$ and spectral bandwidth $\Delta\lambda \approx 4$ nm (see Section 7.1).

Figure 6.2 presents experimental spectra of t_{PA} for a twisted nematic cell which will be called *cell E1*. This cell has the usual sandwich-type structure (as well as the other cells considered in this chapter) and is electrically controllable (see Figure 6.3). Plates of commercial indium tin oxide (ITO)-coated glass for LC displays were used as substrates. The glass thickness is 1.1 mm. The thickness of the ITO films is about 0.07 μm. The alignment films in this cell are photoaligned layers of sulfonic azo-dye SD-1 [5] with a thickness of the order of 10 nm. The cell was filled with nematic LC ZLI-5700-000 (Merck) (this LC material has $n_\perp = 1.4894$ and $n_\parallel = 1.6122$ for $\lambda = 589.3$ nm at 20°C). The parameters of the LC layer are the following: twist angle $\Phi \approx 87°$, pretilt angles $\approx 0°$, and thickness $d \approx 5.3$ μm. In the experiment the results of which are presented in Figure 6.2, the cell was oriented so that the LC director on the LC layer boundary nearest to the input polarizer was nearly perpendicular to the reference axis for the angles ϑ and ϑ'.

Figure 6.4a shows the spectral dependences of the coefficients B_j computed from the spectra of the polarized transmittances presented in Figure 6.2. In principle, the nine coefficients B_j can be found from nine spectra of the polarized transmittance measured for nine different settings of the polarizer and analyzer. However, in our experiments the error of measurement of the polarized transmittances was relatively large (~2%), in particular, because of fluctuations of the light source intensity. To decrease the effect of measurement errors caused by random factors, we calculated the coefficients B_j from 16 experimental polarized transmittances corresponding to $\vartheta = -45°, 0°, 45°, 90°$ and $\vartheta' = -45°, 0°, 45°, 90°$ by the least-squares method. The standard deviation between the measured values of the polarized transmittance and the t_{PA} values calculated by formula (6.5) with the experimental values of B_j in our experiments was of the order of 3×10^{-3}. This testifies to both a relatively high experimental accuracy and a high degree of correspondence of the experimental data to representation (6.5).

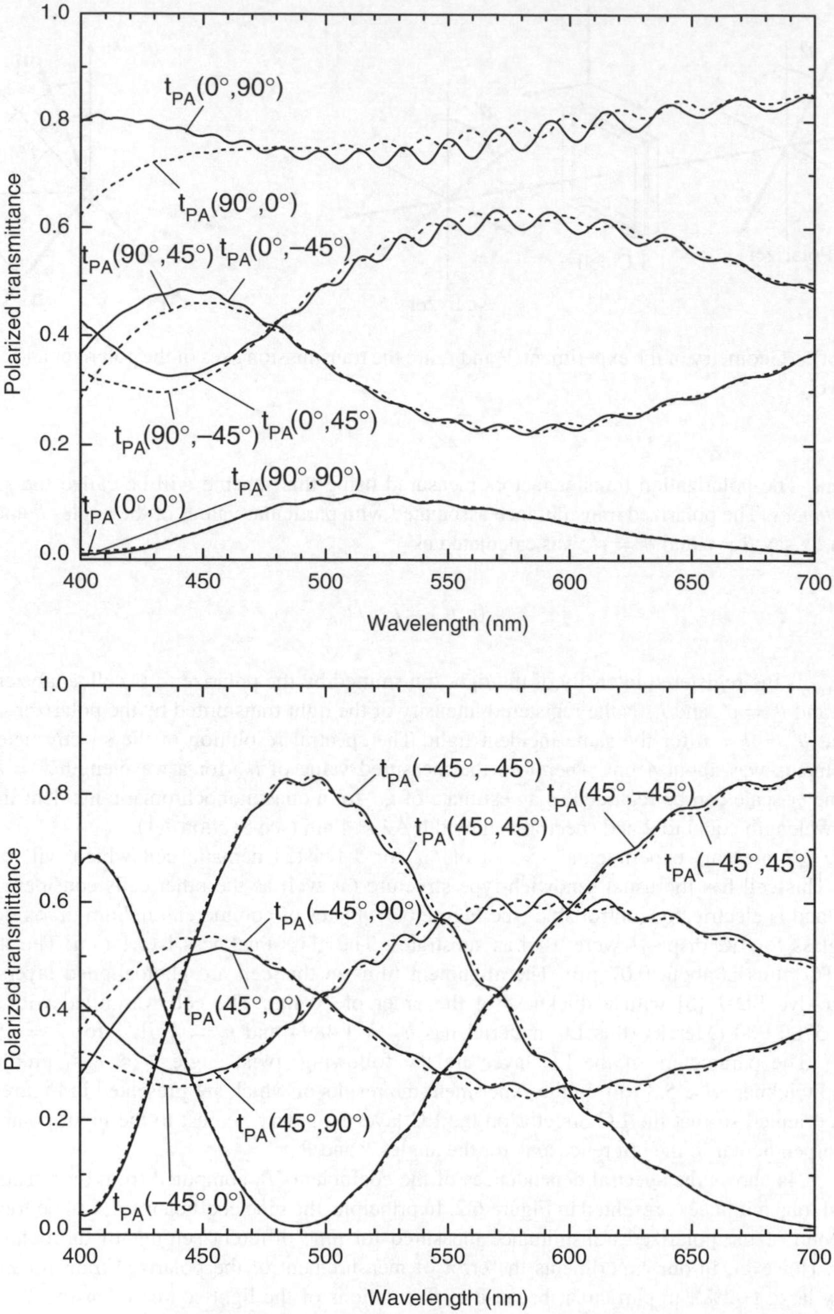

Figure 6.2 Experimental spectra of the polarized transmittances for cell *E1* (with ZLI-5700-000) [4]

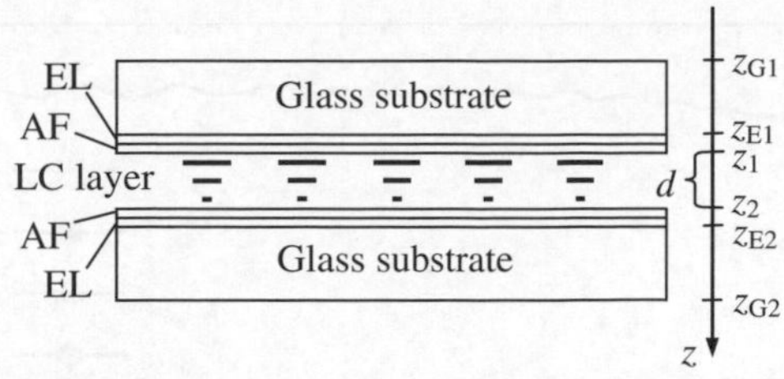

Figure 6.3 Structure of cell *E1*. EL labels the electrodes, and AF the alignment films

6.2.1 Polarization-Dependent Losses and Depolarization. Unpolarized Transmittance

Polarization-Dependent Losses

Figure 6.2 shows eight pairs of the spectra: $\{t_{PA}(0°,0°),\ t_{PA}(90°,90°)\}$, $\{t_{PA}(0°,45°),\ t_{PA}(90°,-45°)\}$, $\{t_{PA}(0°,90°),\ t_{PA}(90°,0°)\}$, $\{t_{PA}(0°,-45°),\ t_{PA}(90°,45°)\}$, $\{t_{PA}(45°,0°),\ t_{PA}(-45°,90°)\}$, $\{t_{PA}(45°,45°),\ t_{PA}(-45°,-45°)\}$, $\{t_{PA}(45°,90°),\ t_{PA}(-45°,0°)\}$, and $\{t_{PA}(45°,-45°),\ t_{PA}(-45°,45°)\}$. In the absence of diattenuation, spectra in each pair would be identical to each other [see (6.15)]. As is seen from the figure, the spectra in each pair are close to each other, but in some pairs, for example, in the pair $\{t_{PA}(0°,90°),\ t_{PA}(90°,0°)\}$, the difference between the spectra is significant, which points to the presence of diattenuation in the cell. The presence of polarization-dependent losses is also evidenced by the noticeable deviation of the coefficients B_5 and B_7 from zero (see Figure 6.4a).

The diattenuation in cell *E1* is determined by the following three factors. The first factor is the multiple-beam interference in the LC layer or, more precisely, the interference of the waves passing through the LC layer once and waves that arise from the multiple reflections from the thin-film systems surrounding the LC layer and pass through the LC layer many times. A feature of this interference is a dependence of the position of the interference extrema on the incident light polarization. The significant effect of this interference is explained by a relatively high reflectivity of the alignment film–electrode systems due to a large difference of the refractive index of ITO ($n_{ITO} \sim 2.0$) from the refractive indices of the glass substrates and alignment layers. This multiple-beam interference appears as relatively fast oscillations in the spectra of the transmittances and B_j. In a real experiment, these oscillations may be somewhat smoothed owing to the finite resolution of the spectrometer used; this effect becomes more pronounced with increasing the LC layer thickness. The second factor leading to the diattenuation is a polarization-dependent transmittance of the liquid crystal–alignment film interfaces. The effect of this factor, as a rule, is smaller than the effect of the first factor, but it is also usually perceptible. In our case, for instance, this factor should be regarded as the reason why in the region from 500 to 700 nm the average value of $t_{PA}(90°,0°)$ is a little higher than the average value of $t_{PA}(0°,90°)$. The significant diattenuation in the spectral region $\lambda < 470$ nm is caused by the action of the third factor; this factor is linear dichroism of the photoalignment films: a polarized absorption band of oriented SD-1 occupies the violet–blue region of the visible spectrum [5]. Another factor of diattenuation in real experiments, more typical than the dichroism of alignment layers, is linear dichroism of the LC material (this factor was absent in the considered experimental example). The diattenuation caused by the first two factors is always present to a greater or lesser extent.

Depolarization

Even if we assume that the LC layer and the other layers of the LC cell in the probed region have no variations in thickness and that the LC layer in this region is strictly 2D-homogeneous, that is, its local

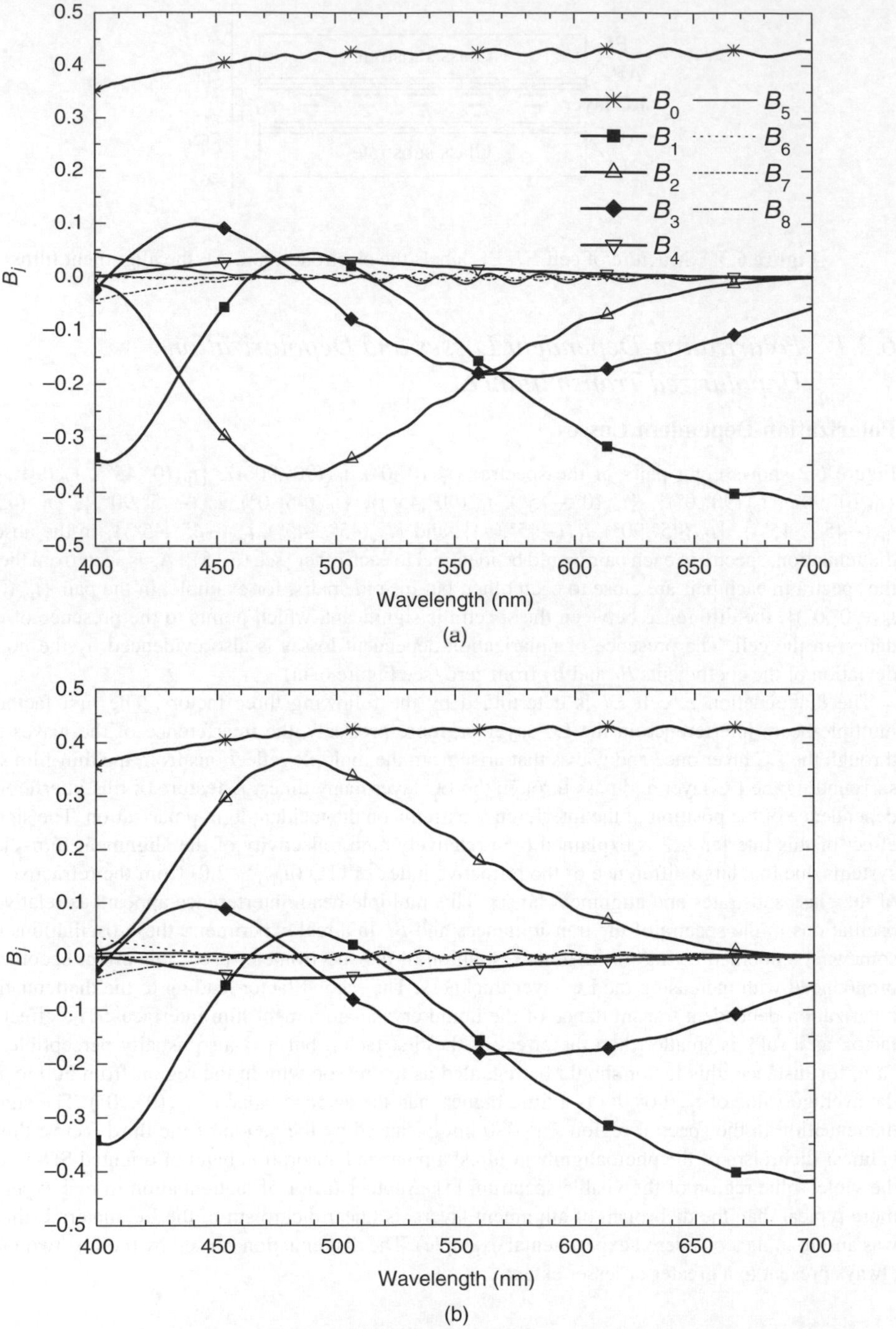

Figure 6.4 Experimental spectra of the coefficients B_j for cell *E1* (a) and their transformation after rotation of the cell by ~45° (b). The symbols are only to mark the lines

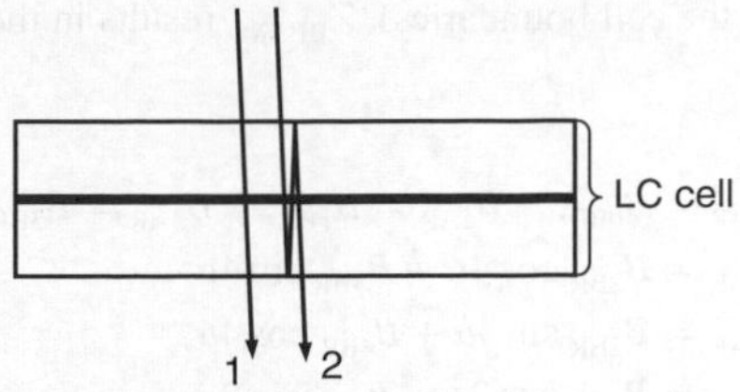

Figure 6.5 To the discussion of the depolarizing action of LC cells

optical parameters depend only on the coordinate along the normal to the interfaces (z), in this experiment, where we measure the overall transmission characteristics of the cell (see Section 7.1), strictly speaking, we cannot consider the LC cell as a nondepolarizing system for quasimonochromatic incident light with bandwidth of the order of 1 nm and larger. Thus, the state of polarization of the fraction of the transmitted beam whose way within the LC cell is shown in Figure 6.5 as way 1 will in general be different from the state of polarization of the fraction whose way is way 2. At the same time, because of a large thickness of the substrates, these fractions are mutually incoherent and add incoherently (see Section 7.1). The incoherent adding of differently polarized components in such situations gives rise to the partial depolarization of the whole transmitted beam. At normal incidence (which we consider in this section), the contribution of the components arising from the multiple reflections at the external surfaces of the LC cell to the net power of the transmitted light is small and often negligible. However, in the case of oblique incidence, the contribution of these components and, consequently, the depolarization may be noticeable. Methods described in Chapter 10 enable a realistic modeling of such situations with due consideration for both coherent and incoherent effects accompanying the propagation of quasimonochromatic light in layered systems.

Unpolarized Transmittance

The transmittance of the sample (the LC cell) for unpolarized incident light (*unpolarized transmittance,* t_{unp}) can be expressed as

$$t_{unp} = 2B_0.$$

The average level and the trend line of the spectrum of t_{unp} in the visible region for cells with the structure shown in Figure 6.3 and negligible absorption, scattering, and reflection losses in the LC layer are determined mainly by the reflection losses at the external interfaces of the cell (~8%) and the losses caused by the multiple-beam interference and absorption in the ITO layer–alignment layer systems (usually of the order of 5–20%). The losses caused by absorption in the glass substrates in the visible region are usually of the order of 1–2%.

6.2.2 *Rotations*

Azimuthal Rotations of the Sample. Rotational Invariants

Analyzing expressions (6.5) as applied to the measurement geometry shown in Figure 6.1, one can easily see that the rotation of the sample by an arbitrary angle α about the axis of the incident beam (parallel to the z_I-axis and perpendicular the cell boundaries), Z_{BEAM}, results in the following transformations of the parameters B_j:

$$\begin{aligned}
B_{0AR} &= B_{0BR}, \quad B_{1AR} = B_{1BR}, \quad B_{3AR} = B_{3BR}, \\
B_{2AR} &= B_{2BR}\cos 4\alpha - B_{4BR}\sin 4\alpha, \\
B_{4AR} &= B_{2BR}\sin 4\alpha + B_{4BR}\cos 4\alpha, \\
B_{5AR} &= B_{5BR}\cos 2\alpha - B_{6BR}\sin 2\alpha, \\
B_{6AR} &= B_{5BR}\sin 2\alpha + B_{6BR}\cos 2\alpha, \\
B_{7AR} &= B_{7BR}\cos 2\alpha - B_{8BR}\sin 2\alpha, \\
B_{8AR} &= B_{7BR}\sin 2\alpha + B_{8BR}\cos 2\alpha,
\end{aligned} \tag{6.34}$$

where B_{jBR} (j = 0,1,…, 8) is the value of the coefficient B_j before the rotation (BR—Before Rotation), and B_{jAR} is the value of B_j after the rotation (AR—After Rotation).

Relations (6.34) indicate that three coefficients, B_0, B_1, and B_3, retain their values upon arbitrary azimuthal rotations of the sample; in other words, these coefficients are rotational invariants. The other coefficients are not invariant with respect to arbitrary azimuthal rotations; however, some of their combinations are also rotational invariants. For example, one can easily see from (6.34) that the quantities

$$\sqrt{B_2^2+B_4^2}, \quad \sqrt{B_5^2+B_6^2}, \text{ and } \quad \sqrt{B_7^2+B_8^2} \tag{6.35}$$

are rotational invariants of this kind. It is also seen that the coefficients B_2 and B_4 are invariant with respect to the rotation by 90° (α = 90°) and assume the opposite values under rotation by 45°, while the coefficients B_5, B_6, B_7, and B_8 do not change under rotation by 180° and assume the opposite values under rotation by 90°.

In Figure 6.4b we show, for illustration, the experimental spectra of the coefficients B_j for the cell *E1* rotated by ~45° with respect to its position for which the spectra presented in Figure 6.4a were obtained. We see that the results of the measurements are in very good agreement with formulas (6.34).

Rotations of the Coordinate System

The rules of transformation of the parameters B_j under rotations of the reference axis for the angles ϑ and ϑ' (x_I in the geometry under consideration) are similar to (6.34). If we denote the value of B_j before rotation by B_{jBR} and the value of B_j after the rotation of the reference frame (x_I, y_I, z_I) by an angle α about the z_I-axis by $B_{jAR,}$ these transformation rules can be written as follows:

$$\begin{aligned}
B_{0AR} &= B_{0BR}, \quad B_{1AR} = B_{1BR}, \quad B_{3AR} = B_{3BR}, \\
B_{2AR} &= B_{2BR}\cos 4\alpha + B_{4BR}\sin 4\alpha, \\
B_{4AR} &= -B_{2BR}\sin 4\alpha + B_{4BR}\cos 4\alpha, \\
B_{5AR} &= B_{5BR}\cos 2\alpha + B_{6BR}\sin 2\alpha, \\
B_{6AR} &= -B_{5BR}\sin 2\alpha + B_{6BR}\cos 2\alpha, \\
B_{7AR} &= B_{7BR}\cos 2\alpha + B_{8BR}\sin 2\alpha, \\
B_{8AR} &= -B_{7BR}\sin 2\alpha + B_{8BR}\cos 2\alpha.
\end{aligned} \tag{6.36}$$

It is clear that the relations (6.36) can be obtained from (6.34) by the replacement $\alpha \to -\alpha$.

Turning Over of the Sample

In the experiments under consideration, we characterize the overall transmission of the sample for the light incident on the sample in the positive direction of the z_I-axis. The LC cells consist of only optically reciprocal materials (Section 8.1.1), so that the operators characterizing the overall transmission of the cell for the beam incident on the cell in the $+z_I$-direction and for an identical beam but incident on the cell in $-z_I$-direction obey certain reciprocity relations. Let $M_D = [m_{Djk}]$ and $M_R = [m_{Rjk}]$ be the transmission Mueller matrices of the sample for the beams incident on it in the $+z_I$- and $-z_I$-directions, respectively. Recall that we use the system (x_I, y_I, z_I) as the input and output reference system for the matrix M_D (Figure 6.1). Let a system (x_R, y_R, z_R) chosen as shown in Figure 6.6a or 6.6b be used as the input and output reference system for M_R. With any of the indicated choices, the matrix M_R can be composed of the elements of the matrix M_D as follows:

$$M_R = \begin{pmatrix} m_{D11} & m_{D21} & -m_{D31} & m_{D41} \\ m_{D12} & m_{D22} & -m_{D32} & m_{D42} \\ -m_{D13} & -m_{D23} & m_{D33} & -m_{D43} \\ m_{D14} & m_{D24} & -m_{D34} & m_{D44} \end{pmatrix}. \tag{6.37}$$

This expression is consistent with (1.256). The rotation of the system (x_R, y_R, z_R) shown in Figure 6.6b by 180° about the x_R-axis makes it coincident with the system (x_I, y_I, z_I). In view of this it is clear that the matrix M_R is equal to the Mueller matrix of the sample rotated by 180° about an axis parallel to the x_I-axis for the beam incident on the sample in the $+z_I$-direction in the system (x_I, y_I, z_I). This gives the following rules of transformations of the coefficients B_j under the rotation of the sample by 180° about an axis parallel to the x_I-axis:

$$\begin{aligned} &B_{0AR} = B_{0BR},\ B_{1AR} = B_{1BR},\ B_{2AR} = B_{2BR},\ B_{3AR} = B_{3BR},\ B_{4AR} = -B_{4BR}, \\ &B_{5AR} = B_{7BR},\ B_{6AR} = -B_{8BR},\ B_{7AR} = B_{5BR},\ B_{8AR} = -B_{6BR}, \end{aligned} \tag{6.38}$$

where, as in (6.34), B_{jBR} and B_{jAR} ($j = 0,1,\ldots,8$) are the values of B_j before and after the rotation of the sample. The same rules are valid for thc rotation of the sample by 180° about an axis parallel to the y_I-axis. If the axis about which the sample is rotated by 180° makes an angle β with the positive x_I-axis, this rotation can be represented as the sequence of the rotation of the sample by 180° about an

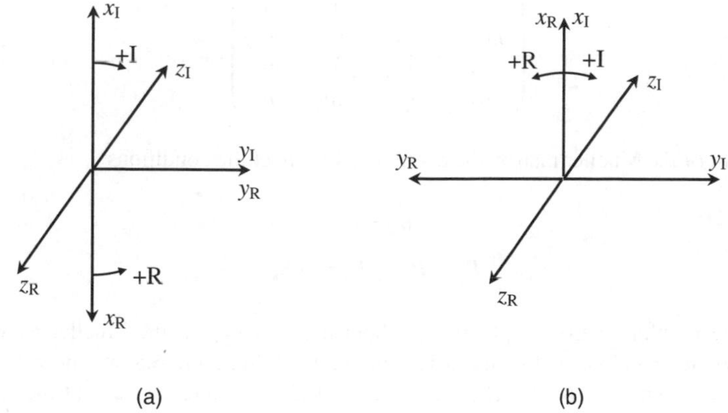

Figure 6.6 Two variants of the reference system for the backward propagating light. The arrows +I and +R indicate positive directions for the angles measured from the x_I-axis and x_R-axis, respectively

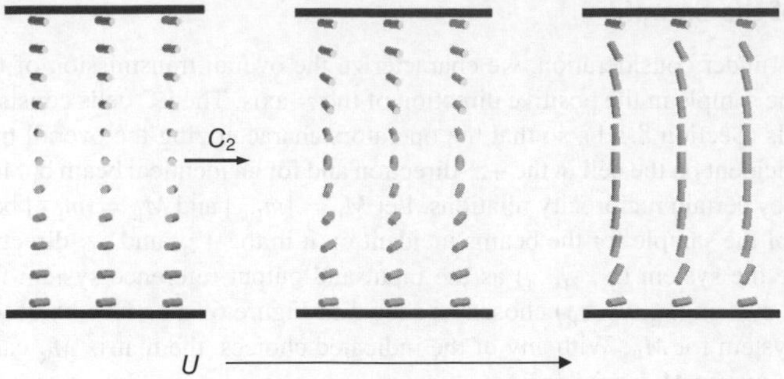

Figure 6.7 Fréedericksz transition in a twisted nematic layer, retaining the symmetry axis C_2

axis parallel to the x_{I}-axis and the rotation by the angle 2β about the axis Z_{BEAM} [see (6.34)]. As is seen from (6.38) and (6.34), under such rotations the coefficients B_0, B_1, and B_3 retain their values.

6.2.3 Symmetry of the Sample

Here we will restrict our consideration to two cases of symmetry, which are most interesting for LCD optics.

LC Cells with the LC Layer Structure Invariant with Respect to the Rotation by 180° About an Axis Parallel to the Layer Boundaries

This kind of symmetry is inherent to TN and STN layers with symmetrical boundary conditions. The second-order symmetry axis C_2 in such layers is perpendicular to the bisector of the twist angle (see Figure 6.7). This symmetry is usually not broken after applying voltage to the LC layer. Let us assume that the properties of a sample are invariant with respect to its rotation by 180° about an axis parallel to the x_{I}-axis. Then, according to (6.37), the transmission Mueller matrix of the sample must have the following structure:

$$\begin{pmatrix} m_{11} & m_{12} & m_{13} & m_{14} \\ m_{12} & m_{22} & m_{23} & m_{24} \\ -m_{13} & -m_{23} & m_{33} & m_{34} \\ m_{14} & m_{24} & -m_{34} & m_{44} \end{pmatrix}. \tag{6.39}$$

With such a form of the Mueller matrix, the coefficients B_j meet the conditions:

$$B_4 = 0, \tag{6.40}$$

$$B_5 = B_7, \quad B_6 = -B_8 \tag{6.41}$$

[see (6.7)]. After rotation of the sample by 90° about the axis Z_{BEAM}, the Mueller matrix retains the symmetry (6.39), and relations (6.40) and (6.41) remain valid. In Figure 6.8, we show the experimental spectra of the coefficients B_0, B_1, B_2, B_3, and B_4 for cell *E1* at different values of the applied voltage (U). The cell is oriented so that the bisector of the twist angle is nearly parallel to the x_{I}-axis. With this orientation, we can expect to observe zero values of B_4 at any wavelength both for the field-off state and the field-on states of the LC cell. As is seen from the figure, B_4 is really very close to zero throughout the

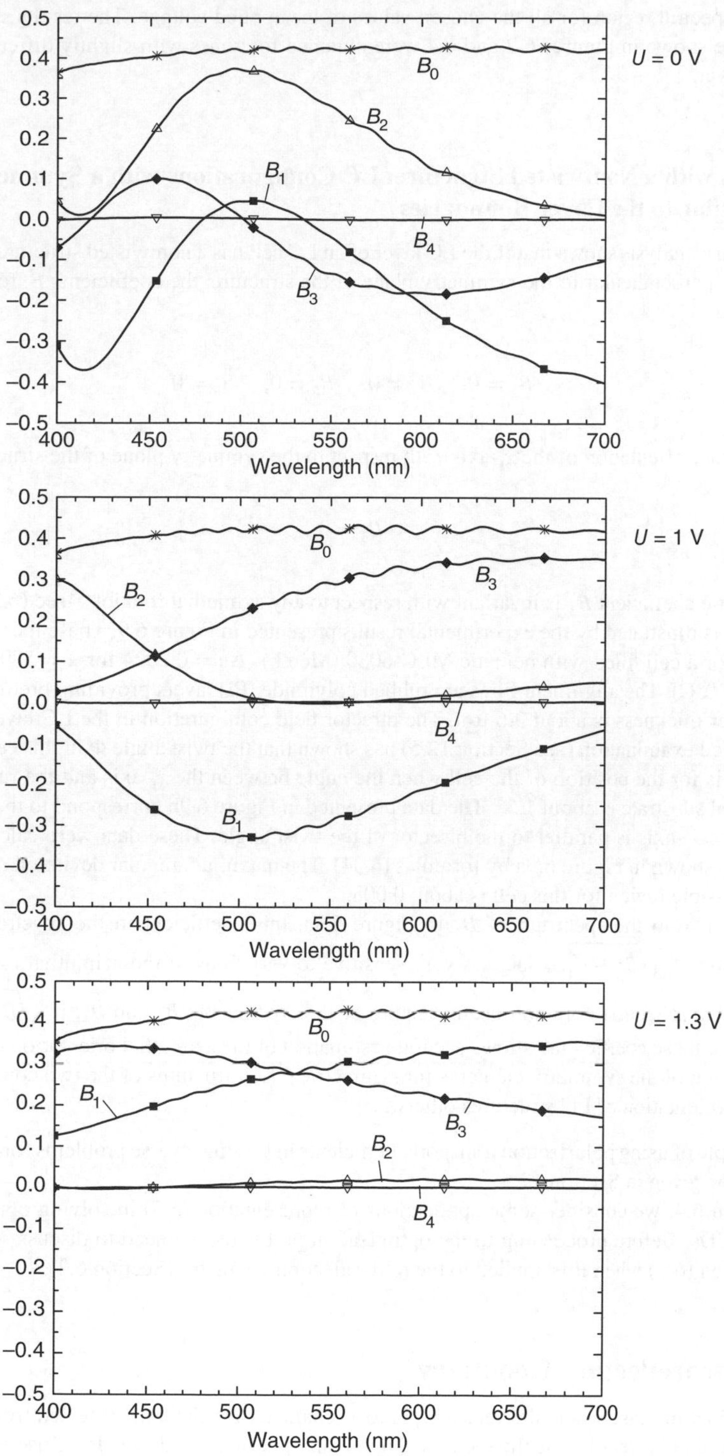

Figure 6.8 Experimental spectra of coefficients B_j for cell *E1* at different values of applied voltage

considered spectral region for all the chosen values of the applied voltage. The results shown in Figure 6.8 and those shown in Figures 6.2 and 6.4 were obtained for zones with slightly different thicknesses of the LC layer.

LC Layers with a Nontwisted Structure: LC Configurations with a Symmetry Plane Perpendicular to the Layer Boundaries

An elementary analysis shows that if the LC layer of an LC cell has a nontwisted structure and the x_1-axis is parallel or perpendicular to the symmetry plane of the structure, the coefficients B_j for this cell meet the conditions

$$B_3 = 0, \quad B_4 = 0, \quad B_6 = 0, \quad B_8 = 0. \tag{6.42}$$

For an arbitrary orientation of the x_1-axis with respect to the symmetry plane of the structure,

$$B_3 = 0. \tag{6.43}$$

Recall that the coefficient B_3 is invariant with respect to any azimuthal rotations [see (6.34) and (6.36)].

This case is illustrated by the experimental results presented in Figure 6.9. These measurements were performed for a cell filled with nematic MLC-6080 (Merck) ($\Delta n = 0.2024$ for $\lambda = 589.3$ nm at a temperature of 20°C). The alignment films are rubbed polyimide (PI) layers providing pretilt angles ≈2–4°. The LC layer thickness is about 3.6 μm. The director field configuration in the LC layer is nearly uniform. Detailed examination (see Section 12.5) has shown that the twist angle Φ in this cell is about 0.4°. Figure 6.9a is for the position of the cell when the angle between the x_1-axis and the rubbing direction on the frontal substrate is about 1.5°. The data presented in Figure 6.9b correspond to the position of the cell when the x_1-axis is parallel to the bisector of the twist angle. These data were calculated from the spectra of B_i shown in Figure 6.9a by formulas (6.34). The maximum regular deviation of $|B_3|$ from zero within the visible region for this cell is about 0.008.

As is seen from the spectrum of B_4 in Figure 6.9a, this coefficient in the spectral region where the magnitude of $\sqrt{B_2^2 + B_4^2}$ is large is very sensitive to variations of the azimuthal orientation of the LC layer. This reflects a common useful feature of the coefficients B_2 and B_4; if $\sqrt{B_2^2 + B_4^2}$; is large, knowledge of these coefficients allows accurate estimation of the azimuthal orientation of LC structures (the orientation of the symmetry elements for symmetrical LC structures of the two considered kinds or the relative orientation of LC structures otherwise).

An example of using polarization transport coefficients in solving inverse problems for twisted nematic layers will be given in Section 12.5.

In Section 6.4, we consider some applications of representation (6.5) in solving optimization problems for LCDs. Before proceeding to the optimization problems, we need to discuss some features of representation (6.5) when it is applied to the retroreflection geometry (Section 6.3).

6.3 Retroreflection Geometry

Let us consider the case where the vector $\mathbf{S}_\mathrm{O}$ in (6.1) characterizes the light reflected from a layered system consisting of layers of optically reciprocal materials at normal incidence. If we use one of the frames

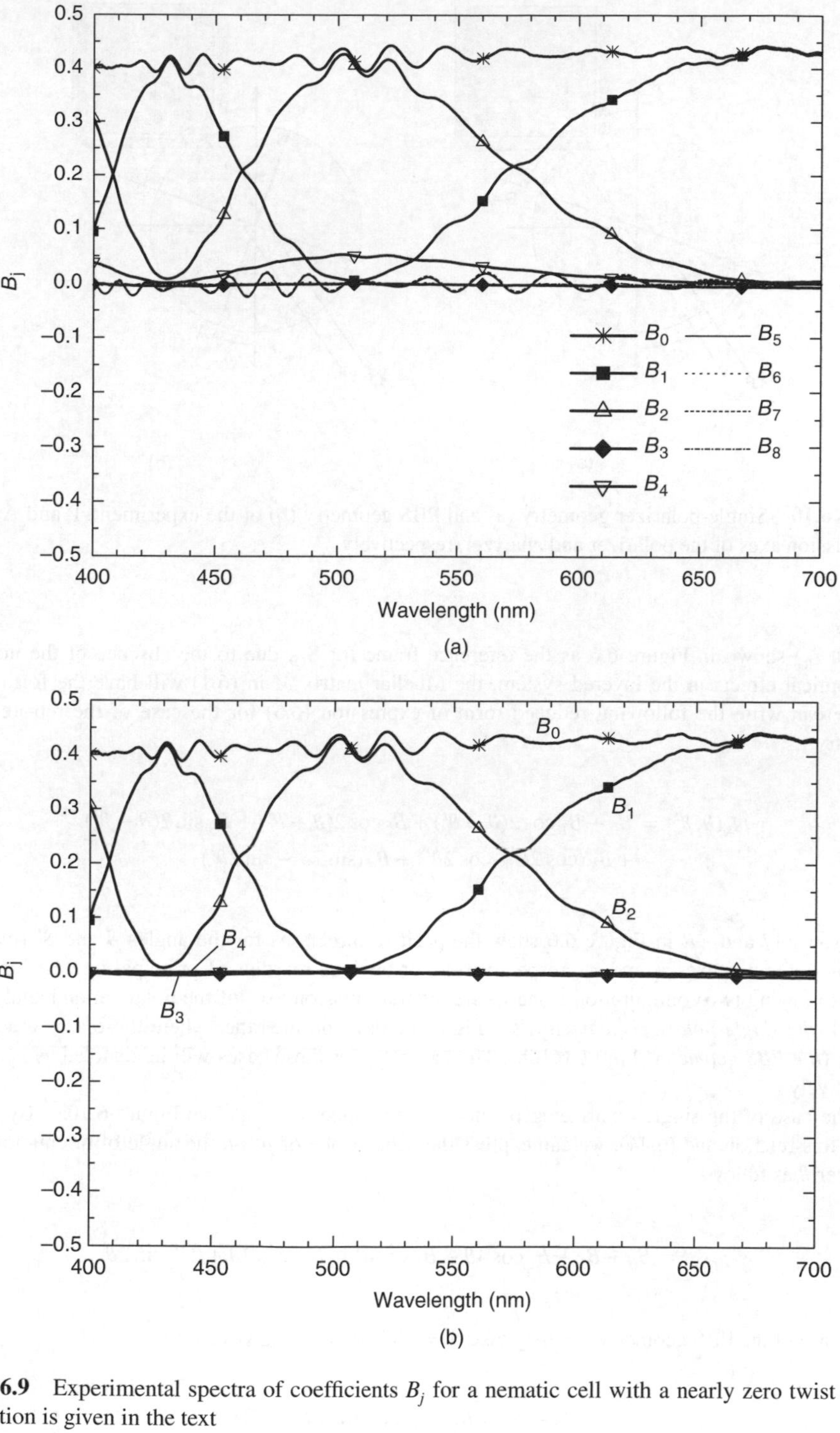

Figure 6.9 Experimental spectra of coefficients B_j for a nematic cell with a nearly zero twist angle. Description is given in the text

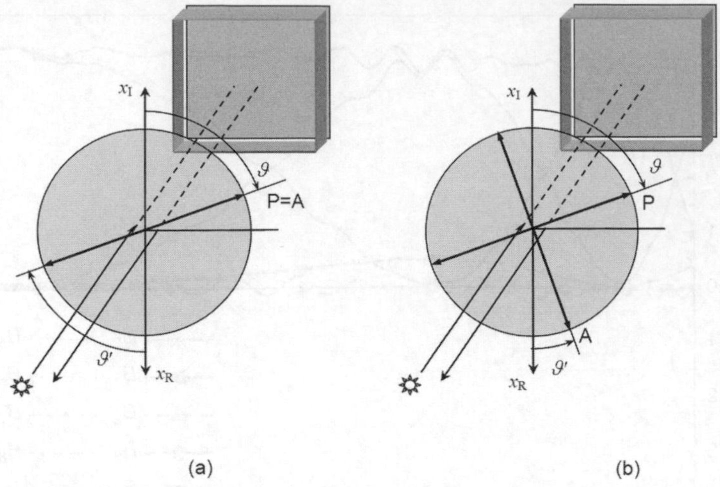

Figure 6.10 Single-polarizer geometry (a) and PBS geometry (b) of the experiment. P and A are the transmission axes of the polarizer and analyzer, respectively

(x_R, y_R, z_R) shown in Figure 6.6 as the reference frame for $\mathbf{S}_O$, due to the absence of the nonreciprocal optical effects in the layered system, the Mueller matrix M in (6.1) will have the form (6.39), and we can write the following reduced form of expression (6.5) for the case of the retroreflection geometry:

$$\begin{aligned} t_{PA}(\vartheta, \vartheta') = B_0 + B_1 \cos 2(\vartheta - \vartheta') + B_2 \cos 2(\vartheta + \vartheta') + B_3 \sin 2(\vartheta - \vartheta') \\ + B_5 (\cos 2\vartheta + \cos 2\vartheta') + B_6 (\sin 2\vartheta - \sin 2\vartheta'). \end{aligned} \tag{6.44}$$

The arrows $+I$ and $+R$ in Figure 6.6 show the positive directions for the angles ϑ and ϑ' (measured from the reference axes x_I and x_R, respectively). For obvious reasons, of principal interest for us will be the following two configurations: one where the transmission axes of the polarizer and analyzer are parallel (*the single-polarizer geometry*, see Figure 6.10a), and the other where these axes are perpendicular (*the PBS geometry*, Figure 6.10b). The factor t_{PA} for these cases will be denoted as $\rho_{\parallel}$ and $\rho_{\perp}$, respectively.

In the case of the single-polarizer geometry, we may take $\vartheta' = -\vartheta$ (see Figure 6.10a). By making use of this relation and (6.44), we can express the dependence of $\rho_{\parallel}$ on the angle of orientation of the polarizer ϑ as follows:

$$\rho_{\parallel}(\vartheta) = B_0 + B_2 + B_1 \cos 4\vartheta + B_3 \sin 4\vartheta + 2B_5 \cos 2\vartheta + 2B_6 \sin 2\vartheta. \tag{6.45}$$

In the case of the PBS geometry, we may take $\vartheta' = 90° - \vartheta$, which gives

$$\rho_{\perp}(\vartheta) = B_0 - B_2 - B_1 \cos 4\vartheta - B_3 \sin 4\vartheta. \tag{6.46}$$

Azimuthal Rotations and Rotational Invariants

The rules of transformation of the polarization transport coefficients under the rotation of the system by an angle α about the axis Z_{BEAM} are

$$\begin{aligned} B_{0\mathrm{AR}} &= B_{0\mathrm{BR}}, \quad B_{2\mathrm{AR}} = B_{2\mathrm{BR}}, \\ B_{1\mathrm{AR}} &= B_{1\mathrm{BR}} \cos 4\alpha - B_{3\mathrm{BR}} \sin 4\alpha, \\ B_{3\mathrm{AR}} &= B_{1\mathrm{BR}} \sin 4\alpha + B_{3\mathrm{BR}} \cos 4\alpha, \\ B_{5\mathrm{AR}} &= B_{5\mathrm{BR}} \cos 2\alpha - B_{6\mathrm{BR}} \sin 2\alpha, \\ B_{6\mathrm{AR}} &= B_{5\mathrm{BR}} \sin 2\alpha + B_{6\mathrm{BR}} \cos 2\alpha, \end{aligned} \tag{6.47}$$

where we use the same notation as in (6.34).

As is seen from (6.47), the quantities

$$B_0,\ B_2,\ \sqrt{B_1^2 + B_3^2},\ \text{and}\ \sqrt{B_5^2 + B_6^2} \tag{6.48}$$

are invariant with respect to arbitrary azimuthal rotations for the retroreflection geometry.

6.4 Applications of Polarization Transport Coefficients in Optimization of LC Devices

To begin with, we consider the following experimental example.

Search for Optimal Polarizer Orientation for an Experimental SSFLC Cell

Figure 6.11 shows experimental spectra of the coefficients B_0 through B_4 for two stable states (states 1 and 2) of an SSFLC cell. The coefficients B_5 through B_8, which are not shown in this figure, for both states fluctuate near zero; the absolute values of these coefficients were less than 0.025 throughout the visible region. The cell exhibited an imperfect bookshelf switching. The structures realized in the LC layer in both stable states were inhomogeneous and chiral; this is evidenced by significant deviations of the experimental values of the coefficient B_3 from zero (see Figure 6.11).

Suppose that we wish to use this cell in an optical shutter, with the polarizer–LC cell–polarizer scheme, for white light. Knowledge of the wavelength dependences of the polarization transport coefficients can help us to find an optimal polarizer orientation and estimate how good the optical characteristics of the device can be. In particular, we can find the polarizer orientation providing the maximum contrast ratio as well as the polarizer orientation providing the maximum contrast ratio attainable at a given level of a wavelength-averaged transmittance in the bright state. In calculations the results of which are presented below, we used the quantity

$$\bar{t}_{\mathrm{PA}} = \frac{1}{\lambda_2 - \lambda_1} \int_{\lambda_1}^{\lambda_2} t_{\mathrm{PA}} \mathrm{d}\lambda, \tag{6.49}$$

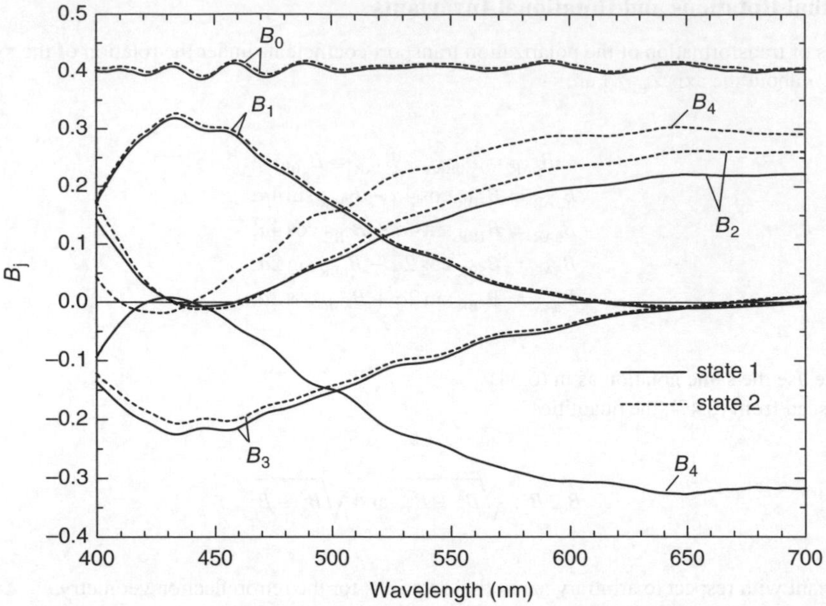

Figure 6.11 Measured spectra of coefficients B_j for two stable states of an SSFLC cell

where $\lambda_1 = 400$ nm and $\lambda_2 = 700$ nm, to characterize the average transmittance of the shutter and the quantity

$$C_a = \frac{\bar{t}_{\text{PA-bright}}}{\bar{t}_{\text{PA-dark}}}, \tag{6.50}$$

where $\bar{t}_{\text{PA-bright}}$ and $\bar{t}_{\text{PA-dark}}$ are the values of $\bar{t}_{\text{PA}}$ in the bright and dark states, respectively, to characterize the contrast. State 1 was taken as the dark state. It was found that the unconditional maximum of the C_a is achieved at $\vartheta = -85.91°$ and $\vartheta' = 23.51°$. For this variant of polarizer orientation, called *variant A*, $C_a = 20.51$ and $\bar{t}_{\text{PA-bright}} = 0.367$. The spectra of t_{PA} of the cell in the dark and bright states for this variant are shown in Figure 6.12. Figures 6.13 and 6.14 demonstrate the results of the conditional optimization of the contrast ratio. Figure 6.13 shows the maximum values of C_a reachable at different values of $\bar{t}_{\text{PA-bright}}$ greater than 0.367. Figure 6.14 shows the values of ϑ and ϑ' providing the maximum C_a for different $\bar{t}_{\text{PA-bright}}$ values, that is, corresponding to the curve C_a ($\bar{t}_{\text{PA-bright}}$) presented in Figure 6.13. In Figure 6.12, along with the spectra corresponding to the unconditional maximum of the C_a, for comparison, we show the spectra corresponding to the maximum C_a ($C_a = 14.9$) attainable at $\bar{t}_{\text{PA-bright}} = 0.42$ ($\vartheta = -81.88°$, $\vartheta' = 26.49°$, *variant B*).

Representation (6.5) is very convenient in solving problems of this kind because it enables one to express the dependences of some characteristics of LCDs defined through integrals over a spectral region on the orientation angles of the polarizers in a simple analytical form. Thus, if a characteristic to be analyzed can be expressed as

$$\bar{t}_f = \int_{\lambda_a}^{\lambda_b} f(\lambda) t_{\text{PA}}(\lambda) \text{d}\lambda, \tag{6.51}$$

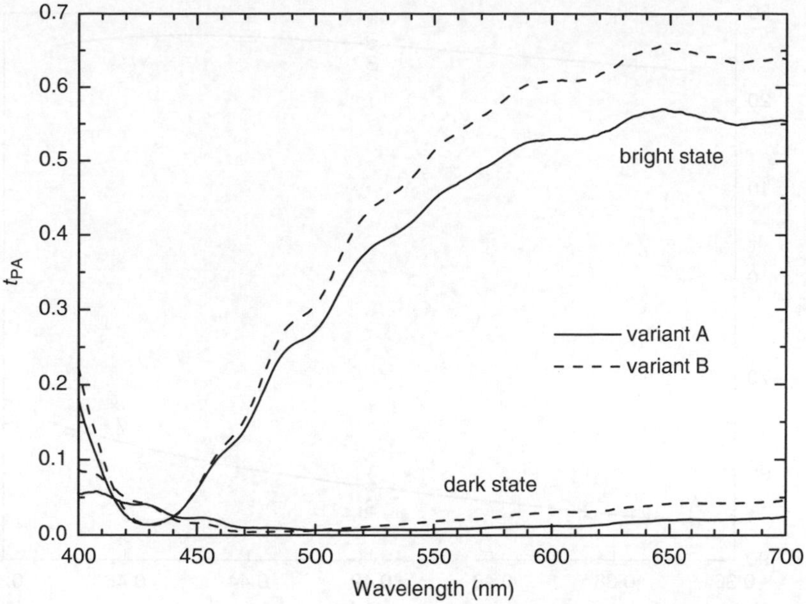

Figure 6.12 Wavelength dependences of t_{PA} of the SSFLC cell in the dark and bright states for two optimized variants of polarizer orientation

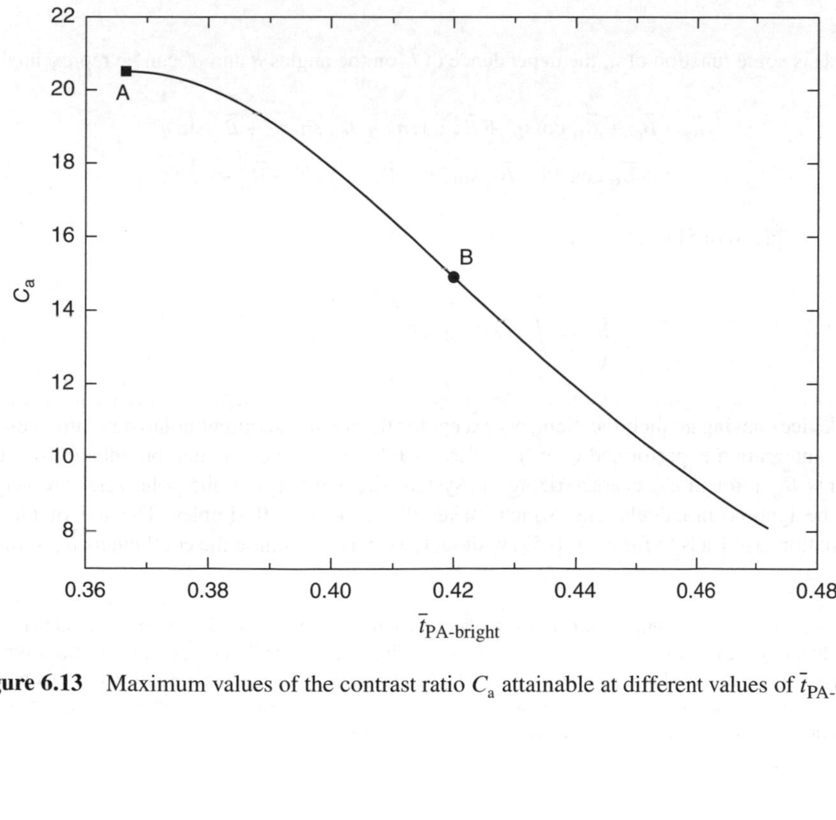

Figure 6.13 Maximum values of the contrast ratio C_a attainable at different values of $\bar{t}_{PA\text{-bright}}$

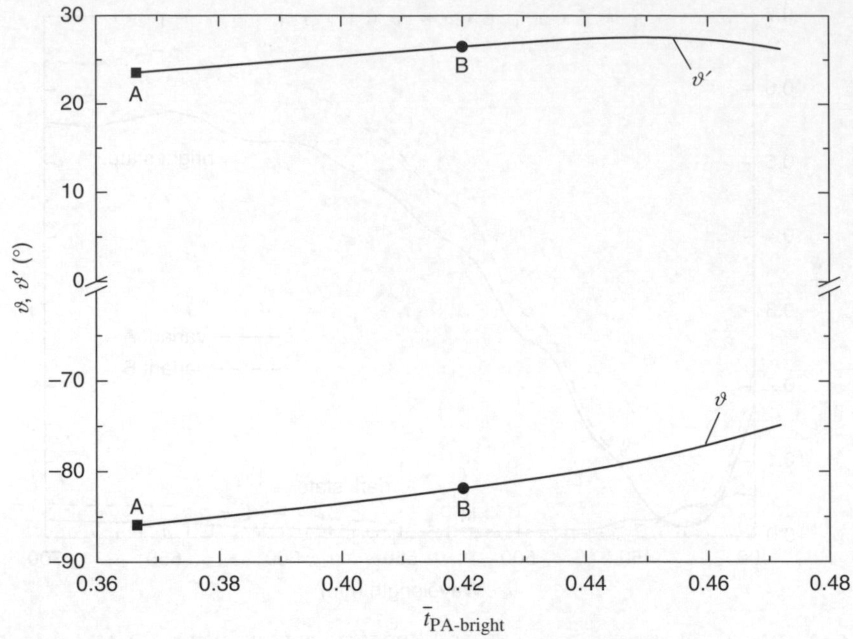

Figure 6.14 The orientation of the polarizers providing the maximum contrast ratio (C_a) at different values of $\bar{t}_{\text{PA-bright}}$

where $f(\lambda)$ is some function of λ, the dependence of $\bar{t}_f$ on the angles ϑ and ϑ' can be represented as

$$\begin{aligned}\bar{t}_f = \overline{B}_{0f} + \overline{B}_{1f}\cos\eta^- + \overline{B}_{2f}\cos\eta^+ + \overline{B}_{3f}\sin\eta^- + \overline{B}_{4f}\sin\eta^+ \\ + \overline{B}_{5f}\cos 2\vartheta + \overline{B}_{6f}\sin 2\vartheta + \overline{B}_{7f}\cos 2\vartheta' + \overline{B}_{8f}\sin 2\vartheta',\end{aligned} \tag{6.52}$$

where, according to (6.51) and (6.5),

$$\overline{B}_{jf} = \int_{\lambda_a}^{\lambda_b} f(\lambda)B_j(\lambda)\mathrm{d}\lambda \quad j = 0, 1, \ldots, 8. \tag{6.53}$$

For LC devices having no dichroic elements except for the input and output polarizers, often, especially when the integration is performed over the entire visible region and $f(\lambda)$ is a broadband function, the coefficients $\overline{B}_{5f}$ through $\overline{B}_{8f}$ characterizing the system situated between the polarizers[1] are very small and can be ignored in calculations, which makes the analysis still simpler. The use of the unitary approximation also leads to formula (6.52) without the terms containing the coefficients $\overline{B}_{5f}$ through $\overline{B}_{8f}$

[1] In the above experimental example, the coefficients B_j characterize the cell surrounded by air. It is clear that in solving such a problem by means of numerical modeling, corresponding coefficients B_j must describe the transformation of the light on its path from the input polarizer to the output polarizer with the medium just after the input polarizer as the input one and the medium preceding the output polarizer as the output one. The model of the device should be chosen so that these input and output media are isotropic and nonabsorbing.

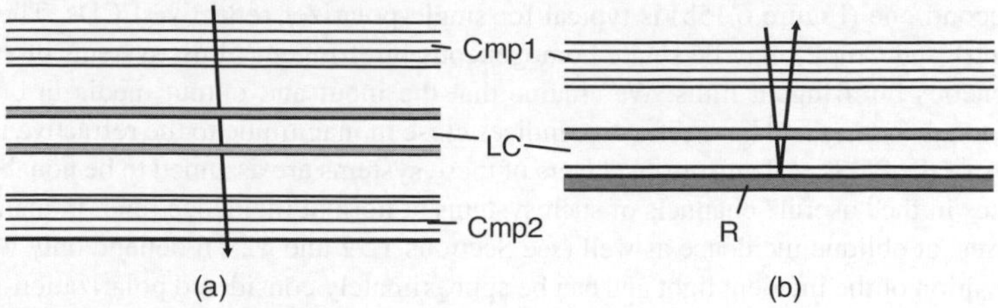

Figure 6.15 PCSs of the transmissive (a) and reflective (b) LC devices under consideration and the "useful" channels of these PCSs. Cmp1, Cmp2—compensators, LC—LC layer, R—reflector

[see (6.30)]. The rest of this section is devoted to examples of application of formulas derived from (6.5) with the use of the unitary approximation in solving optimization problems for LCDs.

"Useful" Channel and Unitary Approximation

Polarization-optical devices that we consider here have the common scheme: input polarizer–*polarization-converting system* (*PCS*)–output polarizer (analyzer). Standard polarization elements of PCSs of LCDs are the LC layer and phase compensators (retarders). The SSFLC cell in the above example was regarded as the PCS of the shutter. The coefficients B_j described the overall transmission of that PCS. In solving optimization problems, when the optics of a PCS is modeled, as a rule, it is more reasonable to deal with the coefficients B_j characterizing a "useful" channel of light propagation in this PCS rather than its overall transmission or overall reflection (see Section 7.1). In considering such optical elements as dichroic polarizers, retarders, LC layers in TN and STN LCDs, the useful effect of the element, in an act of interaction of light with it, is associated with a single passage of the light through the bulk of the element. Light beams arising from reflections at surfaces of such elements and circulating within and between them are considered as parasitic. In this context, considering the optical action of an optical system without taking the light circulation within polarization elements and between them into account is, in fact, considering a "useful" channel of light propagation in this optical system. Consideration of "useful" channels of optical systems is inherent in the classical JC [3] and is usual in polarization optics. The analysis and optimization of transfer characteristics of the "useful" channels of LCDs at normal incidence are major problems that are solved with the help of the classical JC for LCDs. The improved general variant of the Jones matrix method considered in Chapters 8, 11, and 12 enables one to solve these problems more accurately and for the case of oblique incidence.

As for the PCSs of LC displays, the Mueller matrices of their "useful" channels for monochromatic light are relatively slow functions of the wavelength. This allows one to rather accurately represent the Mueller matrix of the "useful" channel of such a PCS for incident quasimonochromatic light with mean wavelength λ_c and bandwidth of the order of a few nanometers as follows:

$$M = \mathbf{L}\left(\mathbf{t}(\lambda_c) \otimes \mathbf{t}(\lambda_c)^*\right)\mathbf{L}^{-1} \tag{6.54}$$

[see (6.16) and Sections 7.1 and 11.1], where $\mathbf{t}(\lambda_c)$ is the corresponding Jones matrix of this channel for $\lambda = \lambda_c$ relating FI Jones vectors (see Section 5.4.2). Note that using (6.54), we assume that the "useful" channel is nondepolarizing. The assumption of the absence of depolarization is, as a rule, justified for the "useful" channels of layered polarization-optical systems, although may be inadequate in considering their overall transmission and reflection (see the above example illustrated by Figure 6.5).

In what follows, we will consider two general schemes of PCSs of LCDs shown in Figure 6.15. The first scheme (Figure 6.15a) is the most general one in the case of double-polarizer transmissive

LCDs. The second one (Figure 6.15b) is typical for single-polarizer reflective LCDs. The phase compensators (Cmp1 and Cmp2) may be single homogeneous birefringent films, systems of such films, or 1D-inhomogeneous birefringent films. We assume that the input and output media in both cases are isotropic and nonabsorbing and have refractive indices close in magnitude to the refractive indices of the adjacent layers of the PCSs. All anisotropic layers of these systems are assumed to be nonabsorbing. The reflection losses in the "useful" channels of such systems at normal incidence [and, in many practically interesting cases, at oblique incidence as well (see Sections 12.2 and 12.4)] depend only weakly on the state of polarization of the incident light and can be approximately considered polarization-independent. Note that considering the "useful" channels, we ignore the multiple-beam interference in the LC layer and thereby exclude one of the main sources of the polarization-dependent losses from consideration. A negligibly small diattenuation in the channels allows one to use the factorization (6.18) for their Jones matrices. As has been noted, in many cases the effects described by the scalar loss factor and the polarization Jones matrix in (6.18) can be estimated and analyzed separately. Sometimes, to solve a problem, the evaluation of the loss factor is not required or rough estimates of this factor can be used. At the same time, an accurate estimation of the polarization Jones matrix can be performed by using a simplified unitary model of the PCS, which includes only polarization-converting elements (LC layer, phase compensators, reflector) which are represented in the calculations by their polarization Jones matrices. The separate consideration of the polarization effects in terms of polarization Jones matrices (the consideration of associated unitary systems) is customary for the classical JC (see Section 1.4.1). In Chapter 12, we will show that this approach can be effectively used within the framework of a more rigorous and general theory as well.

Transmissive Devices

Let us consider a double-polarizer LC device whose PCS is as in Figure 6.15a or simpler, without one or both of the phase compensators. Taking into account real properties of film polarizers, the transmittance of the "useful" channel of this device for quasimonochromatic incident light linearly polarized along the transmission axis of the input polarizer can be expressed in terms of the factors t_{PA} of the "useful" channel of the PCS as follows:

$$t_{||} = C_p t_{||p1} \left(t_{||p2} t_{PA}(\upsilon, \upsilon') + t_{\perp p2} t_{PA}(\upsilon, \upsilon' + 90°) \right), \tag{6.55}$$

where $t_{||pj}$ and $t_{\perp pj}$ are the principal (bulk) transmittances, respectively the maximum and minimum ones, of the input ($j = 1$) and output ($j = 2$) polarizers, υ and υ' are the orientation angles of the input and output polarizers, respectively, and C_p is a coefficient taking into account reflection losses at the surfaces of the polarizers. Analogously, the transmittance of the device for an unpolarized quasimonochromatic incident light can be expressed in terms of t_{PA} as

$$\begin{aligned} t_{unp} = \frac{C_p}{2} & \left[t_{||p1} \left(t_{||p2} t_{PA}(\upsilon, \upsilon') + t_{\perp p2} t_{PA}(\upsilon, \upsilon' + 90°) \right) \right. \\ & \left. + t_{\perp p1} \left(t_{||p2} t_{PA}(\upsilon + 90°, \upsilon') + t_{\perp p2} t_{PA}(\upsilon + 90°, \upsilon' + 90°) \right) \right]. \end{aligned} \tag{6.56}$$

Suppose that the diattenuation in the PCS is negligible. Then, using (6.12), we may represent the function $t_{PA}(\vartheta, \vartheta')$ as follows:

$$t_{PA}(\vartheta, \vartheta') = B_0 + B(\vartheta, \vartheta'), \tag{6.57}$$

where

$$B(\vartheta, \vartheta') = B_1 \cos 2(\vartheta - \vartheta') + B_2 \cos 2(\vartheta + \vartheta') + B_3 \sin 2(\vartheta - \vartheta') + B_4 \sin 2(\vartheta + \vartheta'), \tag{6.58}$$

with coefficients B_j characterizing the "useful" channel of the PCS. Note that, according to (6.58),

$$B(\upsilon, \upsilon') = B(\upsilon + 90°, \upsilon' + 90°) = -B(\upsilon, \upsilon' + 90°) = -B(\upsilon + 90°, \upsilon'). \tag{6.59}$$

Substituting (6.57) into (6.55) and (6.56) and using (6.59), we obtain the following expressions for $t_{||}$ and t_{unp}:

$$t_{||} = C_{\text{p}} t_{||\text{p}1}[(t_{||\text{p}2} + t_{\perp\text{p}2})B_0 + (t_{||\text{p}2} - t_{\perp\text{p}2})B(\upsilon, \upsilon')] \tag{6.60}$$

and

$$t_{\text{unp}} = \frac{C_{\text{p}}}{2}[(t_{||\text{p}1} + t_{\perp\text{p}1})(t_{||\text{p}2} + t_{\perp\text{p}2})B_0 + (t_{||\text{p}1} - t_{\perp\text{p}1})(t_{||\text{p}2} - t_{\perp\text{p}2})B(\upsilon, \upsilon')]. \tag{6.61}$$

It is convenient to rewrite these expressions as follows:

$$t_{||} = 2C_{\text{p}} t_{||\text{p}1} t_{\text{p}2}(B_0 + p_{\text{p}2} B(\upsilon, \upsilon')), \tag{6.62}$$

$$t_{\text{unp}} = 2C_{\text{p}} t_{\text{p}1} t_{\text{p}2}(B_0 + p_{\text{p}1} p_{\text{p}2} B(\upsilon, \upsilon')), \tag{6.63}$$

or

$$t_{||} = 2C_{\text{p}} t_{||\text{p}1} t_{\text{p}2}\left[B_0 + p_{\text{p}2}\left(B_1 \cos\eta^- + B_2 \cos\eta^+ + B_3 \sin\eta^- + B_4 \sin\eta^+\right)\right], \tag{6.64}$$

$$t_{\text{unp}} = 2C_{\text{p}} t_{\text{p}1} t_{\text{p}2}\left[B_0 + p_{\text{p}1} p_{\text{p}2}\left(B_1 \cos\eta^- + B_2 \cos\eta^+ + B_3 \sin\eta^- + B_4 \sin\eta^+\right)\right], \tag{6.65}$$

$$\eta^- \equiv 2(\upsilon - \upsilon'), \quad \eta^+ \equiv 2(\upsilon + \upsilon'), \tag{6.66}$$

where $t_{\text{p}j}$ and $p_{\text{p}j}$ are respectively the average (bulk) transmittance and polarizing efficiency of the jth polarizer, which are defined as

$$t_{\text{p}j} = \frac{t_{||\text{p}j} + t_{\perp\text{p}j}}{2}, \quad p_{\text{p}j} = \frac{t_{||\text{p}j} - t_{\perp\text{p}j}}{t_{||\text{p}j} + t_{\perp\text{p}j}}. \tag{6.67}$$

In the ideal case of the polarizers of unit polarizing efficiency ($p_{\text{p}j} = 1$),

$$t_{||} = 2C_{\text{p}} t_{||\text{p}1} t_{\text{p}2}\left(B_0 + B_1 \cos\eta^- + B_2 \cos\eta^+ + B_3 \sin\eta^- + B_4 \sin\eta^+\right), \tag{6.68}$$

$$t_{\text{unp}} = 2C_{\text{p}} t_{\text{p}1} t_{\text{p}2}\left(B_0 + B_1 \cos\eta^- + B_2 \cos\eta^+ + B_3 \sin\eta^- + B_4 \sin\eta^+\right). \tag{6.69}$$

The coefficients B_j entering into the above expressions can be calculated from the Jones matrix, corresponding to the mean wavelength of the incident light and relating FI Jones vectors, of the "useful" channel of the PCS by formulas (6.17) or (6.19)–(6.24).

The quantities $t_{||}$ and t_{unp} are transmittances of the device for quasimonochromatic light (quasimonochromatic transmittances). For LCDs, which work with broadband light sources, as a rule, it is desired to estimate a set of integral transmission characteristics defined as weighted means of a quasi-monochromatic transmittance ($t_\parallel$, t_{unp}, or a linear combination of $t_\parallel$ and t_{unp}) over the visible region with appropriate weighting functions. In particular, such a set may include the following quantities:

$$
\begin{aligned}
t_X &= \frac{1}{S_Y}\int \bar{x}_\lambda(\lambda)S(\lambda)t(\lambda)\mathrm{d}\lambda,\\
t_Y &= \frac{1}{S_Y}\int \bar{y}_\lambda(\lambda)S(\lambda)t(\lambda)\mathrm{d}\lambda,\\
t_Z &= \frac{1}{S_Y}\int \bar{z}_\lambda(\lambda)S(\lambda)t(\lambda)\mathrm{d}\lambda,
\end{aligned}
\tag{6.70}
$$

where

$$S_Y = \int \bar{y}_\lambda(\lambda)S(\lambda)\mathrm{d}\lambda \tag{6.71}$$

with $t(\lambda)$ being the quasimonochromatic transmittance, $S(\lambda)$ the spectral distribution of the incident light, and $\bar{x}_\lambda(\lambda)$, $\bar{y}_\lambda(\lambda)$, and $\bar{z}_\lambda(\lambda)$ the color matching functions; the function $\bar{y}_\lambda(\lambda)$ is called the photopic luminous efficiency function for the standard observer. The parameter t_Y is usually treated as the average transmission of the device [6–8]. For displays with primary color (RGB) filters, to estimate transmission properties of the device taking into account the effect of any of the filters, the function $S(\lambda)$ may be taken as

$$S(\lambda) = t_{\mathrm{F}}(\lambda)S_0(\lambda), \tag{6.72}$$

where $S_0(\lambda)$ is the spectral distribution of the light from the light source and $t_{\mathrm{F}}(\lambda)$ is the transmittance of the filter. It is clear that in this case, the transmittance $t(\lambda)$ should be calculated for the LCD model without the filters. This trick is fully justified when the "useful" channel of the LCD is considered. The profit from the use of it is evident: the same function $t(\lambda)$ can be used to obtain the results for all the filters.

Substitution of the above expressions for the transmittances $t_\parallel$ and t_{unp} [(6.64), (6.65), (6.68), or (6.69)] into (6.70) gives explicit expressions for the dependences of t_X, t_Y, and t_Z on the polarizer orientation angles. Thus, for example, in the case $t = t_{\mathrm{unp}}$, using (6.65), we obtain the following expression for the average transmission t_Y:

$$t_Y = \overline{B}_0 + \overline{B}_1 \cos\eta^- + \overline{B}_2 \cos\eta^+ + \overline{B}_3 \sin\eta^- + \overline{B}_4 \sin\eta^+, \tag{6.73}$$

where

$$
\begin{aligned}
\overline{B}_0 &= \frac{2}{S_Y}\int C_{\mathrm{p}}t_{\mathrm{p1}}t_{\mathrm{p2}}B_0\bar{y}_\lambda S\mathrm{d}\lambda,\\
\overline{B}_j &= \frac{2}{S_Y}\int C_{\mathrm{p}}t_{\mathrm{p1}}t_{\mathrm{p2}}P_{\mathrm{p1}}P_{\mathrm{p2}}B_j\bar{y}_\lambda S\mathrm{d}\lambda \qquad j = 1,\ 2,\ 3,\ 4.
\end{aligned}
\tag{6.74}
$$

In many cases, with the help of expressions of this kind, it is easy to estimate the effect of polarizer orientation, ascertain to what extent the PCS at given values of its parameters is appropriate in view of stated goals and criteria, and find optimal polarizer orientation. Some such problems can be solved without scanning the polarizer orientation. Some benefits from using expressions like (6.73) in LCD optimization can be seen from the following example.

Suppose that we are optimizing an LCD whose average transmission t_Y can be expressed by (6.73), considering as principal characteristics of the LCD the values of t_Y in the bright ($\bar{t}_B$) and dark ($\bar{t}_D$) states and the contrast ratio

$$C \equiv \frac{\bar{t}_B}{\bar{t}_D}, \tag{6.75}$$

and have calculated the coefficients $\overline{B}_j$ for the bright state ($\overline{B}_{jB}$) and dark state ($\overline{B}_{jD}$) for a given set of values of the parameters of PCS elements. First of all, we can estimate the minimum dark-state transmittance

$$\bar{t}_{D\,min} \equiv \min \bar{t}_D(\eta^-, \eta^+) = \bar{t}_D\left(\eta^-_{D\,min}, \eta^+_{D\,min}\right) \tag{6.76}$$

that can be attained at the given $\overline{B}_{jD}$. According to (6.73),

$$\bar{t}_{D\,min} = \overline{B}_{0D} - \overline{B}_{13D} - \overline{B}_{24D}, \tag{6.77}$$

where

$$\overline{B}_{13D} = \sqrt{\overline{B}^2_{1D} + \overline{B}^2_{3D}}, \quad \overline{B}_{24D} = \sqrt{\overline{B}^2_{2D} + \overline{B}^2_{4D}}.$$

It is clear that this estimate may be sufficient to reject the given set of values of PCS parameters. The condition

$$\overline{B}_{0D} - \overline{B}_{13D} - \overline{B}_{24D} \leq t_{D\,acceptable}, \tag{6.78}$$

where $t_{D\,acceptable}$ is a maximum acceptable value of $\bar{t}_D$, can be used as a primary criterion in an algorithm of automatic optimization of the PCSs. If condition (6.78) is satisfied, it is reasonable to estimate the bright-state transmittance for the polarizer orientation giving $\bar{t}_D = \bar{t}_{D\,min}$. According to (6.73), the values of η^- (if $\overline{B}_{13D} \neq 0$) and η^+ (if $\overline{B}_{24D} \neq 0$) corresponding to the minimum of $\bar{t}_D$ [see (6.76)] satisfy the relations

$$\begin{aligned} \cos\eta^-_{D\,min} &= -\overline{B}_{1D}\overline{B}^{-1}_{13D}, \quad \sin\eta^-_{D\,min} = -\overline{B}_{3D}\overline{B}^{-1}_{13D}, \\ \cos\eta^+_{D\,min} &= -\overline{B}_{2D}\overline{B}^{-1}_{24D}, \quad \sin\eta^+_{D\,min} = -\overline{B}_{4D}\overline{B}^{-1}_{24D}. \end{aligned} \tag{6.79}$$

Hence, $\bar{t}_B(\eta^-_{D\,min}, \eta^+_{D\,min})$ can be calculated by the formula

$$\bar{t}_B\left(\eta^-_{D\,min}, \eta^+_{D\,min}\right) = \overline{B}_{0B} - (\overline{B}_{1B}\overline{B}_{1D} + \overline{B}_{3B}\overline{B}_{3D})\overline{B}^{-1}_{13D} - (\overline{B}_{2B}\overline{B}_{2D} + \overline{B}_{4B}\overline{B}_{4D})\overline{B}^{-1}_{24D}. \tag{6.80}$$

Using the obtained values of $\bar{t}_{D\,min} = \bar{t}_D(\eta^-_{D\,min}, \eta^+_{D\,min})$ and $\bar{t}_B(\eta^-_{D\,min}, \eta^+_{D\,min})$, we can calculate the contrast ratio corresponding to $\bar{t}_D = \bar{t}_{D\,min}$:

$$C_{D\,min} \equiv \frac{\bar{t}_B\left(\eta^-_{D\,min}, \eta^+_{D\,min}\right)}{\bar{t}_D\left(\eta^-_{D\,min}, \eta^+_{D\,min}\right)}. \tag{6.81}$$

Let $C_{\text{acceptable}}$ be an acceptable level of the contrast ratio C. A simple way to ascertain whether the condition

$$C \geq C_{\text{acceptable}} \tag{6.82}$$

can be satisfied given $\overline{B}_{j\text{B}}$ and $\overline{B}_{j\text{D}}$ or not is the following. It is obvious that condition (6.82) can be satisfied only if for some values of the variables η^- and η^+, the function

$$\Delta t_{\text{s}}(\eta^-, \eta^+, s) = \bar{t}_{\text{B}}(\eta^-, \eta^+) - s\bar{t}_{\text{D}}(\eta^-, \eta^+) \tag{6.83}$$

at $s = C_{\text{acceptable}}$ has positive values, that is, if the maximum value of the function $\Delta t_{\text{s}}(\eta^-, \eta^+, C_{\text{acceptable}})$ is positive. Using (6.73), we can represent the function $\Delta t_{\text{s}}(\eta^-, \eta^+, s)$ as follows:

$$\begin{aligned}\Delta t_{\text{s}}(\eta^-, \eta^+, s) = \overline{B}_{0\text{B}} - s\overline{B}_{0\text{D}} + (\overline{B}_{1\text{B}} - s\overline{B}_{1\text{D}})\cos\eta^- + (\overline{B}_{2\text{B}} - s\overline{B}_{2\text{D}})\cos\eta^+ \\ +(\overline{B}_{3\text{B}} - s\overline{B}_{3\text{D}})\sin\eta^- + (\overline{B}_{4\text{B}} - s\overline{B}_{4\text{D}})\sin\eta^+.\end{aligned} \tag{6.84}$$

According to (6.84), the greatest value of $\Delta t_{\text{s}}(\eta^-, \eta^+, s)$ at a fixed s is equal to

$$\Delta t_{\text{s max}}(s) = \Delta_0(s) + \Delta_{13}(s) + \Delta_{24}(s), \tag{6.85}$$

where

$$\begin{aligned}\Delta_{13}(s) = \sqrt{\Delta_1(s)^2 + \Delta_3(s)^2}, \quad \Delta_{24}(s) = \sqrt{\Delta_2(s)^2 + \Delta_4(s)^2}, \\ \Delta_j(s) = \overline{B}_{j\text{B}} - s\overline{B}_{j\text{D}}, \quad j = 0, 1, 2, 3, 4.\end{aligned} \tag{6.86}$$

Thus, in order to examine whether condition (6.82) can be satisfied or not at the given $\overline{B}_{j\text{B}}$ and $\overline{B}_{j\text{D}}$, it suffices to calculate $\Delta t_{\text{s max}}(C_{\text{acceptable}})$: $\Delta t_{\text{s max}}(C_{\text{acceptable}}) \geq 0$ means that condition (6.82) can be met. The condition

$$\Delta t_{\text{s max}}(s) = 0 \tag{6.87}$$

is satisfied at $s = C_{\text{max}}$, where C_{max} is the maximum value of the contrast ratio C that can be attained at the given $\overline{B}_{j\text{B}}$ and $\overline{B}_{j\text{D}}$. Therefore, one of the possible ways to find C_{max} is to solve (6.87). The root of this equation can readily be found by numerical methods.

Denote values of η^- and η^+ at which $\Delta t_{\text{s}}(\eta^-, \eta^+, s) = \Delta t_{\text{s max}}(s)$ by $\eta_{\text{s}}^-(s)$ and $\eta_{\text{s}}^+(s)$, respectively. It follows from (6.84) that $\eta_{\text{s}}^-(s)$ and $\eta_{\text{s}}^+(s)$ can be found from the equations

$$\cos\eta_{\text{s}}^-(s) = \Delta_1(s)\Delta_{13}(s)^{-1}, \quad \sin\eta_{\text{s}}^-(s) = \Delta_3(s)\Delta_{13}(s)^{-1}, \tag{6.88}$$

$$\cos\eta_{\text{s}}^+(s) = \Delta_2(s)\Delta_{24}(s)^{-1}, \quad \sin\eta_{\text{s}}^+(s) = \Delta_4(s)\Delta_{24}(s)^{-1} \tag{6.89}$$

provided that $\Delta_{13}(s) \neq 0$ and $\Delta_{24}(s) \neq 0$. We denote

$$\bar{t}_{\text{Bs}}(s) \equiv \bar{t}_{\text{B}}\left(\eta_{\text{s}}^-(s), \eta_{\text{s}}^+(s)\right), \quad \bar{t}_{\text{Ds}}(s) \equiv \bar{t}_{\text{D}}\left(\eta_{\text{s}}^-(s), \eta_{\text{s}}^+(s)\right), \tag{6.90}$$

and

$$C_{\text{s}}(s) \equiv \frac{\bar{t}_{\text{Bs}}(s)}{\bar{t}_{\text{Ds}}(s)} \equiv \frac{\bar{t}_{\text{B}}\left(\eta_{\text{s}}^-(s), \eta_{\text{s}}^+(s)\right)}{\bar{t}_{\text{D}}\left(\eta_{\text{s}}^-(s), \eta_{\text{s}}^+(s)\right)}. \tag{6.91}$$

According to (6.88), (6.89), and (6.73),

$$\bar{t}_{Bs}(s) = \overline{B}_{0B} + \frac{\overline{B}_{1B}\Delta_1(s) + \overline{B}_{3B}\Delta_3(s)}{\Delta_{13}(s)} + \frac{\overline{B}_{2B}\Delta_2(s) + \overline{B}_{4B}\Delta_4(s)}{\Delta_{24}(s)}, \tag{6.92}$$

$$\bar{t}_{Ds}(s) = \overline{B}_{0D} + \frac{\overline{B}_{1D}\Delta_1(s) + \overline{B}_{3D}\Delta_3(s)}{\Delta_{13}(s)} + \frac{\overline{B}_{2D}\Delta_2(s) + \overline{B}_{4D}\Delta_4(s)}{\Delta_{24}(s)}. \tag{6.93}$$

It can be proved that at any value of s from the range $0 < s \leq C_{max}$, $\bar{t}_{Bs}(s)$ is the maximum value of $\bar{t}_B(\eta^-, \eta^+)$ attainable at $C = C_s(s)$, which allows one to easily obtain solutions for trade-off choice, like that presented in Figures 6.13 and 6.14.

In calculations of this kind, the symmetry properties of the PCS can be used and sometimes must be taken into account. Thus, if the optical properties of the PCS are invariant with respect to the rotation of it by 180° about an axis **C** parallel to its boundaries [it will take place, for instance, when the PCS is a TN or STN cell with a symmetrical boundary conditions (see Section 6.2.3)] and the x_I-axis [the input and output polarization bases for the PCS, (x_I, y_I, z_I) and (x_O, y_O, z_O), are assumed to be equivalent] is parallel or perpendicular to the axis **C**, the Mueller matrix of the "useful" channel of the PCS will have the form (6.39), and the Jones matrix of this channel, **t**, will have the following form:

$$\begin{pmatrix} t_{11} & t_{12} \\ -t_{12} & t_{22} \end{pmatrix} \tag{6.94}$$

[see (1.252)][2]. In this case,

$$B_2 \geq 0, \quad B_4 = 0 \tag{6.95}$$

[see (6.17)]. With such B_2 and B_4, in (6.73), $\overline{B}_2 \geq 0$ and $\overline{B}_4 = 0$. According to (6.73), the maximum and minimum values of t_Y in this case are reached at $\eta^+ = 2\pi j$ (j = 0, ±1, ±2,...) and $\eta^+ = \pi + 2\pi j$ (j = 0, ±1, ±2,...), respectively. For a TN or STN LCD, the axis **C** will be common for dark and bright states and we will have $\overline{B}_{4D} = \overline{B}_{4B} = 0$ when the x_I-axis is parallel or perpendicular to **C**, and, according to (6.36),

$$\overline{B}_{2B}\overline{B}_{4D} - \overline{B}_{4B}\overline{B}_{2D} = 0 \tag{6.96}$$

for an arbitrary orientation of x_I with respect to **C**. If relation (6.96) holds, $\Delta_{24}(s)$ [see (6.86), (6.89), and (6.93)] at a certain value of s may be equal to zero, which should be taken into account in calculations.

In optimizing PCSs, it is often reasonable to consider a reduced quasimonochromatic transmittance of the "useful" channel of the LC device. In Section 1.4.5, we have defined the reduced transmittance by (1.271) in terms of intensities used in the classical JC. Dealing with more rigorous methods, like that considered in Chapter 12, it is convenient to define the reduced quasimonochromatic transmittance of an LC device by (1.272) with $\boldsymbol{t}_{PCS}$ being the *polarization* Jones matrix of the "useful" channel of the PCS of this device. In what follows, we will deal only with thus defined reduced transmittance, which will be denoted by t_U. For a model of an LC device in which the polarizers are taken to be ideal (C_p = 1,

[2] This form of the matrix **t** is determined by the symmetry of the PCS and the reciprocity properties of its "useful" channel. In Section 1.4.2, we arrived at (1.252) considering a particular system within the framework of the classical JC. An analogous general relation for exact Jones matrices, relating FI Jones vectors, of transfer channels in layered systems with the symmetry in question can be obtained using the reciprocity relations of Section 8.6.2.

$t_{\|pj} = 1$, $t_{\perp pj} = 0, j = 1,2$), $t_{\mathrm{U}} = t_{\|} = 2t_{\mathrm{unp}}$. t_{U} is equal to 1 for a perfect bright state and to 0 for a perfect dark state. The reduced transmittance of a transmissive LC device can be expressed as follows:

$$t_{\mathrm{U}} = 0.5 + B_{\mathrm{U1}} \cos \eta^{-} + B_{\mathrm{U2}} \cos \eta^{+} + B_{\mathrm{U3}} \sin \eta^{-} + B_{\mathrm{U4}} \sin \eta^{+}, \tag{6.97}$$

where $B_{\mathrm{U}j}$ are the unitary polarization transport coefficients [see (6.30)–(6.32)] for the "useful" channel of the PCS of this device. In terms of t_{U}, corresponding reduced integral characteristics are defined which are used in optimization procedures.

Reflective Devices

The Jones matrix of the "useful" channel of the PCS shown in Figure 6.15b can be expressed as follows:

$$\mathbf{t} = \mathbf{t}_{\mathrm{B}}\mathbf{r}_{\mathrm{R}}\mathbf{t}_{\mathrm{F}}, \tag{6.98}$$

where $\mathbf{t}_{\mathrm{F}}$ is the Jones matrix relating the FI Jones vectors of the light incident on the PCS ($\mathbf{J}_{\mathrm{inc}}$) and the light incident on the reflector ($\mathbf{J}_{\mathrm{iR}}$; $\mathbf{J}_{\mathrm{iR}} = \mathbf{t}_{\mathrm{F}}\mathbf{J}_{\mathrm{inc}}$), $\mathbf{t}_{\mathrm{B}}$ is the Jones matrix linking the FI Jones vectors of the light reflected from the reflector ($\mathbf{J}_{\mathrm{rR}}$) and the light emerging from the PCS ($\mathbf{J}_{\mathrm{out}}$; $\mathbf{J}_{\mathrm{out}} = \mathbf{t}_{\mathrm{B}}\mathbf{J}_{\mathrm{rR}}$), and $\mathbf{r}_{\mathrm{R}}$ is the Jones matrix characterizing reflection from the reflector ($\mathbf{J}_{\mathrm{rR}} = \mathbf{r}_{\mathrm{R}}\mathbf{J}_{\mathrm{iR}}$). We assume that there is an isotropic nonabsorbing layer between the LC layer and the reflector (see Figure 6.15b). This layer is considered as the medium of incidence for the reflector. In the case of normal incidence, using the frame $(x_{\mathrm{I}}, y_{\mathrm{I}})$ as the reference frame for the vectors $\mathbf{J}_{\mathrm{inc}}$ and $\mathbf{J}_{\mathrm{iR}}$ and the frame $(x_{\mathrm{R}}, y_{\mathrm{R}})$ oriented with respect to the frame $(x_{\mathrm{I}}, y_{\mathrm{I}})$ as shown in Figure 6.6a or 6.6b as the reference frame for the vectors $\mathbf{J}_{\mathrm{rR}}$ and $\mathbf{J}_{\mathrm{out}}$, we may express the matrices $\mathbf{r}_{\mathrm{R}}$ and $\mathbf{t}_{\mathrm{B}}$ as

$$\mathbf{r}_{\mathrm{R}} = R_{\mathrm{R}} \begin{pmatrix} \mp 1 & 0 \\ 0 & \pm 1 \end{pmatrix}, \tag{6.99}$$

$$\mathbf{t}_{\mathrm{B}} = \begin{pmatrix} t_{\mathrm{F11}} & -t_{\mathrm{F21}} \\ -t_{\mathrm{F12}} & t_{\mathrm{F22}} \end{pmatrix}. \tag{6.100}$$

In (6.99), R_{R} is the amplitude reflection coefficient of the reflector; the upper signs correspond to the orientation of the frame $(x_{\mathrm{R}}, y_{\mathrm{R}})$ shown in Figure 6.6a, and the lower signs to that shown in Figure 6.6b. In (6.100), $t_{\mathrm{F}jk}$ are the elements of the matrix

$$\mathbf{t}_{\mathrm{F}} = \begin{pmatrix} t_{\mathrm{F11}} & t_{\mathrm{F12}} \\ t_{\mathrm{F21}} & t_{\mathrm{F22}} \end{pmatrix}. \tag{6.101}$$

Equation (6.100) holds due to the absence of nonreciprocal optical effects in the system (see Sections 1.4.2 and 8.6.2). It follows from (6.98) that with $\mathbf{r}_{\mathrm{R}}$ and $\mathbf{t}_{\mathrm{B}}$ expressed by (6.99) and (6.100), the matrix $\mathbf{t}$ has the form (6.94) [the corresponding Mueller matrix of the channel will have the form (6.39)]. Neglecting the diattenuation in the "useful" channel of the PCS, we obtain the following expressions for the factors $\rho_{\|}$ and $\rho_{\perp}$ (see Section 6.3) associated with this channel:

$$\rho_{\|}(\vartheta) = B_0 + B_2 + B_1 \cos 4\vartheta + B_3 \sin 4\vartheta, \tag{6.102}$$

$$\rho_{\perp}(\vartheta) = B_0 - B_2 - B_1 \cos 4\vartheta - B_3 \sin 4\vartheta, \tag{6.103}$$

with B_j expressed by (6.17). In this case, B_j satisfy the relations

$$B_0 = B_2 + \sqrt{B_1^2 + B_3^2}, \tag{6.104}$$

$$B_2 \geq 0. \tag{6.105}$$

How to use expressions (6.102) and (6.103) in optimization procedures is obvious from the above consideration of transmissive devices. The reduced quasimonochromatic transmittance of the "useful" channel of a single-polarizer RLCD can be expressed as follows:

$$t_U = 0.5 + B_{U2} + B_{U1}\cos 4\vartheta + B_{U3}\sin 4\vartheta. \tag{6.106}$$

In the case of an RLCD with PBS, the reduced transmittance can be expressed as

$$t_U = 0.5 - B_{U2} - B_{U1}\cos 4\vartheta - B_{U3}\sin 4\vartheta. \tag{6.107}$$

The unitary polarization transport coefficients of the PCS entering into these expressions can be computed by formulas (6.31) from the polarization Jones matrix of this channel, $\mathbf{t}_{UM}$ [see (6.21)], calculated as follows:

$$\mathbf{t}_{UM} = \mathbf{t}_{UB}\begin{pmatrix} -i & 0 \\ 0 & i \end{pmatrix}\mathbf{t}_{UF}, \tag{6.108}$$

where $\mathbf{t}_{UF}$ and $\mathbf{t}_{UB}$ are the unimodular polarization Jones matrices of determinant 1 characterizing the sections of the "useful" channel described by the matrices $\mathbf{t}_F$ and $\mathbf{t}_B$, respectively.

6.5 Evaluation of Ultimate Characteristics of an LCD that can be Attained by Fitting the Compensation System. Modulation Efficiency of LC Layers

For many types of LCDs, the best performance is achieved by using appropriate compensation systems. The usual compensation system is a set of homogeneous or inhomogeneous birefringent films situated between the LC layer and the polarizers. In most cases, the function of the compensation system is to ensure a black–white switching, as good as possible, for a given pair of working states of the LC layer corresponding to the dark state and the bright state of the LCD. In the case of 90° TN LCDs, where compensators are generally not required to obtain good black–white switching for the normal viewing direction, the primary purpose of compensation systems is to ensure satisfactory dark and bright states for a wide range of viewing angles. In the case of STN LCDs, in practically interesting situations the "bright" and "dark" states of the LC layer are not suited for a good black–white switching without compensators even for the normal viewing direction. In this case, the primary purpose of the compensation system is to improve the transmission spectra of the LCD panel in the dark and bright states for the normal viewing direction.

Almost always, with the help of an appropriate compensation system, it is possible to obtain very good dark-state performance. However, the improvement of the dark state is often accompanied by unacceptable deterioration or insufficient improvement of the bright-state performance. It is clear that the possibility to obtain a good dark state and a good bright state with the same compensation system is fully determined by properties of the LC layer in the "bright" and "dark" states. We will call the ability of an LC layer, in two specified states, to give a good bright state along with a good dark state, with a proper compensation system, the *modulation efficiency*. Two different, but closely related, parameters characterizing the modulation efficiency of LC layers were proposed in [9, 10]. In this section, we will consider in detail one of them, introduced in [10]. This parameter, called the *modulation efficiency factor* (*MEF*), is proportional to the maximum transmittance of the "useful" channel of the LCD in the bright state that can be obtained with the help of polarization–compensation systems (polarizers and compensators) providing a zero dark-state transmittance. As everywhere in this chapter, we restrict our consideration to the case of normal incidence. The case of oblique incidence in this context will be considered in Section 12.6.

Transmissive Devices

Let us consider a transmissive device with the structure: polarizer–compensator–LC layer–compensator–polarizer. The polarizer and compensator that are passed by the light first will be called polarizer 1 and compensator 1. The second polarizer and second compensator will be called polarizer 2 and compensator 2. The polarizers are assumed to be ideal. We consider the light propagation through the "useful" channel of this device. The diattenuation in the compensators and LC layer is assumed to be negligible. We denote the polarization Jones matrices of the LC layer in the "dark" and "bright" states by $\boldsymbol{t}_{\text{LC-D}}$ and $\boldsymbol{t}_{\text{LC-B}}$, respectively, the polarization Jones matrix of compensator 2 by $\boldsymbol{t}_{\text{C2}}$, the polarization Jones vector of the light transmitted by polarizer 1 and compensator 1 and incident on the LC layer by $\boldsymbol{j}_{\text{i}}$, and the polarization Jones vectors of the light emerging from the PCS of the device and incident on polarizer 2 for the "dark" and "bright" states of the LC layer by $\boldsymbol{j}_{\text{tD}}$ and $\boldsymbol{j}_{\text{tB}}$, respectively. The vectors $\boldsymbol{j}_{\text{tD}}$ and $\boldsymbol{j}_{\text{tB}}$ are related to $\boldsymbol{j}_{\text{i}}$ by

$$\boldsymbol{j}_{\text{tD}} = \boldsymbol{t}_{\text{2D}}\boldsymbol{j}_{\text{i}}, \tag{6.109}$$

$$\boldsymbol{j}_{\text{tB}} = \boldsymbol{t}_{\text{2B}}\boldsymbol{j}_{\text{i}}, \tag{6.110}$$

where

$$\boldsymbol{t}_{\text{2D}} = \boldsymbol{t}_{\text{C2}}\boldsymbol{t}_{\text{LC-D}}, \tag{6.111}$$

$$\boldsymbol{t}_{\text{2B}} = \boldsymbol{t}_{\text{C2}}\boldsymbol{t}_{\text{LC-B}}. \tag{6.112}$$

We denote the polarization Jones vectors of the polarizations fully transmitted and fully extinguished by polarizer 2 by $\boldsymbol{P}$ and $\tilde{\boldsymbol{P}}$, respectively ($\tilde{\boldsymbol{P}}^{\dagger}\boldsymbol{P} = 0$). The transmittance of this polarizer for a wave with polarization Jones vector $\boldsymbol{j}$, incident on it, can be expressed as

$$t = |\boldsymbol{P}^{\dagger}\boldsymbol{j}|^2 \tag{6.113}$$

and

$$t = 1 - |\tilde{\boldsymbol{P}}^{\dagger}\boldsymbol{j}|^2. \tag{6.114}$$

In view of this, we may express the reduced transmittances of the device in the dark ($t_{\text{U-D}}$) and bright ($t_{\text{U-B}}$) states as follows:

$$t_{\text{U-D}} = \left|\boldsymbol{P}^{\dagger}\boldsymbol{j}_{\text{tD}}\right|^2, \tag{6.115}$$

$$t_{\text{U-B}} = \left|\boldsymbol{P}^{\dagger}\boldsymbol{j}_{\text{tB}}\right|^2 = 1 - \left|\tilde{\boldsymbol{P}}^{\dagger}\boldsymbol{j}_{\text{tB}}\right|^2. \tag{6.116}$$

The transmittance of the device in the dark state will be zero, that is,

$$t_{\text{U-D}} = \left|\boldsymbol{P}^{\dagger}\boldsymbol{j}_{\text{tD}}\right|^2 = 0, \tag{6.117}$$

when $\boldsymbol{j}_{\text{tD}}$ and $\tilde{\boldsymbol{P}}$ are such that

$$\tilde{\boldsymbol{P}} = k\boldsymbol{j}_{\text{tD}}, \tag{6.118}$$

where k is a scalar factor with $|k|^2 = 1$ (since, by definition, $|\boldsymbol{j}_{\text{tD}}|^2 = 1$ and $|\tilde{\boldsymbol{P}}|^2 = 1$). Assume that parameters of the polarizers and compensators are chosen so that the transmittance of the device in the dark state is zero. We denote the reduced transmittance of the device in the bright state under these conditions by $t_{\text{U-B0}}$. It follows from (6.116) and (6.118) that

$$t_{\text{U-B0}} = 1 - \left| \boldsymbol{j}_{t\text{D}}^{\dagger} \boldsymbol{j}_{t\text{B}} \right|^2. \tag{6.119}$$

Substituting (6.109)–(6.112) into (6.119), using the matrix identity (5.15), and taking into account that the matrix $\boldsymbol{t}_{\text{C2}}$ is unitary (i.e., $\boldsymbol{t}_{\text{C2}}^{\dagger} \boldsymbol{t}_{\text{C2}} = \mathbf{U}$, where $\mathbf{U}$ is the unit matrix), we obtain

$$t_{\text{U-B0}} = 1 - \left| \boldsymbol{j}_{\text{i}}^{\dagger} \boldsymbol{t}_{2\text{D}}^{\dagger} \boldsymbol{t}_{2\text{B}} \boldsymbol{j}_{\text{i}} \right|^2 = 1 - \left| \boldsymbol{j}_{\text{i}}^{\dagger} \boldsymbol{t}_{\text{LC-D}}^{\dagger} \boldsymbol{t}_{\text{C2}}^{\dagger} \boldsymbol{t}_{\text{C2}} \boldsymbol{t}_{\text{LC-B}} \boldsymbol{j}_{\text{i}} \right|^2 = 1 - \left| \boldsymbol{j}_{\text{i}}^{\dagger} \boldsymbol{t}_{\text{LC-D}}^{\dagger} \boldsymbol{t}_{\text{LC-B}} \boldsymbol{j}_{\text{i}} \right|^2. \tag{6.120}$$

We may therefore express the transmittance $t_{\text{U-B0}}$ as follows:

$$t_{\text{U-B0}} = 1 - \left| \boldsymbol{j}_{\text{i}}^{\dagger} \boldsymbol{t}_{\text{D-B}} \boldsymbol{j}_{\text{i}} \right|^2, \tag{6.121}$$

where

$$\boldsymbol{t}_{\text{D-B}} = \boldsymbol{t}_{\text{LC-D}}^{\dagger} \boldsymbol{t}_{\text{LC-B}}. \tag{6.122}$$

We see that $t_{\text{U-B0}}$ depends only on properties of the LC layer and the state of polarization of the light incident on the LC layer. Using suitable compensator 1, and properly orienting polarizer 1, one can set any desired polarization of the light incident on the LC layer. The largest value of the transmittance $t_{\text{U-B0}}$ as a function of $\boldsymbol{j}_{\text{i}}$ will be referred to as the *modulation efficiency factor* (*MEF*) of the LC layer for the given pair of its states. In the case under consideration, the MEF can be estimated using the following mathematics.

Let $\mathbf{A}$ be a unitary 2×2 matrix of the form

$$\mathbf{A} = \begin{pmatrix} a' + ia'' & b' + ib'' \\ -b' + ib'' & a' - ia'' \end{pmatrix}, \tag{6.123}$$

where a', a'', b', and b'' are real numbers, and let

$$\boldsymbol{X}(\alpha, \beta) = \begin{pmatrix} \cos\alpha \\ e^{i\beta} \sin\alpha \end{pmatrix}, \tag{6.124}$$

where α and β are real variables. It is easy to find that the minimum value of the function

$$f(\alpha, \beta) = |\boldsymbol{X}(\alpha, \beta)^{\dagger} \mathbf{A} \boldsymbol{X}(\alpha, \beta)|^2 \tag{6.125}$$

is equal to a'^2. Actually, substitution of (6.123) and (6.124) into (6.125) gives the following expression:

$$f(\alpha, \beta) = a'^2 + [a'' \cos 2\alpha + (b' \sin\beta + b'' \cos\beta) \sin 2\alpha]^2. \tag{6.126}$$

As can be seen from this expression, at any given value of β, there exist values of α such that the term in the square brackets in (6.126) is equal to zero and hence $f(\alpha, \beta) = a'^2$. By making use of this property, we can easily find the maximum value of the function $t_{\text{U-B0}}(\boldsymbol{j}_{\text{i}})$ (6.121). Any polarization state of polarized light can be represented by a polarization Jones vector of the form (6.124). At the same time, the matrix $t_{\text{D-B}}$, being the product of two unitary matrices [see (6.122)], is unitary and, consequently, can be represented as

$$t_{\text{D-B}} = (\det t_{\text{D-B}})^{1/2} t_{\text{D-B(UM)}}, \tag{6.127}$$

where the matrix $t_{\text{D-B(UM)}} = (\det t_{\text{D-B}})^{-1/2} t_{\text{D-B}}$ is unitary and of determinant 1 and, consequently, has the form (6.123) (see Section 5.1.3). Since the matrix $t_{\text{D-B}}$ is unitary, $|\det t_{\text{D-B}}| = 1$, and hence

$$\left| j_{\text{i}}^{\dagger} \mathbf{t}_{\text{D-B}} j_{\text{i}} \right|^2 = \left| j_{\text{i}}^{\dagger} \mathbf{t}_{\text{D-B(UM)}} j_{\text{i}} \right|^2. \tag{6.128}$$

From the aforesaid, the minimum of $|j_{\text{i}}^{\dagger} \mathbf{t}_{\text{D-B(UM)}} j_{\text{i}}|^2$ as function of j_{i} is equal to $(\text{Re}[\mathbf{t}_{\text{D-B(UM)}}]_{11})^2$ ($[\mathbf{t}]_{kj}$ stands for the element (k, j) of a matrix $\mathbf{t}$). Therefore, the MEF of the LC layer,

$$Q \equiv \max t_{\text{U-B0}}(j_{\text{i}}), \tag{6.129}$$

can be expressed as follows:

$$Q = 1 - (\text{Re}[\mathbf{t}_{\text{D-B(UM)}}]_{11})^2, \tag{6.130}$$

or, equivalently,

$$Q = 1 - \left[\frac{1}{2} \text{Tr} \left(\frac{1}{\sqrt{\det \mathbf{t}_{\text{D-B}}}} \mathbf{t}_{\text{D-B}} \right) \right]^2. \tag{6.131}$$

The maximum modulation of the transmittance, that is, an ideal switching, can be attained when the modulation efficiency factor Q is equal to 1. If $Q \ll 1$, it is impossible to obtain a good switching in principle: under conditions providing a good dark state, the bright-state transmittance will be small, whatever compensators are used. The minimum fraction of the light that will be lost in the bright state with polarizers and compensators maintaining zero transmittance in the dark state is equal to $1 - Q$. Note that, according to (6.131) and (6.122), for two arbitrary states of an LC layer, state 1 and state 2, considered as the dark and bright states, the choice of state 1 as the dark state and state 2 as the bright state and the choice of state 2 as the dark state and state 1 as the bright state will give the same values of the MEF. One can show that the maximum difference of the reduced transmittances in the bright and dark states, $t_{\text{U-B}} - t_{\text{U-D}}$, that can be reached by variation of parameters of polarizers and compensators is equal to $\sqrt{Q}$. A characteristic of the modulation efficiency of LC layers, derived by maximization of the difference $t_{\text{U-B}} - t_{\text{U-D}}$ and equivalent to that which we could define here as $Q_{\Delta} \equiv \sqrt{Q}$, was introduced in [9].

An arbitrary linear polarization can be represented by a polarization Jones vector of the form (6.124) with $\beta = 0$. We have noted that the least value of the function $f(\alpha, \beta)$ [see (6.126)], equal to a'^2, can be reached at any β. From this it is clear that the limit $t_{\text{U-B0}}(j_{\text{i}}) = Q$ can be attained, in particular, for certain linear polarizations of the incident light, that is, in the absence of compensator 1. This limit can also be attained with compensator 1 only (without compensator 2), which is evident in view of the reciprocity of the "useful" channel transmission.

The factor Q in general varies with wavelength. For example, for a nontwisted nematic LC layer, the wavelength dependence of Q can be described by the formula

$$Q = \sin^2 \left[\frac{\pi d}{\lambda} (\overline{n}_{\text{e(B)}}(\lambda) - \overline{n}_{\text{e(D)}}(\lambda)) \right], \tag{6.132}$$

where $\overline{n}_{e(B)}$ and $\overline{n}_{e(D)}$ are the values of the effective extraordinary refractive index of the LC layer

$$\overline{n}_e = \int_{z_1}^{z_2} \frac{n_{\parallel} n_{\perp}}{\sqrt{n_{\perp}^2 \cos^2 \theta(z) + n_{\parallel}^2 \sin^2 \theta(z)}} dz \tag{6.133}$$

in the bright and dark states, respectively. In (6.133), $n_{\parallel}$ and $n_{\perp}$ are the principal refractive indices of the liquid crystal, and θ is the tilt angle of the LC director; $z = z_1$ and $z = z_2$ are the planes of the boundaries of the LC layer (see Figure 6.3).

A good black–white switching can be obtained if the modulation efficiency factor Q is close to 1 throughout the visible region. For the dark and black states of a typical 90° TN LCD, the values of Q in the visible region lie between about 0.8 and 1. For STN LCDs, there exist combinations of the values of parameters of the LC layer at which satisfactory black–white switching can be reached as well. However, in this case, in contrast to the case of TN LCDs, it is impossible to obtain sufficiently high and achromatic bright-state transmission in combination with sufficiently small dark-state transmission throughout the visible region without compensators. Moreover, domains in the parameter space of the LC layer (including the working voltages) that are suitable for good black–white switching for STN LCDs are much narrower than for 90° TN LCDs. The assessment of appropriateness of LC layers and working voltages for black–white switching by examination of the MEF spectrum significantly simplifies finding such domains.

If the wavelength dependence of Q for two working states of the LC layer is suitable for black–white switching, by using optimization methods, it is possible to find parameters of compensation systems providing a small dark-state transmittance ($t_{U\text{-}D} \approx 0$) and $t_{U\text{-}B}(\lambda) \approx Q(\lambda)$ throughout the visible region (optimization engines of our program MOUSE-LCD successfully and rather quickly solve problems of this kind). As an illustration, let us consider the following example. Figure 6.16 shows results of solving the deformation problem for an STN layer with the following parameters: $K_{11} = 1.28 \times 10^{-6}$ dyn, $K_{22} = 7.25 \times 10^{-7}$ dyn, $K_{33} = 2.06 \times 10^{-6}$ dyn, $\varepsilon_{\parallel} = 11.07$, $\varepsilon_{\perp} = 3.61$, twist angle $\Phi = 240°$, pretilt angles $\theta_1 = \theta_2 = 4°$, and $d/p_0 = 0.5$, where d is the thickness of the LC layer and p_0 is the natural helix pitch of the LC; the anchoring is assumed to be infinitely strong. The drastic changes of the LC director field configuration with voltage in the narrow voltage range 2.35–2.5 V allow us to choose the dark-state voltage and bright-state voltage rather close to each other to have a high multiplex ratio N. Our aim was to find a thickness of the LC layer, dark- and bright-state voltages, a compensation system, and orientation angles of the polarizers providing good black–white switching at $N \approx 400$. The wavelength dependences of the principal refractive indices of the LC were represented by the three-coefficient Cauchy equations[3] with coefficients calculated from the following data:

Wavelength (nm)	$n_{\perp}$	$n_{\parallel}$
400	1.5012	1.6086
550	1.4816	1.5805
700	1.4752	1.5696

By examination of the dependence of the MEF spectrum on the LC layer thickness and working voltages, we found that the situation most suitable for black–white switching under the given conditions is realized at $d = 8.17$ μm and working voltages $U_1 = 2.35$ V and $U_2 = 2.47$ V ($N \approx 400$). The LC

[3] We use this kind of approximation of wavelength dependences of refractive indices in all numerical examples in this and following chapters where the spectrum of a refractive index for a nonabsorbing medium is specified by values of this index for three wavelength values.

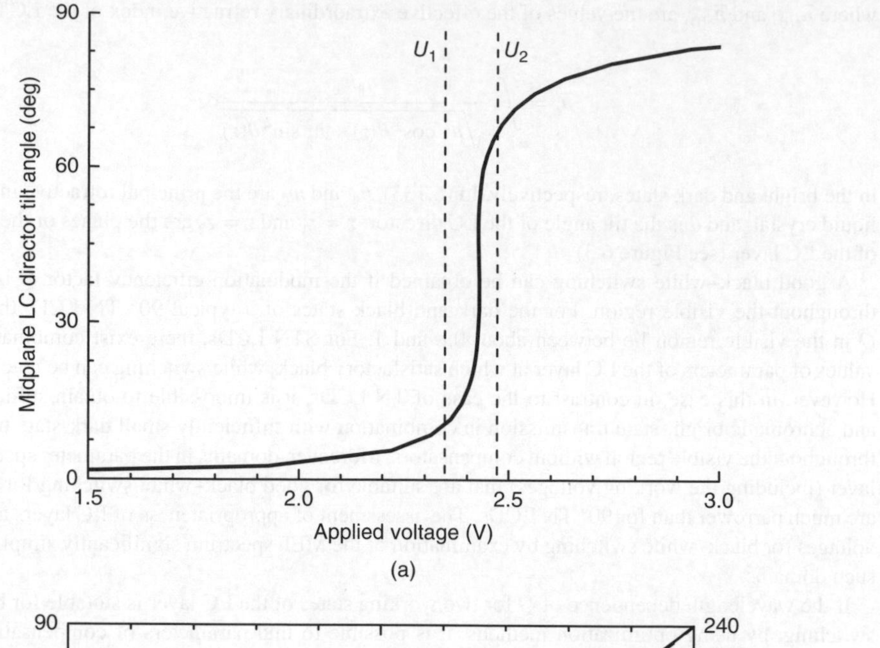

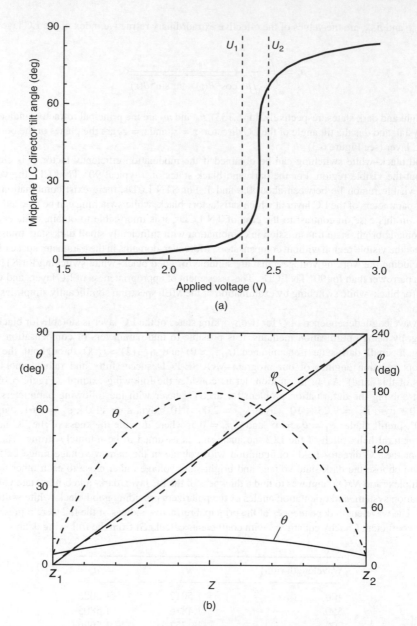

Figure 6.16 (a) Voltage dependence of the midplane tilt angle of the LC director in the LC layer under consideration. (b) LC director profiles in this layer at voltages 2.35 V (U_1, solid lines) and 2.47 V (U_2, dashed lines); θ and φ are respectively the tilt angle and azimuthal angle of the LC director

director field configurations for these voltages are shown in Figure 6.16b. We denote the dark-state and bright-state voltages by U_D and U_B, respectively. The spectrum of Q corresponding to the choice (U_B = 2.35 V, U_D = 2.47 V) or (U_B = 2.47 V, U_D = 2.35 V) at d = 8.17 μm is shown in Figures 6.17–6.19 along with the reduced transmittance spectra for four optimized variants of the device in the dark and bright states. The structure and parameters of the optimized variants of the device are shown in Table 6.1.

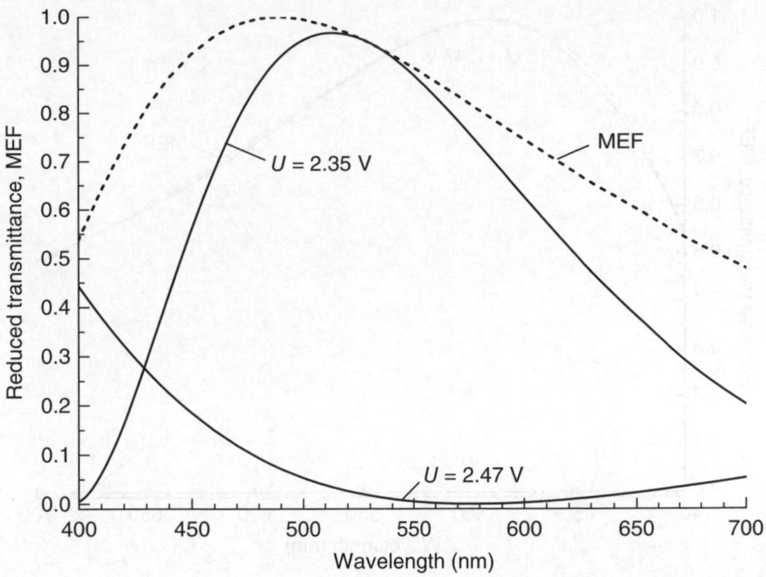

Figure 6.17 Spectrum of the modulation efficiency factor Q (MEF, dashed line) of the LC layer at working voltages $U_1 = 2.35$ V and $U_2 = 2.47$ V and the reduced transmittance spectra for variant 1 (no compensators) of the LCD at these voltages (solid lines)

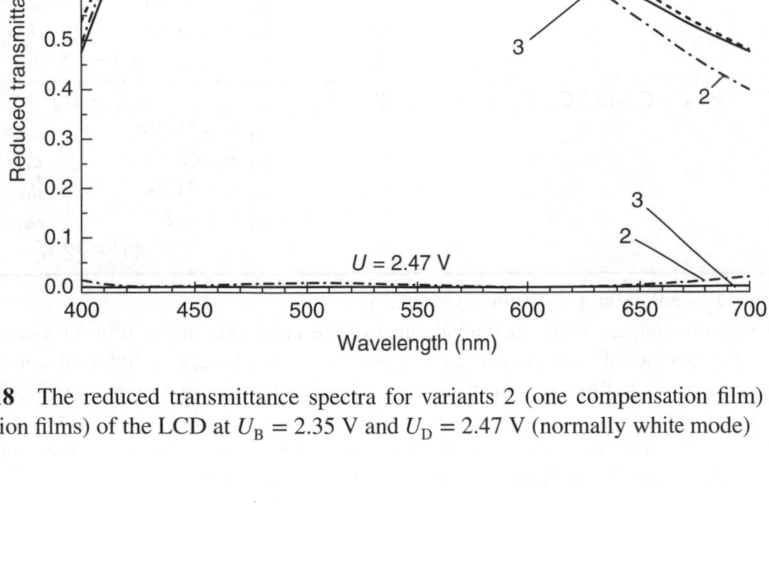

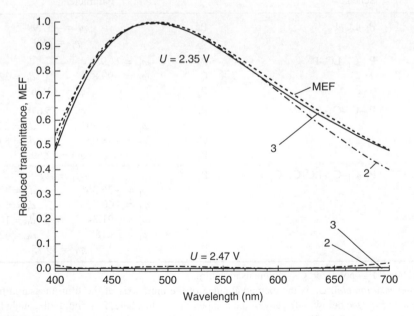

Figure 6.18 The reduced transmittance spectra for variants 2 (one compensation film) and 3 (two compensation films) of the LCD at $U_B = 2.35$ V and $U_D = 2.47$ V (normally white mode)

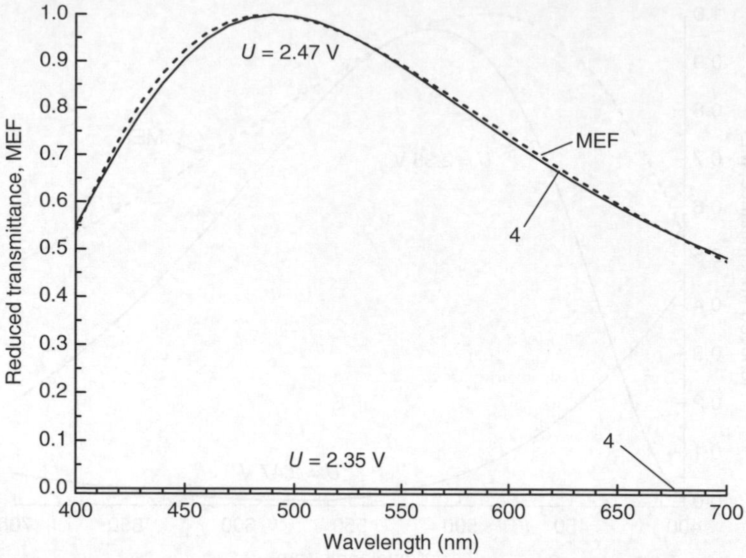

Figure 6.19 The reduced transmittance spectra for variant 4 (four compensation films, normally black mode) of the LCD at $U_D = 2.35$ V and $U_B = 2.47$ V

Table 6.1 Optimized variants of the STN LCD under consideration

Variant	Scheme[a]		Parameters[b]	
1	P_1–LC–P_2	P_1	$\varphi_L = -37.99°$	
		P_2	$\varphi_L = 7.99°$	
2	P_1–C_1–LC–P_2	P_1	$\varphi_L = 21.98°$	
		C_1	$\varphi_L = -99.82°$	$d_L = 32.97$ μm
		P_2	$\varphi_L = 11.36°$	
3	P_1–C_1–C_2–LC–P_2	P_1	$\varphi_L = 85.67°$	
		C_1	$\varphi_L = -64.39°$	$d_L = 22.93$ μm
		C_2	$\varphi_L = -88.05°$	$d_L = 8.69$ μm
		P_2	$\varphi_L = 12.69°$	
4	P_1–C_1–C_2–LC–C_3–C_4–P_2	P_1	$\varphi_L = 1.97°$	
		C_1	$\varphi_L = -26.36°$	$d_L = 42.36$ μm
		C_2	$\varphi_L = -60.62°$	$d_L = 16.73$ μm
		C_3	$\varphi_L = -91.34°$	$d_L = 16.73$ μm
		C_4	$\varphi_L = 28.18°$	$d_L = 42.36$ μm
		P_2	$\varphi_L = 89.75°$	

[a]LC—LC layer; P_j—polarizer; C_j—compensation film.
[b]For *a compensation film*, φ_L is the azimuthal angle of the optic axis of the film, measured from the reference axis P_{LC} (see below)—for any of the compensation films here, it is simply the angle between P_{LC} and the optic axis of the film, d_L is the thickness of the film. For *a polarizer*, φ_L is the angle between the axis P_{LC} and transmission axis of the polarizer. The reference axis P_{LC} is parallel to the boundaries and oriented so that this axis, the easy axis at the frontal boundary of the LC layer, and the normal to this boundary are coplanar and the angle between P_{LC} and the easy axis is acute.

Parameters of the compensation systems and polarizers were calculated with the help of the optimization instruments of MOUSE-LCD. Figure 6.17 shows the spectra for a variant with no compensation system (variant 1) at $U_B = 2.35$ V and $U_D = 2.47$ V. In this case, the polarizers are oriented in such a way as to maximize the contrast ratio $\overline{C}$ calculated as

$$\overline{C} = \int_{400\,\text{nm}}^{700\,\text{nm}} t_{\text{U-B}} d\lambda \Bigg/ \int_{400\,\text{nm}}^{700\,\text{nm}} t_{\text{U-D}} d\lambda. \tag{6.134}$$

As can be seen from Figure 6.17, black–white switching is not attained in this case: the bright-state transmission of the LCD is too low in the blue and red regions, while the dark-state transmission is too high in the blue region. The contrast ratio $\overline{C}$ in this case is as low as 5.5. All these defects can be easily corrected by using compensation films. The bright- and dark-state spectra which can be obtained by introducing compensation films are shown in Figure 6.18 (variants 2 and 3). In all examples presented here, compensation systems are composed of homogeneous uniaxial films with optic axis parallel to the film surfaces and the following principal refractive indices:

Wavelength (nm)	$n_\perp$	$n_\parallel$
400	1.5578	1.5673
550	1.5461	1.5554
700	1.5407	1.5554

In variant 2, only one film is employed for compensation. As is seen from Figure 6.18, the use of this film significantly improves both the dark- and bright-state spectra. The contrast ratio $\overline{C}$ for this variant is about 98. The double-layer compensation system used in variant 3 increases $\overline{C}$ up to 818 and makes the bright-state spectrum rather close to the spectrum of the MEF of the LC layer. In variants 1, 2, and 3, the bright-state voltage is less than the dark-state voltage ($U_B = 2.35$ V, $U_D = 2.47$ V). As has been noted, MEF is invariant with respect to the swap of working states, that is, the ultimate characteristics for the cases ($U_B = U_1$, $U_D = U_2$) and ($U_B = U_2$, $U_D = U_1$) are equivalent. As an illustration, we took $U_B = 2.47$ V and $U_D = 2.35$ V in variant 4 (see Table 6.1 and Figure 6.19). In this case, four compensation films are used, which provides $\overline{C} > 1000$ and, again, $t_{\text{U-B}}(\lambda) \approx Q(\lambda)$ throughout the visible region. We should note that compensation systems providing such good correction of the spectra for the normal direction as for variants 3 and 4 often give viewing angle characteristics that are far from the best ones attainable for the given LC layer and working voltages. This is the case for all the variants presented. Examples of compensation systems for the LC layer and working voltages taken in the above examples that improve the viewing angle characteristics of the LCD will be given in Section 12.6.

Reflective Devices

In an analogous way, one can show that for reflective devices with PCSs having the structure shown in Figure 6.15b, the maximum reduced transmittance of the "useful" channel of the device in the bright state ($t_{\text{U-B}}$) attainable at zero transmittance of this channel in the dark state (i.e., at $t_{\text{U-D}} = 0$) can be expressed as follows:

$$Q = 1 - \left[\frac{1}{2}\text{Tr}\left(\frac{1}{\sqrt{\det \mathbf{t}'_{\text{D-B}}}}\mathbf{t}'_{\text{D-B}}\right)\right]^2, \tag{6.135}$$

where

$$t'_{\text{D-B}} = t^{\dagger}_{\text{LCR-D}} t_{\text{LCR-B}} \tag{6.136}$$

with $t_{\text{LCR-B}}$ and $t_{\text{LCR-D}}$ being the values of the polarization Jones matrix t_{LCR} characterizing the chain of operations {(forward) transmission of the LC layer–reflection from the reflector–(backward) transmission of the LC layer} (see Figure 6.15b) in the bright state and dark state, respectively. The matrix t_{LCR} can be expressed as

$$t_{\text{LCR}} = t^{\uparrow}_{\text{LC}} r t^{\downarrow}_{\text{LC}}, \tag{6.137}$$

where $t^{\downarrow}_{\text{LC}}$ and $t^{\uparrow}_{\text{LC}}$ are the polarization Jones matrices characterizing respectively the forward and backward transmission of the LC layer and r is the polarization Jones matrix of the reflector. In the case of normal incidence, which we consider here, the matrix t_{LCR} can be expressed in terms of the elements of the matrix $t^{\downarrow}_{\text{LC}} = \left[t^{\downarrow}_{ij}\right]$ as follows

$$t_{\text{LCR}} = c_i \begin{pmatrix} -t^{\downarrow 2}_{11} - t^{\downarrow 2}_{21} & -t^{\downarrow}_{11} t^{\downarrow}_{12} - t^{\downarrow}_{21} t^{\downarrow}_{22} \\ t^{\downarrow}_{11} t^{\downarrow}_{12} + t^{\downarrow}_{21} t^{\downarrow}_{22} & t^{\downarrow 2}_{12} + t^{\downarrow 2}_{22} \end{pmatrix}, \tag{6.138}$$

where we have used the reciprocity relation for the matrices $t^{\downarrow}_{\text{LC}}$ and $t^{\uparrow}_{\text{LC}}$ (see Sections 1.4.3 and 8.6.2). Depending on the choice of the polarization reference system for backward propagating light (see Figure 6.6) and the matrix r, the coefficient c_i in (6.138) may be equal to 1, –1, i, $-i$, or other complex number of magnitude 1. In any case, the coefficients c_i for the matrices $t_{\text{LCR-B}}$ and $t_{\text{LCR-D}}$ are the same, and they cancel out in (6.136). One can show that both in the case of a single-polarizer RLCD and in the case of an RLCD with PBS, the polarization–compensation system can be chosen such that the conditions $t_{\text{U-B}} = Q$ and $t_{\text{U-D}} = 0$ are satisfied. Furthermore, one can prove that in the former case, to obtain $t_{\text{U-B}} = Q$ it suffices to satisfy the condition $t_{\text{U-D}} = 0$.

Examples of using estimates of MEF in optimization of RLCDs can be found in [10].

References

[1] R. M. A. Azzam and N. M. Bashara, *Ellipsometry and Polarized Light* (North-Holland, Amsterdam, 1977).

[2] R.A. Chipman, "Mueller matrices," in *Handbook of Optics, Vol. 1. Geometrical and Physical Optics, Polarized Light, Components and Instruments*, edited by M. Bass and V. N. Mahajan, 3rd ed. (McGraw-Hill, New York, 2010).

[3] R. C. Jones, "A new calculus for the treatment of optical systems I. Description and discussion of the calculus," *J. Opt. Soc. Am.* **31**, 488 (1941).

[4] D. A. Yakovlev and V. G. Chigrinov, "A robust polarization-spectral method for determination of twisted liquid crystal layer parameters," *J. Appl. Phys.* **102**, 023510 (2007).

[5] V. G. Chigrinov, V. M. Kozenkov, and H. S. Kwok, *Photoalignment of Liquid Crystalline Materials: Physics and Applications* (Wiley, Chichester, 2008).

[6] T. J. Scheffer and J. Nehring, "Optimization of contrast ratio in reversed-polarizer, transmissive-type twisted nematic displays," *J. Appl. Phys.* **56**, 908 (1984).

[7] T. J. Scheffer and J. Nehring, "Investigation of the electro-optical properties of 270° chiral nematic layers in the birefringence mode," *J. Appl. Phys.* **58**, 3022 (1985).

[8] F. Leenhouts and M. Schadt, "Optics of twisted nematic and supertwisted nematic liquid-crystal displays," *J. Appl. Phys.* **60**, 3275 (1986).

[9] S. Stallinga, J. M. A. van den Eerenbeemd, and J. A. M. M. van Haaren, "Fundamental limitations to the viewing angle of liquid crystal displays with birefringent compensators," *Jpn. J. Appl. Phys.* **37**, 560 (1998).

[10] D. A. Yakovlev, V. G. Chigrinov, and H. S. Kwok, "Optimization of one-polarizer reflective LCDs with phase compensator," *Mol. Cryst. Liq. Cryst.* **410**, 351 (2004).

7

Some Physical Models and Mathematical Algorithms Used in Modeling the Optical Performance of LCDs

A key point in the modeling of optical characteristics of an optical system is the choice of the physical model of the interaction of the optical system and light. The physical model is a set of physical laws, assumptions, and approximations used in constructing the corresponding mathematical model. One of few optical problems that can be solved exactly starting from the Maxwell equations is the problem on interaction of a plane monochromatic wave and an *ideal stratified medium*, a medium that is 2D-homogeneous and infinitely extended in the transverse directions. Known methods, some of which are considered in detail in Chapter 8, provide an accurate and fast estimation of various transmission and reflection characteristics of ideal stratified media for incident plane monochromatic waves. In most cases, the physical model for estimating an optical characteristic of a real layered system is what prescribes how to use such easily obtainable solutions to get the desired estimates. Physical models of this sort are of primary importance in LCD optics. In Section 7.1, we discuss some standard approximations used in such models.

In many cases, the accurate modeling of the optical behavior of LC devices involves taking account of multiple reflections. In Section 7.2, we consider two general algorithms used for calculating transmission and reflection characteristics of layered structures with allowance for multiple reflections.

In Section 7.3, optical models of some basic elements of LCDs are considered.

7.1 Physical Models of the Light–Layered System Interaction Used in Modeling the Optical Behavior of LC Devices. Plane-Wave Approximations. Transfer Channel Approach

Usually, considering a real optical layered system such as an LC cell or an LCD panel, one starts from a realistic model of this system, which takes into account the real geometry of elements of the system. One of the basic tricks used in modeling is a direct approximation of the values of a transmission or reflection

Modeling and Optimization of LCD Optical Performance, First Edition.
Dmitry A. Yakovlev, Vladimir G. Chigrinov and Hoi-Sing Kwok.

Website Companion: www.wiley.com/go/yakovlev/modelinglcd

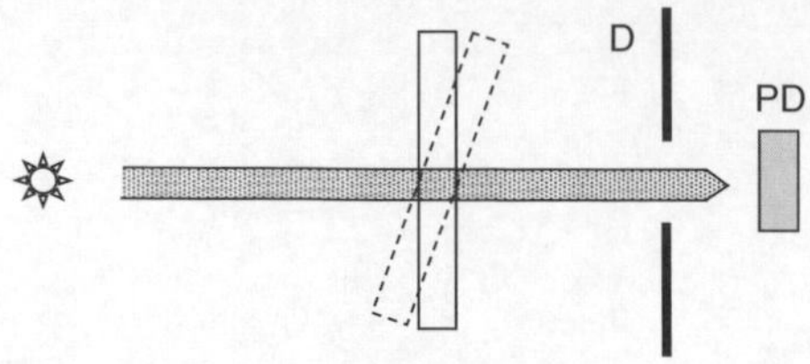

Figure 7.1 Scheme of the experiment considered in problems P1 and P2. D is a diaphragm; PD is a photodetector

characteristic of such a realistic model system by the values of an allied characteristic describing the interaction of a plane monochromatic or quasimonochromatic wave (*approximating incident field*) and an ideal stratified medium whose optical properties are identical or close to those of the realistic model system in the region being considered; we will call this ideal medium the *approximating medium*. Thus, the use of this approximation involves choosing the approximating medium, approximating incident field, and approximating characteristic. As we will see below, depending on the conditions of the problem, the optimal choice of these items may be very different even for the same layered system.

Power-Based Plane-Wave Approximations

Most often, the direct optical problem consists in evaluating *measurable* optical characteristics of the optical system under consideration (e.g., transmittance, reflectance). Therefore, when choosing the physical model, one should take into consideration the experimental conditions under which these characteristics are measured. In the present discussion, we will turn repeatedly to the following two problems: Let us need to estimate theoretically the values of the transmittance of a homogeneous layer (problem P1) or a multilayer system including anisotropic layers (problem P2) that might be obtained using the experimental scheme shown in Figure 7.1. The incident beam is well-collimated and quasimonochromatic. Its power (radiant flux) is assumed to be known. The photodetector measures the net power of the radiation passing through the diaphragm D. The diameter of the opening of the diaphragm D is somewhat larger than the diameter of the probe beam in the plane of the diaphragm and hence the net transmitted power can be measured at least in the case of normal incidence.

In these problems, we deal with finite light beams, and the main characteristic of a beam for us is its power. Our aim is to estimate the ratio of the power of the detected part of the transmitted light to the power of the incident beam by making use of the knowledge on the interaction of plane waves, of infinite extension and infinite power, and ideal stratified media. The problem P1 (a homogeneous layer), in most cases, can be successfully solved using the geometrical optics approximation: by considering the incident beam and the derivative beams as bundles of rays and using the known laws of transformation of the amplitudes and phases of plane waves inside homogeneous layers and at interfaces to calculate the amplitudes and phases associated with the rays (solutions of this kind may be found in many textbooks on optics). In the case of a multilayer structure (problem P2), especially when oblique incidence is considered, this approach is usually not efficient or not applicable at all.

A more general method of solving such problems is based on considering the incident beam as a superposition of plane waves with different propagation directions, whose wave normals $\mathbf{l}$ are confined within a narrow cone around the unit vector $\mathbf{l}_c$ indicating the nominal propagation direction of the beam. A quasimonochromatic beam[1] is regarded as a superposition of monochromatic beams with wavelengths

[1] Any of the quasimonochromatic light fields considered in this book is assumed to allow treating it as a realization of an ergodic stationary random process [2–4].

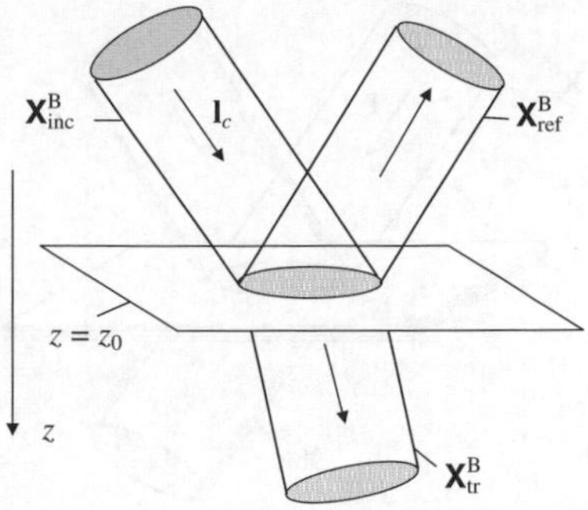

Figure 7.2 To approximation 1: situation 1

λ filling a narrow spectral interval around a mean wavelength λ_c. Using known transfer functions of the layered system for plane monochromatic waves, one calculates the characteristics of the emergent wave fields of interest. It is a method of Fourier optics [1]. The computation of spatial distributions of local characteristics, for example, irradiance, of the emergent fields by this method is rather laborious. But when such a structural detailing of the fields is not required and the problem is stated in terms of nonlocal characteristics of radiation, such as radiant power, as in the problems P1 and P2, the Fourier optics approach may lead to very simple, reliable, and accurate methods of solution—methods based on the approximation formulated above in terms of "approximating characteristic," "approximating medium," and "approximating incident field." Methods of this kind are successfully used for accurate modeling of the optical performance of LCDs. In this section, we start presenting these methods, focusing attention on physical aspects of the modeling. The consideration of these methods will be continued in Chapters 8–11.

Before we proceed we need to define the notions of the "effective spectral range" and "effective angular spectrum" of a beam. The *effective spectral range* of a quasimonochromatic beam is the wavelength (or frequency) range filled by the wavelengths (frequencies) of the monochromatic components giving a significant contribution to the radiant power of this beam. Analogously, the *effective angular spectrum* of a beam is the set of the wave normals of the plane-wave components of the beam that give a significant contribution to the radiant power of this beam.

To begin with, we consider the following two approximations:

Approximation 1 (an interface) Let us consider two situations. *Situation* 1: A monochromatic, uniformly polarized light beam with a wavelength λ and nominal propagation direction $\mathbf{l}_c$ impinges on a plane interface between two nonabsorbing homogeneous isotropic media (Figure 7.2). The interface coincides with the plane $z = z_0$ in a Cartesian coordinate system (x, y, z). We denote the incident beam by $\mathbf{X}^{B}_{inc}$, the transmitted beam by $\mathbf{X}^{B}_{tr}$, and the reflected beam by $\mathbf{X}^{B}_{ref}$ (the superscript ^{B} serves as a reminder that we are dealing with a beam). *Situation* 2: A plane monochromatic wave whose wave normal $\mathbf{l} = \mathbf{l}_c$ and whose wavelength and polarization state are identical to the incident beam in situation 1 falls on the same interface. We denote the incident wave as $\mathbf{X}^{MPW}_{inc}$, transmitted wave as $\mathbf{X}^{MPW}_{tr}$, and reflected wave as $\mathbf{X}^{MPW}_{ref}$, where ^{MPW} reminds that we are dealing with a monochromatic plane wave. Almost always, to a good approximation,

$$\frac{\Phi\left\{\mathbf{X}^{B}_{tr}(z_0+0)\right\}}{\Phi\left\{\mathbf{X}^{B}_{inc}(z_0-0)\right\}} \approx \frac{E\left\{\mathbf{X}^{MPW}_{tr}(z_0+0)\right\}}{E\left\{\mathbf{X}^{MPW}_{inc}(z_0-0)\right\}}, \tag{7.1a}$$

$$\frac{\Phi\left\{\mathbf{X}^{B}_{ref}(z_0-0)\right\}}{\Phi\left\{\mathbf{X}^{B}_{inc}(z_0-0)\right\}} \approx \frac{E\left\{\mathbf{X}^{MPW}_{ref}(z_0-0)\right\}}{E\left\{\mathbf{X}^{MPW}_{inc}(z_0-0)\right\}}, \tag{7.1b}$$

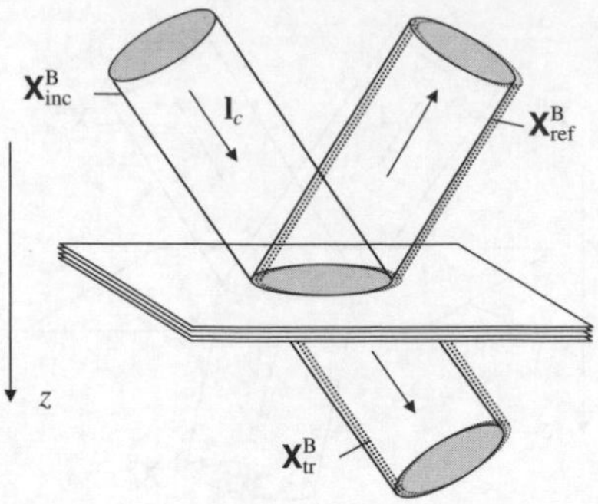

Figure 7.3 To approximation 2: situation 1

where $\Phi\{\mathbf{X}(z')\}$ denotes the radiant flux of a wave field $\mathbf{X}$ across the plane $z = z'$, and $E\{\mathbf{X}^{PW}(z')\}$ denotes the irradiance produced by a plane wave $\mathbf{X}^{PW}$ at points of the plane $z = z'$. Since the media are nonabsorbing, these relations may be rewritten as follows:

$$\frac{P\{\mathbf{X}^{B}_{tr}\}}{P\{\mathbf{X}^{B}_{inc}\}} \approx \frac{E\{\mathbf{X}^{MPW}_{tr}(z_0+0)\}}{E\{\mathbf{X}^{MPW}_{inc}(z_0-0)\}}, \tag{7.2a}$$

$$\frac{P\{\mathbf{X}^{B}_{ref}\}}{P\{\mathbf{X}^{B}_{inc}\}} \approx \frac{E\{\mathbf{X}^{MPW}_{ref}(z_0-0)\}}{E\{\mathbf{X}^{MPW}_{inc}(z_0-0)\}}, \tag{7.2b}$$

where $P\{\mathbf{X}\}$ denotes the power of a beam $\mathbf{X}$ in the traditional engineering sense. This is a well-known approximation, which is mentioned in some textbooks on optics (see, e.g., Reference 5). In [5], this approximation is treated using representations of geometrical optics.

Approximation 2 (a thin layered system) Again we will consider two similar situations differing in illumination conditions only. *Situation* 1: A monochromatic, uniformly polarized light beam $\mathbf{X}^{B}_{inc}$ of wavelength λ and nominal propagation direction $\mathbf{l}_c$ falls on an ideal stratified medium sandwiched between two homogeneous nonabsorbing isotropic media (Figure 7.3). The external boundaries of the stratified medium coincide with the planes $z = z_0$ and $z = z_n$. As in the previous case, we denote the incident beam by $\mathbf{X}^{B}_{inc}$. The total transmitted field and total reflected field, $\mathbf{X}^{B}_{tr}$ and $\mathbf{X}^{B}_{ref}$, at oblique incidence will have a rather complicated structure, each being a set of beams spatially shifted with respect to each other. *Situation* 2: The same ideal stratified medium is illuminated by a plane monochromatic wave $\mathbf{X}^{MPW}_{inc}$ with wave normal $\mathbf{l} = \mathbf{l}_c$ and wavelength and polarization state identical to the incident beam in situation 1. The symbols $\mathbf{X}^{MPW}_{tr}$ and $\mathbf{X}^{MPW}_{ref}$ in this case will denote respectively the total transmitted and total reflected fields. Although each of the fields $\mathbf{X}^{MPW}_{tr}$ and $\mathbf{X}^{MPW}_{ref}$ is a superposition of an infinite number of waves, both $\mathbf{X}^{MPW}_{tr}$ and $\mathbf{X}^{MPW}_{ref}$ are plane waves. It can be shown that under certain rather general conditions, the following relations are valid:

$$\frac{\Phi\{\mathbf{X}^{B}_{tr}(z_n+0)\}}{\Phi\{\mathbf{X}^{B}_{inc}(z_0-0)\}} \approx \frac{E\{\mathbf{X}^{MPW}_{tr}(z_n+0)\}}{E\{\mathbf{X}^{MPW}_{inc}(z_0-0)\}}, \tag{7.3a}$$

$$\frac{\Phi\{\mathbf{X}^{B}_{ref}(z_0-0)\}}{\Phi\{\mathbf{X}^{B}_{inc}(z_0-0)\}} \approx \frac{E\{\mathbf{X}^{MPW}_{ref}(z_0-0)\}}{E\{\mathbf{X}^{MPW}_{inc}(z_0-0)\}}, \tag{7.3b}$$

and, since the surrounding media are nonabsorbing,

$$\frac{P\left\{\mathbf{X}_{\rm tr}^{\rm B}\right\}}{P\left\{\mathbf{X}_{\rm inc}^{\rm B}\right\}} \approx \frac{E\left\{\mathbf{X}_{\rm tr}^{\rm MPW}(z_n+0)\right\}}{E\left\{\mathbf{X}_{\rm inc}^{\rm MPW}(z_0-0)\right\}}, \quad \frac{P\left\{\mathbf{X}_{\rm ref}^{\rm B}\right\}}{P\left\{\mathbf{X}_{\rm inc}^{\rm B}\right\}} \approx \frac{E\left\{\mathbf{X}_{\rm ref}^{\rm MPW}(z_0-0)\right\}}{E\left\{\mathbf{X}_{\rm inc}^{\rm MPW}(z_0-0)\right\}}. \tag{7.4}$$

Relations (7.1)–(7.4) can be derived using the plane-wave decomposition of the incident beam on the assumption that the angular spectrum of the incident beam is so narrow that the transmissivities (reflectivities) of the layered system for different plane-wave components of the effective angular spectrum of the incident beam are almost the same. This condition may be satisfied if $a_{\rm B} \gg d_{\rm sys}|\sin\beta_c|$, where $a_{\rm B}$ is the minimum diameter of the incident beam, $d_{\rm sys}$ is the net thickness of the layered system, and β_c is the angle of incidence (the angle between the z-axis and the vector $\mathbf{l}_c$). Regarding a plane interface as the simplest variant of the stratified medium, we will consider approximation 1 as a particular case of approximation 2.

The ratios on the left-hand side of relations (7.1)–(7.4) are transmittances and reflectances defined for beams. The ratios on the right-hand side of these relations are quantities usually called "transmissivity" and "reflectivity" [2], which are defined in terms of irradiances for plane waves. Thus, these relations allow us to estimate the transmittances and reflectances by approximating their values by the values of the corresponding transmissivities and reflectivities.

In situation 1 of approximation 2, we have assumed that the layered system is an ideal stratified medium. But it is obvious that relations (7.3) and (7.4) will remain valid if in situation 1 we replace the ideal stratified medium by a more realistic model medium, supposing that this medium is 2D-homogeneous at least throughout the region where significant parts of the reflected and transmitted wave fields are formed, while in situation 2 we take the *approximating medium*, an ideal stratified medium whose parameters are identical to the realistic model of situation 1 in the 2D-homogeneous region. Further, relations (7.3) and (7.4) may remain valid if in situation 1 we replace the monochromatic incident beam by a quasimonochromatic one with a mean wavelength λ_c, while in situation 2 we take the incident wave with $\lambda = \lambda_c$. The conditions of validity of relations (7.3) and (7.4) in this case are

$$\tilde{t}(\lambda) \approx \tilde{t}(\lambda_c), \quad \tilde{r}(\lambda) \approx \tilde{r}(\lambda_c) \quad \forall \lambda \in \Omega_\lambda, \tag{7.5}$$

where $\tilde{t}(\lambda')$ and $\tilde{r}(\lambda')$ denote respectively the transmissivity and reflectivity in situation 2 for the incident wave with $\lambda = \lambda'$, and $\Omega_\lambda = [\lambda_c - \Delta\lambda/2,\ \lambda_c + \Delta\lambda/2]$ $(\Delta\lambda \ll \lambda_c)$ is the effective spectral range of the incident beam in situation 1. In other words, relations (7.3) and (7.4) hold if variations of the transmissivity and reflectivity within the spectral range Ω_λ are negligible. After the replacements, situation 1 becomes quite realistic. At the same time, relations (7.3) enable us to evaluate the transmittance and reflectance for this realistic situation using estimates of the transmissivity and reflectivity (*approximating characteristics*) for the incident plane monochromatic wave (*approximating incident field*) and the *approximating* ideal stratified *medium* as follows:

$$t_\Phi \approx \tilde{t}(\mathbf{l}_c, \lambda_c), \quad r_\Phi \approx \tilde{r}(\mathbf{l}_c, \lambda_c), \tag{7.6}$$

where

$$t_\Phi \equiv \frac{\Phi\left\{\mathbf{X}_{\rm tr}^{\rm B}(z_n+0)\right\}}{\Phi\left\{\mathbf{X}_{\rm inc}^{\rm B}(z_0-0)\right\}}, \quad r_\Phi \equiv \frac{\Phi\left\{\mathbf{X}_{\rm ref}^{\rm B}(z_0-0)\right\}}{\Phi\left\{\mathbf{X}_{\rm inc}^{\rm B}(z_0-0)\right\}}, \tag{7.7}$$

$$\tilde{t} \equiv \frac{E\left\{\mathbf{X}_{\rm tr}^{\rm MPW}(z_n+0)\right\}}{E\left\{\mathbf{X}_{\rm inc}^{\rm MPW}(z_0-0)\right\}}, \quad \tilde{r} \equiv \frac{E\left\{\mathbf{X}_{\rm ref}^{\rm MPW}(z_0-0)\right\}}{E\left\{\mathbf{X}_{\rm inc}^{\rm MPW}(z_0-0)\right\}}. \tag{7.8}$$

The same approximation but in terms of Mueller matrices is

$$\mathbf{M}_{\mathrm{tr}(\Phi)} \approx \tilde{\mathbf{M}}_{\mathrm{tr}(E)}(\mathbf{l}_c, \lambda_c), \quad \mathbf{M}_{\mathrm{ref}(\Phi)} \approx \tilde{\mathbf{M}}_{\mathrm{ref}(E)}(\mathbf{l}_c, \lambda_c), \tag{7.9}$$

where $\mathbf{M}_{\mathrm{tr}(\Phi)}$, $\tilde{\mathbf{M}}_{\mathrm{tr}(E)}$, $\mathbf{M}_{\mathrm{ref}(\Phi)}$, and $\tilde{\mathbf{M}}_{\mathrm{ref}(E)}$ are the transmission and reflection Mueller matrices such that

$$S_{(\Phi)}\left\{\mathbf{X}^{\mathrm{B}}_{\mathrm{tr}}(z_n+0)\right\} = \mathbf{M}_{\mathrm{tr}(\Phi)} S_{(\Phi)}\left\{\mathbf{X}^{\mathrm{B}}_{\mathrm{inc}}(z_0-0)\right\}, \tag{7.10}$$

$$S_{(\Phi)}\left\{\mathbf{X}^{\mathrm{B}}_{\mathrm{ref}}(z_0-0)\right\} = \mathbf{M}_{\mathrm{ref}(\Phi)} S_{(\Phi)}\left\{\mathbf{X}^{\mathrm{B}}_{\mathrm{inc}}(z_0-0)\right\}, \tag{7.11}$$

$$S_{(E)}\left\{\mathbf{X}^{\mathrm{MPW}}_{\mathrm{tr}}(z_n+0)\right\} = \tilde{\mathbf{M}}_{\mathrm{tr}(E)} S_{(E)}\left\{\mathbf{X}^{\mathrm{MPW}}_{\mathrm{inc}}(z_0-0)\right\}, \tag{7.12}$$

$$S_{(E)}\left\{\mathbf{X}^{\mathrm{MPW}}_{\mathrm{ref}}(z_0-0)\right\} = \tilde{\mathbf{M}}_{\mathrm{ref}(E)} S_{(E)}\left\{\mathbf{X}^{\mathrm{MPW}}_{\mathrm{inc}}(z_0-0)\right\}, \tag{7.13}$$

where $S_{(\Phi)}\{\mathbf{X}^{\mathrm{B}}(z')\}$ denotes the flux-based Stokes vector for a beam $\mathbf{X}^{\mathrm{B}}$ and the plane $z = z'$, and $S_{(E)}\{\mathbf{X}^{\mathrm{PW}}(z')\}$ the irradiance-based Stokes vector of a plane wave $\mathbf{X}^{\mathrm{PW}}$ at points of the plane $z = z'$ (see Section 5.3). Approximations (7.6) and (7.9) are widely used in modeling optical characteristics of thin films and thin-film systems (AR-coatings, multilayer dielectric mirrors, multilayer transflective polarizers, etc.), optical systems in which multiple-beam (Fabry–Perot, FP) interference plays a key role.

The class of approximations that use a monochromatic plane wave (MPW) as the approximating incident field will be called the *MPW approximation*. We have considered a variant of the application of the MPW approximation in the problem on the overall transmittance and overall reflectance of a layered system. This approximation may also be used in calculating the characteristics of partial transmission and reflection (see below). A general formula of this approximation may be written as

$$\mathbf{M}_{(\Phi)} \approx \tilde{\mathbf{M}}_{(E)}(\mathbf{l}_c, \lambda_c), \tag{7.14}$$

where $\mathbf{M}_{(\Phi)}$ and $\tilde{\mathbf{M}}_{(E)}$ are the Mueller matrices characterizing the same operation (overall or partial transmission or overall or partial reflection) performed by a realistic layered medium and the corresponding approximating medium, respectively. The matrix $\mathbf{M}_{(\Phi)}$ relates the flux-based Stokes vectors of the incident and emergent beams and corresponds to incident quasimonochromatic beams with nominal propagation direction $\mathbf{l}_c$ and mean wavelength λ_c. The matrix $\tilde{\mathbf{M}}_{(E)}(\mathbf{l}, \lambda)$ relates the irradiance-based Stokes vectors of the incident and emergent plane waves and corresponds to incident monochromatic waves with wave normal $\mathbf{l}$ and wavelength λ. The condition of applicability of approximation (7.14) may be expressed as follows:

$$\tilde{\mathbf{M}}_{(E)}(\mathbf{l}, \lambda) \approx \tilde{\mathbf{M}}_{(E)}(\mathbf{l}_c, \lambda_c) \quad \forall \lambda \in \Omega_\lambda, \mathbf{l} \in \Omega_{\mathbf{l}}, \tag{7.15}$$

where Ω_λ and $\Omega_{\mathbf{l}}$ are respectively the effective spectral range and effective angular spectrum of the incident beam.

Let us consider the following example. The transmissivity of an ideal homogeneous nonabsorbing isotropic layer surrounded by isotropic media with refractive index 1 for an incident monochromatic linearly polarized plane wave with an arbitrary orientation of the polarization plane in the case of normal incidence and with polarization plane parallel (p-polarization) or perpendicular (s-polarization) to the plane of incidence at oblique incidence may be expressed as follows:

$$\tilde{t} = \frac{T_I^2}{1 + R_I^2 - 2R_I \cos\left(\dfrac{4\pi d}{\lambda}\sqrt{n^2 - \sin^2\beta}\right)}, \tag{7.16}$$

where β is the angle of incidence, n and d are respectively the refractive index and thickness of the layer, and T_I and R_I are respectively the transmissivity and reflectivity of the frontal interface for the incident wave (in this example, both the transmissivities and reflectivities of the two interfaces are identical; $T_I + R_I = 1$). Let this layer be considered as the approximating medium in solving the problem P1 (Figure 7.1). Assume that the incident beam in the experiment is perfectly collimated and its effective angular spectrum is bounded by a cone with a half-angle of the order of λ_c/a_B, where, as before, a_B is the minimum diameter of the beam ($a_B \gg \lambda_c$). T_I and R_I usually vary relatively slowly with β and λ for a wide domain of values of β and λ, and, as a rule, the neglect of the variation of T_I and R_I in the ranges $\lambda \in \Omega_\lambda$ and $\mathbf{l} \in \Omega_\mathbf{l}$ is fully justified. In such cases, the rate of change of $\tilde{t}$ with β and λ is determined by the cosine term in (7.16). Representing β as $\beta = \beta_c + \Delta\beta$, where, as before, β_c is the nominal angle of incidence of the beam and $\Delta\beta$ is a small angle, we find that

$$\sqrt{n^2 - \sin^2(\beta_c + \Delta\beta)} = \sigma_c - \frac{\sin\beta_c \cos\beta_c}{\sigma_c}\Delta\beta + O(\Delta\beta^2), \tag{7.17}$$

where

$$\sigma_c = \sqrt{n^2 - \sin^2\beta_c}. \tag{7.18}$$

Taking $\Delta\beta = \lambda_c/a_B$, it is easy to see from (7.16) and (7.17) that the changes in $\tilde{t}$ over the range $\Omega_\mathbf{l}$ will be small if

$$a_B \gg \frac{d|\sin\beta_c|}{n}. \tag{7.19}$$

From (7.16), we also see that the variation of $\tilde{t}$ with λ in the range $\Omega_\lambda = [\lambda_c - \Delta\lambda/2, \lambda_c + \Delta\lambda/2]$ may be neglected if

$$2\sigma_c d \ll \frac{\lambda_c^2}{\Delta\lambda}. \tag{7.20}$$

Hence, provided that in the experiment (Figure 7.1) all significant components of the transmitted field fall into the detector, we may expect to obtain a good agreement of the theoretical estimates of t_Φ obtained according to (7.6), that is, by using the MPW approximation, and the measured values of t_Φ if conditions (7.19) and (7.20) are satisfied in the experiment.

In most cases, condition (7.20) is essentially equivalent to the following one: the coherence length l_{coh} ($l_{coh} \sim \lambda_c^2/\Delta\lambda$) of the incident beam is many times larger than the thickness of the layer.

Real light has a random nature and cannot be perfectly monochromatic in principle. Irradiance produced by the superposition of two real light fields depends on the degree of their mutual coherence [2–5]. If these fields, call them field $\mathbf{X}_1$ and field $\mathbf{X}_2$, are fully or partially coherent, the irradiance produced by their superposition, $E\{\mathbf{X}_1 + \mathbf{X}_2\}$, is in general not equal to the sum of the irradiances produced by each of these fields individually, $E\{\mathbf{X}_1\} + E\{\mathbf{X}_2\}$, that is, the fields $\mathbf{X}_1$ and $\mathbf{X}_2$ interfere with each other. If the fields are incoherent, $E\{\mathbf{X}_1 + \mathbf{X}_2\} = E\{\mathbf{X}_1\} + E\{\mathbf{X}_2\}$, that is, there is no interference. Two perfectly monochromatic fields of the same frequency are completely mutually coherent.

In solving the practical optical problems entering into the scope of this book, as a rule, one can rely on the following simple concepts of the interaction of quasimonochromatic light with optical systems.

Wavetrain Representation

Quasimonochromatic light incident on an optical system is considered as a swarm of mutually incoherent wavetrains within each of which the electromagnetic field is similar to the field of a monochromatic

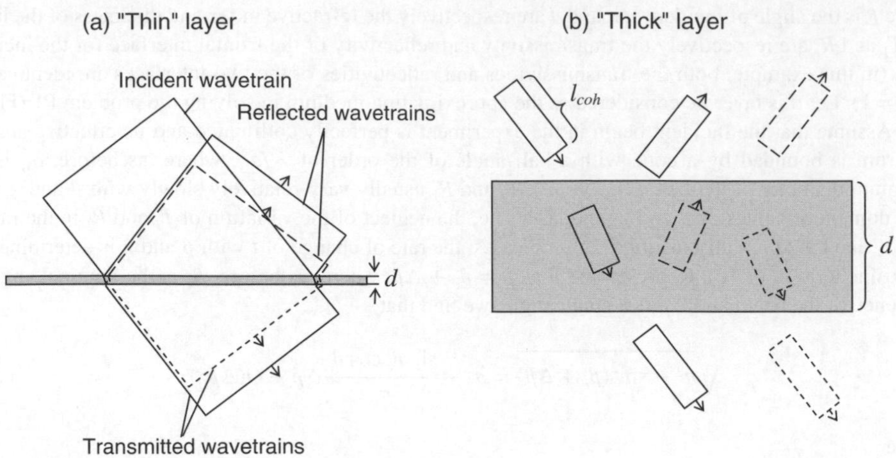

Figure 7.4 Interaction of a quasimonochromatic beam with a "thin" layer (a) and a "thick" layer (b)

wave. The length of each wavetrain is approximately equal to the coherence length l_{coh}. When the light propagates in the optical system, only derivatives of the same wavetrain are able to interfere with each other and interfere more effectively at smaller spatial shift with respect to each other.

In Figure 7.4 we try, based on this concept, to illustrate the fact that for a layer with a thickness $d \ll l_{\mathrm{coh}}$, the conditions for interference of components of the emergent fields that undergo a different number of reflections from the interfaces are fulfilled, while for a layer with a thickness $d > l_{\mathrm{coh}}$ such components are mutually incoherent. In engineering optics, layers with $d \ll l_{\mathrm{coh}}$ are customarily called "thin," and layers with $d > l_{\mathrm{coh}}$ "thick." Hereinafter, where we use the terms "thin" and "thick" in this sense, we retain the quotes. Thus, for example, layered systems with a net thickness many times smaller than l_{coh} will be referred to as "thin" layered systems. The classification "thin"/"thick" is very important for us because, supposing the usual conditions of measuring optical characteristics of LCD panels, the LCD panel should be considered as a system including both "thin" layered systems and "thick" layers and, in most cases, can be regarded as a system consisting only of elements of these classes. It is needless to say that in such situations, real statistical properties of the light must be taken into account in modeling.

Spectral Representation

Another fruitful concept of the interaction of quasimonochromatic light with optical systems is based on the spectral representation: The incident light is regarded as a composition of monochromatic components of different frequencies, each of which interacts with the optical system independently of the others. The validity of this representation is rigorously proved in the theory of coherence [3, 4]. Considering a wave field emerging from the system as a mixture of monochromatic components produced by the corresponding monochromatic components of the incident light, one can calculate characteristics of the system for incident quasimonochromatic light in terms of those for incident monochromatic light. For instance, let t_λ, in the scalar variant of the problem, be the transmittance of a system for monochromatic light, Φ_{inc} and $\Phi_{\lambda\,\mathrm{inc}}$ respectively the radiant power and spectral radiant power of the incident light, and Φ_{tr} and $\Phi_{\lambda\,\mathrm{tr}}$ the radiant power and spectral radiant power of the transmitted light. By definition,

$$\Phi_{\mathrm{inc}} = \int \Phi_{\lambda\,\mathrm{inc}}(\lambda)\mathrm{d}\lambda, \tag{7.21}$$

$$\Phi_{\mathrm{tr}} = \int \Phi_{\lambda\,\mathrm{tr}}(\lambda)\mathrm{d}\lambda. \tag{7.22}$$

On substituting the expression

$$\Phi_{\lambda\,\mathrm{tr}}(\lambda) = t_\lambda(\lambda)\Phi_{\lambda\,\mathrm{inc}}(\lambda) \tag{7.23}$$

into (7.22), we obtain

$$\Phi_{\mathrm{tr}} = \int t_\lambda(\lambda)\Phi_{\lambda\,\mathrm{inc}}(\lambda)\mathrm{d}\lambda. \tag{7.24}$$

Using the spectral form-factor of the incident light

$$g(\lambda) \equiv \Phi_{\lambda\,\mathrm{inc}}(\lambda)/\Phi_{\mathrm{inc}}, \tag{7.25}$$

we may rewrite (7.24) as follows:

$$\Phi_{\mathrm{tr}} = \left[\int t_\lambda(\lambda)g(\lambda)\mathrm{d}\lambda\right]\Phi_{\mathrm{inc}}.$$

We see from this equation that the transmittance of the system for the quasimonochromatic light, $t \equiv \Phi_{\mathrm{tr}}/\Phi_{\mathrm{inc}}$, may be calculated as

$$t = \int t_\lambda(\lambda)g(\lambda)\mathrm{d}\lambda. \tag{7.26}$$

Provided that the spectral density $S_{\lambda\,\mathrm{inc}}$ of the Stokes vector of the incident light, S_{inc}, can be represented in the form

$$S_{\lambda\,\mathrm{inc}}(\lambda) = g(\lambda)S_{\mathrm{inc}}, \tag{7.27}$$

an analogous expression can be written for the Mueller matrix describing the interaction of the system with the quasimonochromatic light:

$$\mathbf{M} = \int \mathbf{M}_\lambda(\lambda)g(\lambda)\mathrm{d}\lambda \tag{7.28}$$

with obvious notation. Quantities that characterize the interaction of an optical system with monochromatic light, such as t_λ and $\mathbf{M}_\lambda$, will be called *monochromatic*, and quantities characterizing the interaction of a system with *quasimonochromatic* light, such as t and $\mathbf{M}$, *quasimonochromatic*.

If the variations of t_λ and $\mathbf{M}_\lambda$ within the effective spectral range are small, one may use the approximations

$$t \approx t_\lambda(\lambda_c), \tag{7.29}$$

$$\mathbf{M} \approx \mathbf{M}_\lambda(\lambda_c). \tag{7.30}$$

This kind of approximations will be referred to as *the monochromatic approximation*. The monochromatic approximation is one of those involved in the MPW approximation.

In general, the thicker a layer, the closer peaks of the FP interference pattern in its spectra of monochromatic transmittance (transmissivity) and reflectance (reflectivity). For "thin" layers, distances between neighbor interference peaks in the monochromatic spectra are much greater than the bandwidth of the incident quasimonochromatic light and variations of the "monochromatic" characteristics within the effective spectral range are very small, which justifies the use of the monochromatic approximation. For "thick" layers, many FP interference extrema of the monochromatic spectra fall at the effective spectral range of the incident quasimonochromatic light and consequently the monochromatic approximation is not applicable. The number of periods of the FP interference oscillations falling at the effective spectral range of quasimonochromatic light with coherence length l_{coh} for a "thick" homogeneous isotropic layer of thickness d and refractive index n at normal incidence is approximately equal to $2nd/l_{coh}$.

Experimental Transmission Spectra of "Thin" and "Thick" Layers

One of the problems that require the nonmonochromaticity of the incident light to be allowed for is the modeling of the transmission spectra of LCD panels. As a rule, it is necessary to calculate them so that they correspond to those that might be obtained in a real experiment under typical measurement conditions.

With some idealization, the relationship between the measured spectrum of the transmittance[2] of an optical system, t_{meas}, obtained using a wide-band light source and a spectrometer, and the actual transmittance spectrum of this system for monochromatic light $t_{act}(\lambda)$ may be expressed as follows:

$$t_{meas}(\lambda_m) = \int f_{app}(\lambda_m, \lambda) t_{act}(\lambda) d\lambda, \tag{7.31}$$

where $f_{app}(\lambda_m, \lambda)$ is an instrument function of the spectrometer, such that at any λ_m

$$\int f_{app}(\lambda_m, \lambda) d\lambda = 1,$$

the variable λ_m represents the nominal values of wavelength which the measured transmittance t_{meas} is associated with. At a given λ_m, $f_{app}(\lambda_m, \lambda)$ as function of λ assumes its maximum value at $\lambda = \lambda_m$ and has values significantly different from zero only for λ values from a small neighborhood of λ_m of width $\Delta\lambda_{app}$ ($\Delta\lambda_{app} \ll \lambda_m$). In the simplest case, $f_{app}(\lambda_m, \lambda)$ is the rectangular function:

$$f_{app}(\lambda_m, \lambda) = \begin{cases} 1/\Delta\lambda_{app} & \text{if } |\lambda - \lambda_m| \leq \Delta\lambda_{app}/2 \\ 0 & \text{if } |\lambda - \lambda_m| > \Delta\lambda_{app}/2. \end{cases} \tag{7.32}$$

With such a function $f_{app}(\lambda_m, \lambda)$, $t_{meas}(\lambda_m)$ is equal to the average of $t_{act}(\lambda)$ over the interval $\lambda_m - \Delta\lambda_{app}/2 \leq \lambda \leq \lambda_m + \Delta\lambda_{app}/2$. With other functions $f_{app}(\lambda_m, \lambda)$, $t_{meas}(\lambda_m)$ is a weighted average of $t_{act}(\lambda)$ over a small interval including λ_m. In any case, the spectrum $t_{meas}(\lambda_m)$ is a result of smoothing the spectrum $t_{act}(\lambda)$. If $t_{act}(\lambda)$ has fast interference oscillations, with a large number of interference maxima falling at a spectral interval of width $\Delta\lambda_{app}$, such oscillations will be absent in the spectrum $t_{meas}(\lambda_m)$. Comparing expressions (7.31) and (7.26), we see that the measured transmittance $t_{meas}(\lambda_m)$ may be treated as the true transmittance of the system for incident quasimonochromatic light with spectral form-factor $g(\lambda) = f_{app}(\lambda_m, \lambda)$, or, a little rougher, for light with $\lambda_c = \lambda_m$ and $l_{coh} \approx \lambda_m{}^2/\Delta\lambda_{app}$. If we use such an estimation of l_{coh} to sort the layers of a layered system into "thick" ones and "thin" ones, we may be sure that we will not see the pattern of the FP interference in the "thick" layers or this pattern will be greatly smoothed in the experimental spectra of this system (see an example in Figure 7.5). We should note that in real situations, the absence of FP interference in relatively thick layers may be caused not only by the nonmonochromaticity of the incident light, but also by other factors, for instance, by spatial shift of emergent beams, components of the transmitted (reflected) wave field, with respect to each other at oblique incidence (as in Figure 7.6).

---

[2] The same might be said about reflectance.

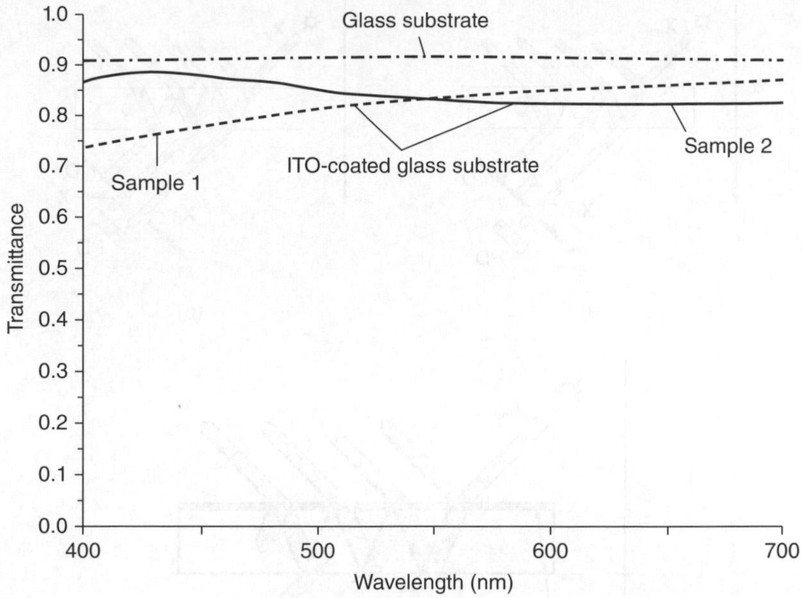

Figure 7.5 Experimental transmittance spectra of an uncoated glass substrate and ITO-coated glass substrates with different thicknesses of ITO layer (samples 1 and 2) (Merck). In all the cases, the glass (soda lime) substrate has a thickness of 1.1 mm. The ITO layer thickness in samples 1 and 2 is about 45 nm and 100 nm, respectively. There is a barrier SiO_2 layer with a thickness of about 20 nm and a refractive index of about 1.55 between the glass substrate and ITO layer in both samples with ITO coating. The optical effect of the barrier layer is relatively small since its refractive index is close to that of soda lime glass (~1.52). The glass substrate is a typical example of a "thick" layer in spectral measurements. ITO and barrier layers are typical "thin" layers. There are no manifestations of the FP interference in the glass plates in the presented experimental spectra. At the same time, we clearly see the effect of FP interference in the "thin" layered systems consisting of the barrier layer and ITO layer

Usually, spectra of LCDs are measured with $\Delta\lambda_{\mathrm{app}}$ of the order of 4–8 nm and consequently the measured spectra correspond to incident light with a coherence length of the order of 30–150 μm. Based on this estimate, considering real sizes of LCD components and their optical action, one may conclude that in most cases it is possible to approximate the LCD panel, when modeling its transmission spectra, by a stratified medium consisting only of "thin" and "thick" layers.

When estimating the overall transmittance or overall reflectance of a layered system including "thin" and "thick" layers, both coherent and incoherent interactions between different waves propagating in the layered system must be taken into account. This may be made by using the spectral averaging, as in (7.26) and (7.28). However, the spectral averaging is very time-consuming when applied to such complicated and relatively thick layered systems as LCD panels, because in this case, to obtain accurate results, it is necessary to compute the spectrum of the monochromatic transmittance (reflectance) of the approximating stratified medium with a very small step. Dealing with layered systems consisting only of "thin" and "thick" layers, one may use significantly more efficient methods of calculating the overall transmittance and overall reflectance than the spectral averaging. Such methods are considered in Section 10.2. In substance, these are methods of calculating the transmission ($\mathbf{M}_{\mathrm{tr}(E)}$) and reflection ($\mathbf{M}_{\mathrm{ref}(E)}$) Mueller matrices characterizing the interaction of a quasimonochromatic plane wave and an

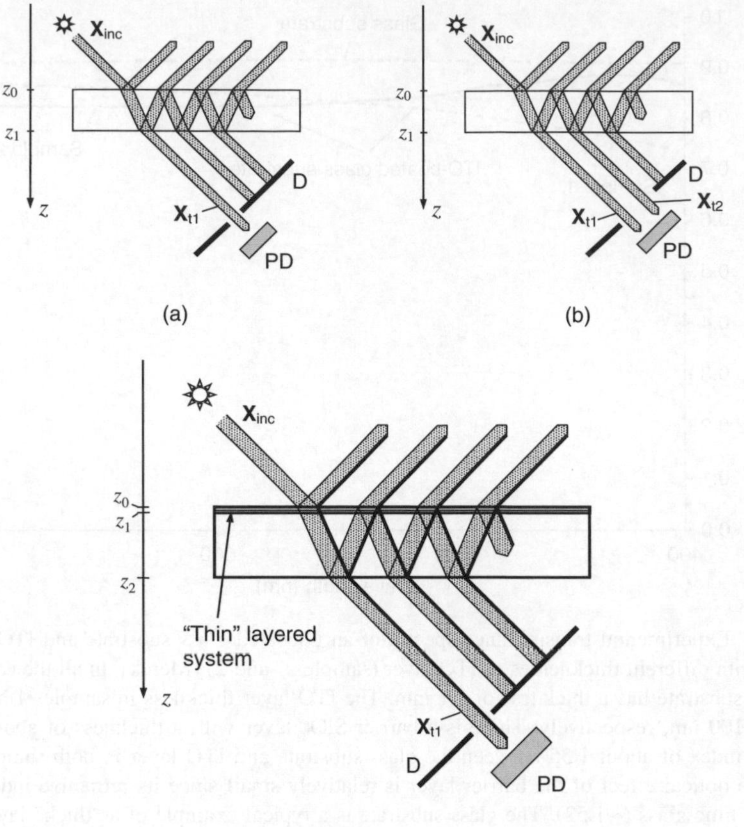

Figure 7.6 Examples of situations that require considering a partial transmission

ideal stratified medium consisting of "thin" and "thick" layers. The spectral averaging method, provided (7.27), implies computing these Mueller matrices by the formulas

$$\mathbf{M}_{\mathrm{tr}(E)} = \int \tilde{\mathbf{M}}_{\mathrm{tr}(E)}(\lambda)g(\lambda)\mathrm{d}\lambda, \tag{7.33}$$

$$\mathbf{M}_{\mathrm{ref}(E)} = \int \tilde{\mathbf{M}}_{\mathrm{ref}(E)}(\lambda)g(\lambda)\mathrm{d}\lambda \tag{7.34}$$

[see (7.12) and (7.13)]. The methods presented in Section 10.2 do not require computing $\tilde{\mathbf{M}}_{\mathrm{tr}(E)}(\lambda)$ and $\tilde{\mathbf{M}}_{\mathrm{ref}(E)}(\lambda)$. In the "realistic" modeling in terms of beams, the matrices $\mathbf{M}_{\mathrm{tr}(E)}$ and $\mathbf{M}_{\mathrm{ref}(E)}$ are used to approximate the matrices $\mathbf{M}_{\mathrm{tr}(\Phi)}$ and $\mathbf{M}_{\mathrm{ref}(\Phi)}$ [see (7.10) and (7.11)]:

$$\mathbf{M}_{\mathrm{tr}(\Phi)} \approx \mathbf{M}_{\mathrm{tr}(E)}, \quad \mathbf{M}_{\mathrm{ref}(\Phi)} \approx \mathbf{M}_{\mathrm{ref}(E)} \tag{7.35}$$

[cf. (7.9)]. The approximating incident field in this case is a quasimonochromatic plane wave.

The class of approximations where a quasimonochromatic plane wave (QMPW) is used as the approximating incident field will be referred to as the *QMPW approximation*. By analogy with (7.14), we write the principal formula of the QMPW approximation as follows:

$$\mathbf{M}_{(\Phi)} \approx \mathbf{M}_{(E)}(\mathbf{l}_c), \tag{7.36}$$

where $\mathbf{M}_{(E)}(\mathbf{l})$ is the Mueller matrix describing the operation under consideration when it is performed by the approximating medium with respect to incident quasimonochromatic plane waves with wave normal $\mathbf{l}$ and mean wavelength and coherence length equal to the mean wavelength and coherence length of the incident beam in the realistic situation described by the matrix $\mathbf{M}_{(\Phi)}$. The general condition of applicability of (7.36) is

$$\mathbf{M}_{(E)}(\mathbf{l}) \approx \mathbf{M}_{(E)}(\mathbf{l}_c) \quad \forall \mathbf{l} \in \Omega_l \tag{7.37}$$

[cf. (7.15)].

The MPW and QMPW approximations allow effective solving of a very large number of practical problems of LCD optics, and we confine our attention to these kinds of problems in this book.

As examples of solutions for characteristics of the interaction of a quasimonochromatic plane wave and an ideal layered medium, different from those that can be obtained by immediate use of the MPW approximation, we give a few useful simple expressions for the overall transmissivity t_E of different "thick" layers. The surrounding medium in all the cases is assumed to be isotropic and of refractive index 1.

1. An isotropic nonabsorbing "thick" layer (any polarization in the case of normal incidence, s- or p-polarization at oblique incidence):

$$t_E = \frac{T_I^2}{1\ R_I^2} \tag{7.38}$$

with the same notation as in (7.16)

2. An isotropic weakly absorbing "thick" layer (any polarization in the case of normal incidence, s- or p-polarization at oblique incidence):

$$t_E = \frac{T_I^2 T_B}{1 - R_I^2 T_B^2}, \tag{7.39}$$

where

$$T_B = \exp\left(-\frac{4\pi d}{\lambda_c} \mathrm{Im}\sqrt{n^2 - \sin^2\beta} \right) \approx \exp\left(-\frac{4\pi d n''}{\lambda_c \cos\beta_L} \right) \tag{7.40}$$

is the transmissivity of the bulk of the layer. In (7.40), $n = n' + in''$ is the complex refractive index of the layer (n' and n'' are real), and

$$\cos\beta_L = \frac{\sqrt{n'^2 - \sin^2\beta}}{n'}. \tag{7.41}$$

3. A uniaxial dichroic "thick" layer with optic axis parallel to the layer boundaries (e.g., a model of a polarizer) at normal incidence of a linearly polarized wave with polarization plane

(i) parallel to the optic axis:

$$t_{E\parallel} = \frac{T_{I\parallel}^2 T_{B\parallel}}{1 - R_{I\parallel}^2 T_{B\parallel}^2} \quad \text{with} \quad T_{B\parallel} = \exp\left(-\frac{4\pi d n''_{\parallel}}{\lambda_c}\right), \tag{7.42}$$

(ii) perpendicular to the optic axis:

$$t_{E\perp} = \frac{T_{I\perp}^2 T_{B\perp}}{1 - R_{I\perp}^2 T_{B\perp}^2} \quad \text{with} \quad T_{B\perp} = \exp\left(-\frac{4\pi d n''_{\perp}}{\lambda_c}\right), \tag{7.43}$$

where $n''_{\parallel}$ and $n''_{\perp}$ are the imaginary parts of the principal refractive indices of the layer, $T_{I\parallel}$, $T_{I\perp}$, $R_{I\parallel}$, and $R_{I\perp}$ are the transmissivities and reflectivities of the frontal interface for the corresponding polarizations of the incident wave, and $t_{E\parallel}$ and $t_{E\perp}$ denote t_E in the cases (i) and (ii), respectively.

Partial Transmission and Partial Reflection. Transfer Channel Approach

In the above discussion, we considered mainly characteristics of the overall transmission and overall reflection of layered media. But often of primary interest are characteristics of a partial transmission or a partial reflection of a layered system. Such situations are typical in LCD optics. In many cases, characteristics of partial transmission and reflection are much easier to analyze than those of overall transmission and reflection.

Characteristics[3] of a partial transmission (reflection) of a layered medium describe the relation between incident light parameters and parameters of a certain part of the transmitted (reflected) light. The part of the transmitted (reflected) light of interest can often be specified by indicating the chain of operations (transmission, reflection) performed by elements of the layered medium that produce this part of the emergent light. For instance, in the situation shown in Figure 7.6a, the detected component of the transmitted light corresponds to the following chain of operations: the transmission of the frontal interface of the layer → the transmission of the bulk of the layer → the transmission of the rear interface of the layer. The way of radiation transfer defined by such a chain of operations will be referred to as an *elementary transfer channel*. The formulation of the optical problem in terms of transmission and reflection operations performed by units of the optical system being considered and solving it by using operators describing these operations are among the main features of a general approach called here the *transfer channel approach*.

Let us introduce some notions and notation. First we define the notions of a TR unit and an OTR unit.

The *TR unit* is a separate domain of a stratified medium about which we can say that this domain transmits and/or reflects light. It may be the interface between layers, the bulk of a homogeneous or an inhomogeneous layer, a layer (interface + bulk + interface), or a system of layers. Generally, we may associate with a TR unit four types of operations (see Figure 7.7): two operations of transmission $\mathbf{T}^{\downarrow}$ and $\mathbf{T}^{\uparrow}$ and two operations of reflection $\mathbf{R}^{\downarrow}$ and $\mathbf{R}^{\uparrow}$. The "operand" of the operations $\mathbf{T}^{\downarrow}$ and $\mathbf{R}^{\downarrow}$ is the forward propagating light incident on this unit; that of the operations $\mathbf{T}^{\uparrow}$ and $\mathbf{R}^{\uparrow}$ is the backward propagating incident light. If necessary, we will write down in the brackets following the symbols of operations and operators for a TR unit the coordinates of the boundaries of this TR unit or the symbol denoting this unit.

The *OTR unit* is a TR unit that is characterized in calculations by operators of the *overall* transmission and/or *overall* reflection.

[3] Jones matrix, Mueller matrix, transmittance (reflectance), and so on.

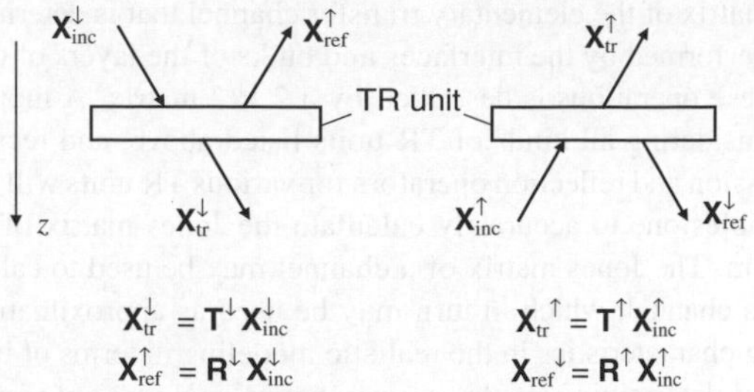

Figure 7.7 Transmission ($\mathbf{T}^{\downarrow}$, $\mathbf{T}^{\uparrow}$) and reflection ($\mathbf{R}^{\downarrow}$, $\mathbf{R}^{\uparrow}$) operations performed by a TR unit. *Convention*: If, as in this figure, a coordinate axis z attached to the layered system under consideration and perpendicular to its interfaces is introduced, we will call wave fields traveling to greater ones z *forward propagating*, and those going to lesser ones z *backward propagating*

For brevity, we will use the following notation. The fact that a wave field $\mathbf{X}''$ is the result of an operation $\mathbf{O}$ performed on a wave field $\mathbf{X}'$ may be expressed by the symbolic relation $\mathbf{X}'' = \mathbf{O}\mathbf{X}'$. If a wave field $\mathbf{X}$ is the result of an operation $\mathbf{O}_2$ on a wave field $\mathbf{X}''$ (i.e., $\mathbf{X} = \mathbf{O}_2\mathbf{X}''$), and the field $\mathbf{X}''$ is the result of an operation $\mathbf{O}_1$ on a wave field $\mathbf{X}'$ ($\mathbf{X}'' = \mathbf{O}_1\mathbf{X}'$), the field $\mathbf{X}$ may be represented as the result of the sequence of operations ($\mathbf{O}_1$, $\mathbf{O}_2$) on $\mathbf{X}'$ ($\mathbf{X} = \mathbf{O}_2\mathbf{O}_1\mathbf{X}'$) or as a result of the operation $\mathbf{O}$ ($\mathbf{X} = \mathbf{O}\mathbf{X}'$) being the product of the operations $\mathbf{O}_1$ and $\mathbf{O}_2$ ($\mathbf{O} = \mathbf{O}_2\mathbf{O}_1$). If a wave field $\mathbf{X}$ is a composition of wave fields $\mathbf{X}_1$ and $\mathbf{X}_2$ ($\mathbf{X} = \mathbf{X}_1 + \mathbf{X}_2$) such that $\mathbf{X}_1 = \mathbf{O}_1'\mathbf{X}'$ and $\mathbf{X}_2' = \mathbf{O}_2'\mathbf{X}'$, then the operation $\mathbf{O}$ that being performed on $\mathbf{X}'$ gives $\mathbf{X}$ may be represented as the sum of $\mathbf{O}_1'$ and $\mathbf{O}_2'$ ($\mathbf{O} = \mathbf{O}_1' + \mathbf{O}_2'$). Transfer channels will be specified by indicating the corresponding operations.

In the situation shown in Figure 7.6a, the channel $\mathbf{M}_{t1}$ giving rise to the detected component $\mathbf{X}_{t1}$ ($\mathbf{X}_{t1} = \mathbf{M}_{t1}\mathbf{X}_{inc}$) can be represented as

$$\mathbf{M}_{t1} = \mathbf{T}^{\downarrow}(z_1 - 0, z_1 + 0)\mathbf{T}^{\downarrow}(z_0 + 0, z_1 - 0)\mathbf{T}^{\downarrow}(z_0 - 0, z_0 + 0). \tag{7.44}$$

In this case, we deal with three TR units: the interface ($z_0 - 0$, $z_0 + 0$), the bulk of the layer ($z_0 + 0$, $z_1 - 0$), and the interface ($z_1 - 0$, $z_1 + 0$).

If an emergent wave field is the result of action of two or more elementary transfer channels, this set of channels can be regarded as a combined transfer channel. Thus, for example, in the situation shown in Figure 7.6b, it is reasonable to consider the combined channel $\mathbf{M}_{t1} + \mathbf{M}_{t2}$, where

$$\mathbf{M}_{t2} = \mathbf{T}^{\downarrow}(z_1 - 0, z_1 + 0)\mathbf{T}^{\downarrow}(z_0 + 0, z_1 - 0)\mathbf{R}^{\uparrow}(z_0 - 0, z_0 + 0)\mathbf{T}^{\uparrow}(z_0 + 0, z_1 - 0)$$
$$\cdot \mathbf{R}^{\downarrow}(z_1 - 0, z_1 + 0)\mathbf{T}^{\downarrow}(z_0 + 0, z_1 - 0)\mathbf{T}^{\downarrow}(z_0 - 0, z_0 + 0)$$

($\mathbf{X}_{t2} = \mathbf{M}_{t2}\mathbf{X}_{inc}$).

In the case shown in Figure 7.6c, the layered system consists of a "thin" layered system with external interfaces at $z = z_0$ and $z = z_1$ and a thick layer. Since the optical action of "thin" layered systems is determined, to a greater or smaller extent, by FP interference, we must take into account the multiple reflections that cause it. Therefore, in this case, it is natural to regard the domain ($z_0 - 0$, $z_1 + 0$), to some approximation, as an OTR unit and to define the channel $\mathbf{M}_{t1'}$ to be considered as follows:

$$\mathbf{M}_{t1'} = \mathbf{T}^{\downarrow}(z_2 - 0, z_2 + 0)\mathbf{T}^{\downarrow}(z_1 + 0, z_2 - 0)\mathbf{T}^{\downarrow}(z_0 - 0, z_1 + 0). \tag{7.45}$$

The transfer channel approach is widely used in the framework of the ray-tracing methods. Some matrix methods of polarization optics of layered systems also use this approach. A well-known example of such a method is the extended Jones matrix method developed by Yeh [6]. In essence, this is a method of calculating the Jones matrix of the elementary transfer channel that is determined by the chain of the transmission operations performed by the interfaces and bulks of the layers of the layered system under consideration. Each of these operations is described by a 2×2 matrix. A more general variant of this method, which allows considering all kinds of TR units listed above, and recipes of fast and accurate calculation of the transmission and reflection operators for various TR units will be presented in Chapters 8–11. This technique enables one to accurately calculate the Jones matrix of any transfer channel of any ideal stratified medium. The Jones matrix of a channel may be used to calculate the transmissivity or Mueller matrix for this channel, which in turn may be used as approximating characteristics or for calculating approximating characteristics in the realistic modeling in terms of beams.

If the chain of operations determining an *elementary* channel includes only operations of the following kinds:

(i) transmission and reflection of an interface,
(ii) transmission and reflection of a "thin" layered system[4],
(iii) transmission of the bulk of a homogeneous layer,
(iv) transmission of the bulk of a "thin" layer with continuously varying parameters,

the Mueller matrix of this channel for plane monochromatic waves is, as a rule, a relatively slow function of λ and $\mathbf{l}$. This is almost always the case for approximating media for LCD panels. The slow change of the monochromatic Mueller matrix allows using the MPW approximation (7.14), which always greatly simplifies the calculations.

The transfer channel method, as a general method of optics of layered systems, is based on tracing the transformation of a basic state characteristic of light (*traced characteristic*)—a characteristic suitable for individual description of the wave fields that are operands and results of the transmission and reflection operations performed by TR units—as the light propagates in the system, interacting with its elements (TR units). The role of the traced characteristic in different cases may be played by a scalar amplitude, irradiance, Jones vector, Stokes vector, and so on. The operations performed by TR units are described by the corresponding operators. If the operators of transmission and reflection are represented by scalars (transmittance, reflectance, etc.) or matrices (Jones matrix, Mueller matrix, etc.), and $\boldsymbol{O}$ is such an operator, defined for a traced state characteristic X and describing an operation O, the corresponding transformation of the traced characteristic is represented as follows:

$$X\{\mathsf{OX}\} = \boldsymbol{O}X\{\mathsf{X}\}, \tag{7.46}$$

where X is a wave field, an operand of O; $X\{\mathsf{X}'\}$ denotes the state characteristic X of a wave field X'. The operator $\boldsymbol{O}_{\mathsf{M}}$ characterizing an elementary channel

$$\mathsf{M} = \mathsf{O}_N \mathsf{O}_{N-1} \cdots \mathsf{O}_2 \mathsf{O}_1$$

is calculated as

$$\boldsymbol{O}_{\mathsf{M}} = \boldsymbol{O}_N \boldsymbol{O}_{N-1} \ldots \boldsymbol{O}_2 \boldsymbol{O}_1, \tag{7.47}$$

where $\boldsymbol{O}_j$ is the operator of the operation O_j. If M is the combined channel consisting of channels M_1 and M_2, and state characteristic X for the components $\mathsf{M}_1\mathsf{X}_{\mathrm{inc}}$ and $\mathsf{M}_2\mathsf{X}_{\mathrm{inc}}$ ($\mathsf{X}_{\mathrm{inc}}$ is the incident light) is additive, that is,

$$X\{\mathsf{M}_1\mathsf{X}_{\mathrm{inc}} + \mathsf{M}_2\mathsf{X}_{\mathrm{inc}}\} = X\{\mathsf{M}_1\mathsf{X}_{\mathrm{inc}}\} + X\{\mathsf{M}_2\mathsf{X}_{\mathrm{inc}}\}, \tag{7.48}$$

[4] Individual "thin" layers are included in this category.

the operator $\boldsymbol{O}_{\mathsf{M}}$ of the channel M can by calculated as follows:

$$\boldsymbol{O}_{\mathsf{M}} = \boldsymbol{O}_{\mathsf{M}1} + \boldsymbol{O}_{\mathsf{M}2}, \quad (7.49)$$

where $\boldsymbol{O}_{\mathsf{M}1}$ and $\boldsymbol{O}_{\mathsf{M}2}$ are the operators characterizing the channels M_1 and M_2, respectively. Whether a state characteristic is additive or not depends on particular conditions. For example, let the layered system under consideration be an ideal stratified medium, and let $\mathbf{X}_{\text{inc}}$ be a plane quasimonochromatic wave. Such characteristics as irradiance and Stokes vector will be additive for the fields $\mathsf{M}_1\mathbf{X}_{\text{inc}}$ and $\mathsf{M}_2\mathbf{X}_{\text{inc}}$ only if these fields are mutually incoherent. If $\mathbf{X}_{\text{inc}}$ is a plane monochromatic wave, a properly defined Jones vector will be additive characteristic for the fields $\mathsf{M}_1\mathbf{X}_{\text{inc}}$ and $\mathsf{M}_2\mathbf{X}_{\text{inc}}$. In the situation shown in Figure 7.6b, radiant power is additive characteristic for the beams $\mathbf{X}_{t1}$ and $\mathbf{X}_{t2}$, irrespective of whether the incident beam is nonmonochromatic or monochromatic.

Using Power-Based Plane-Wave Approximations in Considering Optical Systems Consisting of a Layered System and a Scattering Reflector

When considering direct-view reflective and transflective LCDs with bumpy reflectors [7], a problem of importance is to estimate the optical performance of the LCD under given conditions of the ambient illumination. Figure 7.8 illustrates a subproblem involved, directly or indirectly, in modeling in such cases. A quasimonochromatic beam $\mathbf{X}_{\text{inc}}$ with narrow angular spectrum and nominal direction $\mathbf{l}_i$ falls on a layered system LS behind which a scattering reflector R is situated. The light leaving this combined system (LS + R) consists of a specular component and a diffuse one. The specular component (not shown in Figure 7.8) is caused by reflections from interfaces of the layered system LS. The diffuse component occurs due to the reflector R. The diffuse component, as well as the specular one, has a complicated structure due to multiple reflections. Let us consider the part of the diffuse component that can be expressed as $\mathbf{X}_{\text{dif}} = \mathsf{T}^{\uparrow}_{\mathsf{LS}}\mathsf{R}_{\mathsf{R}}\mathsf{T}^{\downarrow}_{\mathsf{LS}}\mathbf{X}_{\text{inc}}$, where $\mathsf{T}^{\downarrow}_{\mathsf{LS}}$ and $\mathsf{T}^{\uparrow}_{\mathsf{LS}}$ are the operations of transmission (overall or partial) of the system LS for light incident in a forward direction (as $\mathbf{X}_{\text{inc}}$) and for light incident in

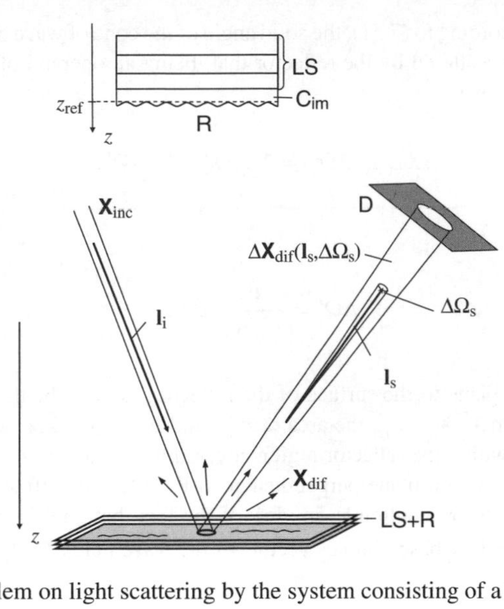

Figure 7.8 To the problem on light scattering by the system consisting of a layered system (LS) and a scattering reflector (R)

a backward direction (as the light scattered from R), respectively, and $\mathbf{R}_\mathsf{R}$ is the reflection from R. Let $\Delta\mathbf{X}_{\mathrm{dif}}(\mathbf{l}_\mathrm{s}, \Delta\Omega_\mathrm{s})$ be the portion of $\mathbf{X}_{\mathrm{dif}}$ whose angular spectrum is bounded by a narrow cone with solid angle $\Delta\Omega_\mathrm{s}$ around a direction $\mathbf{l}_\mathrm{s}$. To some degree of approximation, the beam $\Delta\mathbf{X}_{\mathrm{dif}}(\mathbf{l}_\mathrm{s}, \Delta\Omega_\mathrm{s})$ may be thought of as a part of $\mathbf{X}_{\mathrm{dif}}$ falling into a distant detector whose input diaphragm D is arranged as in Figure 7.8 (cf. Figure 7.1). The problem is to estimate the ratio of the radiant power of $\Delta\mathbf{X}_{\mathrm{dif}}(\mathbf{l}_\mathrm{s}, \Delta\Omega_\mathrm{s})$ to the radiant power of $\mathbf{X}_{\mathrm{inc}}$. Commonly, the optical action of bumpy reflectors and transflectors of LCDs can be described using the microfacet model approach [8–10]. In this approach, the scattering surface is regarded as a collection of differently oriented reflecting microfacets. Each microfacet, an interface between a surrounding substance (the medium of incidence) and the substance of the reflector, is assumed to reflect and transmit light in accordance with Snell's law and Fresnel's formulas for interfaces. Following this approach, we assume that the rear endpoint medium for the system LS, the medium of incidence for the reflector R (see Figure 7.8), is an isotropic nonabsorbing substance with refractive index n_{im}, call it C_{im}, and that the material of the reflector has a complex refractive index n_R. The refractive index of the external medium (where $\mathbf{X}_{\mathrm{inc}}$ and $\mathbf{X}_{\mathrm{dif}}$ propagate) is assumed to be equal to 1.

According to Snell's law (see Sections 1.2.1 and 8.1.3), the result of the operation $\mathbf{T}^{\downarrow}_{\mathsf{LS}}$ performed by an ideal stratified medium approximating LS on an incident plane wave with wave normal $\mathbf{l}_\mathrm{i}$ will be a plane wave with the wave normal

$$\mathbf{l}'_\mathrm{i} = \frac{1}{n_{\mathrm{im}}}\mathbf{l}_{\mathrm{i}\perp} + \mathbf{z}\sqrt{1 - \frac{\mathbf{l}_{\mathrm{i}\perp}\cdot\mathbf{l}_{\mathrm{i}\perp}}{n_{\mathrm{im}}^2}}, \tag{7.50}$$

where $\mathbf{l}_{\mathrm{i}\perp} = \mathbf{l}_\mathrm{i} - \mathbf{z}(\mathbf{z}\mathbf{l}_\mathrm{i})$ and $\mathbf{z}$ is a unit vector codirectional with the positive z axis (see Figure 7.8), in C_{im}. To give an output wave with the wave normal $\mathbf{l}_\mathrm{s}$, the incident wave being the operand of the operation $\mathbf{T}^{\uparrow}_{\mathsf{LS}}$ must have the wave normal

$$\mathbf{l}'_\mathrm{s} = \frac{1}{n_{\mathrm{im}}}\mathbf{l}_{\mathrm{s}\perp} - \mathbf{z}\sqrt{1 - \frac{\mathbf{l}_{\mathrm{s}\perp}\cdot\mathbf{l}_{\mathrm{s}\perp}}{n_{\mathrm{im}}^2}}, \tag{7.51}$$

where $\mathbf{l}_{\mathrm{s}\perp} = \mathbf{l}_\mathrm{s} - \mathbf{z}(\mathbf{z}\mathbf{l}_\mathrm{s})$. According to (7.51), the solid angle of the cone of wave normals of the component $\Delta\mathbf{X}'_{\mathrm{dif}}(\mathbf{l}'_\mathrm{s}, \Delta\Omega'_\mathrm{s})$ of the light scattered by the reflector that, being an operand of the operation $\mathbf{T}^{\uparrow}_{\mathsf{LS}}$, gives $\Delta\mathbf{X}_{\mathrm{dif}}(\mathbf{l}_\mathrm{s}, \Delta\Omega_\mathrm{s})$, that is, such that

$$\Delta\mathbf{X}_{\mathrm{dif}}(\mathbf{l}_\mathrm{s}, \Delta\Omega_\mathrm{s}) = \mathbf{T}^{\uparrow}_{\mathsf{LS}}\Delta\mathbf{X}'_{\mathrm{dif}}(\mathbf{l}'_\mathrm{s}, \Delta\Omega'_\mathrm{s}), \tag{7.52}$$

satisfies the relation

$$\Delta\Omega'_\mathrm{s} = \frac{\mathbf{z}\mathbf{l}_\mathrm{s}}{n_{\mathrm{im}}^2\,\mathbf{z}\mathbf{l}'_\mathrm{s}}\Delta\Omega_\mathrm{s}. \tag{7.53}$$

Let $z = z_{\mathrm{ref}}$ be the tangent plane to the surface of the reflector. Let $\boldsymbol{\sigma}_{\mathrm{ref}}$ be the region of the plane $z = z_{\mathrm{ref}}$ illuminated by the beam $\mathbf{T}^{\downarrow}_{\mathsf{LS}}\mathbf{X}_{\mathrm{inc}}$, σ_{ref} the area of $\boldsymbol{\sigma}_{\mathrm{ref}}$, and $\Delta S(\mathbf{l}'_\mathrm{i}, \mathbf{l}'_\mathrm{s}, \Delta\Omega'_\mathrm{s})$ the area of the projection of the microfacets that lie within the reflector region covered by $\boldsymbol{\sigma}_{\mathrm{ref}}$ and reflect the incident light to the solid angle $\Delta\Omega'_\mathrm{s}$ around $\mathbf{l}'_\mathrm{s}$ onto a plane perpendicular to $\mathbf{l}'_\mathrm{i}$. Then, if diffraction effects and multiple scattering among microfacets are negligible, in view of the fact that $\Delta\Omega'_\mathrm{s}$ is small, the Mueller matrix $\mathbf{M}_\mathsf{R}(\mathbf{l}'_\mathrm{i}, \mathbf{l}'_\mathrm{s}, \Delta\Omega'_\mathrm{s})$ relating the flux-based Stokes vectors of the wave fields $\mathbf{T}^{\downarrow}_{\mathsf{LS}}\mathbf{X}_{\mathrm{inc}}$ and $\Delta\mathbf{X}'_{\mathrm{dif}}(\mathbf{l}'_\mathrm{s}, \Delta\Omega'_\mathrm{s})$,

$$S_{(\Phi)}\left\{\Delta\mathbf{X}'_{\mathrm{dif}}(\mathbf{l}'_\mathrm{s}, \Delta\Omega'_\mathrm{s})\right\} = \mathbf{M}_\mathsf{R}(\mathbf{l}'_\mathrm{i}, \mathbf{l}'_\mathrm{s}, \Delta\Omega'_\mathrm{s})S_{(\Phi)}\left\{\mathbf{T}^{\downarrow}_{\mathsf{LS}}\mathbf{X}_{\mathrm{inc}}\right\}, \tag{7.54}$$

may be represented as follows (see, e.g., Reference 10):

$$\mathbf{M}_{\mathsf{R}}(\mathbf{l}'_{\mathrm{i}}, \mathbf{l}'_{\mathrm{s}}, \Delta\Omega'_{\mathrm{s}}) = f_F(\mathbf{l}'_{\mathrm{i}}, \mathbf{l}'_{\mathrm{s}}, \Delta\Omega'_{\mathrm{s}})\mathbf{M}_{FR}(\mathbf{l}'_{\mathrm{i}}, \mathbf{l}'_{\mathrm{s}}), \tag{7.55}$$

where

$$f_F(\mathbf{l}'_{\mathrm{i}}, \mathbf{l}'_{\mathrm{s}}, \Delta\Omega'_{\mathrm{s}}) = \frac{\Delta S(\mathbf{l}'_{\mathrm{i}}, \mathbf{l}'_{\mathrm{s}}, \Delta\Omega'_{\mathrm{s}})}{\sigma_{\mathrm{ref}}\mathbf{z}\mathbf{l}'_{\mathrm{i}}} \tag{7.56}$$

and $\mathbf{M}_{FR}(\mathbf{l}'_{\mathrm{i}}, \mathbf{l}'_{\mathrm{s}})$ is the Mueller–Jones matrix[5] describing the Fresnel reflection from a plane interface between media with refractive indices n_{im} and n_{R} perpendicular to $\mathbf{l}'_{\mathrm{i}} - \mathbf{l}'_{\mathrm{s}}$.

In common situations, in view of the narrowness of the angular spectra of $\mathbf{X}_{\mathrm{inc}}$ and $\Delta\mathbf{X}_{\mathrm{dif}}(\mathbf{l}_{\mathrm{s}}, \Delta\Omega_{\mathrm{s}})$, one can calculate the Mueller matrix $\mathbf{M}^{\downarrow}_{\mathsf{LS}}(\mathbf{l}_{\mathrm{i}})$ relating $S_{(\Phi)}\{\mathbf{X}_{\mathrm{inc}}\}$ and $S_{(\Phi)}\{\mathbf{T}^{\downarrow}_{\mathsf{LS}}\mathbf{X}_{\mathrm{inc}}\}$,

$$S_{(\Phi)}\{\mathbf{T}^{\downarrow}_{\mathsf{LS}}\mathbf{X}_{\mathrm{inc}}\} = \mathbf{M}^{\downarrow}_{\mathsf{LS}}(\mathbf{l}_{\mathrm{i}})S_{(\Phi)}\{\mathbf{X}_{\mathrm{inc}}\}, \tag{7.57}$$

and the Mueller matrix $\mathbf{M}^{\uparrow}_{\mathsf{LS}}(\mathbf{l}_{\mathrm{s}})$ linking $S_{(\Phi)}\{\Delta\mathbf{X}'_{\mathrm{dif}}(\mathbf{l}'_{\mathrm{s}}, \Delta\Omega'_{\mathrm{s}})\}$ and $S_{(\Phi)}\{\Delta\mathbf{X}_{\mathrm{dif}}(\mathbf{l}_{\mathrm{s}}, \Delta\Omega_{\mathrm{s}})\}$,

$$S_{(\Phi)}\{\Delta\mathbf{X}_{\mathrm{dif}}(\mathbf{l}_{\mathrm{s}}, \Delta\Omega_{\mathrm{s}})\} = \mathbf{M}^{\uparrow}_{\mathsf{LS}}(\mathbf{l}_{\mathrm{s}})S_{(\Phi)}\{\Delta\mathbf{X}'_{\mathrm{dif}}(\mathbf{l}'_{\mathrm{s}}, \Delta\Omega'_{\mathrm{s}})\}, \tag{7.58}$$

using the QMPW approximation.

From (7.54), (7.55), (7.57), and (7.58), we have

$$S_{(\Phi)}\{\Delta\mathbf{X}_{\mathrm{dif}}(\mathbf{l}_{\mathrm{s}}, \Delta\Omega_{\mathrm{s}})\} = f_F(\mathbf{l}'_{\mathrm{i}}, \mathbf{l}'_{\mathrm{s}}, \Delta\Omega'_{\mathrm{s}})\mathbf{M}^{\uparrow}_{\mathsf{LS}}(\mathbf{l}_{\mathrm{s}})\mathbf{M}_{FR}(\mathbf{l}'_{\mathrm{i}}, \mathbf{l}'_{\mathrm{s}})\mathbf{M}^{\downarrow}_{\mathsf{LS}}(\mathbf{l}_{\mathrm{i}})S_{(\Phi)}\{\mathbf{X}_{\mathrm{inc}}\}. \tag{7.59}$$

As can be seen from (7.59), the problem under consideration may be split into two problems. The first problem is to calculate $f_F(\mathbf{l}'_{\mathrm{i}}, \mathbf{l}'_{\mathrm{s}}, \Delta\Omega'_{\mathrm{s}})$. As a rule, this problem can easily be solved by using approaches described in References 8–10. The second problem is the calculation of the matrix

$$\mathbf{M}_{\mathsf{LS}\text{-}\mathsf{R}}(\mathbf{l}_{\mathrm{i}}, \mathbf{l}_{\mathrm{s}}) = \mathbf{M}^{\uparrow}_{\mathsf{LS}}(\mathbf{l}_{\mathrm{s}})\mathbf{M}_{FR}(\mathbf{l}'_{\mathrm{i}}, \mathbf{l}'_{\mathrm{s}})\mathbf{M}^{\downarrow}_{\mathsf{LS}}(\mathbf{l}_{\mathrm{i}}). \tag{7.60}$$

This matrix may be interpreted as that describing the corresponding transfer channel for a system consisting of LS and a tilted plane interface between media with refractive indices n_{im} and n_{R} perpendicular to $\mathbf{l}'_{\mathrm{i}} - \mathbf{l}'_{\mathrm{s}}$, which is characterized by the matrix $\mathbf{M}_{FR}(\mathbf{l}'_{\mathrm{i}}, \mathbf{l}'_{\mathrm{s}})$. An example of using such an approach for modeling direct-view reflective LCDs can be found in Reference 7.

Field-Based Plane-Wave Approximations. Direct-Ray Approximation

The power-based plane-wave approximations considered above are fully justified from the standpoint of the rigorous theory and their applicability is restricted only by the aforementioned rather general conditions. Among other things, these conditions allow one to solve the problem ignoring diffraction effects, which is used in the power-based plane-wave approximations. When dealing with LCDs having a fine transverse structure where parameters of the LC layer vary significantly along the transverse directions over a distance of the order of a light wavelength, the corresponding diffraction effects must be taken into account and the above power-based plane-wave approximations are not applicable, at least to the LC layer. Relatively accurate modeling of the optical action of such LC layers with allowance

[5] See Section 10.1.

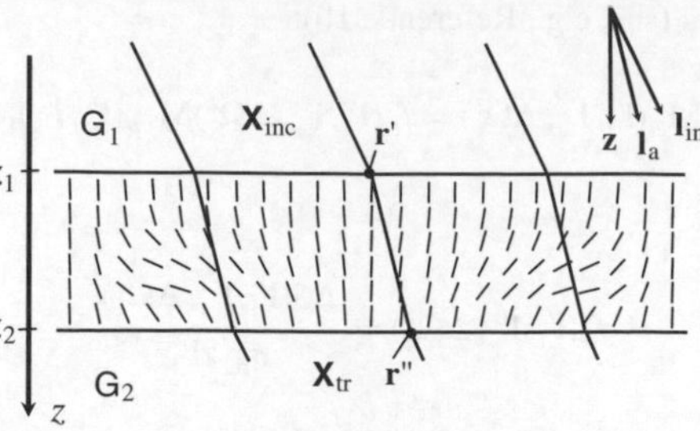

Figure 7.9 An application of the direct-ray approximation

for the diffraction effects can be carried out using finite-difference methods or grating methods (see references in Chapter 13). However, these methods are very time-consuming. For this reason, in practice, developers prefer to use, instead of these methods, simpler but less accurate methods utilizing techniques developed for ideal stratified media. These simple methods are derived with a serious departure from the rigorous theory and are almost always based on a heuristic application of the picture of light propagation in very weakly birefringent inhomogeneous media to media with stronger birefringence. One of such methods is the direct-ray approximation (DRA) method. Different variants of this method are presented and used in many works (see, e.g., References 11 and 12).

To show the basic idea of the DRA method, we consider the following problem: Let a layer of a uniaxial liquid crystal with LC director field $\mathbf{n}(\mathbf{r})$ inhomogeneous in more than one dimension and principal refractive indices $n_{\parallel}$ and $n_{\perp}$ be sandwiched between nonabsorbing isotropic media G_1 and G_2, and let the planes $z = z_1$ and $z = z_2$ be the planes of the interfaces between the LC layer and the media G_1 and G_2, respectively (Figure 7.9). Let a well-collimated monochromatic beam $\mathbf{X}_{\text{inc}}$ with nominal propagation direction along a unit vector $\mathbf{l}_{\text{inc}}$ fall on the LC layer from the medium G_1. Neglecting multiple reflections, the transmitted field $\mathbf{X}_{\text{tr}}$ propagating in G_2 is considered to be a result of the sequence of the transmission operations carried out by the frontal interface ($z = z_1$), the LC layer bulk, and the rear interface ($z = z_2$). The direct-ray approximation as applied to this problem includes the assumption that the state of the field $\mathbf{X}_{\text{tr}}$ at an arbitrary point $\mathbf{r}'' = (x'', y'', z_2 + 0)$ at the interface $z = z_2$ is fully determined by (i) the state of the field $\mathbf{X}_{\text{inc}}$ at the point $\mathbf{r}' = (x', y', z_1 - 0)$ that lies on the straight line passing through the point $\mathbf{r}''$ and parallel to a unit vector $\mathbf{l}_{\text{a}}$ specifying a conventional dominant direction of light propagation in the LC layer and (ii) parameters of the medium at the points of the straight-line segment $(\mathbf{r}', \mathbf{r}'')$. The vector $\mathbf{l}_{\text{a}}$ is usually chosen to be parallel to the wave normal of the ordinary wave in the LC layer [12] or to the wave normal of the wave that would propagate in the LC layer if this layer were isotropic and had a refractive index $n_m = (n_{\parallel} + n_{\perp})/2$ [11]; in both cases, when determining $\mathbf{l}_{\text{a}}$, the incident field $\mathbf{X}_{\text{inc}}$ is assumed to be a plane wave with wave normal $\mathbf{l}_{\text{inc}}$. In the former case,

$$\mathbf{l}_{\text{a}} = \frac{n_g}{n_{\perp}} \mathbf{l}_{\text{inc}\perp} + \mathbf{z}\sqrt{1 - \frac{n_g^2 \mathbf{l}_{\text{inc}\perp} \cdot \mathbf{l}_{\text{inc}\perp}}{n_{\perp}^2}}, \tag{7.61}$$

in the latter,

$$\mathbf{l}_{\text{a}} = \frac{n_g}{n_m} \mathbf{l}_{\text{inc}\perp} + \mathbf{z}\sqrt{1 - \frac{n_g^2 \mathbf{l}_{\text{inc}\perp} \cdot \mathbf{l}_{\text{inc}\perp}}{n_m^2}}, \tag{7.62}$$

where $\mathbf{l}_{\text{inc}\perp} = \mathbf{l}_{\text{inc}} - \mathbf{z}(\mathbf{z}\mathbf{l}_{\text{inc}})$, and n_g is the refractive index of the medium G_1. The Jones matrix $\boldsymbol{t}_J(\mathbf{r}', \mathbf{r}'')$ linking the Jones vectors $\boldsymbol{J}\{\mathbf{X}_{\text{inc}}(\mathbf{r}')\}$ and $\boldsymbol{J}\{\mathbf{X}_{\text{tr}}(\mathbf{r}'')\}$ as

$$\boldsymbol{J}\{\mathbf{X}_{\text{tr}}(\mathbf{r}'')\} = \boldsymbol{t}_J(\mathbf{r}', \mathbf{r}'')\boldsymbol{J}\{\mathbf{X}_{\text{inc}}(\mathbf{r}')\} \tag{7.63}$$

is approximated by the transmission Jones matrix $\boldsymbol{t}_{J1\text{D}}(\mathbf{r}', \mathbf{r}'')$ of a fictitious ideal stratified medium occupying the region $(z_1 + 0, z_2 - 0)$ whose local optical parameters at any given z, say $z = \xi$, are identical to those of the original medium at the point of intersection of the straight-line segment $(\mathbf{r}', \mathbf{r}'')$ and the plane $z = \xi$. The spatial dependence of the local optic axis $\mathbf{c}$ in this ideal stratified medium may be expressed as follows:

$$\mathbf{c}(z) = \mathbf{n}\left(\mathbf{r}' + \mathbf{l}_a \frac{(z - z')}{\mathbf{z}\mathbf{l}_a}\right), \tag{7.64}$$

where $\mathbf{n}(\mathbf{r})$ is the LC director field in the original LC layer. Calculation of the matrix $\boldsymbol{t}_{J1\text{D}}(\mathbf{r}', \mathbf{r}'')$ is usually performed with the aid of some variant of the Jones matrix method. Techniques providing fast and accurate estimation of $\boldsymbol{t}_{J1\text{D}}(\mathbf{r}', \mathbf{r}'')$ are considered in Chapters 8 and 11.

Given $\boldsymbol{J}\{\mathbf{X}_{\text{inc}}(\mathbf{r})\}$ at the plane $z = z_1$ (at $z = z_1 - 0$), one calculates $\boldsymbol{J}\{\mathbf{X}_{\text{tr}}(\mathbf{r})\}$ at the plane $z = z_2$ (at $z = z_2 + 0$). Further, by using standard approaches of Fourier optics, the field $\mathbf{X}_{\text{tr}}(\mathbf{r})$ may be decomposed into plane-wave components. Considering the interaction of these plane-wave components with following elements of the device being modeled, one may find the parameters of plane-wave components of the output light. Finally one may calculate, for instance, the distribution of the output light in the far-field region (see, e.g., Reference 12).

There are simpler variants of application of DRA in modeling LCDs (see, e.g., Reference 11). One of them is considered in Chapter 13. Though rough, DRA methods often provide very useful estimates. DRA is widely used in modeling many kinds of LC devices, including PDLC (polymer-dispersed liquid crystal) devices. In the case of PDLCs, DRA is used in calculations of the scattering cross-sections of LC droplets in the anomalous diffraction approximation [13, 14].

DRA as presented above and other approximations that involve plane-wave approximations of wave fields on the level of local amplitude characteristics belong to the class of field-based plane-wave approximations.

7.2 Transfer Matrix Technique and Adding Technique

In this section, we consider two general techniques, the transfer matrix technique and the adding technique, which underlie many practical methods of calculating transmission and reflection characteristics of layered systems with allowance for multiple reflections. Like the channel technique described in the previous section, the adding technique is an algorithm of calculating transmission and reflection operators (Jones matrix, Mueller matrix, etc.) of the layered system using known transmission and reflection operators (of the same kind) of its TR units. A flexible variant of the adding technique, which will be presented here, can be used for calculating characteristics of both overall and partial transmission and reflection of layered systems. The transfer matrix technique is intended for calculating operators of the overall reflection and transmission of layered systems. Like the channel technique, the transfer matrix technique and adding technique are not restricted by any particular choice of the traced characteristic of light. For this reason, as in the above description of the channel technique, we will not specify the traced characteristic, denoting it simply by X and assuming that it may be Jones vector, Stokes vector, irradiance, or any other suitable characteristic of light.

7.2.1 *Transfer Matrix Technique*

TR-Additivity

Let us consider a TR unit (z', z'') of a layered medium. Let $\mathbf{X}^{\downarrow}(z')$ and $\mathbf{X}^{\uparrow}(z'')$ be wave fields incident on this unit from the regions $z < z'$ and $z > z''$, respectively. The arrow $^{\downarrow}$ ($^{\uparrow}$) at a symbol or a characteristic of a wave field indicates that this wave field propagates to greater (lesser) z. The outgoing fields $\mathbf{X}^{\uparrow}(z')$ and $\mathbf{X}^{\downarrow}(z'')$ generated by the fields $\mathbf{X}^{\downarrow}(z')$ и $\mathbf{X}^{\uparrow}(z'')$ may be represented as

$$\mathbf{X}^{\downarrow}(z'') = \mathbf{X}^{\downarrow}_{\mathrm{tr}}(z'') + \mathbf{X}^{\downarrow}_{\mathrm{ref}}(z''), \quad \mathbf{X}^{\uparrow}(z') = \mathbf{X}^{\uparrow}_{\mathrm{tr}}(z') + \mathbf{X}^{\uparrow}_{\mathrm{ref}}(z'), \tag{7.65}$$

where

$$\mathbf{X}^{\downarrow}_{\mathrm{tr}}(z'') \equiv \mathbf{T}^{\downarrow}(z', z'')\mathbf{X}^{\downarrow}(z'), \quad \mathbf{X}^{\downarrow}_{\mathrm{ref}}(z'') \equiv \mathbf{R}^{\uparrow}(z', z'')\mathbf{X}^{\uparrow}(z''),$$
$$\mathbf{X}^{\uparrow}_{\mathrm{tr}}(z') \equiv \mathbf{T}^{\uparrow}(z', z'')\mathbf{X}^{\uparrow}(z''), \quad \mathbf{X}^{\uparrow}_{\mathrm{ref}}(z') \equiv \mathbf{R}^{\downarrow}(z', z'')\mathbf{X}^{\downarrow}(z')$$

(see Figure 7.10), that is, we can write

$$\mathbf{X}^{\downarrow}(z'') = \mathbf{T}^{\downarrow}(z', z'')\mathbf{X}^{\downarrow}(z') + \mathbf{R}^{\uparrow}(z', z'')\mathbf{X}^{\uparrow}(z''), \tag{7.66}$$
$$\mathbf{X}^{\uparrow}(z') = \mathbf{T}^{\uparrow}(z', z'')\mathbf{X}^{\uparrow}(z'') + \mathbf{R}^{\downarrow}(z', z'')\mathbf{X}^{\downarrow}(z'). \tag{7.67}$$

Let the chosen traced characteristic X be a scalar quantity or a quantity represented by a column vector, and let the transmission and reflection operators be correspondingly represented by scalars or square matrices. We will say that characteristic X is *TR-additive* for the unit (z', z'') if the interaction of the outgoing wave fields is such that

$$X^{\uparrow}(z') = X^{\uparrow}_{\mathrm{tr}}(z') + X^{\uparrow}_{\mathrm{ref}}(z'), \quad X^{\downarrow}(z'') = X^{\downarrow}_{\mathrm{tr}}(z'') + X^{\downarrow}_{\mathrm{ref}}(z''), \tag{7.68}$$

where $X^{\downarrow}_{\mathrm{tr}}(z'')$, $X^{\downarrow}_{\mathrm{ref}}(z'')$, $X^{\uparrow}_{\mathrm{tr}}(z')$, and $X^{\uparrow}_{\mathrm{ref}}(z')$ are the X-characteristics of the wave fields $\mathbf{X}^{\downarrow}_{\mathrm{tr}}(z'')$, $\mathbf{X}^{\downarrow}_{\mathrm{ref}}(z'')$, $\mathbf{X}^{\uparrow}_{\mathrm{tr}}(z')$, and $\mathbf{X}^{\uparrow}_{\mathrm{ref}}(z')$, respectively. For example, irradiance can be considered as a TR-additive characteristic

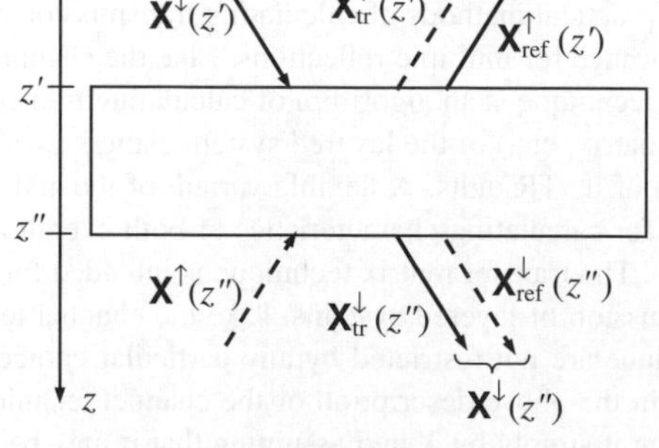

Figure 7.10 A TR unit of a layered medium

for the unit (z', z'') only if the superposed outgoing fields, $\mathbf{X}^{\downarrow}_{tr}(z'')$ and $\mathbf{X}^{\downarrow}_{ref}(z'')$ as well as $\mathbf{X}^{\uparrow}_{tr}(z')$ and $\mathbf{X}^{\uparrow}_{ref}(z')$, are mutually incoherent. It follows from (7.65)–(7.68) that

$$\begin{aligned} \boldsymbol{X}^{\downarrow}(z'') &= \boldsymbol{T}^{\downarrow}(z',z'')\boldsymbol{X}^{\downarrow}(z') + \boldsymbol{R}^{\uparrow}(z',z'')\boldsymbol{X}^{\uparrow}(z''), \\ \boldsymbol{X}^{\uparrow}(z') &= \boldsymbol{T}^{\uparrow}(z',z'')\boldsymbol{X}^{\uparrow}(z'') + \boldsymbol{R}^{\downarrow}(z',z'')\boldsymbol{X}^{\downarrow}(z'), \end{aligned} \tag{7.69}$$

where $\boldsymbol{T}^{\downarrow}(z', z'')$, $\boldsymbol{T}^{\uparrow}(z', z'')$, $\boldsymbol{R}^{\uparrow}(z', z'')$, and $\boldsymbol{R}^{\downarrow}(z', z'')$ are the operators of the operations $\mathbf{T}^{\downarrow}(z', z'')$, $\mathbf{T}^{\uparrow}(z', z'')$, $\mathbf{R}^{\uparrow}(z', z'')$, and $\mathbf{R}^{\downarrow}(z', z'')$.

Transfer Matrix Approach

Consider a layered system consisting of n OTR units for each of which the chosen traced characteristic $\boldsymbol{X}$ is TR-additive. The TR-additivity allows one to link the $\boldsymbol{X}$-characteristics of the incoming and outgoing fields for each unit by relations of the form (7.69):

$$\begin{aligned} \boldsymbol{X}^{\downarrow}(\bar{z}_j) &= \boldsymbol{T}^{\downarrow}(\bar{z}_{j-1},\bar{z}_j)\boldsymbol{X}^{\downarrow}(\bar{z}_{j-1}) + \boldsymbol{R}^{\uparrow}(\bar{z}_{j-1},\bar{z}_j)\boldsymbol{X}^{\uparrow}(\bar{z}_j), \\ \boldsymbol{X}^{\uparrow}(\bar{z}_{j-1}) &= \boldsymbol{T}^{\uparrow}(\bar{z}_{j-1},\bar{z}_j)\boldsymbol{X}^{\uparrow}(\bar{z}_j) + \boldsymbol{R}^{\downarrow}(\bar{z}_{j-1},\bar{z}_j)\boldsymbol{X}^{\downarrow}(\bar{z}_{j-1}), \\ j &= 1,2,\ldots,n, \end{aligned} \tag{7.70}$$

where $\bar{z}_{j-1}$ and $\bar{z}_j$ ($\bar{z}_{j-1} < \bar{z}_j$) are the z-coordinates of the boundaries of the jth OTR unit. Relations (7.70) may be treated as a basic set of equations for finding the operators of the overall transmission and reflection of the layered system when the transmission and reflection operators for each of the OTR units are known. Here is one of simple ways to find the overall transmission and reflection operators of the layered system, $\boldsymbol{T}^{\downarrow}(\bar{z}_0, \bar{z}_n)$, $\boldsymbol{R}^{\downarrow}(\bar{z}_0, \bar{z}_n)$, $\boldsymbol{T}^{\uparrow}(\bar{z}_0, \bar{z}_n)$, and $\boldsymbol{R}^{\uparrow}(\bar{z}_0, \bar{z}_n)$, starting from (7.70).

If the operator $\boldsymbol{T}^{\uparrow}(z', z'')^{-1}$ exists, the system of equations (7.69) is equivalent to the following one:

$$\begin{aligned} \boldsymbol{X}^{\downarrow}(z'') &= \boldsymbol{N}_{11}(z'',z')\boldsymbol{X}^{\downarrow}(z') + \boldsymbol{N}_{12}(z'',z')\boldsymbol{X}^{\uparrow}(z'), \\ \boldsymbol{X}^{\uparrow}(z'') &= \boldsymbol{N}_{21}(z'',z')\boldsymbol{X}^{\downarrow}(z') + \boldsymbol{N}_{22}(z'',z')\boldsymbol{X}^{\uparrow}(z'), \end{aligned} \tag{7.71}$$

where

$$\begin{aligned} \boldsymbol{N}_{11}(z'',z') &= \boldsymbol{T}^{\downarrow}(z',z'') - \boldsymbol{R}^{\uparrow}(z',z'')\boldsymbol{T}^{\uparrow}(z',z'')^{-1}\boldsymbol{R}^{\downarrow}(z',z''), \\ \boldsymbol{N}_{12}(z'',z') &= \boldsymbol{R}^{\uparrow}(z',z'')\boldsymbol{T}^{\uparrow}(z',z'')^{-1}, \\ \boldsymbol{N}_{21}(z'',z') &= -\boldsymbol{T}^{\uparrow}(z',z'')^{-1}\boldsymbol{R}^{\downarrow}(z',z''), \quad \boldsymbol{N}_{22}(z'',z') = \boldsymbol{T}^{\uparrow}(z',z'')^{-1}. \end{aligned} \tag{7.72}$$

Introducing the state vector

$$\overleftrightarrow{\boldsymbol{X}}(z) = \begin{pmatrix} \boldsymbol{X}^{\downarrow}(z) \\ \boldsymbol{X}^{\uparrow}(z) \end{pmatrix} \tag{7.73}$$

and the matrix

$$\bar{\boldsymbol{N}}(z'',z') = \begin{pmatrix} \boldsymbol{N}_{11}(z'',z') & \boldsymbol{N}_{12}(z'',z') \\ \boldsymbol{N}_{21}(z'',z') & \boldsymbol{N}_{22}(z'',z') \end{pmatrix},$$

we may write the system (7.71) as follows:

$$\overset{\updownarrow}{\boldsymbol{X}}(z'') = \bar{\boldsymbol{N}}(z'', z')\overset{\updownarrow}{\boldsymbol{X}}(z'). \tag{7.74}$$

This relation represents the vector $\overset{\updownarrow}{\boldsymbol{X}}(z'')$ characterizing the state of the radiation in the plane $z = z''$ as the result of a linear transformation of the vector $\overset{\updownarrow}{\boldsymbol{X}}(z')$ describing the state of the radiation in the plane $z = z'$. The operator $\bar{\boldsymbol{N}}(z'', z')$ may be called the *transfer matrix* of state vector $\overset{\updownarrow}{\boldsymbol{X}}$ for the fragment (z', z''). Note that here and in what follows, specifying transfer matrices, we write down the arguments in the reverse order.

If each of the units $(\bar{z}_{j-1}, \bar{z}_j)$ (j = 1, 2,…, n) of the layered structure under consideration can be described by the corresponding transfer matrix, $\bar{\boldsymbol{N}}(\bar{z}_j, \bar{z}_{j-1})$, we may replace the set of equations (7.70) by the following one:

$$\overset{\updownarrow}{\boldsymbol{X}}(\bar{z}_j) = \bar{\boldsymbol{N}}(\bar{z}_j, \bar{z}_{j-1})\overset{\updownarrow}{\boldsymbol{X}}(\bar{z}_{j-1}) \qquad j = 1, 2, \ldots, n. \tag{7.75}$$

According to (7.75), the vectors

$$\overset{\updownarrow}{\boldsymbol{X}}(\bar{z}_0) = \begin{pmatrix} \boldsymbol{X}^{\downarrow}(\bar{z}_0) \\ \boldsymbol{X}^{\uparrow}(\bar{z}_0) \end{pmatrix} \quad \text{and} \quad \overset{\updownarrow}{\boldsymbol{X}}(\bar{z}_n) = \begin{pmatrix} \boldsymbol{X}^{\downarrow}(\bar{z}_n) \\ \boldsymbol{X}^{\uparrow}(\bar{z}_n) \end{pmatrix},$$

must satisfy the equation

$$\overset{\updownarrow}{\boldsymbol{X}}(\bar{z}_n) = \bar{\boldsymbol{N}}(\bar{z}_n, \bar{z}_0)\overset{\updownarrow}{\boldsymbol{X}}(\bar{z}_0), \tag{7.76}$$

where

$$\bar{\boldsymbol{N}}(\bar{z}_n, \bar{z}_0) = \begin{pmatrix} \widehat{\boldsymbol{N}}_{11} & \widehat{\boldsymbol{N}}_{12} \\ \widehat{\boldsymbol{N}}_{21} & \widehat{\boldsymbol{N}}_{22} \end{pmatrix} = \bar{\boldsymbol{N}}_n \bar{\boldsymbol{N}}_{n-1} \cdot \ldots \cdot \bar{\boldsymbol{N}}_2 \bar{\boldsymbol{N}}_1, \quad \bar{\boldsymbol{N}}_j \equiv \bar{\boldsymbol{N}}(\bar{z}_j, \bar{z}_{j-1}). \tag{7.77}$$

Thus, the set of equations (7.70) is reduced to the following one

$$\boldsymbol{X}^{\downarrow}(\bar{z}_n) = \widehat{\boldsymbol{N}}_{11}\boldsymbol{X}^{\downarrow}(\bar{z}_0) + \widehat{\boldsymbol{N}}_{12}\boldsymbol{X}^{\uparrow}(\bar{z}_0), \tag{7.78a}$$

$$\boldsymbol{X}^{\uparrow}(\bar{z}_n) = \widehat{\boldsymbol{N}}_{21}\boldsymbol{X}^{\downarrow}(\bar{z}_0) + \widehat{\boldsymbol{N}}_{22}\boldsymbol{X}^{\uparrow}(\bar{z}_0). \tag{7.78b}$$

The sought overall transmission and reflection operators of the layered system can readily be expressed in terms of $\widehat{\boldsymbol{N}}_{kl}$. Thus, to find the operators $\boldsymbol{T}^{\downarrow}(\bar{z}_0, \bar{z}_n)$ and $\boldsymbol{R}^{\downarrow}(\bar{z}_0, \bar{z}_n)$, we should put $\boldsymbol{X}^{\uparrow}(\bar{z}_n) = \boldsymbol{0}$, where $\boldsymbol{0}$ is zero or the zero vector, in (7.78b), which gives

$$\widehat{\boldsymbol{N}}_{22}\boldsymbol{X}^{\uparrow}(\bar{z}_0) = -\widehat{\boldsymbol{N}}_{21}\boldsymbol{X}^{\downarrow}(\bar{z}_0). \tag{7.79}$$

In this case, $\boldsymbol{X}^{\downarrow}(\bar{z}_n)$ characterizes the field transmitted by the system, $\boldsymbol{X}^{\uparrow}(\bar{z}_0)$ describes the reflected field, and, by definition,

$$\boldsymbol{X}^{\downarrow}(\bar{z}_n) = \boldsymbol{T}^{\downarrow}(\bar{z}_0, \bar{z}_n)\boldsymbol{X}^{\downarrow}(\bar{z}_0), \tag{7.80a}$$

$$\boldsymbol{X}^{\uparrow}(\bar{z}_0) = \boldsymbol{R}^{\downarrow}(\bar{z}_0, \bar{z}_n)\boldsymbol{X}^{\downarrow}(\bar{z}_0). \tag{7.80b}$$

Premultiplying (7.79) by $\widehat{\boldsymbol{N}}_{22}^{-1}$ we have

$$\boldsymbol{X}^{\uparrow}(\bar{z}_0) = -\widehat{\boldsymbol{N}}_{22}^{-1}\widehat{\boldsymbol{N}}_{21}\boldsymbol{X}^{\downarrow}(\bar{z}_0). \tag{7.81}$$

Comparing (7.81) with (7.80b), we see that

$$\boldsymbol{R}^{\downarrow}(\bar{z}_0, \bar{z}_n) = -\widehat{\boldsymbol{N}}_{22}^{-1}\widehat{\boldsymbol{N}}_{21}. \tag{7.82a}$$

Substituting (7.81) into (7.78a) gives

$$\boldsymbol{X}^{\downarrow}(\bar{z}_n) = \left(\widehat{\boldsymbol{N}}_{11} - \widehat{\boldsymbol{N}}_{12}\widehat{\boldsymbol{N}}_{22}^{-1}\widehat{\boldsymbol{N}}_{21}\right)\boldsymbol{X}^{\downarrow}(\bar{z}_0),$$

that is,

$$\boldsymbol{T}^{\downarrow}(\bar{z}_0, \bar{z}_n) = \widehat{\boldsymbol{N}}_{11} - \widehat{\boldsymbol{N}}_{12}\widehat{\boldsymbol{N}}_{22}^{-1}\widehat{\boldsymbol{N}}_{21}. \tag{7.82b}$$

The expressions for the operators $\boldsymbol{T}^{\uparrow}(\bar{z}_0, \bar{z}_n)$ and $\boldsymbol{R}^{\uparrow}(\bar{z}_0, \bar{z}_n)$, obtained on putting $\boldsymbol{X}^{\downarrow}(\bar{z}_0) = \mathbf{0}$ in (7.78), are

$$\boldsymbol{T}^{\uparrow}(\bar{z}_0, \bar{z}_n) = \widehat{\boldsymbol{N}}_{22}^{-1}, \tag{7.82c}$$

$$\boldsymbol{R}^{\uparrow}(\bar{z}_0, \bar{z}_n) = \widehat{\boldsymbol{N}}_{12}\widehat{\boldsymbol{N}}_{22}^{-1}. \tag{7.82d}$$

It is convenient to introduce the following notation. Let $\mathbf{T}$ be a $2m \times 2m$ matrix with $m \times m$ blocks $\mathbf{t}_{11}$, $\mathbf{t}_{12}$, $\mathbf{t}_{21}$, and $\mathbf{t}_{22}$,

$$\mathbf{T} = \begin{pmatrix} \mathbf{t}_{11} & \mathbf{t}_{12} \\ \mathbf{t}_{21} & \mathbf{t}_{22} \end{pmatrix}. \tag{7.83}$$

By $\mathtt{t}^{\downarrow}\{\mathbf{T}\}$, $\mathtt{r}^{\downarrow}\{\mathbf{T}\}$, $\mathtt{t}^{\uparrow}\{\mathbf{T}\}$, and $\mathtt{r}^{\uparrow}\{\mathbf{T}\}$ we will denote the $m \times m$ matrices calculated as

$$\begin{gathered} \mathtt{t}^{\downarrow}\{\mathbf{T}\} = \mathbf{t}_{11} - \mathbf{t}_{12}\mathbf{t}_{22}^{-1}\mathbf{t}_{21}, \quad \mathtt{r}^{\downarrow}\{\mathbf{T}\} = -\mathbf{t}_{22}^{-1}\mathbf{t}_{21}, \\ \mathtt{t}^{\uparrow}\{\mathbf{T}\} = \mathbf{t}_{22}^{-1}, \quad \mathtt{r}^{\uparrow}\{\mathbf{T}\} = \mathbf{t}_{12}\mathbf{t}_{22}^{-1}. \end{gathered} \tag{7.84}$$

Using this notation, we may rewrite expressions (7.82) as follows:

$$\begin{aligned} \boldsymbol{T}^{\downarrow}(\bar{z}_0, \bar{z}_n) &= \mathtt{t}^{\downarrow}\left\{\bar{\boldsymbol{N}}(\bar{z}_n, \bar{z}_0)\right\}, \quad \boldsymbol{R}^{\downarrow}(\bar{z}_0, \bar{z}_n) = \mathtt{r}^{\downarrow}\left\{\bar{\boldsymbol{N}}(\bar{z}_n, \bar{z}_0)\right\}, \\ \boldsymbol{T}^{\uparrow}(\bar{z}_0, \bar{z}_n) &= \mathtt{t}^{\uparrow}\left\{\bar{\boldsymbol{N}}(\bar{z}_n, \bar{z}_0)\right\}, \quad \boldsymbol{R}^{\uparrow}(\bar{z}_0, \bar{z}_n) = \mathtt{r}^{\uparrow}\left\{\bar{\boldsymbol{N}}(\bar{z}_n, \bar{z}_0)\right\}. \end{aligned}$$

It is worth mentioning another example of application of the transfer matrix $\bar{\boldsymbol{N}}(\bar{z}_n, \bar{z}_0)$. Assume that we deal with a combined system (see Figure 7.11) consisting of the layered medium considered above and a reflector characterized by the reflection operator $\boldsymbol{R}^{\downarrow}(\bar{z}_n, \infty)$ such that

$$\boldsymbol{X}^{\uparrow}(\bar{z}_n) = \boldsymbol{R}^{\downarrow}(\bar{z}_n, \infty)\boldsymbol{X}^{\downarrow}(\bar{z}_n). \tag{7.85}$$

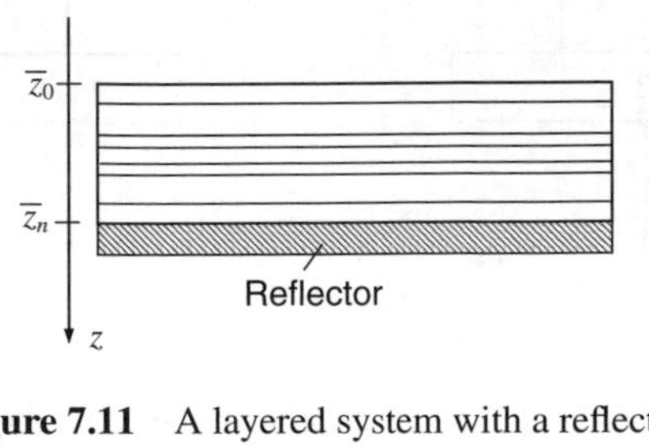

Figure 7.11 A layered system with a reflector

According to (7.78) and (7.85), the operator of the overall reflection for this combined system, $\boldsymbol{R}^{\downarrow}(\bar{z}_0, \infty)$, can be calculated as follows:

$$\boldsymbol{R}^{\downarrow}(\bar{z}_0, \infty) = -\left(\widehat{\boldsymbol{N}}_{22} - \boldsymbol{R}^{\downarrow}(\bar{z}_n, \infty)\widehat{\boldsymbol{N}}_{12}\right)^{-1}\left(\widehat{\boldsymbol{N}}_{21} - \boldsymbol{R}^{\downarrow}(\bar{z}_n, \infty)\widehat{\boldsymbol{N}}_{11}\right). \tag{7.86}$$

Relations (7.77) and (7.72) show a way of calculating $\widehat{\boldsymbol{N}}_{kl}$ $(k, l = 1, 2)$ when we know the transmission and reflection operators of the units of the layered structure. In many cases, it is possible to calculate the transfer matrices $\bar{\boldsymbol{N}}(\bar{z}_j, \bar{z}_{j-1})$ more efficiently than via transmission and reflection operators of the units. We will see examples of this in the next chapters.

The feature of the vector $\overleftrightarrow{\boldsymbol{X}}$ is that it describes the state of both the forward and backward propagating fields, that is, the total field, in a given plane. In the literature, methods in which transmission and reflection characteristics of the layered structure are calculated from operators linking state vectors of the total fields, such as $\overleftrightarrow{\boldsymbol{X}}$ or reducible to such as $\overleftrightarrow{\boldsymbol{X}}$ by a linear transformation, at the outer boundaries of the structure are customarily referred to the class of *transfer matrix methods*. Well-known representatives of this class are Abelès's method [15, 3] and the method of Hayfield and White [16, 17] for isotropic layered systems as well as the more general Berreman's method [18] and Yeh's method [19].

The main disadvantage of the transfer matrix methods is their numerical instability in situations when strong attenuation (because of absorption or realization of TIR mode) takes place in a layer or layers of the system characterized by the resulting transfer matrix. This instability is connected with the presence of very large components in certain transfer matrices and the necessity to compute small quantities as the difference of very large numbers; computations of this kind are known to lead to critical loss of accuracy. In the above algorithm, the appearance of large elements in the matrix $\bar{\boldsymbol{N}}(z'', z')$ is connected with the presence of the term $\boldsymbol{T}^{\uparrow}(z', z'')^{-1}$ in (7.72). This can easily be seen from (7.72) on taking, for simplicity, radiant flux as the traced characteristic. In this case, the operators $\boldsymbol{T}^{\downarrow}$, $\boldsymbol{T}^{\uparrow}$, $\boldsymbol{R}^{\downarrow}$, and $\boldsymbol{R}^{\uparrow}$ are transmittances and reflectances, and, as seen from (7.72), some or all of the elements of the matrix $\bar{\boldsymbol{N}}(z'', z')$ tend to infinity as $\boldsymbol{T}^{\uparrow}(z', z'') \to 0$. The adding technique (also called *the S-matrix algorithm* [20]) described in the next section and some other allied techniques, such as the scattering matrix technique [21], are numerically stable in such situations.

7.2.2 *Adding Technique*

Let us consider two adjacent fragments (TR units) **A** and **B** of a layered medium (Figure 7.12). Denote the transmission and reflection operators of the fragment **A** by $\boldsymbol{T}^{\downarrow}_{\mathrm{A}}$, $\boldsymbol{R}^{\downarrow}_{\mathrm{A}}$, $\boldsymbol{T}^{\uparrow}_{\mathrm{A}}$, and $\boldsymbol{R}^{\uparrow}_{\mathrm{A}}$, and those of the

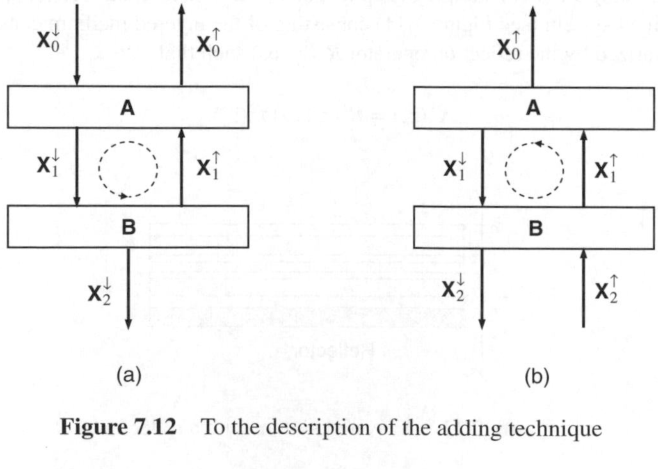

Figure 7.12 To the description of the adding technique

fragment **B** by $\boldsymbol{T}^{\downarrow}_{\mathrm{B}}$, $\boldsymbol{R}^{\downarrow}_{\mathrm{B}}$, $\boldsymbol{T}^{\uparrow}_{\mathrm{B}}$, and $\boldsymbol{R}^{\uparrow}_{\mathrm{B}}$. The corresponding state characteristic $\boldsymbol{X}$ is assumed to be TR-additive for both fragments. One of ways to calculate the transmission and reflection operators of the system **A** + **B**, $\boldsymbol{T}^{\downarrow}_{\mathrm{A+B}}$, $\boldsymbol{R}^{\downarrow}_{\mathrm{A+B}}$, $\boldsymbol{T}^{\uparrow}_{\mathrm{A+B}}$, and $\boldsymbol{R}^{\uparrow}_{\mathrm{A+B}}$, when we know the transmission and reflection operators of the fragments **A** and **B** is the following.

In order to find the operators $\boldsymbol{T}^{\downarrow}_{\mathrm{A+B}}$ and $\boldsymbol{R}^{\downarrow}_{\mathrm{A+B}}$, consider the fields generated in the system **A** + **B** by the field $\boldsymbol{X}^{\downarrow}_{0}$ (here, to refer to a particular field we use the symbol denoting the state vector of this field) incident on the fragment **A** (see Figure 7.12a; there is no gap between the fragments **A** and **B**; we disjoined these fragments in Figure 7.12 only to make this figure clearer). Using the notation introduced in Figure 7.12a, we may write the following relations for the state vectors of the fields involved in the process:

$$\boldsymbol{X}^{\uparrow}_{0} = \boldsymbol{R}^{\downarrow}_{\mathrm{A}}\boldsymbol{X}^{\downarrow}_{0} + \boldsymbol{T}^{\uparrow}_{\mathrm{A}}\boldsymbol{X}^{\uparrow}_{1}, \tag{7.87}$$

$$\boldsymbol{X}^{\downarrow}_{1} = \boldsymbol{T}^{\downarrow}_{\mathrm{A}}\boldsymbol{X}^{\downarrow}_{0} + \boldsymbol{R}^{\uparrow}_{\mathrm{A}}\boldsymbol{X}^{\uparrow}_{1} = \boldsymbol{T}^{\downarrow}_{\mathrm{A}}\boldsymbol{X}^{\downarrow}_{0} + \boldsymbol{R}^{\uparrow}_{A}\boldsymbol{R}^{\downarrow}_{B}\boldsymbol{X}^{\downarrow}_{1}, \tag{7.88}$$

$$\boldsymbol{X}^{\downarrow}_{2} = \boldsymbol{T}^{\downarrow}_{\mathrm{B}}\boldsymbol{X}^{\downarrow}_{1}. \tag{7.89}$$

From (7.88), we have

$$(\mathbf{U} - \boldsymbol{R}^{\uparrow}_{\mathrm{A}}\boldsymbol{R}^{\downarrow}_{\mathrm{B}})\boldsymbol{X}^{\downarrow}_{1} = \boldsymbol{T}^{\downarrow}_{A}\boldsymbol{X}^{\downarrow}_{0}, \tag{7.90}$$

where **U** is the unit matrix. Therefore, we may express the vector $\boldsymbol{X}^{\downarrow}_{1}$ as follows:

$$\boldsymbol{X}^{\downarrow}_{1} = (\mathbf{U} - \boldsymbol{R}^{\uparrow}_{\mathrm{A}}\boldsymbol{R}^{\downarrow}_{\mathrm{B}})^{-1}\boldsymbol{T}^{\downarrow}_{A}\boldsymbol{X}^{\downarrow}_{0}. \tag{7.91}$$

Since $\boldsymbol{X}^{\uparrow}_{1} = \boldsymbol{R}^{\downarrow}_{\mathrm{B}}\boldsymbol{X}^{\downarrow}_{1}$, using (7.91), we may write

$$\boldsymbol{X}^{\uparrow}_{1} = \boldsymbol{R}^{\downarrow}_{\mathrm{B}}(\mathbf{U} - \boldsymbol{R}^{\uparrow}_{\mathrm{A}}\boldsymbol{R}^{\downarrow}_{\mathrm{B}})^{-1}\boldsymbol{T}^{\downarrow}_{\mathrm{A}}\boldsymbol{X}^{\downarrow}_{0}. \tag{7.92}$$

On substituting (7.91) and (7.92) into (7.89) and (7.87), respectively, we obtain

$$\boldsymbol{X}^{\downarrow}_{2} = \boldsymbol{T}^{\downarrow}_{\mathrm{B}}(\mathbf{U} - \boldsymbol{R}^{\uparrow}_{\mathrm{A}}\boldsymbol{R}^{\downarrow}_{\mathrm{B}})^{-1}\boldsymbol{T}^{\downarrow}_{\mathrm{A}}\boldsymbol{X}^{\downarrow}_{0}, \tag{7.93}$$

$$\boldsymbol{X}^{\uparrow}_{0} = [\boldsymbol{R}^{\downarrow}_{\mathrm{A}} + \boldsymbol{T}^{\uparrow}_{\mathrm{A}}\boldsymbol{R}^{\downarrow}_{\mathrm{B}}(\mathbf{U} - \boldsymbol{R}^{\uparrow}_{\mathrm{A}}\boldsymbol{R}^{\downarrow}_{\mathrm{B}})^{-1}\boldsymbol{T}^{\downarrow}_{\mathrm{A}}]\boldsymbol{X}^{\downarrow}_{0}. \tag{7.94}$$

From these relations we see that

$$\boldsymbol{T}^{\downarrow}_{\mathrm{A+B}} = \boldsymbol{T}^{\downarrow}_{\mathrm{B}}(\mathbf{U} - \boldsymbol{R}^{\uparrow}_{\mathrm{A}}\boldsymbol{R}^{\downarrow}_{\mathrm{B}})^{-1}\boldsymbol{T}^{\downarrow}_{\mathrm{A}}, \tag{7.95}$$

$$\boldsymbol{R}^{\downarrow}_{\mathrm{A+B}} = \boldsymbol{R}^{\downarrow}_{\mathrm{A}} + \boldsymbol{T}^{\uparrow}_{\mathrm{A}}\boldsymbol{R}^{\downarrow}_{\mathrm{B}}(\mathbf{U} - \boldsymbol{R}^{\uparrow}_{\mathrm{A}}\boldsymbol{R}^{\downarrow}_{\mathrm{B}})^{-1}\boldsymbol{T}^{\downarrow}_{\mathrm{A}}. \tag{7.96}$$

The operators $\boldsymbol{T}^{\downarrow}_{\mathrm{A+B}}$ and $\boldsymbol{R}^{\downarrow}_{\mathrm{A+B}}$ expressed by formulas (7.95) and (7.96) allow for all re-reflections between the fragments **A** and **B**. If we want to ignore these re-reflections at all, we should use, instead of (7.91), the following expression for $\boldsymbol{X}^{\downarrow}_{1}$,

$$\boldsymbol{X}^{\downarrow}_{1} = \boldsymbol{T}^{\downarrow}_{\mathrm{A}}\boldsymbol{X}^{\downarrow}_{0}, \tag{7.97}$$

which leads to the following expressions for the operators $\boldsymbol{T}^{\downarrow}_{\mathrm{A+B}}$ and $\boldsymbol{R}^{\downarrow}_{\mathrm{A+B}}$

$$\boldsymbol{T}^{\downarrow}_{\mathrm{A+B}} = \boldsymbol{T}^{\downarrow}_{\mathrm{B}}\boldsymbol{T}^{\downarrow}_{\mathrm{A}}, \tag{7.98}$$

$$\boldsymbol{R}^{\downarrow}_{\mathrm{A+B}} = \boldsymbol{R}^{\downarrow}_{\mathrm{A}} + \boldsymbol{T}^{\uparrow}_{\mathrm{A}}\boldsymbol{R}^{\downarrow}_{\mathrm{B}}\boldsymbol{T}^{\downarrow}_{\mathrm{A}}. \tag{7.99}$$

We note that in this case, we ignore only re-reflections, that is, the sequences $\mathbf{R}^{\uparrow}_{\mathrm{A}}\mathbf{R}^{\downarrow}_{\mathrm{B}}$. As seen from (7.99), the contribution of reflection from **B** to the field $X^{\uparrow}_0$ is taken into account. Neglecting this contribution, we have simply

$$R^{\downarrow}_{\mathrm{A+B}} = R^{\downarrow}_{\mathrm{A}}.$$

If we want to take account of the re-reflections up to n-fold ones, we should express $X^{\downarrow}_1$ as follows:

$$X^{\downarrow}_1 = \left[\mathrm{U} + \sum_{k=1}^{n} \left(R^{\uparrow}_{\mathrm{A}} R^{\downarrow}_{\mathrm{B}}\right)^k\right] T^{\downarrow}_{\mathrm{A}} X^{\downarrow}_0. \tag{7.100}$$

This gives the following expressions for $T^{\downarrow}_{\mathrm{A+B}}$ and $R^{\downarrow}_{\mathrm{A+B}}$:

$$T^{\downarrow}_{\mathrm{A+B}} = T^{\downarrow}_{\mathrm{B}} \left[\mathrm{U} + \sum_{k=1}^{n} \left(R^{\uparrow}_{\mathrm{A}} R^{\downarrow}_{\mathrm{B}}\right)^k\right] T^{\downarrow}_{\mathrm{A}}, \tag{7.101}$$

$$R^{\downarrow}_{\mathrm{A+B}} = R^{\downarrow}_{\mathrm{A}} + T^{\uparrow}_{\mathrm{A}} R^{\downarrow}_{\mathrm{B}} \left[\mathrm{U} + \sum_{k=1}^{n} \left(R^{\uparrow}_{\mathrm{A}} R^{\downarrow}_{\mathrm{B}}\right)^k\right] T^{\downarrow}_{\mathrm{A}}. \tag{7.102}$$

Formulas (7.101) and (7.102) are consistent with formulas (7.95) and (7.96), because

$$\lim_{n\to\infty} \left[\mathrm{U} + \sum_{k=1}^{n} \left(R^{\uparrow}_{\mathrm{A}} R^{\downarrow}_{\mathrm{B}}\right)^k\right] = \left(\mathrm{U} - R^{\uparrow}_{\mathrm{A}} R^{\downarrow}_{\mathrm{B}}\right)^{-1} \tag{7.103}$$

[see (5.71)].

In a similar way, considering the situation shown in Figure 7.12b, one may obtain the following expressions for the operators $T^{\uparrow}_{\mathrm{A+B}}$ and $R^{\uparrow}_{\mathrm{A+B}}$:

(i) taking account of all re-reflections between **A** and **B**,

$$T^{\uparrow}_{\mathrm{A+B}} = T^{\uparrow}_{\mathrm{A}}(\mathrm{U} - R^{\downarrow}_{\mathrm{B}} R^{\uparrow}_{\mathrm{A}})^{-1} T^{\uparrow}_{\mathrm{B}}, \tag{7.104}$$

$$R^{\uparrow}_{\mathrm{A+B}} = R^{\uparrow}_{\mathrm{B}} + T^{\downarrow}_{\mathrm{B}} R^{\uparrow}_{\mathrm{A}}(\mathrm{U} - R^{\downarrow}_{\mathrm{B}} R^{\uparrow}_{\mathrm{A}})^{-1} T^{\uparrow}_{\mathrm{B}}; \tag{7.105}$$

(ii) ignoring all re-reflections between **A** and **B**,

$$T^{\uparrow}_{\mathrm{A+B}} = T^{\uparrow}_{\mathrm{A}} T^{\uparrow}_{\mathrm{B}}, \tag{7.106}$$

$$R^{\uparrow}_{\mathrm{A+B}} = R^{\uparrow}_{\mathrm{B}} + T^{\downarrow}_{\mathrm{B}} R^{\uparrow}_{\mathrm{A}} T^{\uparrow}_{\mathrm{B}}; \tag{7.107}$$

(iii) taking account of all the re-reflections up to n-fold ones,

$$T^{\uparrow}_{\mathrm{A+B}} = T^{\uparrow}_{\mathrm{A}} \left[\mathrm{U} + \sum_{k=1}^{n} \left(R^{\downarrow}_{\mathrm{B}} R^{\uparrow}_{\mathrm{A}}\right)^k\right] T^{\uparrow}_{\mathrm{B}}, \tag{7.108}$$

$$R^{\uparrow}_{\mathrm{A+B}} = R^{\uparrow}_{\mathrm{B}} + T^{\downarrow}_{\mathrm{B}} R^{\uparrow}_{\mathrm{A}} \left[\mathrm{U} + \sum_{k=1}^{n} \left(R^{\downarrow}_{\mathrm{B}} R^{\uparrow}_{\mathrm{A}}\right)^k\right] T^{\uparrow}_{\mathrm{B}}. \tag{7.109}$$

Flexible Adding Algorithm

The above consideration suggests the following flexible recursion technique for calculating transmission and reflection operators of multicomponent systems. Let us deal with a system $\mathbf{A}_{(N)}$ consisting of N fragments (TR units) $\mathbf{A}_j$ characterized by their reflection and transmission operators $\boldsymbol{T}_j^{\downarrow}$, $\boldsymbol{R}_j^{\downarrow}$, $\boldsymbol{T}_j^{\uparrow}$, and $\boldsymbol{R}_j^{\uparrow}$ ($j = 1,2,\ldots,N$). Denote the part of the system $\mathbf{A}_{(N)}$ that includes the first j fragments of this system by $\mathbf{A}_{(j)}$ and the reflection and transmission operators of $\mathbf{A}_{(j)}$ by $\boldsymbol{T}_{(j)}^{\downarrow}$, $\boldsymbol{R}_{(j)}^{\downarrow}$, $\boldsymbol{T}_{(j)}^{\uparrow}$, and $\boldsymbol{R}_{(j)}^{\uparrow}$ ($j = 1,2,\ldots,N-1$). We may use the following variants of calculation of the reflection and transmission operators for the system $\mathbf{A}_{(j+1)}$ from those for the system $\mathbf{A}_{(j)}$ and fragment $\mathbf{A}_{j+1}$:

(i) if we want to take all re-reflections between $\mathbf{A}_{(j)}$ and $\mathbf{A}_{j+1}$ into account,

$$\boldsymbol{T}_{(j+1)}^{\downarrow} = \boldsymbol{T}_{j+1}^{\downarrow}\left(\mathbf{U} - \boldsymbol{R}_{(j)}^{\uparrow}\boldsymbol{R}_{j+1}^{\downarrow}\right)^{-1}\boldsymbol{T}_{(j)}^{\downarrow}, \tag{7.110a}$$

$$\boldsymbol{R}_{(j+1)}^{\downarrow} = \boldsymbol{R}_{(j)}^{\downarrow} + \boldsymbol{T}_{(j)}^{\uparrow}\boldsymbol{R}_{j+1}^{\downarrow}\left(\mathbf{U} - \boldsymbol{R}_{(j)}^{\uparrow}\boldsymbol{R}_{j+1}^{\downarrow}\right)^{-1}\boldsymbol{T}_{(j)}^{\downarrow}, \tag{7.110b}$$

$$\boldsymbol{T}_{(j+1)}^{\uparrow} = \boldsymbol{T}_{(j)}^{\uparrow}\left(\mathbf{U} - \boldsymbol{R}_{j+1}^{\downarrow}\boldsymbol{R}_{(j)}^{\uparrow}\right)^{-1}\boldsymbol{T}_{j+1}^{\uparrow}, \tag{7.110c}$$

$$\boldsymbol{R}_{(j+1)}^{\uparrow} = \boldsymbol{R}_{j+1}^{\uparrow} + \boldsymbol{T}_{j+1}^{\downarrow}\boldsymbol{R}_{(j)}^{\uparrow}\left(\mathbf{U} - \boldsymbol{R}_{j+1}^{\downarrow}\boldsymbol{R}_{(j)}^{\uparrow}\right)^{-1}\boldsymbol{T}_{j+1}^{\uparrow}; \tag{7.110d}$$

(ii) if we want to ignore all re-reflections between $\mathbf{A}_{(j)}$ and $\mathbf{A}_{j+1}$,

$$\boldsymbol{T}_{(j+1)}^{\downarrow} = \boldsymbol{T}_{j+1}^{\downarrow}\boldsymbol{T}_{(j)}^{\downarrow}, \tag{7.111a}$$

$$\boldsymbol{R}_{(j+1)}^{\downarrow} = \boldsymbol{R}_{(j)}^{\downarrow} + \boldsymbol{T}_{(j)}^{\uparrow}\boldsymbol{R}_{j+1}^{\downarrow}\boldsymbol{T}_{(j)}^{\downarrow}, \tag{7.111b}$$

$$\boldsymbol{T}_{(j+1)}^{\uparrow} = \boldsymbol{T}_{(j)}^{\uparrow}\boldsymbol{T}_{j+1}^{\uparrow}, \tag{7.111c}$$

$$\boldsymbol{R}_{(j+1)}^{\uparrow} = \boldsymbol{R}_{j+1}^{\uparrow} + \boldsymbol{T}_{j+1}^{\downarrow}\boldsymbol{R}_{(j)}^{\uparrow}\boldsymbol{T}_{j+1}^{\uparrow}; \tag{7.111d}$$

(iii) if we want to ignore all re-reflections between $\mathbf{A}_{(j)}$ and $\mathbf{A}_{j+1}$ and contributions of reflections $\mathbf{R}_{j+1}^{\downarrow}$ and $\mathbf{R}_{(j)}^{\uparrow}$ to the fields emerging from the system $\mathbf{A}_{(N)}$,

$$\boldsymbol{T}_{(j+1)}^{\downarrow} = \boldsymbol{T}_{j+1}^{\downarrow}\boldsymbol{T}_{(j)}^{\downarrow}, \tag{7.112a}$$

$$\boldsymbol{R}_{(j+1)}^{\downarrow} = \boldsymbol{R}_{(j)}^{\downarrow}, \tag{7.112b}$$

$$\boldsymbol{T}_{(j+1)}^{\uparrow} = \boldsymbol{T}_{(j)}^{\uparrow}\boldsymbol{T}_{j+1}^{\uparrow}, \tag{7.112c}$$

$$\boldsymbol{R}_{(j+1)}^{\uparrow} = \boldsymbol{R}_{j+1}^{\uparrow}; \tag{7.112d}$$

(iv) if we want to take account of only one-fold re-reflection between $\mathbf{A}_{(j)}$ and $\mathbf{A}_{j+1}$,

$$\boldsymbol{T}_{(j+1)}^{\downarrow} = \boldsymbol{T}_{j+1}^{\downarrow}\left(\mathbf{U} + \boldsymbol{R}_{(j)}^{\uparrow}\boldsymbol{R}_{j+1}^{\downarrow}\right)\boldsymbol{T}_{(j)}^{\downarrow}, \tag{7.113a}$$

$$\boldsymbol{R}_{(j+1)}^{\downarrow} = \boldsymbol{R}_{(j)}^{\downarrow} + \boldsymbol{T}_{(j)}^{\uparrow}\boldsymbol{R}_{j+1}^{\downarrow}\left(\mathbf{U} + \boldsymbol{R}_{(j)}^{\uparrow}\boldsymbol{R}_{j+1}^{\downarrow}\right)\boldsymbol{T}_{(j)}^{\downarrow}, \tag{7.113b}$$

$$\boldsymbol{T}_{(j+1)}^{\uparrow} = \boldsymbol{T}_{(j)}^{\uparrow}\left(\mathbf{U} + \boldsymbol{R}_{j+1}^{\downarrow}\boldsymbol{R}_{(j)}^{\uparrow}\right)\boldsymbol{T}_{j+1}^{\uparrow}, \tag{7.113c}$$

$$\boldsymbol{R}_{(j+1)}^{\uparrow} = \boldsymbol{R}_{j+1}^{\uparrow} + \boldsymbol{T}_{j+1}^{\downarrow}\boldsymbol{R}_{(j)}^{\uparrow}\left(\mathbf{U} + \boldsymbol{R}_{j+1}^{\downarrow}\boldsymbol{R}_{(j)}^{\uparrow}\right)\boldsymbol{T}_{j+1}^{\uparrow}; \tag{7.113d}$$

and so on. There are many other, rather obvious, variants of involving a fragment.

Starting from

$$T^{\downarrow}_{(1)} = T^{\downarrow}_{1},\ R^{\downarrow}_{(1)} = R^{\downarrow}_{1},\ T^{\uparrow}_{(1)} = T^{\uparrow}_{1},\ R^{\uparrow}_{(1)} = T^{\uparrow}_{1} \tag{7.114}$$

and using the above formulas successively for $j = 2, \ldots, N-1$, we may find the operators $T^{\downarrow}_{(N)}$, $R^{\downarrow}_{(N)}$, $T^{\uparrow}_{(N)}$, and $R^{\uparrow}_{(N)}$, which characterize the whole system $\mathbf{A}_{(N)}$.

We have successfully used this very convenient and reliable algorithm within different methods with various state vectors. Some particular examples of application of this algorithm may be found in the next chapters.

7.3 Optical Models of Some Elements of LCDs

In examples given in this book, we, as a rule, consider standard models of LCD elements. In this section, giving a summary on the standard models, we will mention some more complicated models (for polarizers, glass plates, ITO layers) which are used in practice.

LC Layers

In this book, we deal with three classes of liquid crystals: nematics, cholesterics (chiral nematics), and chiral smectics C^* (FLC). Nematics and cholesterics are considered to be locally uniaxial media with local optic axis **c** directed along the LC director **n**. When considering a nematic or cholesteric LC layer of an LCD, it is usually assumed that throughout the volume occupied by the liquid crystal, the principal refractive indices and, consequently, the principal values of the permittivity tensor of the LC at optical frequencies are spatially invariant. The spatial dependence of the complex permittivity tensor of such a medium at optical frequencies can be represented as follows:

$$\boldsymbol{\varepsilon}(\mathbf{r}) = \varepsilon_{\perp}\mathbf{U} + (\varepsilon_{||} - \varepsilon_{\perp})\mathbf{n}(\mathbf{r}) \otimes \mathbf{n}(\mathbf{r}),$$
$$\varepsilon_{||} = n_{||}^2,\ \varepsilon_{\perp} = n_{\perp}^2,$$

where $n_{||}$ and $n_{\perp}$ are the principal refractive indices of the medium, $\varepsilon_{||}$ and $\varepsilon_{\perp}$ are the principal values of the permittivity tensor, and **U** is the unit matrix [see also (5.66)]. Therefore, specification of a nematic or cholesteric layer in a typical optical problem is a specification of the principal refractive indices of the LC, the layer geometry, and the LC director field **n**(**r**).

Smectics C^* are locally biaxial media, and the real permittivity tensor of a smectic C^* at optical frequencies has in general three different principal values (ε_1, ε_2, ε_3) corresponding to three mutually orthogonal principal axes (see Section 9.4). One of the principal axes is perpendicular to the LC director and parallel to boundaries of smectic monomolecular layers. For most smectics C^*, the principal axis corresponding to the largest principal value of the tensor ε (let it be ε_3) is approximately parallel to the LC director **n**, and ε_2 is very close to ε_1 (the difference of the corresponding principal refractive indices does not exceed 10^{-3}), that is, the optical properties of a typical smectic C^* are very close to those of a locally uniaxial medium with $n_{||} = \varepsilon_3^{1/2}$, $n_{\perp} = \varepsilon_1^{1/2}$, and optic axis directed along **n** [22]. For this reason, in modeling optical characteristics of thin (with a thickness less than 10 μm) layers of smectics C^*, the LC medium is usually considered as optically locally uniaxial. When an inhomogeneous smectic layer is considered as an optically locally biaxial medium, the principal values of its permittivity tensor at optical frequencies are usually assumed to be spatially invariant, just as in the case of nematics and cholesterics.

Film Polarizers

The simplest model of a film polarizer is a uniaxial absorbing layer whose optic axis is parallel to the layer boundary. In calculations, this layer is specified by the azimuth of the optic axis and the complex principal refractive indices, $n_{||}$ and $n_{\perp}$, of the layer. Polarizers of *o*-type are modeled by layers with $|\mathrm{Im}(n_{||})| \gg |\mathrm{Im}(n_{\perp})|$, and polarizers of *e*-type by layers with $|\mathrm{Im}(n_{||})| \ll |\mathrm{Im}(n_{\perp})|$. The real parts of $n_{||}$ and $n_{\perp}$ are usually chosen close to 1.5 and equal or almost equal to each other, for example, $\mathrm{Re}(n_{||}) = 1.50001$ and $\mathrm{Re}(n_{\perp}) = 1.5$. When it is necessary to model a particular polarizer with given spectra of the principal transmittances at normal incidence, the imaginary parts of the principal refractive indices may be calculated from the given principal transmittance spectra by using formulas (7.42) and (7.43); an alternative way is described in Reference 23. The simplest model reflects only the main property of polarizers, namely, diattenuation. Real film polarizers include a few optically anisotropic layers and, in many cases, the phase retardation introduced by the elements of real polarizers must be taken into account. This leads to more realistic layered models of polarizers. For example, the usual polarization film for LCDs is a stretched iodine(or dye)-doped PVA (polyvinyl alcohol; the refractive indices are ~1.48 to 1.52) film—this film plays the role of a polarizing element—sandwiched between two protective TAC (cellulose triacetate; the refractive indices are ~1.48 to 1.51) films. The protective TAC films of usual polarizers behave as uniaxial layers with negative birefringence and optic axis perpendicular to the layer boundaries [24–26]. The optical anisotropy of TAC films is taken into account in modeling and optimization of LCDs with phase compensators (retarders) improving their viewing angle characteristics. As shown in Reference 25, in some applications, TAC films of polarizers are able to act as protective layers and compensators simultaneously. The stretched PVA film may also introduce a nonzero retardation, having significantly different $\mathrm{Re}(n_{||})$ and $\mathrm{Re}(n_{\perp})$. There are some situations where this retardation has to be taken into account as well (mainly, for polarizers with weak absorbance for both principal components in a part of the visible region).

Compensation Films

There are many kinds of homogeneous compensation films: uniaxial films of positive and negative birefringence with various polar orientation of the optic axis, various biaxial films [26–28]. The choice of optical models for such films does not require any comments. Some variants of combined compensators containing anisotropic layers with continuously varying optical parameters along the stratification direction are described in References 29–34. The most-used compensators of this kind are Fuji wide view (WV) films [29, 30]. The Fuji WV film consists of a polymerized discotic layer with a splay-bend hybrid alignment and a TAC substrate. In most cases, smoothly inhomogeneous layers of such combined compensators are modeled in the same way as inhomogeneous liquid crystal layers. Experimental estimates of parameters of such models for Fuji WV films can be found, for example, in works [29, 35, 36]. In References 29 and 35 the discotic layer is regarded as a locally uniaxial medium, and in Reference 36 as a locally biaxial one. As shown in Reference 36, the biaxial model better agrees with the experimental data presented in this work than the uniaxial one.

Glass Substrates

The standard model of a glass plate of an LCD is a homogeneous isotropic nonabsorbing layer. Real glass substrates used in LCDs slightly absorb the light, usually, to a greater extent at the edges of the visible region (~1% to 2% for 1.1-mm thick soda-lime glass plates). Sometimes these absorption losses are required to be taken into account. In such cases, in models, glass plates are specified by the wavelength-dependent complex refractive index. The usual float glass plate has a thin (~6 to 10 μm) surface layer with a significant gradient of the refractive index on the so-called tin side of the plate [37]. If necessary, this surface layer can be modeled by a pile of thin homogeneous sublayers of different refractive indices (see Section 8.3.3).

ITO Layers

The simplest model of an ITO layer is a homogeneous nonabsorbing isotropic layer with a refractive index of the order of 2. This model does not provide an accurate description of the optical effect of ITO layers. Real ITO layers are absorbing and have a complicated (graded) microstructure. The optical constants of ITO layers significantly vary along the axis normal to the boundaries and strongly depend on the wavelength. Description of models taking account of the grading and experimental data on the optical constants of ITO layers can be found in References 38 and 39.

Alignment Layers

Alignment layers are commonly modeled by homogeneous nonabsorbing isotropic layers. Optical manifestations of the inhomogeneity and anisotropy of real alignment layers are, as a rule, very weak. But sometimes, as we saw in an experimental example considered in Section 6.2.1 [the cell with photoalignment (SD-1) films], the optical anisotropy of alignment layers and absorption losses in them must be taken into account.

References

[1] J. W. Goodman, *Introduction to Fourier Optics*, 2nd ed. (McGraw-Hill, New York, 1996).
[2] M. Born and E. Wolf, *Principles of Optics*, 7th ed. (Pergamon Press, New York, 1999).
[3] L. Mandel and E. Wolf, *Optical Coherence and Quantum Optics* (Cambridge University Press, Cambridge, 1995).
[4] J. W. Goodman, *Statistical Optics* (John Wiley & Sons, Inc., New York, 2000).
[5] E. Hecht, *Optics*, 4th ed. (Addison Wesley, San Francisco, 2002).
[6] P. Yeh, "Extended Jones matrix method," *J. Opt. Soc. Am.* **72**, 507 (1982).
[7] Z. Ge, T. X. Wu, X. Zhu, and S.-T. Wu, "Reflective liquid-crystal displays with asymmetric incident and exit angles," *J. Opt. Soc. Am. A* **22**, 966 (2005).
[8] K. E. Torrance and E. M. Sparrow, "Theory for off-specular reflection from roughened surfaces," *J. Opt. Soc. Am.* **57**, 1105 (1967).
[9] B. Walter, S. R. Marschner, H. Li, and K. E. Torrance, "Microfacet models for refraction through rough surfaces," in *Eurographics Symposium on Rendering*, edited by J. Kautz and S. Pattanaik (The Eurographics Association, 2007).
[10] R. G. Priest and S. R. Meier, "Polarimetric microfacet scattering theory with applications to absorptive and reflective surfaces," *Opt. Eng.* **41**, 988 (2002).
[11] W. Liu and J. Kelly, "Multidimensional modeling of liquid crystal optics using a ray-tracing technique," *SID Dig.* **31**, 847 (2000).
[12] C. Desimpel, K. Neyts, D. Olivero, C. Oldano, D. K. G. de Boer, and R. Cortie, "Optical transmission model for thin two-dimensional layers," *Mol. Cryst. Liq. Cryst.* **422**, 185 (2004).
[13] S. Zumer, "Light scattering from nematic droplets: anomalous-diffraction approach," *Phys. Rev. A* **37**, 4006 (1988).
[14] D. A. Yakovlev and O. A. Afonin, "Method for calculating the amplitude scattering matrix for nonuniform anisotropic particles in the approximation of anomalous diffraction," *Opt. Spectr.* **82**, 86 (1997) (in Russian).
[15] F. Abelès, "Sur la propagation des ondes electromagnetiques dans les milieux statifiés," *Ann. Phys. (Paris)* **3**, 504 (1948).
[16] P. C. S. Hayfield and G. W. T. White, "An assessment of the suitability of the Drude-Tronstad polarized light method for the study of film growth on polycrystalline metals," in *Ellipsometry in the Measurement of Surfaces and Thin Films*, edited by E. Passaglia, R. R. Stromberg, and J. Kruger, *Natl. Bur. Stand. Misc. Publ.* 256 (U.S. Government Printing Office, Washington, DC, 1964), p. 157.
[17] R. M. A. Azzam and N. M. Bashara, *Ellipsometry and Polarized Light* (North-Holland, Amsterdam, 1977).
[18] D. W. Berreman, "Optics in stratified and anisotropic media: 4x4 matrix formulation," *J. Opt. Soc. Am.* **62**, 502 (1972).
[19] P. Yeh, "Electromagnetic propagation in birefringent layered media," *J. Opt. Soc. Am.* **69**, 742 (1979).

[20] L. Li, "Formulation and comparison of two recursive matrix algorithms for modeling layered diffraction gratings," *J. Opt. Soc. Am. A* **13**, 1024 (1996).

[21] D. Y. K. Ko and J. R. Sambles, "Scattering matrix method for propagation of radiation in stratified media: attenuated total reflection studies of liquid crystals," *J. Opt. Soc. Am. A* **5**, 1863 (1988).

[22] G. S. Chilaya and V. G. Chigrinov, "Optics and electro-optics of chiral smectic *C* liquid crystals," *Uspechi Fiz. Nauk* **163**, 1 (1993).

[23] G. Haas, H. H. Wöhler, M. Fritsch, and D. A. Mlynski, "Polarizer model for liquid-crystal devices," *J. Opt. Soc. Am. A* **5**, 1571 (1988).

[24] H. Mori and P. J. Bos, "Application of a negative birefringence film to various LCD modes," in Proceedings of the International Display Research Conference, Toronto, ON, Canada, M88 (1997).

[25] H. Mori, "High performance TAC film for LCDs," *Proc. SPIE.* **6135**, 16 (2006).

[26] M. G. Robinson, J. Chen, and G. D. Sharp, *Polarization Engineering for LCD Projection* (John Wiley & Sons, Ltd, Chichester, 2005).

[27] D.-K. Yang and S.-T. Wu, *Fundamentals of Liquid Crystal Devices* (John Wiley & Sons, Ltd, Chichester, 2006).

[28] *Mobile Displays: Technology and Applications*, edited by A. Bhowmik, Z. Li, and P. Bos (John Wiley & Sons, Ltd, Chichester, 2008).

[29] H. Mori, Y. Itoh, Y. Nishiura, T. Nakamura, and Y. Shinagawa, "Performance of a novel optical compensation film based on negative birefringence of discotic compound for wide-viewing-angle twisted-nematic liquid-crystal displays," *Jpn. J. Appl. Phys.* **36**, 143 (1997).

[30] H. Mori, "Novel optical compensators of negative birefringence for wide-viewing-angle twisted-nematic liquid-crystal displays," *Jpn. J. Appl. Phys.* **36**, 1068 (1997).

[31] P. van de Witte, J. van Haaren, J. Tuijtelaars, S. Stallinga, and J. Lub, "Preparation of retarders with a tilted optic axis," *Jpn. J. Appl. Phys.* **38**, 748 (1999).

[32] P. van de Witte, S. Stallinga, and J. A. M. M. van Haaren, "Viewing angle compensators for liquid crystal displays based on layers with a positive birefrigence," *Jpn. J. Appl. Phys.* **39**, 101 (2000).

[33] T. Bachels, J. Fünfschilling, H. Seiberle, and M. Schadt "Novel photo-aligned LC-polymer wide-view film for TN displays," in Proceedings of Eurodisplay'02, p. 183 (2002).

[34] H. Seiberle, C. Benecke, and T. Bachels, "Photo-aligned anisotropic optical thin films," *SID Digest.* 1162 (2003).

[35] T. A. Sergan, S. H. Jamal, and J. R. Kelly, "Polymer negative birefringence films for compensation of twisted nematic devices," *Displays* **20**, 259 (1999).

[36] K. Vermeirsch, K. D'havé, and B. Verweire "Characterization of the biaxiality of the Fuji-Wide View film," *SID Digest.* 677 (1999).

[37] B. Yang, P. D. Townsend, and S. A. Holgate, "Cathodoluminescence and depth profiles of tin in float glass," *J. Phys. D Appl. Phys.* **27**, 1757 (1994).

[38] R. A. Synowicki, "Spectroscopic ellipsometry characterization of indium tin oxide film microstructure and optical constants," *Thin Solid Films* **313–314**, 394 (1998).

[39] Y. S. Jung, "Spectroscopic ellipsometry studies on the optical constants of indium tin oxide films deposited under various sputtering conditions," *Thin Solid Films* **467**, 362 (2004).

8

Modeling Methods Based on the Rigorous Theory of the Interaction of a Plane Monochromatic Wave with an Ideal Stratified Medium. Eigenwave (EW) Methods. EW Jones Matrix Method

A firm foundation for an accurate modeling of the optical performance of LCDs is the classical electromagnetic theory. This chapter is devoted to modeling methods that are rigorously derived from the basic equations of electromagnetic theory, considering the interaction of a plane monochromatic wave and a 1D-inhomogeneous medium. Such methods, in different variants, were used for solving a huge number of problems of LCD optics. It cannot but be mentioned that some of these methods were developed in solving problems of optics of liquid crystals and LCDs.

In this chapter, starting from Maxwell's equations (Section 8.1.1) and basic notions of crystal optics (Section 8.1.2), we arrive at classical transfer matrix methods (Sections 8.1.3, 8.2.4, and 8.3) and a method based on the use of 2×2 transmission and reflection matrices (Sections 8.2.1 and 8.4) for describing the optical effect of constituents of the layered system. The latter method exploits the former ones along with other tools for calculation of the transmission and reflection operators. This 2×2 matrix method, called in this book *the eigenwave (EW) Jones matrix method*, may be considered as an implementation of basic ideas of the classical Jones matrix method within the framework of the rigorous electromagnetic theory. A sufficiently complete description of this method is given for the first time. In Section 8.2.1, we present the basic concepts of the EW Jones matrix method and briefly outline the capabilities of this method in the context of LCD modeling.

In Section 8.3, devoted mainly to Berreman's transfer matrix method, we also discuss a staircase approximation of smoothly inhomogeneous media. An understanding of the nature of this approximation is very important for us, because in most cases, the optical properties of inhomogeneous liquid crystal layers are modeled with the use of this approximation.

Modeling and Optimization of LCD Optical Performance, First Edition.
Dmitry A. Yakovlev, Vladimir G. Chigrinov and Hoi-Sing Kwok.

Website Companion: www.wiley.com/go/yakovlev/modelinglcd

In Section 8.1.3, we introduce the notion of an eigenwave basis as used in the methods described in this and following chapters. The eigenwave representation allows a versatile and flexible use of solutions obtained with the aid of electromagnetic theory in modeling the optical properties of complicated layered systems such as LCDs, in particular enabling use of the transfer channel technique and other general techniques described in Chapter 7. The eigenwave decomposition is exploited in most matrix techniques used for modeling the optical behavior of LCDs. In Section 8.4.1, we consider some specific properties of eigenwave bases, the use of which greatly simplifies calculations and analysis. Sections 8.2.2, 8.4.2, and 8.4.3 are devoted to the methods of calculation of the transmission and reflection operators for different types of elements of layered structures. In Section 8.5, we discuss the calculation of transmissivities and reflectivities in the EW Jones matrix method. In Section 8.6, we consider mathematical properties of the transfer matrices as well as transmission and reflection matrices peculiar to nonabsorbing media and reciprocal (see Section 8.1.1) media. Knowledge of these properties helps in analysis, algorithm optimization, and testing computer programs. In Sections 8.7 and 8.8, we discuss some special topics connected with application of methods presented in this chapter.

Where convenient, we give references to the literature at the end of sections, as "Relevant sources."

8.1 General Properties of the Electromagnetic Field Induced by a Plane Monochromatic Wave in a Linear Stratified Medium

8.1.1 Maxwell's Equations and Constitutive Relations

The starting point of considering the propagation of time-harmonic (monochromatic) electromagnetic fields in stratified media in the framework of the classical electromagnetic theory is the Maxwell equations for curls of electric and magnetic fields with appropriate constitutive relations.

Maxwell's Equations

Using the complex representation of the electromagnetic fields, the Maxwell equations for curls for a source-free region may be written in the following form:

$$\nabla \times \mathbf{H} = -ik_0\mathbf{D}, \tag{8.1}$$

$$\nabla \times \mathbf{E} = ik_0\mathbf{B}, \tag{8.2}$$

where **E** is the electric field strength vector, **H** the magnetic field strength vector, **D** the electric displacement vector, and **B** the magnetic induction vector; $\nabla\times$ is the curl operator (∇ is the nabla operator); $k_0 \equiv \omega/c$ with c being the velocity of light in free space and ω being the angular frequency. The complex electric displacement **D** here is defined so that it involves the effects of the conduction current (if an absorbing medium is considered) [1]. Therefore, the term with the conduction current density is absent in (8.1). The Maxwell equations for the divergences of **D** and **B**,

$$\nabla \cdot \mathbf{D} = 0, \tag{8.3}$$

$$\nabla \cdot \mathbf{B} = 0, \tag{8.4}$$

where $\nabla\cdot$ is the divergence operator, in the case of harmonic fields cannot be treated as independent, because they can be derived from equations (8.1) and (8.2).

Constitutive Relations

The constitutive relations describe the effect of the medium on the electromagnetic field by establishing the relationship between **E**, **H**, **D**, and **B** in terms of material parameters of the medium. One of the most general forms of constitutive relations for linear media is

$$\mathbf{D} = \boldsymbol{\varepsilon}\mathbf{E} + \boldsymbol{\rho}\mathbf{H}, \\ \mathbf{B} = \boldsymbol{\rho}'\mathbf{E} + \boldsymbol{\mu}\mathbf{H}, \tag{8.5}$$

where $\boldsymbol{\varepsilon}$ is the permittivity tensor, $\boldsymbol{\mu}$ is the permeability tensor, $\boldsymbol{\rho}$ and $\boldsymbol{\rho}'$ are tensors characterizing the optical activity of the medium. These tensors are assumed to be independent of the amplitudes and directions of the vectors **E** and **H** but in general dependent on the frequency of the field. The constitutive relations of the form (8.5) were adopted in the classical Berreman's paper on the optics of stratified media [2]. When modeling the LCD optics, we always deal with media that are specified with much simpler constitutive relations, namely

$$\mathbf{D} = \boldsymbol{\varepsilon}\mathbf{E}, \\ \mathbf{B} = \mathbf{H}, \tag{8.6}$$

where $\boldsymbol{\varepsilon}$ is a symmetric tensor, that is,

$$\boldsymbol{\varepsilon} = \boldsymbol{\varepsilon}^{\mathrm{T}}, \tag{8.7}$$

where the symbol ^{T}, as usual, denotes the matrix transpose operation. Nevertheless, before we concentrate our attention on such media, let us mention some fundamental classes of optical media and general restrictions on the form of the material tensors involved in the constitutive relations (8.5) for media of these classes. First of all, we note the classes of *reciprocal* media and *nonreciprocal* media. In contrast to *nonreciprocal* media, the material tensors of reciprocal media satisfy the following relations:

$$\boldsymbol{\varepsilon} = \boldsymbol{\varepsilon}^{\mathrm{T}},\ \boldsymbol{\mu} = \boldsymbol{\mu}^{\mathrm{T}},\ \boldsymbol{\rho}' = -\boldsymbol{\rho}^{\mathrm{T}} \tag{8.8}$$

[3]. Most optical media, including all those that we see in LC displays, are reciprocal. With condition (8.8), the constitutive relations (8.5) allow consideration of natural optical activity (which is a reciprocal phenomenon) but cannot describe nonreciprocal optical phenomena such as the Faraday rotation.

Next we should mention the class of optically *locally centrosymmetric* media, for which at optical frequencies

$$\boldsymbol{\rho}' = \boldsymbol{\rho} = \mathbf{O}, \tag{8.9}$$

where **O** is the zero tensor. It is clear that with this restriction, relations (8.5) cannot describe natural optical activity.

The material tensors of *nonabsorbing* media meet the following conditions (see, e.g., Reference 3):

$$\boldsymbol{\varepsilon} = \boldsymbol{\varepsilon}^{\dagger},\quad \boldsymbol{\mu} = \boldsymbol{\mu}^{\dagger},\quad \boldsymbol{\rho}' = \boldsymbol{\rho}^{\dagger}, \tag{8.10}$$

where, as before, † is the symbol of Hermitian conjugation. For absorbing media, one or more of the conditions (8.10) are violated.

The great majority of optical materials, including all optical materials of LC displays, are nonmagnetic at optical frequencies, that is, for them

$$\boldsymbol{\mu} = \mathbf{U}, \tag{8.11}$$

where **U** is the unit tensor.

Thus, adopting the constitutive relations in the form (8.6)–(8.7), we accept that the medium under consideration, when interacting with optical electromagnetic fields, behaves as nonmagnetic, reciprocal, and locally centrosymmetric. From now on we consider only such media unless otherwise stated. In view of (8.11), we will sometimes omit equations for **B**, assuming that **B** = **H**.

In the case of a *nonabsorbing* medium, the permittivity tensor in (8.6)–(8.7) must be real [according to conditions (8.7) and $\varepsilon = \varepsilon^{\dagger}$ (8.10)], that is,

$$\varepsilon = \mathrm{Re}(\varepsilon) = \varepsilon_{\mathrm{R}}, \tag{8.12}$$

where ε_{R} is the real permittivity tensor which is used in the constitutive relations corresponding to the real representation of electromagnetic fields. For *absorbing* media,

$$\varepsilon = \varepsilon_{\mathrm{R}} + i\varepsilon_{\mathrm{I}} \tag{8.13}$$

with

$$\varepsilon_{\mathrm{I}} = (4\pi/\omega)\sigma, \tag{8.14}$$

where σ is the conductivity tensor. In the case of an isotropic medium,

$$\varepsilon = \varepsilon\,\mathbf{U} = n^2\mathbf{U}, \tag{8.15}$$

where n and $\varepsilon = n^2$ are respectively the complex refractive index and scalar complex permittivity of the medium. For any nonabsorbing anisotropic medium, any uniaxial absorbing medium, and any biaxial absorbing medium of orthorhombic symmetry, the tensor ε can be expressed in terms of the principal refractive indices of the medium (see Sections 9.3 and 9.4 of the next chapter). Such a representation cannot in general be used for biaxial absorbing media of lower symmetry, because for such media the system of principal axes of the tensor σ may not coincide with the system of principal axes of the tensor ε_{R}. The system of principal axes of a tensor includes three mutually orthogonal axes that show the directions of the axes of Cartesian coordinate systems in which the matrix of this tensor is diagonal. If the systems of principal axes of σ and ε_{R} are different, the tensor ε cannot be diagonalized and, consequently, the principal complex refractive indices cannot be defined. Note that the material symmetry of some liquid crystals, for example, smectics C^*, allows such a situation.

Poynting Vector

The quantity

$$\langle \mathbf{S} \rangle = \frac{c}{8\pi}\mathrm{Re}\,(\mathbf{E}\times\mathbf{H}^*) \tag{8.16}$$

represents the time-averaged Poynting vector of a field $\{\mathbf{E}, \mathbf{H}\}$. In optics, time-averaged Poynting vector is interpreted, with some reservations [4], as time-averaged energy flux density vector [1, 4–6] and serves as a connecting link between electromagnetic theory and radiometry, allowing one to express radiometric quantities, such as irradiance, in terms of quantities of electromagnetic theory (see, e.g., Sections 5.4.2 and 8.5).

Energy Flux in Nonabsorbing Media

According to Poynting's theorem [3, 4], the time-averaged Poynting vector of a field in a nonabsorbing medium satisfies the equation

$$\nabla \cdot \langle \mathbf{S} \rangle = 0. \tag{8.17}$$

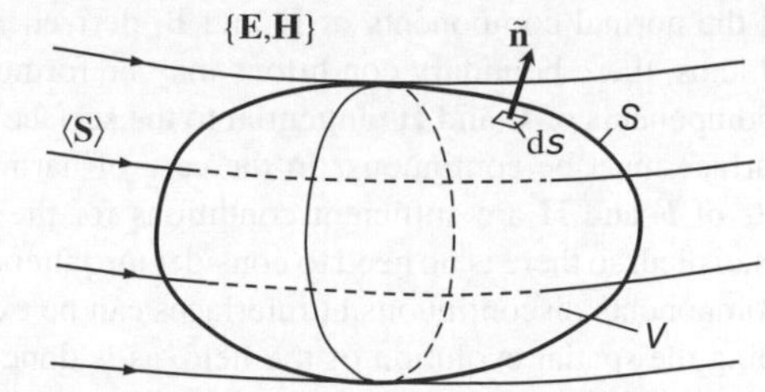

Figure 8.1 To relation (8.18). Flux of the time-averaged Poynting vector ⟨**S**⟩ through a closed surface *S*

Properties (8.10) of material tensors as conditions for this equation to be satisfied are derived in Appendix B.1. On the assumption that (8.17) holds throughout some volume region *V* of the medium (possibly inhomogeneous), integrating $\nabla \cdot \langle \mathbf{S} \rangle$ over *V* and applying the divergence theorem give the following equation:

$$\oint\!\!\!\oint_S \langle \mathbf{S} \rangle \, \tilde{\mathbf{n}} ds = 0, \tag{8.18}$$

where *S* is the surface bounding the region *V*, and $\tilde{\mathbf{n}}$ is the unit outward normal to *S* (see Figure 8.1).

Lorentz's Lemma for Reciprocal Media

This lemma is a version of the Lorentz reciprocity theorem [3, 7] for the case of a source-free region. Let $\{\mathbf{E}_1, \mathbf{H}_1\}$ and $\{\mathbf{E}_2, \mathbf{H}_2\}$ be arbitrary fields each satisfying the Maxwell equations (8.1)–(8.2) in some volume region *V*; the medium in *V* may be inhomogeneous and absorbing. The Lorentz lemma states that

$$\oint\!\!\!\oint_S (\mathbf{E}_1 \times \mathbf{H}_2 - \mathbf{E}_2 \times \mathbf{H}_1) \tilde{\mathbf{n}} ds = 0, \tag{8.19}$$

where, as in (8.18), *S* is the boundary of *V*, and $\tilde{\mathbf{n}}$ is the unit outward normal. The differential form of this lemma is

$$\nabla \cdot (\mathbf{E}_1 \times \mathbf{H}_2 - \mathbf{E}_2 \times \mathbf{H}_1) = 0. \tag{8.20}$$

In this case, too, the integral form (8.19) can be derived from the differential one (8.20) by integrating (8.20) over *V* and applying the divergence theorem. The Lorentz lemma may be violated if the material tensors of the medium do not satisfy (8.8) (see Appendix B.2).

We will see in later sections of this chapter that relations (8.17)–(8.20) are very useful in optics of stratified media.

Boundary Conditions at Interfaces

The differential equations (8.1) and (8.2) hold only for points at which the fields are continuous functions of the spatial coordinates. At surfaces across which the properties of the medium change stepwise, some of the components of the field vectors exhibit discontinuities. Transition of the fields across such surfaces of discontinuity is usually described using the familiar boundary conditions for the tangential components of **E** and **H** and the normal components of **D** and **B**, derived from the integral Maxwell equations [4, 7]. For our problems, these boundary conditions may be formulated as follows: Across a surface of discontinuity, the components of **E** and **H** tangential to the surface as well as the components of **D** and **B** normal to the surface must be continuous. In the case of harmonic fields, the conditions for the tangential components of **E** and **H** are sufficient conditions for the conditions for the normal components of **D** and **B** to be satisfied, so there is no need to consider the latter ones. In the case of an ideal stratified medium, the field components discontinuous at interfaces can be excluded from simultaneous differential equations describing the spatial evolution of the field, as is done in Berreman's formalism, the property of continuity of the tangential components being directly expressed by those equations (see Section 8.1.3). Let us note another, more physical, interpretation of the boundary conditions. A stepwise change in ε is nothing more than a mathematical idealization. A more physically justified model of a region with a sharp change in ε, say, from ε_A to ε_B, is a transition layer in which ε varies rapidly but continuously from ε_A to ε_B. Almost always the transition layer can be treated as infinitesimally thin in the physical (but not mathematical) sense. Considering the interfaces as infinitesimal transition layers, we extend the region of validity of equations (8.1) and (8.2) to the whole space. The boundary conditions are then used as a means for describing the transition of the fields across transition layers.

8.1.2 Plane Waves

A key role in optics of stratified media is played by plane waves. By a common convention, the term "plane wave" can be used for any field of the form

$$\mathbf{E}(\mathbf{r},t) = \underline{\mathbf{E}}e^{i(\mathbf{kr}-\omega t)},\ \mathbf{H}(\mathbf{r},t) = \underline{\mathbf{H}}e^{i(\mathbf{kr}-\omega t)},\ \mathbf{D}(\mathbf{r},t) = \underline{\mathbf{D}}e^{i(\mathbf{kr}-\omega t)} \tag{8.21}$$

(see Section 1.1.1). Substituting (8.21) into the Maxwell equations (8.1)–(8.2) and carrying out the differentiation with respect to the coordinates, we have

$$\mathbf{k}\times\underline{\mathbf{H}} = -k_0\underline{\mathbf{D}}, \tag{8.22}$$

$$\mathbf{k}\times\underline{\mathbf{E}} = k_0\underline{\mathbf{H}}. \tag{8.23}$$

Eliminating from these equations **H** and **D**, the latter by using (8.6), we can obtain the well-known wave equation

$$\mathbf{k}\times\left(\mathbf{k}\times\underline{\mathbf{E}}\right) = -k_0^2\varepsilon\underline{\mathbf{E}} \tag{8.24}$$

[8,9]. The left-hand side of this equation can be rewritten as follows:

$$\mathbf{k}\times\left(\mathbf{k}\times\underline{\mathbf{E}}\right) = \mathbf{k}\left(\mathbf{k}\cdot\underline{\mathbf{E}}\right) - (\mathbf{k}\cdot\mathbf{k})\,\underline{\mathbf{E}} = \mathbf{k}\otimes\mathbf{k}\underline{\mathbf{E}} - \mathbf{k}^2\underline{\mathbf{E}} = \left(\mathbf{k}\otimes\mathbf{k} - \mathbf{k}^2\mathbf{U}\right)\underline{\mathbf{E}}, \tag{8.25}$$

where $\mathbf{k}^2 \equiv \mathbf{k}\cdot\mathbf{k}$. In (8.25), the so-called *dyadic product* is used [see equation (5.66)]. For arbitrary vectors **a** and **b**, the *dyadic product* **a**⊗**b** is the tensor whose components are

$$[\mathbf{a}\otimes\mathbf{b}]_{kj} = [\mathbf{a}]_k\,[\mathbf{b}]_j\,. \tag{8.26}$$

For arbitrary vectors **a**, **b**, and **c**,

$$\mathbf{a}(\mathbf{b}\cdot\mathbf{c}) = (\mathbf{a}\otimes\mathbf{b})\mathbf{c}. \tag{8.27}$$

This property has been employed in (8.25). Now, using (8.25), we can rewrite (8.24) in the form

$$\left(\mathbf{k}\otimes\mathbf{k} - \mathbf{k}^2\mathbf{U} + k_0^2\varepsilon\right)\underline{\mathbf{E}} = \widehat{\mathbf{0}}, \tag{8.28}$$

where $\widehat{\mathbf{0}}$ is the zero vector. Equation (8.28) has nontrivial solutions ($\underline{\mathbf{E}} \neq \widehat{\mathbf{0}}$) only if

$$\det\left(\mathbf{k} \otimes \mathbf{k} - \mathbf{k}^2\mathbf{U} + k_0^2\varepsilon\right) = 0. \tag{8.29}$$

This equation defines the set of all wave vectors allowed at a given ε.

Refraction Vector

For our purposes, it is convenient to represent the wave vector **k** as follows

$$\mathbf{k} = k_0\mathbf{m} \tag{8.30}$$

and consider the vector **m** rather than **k** as a basic parameter describing the spatial dependence of the fields in a wave. This simplifies many important computational formulas, making them free of the factor k_0. For example, equations (8.22) and (8.23) in terms of **m** are simply

$$\mathbf{m} \times \underline{\mathbf{H}} = -\underline{\mathbf{D}}, \tag{8.31}$$

$$\mathbf{m} \times \underline{\mathbf{E}} = \underline{\mathbf{H}}. \tag{8.32}$$

Following Reference 10, we will call **m** the *refraction vector*. The appropriateness of this term will be seen from our discussion of homogeneous plane waves below in this section. Equations (8.24), (8.28), and (8.29) in terms of **m** are respectively

$$\mathbf{m} \times \left(\mathbf{m} \times \underline{\mathbf{E}}\right) = -\varepsilon\underline{\mathbf{E}}, \tag{8.33}$$

$$\left(\mathbf{m} \otimes \mathbf{m} - \mathbf{m}^2\mathbf{U} + \varepsilon\right)\underline{\mathbf{E}} = \widehat{\mathbf{0}}, \tag{8.34}$$

and

$$\det\left(\mathbf{m} \otimes \mathbf{m} - \mathbf{m}^2\mathbf{U} + \varepsilon\right) = 0, \tag{8.35}$$

where $\mathbf{m}^2 \equiv \mathbf{m}\cdot\mathbf{m}$. Equation (8.35) can be cast into a more convenient form [10], namely,

$$\mathbf{m}^2(\mathbf{m}\varepsilon\mathbf{m}) - \mathbf{m}[\mathrm{Tr}(\varepsilon^{\mathrm{adj}})\mathbf{U} - \varepsilon^{\mathrm{adj}}]\mathbf{m} + \det(\varepsilon) = 0 \tag{8.36}$$

or, equivalently,

$$\mathbf{m}^2\mathbf{m}\varepsilon\mathbf{m} - \mathbf{m}[\mathrm{Tr}(\varepsilon)\varepsilon - \varepsilon^2]\mathbf{m} + \det(\varepsilon) = 0, \tag{8.37}$$

where $\mathrm{Tr}(\mathbf{t})$ denotes the trace of a matrix **t**. In (8.36), $\varepsilon^{\mathrm{adj}}$ is the adjugate of ε (see Section 5.1.1),

$$\varepsilon^{\mathrm{adj}} = \begin{pmatrix} \varepsilon_{22}\varepsilon_{33} - \varepsilon_{23}\varepsilon_{32} & \varepsilon_{32}\varepsilon_{13} - \varepsilon_{12}\varepsilon_{33} & \varepsilon_{12}\varepsilon_{23} - \varepsilon_{13}\varepsilon_{22} \\ \varepsilon_{23}\varepsilon_{31} - \varepsilon_{21}\varepsilon_{33} & \varepsilon_{11}\varepsilon_{33} - \varepsilon_{13}\varepsilon_{31} & \varepsilon_{13}\varepsilon_{21} - \varepsilon_{11}\varepsilon_{23} \\ \varepsilon_{21}\varepsilon_{32} - \varepsilon_{31}\varepsilon_{22} & \varepsilon_{12}\varepsilon_{31} - \varepsilon_{11}\varepsilon_{32} & \varepsilon_{11}\varepsilon_{22} - \varepsilon_{21}\varepsilon_{12} \end{pmatrix},$$

where ε_{kj} are components of ε. Equation (8.36) is a fundamental equation in optics of nongyrotropic media [10, 11]; it is sometimes called *the generalized Fresnel equation*. The classical Fresnel equation for transparent crystals [4] can easily be derived from (8.36).

Scalar Amplitude and Vibration Vectors

It is also convenient to introduce a common scalar amplitude characteristic of the wave. Equations (8.21) can be rewritten in the form

$$\mathbf{E}(\mathbf{r},\ t) = A(\mathbf{r},\ t)\mathbf{e},\quad \mathbf{H}(\mathbf{r},\ t) = A(\mathbf{r},\ t)\mathbf{h},\quad \mathbf{D}(\mathbf{r},\ t) = A(\mathbf{r},\ t)\mathbf{d}, \tag{8.38}$$

where $A(\mathbf{r},t)$ is a scalar complex function such that

$$A(\mathbf{r}, t) = A(\mathbf{r}', t')e^{i[k_0\mathbf{m}(\mathbf{r}-\mathbf{r}')-\omega(t-t')]} \tag{8.39}$$

at any $\mathbf{r}$ and $\mathbf{r}'$ within the region where the wave exists and at any t and t', and $\mathbf{e}$, $\mathbf{h}$, and $\mathbf{d}$ are vectors satisfying the equations

$$\mathbf{m} \times \mathbf{h} = -\mathbf{d}, \tag{8.40}$$

$$\mathbf{m} \times \mathbf{e} = \mathbf{h}, \tag{8.41}$$

$$\mathbf{d} = \varepsilon\mathbf{e}, \tag{8.42}$$

and a normalization condition, for example,

$$\mathbf{e}^* \cdot \mathbf{e} = 1. \tag{8.43}$$

The quantity $A(\mathbf{r},\ 0)$ will be referred to as *the scalar amplitude of the wave*. The vectors **e**, **h**, and **d** will be called, respectively, the *electric vibration vector*, *magnetic vibration vector*, and *displacement vibration vector*.

Notation: In this and the following chapters, we consider many quantities which are functions of position and time. To simplify notation, when indicating explicitly arguments of such functions in brackets, we will drop the arguments that are assumed to be zero. For example, for a function $f(\mathbf{r},\ t)$, where $\mathbf{r} = (x,\ y,\ z)$, or, what is the same, $f(x,\ y,\ z,\ t)$,

$$f(\mathbf{r}) \equiv f(\mathbf{r},\ 0), \tag{8.44}$$

$$f(z) \equiv f(0,\ 0,\ z,\ 0). \tag{8.45}$$

Using this convention, we may describe the spatial evolution of the local scalar complex amplitude of a wave in a homogeneous medium by the equation

$$A(\mathbf{r}) = A(\mathbf{r}')e^{ik_0\mathbf{m}(\mathbf{r}-\mathbf{r}')}. \tag{8.46}$$

Homogeneous and Inhomogeneous Plane Waves

Waves of the form (8.21) for which (i) **m** is real or (ii) **m** is complex but Im(**m**) is parallel to Re(**m**) are called *homogeneous plane waves*. All other waves of the form (8.21), that is, waves having a complex **m** with Im(**m**) nonparallel to Re(**m**), are called *inhomogeneous plane waves*. These definitions are illustrated by Figure 8.2. For a homogeneous plane wave, the surfaces on which the field is constant are planes perpendicular to the wave normal, that is, the wave is literally plane. The refraction vector of such a wave can be represented as

$$\mathbf{m} = n_w\mathbf{l}, \tag{8.47}$$

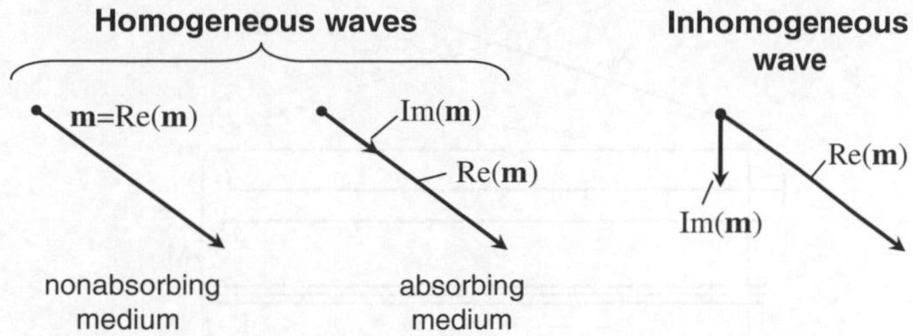

Figure 8.2 Refraction vectors of homogeneous and inhomogeneous plane waves

where **l** is the wave normal, and n_w is the complex refractive index of the medium for this wave. If we define the length of **m** as

$$m = \sqrt{\mathbf{m}^2}, \tag{8.48}$$

we have

$$m = n_w. \tag{8.49}$$

Thus, in the case of a homogeneous plane wave, the refraction vector is a vector directed along the wave normal and having a length m [see (8.48)] equal to the complex refractive index of the medium for this wave. In the case of an inhomogeneous plane wave, it is impossible to define "complex refractive index for the wave"; therefore, this simple interpretation of the refraction vector is inapplicable. It should be noted that inhomogeneous plane waves are called "plane waves" purely conventionally, because the surfaces of constant field in them are not plane.

Mutual Orientation of the Vibration Vectors and Refraction Vector of a Wave

Taking the scalar products of (8.40) and (8.41) with **m**, (8.40) with **h**, and (8.41) with **e**, we find that

$$\mathbf{m} \cdot \mathbf{d} = 0, \tag{8.50}$$

$$\mathbf{m} \cdot \mathbf{h} = 0, \tag{8.51}$$

$$\mathbf{d} \cdot \mathbf{h} = 0, \tag{8.52}$$

$$\mathbf{e} \cdot \mathbf{h} = 0. \tag{8.53}$$

In the case of a homogeneous wave, these relations have a clear geometrical meaning, indicating that **m**, **d**, and **h** are mutually perpendicular and that **e** is perpendicular to **h**. In the case of an inhomogeneous wave, the real and imaginary constituents of any of these vectors may be nonparallel, and relations (8.50)–(8.53) therefore do not have such a simple interpretation.

8.1.3 Field Geometry

In this section, we consider the general properties of the wave fields induced in an arbitrary 1D-inhomogeneous medium by an incident plane monochromatic wave, properties which make the analysis of the spatial evolution of such fields relatively simple and are of fundamental importance in the context of the modeling methods presented in this chapter.

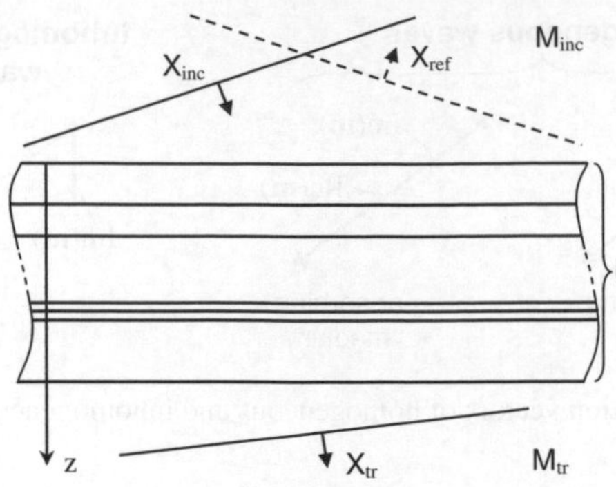

Figure 8.3 Geometry of the problem

Translational Invariance

Let us consider a 1D-inhomogeneous medium M_{SM} sandwiched between two semiinfinite homogeneous media M_{inc} and M_{tr} (Figure 8.3). Let the medium M_{inc} be nonabsorbing and isotropic, and let a homogeneous plane monochromatic wave X_{inc} fall on M_{SM} from M_{inc}. What can we say about the resulting field based on these conditions?

Let us introduce a rectangular Cartesian coordinate system (x, y, z) with the z-axis parallel to the direction of stratification, so that $\varepsilon(\mathbf{r}) = \varepsilon(z)$, and directed from M_{inc} to M_{tr} and represent the refraction vector of the incident wave, $\mathbf{m}_{inc}$, as

$$\mathbf{m}_{inc} = \mathbf{b} + \mathbf{z}\sigma_{inc}, \tag{8.54}$$

where $\mathbf{z}$ is the unit vector along the z-axis and $\mathbf{b}$ is the component of $\mathbf{m}_{inc}$ perpendicular to $\mathbf{z}$ (see Figure 8.4a). Since the medium M_{inc} is assumed to be nonabsorbing and the incident wave to be homogeneous, $\mathbf{m}_{inc}$ and $\mathbf{b}$ are real vectors. Using (8.54), we may write the field of the incident wave in the form

$$\begin{pmatrix} \mathbf{E}_{inc}(\mathbf{r},t) \\ \mathbf{H}_{inc}(\mathbf{r},t) \end{pmatrix} = \begin{pmatrix} \mathbf{E}_{inc}(z) \\ \mathbf{H}_{inc}(z) \end{pmatrix} \exp\left[i\omega\left(\frac{\mathbf{br}}{c} - t\right)\right]. \tag{8.55}$$

Because $\mathbf{b}$ is real, the incident field, as may be seen from (8.55), is invariant with respect to the translations

$$(\mathbf{r},\ t) \rightarrow \left(\mathbf{r} + \Delta\mathbf{r}_{\perp},\ t + \frac{\mathbf{b}\Delta\mathbf{r}_{\perp}}{c}\right), \tag{8.56}$$

where $\Delta\mathbf{r}_{\perp}$ is an arbitrary vector perpendicular to the z-axis. We illustrate this statement by Figure 8.4b where we use, instead of the system (x, y, z) or as a particular choice of this system, a coordinate system (x, y, z) with basis vectors $(\boldsymbol{x}, \boldsymbol{y}, \boldsymbol{z})$, oriented so that $\boldsymbol{x} \| \mathbf{b}$ with $\zeta \equiv \boldsymbol{x}\cdot\mathbf{b}$ being nonnegative. The material properties of the media M_{inc}, M_{SM}, and M_{tr} are independent of t and invariant under any translations in directions perpendicular to the z-axis. Therefore, the resulting field must be invariant under translations (8.56). According to (8.56), we have

$$\begin{pmatrix} \mathbf{E}(\mathbf{r}) \\ \mathbf{H}(\mathbf{r}) \end{pmatrix} \exp(-i\omega t) = \begin{pmatrix} \mathbf{E}(\mathbf{r} + \Delta\mathbf{r}_{\perp}) \\ \mathbf{H}(\mathbf{r} + \Delta\mathbf{r}_{\perp}) \end{pmatrix} \exp\left[-i\omega\left(t + \frac{\mathbf{b}\Delta\mathbf{r}_{\perp}}{c}\right)\right],$$

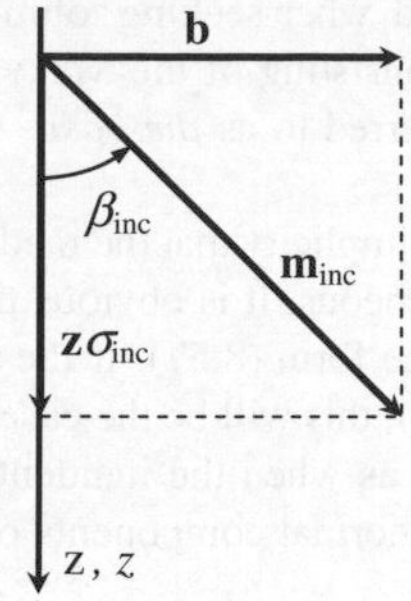

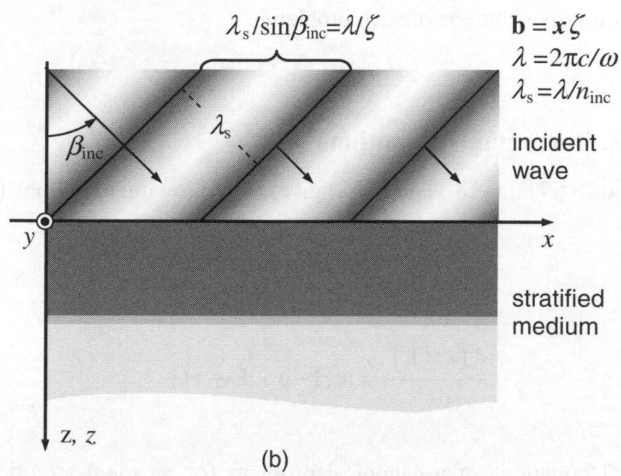

Figure 8.4 (a) Refraction vector of the incident wave. (b) Illustration of the translational invariance of the incident wave–stratified medium system. The field inside the stratified medium is not shown

that is,

$$\begin{pmatrix} \mathbf{E}(\mathbf{r}+\Delta\mathbf{r}_\perp) \\ \mathbf{H}(\mathbf{r}+\Delta\mathbf{r}_\perp) \end{pmatrix} = \begin{pmatrix} \mathbf{E}(\mathbf{r}) \\ \mathbf{H}(\mathbf{r}) \end{pmatrix} \exp\left(ik_0\mathbf{b}\Delta\mathbf{r}_\perp\right).$$

Thus, the translational symmetry of the system determines the following form of the resulting field

$$\mathbf{E}(\mathbf{r},t) = \tilde{\mathbf{E}}(\mathrm{z})\exp\left\{i\left(k_0\mathbf{br}-\omega t\right)\right\},\ \mathbf{H}(\mathbf{r},t) = \tilde{\mathbf{H}}(\mathrm{z})\exp\left\{i\left(k_0\mathbf{br}-\omega t\right)\right\}. \tag{8.57}$$

This means in particular that the refraction vectors of all plane waves induced by the incident wave in any homogeneous layer of the medium M_{SM} and in the media M_{inc} and M_{tr} have the same tangential component, equal to the tangential component **b** of the refraction vector of the incident wave. It is easy to see a connection between this conclusion and Snell's law remembering that, according to (8.54) and (8.47),

$$\zeta \equiv \boldsymbol{x}\cdot\mathbf{b} \equiv |\mathbf{b}| = n_{inc}\sin\beta_{inc},$$

where n_{inc} is the refractive index of the medium M_{SM}, and β_{inc} is the angle of incidence (see Figure 8.4). The special form of the fields (8.57) greatly simplifies the problem, as the dependence of the fields on the transverse coordinates, x and y, is known, and, when seeking solutions, we can deal with functions of only one variable z. The system of equations consisting of the Maxwell equations (8.1)–(8.2), constitutive relations, and restriction (8.57) will be referred to as *the basic equation system* of optics of stratified media.

In the above discussion, we assumed for simplicity that the medium M_{inc} was isotropic and nonabsorbing and that the incident wave was homogeneous. It is obvious that those conditions are not necessary conditions for the resulting field to be of the form (8.57). If the incident field has the form (8.57), the resulting field will also be of the form (8.57); this will be the case, in particular, when the incident wave is inhomogeneous but has a real **b** as well as when the incident field is a superposition of two plane waves having the same real **b** but different normal components of their refraction vectors. The latter is possible when M_{inc} is anisotropic.

The problems of optics of stratified media where all wave fields to be considered have the form (8.57) will be called *canonical problems*. Optical methods presented in this chapter are methods of solving canonical problems or employ solutions of such problems.

Principal Equations. Berreman's Equation

Substitution of (8.6) and (8.57) into (8.1) and (8.2) gives the following equations for $\tilde{\mathbf{E}}(z)$ and $\tilde{\mathbf{H}}(z)$

$$\begin{aligned}\frac{d\left(\mathbf{z}\times\tilde{\mathbf{H}}\right)}{dz} &= -ik_0\left(\boldsymbol{\varepsilon}\tilde{\mathbf{E}}+\mathbf{b}\times\tilde{\mathbf{H}}\right),\\ \frac{d\left(\mathbf{z}\times\tilde{\mathbf{E}}\right)}{dz} &= ik_0\left(-\mathbf{b}\times\tilde{\mathbf{E}}+\tilde{\mathbf{H}}\right).\end{aligned} \tag{8.58}$$

These are the principal equations of a canonical problem for an ideal stratified medium described by constitutive relations (8.6). The complete set of equations describing spatial evolution of the field $\{\mathbf{E}, \mathbf{H}\}$ includes equations (8.58) and (8.57). Equations (8.58) written in the coordinate form are

$$-\frac{d\tilde{H}_y}{dz} = -ik_0\left(\varepsilon_{xx}\tilde{E}_x+\varepsilon_{xy}\tilde{E}_y+\varepsilon_{xz}\tilde{E}_z+b_y\tilde{H}_z\right), \tag{8.59a}$$

$$\frac{d\tilde{H}_x}{dz} = -ik_0\left(\varepsilon_{yx}\tilde{E}_x+\varepsilon_{yy}\tilde{E}_y+\varepsilon_{yz}\tilde{E}_z-b_x\tilde{H}_z\right), \tag{8.59b}$$

$$-ik_0\left(\varepsilon_{zx}\tilde{E}_x+\varepsilon_{zy}\tilde{E}_y+\varepsilon_{zz}\tilde{E}_z+b_x\tilde{H}_y-b_y\tilde{H}_x\right)=0, \tag{8.59c}$$

$$-\frac{d\tilde{E}_y}{dz} = ik_0\left(-b_y\tilde{E}_z+\tilde{H}_x\right),\quad \frac{d\tilde{E}_x}{dz} = ik_0\left(b_x\tilde{E}_z+\tilde{H}_y\right), \tag{8.59d,e}$$

$$ik_0\left(-b_x\tilde{E}_y+b_y\tilde{E}_x+\tilde{H}_z\right)=0, \tag{8.59f}$$

where $\tilde{E}_j$, $\tilde{H}_j$, b_j, and ε_{jk} (j, k = x, y) are components of $\tilde{\mathbf{E}}$, $\tilde{\mathbf{H}}$, **b**, and $\boldsymbol{\varepsilon}$. In (8.58) there are no derivatives of components of the field vectors that are discontinuous at interfaces, so that these equations are valid over the whole space. In the more general case of constitutive relations (8.5), the principal equations have the same structure as (8.58): they consist of two linear homogeneous algebraic equations and four linear homogeneous first-order differential equations in the Cartesian components of $\tilde{\mathbf{E}}(z)$ and $\tilde{\mathbf{H}}(z)$ (see, e.g., Reference 2). Using the algebraic equations, one can express two of the six field variables in terms of the other four. Thus, from (8.59c) and (8.59f) we have

$$\tilde{E}_z = -\frac{1}{\varepsilon_{zz}}\left(\varepsilon_{zx}\tilde{E}_x + \varepsilon_{zy}\tilde{E}_y + b_x\tilde{H}_y - b_y\tilde{H}_x\right), \tag{8.60a}$$

$$\tilde{H}_z = b_x\tilde{E}_y - b_y\tilde{E}_x. \tag{8.60b}$$

Substituting (8.60) into (8.59a,b,d,e) gives four linear homogeneous first-order differential equations in four field variables:

$$\begin{aligned}
\frac{d\tilde{E}_x}{dz} &= ik_0\left[\left(-\frac{b_x\varepsilon_{zx}}{\varepsilon_{zz}}\right)\tilde{E}_x + \left(-\frac{b_x\varepsilon_{zy}}{\varepsilon_{zz}}\right)\tilde{E}_y + \left(\frac{b_xb_y}{\varepsilon_{zz}}\right)\tilde{H}_x + \left(1-\frac{b_x^2}{\varepsilon_{zz}}\right)\tilde{H}_y\right],\\
\frac{d\tilde{E}_y}{dz} &= ik_0\left[\left(-\frac{b_y\varepsilon_{zx}}{\varepsilon_{zz}}\right)\tilde{E}_x + \left(-\frac{b_y\varepsilon_{zy}}{\varepsilon_{zz}}\right)\tilde{E}_y + \left(\frac{b_y^2}{\varepsilon_{zz}}-1\right)\tilde{H}_x + \left(-\frac{b_xb_y}{\varepsilon_{zz}}\right)\tilde{H}_y\right],\\
\frac{d\tilde{H}_x}{dz} &= ik_0\left[\left(\frac{\varepsilon_{yz}\varepsilon_{zx}}{\varepsilon_{zz}}-\varepsilon_{yx}-b_xb_y\right)\tilde{E}_x + \left(\frac{\varepsilon_{yz}\varepsilon_{zy}}{\varepsilon_{zz}}-\varepsilon_{yy}+b_x^2\right)\tilde{E}_y + \left(-\frac{b_y\varepsilon_{yz}}{\varepsilon_{zz}}\right)\tilde{H}_x + \left(\frac{b_x\varepsilon_{yz}}{\varepsilon_{zz}}\right)\tilde{H}_y\right],\\
\frac{d\tilde{H}_y}{dz} &= ik_0\left[\left(\varepsilon_{xx}-\frac{\varepsilon_{zx}\varepsilon_{xz}}{\varepsilon_{zz}}-b_y^2\right)\tilde{E}_x + \left(\varepsilon_{xy}-\frac{\varepsilon_{zy}\varepsilon_{xz}}{\varepsilon_{zz}}+b_yb_x\right)\tilde{E}_y + \left(\frac{b_y\varepsilon_{xz}}{\varepsilon_{zz}}\right)\tilde{H}_x + \left(-\frac{b_x\varepsilon_{xz}}{\varepsilon_{zz}}\right)\tilde{H}_y\right].
\end{aligned} \tag{8.61}$$

This system of equations can be cast into the following form:

$$\frac{d\Psi}{dz} = ik_0\mathbf{\Delta}(z)\Psi, \tag{8.62}$$

where

$$\Psi(z) \equiv \begin{pmatrix} \tilde{E}_x(z) \\ \tilde{H}_y(z) \\ \tilde{E}_y(z) \\ -\tilde{H}_x(z) \end{pmatrix}_{xyz} \tag{8.63}$$

is the so-called Berreman vector and

$$\Delta = \begin{pmatrix}
-\frac{b_x\varepsilon_{zx}}{\varepsilon_{zz}} & 1-\frac{b_x^2}{\varepsilon_{zz}} & -\frac{b_x\varepsilon_{zy}}{\varepsilon_{zz}} & -\frac{b_xb_y}{\varepsilon_{zz}} \\
\varepsilon_{xx}-\frac{\varepsilon_{zx}\varepsilon_{xz}}{\varepsilon_{zz}}-b_y^2 & -\frac{b_x\varepsilon_{xz}}{\varepsilon_{zz}} & \varepsilon_{xy}-\frac{\varepsilon_{zy}\varepsilon_{xz}}{\varepsilon_{zz}}+b_yb_x & -\frac{b_y\varepsilon_{xz}}{\varepsilon_{zz}} \\
-\frac{b_y\varepsilon_{zx}}{\varepsilon_{zz}} & -\frac{b_xb_y}{\varepsilon_{zz}} & -\frac{b_y\varepsilon_{zy}}{\varepsilon_{zz}} & 1-\frac{b_y^2}{\varepsilon_{zz}} \\
\varepsilon_{yx}-\frac{\varepsilon_{yz}\varepsilon_{zx}}{\varepsilon_{zz}}+b_xb_y & -\frac{b_x\varepsilon_{yz}}{\varepsilon_{zz}} & \varepsilon_{yy}-\frac{\varepsilon_{yz}\varepsilon_{zy}}{\varepsilon_{zz}}-b_x^2 & -\frac{b_y\varepsilon_{yz}}{\varepsilon_{zz}}
\end{pmatrix}_{xyz} \tag{8.64}$$

is the so-called differential propagation matrix [the notation $(\cdot)_{xyz}$ is explained in Section 9.1]. Expression (8.64) for the matrix $\mathbf{\Delta}$ corresponds to constitutive relations (8.6). The most general expressions for the components of **Δ**, corresponding to constitutive relations (8.5), were given by Berreman in [2]. It should be noted that in [2], Berreman used the coordinate system with the x-axis parallel to **b**. This choice simplifies computational formulas but, as we saw above, is not necessary for obtaining an equation of the form (8.62) for the Berreman vector (8.63) (see also Section 8.3.4).

The Berreman equation is a shortened form of the principal equations. In practical calculations, the algebraic equations of the principal equations are invoked only in rare cases when there is a need to calculate $\tilde{E}_z$ and $\tilde{H}_z$ from the Berreman vector.

Modal Representation

It follows from the theory of linear ordinary differential equations that a fundamental system of solutions for (8.62) can be composed of four linearly independent particular solutions of this equation and that such a set of four linearly independent solutions can always be found. By definition, given a fundamental system ($\Psi_1(z)$, $\Psi_2(z)$, $\Psi_3(z)$, $\Psi_4(z)$), any particular solution of (8.62) can be represented as a linear combination of the fundamental solutions

$$\Psi(z) = c_1\Psi_1(z) + c_2\Psi_2(z) + c_3\Psi_3(z) + c_4\Psi_4(z). \tag{8.65}$$

According to (8.65), (8.59), and (8.57), the field described by the function $\Psi(z)$ can be expressed as

$$\begin{pmatrix} \mathbf{E}(\mathbf{r},t) \\ \mathbf{H}(\mathbf{r},t) \end{pmatrix} = \sum_{l=1}^{4} c_l \begin{pmatrix} \mathbf{E}_l(\mathbf{r},t) \\ \mathbf{H}_l(\mathbf{r},t) \end{pmatrix}, \tag{8.66}$$

where $\{\mathbf{E}_l(\mathbf{r}, t), \mathbf{H}_l(\mathbf{r}, t)\}$ ($l = 1, 2, 3, 4$) are fields characterized by functions $\Psi_l(z)$. Thus, the set of the four fields $\{\mathbf{E}_l(\mathbf{r}, t), \mathbf{H}_l(\mathbf{r}, t)\}$ ($l = 1, 2, 3, 4$) is a fundamental system of solutions of the basic equation system [Maxwell's equations + constitutive relations + equations (8.57)].

Eigenwaves of a Homogeneous Layer. Eigenwave Representation

The fundamental system of solutions of the basic equation system for a homogeneous layer can almost always be composed of four *eigenwaves* of the layer, four waves of the form (8.21). The representation of a field in a homogeneous layer, or in an infinitesimally thin layer of an inhomogeneous medium with continuously varying parameters, as a superposition of eigenwaves of this layer is referred to in this book as *the eigenwave representation*. The set of basis functions used for eigenwave decomposition will be referred to as *the eigenwave basis*. Eigenwave representation is used in most optical methods considered in this book, including all those presented in the program library LMOPTICS (see Chapters 9 and 10 and the companion website). In this section we touch upon only key aspects of this representation. Much useful information about eigenwave bases may be found in Section 8.4.1 and Chapter 9.

Using quantities defined in the previous section, we may write the basic formula of eigenwave representation as follows:

$$\begin{pmatrix} \mathbf{E}(\mathbf{r},t) \\ \mathbf{H}(\mathbf{r},t) \end{pmatrix} = \sum_{l=1}^{4} A_l(z)\mathbf{Q}_l(z)\exp\left\{i\left(k_0\mathbf{br} - \omega t\right)\right\}, \quad \mathbf{Q}_l(z) \equiv \begin{pmatrix} \mathbf{e}_l(z) \\ \mathbf{h}_l(z) \end{pmatrix}, \tag{8.67}$$

where $\mathbf{e}_l$ and $\mathbf{h}_l$ are respectively the electric and magnetic vibration vectors of the lth basis eigenwave. The functions $\mathbf{e}_l(z)$ and $\mathbf{h}_l(z)$ are assumed to have constant values within homogeneous layers.

From (8.67), (8.57), and (8.63), one may see that the four-component column-vector composed of the scalar amplitudes of the eigenwave components,

$$\mathbf{A}(z) = \begin{pmatrix} A_1(z) \\ A_2(z) \\ A_3(z) \\ A_4(z) \end{pmatrix}, \tag{8.68}$$

is related to the vector $\Psi(z)$ characterizing the same field by the following equation

$$\Psi(z) = \mathbf{\Psi}(z)\mathbf{A}(z), \tag{8.69}$$

where

$$\mathbf{\Psi}(z) = \begin{pmatrix} \psi_1(z) & \psi_2(z) & \psi_3(z) & \psi_4(z) \end{pmatrix} \tag{8.70}$$

is a 4×4 matrix with columns

$$\psi_l(z) = \begin{pmatrix} \mathbf{e}_l(z) \cdot \mathbf{x} \\ \mathbf{h}_l(z) \cdot \mathbf{y} \\ \mathbf{e}_l(z) \cdot \mathbf{y} \\ -\mathbf{h}_l(z) \cdot \mathbf{x} \end{pmatrix}, \quad l = 1,\ 2,\ 3, 4. \tag{8.71}$$

According to (8.57) and (8.67), refraction vectors of all eigenwaves used for eigenwave decomposition (*basis waves*) must have the same tangential component, equal to **b**. The z-components of the refraction vectors of the basis waves can be found from a quartic equation that is obtained by substitution of

$$\mathbf{m} = \mathbf{b} + \mathbf{z}\sigma \tag{8.72}$$

into (8.35), (8.36), or (8.37). Thus, substituting (8.72) into (8.37) gives the following equation:

$$d_4\sigma^4 + d_3\sigma^3 + d_2\sigma^2 + d_1\sigma + d_0 = 0, \tag{8.73}$$

where

$$\begin{gathered} d_0 = \left[\mathbf{b}^2 - \mathrm{Tr}(\varepsilon)\right](\mathbf{b}\varepsilon\mathbf{b}) + (\varepsilon\mathbf{b})^2 + \det(\varepsilon), \\ d_1 = 2\left\{\left[\mathbf{b}^2 - \mathrm{Tr}(\varepsilon)\right](\mathbf{b}\varepsilon\mathbf{z}) + (\varepsilon\mathbf{b})(\varepsilon\mathbf{z})\right\}, \\ d_2 = \left[\mathbf{b}^2 - \mathrm{Tr}(\varepsilon)\right](\mathbf{z}\varepsilon\mathbf{z}) + \mathbf{b}\varepsilon\mathbf{b} + (\varepsilon\mathbf{z})^2, \\ d_3 = 2\mathbf{b}\varepsilon\mathbf{z},\ d_4 = \mathbf{z}\varepsilon\mathbf{z}. \end{gathered}$$

This equation has four roots, including possible multiple roots; we denote them by σ_1, σ_2, σ_3, and σ_4. Except some singular situations (see below), it is possible to find four eigenwaves with refraction vectors

$$\mathbf{m}_l = \mathbf{b} + \mathbf{z}\sigma_l \quad (l = 1, 2, 3, 4) \tag{8.74}$$

(see Figure 8.5) that together constitute a fundamental system of solutions of the basic equation system. The vibration vectors $\mathbf{e}_l$ and $\mathbf{h}_l$ of the basis waves can be calculated by using (8.40)–(8.42) and an appropriate normalization condition (see Sections 8.4, 8.5, and Chapter 9).

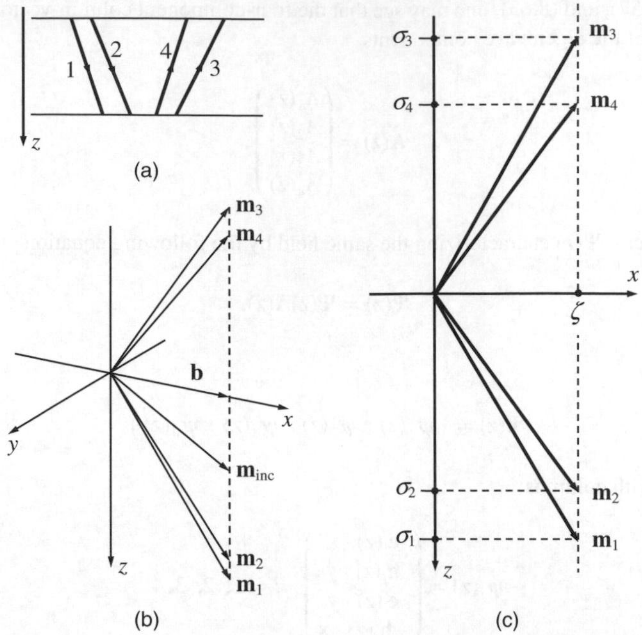

Figure 8.5 (a) Numeration of basis eigenwaves adopted in this book. (b, c) Refraction vectors of the basis eigenwaves for a nonabsorbing anisotropic layer out of TIR mode

The Berreman formalism offers an alternative way of calculating the parameters of the basis eigenwaves. The function Ψ(z) characterizing a plane wave of the form (8.21) may be written as follows:

$$\Psi(\mathrm{z}) = A(\mathrm{z}')\exp\left[ik_0\sigma(\mathrm{z}-\mathrm{z}')\right]\boldsymbol{\psi}, \tag{8.75}$$

where

$$\boldsymbol{\psi} = \begin{pmatrix} \mathbf{e}\cdot\mathbf{x} \\ \mathbf{h}\cdot\mathbf{y} \\ \mathbf{e}\cdot\mathbf{y} \\ -\mathbf{h}\cdot\mathbf{x} \end{pmatrix}. \tag{8.76}$$

Substitution of (8.75) into (8.62) leads to the following equation:

$$(\boldsymbol{\Delta} - \sigma\mathbf{U})\,\boldsymbol{\psi} = \widehat{\mathbf{0}}, \tag{8.77}$$

where $\widehat{\mathbf{0}}$ is the zero column. This equation has nontrivial solutions only if

$$\det(\boldsymbol{\Delta} - \sigma\mathbf{U}) = 0. \tag{8.78}$$

This equation is a quartic in σ and has the same roots as (8.73). With known σ_l (l = 1, 2, 3, 4), vectors $\boldsymbol{\psi}_l$ [see (8.71)] corresponding to the basis eigenwaves can be found from the equation

$$\left(\boldsymbol{\Delta} - \sigma_l\mathbf{U}\right)\boldsymbol{\psi}_l = \widehat{\mathbf{0}}. \tag{8.79}$$

Equation (8.77) indicates that the z-components of refraction vectors of the basis eigenwaves, σ_l, and corresponding vectors $\boldsymbol{\psi}_l$ are respectively eigenvalues and right eigenvectors of the matrix $\boldsymbol{\Delta}$.

Spatial Evolution of the State Vector A in a Stratified Medium

Let $\{\mathbf{E}(\mathbf{r}, t), \mathbf{H}(\mathbf{r}, t)\}$ be a field satisfying the basic equation system in the region $z_a < z < z_b$ of a stratified medium (see Figure 8.6), and let the functions $\mathbf{A}(z)$ and $\Psi(z)$ characterize this field. Let us consider transformation of $\mathbf{A}$ within this region.

A. *Interface*. Let the plane $z = z_j$ $(z_a < z_j < z_b)$ be the plane of the boundary between layers (a surface of discontinuity of ε). In terms of Ψ, the continuity of the tangential components of $\mathbf{E}$ and $\mathbf{H}$ across the surface $z = z_j$ may be expressed by the relation

$$\Psi(z_j - 0) = \Psi(z_j + 0), \tag{8.80}$$

where $z = z_j - 0$ and $z = z_j + 0$, as usual, denote the sides of the plane $z = z_j$ facing the regions $z < z_j$ and $z > z_j$, respectively (see Figure 8.6). By making use of (8.69), we may rewrite (8.80) in terms of $\mathbf{A}$:

$$\mathbf{\Psi}(z_j - 0)\,\mathbf{A}(z_j - 0) = \mathbf{\Psi}(z_j + 0)\,\mathbf{A}(z_j + 0). \tag{8.81}$$

From (8.81), we obtain the relation

$$\mathbf{A}(z_j + 0) = \mathbf{\Psi}(z_j + 0)^{-1}\,\mathbf{\Psi}(z_j - 0)\,\mathbf{A}(z_j - 0), \tag{8.82}$$

which describes the transformation of $\mathbf{A}$ at the interface.

B. *Bulk of a homogeneous layer*. Let $z = z_{n-1}$ and $z = z_n$ $(z_a < z_{n-1} < z_n < z_b)$ be the planes of the boundaries of a homogeneous layer (Figure 8.6). The bulk of this layer occupies the region $(z_{n-1} + 0, z_n - 0)$. Using (8.46), we may write the following relations for the components of $\mathbf{A}$:

$$A_l(z_n - 0) = A_l(z_{n-1} + 0)\exp(ik_0\sigma_l^{(n)}\,h_n) \quad l = 1, 2, 3, 4, \tag{8.83}$$

where $\sigma_l^{(n)}$ is the value of $\sigma_l(z)$ within the layer, and $h_n = z_n - z_{n-1}$ is the thickness of the layer. In matrix form, relations (8.83) can be written as follows:

$$\mathbf{A}(z_n - 0) = \tilde{\mathbf{T}}(\sigma_l^{(n)}, h_n)\mathbf{A}(z_{n-1} + 0), \tag{8.84}$$

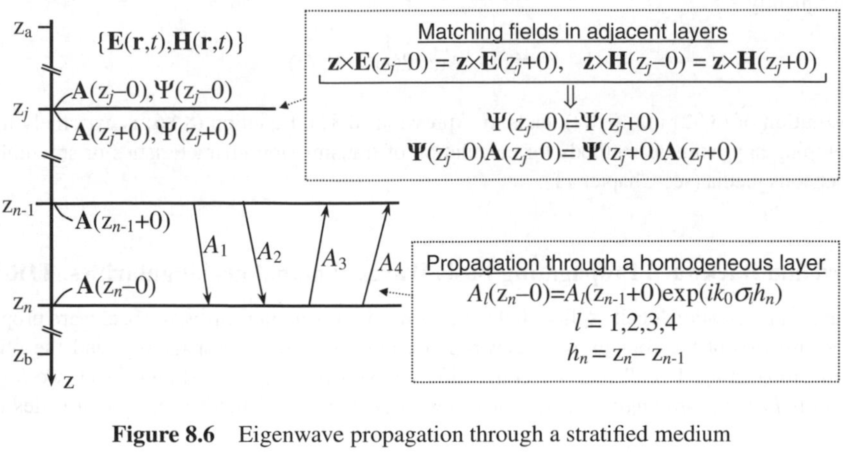

Figure 8.6 Eigenwave propagation through a stratified medium

where

$$\tilde{\mathbf{T}}(\sigma_l, h) \equiv \begin{pmatrix} \exp(ik_0\sigma_1 h) & 0 & 0 & 0 \\ 0 & \exp(ik_0\sigma_2 h) & 0 & 0 \\ 0 & 0 & \exp(ik_0\sigma_3 h) & 0 \\ 0 & 0 & 0 & \exp(ik_0\sigma_4 h) \end{pmatrix}. \tag{8.85}$$

C. *Bulk of a smoothly inhomogeneous layer*. Within a region with a continuously varying tensor ε, the function $\mathbf{A}(z)$ obeys the following equation:

$$\frac{d\mathbf{A}}{dz} = \widehat{\boldsymbol{\Delta}}_{\mathbf{A}}(z)\mathbf{A}(z), \tag{8.86}$$

where

$$\widehat{\boldsymbol{\Delta}}_{\mathbf{A}}(z) = ik_0\boldsymbol{\Delta}_{\mathbf{A}}(z) - \boldsymbol{\Psi}(z)^{-1}\frac{d\boldsymbol{\Psi}}{dz}, \tag{8.87}$$

$$\boldsymbol{\Delta}_{\mathbf{A}}(z) = \begin{pmatrix} \sigma_1(z) & 0 & 0 & 0 \\ 0 & \sigma_2(z) & 0 & 0 \\ 0 & 0 & \sigma_3(z) & 0 \\ 0 & 0 & 0 & \sigma_4(z) \end{pmatrix}. \tag{8.88}$$

Here it is assumed that the functions $\sigma_l(z)$ and $\boldsymbol{\psi}_l(z)$ ($l = 1, 2, 3, 4$) are chosen continuous. Equation (8.86) can easily be derived from (8.62). Substitution of (8.69) into (8.62) gives (8.86) with the following expression for $\widehat{\boldsymbol{\Delta}}_{\mathbf{A}}(z)$:

$$\widehat{\boldsymbol{\Delta}}_{\mathbf{A}}(z) = ik_0\boldsymbol{\Psi}(z)^{-1}\boldsymbol{\Delta}(z)\boldsymbol{\Psi}(z) - \boldsymbol{\Psi}(z)^{-1}\frac{d\boldsymbol{\Psi}}{dz}. \tag{8.89}$$

According to (8.79), we have

$$\boldsymbol{\Delta}(z)\boldsymbol{\psi}_l(z) = \sigma_l(z)\boldsymbol{\psi}_l(z) \quad l = 1, 2, 3, 4. \tag{8.90}$$

These four equations can be written in matrix form as follows:

$$\boldsymbol{\Delta}(z)\boldsymbol{\Psi}(z) = \boldsymbol{\Psi}(z)\boldsymbol{\Delta}_{\mathbf{A}}(z). \tag{8.91}$$

Consequently,

$$\boldsymbol{\Psi}(z)^{-1}\boldsymbol{\Delta}(z)\boldsymbol{\Psi}(z) = \boldsymbol{\Delta}_{\mathbf{A}}(z). \tag{8.92}$$

Substitution of (8.92) into (8.89) leads to expression (8.87). Equation (8.86) is extremely useful in developing approximate methods for calculation of transmission characteristics of smoothly inhomogeneous media (see Chapter 11).

Forward and Backward Propagating Basis Waves. Anomalous Singularities. TIR Mode

Eigenwave representation locally splits a field into forward propagating waves and backward propagating ones, because two of the four basis eigenwaves are always forward propagating, and the other two backward propagating. Recall that, by convention adopted in Chapter 7 (Figure 7.7), we say that a wave field is *forward propagating* if it propagates to greater z, and fields propagating to lesser z are called *backward propagating.* In the context of the general approaches described in Chapter 7, it is important that eigenwave representation allows one to characterize forward propagating and backward propagating fields separately, using convenient state vectors with clear physical meaning, and to introduce corresponding transmission and reflection operators in a natural way (see Section 8.2 and Chapter 10).

In what follows, we will always assign numbers 1 and 2 to forward propagating basis waves, and numbers 3 and 4 to backward propagating ones (see Figure 8.5). With this numeration, the column-vector **A** (8.68) can be written as

$$\mathbf{A} = \begin{pmatrix} \mathbf{a}^{\downarrow} \\ \mathbf{a}^{\uparrow} \end{pmatrix}, \tag{8.93}$$

where two-component column-vectors

$$\mathbf{a}^{\downarrow} = \begin{pmatrix} A_1 \\ A_2 \end{pmatrix} \text{ and } \mathbf{a}^{\uparrow} = \begin{pmatrix} A_3 \\ A_4 \end{pmatrix} \tag{8.94}$$

characterize the forward propagating and backward propagating wave fields, respectively.

There are two main criteria that allow one to determine whether a wave is forward propagating or backward propagating from its σ, **e** and **h** (or $\boldsymbol{\psi}$). The first criterion: when σ has a nonzero imaginary part,

$$\text{the wave is } \begin{cases} \text{forward propagating if } \mathrm{Im}(\sigma) > 0 \\ \text{backward propagating if } \mathrm{Im}(\sigma) < 0. \end{cases} \tag{8.95}$$

To formulate the second criterion, we introduce the quantity

$$s_z \equiv 2\mathrm{Re}[\mathbf{z}(\mathbf{e}\times\mathbf{h}^*)] = \mathbf{z}(\mathbf{e}^*\times\mathbf{h}) + \mathbf{z}(\mathbf{e}\times\mathbf{h}^*) = \psi_1^*\psi_2 + \psi_2^*\psi_1 + \psi_3^*\psi_4 + \psi_4^*\psi_3 = \boldsymbol{\psi}^\dagger \mathbf{I}_0 \boldsymbol{\psi}, \tag{8.96}$$

where ψ_l ($l = 1, 2, 3, 4$) are components of the vector $\boldsymbol{\psi}$ [see (8.76)] and

$$\mathbf{I}_0 = \begin{pmatrix} 0 & 1 & 0 & 0 \\ 1 & 0 & 0 & 0 \\ 0 & 0 & 0 & 1 \\ 0 & 0 & 1 & 0 \end{pmatrix}. \tag{8.97}$$

The second criterion: when σ is real,

$$\text{the wave is } \begin{cases} \text{forward propagating if its } s_z > 0 \\ \text{backward propagating if its } s_z < 0. \end{cases} \tag{8.98}$$

The sign of s_z always coincides with the sign of the z-component of the time-averaged Poynting vector of the wave; to see this, it suffices to substitute (8.38) into (8.16). From this the physical meaning of this criterion is clear.

Figures 8.7–8.11 illustrate different situations encountered in specifying eigenwave bases, showing values of the z-components of refraction vectors of the basis eigenwaves for some media under different illumination conditions. In these examples, illumination conditions are specified by the propagation parameter ζ. Recall that $\mathbf{b} = \zeta \boldsymbol{x}$ and, if the medium M_{inc} (see Figure 8.3) is isotropic, $\zeta = n_{\text{inc}} \sin\beta_{\text{inc}}$, where n_{inc} is the refractive index of M_{inc} and β_{inc} is the angle of incidence (Figure 8.4). Figures 8.7–8.9 show the results for isotropic media: a nonabsorbing medium with refractive index $n = 1$ (air) (Figure 8.7), a nonabsorbing medium with $n = 1.52$ (soda-lime glass) (Figure 8.8), and a strongly

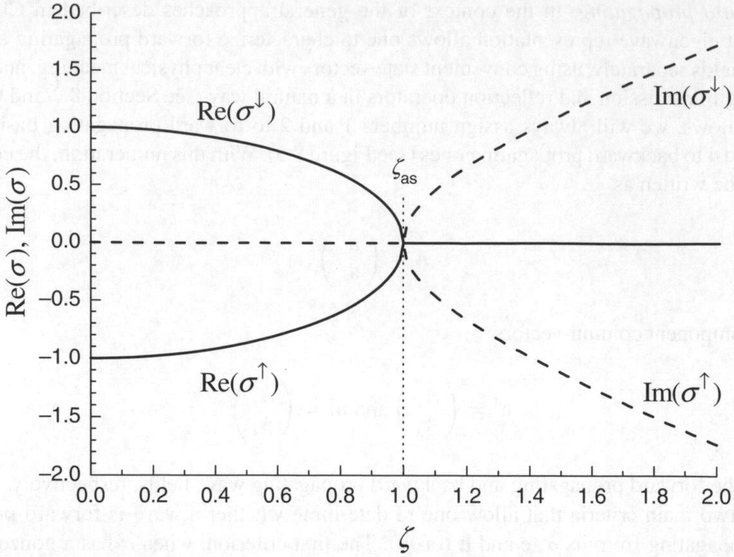

Figure 8.7 The z-components of refraction vectors of the basis eigenwaves for a nonabsorbing medium with refractive index $n = 1$ (air)

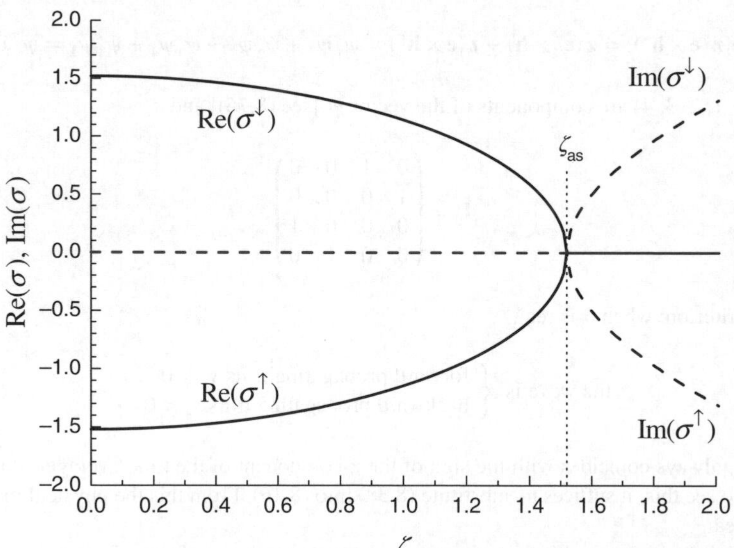

Figure 8.8 The z-components of refraction vectors of the basis eigenwaves for a nonabsorbing medium with refractive index $n = 1.52$ (soda-lime glass)

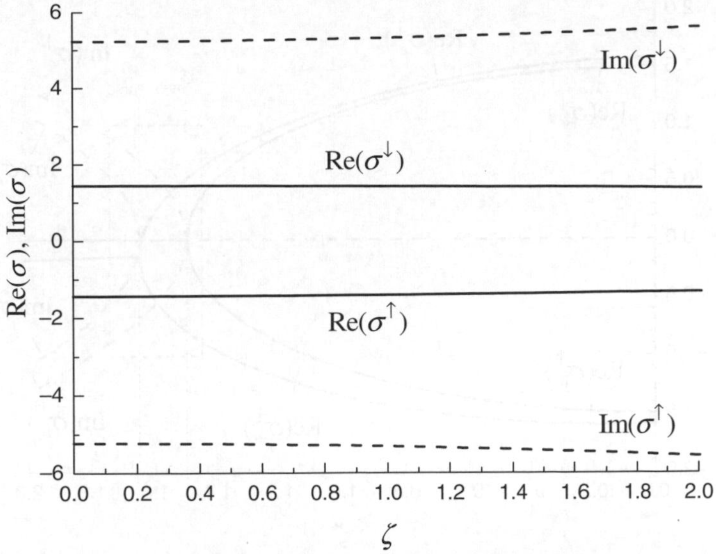

Figure 8.9 The z-components of refraction vectors of the basis eigenwaves for a strongly absorbing isotropic medium with complex refractive index $n = 1.44 + i5.23$ (aluminum)

absorbing medium with complex refractive index $n = 1.44 + i5.23$ (aluminum) (Figure 8.9). In the case of an isotropic medium, equation (8.73) has the following roots:

$$\sigma_1 = \sigma_2 = -\sigma_3 = -\sigma_4 = \sqrt{n^2 - \zeta^2} \tag{8.99}$$

(see Section 9.2). In Figures 8.7–8.9, we use the following notation: $\sigma^{\downarrow} = \sigma_1 = \sigma_2$, $\sigma^{\uparrow} = \sigma_3 = \sigma_4$. Figures 8.10 and 8.11 show the results for nonabsorbing uniaxial media (see Section 9.3). The example shown in Figure 8.10 is for a positive crystal ($n_{\parallel} > n_{\perp}$) with $n_{\parallel} = 1.7$ and $n_{\perp} = 1.5$. Figure 8.11 presents the results for a negative crystal ($n_{\parallel} < n_{\perp}$) with $n_{\parallel} = 1.5$ and $n_{\perp} = 1.7$. In these figures, $\sigma_e^{\downarrow}$ and $\sigma_e^{\uparrow}$ denote the values of σ for the forward and backward propagating extraordinary waves, and $\sigma_o^{\downarrow}$ and $\sigma_o^{\uparrow}$ those for the forward and backward propagating ordinary waves, respectively.

In the case of a nonabsorbing isotropic medium, for $\zeta < n$, all σ_l are real; for $\zeta > n$, all σ_l are imaginary; and at $\zeta = n$,

$$\sigma_1 = \sigma_2 = \sigma_3 = \sigma_4 = 0.$$

In the last case, the eigenwave representation is inapplicable, since here a fundamental system of solutions of the basic equation system cannot be composed of four waves of the form (8.21) (i.e., eigenwaves). The modal representation can still be used in this case, but it is impossible to select among the fundamental modes two forward propagating ones and two backward propagating ones (see Appendix B.3). Situations where a fundamental system of solutions of the basic equation system cannot be formed of four eigenwaves will be referred to as *anomalous singularities*. In the figures, values of ζ corresponding to anomalous singularities are marked as ζ_{as}. In each of the plots for uniaxial nonabsorbing media (Figures 8.10 and 8.11), we see a pair of anomalous singularities. One of them is at the confluence of $\sigma_e^{\downarrow}$ and $\sigma_e^{\uparrow}$, and the other at the confluence of $\sigma_o^{\downarrow}$ and $\sigma_o^{\uparrow}$. Again, on passing a singularity point to greater ζ, values of σ for two corresponding eigenwaves become complex (and complex conjugates, since the coefficients

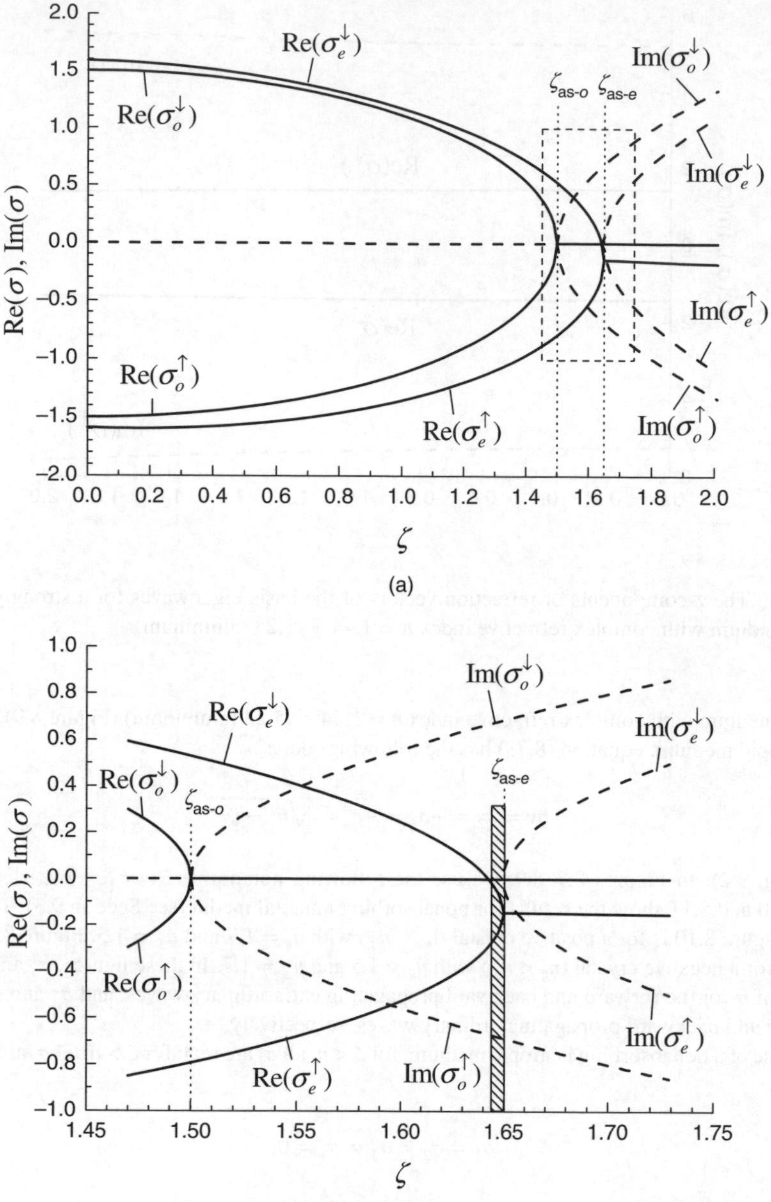

Figure 8.10 The z-components of refraction vectors of the basis eigenwaves for a nonabsorbing positive uniaxial medium with principal refractive indices $n_{\|} = 1.7$ and $n_{\perp} = 1.5$. The optic axis of the medium is parallel to the plane of incidence and makes an angle of 45° with the vector **b**. Plot (b) shows in detail the neighborhood of the anomalous singularities

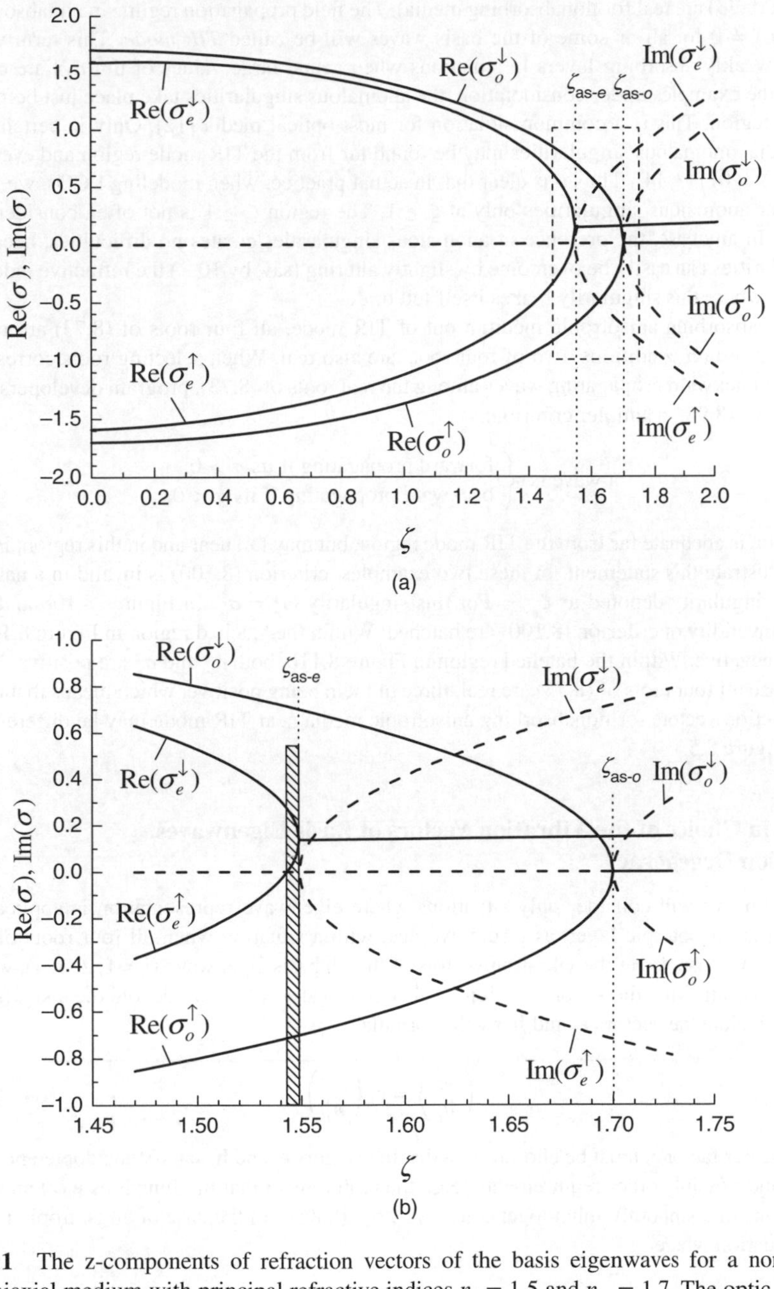

Figure 8.11 The z-components of refraction vectors of the basis eigenwaves for a nonabsorbing negative uniaxial medium with principal refractive indices $n_{\parallel} = 1.5$ and $n_{\perp} = 1.7$. The optic axis of the medium is parallel to the plane of incidence and makes an angle of 45° with the vector **b**. Plot (b) shows in detail the neighborhood of the anomalous singularities

of equation (8.73) are real for nonabsorbing media). The field propagation regime in nonabsorbing layers when $\mathrm{Im}(\sigma_l) \neq 0$ for all or some of the basis waves will be called *TIR mode*. This term will also be applied to weakly absorbing layers in situations when actual large values of $|\mathrm{Im}(\sigma_l)|$ are caused by a large ζ. In the examples under consideration, the anomalous singularities take place just before or in the TIR mode region. This is a common situation for most optical media [12]. Only in certain absorbing biaxial media, anomalous singularities may be found far from the TIR mode region and even at normal incidence ($\zeta = 0$) [13, 14]. Thus it is clear that in actual practice, when modeling LCDs, we may expect to encounter anomalous singularities only at $\zeta \geq 1$. The region $\zeta \geq 1$ is not often considered in LCD modeling.[1] In any case, the presence of anomalous singularities creates no difficulties, because any of such singularities can easily be overcome by slightly altering (say, by 10^{-5}) the refractive index (indices) of the layer where this singularity makes itself felt or ζ.

For a nonabsorbing anisotropic medium out of TIR mode, all four roots of (8.73) are real. In TIR mode, under certain conditions, two of four roots are also real. When selecting roots corresponding to forward and backward propagating waves among the real roots of (8.73), program developers sometimes use, instead of (8.98), a simpler criterion:

$$\text{a wave is } \begin{cases} \text{forward propagating if its } \sigma > 0 \\ \text{backward propagating if its } \sigma < 0. \end{cases} \tag{8.100}$$

This criterion is adequate far from the TIR mode region, but may fail near and in this region. Figures 8.10 and 8.11 illustrate this statement. In these two examples, criterion (8.100) is invalid in a narrow region before the singularity denoted as $\zeta_{\text{as}-e}$. For this singularity, $\sigma_e^{\downarrow} = \sigma_e^{\uparrow}$. In Figures 8.10b and 8.11b, the regions of invalidity of criterion (8.100) are hatched. Within the hatched region in Figure 8.10b, both $\sigma_e^{\downarrow}$ and $\sigma_e^{\uparrow}$ are negative. Within the hatched region in Figure 8.11b, both $\sigma_e^{\downarrow}$ and $\sigma_e^{\uparrow}$ are positive. Note that in the latter case all four roots of (8.73) are real, three of them being positive, which means that the location of the refraction vectors for nonabsorbing anisotropic media near TIR mode may be different from that shown in Figure 8.5.

Freedom in Choice of the Vibration Vectors of Basis Eigenwaves. Polarization Degeneracy

From now on, we will consider only situations where eigenwave representation is applicable. When considering an anisotropic layer, as a rule, we deal with a situation when all four roots of (8.73) are different. In this case, to set the vibration vectors of the jth basis eigenwave (j = 1, 2, 3, 4), we may take any pair ($\mathbf{e}$, $\mathbf{h}$) satisfying the set of equations (8.40)–(8.42) at $\sigma = \sigma_j$—we denote the taken ($\mathbf{e}$, $\mathbf{h}$) by ($\boldsymbol{e}_j$, $\boldsymbol{h}_j$)—and calculate the vectors $\mathbf{e}_j$ and $\mathbf{h}_j$ by the formula

$$\begin{pmatrix} \mathbf{e}_j \\ \mathbf{h}_j \end{pmatrix} = \mathrm{c}_j \begin{pmatrix} \boldsymbol{e}_j \\ \boldsymbol{h}_j \end{pmatrix}, \tag{8.101}$$

where the scalar factor c_j must be chosen such that the vectors $\mathbf{e}_j$ and $\mathbf{h}_j$ satisfy an adopted normalization condition and possibly other requirements [e.g., the requirement that the functions $\mathbf{e}_j(z)$ and $\mathbf{h}_j(z)$ must be continuous in a smoothly inhomogeneous layer; see (8.86)]. In the case of an isotropic medium, we face the situation where

$$\sigma_1 = \sigma_2, \quad \sigma_3 = \sigma_4. \tag{8.102}$$

When dealing with anisotropic media, we may also encounter situations when one or both of these relations are satisfied. This occurs when the refraction vectors of two or all of the four basis waves are

[1] An LCD exploiting TIR mode in the LC layer is considered in Section 4.3.

parallel to the (an) optic axis of the medium. To designate the fact that the refraction vectors of some of the basis waves are identical, we will use the term "*polarization degeneracy*." Literally this term reflects the following feature. When (8.73) gives identical roots for the forward (backward) propagating waves, the set of equations (8.40)–(8.42) admits a much wider variety of solutions than in the case when these roots are different. For example, in the case of an isotropic medium, given $\mathbf{m}$, any $\mathbf{e}$ satisfying the condition $\mathbf{m}\cdot\mathbf{e} = 0$ meets (8.40)–(8.42). In the case of polarization degeneracy, one can always find two linearly independent solutions of the system (8.40)–(8.42) corresponding to the given $\mathbf{m}$. Any linear combination of these solutions will satisfy this system as well. Clearly, it provides a greater freedom in choosing vibration vectors for the basis waves.

Suppose that polarization degeneracy takes place for the eigenwaves with numbers j and $j+1$ ($j = 1$ or 3), that is, $\mathbf{m}_{j+1} = \mathbf{m}_j$. Let the pairs ($\boldsymbol{e}'$, $\boldsymbol{h}'$) and ($\boldsymbol{e}''$, $\boldsymbol{h}''$) be two linearly independent solutions of the system (8.40)–(8.42) at $\mathbf{m} = \mathbf{m}_j$. Then we may write the following general expressions for the vectors $\mathbf{e}_j$, $\mathbf{h}_j$, $\mathbf{e}_{j+1}$, and $\mathbf{h}_{j+1}$:

$$\begin{pmatrix} \mathbf{e}_j \\ \mathbf{h}_j \end{pmatrix} = c'_j \begin{pmatrix} \boldsymbol{e}' \\ \boldsymbol{h}' \end{pmatrix} + c''_j \begin{pmatrix} \boldsymbol{e}'' \\ \boldsymbol{h}'' \end{pmatrix}, \quad \begin{pmatrix} \mathbf{e}_{j+1} \\ \mathbf{h}_{j+1} \end{pmatrix} = c'_{j+1} \begin{pmatrix} \boldsymbol{e}' \\ \boldsymbol{h}' \end{pmatrix} + c''_{j+1} \begin{pmatrix} \boldsymbol{e}'' \\ \boldsymbol{h}'' \end{pmatrix}, \tag{8.103}$$

where c'_j, c''_j, c'_{j+1}, and c''_{j+1}, are scalar coefficients whose region of possible values is determined by the normalization condition and the requirement that the solutions ($\mathbf{e}_j$, $\mathbf{h}_j$) and ($\mathbf{e}_{j+1}$, $\mathbf{h}_{j+1}$) must be linearly independent.

Relevant sources: References 2, 9, and 15–23.

8.2 Transmission and Reflection Operators of Fragments (TR Units) of a Stratified Medium and Their Calculation

8.2.1 EW Jones Vector. EW Jones Matrices. Transmission and Reflection Operators

One of the basic problems that are solved using the theory described in this chapter in the modeling methods presented in this book is the calculation of transmission and reflection operators characterizing TR units of approximating media (see Section 7.1). In this section we introduce an important class of such operators, namely, eigenwave (EW) Jones matrices. State vectors with which operators of this class work will be referred to as EW Jones vectors. We begin with definition of EW Jones vector.

EW Jones Vector

Let a complex vector

$$\mathbf{a}(\xi) = \begin{pmatrix} a_1(\xi) \\ a_2(\xi) \end{pmatrix}, \tag{8.104}$$

where ξ is a given value of z, represent the field

$$\begin{pmatrix} \mathbf{E}(\mathbf{r}_\xi, t) \\ \mathbf{H}(\mathbf{r}_\xi, t) \end{pmatrix} = \left[a_1(\xi) \begin{pmatrix} \mathbf{e}_1(\xi) \\ \mathbf{h}_1(\xi) \end{pmatrix} + a_2(\xi) \begin{pmatrix} \mathbf{e}_2(\xi) \\ \mathbf{h}_2(\xi) \end{pmatrix} \right] \exp\left\{ i \left(k_0 \mathbf{b}\mathbf{r}_\xi - \omega t \right) \right\} \tag{8.105}$$

if this vector characterizes the state of a forward propagating wave field, or the field

$$\begin{pmatrix} \mathbf{E}(\mathbf{r}_\xi, t) \\ \mathbf{H}(\mathbf{r}_\xi, t) \end{pmatrix} = \left[a_1(\xi) \begin{pmatrix} \mathbf{e}_3(\xi) \\ \mathbf{h}_3(\xi) \end{pmatrix} + a_2(\xi) \begin{pmatrix} \mathbf{e}_4(\xi) \\ \mathbf{h}_4(\xi) \end{pmatrix} \right] \exp\left\{ i \left(k_0 \mathbf{b} \mathbf{r}_\xi - \omega t \right) \right\} \tag{8.106}$$

if this vector characterizes the state of a backward propagating wave field. Here, $\mathbf{r}_\xi = (x, y, \xi)$ is a position vector on the plane $z = \xi$. Recall that the indices 1 and 2 are assigned to the forward propagating basis waves, and 3 and 4 to the backward propagating ones. Any forward or backward propagating field of the form (8.57) passing through the plane $z = \xi$ can be represented by the corresponding vector $\mathbf{a}(\xi)$. Such vectors will be called *EW Jones vectors* or **a**-*vectors*. We will mark **a**-vectors for forward and backward propagating fields by the arrows $^\downarrow$ ($\mathbf{a}^\downarrow$) and $^\uparrow$ ($\mathbf{a}^\uparrow$), respectively. We have already met **a**-vectors in expression (8.93). State vectors used in the extended Jones matrix method of Yeh and Gu [24–26] are particular variants of EW Jones vector. Vectors (5.89), (5.90), and (5.104) are examples of EW Jones vectors. Vectors (1.97) and (1.114) taken on the normal coordinate axis also meet the definition of EW Jones vector. One may notice a close similarity of the EW Jones vectors to the canonical Jones vectors (1.63). The conditions under which estimating a canonical Jones vector will be tantamount to estimating a EW Jones vector are evident from the definitions presented. *The EW Jones matrix method* is a method exploiting the general operator approaches described in Chapter 7 and using EW Jones vector as traced characteristic.

EW Jones Matrices of Transmission and Reflection of a TR Unit

Let us consider a domain **F** of a stratified medium (Figure 8.12), whose boundaries, planes $z = z'$ and $z = z''$, do not coincide with interfaces, but, if z_I is the z-coordinate of an interface, z' (or z'') may be equal to $z_I - 0$ or $z_I + 0$, so that this domain can be regarded as a TR unit. We will call the homogeneous layers containing the planes $z = z'$ and $z = z''$ respectively layer $\mathbf{M}_A$ and layer $\mathbf{M}_B$. Layers $\mathbf{M}_A$ and $\mathbf{M}_B$ may be infinitely thin and may be the same layer (if **F** is the bulk of a homogeneous layer). In general, domain **F** may perform four operations: the operations of transmission $\mathbf{T}^\downarrow\{\mathbf{F}\}$ and reflection $\mathbf{R}^\downarrow\{\mathbf{F}\}$ for forward propagating incident light $\mathbf{X}^\downarrow_{inc}$ and the operations of transmission $\mathbf{T}^\uparrow\{\mathbf{F}\}$ and reflection $\mathbf{R}^\uparrow\{\mathbf{F}\}$ for backward propagating incident light $\mathbf{X}^\uparrow_{inc}$ (see also the definition of TR units in Section 7.1). These operations are precisely defined by symbolical relations in Figure 8.12. We define EW Jones matrices $\mathbf{t}^\downarrow(z', z'')$, $\mathbf{r}^\downarrow(z', z'')$, $\mathbf{t}^\uparrow(z', z'')$, and $\mathbf{r}^\uparrow(z', z'')$ corresponding to operations $\mathbf{T}^\downarrow\{\mathbf{F}\}$, $\mathbf{R}^\downarrow\{\mathbf{F}\}$, $\mathbf{T}^\uparrow\{\mathbf{F}\}$, and $\mathbf{R}^\uparrow\{\mathbf{F}\}$ (see the tables in Figure 8.12) by the following relations:

$$\mathbf{a}^\downarrow_{tr}(z'') = \mathbf{t}^\downarrow(z', z'')\, \mathbf{a}^\downarrow_{inc}(z'), \tag{8.107}$$

$$\mathbf{a}^\uparrow_{ref}(z') = \mathbf{r}^\downarrow(z', z'')\mathbf{a}^\downarrow_{inc}(z'), \tag{8.108}$$

$$\mathbf{a}^\uparrow_{tr}(z') = \mathbf{t}^\uparrow(z', z'')\mathbf{a}^\uparrow_{inc}(z''), \tag{8.109}$$

$$\mathbf{a}^\downarrow_{ref}(z'') = \mathbf{r}^\uparrow(z', z'')\, \mathbf{a}^\uparrow_{inc}(z''), \tag{8.110}$$

where $\mathbf{a}^\downarrow_{inc}(z')$ is the **a**-vector of the incident field $\mathbf{X}^\downarrow_{inc}$, $\mathbf{a}^\downarrow_{tr}(z'')$ and $\mathbf{a}^\uparrow_{ref}(z')$ are **a**-vectors characterizing, respectively, the transmitted and reflected fields generated by the field $\mathbf{X}^\downarrow_{inc}$, $\mathbf{a}^\uparrow_{inc}(z'')$ is the **a**-vector of the incident field $\mathbf{X}^\uparrow_{inc}$, and $\mathbf{a}^\uparrow_{tr}(z')$ and $\mathbf{a}^\downarrow_{ref}(z'')$ are the **a**-vectors of the transmitted and reflected fields generated by the field $\mathbf{X}^\uparrow_{inc}$.

In what follows, the symbols $\mathbf{t}^\downarrow$, $\mathbf{t}^\uparrow$, $\mathbf{r}^\downarrow$, and $\mathbf{r}^\uparrow$ will be used as generic symbols for EW Jones matrices of the corresponding types of operations performed by TR units.

Before proceeding to methods of calculation of such transmission and reflection operators for different elements, we wish to present a few typical variants of using such operators in modeling LCDs.

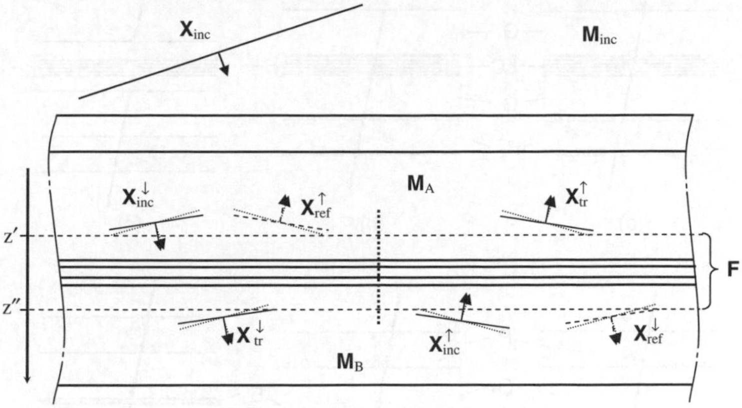

Operations performed by the fragment $\mathbf{F}$

$$\mathbf{X}^{\downarrow}_{tr}(z'') = \mathbf{T}^{\downarrow}\{\mathbf{F}\}\mathbf{X}^{\downarrow}_{inc}(z') \qquad \mathbf{X}^{\uparrow}_{tr}(z') = \mathbf{T}^{\uparrow}\{\mathbf{F}\}\mathbf{X}^{\uparrow}_{inc}(z'')$$

$$\mathbf{X}^{\uparrow}_{ref}(z') = \mathbf{R}^{\downarrow}\{\mathbf{F}\}\mathbf{X}^{\downarrow}_{inc}(z') \qquad \mathbf{X}^{\downarrow}_{ref}(z'') = \mathbf{R}^{\uparrow}\{\mathbf{F}\}\mathbf{X}^{\uparrow}_{inc}(z'')$$

Wave field	State vector
$\mathbf{X}^{\downarrow}_{inc}(z')$	$\mathbf{a}^{\downarrow}_{inc}(z')$
$\mathbf{X}^{\uparrow}_{ref}(z')$	$\mathbf{a}^{\uparrow}_{ref}(z')$
$\mathbf{X}^{\downarrow}_{tr}(z'')$	$\mathbf{a}^{\downarrow}_{tr}(z'')$
$\mathbf{X}^{\uparrow}_{inc}(z'')$	$\mathbf{a}^{\uparrow}_{inc}(z'')$
$\mathbf{X}^{\downarrow}_{ref}(z'')$	$\mathbf{a}^{\downarrow}_{ref}(z'')$
$\mathbf{X}^{\uparrow}_{tr}(z')$	$\mathbf{a}^{\uparrow}_{tr}(z')$

Operation	Operator
$\mathbf{T}^{\downarrow}\{\mathbf{F}\}$	$\mathbf{t}^{\downarrow}(z',z'')$
$\mathbf{R}^{\downarrow}\{\mathbf{F}\}$	$\mathbf{r}^{\downarrow}(z',z'')$
$\mathbf{T}^{\uparrow}\{\mathbf{F}\}$	$\mathbf{t}^{\uparrow}(z',z'')$
$\mathbf{R}^{\uparrow}\{\mathbf{F}\}$	$\mathbf{r}^{\uparrow}(z',z'')$

Figure 8.12 Transmission and reflection of a fragment of a stratified medium. Fields, operations, state vectors, and operators

TR Units and Transfer Channels in Models of LCDs

The simplest way to use EW Jones matrices in the modeling of an LCD panel is suggested by the transfer channel approach (this approach is basic in both the classical Jones matrix method and the extended Jones matrix methods [24–31]): We should divide the stratified medium approximating the LCD panel into elementary fragments so that a sequence of operations performed by these fragments defines a transfer channel such that a characteristic of this channel is a good approximating characteristic for the LCD panel characteristic to be estimated (see Section 7.1). Then we should calculate the corresponding transmission and reflection operators for elementary fragments and, by multiplying these operators, the operator characterizing the channel (and, using this operator, evaluate the approximating characteristic). As an illustration, let us consider several simple examples. In these examples, we deal with (i) a transmissive LCD containing an LC cell and two polarizers (Figure 8.13) and (ii) a single-polarizer LCD including an LC cell with a metal reflector situated just after the LC layer, a retardation film, and a polarizer (Figure 8.14). Polarizers and retardation film are regarded in these examples as homogeneous anisotropic layers. The LC layers are considered to be inhomogeneous.

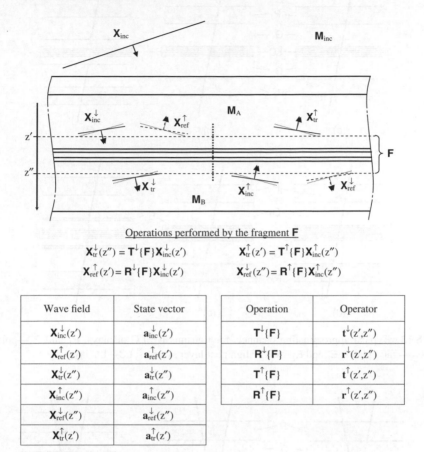

Figures 8.13a and 8.14a show the standard choices of approximating media and approximating channels when calculations in such situations are performed by using the mentioned extended Jones

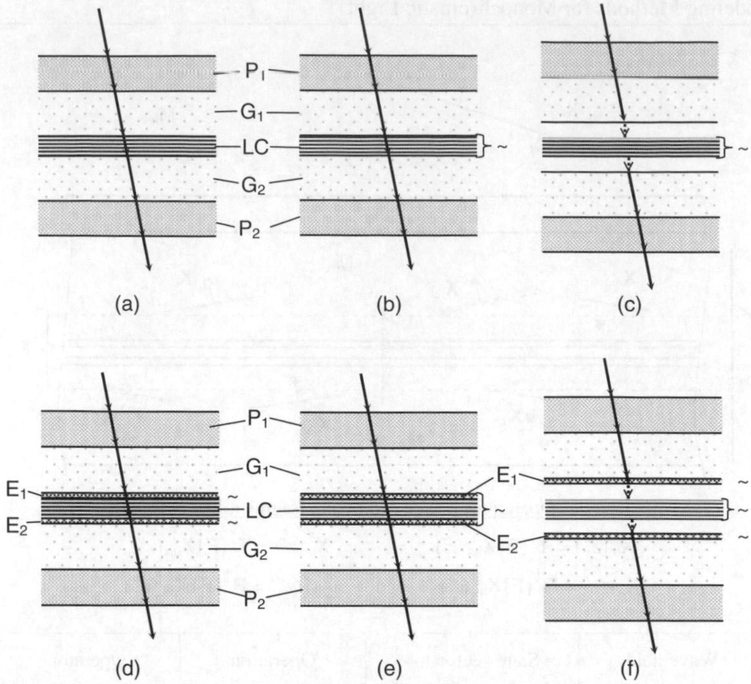

Figure 8.13 Typical approximating channels for transmissive LC displays. P_1 and P_2—polarizers; G_1 and G_2—glass plates; E_1 and E_2—ITO–alignment layer systems; LC—LC layer

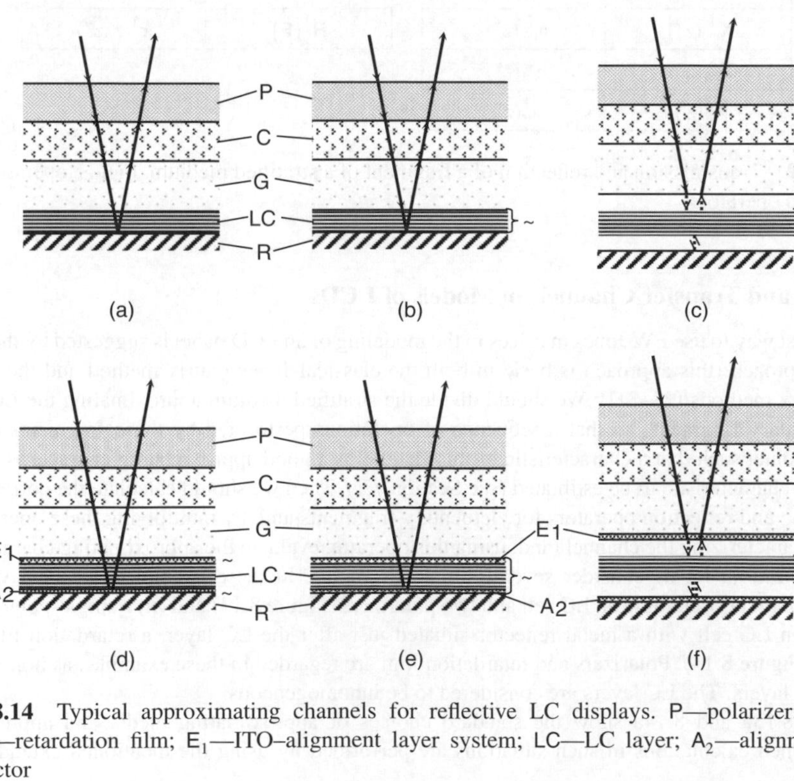

Figure 8.14 Typical approximating channels for reflective LC displays. P—polarizer; G—glass plate; C—retardation film; E_1—ITO–alignment layer system; LC—LC layer; A_2—alignment layer; R—reflector

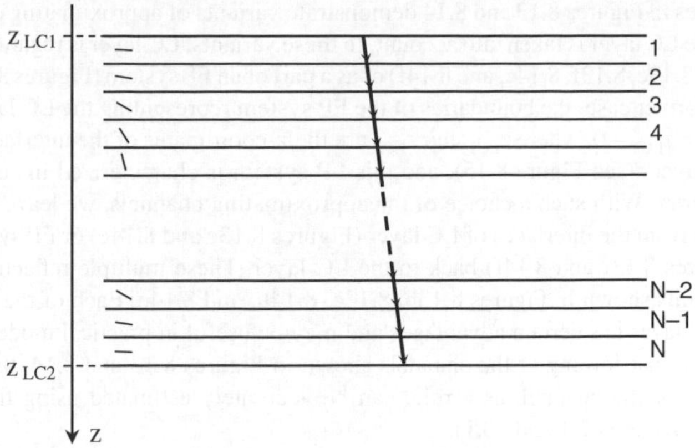

Figure 8.15 Section of the approximating channels shown in Figures 8.13a, 8.13d, 8.14a, and 8.14d within the layered medium approximating the LC layer

matrix methods. Figure 8.15 details a piece of these channels within the media approximating LC layers. In these two cases, as in the other cases, staircase models (see Sections 8.3.3 and 11.3.1) are used to represent inhomogeneous LC layers: the LC layer is regarded as a pile of very thin anisotropic homogeneous layers with differently oriented optic axes. In both cases presented in Figures 8.13a and 8.14a, the elementary fragments (TR units) are the bulks of homogeneous layers and the interfaces. For the transmissive LCD, the approximating channel is the chain of the elementary transmission operations guiding the light from the frontal side of the interface between the air (the entrance medium) and the frontal polarizer (P_1) to the rear side of the interface between the rear polarizer (P_2) and the air (the exit medium). For the reflective LCD, the approximating channel includes the chain of the elementary transmission operations leading the light from the frontal side of the interface [air–polarizer] to the frontal side of the interface [LC–reflector], the operation of reflection from the interface [LC–reflector], and the chain of the elementary transmission operations leading the light back, from the frontal side of the interface [LC–reflector] to the frontal side of the interface [air–polarizer].

Let us wish to incorporate ITO and alignment layers in both models (see Figures 8.13d and 8.14d). We have noted that in many cases, the optical characteristics of thin layers and systems of thin layers (successive thin layers) should be modeled taking into consideration multiple reflections to account for the effect of Fabry–Perot interference. In the case of ITO and alignment layers, we face such a situation. We will refer to single layers and systems of layers in which we intend to take into account the multiple reflections that result in Fabry–Perot interference as *FP systems*. A reasonable way of including an FP system into the models in our case is to regard this whole system as an OTR unit (see Section 7.1), that is, characterize it by operators describing the overall transmission and reflection of this system. This leads to dealing with the approximating channels shown in Figures 8.13d and 8.14d. In the figures, FP systems are marked by tildes. The channel depicted in Figure 8.13d includes, along with transmission operations for interfaces and bulks, the operations of forward transmission performed by the ITO layer–alignment layer system (E_1) and the alignment layer–ITO layer system (E_2). The channel shown in Figure 8.14d contains, among other operations, the operations of forward and backward transmissions for the ITO layer–alignment layer system (E_1) and reflection from the rear alignment layer (A_2)–reflector system. The boundaries of the last TR unit are taken to be $z = z_{LC2} - 0$ and $z = z_{A2\text{-}R} + 0$, where z_{LC2} is the z-coordinate of the interface [LC–alignment layer A_2] and $z_{A2\text{-}R}$ is the z-coordinate of the interface [alignment layer A_2–reflector].

The other sketches in Figures 8.13 and 8.14 demonstrate variants of approximating channels where the bulk reflection in the LC layer is taken into account. In these variants, LC layer is regarded as an individual FP system (Figures 8.13c, 8.13f, 8.14c, and 8.14f) or as a part of an FP system (Figures 8.13b, 8.13e, 8.14b, and 8.14e). In the former case, the boundaries of the FP system representing the LC layer are taken to be $z = z_{LC1} + 0$ and $z = z_{LC2} - 0$, where z_{LC1} and z_{LC2} are the z-coordinates of the interfaces of the LC layer with the adjacent layers (see Figure 8.15), and this FP system is characterized in calculations only by transmission operators. With such a choice of the approximating channels, we leave out of account the multiple reflections from the interfaces of LC layer (Figures 8.13c and 8.14c) or FP systems surrounding the LC layer (Figures 8.13f and 8.14f) back to the LC layer. These multiple reflections are taken into account in the variants shown in Figures 8.13b, 8.13e, 8.14b, and 8.14e. Each of the presented variants of involving the LC layer has certain advantages and may be useful in practical modeling.

It should be noted that for any of the channels shown in Figures 8.13 and 8.14, the quasimonochromatic transmittance of the channel, as a rule, can be accurately estimated using the monochromatic approximation (see Sections 7.1 and 10.1).

Channels similar to those shown in Figures 8.14a, 8.14c, 8.14d, and 8.14f can be used in modeling reflective LCDs with nonspecular reflectors [31] (see also Section 7.1). In this case, the operators characterizing the channel sections leading to the reflector and back in general correspond to different values of **b**.

The above examples have displayed three main kinds of elementary fragments: interfaces, bulks of homogeneous layers, and FP systems.

The calculation of EW Jones matrices for bulks of homogeneous layers does not require special consideration: according to (8.83), for a homogeneous layer

$$\mathbf{t}^{\downarrow}(z_{n-1} + 0, z_n - 0) = \begin{pmatrix} \exp\left(ik_0\sigma_1^{(n)}h_n\right) & 0 \\ 0 & \exp\left(ik_0\sigma_2^{(n)}h_n\right) \end{pmatrix}, \tag{8.111}$$

$$\mathbf{t}^{\uparrow}(z_{n-1} + 0, z_n - 0) = \begin{pmatrix} \exp\left(-ik_0\sigma_3^{(n)}h_n\right) & 0 \\ 0 & \exp\left(-ik_0\sigma_4^{(n)}h_n\right) \end{pmatrix}, \tag{8.112}$$

$$\mathbf{r}^{\downarrow}(z_{n-1} + 0, z_n - 0) = \mathbf{r}^{\uparrow}(z_{n-1} + 0, z_n - 0) = \mathbf{0}, \tag{8.113}$$

where, as in (8.83), z_{n-1} and z_n ($z_{n-1} < z_n$) are the z-coordinates of the interfaces of this layer with the adjacent layers, $h_n = z_n - z_{n-1}$ is the thickness of the layer, $\sigma_l^{(n)}$ is the value of $\sigma_l(z)$ within the layer (l = 1, 2, 3, 4), and **0** is the zero matrix.

Section 8.2.2 shows how the transmission and reflection EW Jones matrices characterizing the overall transmission and overall reflection of a layered system can be calculated with the help of the eigenwave 4×4 transfer matrix method and the Berreman transfer matrix method. In Section 8.4.1, we present a number of relations for eigenwave bases that are useful in calculating EW Jones matrices. Calculation of transmission operators for interfaces is a weak point of the popular variants of the extended Jones matrix method [24–30] (see Section 11.1.2). In Section 8.4.2, we derive simple exact formulas for these operators, which are a very good alternative to the approximate expressions for these matrices used in References 24–30. In Section 8.4.3, we present a technique for calculating the reflection and transmission operators of layered systems that exploits the adding algorithm. For systems with strongly absorbing layers and layers being in TIR mode, this technique is much more reliable (numerically stable) than the transfer matrix approach described in Section 8.2.2. In Chapter 11, we present efficient methods of calculating transmission operators for inhomogeneous LC layers with negligible bulk reflection.

8.2.2 *Calculation of Overall Transmission and Overall Reflection Operators for Layered Systems by Using Transfer Matrices*

Transfer Matrices

To find the operators characterizing the overall transmission and overall reflection of the fragment (z′, z″) (Figure 8.12), we should consider the fields that satisfy the basic equation system throughout the whole region (z′, z″). For any of such fields, the relationship between the values of the function **A**(z) at z = z′ and at z = z″ can be expressed as follows:

$$\mathbf{A}(z'') = \mathbf{T}(z'', z')\mathbf{A}(z'), \tag{8.114}$$

where $\mathbf{T}(z'', z')$ is the 4 × 4 transfer matrix of the fragment (z′, z″) for **A**-vectors in the given eigenwave bases. Such transfer matrices will be referred to as *EW transfer matrices*. The matrix $\mathbf{T}(z'', z')$ characterizes the general solution and hence is common for all the fields under consideration. In Section 8.1.3, we have dealt with a EW transfer matrix for the bulk of a homogeneous layer [see (8.84) and (8.85)]. According to (8.82), the transfer matrix of an interface can be calculated by the formula

$$\mathbf{T}(\underline{z} + 0, \underline{z} - 0) = \mathbf{\Psi}(\underline{z} + 0)^{-1}\mathbf{\Psi}(\underline{z} - 0), \tag{8.115}$$

where $\underline{z}$ is the z-coordinate of this interface. The EW transfer matrix of a fragment consisting of subfragments can be calculated as the product of the EW transfer matrices of the subfragments. For illustration, we return to the problem depicted in Figure 8.12 and suppose that *N* homogeneous layers $\mathsf{F}_1, \mathsf{F}_2, \ldots, \mathsf{F}_N$ are situated between the layers M_A and M_B. Figure 8.16 shows a scheme and formulas for calculation of the matrix $\mathbf{T}(z'', z')$ for this case. The following notation is used:

$\sigma_l[\mathsf{A}]$ and $\mathbf{\Psi}[\mathsf{A}]$ denote the values of σ_l and $\mathbf{\Psi}$ for a layer A (A = $\mathsf{M}_A, \mathsf{F}_1, \mathsf{F}_2, \ldots, \mathsf{F}_N, \mathsf{M}_B$);

z_{Fj-1} and z_{Fj} are the z-coordinates of the interfaces of the layer F_j with the adjacent layers ($i = 1, 2, \ldots, N$);

$h[\mathsf{F}_j]$ is the thickness of the layer F_j ($h[\mathsf{F}_j] = z_{Fj} - z_{Fj-1}$) ($j = 1, 2, \ldots, N$);

h' is the distance between the planes z = z′ and $z = z_{F0}$ ($h' = z_{F0} - z'$);

h'' is the distance between the planes $z = z_{FN}$ and z = z″ ($h'' = z'' - z_{FN}$);

$\mathbf{T}_{[j]} \equiv \mathbf{T}(z_{Fj} - 0, z_{Fj-1} + 0)$ ($j = 1, \ldots, N$) is the EW transfer matrix for the bulk of the layer F_j;

$\mathbf{T}_{[0]} \equiv \mathbf{T}(z_{F0} - 0, z')$ is the EW transfer matrix for a part of the bulk of the layer M_A;

$\mathbf{T}_{[N+1]} \equiv \mathbf{T}(z'', z_{FN} + 0)$ is the EW transfer matrix for a part of the bulk of the layer M_B;

$\mathbf{T}_{|j} \equiv \mathbf{T}(z_{Fj} + 0, z_{Fj} - 0)$ is the EW transfer matrix for the interface $z = z_{Fj}$ ($j = 0, \ldots, N$).

The matrix pattern $\tilde{\mathbf{T}}(\sigma_l, h)$ is given by (8.85).

If the boundaries of the fragment **F** are chosen right at the interfaces $z = z_{F0}$ ($z' = z_{F0} - 0$) and $z = z_{FN}$ ($z'' = z_{FN} + 0$), the matrix $\mathbf{T}(z'', z')$ can be expressed as follows:

$$\mathbf{T}(z'', z') = \mathbf{T}_{|N}\mathbf{T}_{[N]}\mathbf{T}_{|N-1} \cdot \ldots \cdot \mathbf{T}_{|2}\mathbf{T}_{[2]}\mathbf{T}_{|1}\mathbf{T}_{[1]}\mathbf{T}_{|0}. \tag{8.116}$$

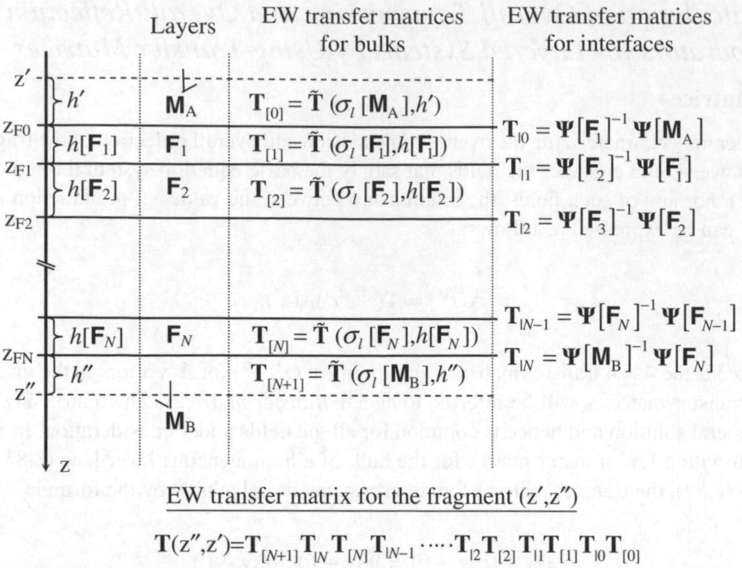

Figure 8.16 Scheme of calculation of the EW transfer matrix for a layered fragment

We should mention another way of calculating $\mathbf{T}(z'', z')$. The relation analogous to (8.114) but written in terms of Berreman vectors is

$$\Psi(z'') = \mathbf{P}(z'', z')\,\Psi(z'), \tag{8.117}$$

where $\mathbf{P}(z'', z')$ is the Berreman transfer matrix of the fragment (z', z''). Calculation of Berreman transfer matrices will be discussed in Section 8.3. Given $\mathbf{P}(z'', z')$, the matrix $\mathbf{T}(z'', z')$ can be calculated by the formula

$$\mathbf{T}(z'', z') = \mathbf{\Psi}(z'')^{-1}\,\mathbf{P}(z'', z')\mathbf{\Psi}(z') \tag{8.118}$$

[see (8.69)].

Calculating the Overall Transmission and Overall Reflection Operators of a Fragment from the EW Transfer Matrix of this Fragment

Given the matrix $\mathbf{T}(z'', z')$, the operators $\mathbf{t}^{\downarrow}(z', z'')$ and $\mathbf{r}^{\downarrow}(z', z'')$ can be found from the equation

$$\begin{pmatrix} \mathbf{a}^{\downarrow}_{tr}(z'') \\ \hat{\mathbf{0}} \end{pmatrix} = \mathbf{T}(z'', z')\begin{pmatrix} \mathbf{a}^{\downarrow}_{inc}(z') \\ \mathbf{a}^{\uparrow}_{ref}(z') \end{pmatrix} \tag{8.119}$$

which models the situation shown in the left part of the sketch in Figure 8.12, and the operators $\mathbf{t}^{\uparrow}(z', z'')$ and $\mathbf{r}^{\uparrow}(z', z'')$ from the equation

$$\begin{pmatrix} \mathbf{a}^{\downarrow}_{ref}(z'') \\ \mathbf{a}^{\uparrow}_{inc}(z'') \end{pmatrix} = \mathbf{T}(z'', z')\begin{pmatrix} \hat{\mathbf{0}} \\ \mathbf{a}^{\uparrow}_{tr}(z') \end{pmatrix} \tag{8.120}$$

which describes the situation shown in the right part of that sketch; here $\widehat{\mathbf{0}}$ is the zero 2×1 column. Such problems have been solved in Section 7.2.1. By using those solutions, we may write the results for the case in question as follows:

$$\begin{aligned}
\mathbf{t}^{\downarrow}(z', z'') &= \mathtt{t}^{\downarrow}\{\mathbf{T}(z'', z')\},\\
\mathbf{r}^{\downarrow}(z', z'') &= \mathtt{r}^{\downarrow}\{\mathbf{T}(z'', z')\},\\
\mathbf{t}^{\uparrow}(z', z'') &= \mathtt{t}^{\uparrow}\{\mathbf{T}(z'', z')\},\\
\mathbf{r}^{\uparrow}(z', z'') &= \mathtt{r}^{\uparrow}\{\mathbf{T}(z'', z')\},
\end{aligned} \tag{8.121}$$

where $\mathtt{t}^{\downarrow}\{\mathbf{T}\}$, $\mathtt{r}^{\downarrow}\{\mathbf{T}\}$, $\mathtt{t}^{\uparrow}\{\mathbf{T}\}$, and $\mathtt{r}^{\uparrow}\{\mathbf{T}\}$ denote the 2×2 matrices calculated from a 4×4 matrix $\mathbf{T}$ by the formulas

$$\begin{aligned}
\mathtt{t}^{\downarrow}\{\mathbf{T}\} &= \mathbf{t}_{11} - \mathbf{t}_{12}\mathbf{t}_{22}^{-1}\mathbf{t}_{21},\\
\mathtt{r}^{\downarrow}\{\mathbf{T}\} &= -\mathbf{t}_{22}^{-1}\mathbf{t}_{21},\\
\mathtt{t}^{\uparrow}\{\mathbf{T}\} &= \mathbf{t}_{22}^{-1},\\
\mathtt{r}^{\uparrow}\{\mathbf{T}\} &= \mathbf{t}_{12}\mathbf{t}_{22}^{-1},
\end{aligned} \tag{8.122}$$

where $\mathbf{t}_{kj}$ $(k, j = 1, 2)$ are 2×2 blocks of the matrix $\mathbf{T}$:

$$\mathbf{T} = \begin{pmatrix} \mathbf{t}_{11} & \mathbf{t}_{12} \\ \mathbf{t}_{21} & \mathbf{t}_{22} \end{pmatrix}. \tag{8.123}$$

In certain situations (see Section 8.4.2), it is convenient to calculate the transmission and reflection operators from the inverse of the transfer matrix $\mathbf{T}(z'', z')$. The computational formulas for this case are

$$\begin{aligned}
\mathbf{t}^{\downarrow}(z', z'') &= \bar{\mathtt{t}}^{\downarrow}\left\{\mathbf{T}(z'', z')^{-1}\right\},\\
\mathbf{r}^{\downarrow}(z', z'') &= \bar{\mathtt{r}}^{\downarrow}\left\{\mathbf{T}(z'', z')^{-1}\right\},\\
\mathbf{t}^{\uparrow}(z', z'') &= \bar{\mathtt{t}}^{\uparrow}\left\{\mathbf{T}(z'', z')^{-1}\right\},\\
\mathbf{r}^{\uparrow}(z', z'') &= \bar{\mathtt{r}}^{\uparrow}\left\{\mathbf{T}(z'', z')^{-1}\right\},
\end{aligned} \tag{8.124}$$

where $\bar{\mathtt{t}}^{\downarrow}\{\mathbf{T}\}$, $\bar{\mathtt{r}}^{\downarrow}\{\mathbf{T}\}$, $\bar{\mathtt{t}}^{\uparrow}\{\mathbf{T}\}$, and $\bar{\mathtt{r}}^{\uparrow}\{\mathbf{T}\}$ denote the 2×2 matrices calculated from a 4×4 matrix $\mathbf{T}$ as follows:

$$\begin{aligned}
\bar{\mathtt{t}}^{\downarrow}\{\mathbf{T}\} &= \mathbf{t}_{11}^{-1},\\
\bar{\mathtt{r}}^{\downarrow}\{\mathbf{T}\} &= \mathbf{t}_{21}\mathbf{t}_{11}^{-1},\\
\bar{\mathtt{t}}^{\uparrow}\{\mathbf{T}\} &= \mathbf{t}_{22} - \mathbf{t}_{21}\mathbf{t}_{11}^{-1}\mathbf{t}_{12},\\
\bar{\mathtt{r}}^{\uparrow}\{\mathbf{T}\} &= -\mathbf{t}_{11}^{-1}\mathbf{t}_{12},
\end{aligned} \tag{8.125}$$

where, as before, $\mathbf{t}_{kj}$ are 2×2 blocks of the matrix $\mathbf{T}$.

8.3 Berreman's Method

8.3.1 Transfer Matrices

In the Berreman transfer matrix method [2, 16], transmission and reflection characteristics of a stratified system—let us denote the z-coordinates of its external boundaries by z_a and z_b $(z_b > z_a)$—are calculated from the transfer matrix $\mathbf{P}(z_b, z_a)$ relating values of functions $\Psi(z)$ [see (8.63)] satisfying equation (8.62) throughout the region (z_a, z_b) at the external boundaries of the system:

$$\Psi(z_b) = \mathbf{P}(z_b, z_a)\Psi(z_a). \tag{8.126}$$

Computational formulas for characteristics of interest are obtained by solving a boundary-value problem; examples may be found in Section 8.2.2. The central problem of this method is to find the transfer matrix. In this section, we consider some basic concepts of Berreman's approach to calculating transfer matrices using his method.

Continuity of the functions $\Psi(z)$ determines the following property of Berreman transfer matrices:

$$\mathbf{P}(z_b, z_a) = \mathbf{P}(z_b, z_{ck})\mathbf{P}(z_{ck}, z_{ck-1}) \cdot \ldots \cdot \mathbf{P}(z_{c2}, z_{c1})\mathbf{P}(z_{c1}, z_a), \tag{8.127}$$

where z_{cj} ($j = 1, \ldots, k$) are arbitrary points on the z-axis satisfying the condition

$$z_a < z_{c1} < z_{c2} < \ldots < z_{ck-1} < z_{ck} < z_b,$$

that is, the Berreman transfer matrix for the interval (z_a, z_b) is equal to the product of the Berreman transfer matrices for any complete series of subintervals between z_a and z_b. This property makes it possible to calculate the matrix $\mathbf{P}(z_b, z_a)$ as the product of the transfer matrices characterizing elementary regions of the region (z_a, z_b). Thus, for a medium being a stack of homogeneous layers, the matrix $\mathbf{P}(z_b, z_a)$ can be calculated as the product of transfer matrices of homogeneous layers, for which relatively simple computational formulas are known (see Section 8.3.2).

An important element of Berreman's approach is a differential equation for transfer matrices. This equation is derived from the basic equation of Berreman's formalism (8.62) as follows. Since equation (8.62) is linear and homogeneous, any function satisfying (8.62) can be represented in the form

$$\Psi(z) = \mathbf{P}(z, z')\Psi(z'), \tag{8.128}$$

where $\mathbf{P}(z, z')$ is a transfer matrix, which is considered here as a function of the coordinate of the point which the column Ψ on the left-hand side of (8.128) is for; z' is a given point on the z-axis within the region being considered. Substitution of (8.128) into (8.62) gives the following equation for $\mathbf{P}(z, z')$:

$$\frac{d\mathbf{P}(z, z')}{dz} = ik_0\mathbf{\Delta}(z)\mathbf{P}(z, z') \qquad [\mathbf{P}(z', z') = \mathbf{U}]. \tag{8.129}$$

Due to the condition $\mathbf{P}(z', z') = \mathbf{U}$, $\mathbf{P}(z, z')$ is a unique solution. $\mathbf{P}(z, z')$ represents the general solution of (8.62): the columns of this matrix are four linearly independent solutions of equation (8.62), which constitute a fundamental system of solutions for this equation. With any given $\Psi(z')$, expression (8.128) yields a solution of the Maxwell equations for the given medium. According to (8.129), the matrix $\mathbf{P}(z'', z')$ can be expressed as follows:

$$\mathbf{P}(z'', z') = \exp\left(ik_0 \int_{z'}^{z''} \mathbf{\Delta}(z)dz\right) \equiv \mathbf{U} + \sum_{j=1}^{\infty} (ik_0)^j \mathbf{\Delta}_{(j)}(z''), \tag{8.130}$$

where

$$\mathbf{\Delta}_{(1)}(\tau) = \int_{z'}^{\tau} \mathbf{\Delta}(z)dz,$$

$$\mathbf{\Delta}_{(j)}(\tau) = \int_{z'}^{\tau} \mathbf{\Delta}(z)\mathbf{\Delta}_{(j-1)}(z)dz \qquad j = 2, 3, 4, \ldots$$

Equation (8.129) and expression (8.130) underlie many practical techniques for calculating Berreman transfer matrices.

8.3.2 Transfer Matrix of a Homogeneous Layer

If the medium between planes z = z′ and z = z″ is homogeneous, the matrix $\mathbf{\Delta}$ in the region (z′, z″) is independent of z. In this case, $\mathbf{P}(z'', z')$ can be expressed in the form of a matrix exponential:

$$\mathbf{P}\left(z' + h, z'\right) = \exp\left(ik_0 h \mathbf{\Delta}\right) \equiv \mathbf{U} + \sum_{j=1}^{\infty} \frac{\left(ik_0 h\right)^j}{j!} \mathbf{\Delta}^j, \tag{8.131}$$

where $h \equiv z'' - z'$ is the thickness of the layer under consideration. There are many mathematical methods for computing matrix exponentials (for a review, see [32]). Some of these methods can be used in the case in question. Here we consider those of them that have already been used in modeling LCDs, having shown themselves efficient; all these methods provide exact closed-form expressions for the transfer matrix.

Eigenvector Method

If the fields in the layer can be decomposed into eigenwaves and if a set of σ_j and $\boldsymbol{\psi}_j$ (j = 1, 2, 3, 4), eigenvalues and eigenvectors of the matrix $\mathbf{\Delta}$, represents an eigenwave basis (see Section 8.1.3), the matrix $\mathbf{P}(z' + h, z')$ can be calculated as follows:

$$\mathbf{P}\left(z' + h,\ z'\right) = \mathbf{\Psi}\mathbf{C}\mathbf{\Psi}^{-1}, \tag{8.132}$$

where

$$\mathbf{\Psi} = \left(\boldsymbol{\psi}_1 \quad \boldsymbol{\psi}_2 \quad \boldsymbol{\psi}_3 \quad \boldsymbol{\psi}_4\right),$$
$$\mathbf{C} = \begin{pmatrix} \exp\left(ik_0\sigma_1 h\right) & 0 & 0 & 0 \\ 0 & \exp\left(ik_0\sigma_2 h\right) & 0 & 0 \\ 0 & 0 & \exp\left(ik_0\sigma_3 h\right) & 0 \\ 0 & 0 & 0 & \exp\left(ik_0\sigma_4 h\right) \end{pmatrix}. \tag{8.133}$$

That this representation is possible is evident from discussion in Section 8.1.3.

Methods Employing a Third-Degree Polynomial in Δ

It follows from the Cayley–Hamilton theorem of matrix algebra that the matrix exponential in (8.131) can be expressed in the form of a third-degree polynomial in $\mathbf{\Delta}$. Several techniques for calculation of Berreman transfer matrices using this possibility were proposed [33–35]. Here are three basic algebraic expressions for the Berreman matrix of a homogeneous layer used in these techniques.

1. Explicit Cayley–Hamilton formula [33]:

$$\mathbf{P}(z' + h,\ z') = \beta_0\mathbf{U} + \beta_1\mathbf{\Delta} + \beta_2\mathbf{\Delta}^2 + \beta_3\mathbf{\Delta}^3, \tag{8.134}$$

where

$$\beta_0 = -\sum_{n=1}^{4} \sigma_j\sigma_k\sigma_l \bar{f}_n,\ \beta_1 = \sum_{n=1}^{4} \left(\sigma_j\sigma_k + \sigma_j\sigma_l + \sigma_k\sigma_l\right)\bar{f}_n,$$
$$\beta_2 = -\sum_{n=1}^{4} \left(\sigma_j + \sigma_k + \sigma_l\right)\bar{f}_n,\ \beta_3 = \sum_{n=1}^{4} \bar{f}_n, \tag{8.135}$$
$$\bar{f}_n = \frac{\exp\left(ik_0\sigma_n h\right)}{(\sigma_n - \sigma_j)(\sigma_n - \sigma_k)(\sigma_n - \sigma_l)},$$
$$n,\ j,\ k,\ l = 1, 2, 3, 4; \quad i \neq j \neq k \neq l.$$

2. Lagrange-interpolation/Sylvester-theorem formula [34, 36]:

$$\mathbf{P}\left(z'+h,\ z'\right)=\sum_{n=1}^{4}\exp\left(ik_0\sigma_n h\right)\prod_{j\neq n}\frac{\left(\boldsymbol{\Delta}-\sigma_j\mathbf{U}\right)}{\left(\sigma_n-\sigma_j\right)}. \tag{8.136}$$

3. Newton-interpolation/Newton–Putzler technique formula [32, 35]:

$$\mathbf{P}(z'+h,\ z')=f_1+l_{12}\boldsymbol{\Delta}_1+l_{123}\boldsymbol{\Delta}_{12}+l_{1234}\boldsymbol{\Delta}_{123}, \tag{8.137}$$

where

$$\begin{gathered}\boldsymbol{\Delta}_{12}=\boldsymbol{\Delta}_1\boldsymbol{\Delta}_2,\quad \boldsymbol{\Delta}_{123}=\boldsymbol{\Delta}_{12}\boldsymbol{\Delta}_3,\\ \boldsymbol{\Delta}_1=(\boldsymbol{\Delta}-\sigma_1\mathbf{U}),\quad \boldsymbol{\Delta}_2=(\boldsymbol{\Delta}-\sigma_2\mathbf{U}),\quad \boldsymbol{\Delta}_3=(\boldsymbol{\Delta}-\sigma_3\mathbf{U}),\\ l_{1234}=(l_{123}-l_{234})/(\sigma_1-\sigma_4),\\ l_{123}=(l_{12}-l_{23})/(\sigma_1-\sigma_3),\quad l_{234}=(l_{23}-l_{34})/(\sigma_2-\sigma_4),\\ l_{12}=(f_1-f_2)/(\sigma_1-\sigma_2),\quad l_{23}=(f_2-f_3)/(\sigma_2-\sigma_3),\quad l_{34}=(f_3-f_4)/(\sigma_3-\sigma_4),\\ f_n=\exp(ik_0\sigma_n h)\qquad n=1,2,3,4.\end{gathered} \tag{8.138}$$

Formulas (8.135), (8.136), and (8.138) can be used for computations only when all the eigenvalues σ_j are distinct; therefore they cannot be directly applied, for example, for isotropic media or for anisotropic media in the presence of polarization degeneracy (see Section 8.1.3). However, from these formulas, using L'Hôspital's rule, one may derive special computational formulas for different cases of coincidence of σ_j. Thus, for example, in the case $\sigma_1=\sigma_2$, l_{12} in (8.137)–(8.138) can be expressed as

$$l_{12}=ik_0h\ \exp(ik_0\sigma_1 h).$$

Furthermore, explicit expressions for the matrix exponentials of 4×4 matrices with coincident eigenvalues are known [37], which can easily be adapted for calculations in such cases, including the case of anomalous singularities (see Section 8.1.3). Therefore, the polynomial representation can be considered absolutely universal.

Due to their generality and efficiency, the polynomial techniques are very popular among LCD modelers. However, we should note that commonly, when used for calculating transfer matrices for LC layers, optimized techniques based on the eigenwave representation (one of them is described in Section 8.8.1) are not inferior to those based on the polynomial representation of Berreman matrices in computational efficiency.

Explicit Expression for the Transfer Matrix of an Isotropic Layer

In the coordinate system (x, y, z), whose xz-plane is parallel to the plane of incidence (Figure 8.4b), the differential propagation matrix of an isotropic layer is

$$\boldsymbol{\Delta}=\begin{pmatrix}0 & 1-\dfrac{\zeta^2}{n^2} & 0 & 0\\ n^2 & 0 & 0 & 0\\ 0 & 0 & 0 & 1\\ 0 & 0 & n^2-\zeta^2 & 0\end{pmatrix}_{xyz}, \tag{8.139}$$

where n is the complex refractive index of the medium. In this case, by using any of the above representations, the following simple expression for the transfer matrix of the layer can be obtained:

$$\mathbf{P}(z'+h,\ z') = \begin{pmatrix} p_c & \dfrac{\sigma_1 p_s}{n^2} & 0 & 0 \\ \dfrac{n^2 p_s}{\sigma_1} & p_c & 0 & 0 \\ 0 & 0 & p_c & \dfrac{p_s}{\sigma_1} \\ 0 & 0 & \sigma_1 p_s & p_c \end{pmatrix}, \tag{8.140}$$

where

$$\sigma_1 = \sqrt{n^2 - \zeta^2},$$

$$p_c = \frac{\exp\left(ik_0\sigma_1 h\right) + \exp\left(-ik_0\sigma_1 h\right)}{2}, \quad p_s = \frac{\exp\left(ik_0\sigma_1 h\right) - \exp\left(-ik_0\sigma_1 h\right)}{2}.$$

If the medium is nonabsorbing, the last two parameters may be expressed as

$$p_c = \cos\left(k_0\sigma_1 h\right), \quad p_s = i\sin\left(k_0\sigma_1 h\right).$$

8.3.3 Transfer Matrix of a Smoothly Inhomogeneous Layer. Staircase Approximation

The basic model of an inhomogeneous liquid crystal layer in LCD modeling is a medium whose local optical parameters smoothly vary with position. Thus, in 1D models, the permittivity tensor ε of the medium being a model of the LC layer is usually considered as a continuous function of the longitudinal coordinate (the z-coordinate in our case). Many numerical techniques—common finite difference techniques, such as the Runge–Kutta methods, and techniques that were specially developed for the problem in question—have been examined to find a reliable and efficient method for calculating the Berreman transfer matrices for inhomogeneous LC layers [2, 16, 38]. In this section, we consider only one technique for calculation of transfer matrices of smoothly inhomogeneous layers, which is the most common, if not the only, one in use now in LCD modeling. In this technique, the 1D-inhomogeneous medium with smoothly varying parameters, which is regarded as the "true" model of an actual layer, is divided into slices so thin that variations of ε within each slice are very small. Then, neglecting these variations, the transfer matrix for the "true" model is calculated as for a system of homogeneous layers. In other words, using this approach, we approximate the transfer matrix of the smoothly inhomogeneous layer by the transfer matrix of a system of homogeneous layers, whose local optical parameters are close to those of the "true" model but change stepwise. This kind of approximation is sometimes called *the staircase approximation*. We will call the approximating layered medium involved in this approximation *the staircase model* of the smoothly inhomogeneous layer, regardless of in what context this model is used (in this book, staircase models are employed within different methods). The validity of the staircase approximation is evident from a physical point of view. Here we will discuss the staircase approximation technique as a mathematical method for integrating (8.129).

According to (8.130), the Berreman transfer matrix of a region $(z_K - h_K/2,\ z_K + h_K/2)$ of a smoothly inhomogeneous medium may be expressed as follows:

$$\begin{aligned}\mathbf{P}\left(z_K + \frac{h_K}{2}, z_K - \frac{h_K}{2}\right) = \mathbf{U} &+ ik_0 \int\limits_{z_K-\frac{h_K}{2}}^{z_K+\frac{h_K}{2}} \mathbf{\Delta}(z_1)dz_1 + (ik_0)^2 \int\limits_{z_K-\frac{h_K}{2}}^{z_K+\frac{h_K}{2}} \mathbf{\Delta}(z_1)dz_1 \int\limits_{z_K-\frac{h_K}{2}}^{z_1} \mathbf{\Delta}(z_2)dz_2 \\ &+ (ik_0)^3 \int\limits_{z_K-\frac{h_K}{2}}^{z_K+\frac{h_K}{2}} \mathbf{\Delta}(z_1)dz_1 \int\limits_{z_K-\frac{h_K}{2}}^{z_1} \mathbf{\Delta}(z_2)dz_2 \int\limits_{z_K-\frac{h_K}{2}}^{z_2} \mathbf{\Delta}(z_3)dz_3 + \ldots\end{aligned} \tag{8.141}$$

Writing the function $\boldsymbol{\Delta}(z)$ in the Taylor-series form

$$\boldsymbol{\Delta}(z) = \boldsymbol{\Delta}(z_K) + \sum_{n=1}^{\infty} \boldsymbol{\Delta}^{(n)}(z_K)\frac{(z - z_K)^n}{n!}, \tag{8.142}$$

where $\boldsymbol{\Delta}^{(n)} = \mathrm{d}^n\boldsymbol{\Delta}/\mathrm{d}z^n$, and substituting (8.142) into (8.141), we find that

$$\mathbf{P}\left(z_K + \frac{h_K}{2}, z_K - \frac{h_K}{2}\right) = \tilde{\mathbf{P}}\left(\boldsymbol{\Delta}(z_K), h_K\right) + O(h_K^3) \quad \text{as } h_K \to 0, \tag{8.143}$$

where

$$\tilde{\mathbf{P}}(\boldsymbol{\Delta}, h) \equiv \mathbf{U} + \sum_{n=1}^{\infty} \frac{(ik_0 h)^n}{n!}\boldsymbol{\Delta}^n. \tag{8.144}$$

Here and in what follows, in matrix formulas, the asymptotic estimate expressed by the term $O(\cdot)$ (big O) is applied to all elements of the matrix being estimated. It is easy to see that the matrix $\tilde{\mathbf{P}}\left(\boldsymbol{\Delta}(z_K), h_K\right)$ is equal to the Berreman transfer matrix of a homogeneous layer of thickness h_K whose $\boldsymbol{\Delta}$ is equal to the $\boldsymbol{\Delta}$ of the inhomogeneous medium at $z = z_K$. Let M_C be a smoothly inhomogeneous medium of thickness d, bounded by planes $z = z'$ and $z = z'' \equiv z' + d$. Denote the permittivity tensor of this medium by $\boldsymbol{\varepsilon}_\mathrm{C}$. Divide the medium M_C into slices by the planes $z = z_{Ij}$, where $z_{Ij} = z' + jd/N$ and $j = 1, 2, \ldots, N-1$. Each slice has thickness $h = d/N$. The midplanes of the slices coincide with the planes $z = z_j$, where

$$z_j = \left(j - \frac{1}{2}\right)\frac{d}{N}. \tag{8.145}$$

According to (8.127), we may write the Berreman transfer matrix of the whole medium M_C as follows

$$\mathbf{P}_\mathrm{C}(z'', z') = \mathbf{P}_\mathrm{C}(z'', z_{IN-1})\mathbf{P}_\mathrm{C}(z_{IN-1}, z_{IN-2}) \cdot \ldots \cdot \mathbf{P}_\mathrm{C}(z_{I2}, z_{I1})\mathbf{P}_\mathrm{C}(z_{I1}, z'), \tag{8.146}$$

where the label C denotes that a matrix pertains to the medium M_C. Now, we introduce a staircase model M_{AN}—let $\boldsymbol{\varepsilon}_\mathrm{A}$ denote the permittivity tensor of this medium—so that $\boldsymbol{\varepsilon}_\mathrm{A}(z) = \boldsymbol{\varepsilon}_\mathrm{C}(z_j)$ for all z lying between z_{Ij-1} and z_{Ij} ($j = 1, 2, \ldots, N$), that is, so that the permittivity tensor within the jth slice of the medium M_{AN} is equal to that of the medium M_C in the midplane of its jth slice (see Figure 8.17). In view of this, by making use of (8.144), we may express the transfer matrix of the medium M_{AN}, $\mathbf{P}_{AN}(z'', z')$, as follows:

$$\mathbf{P}_{AN}(z'', z') = \tilde{\mathbf{P}}\left(\boldsymbol{\Delta}_\mathrm{C}(z_N), h\right)\tilde{\mathbf{P}}\left(\boldsymbol{\Delta}_\mathrm{C}(z_{N-1}), h\right) \cdot \ldots \cdot \tilde{\mathbf{P}}\left(\boldsymbol{\Delta}_\mathrm{C}(z_2), h\right)\tilde{\mathbf{P}}\left(\boldsymbol{\Delta}_\mathrm{C}(z_1), h\right), \tag{8.147}$$

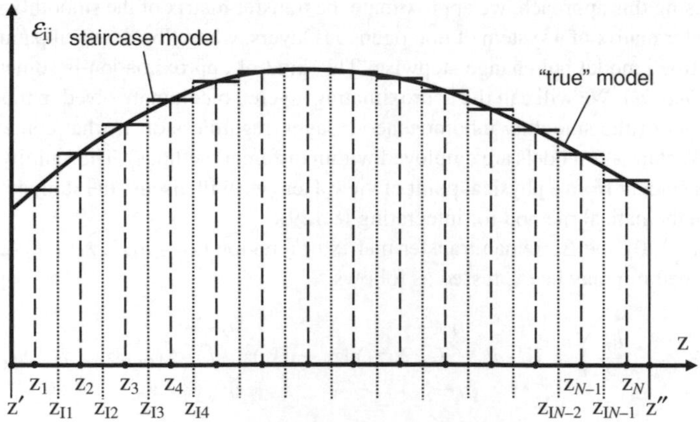

Figure 8.17 Staircase approximation

where $\mathbf{\Delta}_\mathrm{C}$ denotes $\mathbf{\Delta}$ for the medium M_C. On the other hand, according to (8.143), we have

$$\mathbf{P}_\mathrm{C}\left(\mathrm{z}_{\mathrm{I}j}, \mathrm{z}_{\mathrm{I}j-1}\right) = \tilde{\mathbf{P}}\left(\mathbf{\Delta}_\mathrm{C}(\mathrm{z}_j), h\right) + O(h^3). \tag{8.148}$$

Substituting (8.148) into (8.146) and comparing the expression thus obtained with (8.147), considering that $h = d/N$, we find that

$$\mathbf{P}_\mathrm{C}(\mathrm{z}'', \mathrm{z}') = \mathbf{P}_{\mathrm{A}N}(\mathrm{z}'', \mathrm{z}') + O\left(\frac{1}{N^2}\right) \quad \text{as } N \to \infty. \tag{8.149}$$

We see that, as N tends to infinity, the transfer matrix of the staircase model $\mathsf{M}_{\mathrm{A}N}$ does approach the transfer matrix of the smoothly inhomogeneous medium M_C, and rather quickly. According to (8.149), for sufficiently large values of N, doubling N increases the accuracy of the approximation

$$\mathbf{P}_\mathrm{C}(\mathrm{z}'', \mathrm{z}') \approx \mathbf{P}_{\mathrm{A}N}(\mathrm{z}'', \mathrm{z}') \tag{8.150}$$

by a factor of about four. Numerical estimates of the accuracy of different techniques which use the staircase approximation for LC layers will be given in Chapter 11.

8.3.4 *Coordinate Systems*

In Berreman's papers [2, 16], one of the axes of the coordinate system for representation of the field vectors is perpendicular to the plane of incidence. The use of such a coordinate system simplifies the calculations and final computational formulas. For example, in the system (x, y, z), which was introduced (see Section 8.1.3 and Figure 8.4 therein) in the same manner as in [2, 16] (only, for our later purposes, we imposed an additional requirement that the positive x-axis must be in the direction of **b**), the matrix $\mathbf{\Delta}$ has six elements equal to zero:

$$\mathbf{\Delta} = \begin{pmatrix} \Delta_{11} & \Delta_{12} & \Delta_{13} & 0 \\ \Delta_{21} & \Delta_{11} & \Delta_{23} & 0 \\ 0 & 0 & 0 & \Delta_{34} \\ \Delta_{23} & \Delta_{13} & \Delta_{43} & 0 \end{pmatrix}_{xyz}, \tag{8.151}$$

where

$$\Delta_{11} = -\zeta\frac{\varepsilon_{xz}}{\varepsilon_{zz}}, \quad \Delta_{12} = 1 - \frac{\zeta^2}{\varepsilon_{zz}}, \quad \Delta_{13} = -\zeta\frac{\varepsilon_{yz}}{\varepsilon_{zz}},$$

$$\Delta_{21} = \varepsilon_{xx} - \frac{\varepsilon_{xz}^2}{\varepsilon_{zz}}, \quad \Delta_{23} = \varepsilon_{xy} - \frac{\varepsilon_{xz}\varepsilon_{yz}}{\varepsilon_{zz}},$$

$$\Delta_{34} = 1, \quad \Delta_{43} = \varepsilon_{yy} - \frac{\varepsilon_{yz}^2}{\varepsilon_{zz}} - \zeta^2$$

with ε_{xx}, ε_{xy}, ε_{xz}, ε_{yy} ε_{yz}, and ε_{zz} being components of the tensor ε referred to the system (x, y, z). In our consideration, we regard the coordinate system (x, y, z) only as a possible variant of the coordinate system for representation of Berreman vectors. Thus, in Section 8.1.3, we did not attach the coordinate system $(\mathrm{x}, \mathrm{y}, \mathrm{z})$ to the plane of incidence. Such a choice will be convenient in Sections 8.4.1 and 8.6.2. Formulas presented here help to transfer relations and estimates for vectors and matrix operators of Berreman's formalism obtained with the system $(\mathrm{x}, \mathrm{y}, \mathrm{z})$ to the system (x, y, z) and vice versa.

Suppose that the system (x, y, z) may be obtained from a system $(\mathrm{x}, \mathrm{y}, \mathrm{z})$ by rotating the latter through an angle $\alpha_{\mathrm{x}\to x}$ about the z-axis (Figure 8.18). Then the relationship between the column Ψ_{xy} representing

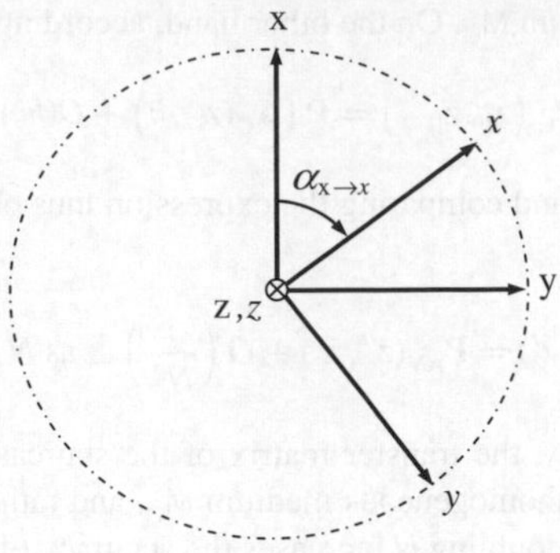

Figure 8.18 To relations (8.152)–(8.159). Coordinate systems under consideration

a Berreman vector Ψ in the system (x, y, z) and the column Ψ_{xy} representing the same Berreman vector in the system (*x*, *y*, *z*) can be expressed as follows:

$$\Psi_{xy} = \mathbf{R}_{\mathrm{BR}}(\alpha_{\mathrm{x}\to x})\Psi_{\mathrm{xy}}, \tag{8.152a}$$

$$\Psi_{\mathrm{xy}} = \mathbf{R}_{\mathrm{BR}}(-\alpha_{\mathrm{x}\to x})\Psi_{xy}, \tag{8.152b}$$

where

$$\mathbf{R}_{\mathrm{BR}}(\alpha) = \begin{pmatrix} \cos\alpha & 0 & \sin\alpha & 0 \\ 0 & \cos\alpha & 0 & \sin\alpha \\ -\sin\alpha & 0 & \cos\alpha & 0 \\ 0 & -\sin\alpha & 0 & \cos\alpha \end{pmatrix}.$$

Similarly, using the same notation, we may write

$$\psi_{jxy}\mathbf{R}_{\mathrm{BR}}(\alpha_{\mathrm{x}\to x})\psi_{j\mathrm{xy}}, \quad \psi_{j\mathrm{xy}} = \mathbf{R}_{\mathrm{BR}}(-\alpha_{\mathrm{x}\to x})\psi_{jxy} \quad j = 1, 2, 3, 4; \tag{8.153}$$

$$\boldsymbol{\Psi}_{xy} = \mathbf{R}_{\mathrm{BR}}(\alpha_{\mathrm{x}\to x})\,\boldsymbol{\Psi}_{\mathrm{xy}}, \quad \boldsymbol{\Psi}_{\mathrm{xy}} = \mathbf{R}_{\mathrm{BR}}(-\alpha_{\mathrm{x}\to x})\,\boldsymbol{\Psi}_{xy}. \tag{8.154}$$

Inverting (8.154) gives

$$\boldsymbol{\Psi}_{xy}^{-1} = \boldsymbol{\Psi}_{\mathrm{xy}}^{-1}\mathbf{R}_{\mathrm{BR}}(-\alpha_{\mathrm{x}\to x}), \quad \boldsymbol{\Psi}_{\mathrm{xy}}^{-1} = \boldsymbol{\Psi}_{xy}^{-1}\mathbf{R}_{\mathrm{BR}}(\alpha_{\mathrm{x}\to x}). \tag{8.155}$$

With the help of (8.152), the following relations for differential propagation matrices and transfer matrices of Berreman's formalism can readily be obtained:

$$\boldsymbol{\Delta}_{xy} = \mathbf{R}_{\mathrm{BR}}(\alpha_{\mathrm{x}\to x})\boldsymbol{\Delta}_{\mathrm{xy}}\mathbf{R}_{\mathrm{BR}}(-\alpha_{\mathrm{x}\to x}), \tag{8.156}$$

$$\mathbf{P}_{xy} = \mathbf{R}_{\mathrm{BR}}(\alpha_{\mathrm{x}\to x})\mathbf{P}_{\mathrm{xy}}\mathbf{R}_{\mathrm{BR}}(-\alpha_{\mathrm{x}\to x}), \tag{8.157}$$

and, conversely,

$$\boldsymbol{\Delta}_{\mathrm{xy}} = \mathbf{R}_{\mathrm{BR}}(-\alpha_{\mathrm{x}\to x})\boldsymbol{\Delta}_{xy}\mathbf{R}_{\mathrm{BR}}(\alpha_{\mathrm{x}\to x}), \tag{8.158}$$

$$\mathbf{P}_{\mathrm{xy}} = \mathbf{R}_{\mathrm{BR}}(-\alpha_{\mathrm{x}\to x})\mathbf{P}_{xy}\mathbf{R}_{\mathrm{BR}}(\alpha_{\mathrm{x}\to x}). \tag{8.159}$$

8.4 Simplifications, Useful Relations, and Advanced Techniques

8.4.1 *Orthogonality Relations and Other Useful Relations for Eigenwave Bases*

In this section, we consider some specific geometrical properties of eigenwave bases determined by the nature of the medium. As we will see in this and subsequent chapters, knowledge of these properties helps in solving many problems.

Orthogonality Relations

One of the problems that we face dealing with eigenwave representation is inverting the matrix $\boldsymbol{\Psi}$. We see the matrix $\boldsymbol{\Psi}^{-1}$ in many key formulas. The problem of calculating $\boldsymbol{\Psi}^{-1}$ is easily solved numerically by standard mathematical methods but is a big obstacle in analysis. Very simple and convenient expressions for the elements of $\boldsymbol{\Psi}^{-1}$ in terms of the elements of $\boldsymbol{\Psi}$ can be obtained by using specific orthogonality properties of eigenwave bases. For many optical media, the vectors $\boldsymbol{\psi}_j$ of the basis eigenwaves are mutually orthogonal in a certain sense (in the absence of polarization degeneracy) or can be chosen to be mutually orthogonal (in the presence of polarization degeneracy). Here we consider two kinds of orthogonality of eigenwave bases. The first kind implies the fulfillment of the following relations for the basis eigenwaves

$$\boldsymbol{\psi}_j^\dagger \mathbf{I}_0 \boldsymbol{\psi}_k \equiv \mathbf{z}(\mathbf{e}_j^* \times \mathbf{h}_k) + \mathbf{z}(\mathbf{e}_k \times \mathbf{h}_j^*) = 0 \quad k \neq j \tag{8.160}$$

[19, 21]; the matrix $\mathbf{I}_0$ is defined by (8.97). In the absence of polarization degeneracy, any eigenwave basis for a nonabsorbing medium out of TIR mode is orthogonal in the sense (8.160). The orthogonality of the second kind is expressed by the following relations

$$\boldsymbol{\psi}_j^{\mathrm{T}} \mathbf{I}_0 \boldsymbol{\psi}_k \equiv \mathbf{z}\left(\mathbf{e}_j \times \mathbf{h}_k\right) + \mathbf{z}\left(\mathbf{e}_k \times \mathbf{h}_j\right) = 0 \quad k \neq j \tag{8.161}$$

[22, 23]. In the absence of polarization degeneracy, relations (8.161) hold for any eigenwave basis in any reciprocal optically locally centrosymmetric medium [i.e., for any medium whose material tensors satisfy conditions (8.8) and (8.9)]. Recall that all optical materials that we deal with in LCD modeling belong to this class of media. We will derive relations (8.160) and (8.161) below. But first we will show how these relations may be used in calculating the matrix $\boldsymbol{\Psi}^{-1}$.

Let us denote the jth row of the matrix $\boldsymbol{\Psi}^{-1}$ by $\bar{\boldsymbol{\psi}}_j$. Then the problem of finding the matrix $\boldsymbol{\Psi}^{-1}$ can be considered as the problem of finding four row-vectors $\bar{\boldsymbol{\psi}}_j$ satisfying the condition

$$\bar{\boldsymbol{\psi}}_j \boldsymbol{\psi}_k = \delta_{jk}, \tag{8.162}$$

where δ_{jk} is the Kronecker delta. Turning back to relations (8.160), one can see that, subject to (8.160), the vectors $\bar{\boldsymbol{\psi}}_j$ may be expressed as follows:

$$\bar{\boldsymbol{\psi}}_j = \frac{1}{\boldsymbol{\psi}_j^\dagger \mathbf{I}_0 \boldsymbol{\psi}_j} \boldsymbol{\psi}_j^\dagger \mathbf{I}_0 = \frac{1}{2\mathrm{Re}\left(\psi_{1j}^* \psi_{2j} + \psi_{3j}^* \psi_{4j}\right)} \begin{pmatrix} \psi_{2j}^* & \psi_{1j}^* & \psi_{4j}^* & \psi_{3j}^* \end{pmatrix}, \quad j = 1,\ 2,\ 3,\ 4, \tag{8.163}$$

where ψ_{kj} are elements of the matrix $\boldsymbol{\Psi}$. If the eigenwave basis is normalized so that

$$\boldsymbol{\psi}_j^\dagger \mathbf{I}_0 \boldsymbol{\psi}_j \equiv 2\mathrm{Re}\left[\mathbf{z}(\mathbf{e}_j^* \times \mathbf{h}_j)\right] = \begin{cases} 1 & j = 1, 2 \\ -1 & j = 3, 4, \end{cases} \tag{8.164}$$

using (8.163), we may express the matrix $\mathbf{\Psi}^{-1}$ in terms of the elements of the matrix $\mathbf{\Psi}$ as follows:

$$\mathbf{\Psi}^{-1} = \begin{pmatrix} \psi_{21}^* & \psi_{11}^* & \psi_{41}^* & \psi_{31}^* \\ \psi_{22}^* & \psi_{12}^* & \psi_{42}^* & \psi_{32}^* \\ -\psi_{23}^* & -\psi_{13}^* & -\psi_{43}^* & -\psi_{33}^* \\ -\psi_{24}^* & -\psi_{14}^* & -\psi_{44}^* & -\psi_{34}^* \end{pmatrix}. \tag{8.165}$$

We will call normalization (8.164) the *flux normalization* or *F-normalization* [see also (8.96)].

Relation (8.161) suggests another representation of the rows $\bar{\psi}_j$, namely,

$$\bar{\psi}_j = \frac{1}{\psi_j^{\mathrm{T}} \mathbf{I}_0 \psi_j} \psi_j^{\mathrm{T}} \mathbf{I}_0 = \frac{1}{2\left(\psi_{1j}\psi_{2j} + \psi_{3j}\psi_{4j}\right)} \left(\psi_{2j} \quad \psi_{1j} \quad \psi_{4j} \quad \psi_{3j}\right), \quad j = 1,\, 2,\, 3,\, 4. \tag{8.166}$$

Using (8.166) together with the following normalization

$$\psi_j^{\mathrm{T}} \mathbf{I}_0 \psi_j \equiv 2\mathbf{z}\left(\mathbf{e}_j \times \mathbf{h}_j\right) = \begin{cases} 1 & j = 1,2 \\ -1 & j = 3,4, \end{cases} \tag{8.167}$$

we obtain the following simplest expression for the matrix $\mathbf{\Psi}^{-1}$:

$$\mathbf{\Psi}^{-1} = \begin{pmatrix} \psi_{21} & \psi_{11} & \psi_{41} & \psi_{31} \\ \psi_{22} & \psi_{12} & \psi_{42} & \psi_{32} \\ -\psi_{23} & -\psi_{13} & -\psi_{43} & -\psi_{33} \\ -\psi_{24} & -\psi_{14} & -\psi_{44} & -\psi_{34} \end{pmatrix}. \tag{8.168}$$

In terms of the vibration vectors $\mathbf{e}_j$ and $\mathbf{h}_j$, this expression can be written as follows

$$\mathbf{\Psi}^{-1} = \begin{pmatrix} \mathbf{h}_1 \cdot \mathbf{y} & \mathbf{e}_1 \cdot \mathbf{x} & -\mathbf{h}_1 \cdot \mathbf{x} & \mathbf{e}_1 \cdot \mathbf{y} \\ \mathbf{h}_2 \cdot \mathbf{y} & \mathbf{e}_2 \cdot \mathbf{x} & -\mathbf{h}_2 \cdot \mathbf{x} & \mathbf{e}_2 \cdot \mathbf{y} \\ -\mathbf{h}_3 \cdot \mathbf{y} & -\mathbf{e}_3 \cdot \mathbf{x} & \mathbf{h}_3 \cdot \mathbf{x} & -\mathbf{e}_3 \cdot \mathbf{y} \\ -\mathbf{h}_4 \cdot \mathbf{y} & -\mathbf{e}_4 \cdot \mathbf{x} & \mathbf{h}_4 \cdot \mathbf{x} & -\mathbf{e}_4 \cdot \mathbf{y} \end{pmatrix}. \tag{8.169}$$

The normalization (8.167) will be referred to as the *symmetrical normalization* or *S-normalization*. In our programs, to calculate $\mathbf{\Psi}^{-1}$, we use only formulas (8.166) and (8.168), because these formulas, being applicable to absorbing media and in the case of TIR mode, are more general for our purposes than (8.163) and (8.165). But we, of course, should mention that in the case of a nonabsorbing medium out of TIR mode, the eigenwave basis can be chosen so that all the vectors ψ_j will be real. Dealing with such "real" bases, there is no sense in distinguishing between relations (8.160) and (8.161), between normalizations (8.164) and (8.167), and between expressions [(8.163), (8.165)] and [(8.166), (8.168)] in view of identity of their implications in this case.

Derivation of Relations (8.161)

In Reference 22, relations (8.161) were derived with consideration of the general constitutive relations (8.5). Here, we will derive them in a simpler way, starting from the simpler constitutive relations (8.6) which are most important for our purposes. Let us consider two arbitrary eigenwaves, a wave a and a wave b, in a medium with a given symmetrical permittivity tensor ε [see (8.7)]. Denote the vectors (**m**, **e**, **h**) of the waves a and b respectively ($\mathbf{m}_a$, $\mathbf{e}_a$, $\mathbf{h}_a$) and ($\mathbf{m}_b$, $\mathbf{e}_b$, $\mathbf{h}_b$). According to (8.40)–(8.42), these vectors must satisfy the equations

$$\mathbf{m}_a \times \mathbf{h}_a = -\varepsilon \mathbf{e}_a, \tag{8.170}$$

$$\mathbf{m}_a \times \mathbf{e}_a = \mathbf{h}_a, \tag{8.171}$$

$$\mathbf{m}_b \times \mathbf{h}_b = -\varepsilon \mathbf{e}_b, \tag{8.172}$$

$$\mathbf{m}_b \times \mathbf{e}_b = \mathbf{h}_b. \tag{8.173}$$

By multiplying (8.170), (8.171), (8.172), and (8.173) scalarly by $\mathbf{e}_b$, $\mathbf{h}_b$, $\mathbf{e}_a$, and $\mathbf{h}_a$, respectively, we obtain

$$\begin{aligned} \mathbf{e}_b(\mathbf{m}_a \times \mathbf{h}_a) &= -\mathbf{e}_b \varepsilon \mathbf{e}_a, \quad \mathbf{h}_b(\mathbf{m}_a \times \mathbf{e}_a) = \mathbf{h}_b \mathbf{h}_a, \\ \mathbf{e}_a(\mathbf{m}_b \times \mathbf{h}_b) &= -\mathbf{e}_a \varepsilon \mathbf{e}_b, \quad \mathbf{h}_a(\mathbf{m}_b \times \mathbf{e}_b) = \mathbf{h}_a \mathbf{h}_b. \end{aligned} \tag{8.174}$$

Using the vector identities

$$\mathbf{a}(\mathbf{b} \times \mathbf{c}) = \mathbf{b}(\mathbf{c} \times \mathbf{a}) = \mathbf{c}(\mathbf{a} \times \mathbf{b}) = -\mathbf{a}(\mathbf{c} \times \mathbf{b}) = -\mathbf{b}(\mathbf{a} \times \mathbf{c}) = -\mathbf{c}(\mathbf{b} \times \mathbf{a}), \tag{8.175}$$

we may rewrite (8.174) as follows:

$$-\mathbf{m}_a(\mathbf{e}_b \times \mathbf{h}_a) = -\mathbf{e}_b \varepsilon \mathbf{e}_a, \tag{8.176}$$

$$\mathbf{m}_a(\mathbf{e}_a \times \mathbf{h}_b) = \mathbf{h}_b \mathbf{h}_a, \tag{8.177}$$

$$\mathbf{m}_b(\mathbf{e}_a \times \mathbf{h}_b) = \mathbf{e}_a \varepsilon \mathbf{e}_b, \tag{8.178}$$

$$-\mathbf{m}_b(\mathbf{e}_b \times \mathbf{h}_a) = -\mathbf{h}_a \mathbf{h}_b. \tag{8.179}$$

On adding (8.176) and (8.178) we obtain

$$-\mathbf{m}_a(\mathbf{e}_b \times \mathbf{h}_a) + \mathbf{m}_b(\mathbf{e}_a \times \mathbf{h}_b) = 0, \tag{8.180}$$

where the fact that for a symmetrical ε ($\varepsilon = \varepsilon^{\mathrm{T}}$) $\mathbf{e}_b \varepsilon \mathbf{e}_a = \mathbf{e}_a \varepsilon \mathbf{e}_b$ has been used. Adding (8.177) and (8.179) gives

$$\mathbf{m}_a(\mathbf{e}_a \times \mathbf{h}_b) - \mathbf{m}_b(\mathbf{e}_b \times \mathbf{h}_a) = 0. \tag{8.181}$$

Finally, by subtracting (8.181) from (8.180) we obtain

$$\left(\mathbf{m}_b - \mathbf{m}_a\right) \left[(\mathbf{e}_a \times \mathbf{h}_b) + (\mathbf{e}_b \times \mathbf{h}_a)\right] = 0. \tag{8.182}$$

Let us consider any two basis waves of an eigenwave basis, say, jth and kth ones. In view of the fact that for these waves

$$\mathbf{m}_j = \mathbf{b} + \mathbf{z}\sigma_j, \quad \mathbf{m}_k = \mathbf{b} + \mathbf{z}\sigma_k, \tag{8.183}$$

application of (8.182) to these waves gives

$$\left(\sigma_k - \sigma_j\right) \mathbf{z} \left[(\mathbf{e}_j \times \mathbf{h}_k) + (\mathbf{e}_k \times \mathbf{h}_j)\right] = 0. \tag{8.184}$$

Hence, at $\sigma_k \neq \sigma_j$, the relation

$$\mathbf{z}\left[(\mathbf{e}_j \times \mathbf{h}_k) + (\mathbf{e}_k \times \mathbf{h}_j)\right] = 0 \tag{8.185}$$

is valid. Relation (8.185) is identical to relation (8.161).

Derivation of Relations (8.160)

In principle, we could derive relations (8.160) in the same manner as relations (8.161). But we will use another way in order to demonstrate some useful resources of the Berreman formalism and show once again the interrelation of this formalism with the eigenwave method. In this derivation, we will exploit the fact that the matrix $\mathbf{\Delta}$ of an arbitrary nonabsorbing medium satisfies the relation

$$\mathbf{\Delta} = \mathbf{I}_0 \mathbf{\Delta}^{\dagger} \mathbf{I}_0 \tag{8.186}$$

[19, 21]. Another form of this relation is

$$\mathbf{I}_0 \mathbf{\Delta} = \mathbf{\Delta}^{\dagger} \mathbf{I}_0 \tag{8.187}$$

($\mathbf{I}_0\mathbf{I}_0 = \mathbf{U}$). For an arbitrary pair of eigenvectors of the matrix $\mathbf{\Delta}$, say, $\boldsymbol{\psi}_j$ and $\boldsymbol{\psi}_k$, according to (8.90), we may write

$$\mathbf{\Delta}\boldsymbol{\psi}_j = \sigma_j \boldsymbol{\psi}_j, \tag{8.188}$$

$$\mathbf{\Delta}\boldsymbol{\psi}_k = \sigma_k \boldsymbol{\psi}_k. \tag{8.189}$$

Premultiply (8.188) and (8.189) respectively by the row-vectors $\boldsymbol{\psi}_k^{\dagger}\mathbf{I}_0$ and $\boldsymbol{\psi}_j^{\dagger}\mathbf{I}_0$,

$$\boldsymbol{\psi}_k^{\dagger}\mathbf{I}_0\mathbf{\Delta}\boldsymbol{\psi}_j = \sigma_j \boldsymbol{\psi}_k^{\dagger}\mathbf{I}_0\boldsymbol{\psi}_j, \tag{8.190}$$

$$\boldsymbol{\psi}_j^{\dagger}\mathbf{I}_0\mathbf{\Delta}\boldsymbol{\psi}_k = \sigma_k \boldsymbol{\psi}_j^{\dagger}\mathbf{I}_0\boldsymbol{\psi}_k, \tag{8.191}$$

and take the Hermitian conjugate of the former equation:

$$\boldsymbol{\psi}_j^{\dagger}\mathbf{\Delta}^{\dagger}\mathbf{I}_0\boldsymbol{\psi}_k = \sigma_j^{*} \boldsymbol{\psi}_j^{\dagger}\mathbf{I}_0\boldsymbol{\psi}_k. \tag{8.192}$$

Subtracting (8.192) from (8.191), we obtain

$$\boldsymbol{\psi}_j^{\dagger}\left(\mathbf{I}_0\mathbf{\Delta} - \mathbf{\Delta}^{\dagger}\mathbf{I}_0\right)\boldsymbol{\psi}_k = (\sigma_k - \sigma_j^{*})\boldsymbol{\psi}_j^{\dagger}\mathbf{I}_0\boldsymbol{\psi}_k. \tag{8.193}$$

According to (8.187) the left-hand side of this equation is equal to zero and consequently

$$(\sigma_k - \sigma_j^{*})\boldsymbol{\psi}_j^{\dagger}\mathbf{I}_0\boldsymbol{\psi}_k = 0. \tag{8.194}$$

Out of TIR mode all σ_l are real, and therefore, according to (8.194), at $\sigma_k \neq \sigma_j$, we have

$$\boldsymbol{\psi}_j^{\dagger}\mathbf{I}_0\boldsymbol{\psi}_k = 0, \tag{8.195}$$

that is, we have arrived at relation (8.160). We could derive in a similar manner the relation (8.161), using the fact that with a symmetrical ε ($\varepsilon = \varepsilon^{\mathrm{T}}$) the matrix $\mathbf{\Delta}$ meets the condition

$$\mathbf{\Delta} = \mathbf{I}_0 \mathbf{\Delta}^{\mathrm{T}} \mathbf{I}_0. \tag{8.196}$$

The Case of Polarization Degeneracy. Optimal Basis

In the case of polarization degeneracy, thanks to (8.103), the vibration vectors of the basis waves with identical refraction vectors can always be chosen such that a desired orthogonality relation [(8.160), or (8.161), or both (in the case of a nonabsorbing medium)] for this pair of waves will be fulfilled. This choice is optimal, because it allows a simple inversion of the matrix $\mathbf{\Psi}$. We will call eigenwave bases satisfying relations (8.161) *optimal*. In Chapter 9, we will present variants of optimal bases for isotropic, uniaxial, and biaxial media. Here, we present some general relations simplifying the choice of the vibration vectors for optimal bases in the case of polarization degeneracy and some other allied relations needed in future discussions. According to (8.171) and (8.173), for an arbitrary pair of waves a and b, the vectors ($\mathbf{m}_a$, $\mathbf{e}_a$, $\mathbf{h}_a$) and ($\mathbf{m}_b$, $\mathbf{e}_b$, $\mathbf{h}_b$) satisfy the following relations:

$$\mathbf{z}\left(\mathbf{e}_b \times \mathbf{h}_a\right) + \mathbf{z}\left(\mathbf{e}_a \times \mathbf{h}_b\right) = \left(\mathbf{zm}_a + \mathbf{zm}_b\right)\left(\mathbf{e}_b\mathbf{e}_a\right) - \left(\mathbf{ze}_a\right)\left(\mathbf{e}_b\mathbf{m}_a\right) - \left(\mathbf{ze}_b\right)\left(\mathbf{e}_a\mathbf{m}_b\right), \quad (8.197)$$

$$\mathbf{z}\left(\mathbf{e}_b^* \times \mathbf{h}_a\right) + \mathbf{z}\left(\mathbf{e}_a \times \mathbf{h}_b^*\right) = \left(\mathbf{zm}_a + \mathbf{zm}_b^*\right)\left(\mathbf{e}_b^*\mathbf{e}_a\right) - \left(\mathbf{ze}_a\right)\left(\mathbf{e}_b^*\mathbf{m}_a\right) - \left(\mathbf{ze}_b^*\right)\left(\mathbf{e}_a\mathbf{m}_b^*\right), \quad (8.198)$$

$$\mathbf{z}\left(\mathbf{e}_b \times \mathbf{h}_a\right) - \mathbf{z}\left(\mathbf{e}_a \times \mathbf{h}_b\right) = \left(\mathbf{zm}_a - \mathbf{zm}_b\right)\left(\mathbf{e}_b\mathbf{e}_a\right) - \left(\mathbf{ze}_a\right)\left(\mathbf{e}_b\mathbf{m}_a\right) + \left(\mathbf{ze}_b\right)\left(\mathbf{e}_a\mathbf{m}_b\right). \quad (8.199)$$

Suppose that polarization degeneracy takes place for the *j*th and *k*th basis waves, that is,

$$\mathbf{m}_j = \mathbf{m}_k. \quad (8.200)$$

In the case of an isotropic or a uniaxial medium, for these waves the conditions

$$\mathbf{m}_j \cdot \mathbf{e}_j = 0 \text{ and } \mathbf{m}_k \cdot \mathbf{e}_k = 0 \quad (8.201)$$

must be satisfied. In view of (8.200) and (8.201), we can write relation (8.197) for these waves as follows:

$$\mathbf{z}\left(\mathbf{e}_j \times \mathbf{h}_k\right) + \mathbf{z}\left(\mathbf{e}_k \times \mathbf{h}_j\right) = 2\mathbf{zm}_j\left(\mathbf{e}_j\mathbf{e}_k\right). \quad (8.202)$$

Hence, in this case, relation (8.161) will be fulfilled for the *j*th and *k*th basis waves if we choose the vectors $\mathbf{e}_j$ and $\mathbf{e}_k$ such that

$$\mathbf{e}_j \cdot \mathbf{e}_k = 0. \quad (8.203)$$

If $\mathbf{m}_j$ and $\mathbf{m}_k$ are real (a nonabsorbing medium out of TIR mode), along with (8.201) the following relations are valid:

$$\mathbf{m}_j \cdot \mathbf{e}_j^* = 0, \ \mathbf{m}_k \cdot \mathbf{e}_k^* = 0. \quad (8.204)$$

Therefore, in this case, we may write relation (8.198) for the *j*th and *k*th basis waves as

$$\mathbf{z}(\mathbf{e}_j^* \times \mathbf{h}_k) + \mathbf{z}(\mathbf{e}_k \times \mathbf{h}_j^*) = 2\mathbf{zm}_j(\mathbf{e}_j^*\mathbf{e}_k). \quad (8.205)$$

We see that in this situation, the choice of $\mathbf{e}_j$ and $\mathbf{e}_k$ in accordance with the condition

$$\mathbf{e}_j^*\mathbf{e}_k = 0 \quad (8.206)$$

guarantees the fulfillment of relation (8.160) for these waves.

Some Consequences of Reciprocity and Symmetry of the Medium

Here we will consider the parameters of basis eigenwaves as functions of the vector **b**. Let us consider quadruples of eigenwaves corresponding to $\mathbf{b} = \tilde{\mathbf{b}}$, where $\tilde{\mathbf{b}}$ is any of the possible values of the vector **b**, and $\mathbf{b} = -\tilde{\mathbf{b}}$. One may notice that the coefficients of the equation (8.73) at $\mathbf{b} = \tilde{\mathbf{b}}$ and at $\mathbf{b} = -\tilde{\mathbf{b}}$ are related as follows:

$$d_0(-\tilde{\mathbf{b}}) = d_0(\tilde{\mathbf{b}}),\ d_1(-\tilde{\mathbf{b}}) = -d_1(\tilde{\mathbf{b}}),\ d_2(-\tilde{\mathbf{b}}) = d_2(\tilde{\mathbf{b}}),$$
$$d_3(-\tilde{\mathbf{b}}) = -d_3(\tilde{\mathbf{b}}),\ d_4(-\tilde{\mathbf{b}}) = d_4(\tilde{\mathbf{b}}).$$

This means that the roots of this equation for $\mathbf{b} = \tilde{\mathbf{b}}$ and $\mathbf{b} = -\tilde{\mathbf{b}}$ differ only in sign. Generally speaking, this is the case for any reciprocal medium [22]. Let us assign these roots to the basis eigenwaves so that the relations

$$\sigma_j(-\tilde{\mathbf{b}}) = -\sigma_k(\tilde{\mathbf{b}}) \tag{8.207}$$

and consequently

$$\mathbf{m}_j(-\tilde{\mathbf{b}}) = -\mathbf{m}_k(\tilde{\mathbf{b}})$$

will hold for the pairs

$$(j,\ k) = (1,\ 3),\ (2,\ 4),\ (3,\ 1),\ (4,\ 2). \tag{8.208}$$

Adding (8.180) and (8.181) gives

$$\left(\mathbf{m}_a + \mathbf{m}_b\right)\left(\mathbf{e}_a \times \mathbf{h}_b - \mathbf{e}_b \times \mathbf{h}_a\right) = 0. \tag{8.209}$$

From (8.209) it follows that for the bases under consideration

$$\left[\sigma_j(\tilde{\mathbf{b}}) + \sigma_k(-\tilde{\mathbf{b}})\right] \mathbf{z} \cdot \left[\mathbf{e}_j(\tilde{\mathbf{b}}) \times \mathbf{h}_k(-\tilde{\mathbf{b}}) - \mathbf{e}_k(-\tilde{\mathbf{b}}) \times \mathbf{h}_j(\tilde{\mathbf{b}})\right] = 0 \tag{8.210}$$

for any pair $(j,\ k)$ of the basis waves. We should note that this relation might be derived directly from Lorentz's lemma (8.20) and holds for any reciprocal medium. As seen from (8.210), (8.207), and (8.208), in the absence of polarization degeneracy the vibration vectors of these bases are such that

$$\begin{gathered}\mathbf{z} \cdot \left[\mathbf{e}_j(\tilde{\mathbf{b}}) \times \mathbf{h}_k(-\tilde{\mathbf{b}}) - \mathbf{e}_k(-\tilde{\mathbf{b}}) \times \mathbf{h}_j(\tilde{\mathbf{b}})\right] = 0 \\ |\,j - k| \neq 2.\end{gathered} \tag{8.211}$$

Equation (8.211) can be written in terms of the $\boldsymbol{\psi}$ vectors as follows:

$$\boldsymbol{\psi}_j(\tilde{\mathbf{b}})^{\mathrm{T}} \mathbf{I}_r \boldsymbol{\psi}_k(-\tilde{\mathbf{b}}) = 0, \tag{8.212}$$

where

$$\mathbf{I}_r = \begin{pmatrix} 0 & 1 & 0 & 0 \\ -1 & 0 & 0 & 0 \\ 0 & 0 & 0 & 1 \\ 0 & 0 & -1 & 0 \end{pmatrix}. \tag{8.213}$$

In (8.212), the vectors $\boldsymbol{\psi}_j(\tilde{\mathbf{b}})$ and $\boldsymbol{\psi}_k(-\tilde{\mathbf{b}})$ are referred to the same coordinate system (see Section 8.3.4). We may notice that the vectors $\mathbf{m}_b = -\mathbf{m}_a$, $\mathbf{e}_b = \mathbf{e}_a$ and $\mathbf{h}_b = -\mathbf{h}_a$, where $\mathbf{m}_a$, $\mathbf{e}_a$, and $\mathbf{h}_a$ satisfy

(8.170)–(8.171), meet (8.172) and (8.173). Therefore, for the pairs (8.208), in the absence of polarization degeneracy,

$$\mathbf{e}_k(-\tilde{\mathbf{b}}) = b_{kj}\mathbf{e}_j(\tilde{\mathbf{b}}), \quad \mathbf{h}_k(-\tilde{\mathbf{b}}) = -b_{kj}\mathbf{h}_j(\tilde{\mathbf{b}}), \tag{8.214}$$

where b_{kj} are scalar factors whose presence in these relations is connected with the freedom in choice of the vibration vectors [see (8.101)]. Let us assume that polarization degeneracy takes place and we have an optimal polarization basis for $\mathbf{b} = \tilde{\mathbf{b}}$. If in this situation we choose the eigenwave basis for $\mathbf{b} = -\tilde{\mathbf{b}}$ so that relations (8.214) are satisfied for all its basis waves, this basis will be optimal and satisfy condition (8.211). If two optimal bases corresponding to opposite values of $\mathbf{b}$ satisfy (8.211), we will call them *optimal reciprocal bases.*

8.4.2 Simple General Formulas for Transmission Operators of Interfaces

One of the most frequently executed operations in numerical calculations for LCDs is the calculation of the transmission operators for interfaces. If the liquid crystal layer in the LCD being modeled is not regarded as an FP system or an element of an FP system (see Section 8.2.1), the number of interfaces whose transmission operators must be calculated is typically hundreds, and most of these interfaces are the interfaces between anisotropic layers of a staircase model of the LC layer (see Figure 8.15). In the well-known variants [24–30] of the extended Jones matrix method, transmission operators for interfaces between anisotropic media are calculated by approximate formulas. However, even in the case of normal incidence, the use of the approximate formulas employed in [24–30] may lead to large errors (see Sections 11.1.2 and 12.2). In our programs, we, as a rule, calculate the transmission operators for interfaces rigorously by using sufficiently simple and general exact formulas obtained in [39]. These formulas are presented in this section.

General Computational Formulas for Transmission Operators of Interfaces

Let us consider the interface between arbitrary layers A and B characterized by symmetrical permittivity tensors. Let the eigenwave bases in these layers be orthogonal in the sense of (8.161). Recall that in the absence of polarization degeneracy the condition (8.161) is satisfied for any choice of the eigenwave basis, while in the presence of polarization degeneracy the polarization basis can always be chosen so that the condition (8.161) will be met. Let $z = \underline{z}$ be the plane of the interface. The transmission operators $\mathbf{t}^{\downarrow}_{AB} \equiv \mathbf{t}^{\downarrow}(\underline{z}-0, \underline{z}+0)$ and $\mathbf{t}^{\uparrow}_{AB} \equiv \mathbf{t}^{\uparrow}(\underline{z}-0, \underline{z}+0)$ can be calculated from the transfer matrix $\mathbf{T}(\underline{z}+0, \underline{z}-0)$ or from the inverse of this matrix [see (8.121) and (8.124)]. The simplest variants of the calculation are

$$\mathbf{t}^{\downarrow}_{AB} = \frac{1}{t^{(BA)}_{11}t^{(BA)}_{22} - t^{(BA)}_{12}t^{(BA)}_{21}} \begin{pmatrix} t^{(BA)}_{22} & -t^{(BA)}_{12} \\ -t^{(BA)}_{21} & t^{(BA)}_{11} \end{pmatrix}, \tag{8.215}$$

$$\mathbf{t}^{\uparrow}_{AB} = \frac{1}{t^{(AB)}_{33}t^{(AB)}_{44} - t^{(AB)}_{34}t^{(AB)}_{43}} \begin{pmatrix} t^{(AB)}_{44} & -t^{(AB)}_{34} \\ -t^{(AB)}_{43} & t^{(AB)}_{33} \end{pmatrix}, \tag{8.216}$$

where $t^{(AB)}_{jk}$ and $t^{(BA)}_{jk}$ are elements of the matrices $\mathbf{T}(\underline{z}+0, \underline{z}-0)$ and $\mathbf{T}(\underline{z}+0, \underline{z}-0)^{-1}$, respectively. According to (8.115), we may represent the matrices $\mathbf{T}(\underline{z}+0, \underline{z}-0)$ and $\mathbf{T}(\underline{z}+0, \underline{z}-0)^{-1}$ as follows:

$$\mathbf{T}(\underline{z}+0,\ \underline{z}-0) = \mathbf{\Psi}(\underline{z}+0)^{-1}\mathbf{\Psi}(\underline{z}-0), \tag{8.217a}$$

$$\mathbf{T}(\underline{z}+0,\ \underline{z}-0)^{-1} = \mathbf{\Psi}(\underline{z}-0)^{-1}\mathbf{\Psi}(\underline{z}+0), \tag{8.217b}$$

where

$$\boldsymbol{\Psi}(\underline{z}-0) = \begin{pmatrix} \boldsymbol{\psi}_{1A} & \boldsymbol{\psi}_{2A} & \boldsymbol{\psi}_{3A} & \boldsymbol{\psi}_{4A} \end{pmatrix},$$
$$\boldsymbol{\Psi}(\underline{z}+0) = \begin{pmatrix} \boldsymbol{\psi}_{1B} & \boldsymbol{\psi}_{2B} & \boldsymbol{\psi}_{3B} & \boldsymbol{\psi}_{4B} \end{pmatrix},$$
$$\boldsymbol{\psi}_{jA} = \begin{pmatrix} \psi_{1jA} \\ \psi_{2jA} \\ \psi_{3jA} \\ \psi_{4jA} \end{pmatrix} = \begin{pmatrix} \mathbf{e}_{jA}\mathbf{x} \\ \mathbf{h}_{jA}\mathbf{y} \\ \mathbf{e}_{jA}\mathbf{y} \\ -\mathbf{h}_{jA}\mathbf{x} \end{pmatrix}, \quad \boldsymbol{\psi}_{jB} = \begin{pmatrix} \psi_{1jB} \\ \psi_{2jB} \\ \psi_{3jB} \\ \psi_{4jB} \end{pmatrix} = \begin{pmatrix} \mathbf{e}_{jB}\mathbf{x} \\ \mathbf{h}_{jB}\mathbf{y} \\ \mathbf{e}_{jB}\mathbf{y} \\ -\mathbf{h}_{jB}\mathbf{x} \end{pmatrix}, \quad j = 1, 2, 3, 4. \tag{8.218}$$

Here and in what follows, the quantities pertaining to the layers $\mathcal{A}$ and $\mathcal{B}$ are labeled by the subscripts A and B, respectively. Let us denote the rows of the matrices $\boldsymbol{\Psi}(\underline{z}-0)^{-1}$ and $\boldsymbol{\Psi}(\underline{z}+0)^{-1}$ by $\bar{\boldsymbol{\psi}}_{jA}$ and $\bar{\boldsymbol{\psi}}_{jB}$, respectively ($j$ = 1, 2, 3, 4). Then, according to (8.217), the components of the matrices $\mathbf{T}(\underline{z}+0, \underline{z}-0)$ and $\mathbf{T}(\underline{z}+0, \underline{z}-0)^{-1}$ can be represented as

$$t_{jk}^{(BA)} = \bar{\boldsymbol{\psi}}_{jA}\boldsymbol{\psi}_{kB}, \quad t_{jk}^{(AB)} = \bar{\boldsymbol{\psi}}_{jB}\boldsymbol{\psi}_{kA}. \tag{8.219}$$

Due to the orthogonality of both eigenwave bases, we may, using (8.166), express the rows $\bar{\boldsymbol{\psi}}_{jA}$ and $\bar{\boldsymbol{\psi}}_{jB}$ as follows:

$$\bar{\boldsymbol{\psi}}_{jA} = \frac{1}{2\left(\psi_{1jA}\psi_{2jA} + \psi_{3jA}\psi_{4jA}\right)}\left(\psi_{2jA} \quad \psi_{1jA} \quad \psi_{4jA} \quad \psi_{3jA}\right), \tag{8.220}$$
$$\bar{\boldsymbol{\psi}}_{jB} = \frac{1}{2\left(\psi_{1jB}\psi_{2jB} + \psi_{3jB}\psi_{4jB}\right)}\left(\psi_{2jB} \quad \psi_{1jB} \quad \psi_{4jB} \quad \psi_{3jB}\right) \tag{8.221}$$

[see (8.218)]. Substitution of (8.220) and (8.221) into (8.219) gives the following simple expressions for $t_{jk}^{(BA)}$ and $t_{jk}^{(AB)}$:

$$t_{jk}^{(BA)} = \frac{\psi_{2jA}\psi_{1kB} + \psi_{1jA}\psi_{2kB} + \psi_{4jA}\psi_{3kB} + \psi_{3jA}\psi_{4kB}}{2\left(\psi_{1jA}\psi_{2jA} + \psi_{3jA}\psi_{4jA}\right)}, \tag{8.222}$$
$$t_{jk}^{(AB)} = \frac{\psi_{2jB}\psi_{1kA} + \psi_{1jB}\psi_{2kA} + \psi_{4jB}\psi_{3kA} + \psi_{3jB}\psi_{4kA}}{2\left(\psi_{1jB}\psi_{2jB} + \psi_{3jB}\psi_{4jB}\right)}. \tag{8.223}$$

These expressions can also be written in the form

$$t_{jk}^{(BA)} = \frac{\mathbf{z}(\mathbf{e}_{jA} \times \mathbf{h}_{kB}) + \mathbf{z}(\mathbf{e}_{kB} \times \mathbf{h}_{jA})}{2\mathbf{z}(\mathbf{e}_{jA} \times \mathbf{h}_{jA})}, \tag{8.224}$$

$$t_{jk}^{(AB)} = \frac{\mathbf{z}(\mathbf{e}_{jB} \times \mathbf{h}_{kA}) + \mathbf{z}(\mathbf{e}_{kA} \times \mathbf{h}_{jB})}{2\mathbf{z}(\mathbf{e}_{jB} \times \mathbf{h}_{jB})}. \tag{8.225}$$

Using these formulas for the elements of the transfer matrices $\mathbf{T}(\underline{z}+0, \underline{z}-0)$ and $\mathbf{T}(\underline{z}+0, \underline{z}-0)^{-1}$ simplifies the calculation of the reflection operators for interfaces by formulas (8.121) and (8.124) as well.

With the symmetrical normalization (8.167) of the eigenwave bases, the expressions for $t_{jk}^{(BA)}$ and $t_{jk}^{(AB)}$ become still simpler. Here we present expressions only for those elements of $\mathbf{T}(\underline{z}+0, \underline{z}-0)$ and $\mathbf{T}(\underline{z}+0, \underline{z}-0)^{-1}$ that are required to calculate $\mathbf{t}_{AB}^{\downarrow}$ and $\mathbf{t}_{AB}^{\uparrow}$ [see (8.215) and (8.216)]. With normalization (8.167), the elements required to calculate $\mathbf{t}_{AB}^{\downarrow}$ are expressed as follows:

$$t_{jk}^{(BA)} = \psi_{2jA}\psi_{1kB} + \psi_{1jA}\psi_{2kB} + \psi_{4jA}\psi_{3kB} + \psi_{3jA}\psi_{4kB}, \tag{8.226a}$$

or, equivalently,

$$t_{jk}^{(\mathrm{BA})} = \mathbf{z}(\mathbf{e}_{j\mathrm{A}} \times \mathbf{h}_{k\mathrm{B}}) + \mathbf{z}(\mathbf{e}_{k\mathrm{B}} \times \mathbf{h}_{j\mathrm{A}}), \quad j = 1,2, \quad k = 1,2; \tag{8.226b}$$

the elements required to calculate $\mathbf{t}_{\mathrm{AB}}^{\uparrow}$:

$$t_{jk}^{(\mathrm{AB})} = -\left(\psi_{2j\mathrm{B}}\psi_{1k\mathrm{A}} + \psi_{1j\mathrm{B}}\psi_{2k\mathrm{A}} + \psi_{4j\mathrm{B}}\psi_{3k\mathrm{A}} + \psi_{3j\mathrm{B}}\psi_{4k\mathrm{A}}\right), \tag{8.227a}$$

or

$$t_{jk}^{(\mathrm{AB})} = -\left(\mathbf{z}(\mathbf{e}_{j\mathrm{B}} \times \mathbf{h}_{k\mathrm{A}}) + \mathbf{z}(\mathbf{e}_{k\mathrm{A}} \times \mathbf{h}_{j\mathrm{B}})\right), \quad j = 3,4, \quad k = 3,4. \tag{8.227b}$$

One may notice that in the above formulas for the transmission operators, the forward transmission operator $\mathbf{t}_{\mathrm{AB}}^{\downarrow}$ is expressed in terms of parameters of only forward propagating basis eigenwaves, and the backward transmission operator $\mathbf{t}_{\mathrm{AB}}^{\uparrow}$ in terms of parameters of only backward propagating ones. Therefore, when simple transmission channels like those shown in Figures 8.13a and 8.15 are considered, parameters of only forward propagating basis eigenwaves are required in calculations, as in the variants [24–30] of the extended Jones matrix method (but we emphasize again that, in contrast to [24–30], in our case transmission operators for interfaces are calculated rigorously).

Analytical Expressions for the Forward Transmission Operator of the Interface Between Uniaxial Media

The above expressions for the transmission operators of interfaces are very convenient for computations, but not for analysis. However, substitution of explicit expressions for parameters of eigenwaves in terms of optical parameters of the media into these expressions may give useful analytical expressions for the transmission operators of interfaces. As an example, we will obtain here such an expression for the transmission operator $\mathbf{t}_{\mathrm{AB}}^{\downarrow}$ of the interface of two uniaxial media with identical principal refractive indices but different orientation of the optic axes (Figure 8.19). We deal with such interfaces when considering staircase models of inhomogeneous LC layers. This example allows us to see some analogy of the exact expression for this operator with approximate ones used in different variants of the Jones matrix method.

In the calculations, we will utilize a template of eigenwave basis for uniaxial media that is presented in Section 9.3. Applying the symmetrical normalization (8.167) to this template, we can set and represent the parameters of the forward propagating basis waves in a uniaxial medium as follows:

$$\begin{gathered}
\sigma_1 = \frac{-\Delta\varepsilon(\mathbf{bc})(\mathbf{zc}) + \sqrt{[\Delta\varepsilon(\mathbf{bc})(\mathbf{zc})]^2 - \left(\varepsilon_\perp + \Delta\varepsilon(\mathbf{zc})^2\right)\left(\varepsilon_\perp\zeta^2 + \Delta\varepsilon(\mathbf{bc})^2 - \varepsilon_\perp\varepsilon_\parallel\right)}}{\varepsilon_\perp + \Delta\varepsilon(\mathbf{zc})^2}, \\
\sigma_2 = \sqrt{\varepsilon_\perp - \zeta^2}, \\
\Delta\varepsilon = \varepsilon_\parallel - \varepsilon_\perp,\ \varepsilon_\parallel = n_\parallel^2,\ \varepsilon_\perp = n_\perp^2, \\
\mathbf{e}_1 = \frac{\boldsymbol{e}_1}{\sqrt{2\mathbf{z}\left(\boldsymbol{e}_1 \times \boldsymbol{h}_1\right)}},\ \mathbf{h}_1 = \frac{\boldsymbol{h}_1}{\sqrt{2\mathbf{z}\left(\boldsymbol{e}_1 \times \boldsymbol{h}_1\right)}},\ \mathbf{e}_2 = \frac{\boldsymbol{e}_2}{\sqrt{2\mathbf{z}\left(\boldsymbol{e}_2 \times \boldsymbol{h}_2\right)}},\ \mathbf{h}_2 = \frac{\boldsymbol{h}_2}{\sqrt{2\mathbf{z}\left(\boldsymbol{e}_2 \times \boldsymbol{h}_2\right)}}, \\
\boldsymbol{e}_1 = \mathbf{c} - \frac{(\mathbf{m}_1\mathbf{c})}{\varepsilon_\perp}\mathbf{m}_1,\ \boldsymbol{h}_1 = \mathbf{m}_1 \times \boldsymbol{e}_1,\ \boldsymbol{e}_2 = \mathbf{m}_2 \times \mathbf{c},\ \boldsymbol{h}_2 = \mathbf{m}_2 \times \boldsymbol{e}_2, \\
\mathbf{m}_1 = \mathbf{b} + \mathbf{z}\sigma_1,\ \mathbf{m}_2 = \mathbf{b} + \mathbf{z}\sigma_2,
\end{gathered} \tag{8.228}$$

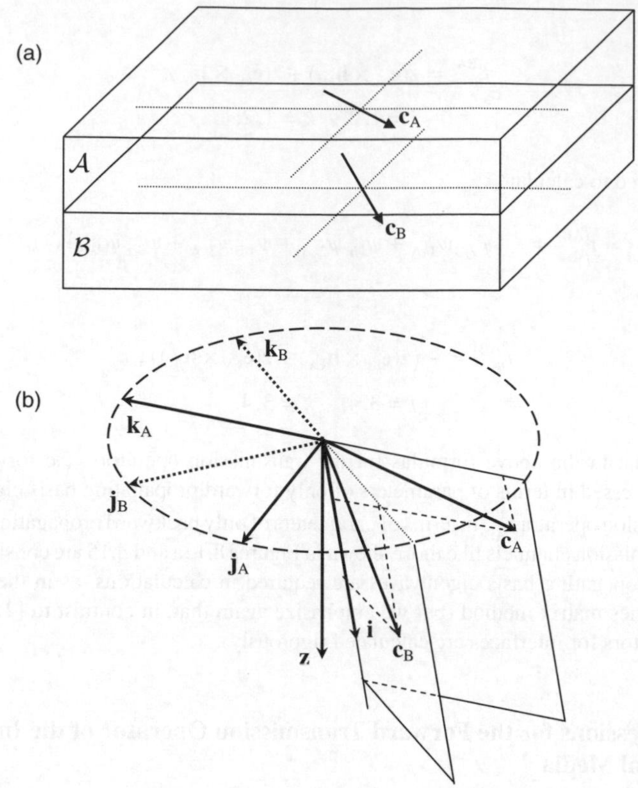

Figure 8.19 To the problem on the transmission characteristics of the interface between uniaxial layers. In (b), we assume, for simplicity, that the layers are nonabsorbing

where $n_{\|}$ and $n_{\perp}$ are the principal complex refractive indices of the medium; **c** is a unit vector parallel to the optic axis. The index 1 is assigned to the extraordinary basis wave, and the index 2 to the ordinary one. Let us introduce the vectors

$$\mathbf{i} = \mathbf{m}_2/M_2, \quad \mathbf{j} = (\mathbf{i}\times\mathbf{c})/p, \quad \mathbf{k} = \mathbf{i}\times\mathbf{j}, \tag{8.229}$$

where

$$M_2 \equiv \sqrt{\mathbf{m}_2^2} = n_{\perp}, \quad p = \sqrt{(\mathbf{i}\times\mathbf{c})^2}. \tag{8.230}$$

The vectors **i**, **j**, and **k** satisfy the conditions

$$\mathbf{i}^2 = \mathbf{j}^2 = \mathbf{k}^2 = 1, \quad \mathbf{ij} = \mathbf{jk} = \mathbf{ik} = 0. \tag{8.231}$$

In the case of a nonabsorbing medium out of TIR mode, the vectors **i**, **j**, and **k** are real and have a simple meaning: **i** is the wave normal of the ordinary basis wave; **j** is a unit vector parallel to the electric vibration vector of the ordinary basis wave—the vector **j** is perpendicular to the optic axis **c** and the wave normal **i**; **k** is a unit vector perpendicular to **i** and **j** and parallel to the magnetic vibration vector of the ordinary basis wave (Figure 8.19). In the general case, using these vectors, we may represent the vibration vectors of the ordinary basis wave as follows:

$$\mathbf{e}_2 = \frac{\mathbf{j}}{\sqrt{2n_\perp \mathbf{iz}}}, \quad \mathbf{h}_2 = \frac{n_\perp \mathbf{k}}{\sqrt{2n_\perp \mathbf{iz}}}. \tag{8.232}$$

Then, representing the refraction vector $\mathbf{m}_1$ as

$$\mathbf{m}_1 = \mathbf{m}_2 + \mathbf{z} n_\perp \gamma, \tag{8.233}$$

where

$$\gamma = (\sigma_1 - \sigma_2)/n_\perp,$$

and substituting (8.233) into expressions for $\boldsymbol{e}_1$ and $\boldsymbol{h}_1$ in (8.228), we obtain

$$\begin{aligned} \boldsymbol{e}_1 &= -\left\{p\mathbf{k} + \gamma\,[\mathbf{z}(\mathbf{ic}) + \mathbf{i}(\mathbf{zc})] + \gamma^2 \mathbf{z}(\mathbf{zc})\right\}, \\ \boldsymbol{h}_1 &= n_\perp\,[p\mathbf{j} + \gamma\,(\mathbf{z} \times \mathbf{c})]. \end{aligned} \tag{8.234}$$

By expressing the vectors $\mathbf{e}_{j\mathrm{A}}$, $\mathbf{h}_{j\mathrm{A}}$, $\mathbf{e}_{j\mathrm{B}}$, and $\mathbf{h}_{j\mathrm{B}}$ (i = 1,2) in accordance with (8.228), (8.232), and (8.234) and substituting the obtained expressions into (8.226), we obtain the following expressions for the parameters $t_{jk}^{(\mathrm{BA})}$:

$$t_{11}^{(\mathrm{BA})} = \frac{C_{11}}{C_\mathrm{A} C_\mathrm{B}}, \quad t_{12}^{(\mathrm{BA})} = \frac{C_{12}}{C_\mathrm{A}}, \quad t_{21}^{(\mathrm{BA})} = \frac{C_{21}}{C_\mathrm{B}}, \quad t_{22}^{(\mathrm{BA})} = \mathbf{j}_\mathrm{B}\mathbf{j}_\mathrm{A}, \tag{8.235a}$$

where

$$C_{11} = \mathbf{j}_\mathrm{B}\mathbf{j}_\mathrm{A}\left(1 + \frac{\gamma_\mathrm{A} + \gamma_\mathrm{B}}{2\mathbf{iz}}\right) + \gamma_\mathrm{A}\gamma_\mathrm{B} \frac{(\mathbf{i} \times \mathbf{z})\left[(\mathbf{z} \times \mathbf{c}_\mathrm{B})(\mathbf{zc}_\mathrm{A}) + (\mathbf{z} \times \mathbf{c}_\mathrm{A})(\mathbf{zc}_\mathrm{B})\right]}{2\mathbf{iz}p_\mathrm{A}p_\mathrm{B}}, \tag{8.235b}$$

$$C_{12} = -\mathbf{j}_\mathrm{B}\mathbf{k}_\mathrm{A}\left(1 + \frac{\gamma_\mathrm{A}}{2\mathbf{iz}}\right), \quad C_{21} = \mathbf{j}_\mathrm{B}\mathbf{k}_\mathrm{A}\left(1 + \frac{\gamma_\mathrm{B}}{2\mathbf{iz}}\right), \tag{8.235c}$$

$$C_{\mathrm{A,B}} = \left(1 + \frac{\gamma_{\mathrm{A,B}}}{\mathbf{iz}} + \frac{\gamma_{\mathrm{A,B}}^2 (\mathbf{i} \times \mathbf{z})(\mathbf{z} \times \mathbf{c}_{\mathrm{A,B}})(\mathbf{zc}_{\mathrm{A,B}})}{(\mathbf{iz})p_{\mathrm{A,B}}^2}\right)^{\frac{1}{2}}. \tag{8.235d}$$

The subscripts A and B mark the values of the parameters $\mathbf{c}$, $\mathbf{j}$, $\mathbf{k}$, γ, and p for the layers $\mathcal{A}$ and $\mathcal{B}$, respectively. From these expressions, a very simple approximate formula for the matrix $\mathbf{t}_{\mathrm{AB}}^{\downarrow}$ may be derived for the case of small $\gamma_\mathrm{A}/\mathbf{iz}$ and $\gamma_\mathrm{B}/\mathbf{iz}$. According to (8.235d),

$$\frac{1}{C_{\mathrm{A,B}}} = 1 - \frac{\gamma_{\mathrm{A,B}}}{2\mathbf{iz}} + \ldots, \tag{8.236}$$

where terms containing the factors $\gamma_{\mathrm{A,B}}^m$ with $m \geq 2$ are omitted. Substituting (8.236), (8.235b), and (8.235c) into (8.235a), one may see that

$$\begin{pmatrix} t_{11}^{(\mathrm{BA})} & t_{12}^{(\mathrm{BA})} \\ t_{21}^{(\mathrm{BA})} & t_{22}^{(\mathrm{BA})} \end{pmatrix} \approx \begin{pmatrix} \mathbf{j}_\mathrm{B}\mathbf{j}_\mathrm{A} & -\mathbf{j}_\mathrm{B}\mathbf{k}_\mathrm{A} \\ \mathbf{j}_\mathrm{B}\mathbf{k}_\mathrm{A} & \mathbf{j}_\mathrm{B}\mathbf{j}_\mathrm{A} \end{pmatrix}$$

to a high accuracy in the case $(\gamma_{A,B}/\mathbf{iz})^2 \ll 1$. Using this approximation, we have

$$\mathbf{t}^{\downarrow}_{AB} \approx \begin{pmatrix} \mathbf{j}_B\mathbf{j}_A & \mathbf{j}_B\mathbf{k}_A \\ -\mathbf{j}_B\mathbf{k}_A & \mathbf{j}_B\mathbf{j}_A \end{pmatrix}. \tag{8.237}$$

In the absence of absorption, the matrix approximating the matrix $\mathbf{t}^{\downarrow}_{AB}$ in (8.237) is equal to the matrix of (coordinate) rotation (1.53) with α equal to the angle between the vectors $\mathbf{j}_A$ and $\mathbf{j}_B$. Such a matrix is used to approximate the interface transmission operator in the variant of the extended Jones matrix method developed by Yeh and Gu [17, 18]. But, in that method, a different normalization of the eigenwave basis is used, namely,

$$\mathbf{e}_j \cdot \mathbf{e}_j = 1 \qquad j = 1,\ 2,\ 3,\ 4. \tag{8.238}$$

The normalization (8.238) will be referred to as *Yeh's normalization.* The exact matrices $\mathbf{t}^{\downarrow}_{AB}$ corresponding to Yeh's normalization in general differ from those corresponding to the symmetrical normalization. One may show that in common situations, the approximation (8.237), being relatively good in the case of the symmetrical normalization, is not so good in the case of Yeh's normalization. We illustrate this by the following numerical example. Table 8.1 compares the values of the matrix

$$\widehat{R}(\mathbf{j}_A, \mathbf{j}_B) = \begin{pmatrix} \mathbf{j}_B\mathbf{j}_A & \mathbf{j}_B\mathbf{k}_A \\ -\mathbf{j}_B\mathbf{k}_A & \mathbf{j}_B\mathbf{j}_A \end{pmatrix}$$

and exactly calculated matrices $\mathbf{t}^{\downarrow}_{AB}$ corresponding to the symmetrical normalization and Yeh's normalization for the interface between nonabsorbing uniaxial layers. The principal refractive indices of both layers are $n_{\parallel} = 1.7$ and $n_{\perp} = 1.5$. The angles of orientation of the optic axis for the layer $\mathcal{A}$ are $\theta = 20°$ and $\varphi = 0°$ (see Figure 9.3 in Section 9.3 of the next chapter), those for the layer $\mathcal{B}$ are $\theta = 30°$ and $\varphi = 10°$. Here, to specify the optical system and illumination conditions, we use the reference system, quantities, and notation adopted in Chapter 9. It is assumed that the light falls on a layered system that contains the layers $\mathcal{A}$ and $\mathcal{B}$ from air at an angle of 45° (β_{inc}). Table 8.1 shows results for three orientations of the plane of incidence: $\alpha_{inc} = 0°, 45°, 90°$ (see Figure 9.1). As seen from this table, in all the cases, the matrix $\mathbf{t}^{\downarrow}_{AB}$ calculated with symmetrical normalization is much closer to the rotation matrix $\widehat{R}(\mathbf{j}_A, \mathbf{j}_B)$ than the matrix $\mathbf{t}^{\downarrow}_{AB}$ calculated with Yeh's normalization. We see also that even at relatively close parameters of the adjacent layers, as in this example, the exact matrix $\mathbf{t}^{\downarrow}_{AB}$ calculated with Yeh's normalization may significantly differ from the rotation matrix. Thus, in principle, we may use approximation (8.237), but we should remember that this approximation is relatively good only for certain normalizations of the eigenwave basis, such as the symmetrical normalization.

Table 8.1 The transmission matrices for interface between uniaxial layers and the rotation matrix $\widehat{R}(\mathbf{j}_A, \mathbf{j}_B)$

α_{inc}	$\widehat{R}(\mathbf{j}_A, \mathbf{j}_B)$	$\mathbf{t}^{\downarrow}_{AB}$, symmetrical normalization	$\mathbf{t}^{\downarrow}_{AB}$, Yeh's normalization
0°	$\begin{pmatrix} 0.960124 & 0.279572 \\ -0.279572 & 0.960125 \end{pmatrix}$	$\begin{pmatrix} 0.959964 & 0.279905 \\ -0.279766 & 0.959958 \end{pmatrix}$	$\begin{pmatrix} 0.966793 & 0.283090 \\ -0.278586 & 0.959958 \end{pmatrix}$
45°	$\begin{pmatrix} 0.995325 & 0.096583 \\ -0.096583 & 0.995325 \end{pmatrix}$	$\begin{pmatrix} 0.995163 & 0.096749 \\ -0.096678 & 0.995298 \end{pmatrix}$	$\begin{pmatrix} 1.015149 & 0.096384 \\ -0.098993 & 0.995298 \end{pmatrix}$
90°	$\begin{pmatrix} 0.998768 & 0.049627 \\ -0.049627 & 0.998768 \end{pmatrix}$	$\begin{pmatrix} 0.998600 & 0.049740 \\ -0.049709 & 0.998758 \end{pmatrix}$	$\begin{pmatrix} 1.016518 & 0.047243 \\ -0.053275 & 0.998758 \end{pmatrix}$

8.4.3 *Calculation of Transmission and Reflection Operators of Layered Systems by Using the Adding Technique*

In Section 8.2.2, we described the technique in which the overall transmission and reflection operators of a layered system are calculated from the EW transfer matrix or Berreman transfer matrix of this system. In the transfer matrix methods, the spatial evolution of damped backward propagating waves is described by exponentially growing functions. Therefore, the transfer matrices of layers where such waves are present may contain very large elements. In numerical calculations, the presence of such elements may cause severe loss of significant digits in the numerical values of elements of the resulting transfer matrix of the layered system. That is why the transfer matrix technique is numerically unstable and may give a huge error when applied to layered systems including strongly absorbing layers or layers in TIR mode.[2] For such systems, the overall transmission and reflection operators can be accurately calculated with the help of an adding-technique-based method that is presented in this section (the adding technique is considered in detail in Section 7.2.2). The well-known method of Ko and Sambles [40] may be considered as a variant of this method.

An Algorithm Using the Transmission and Reflection Matrices of Elements to be Added

In this algorithm, two kinds of elements of the layered system are considered: bulks of layers and interfaces. The elements are presented in the calculations by their transmission and reflection operators (EW Jones matrices). The transmission matrices for the bulks are calculated by formulas (8.111) and (8.112). The transmission and reflection matrices for an interface may be computed by formulas (8.121) from the transfer matrix of this interface (8.115) or somehow else (see Sections 8.2.2 and 8.4.2). Denoting the reflection and transmission operators of the jth element by $\mathbf{t}^{\downarrow}_{j}$, $\mathbf{r}^{\downarrow}_{j}$, $\mathbf{t}^{\uparrow}_{j}$, and $\mathbf{r}^{\uparrow}_{j}$ ($j = 1,2, \ldots, N$, where N is the number of the elements in the layered system), and the reflection and transmission operators of the part of the system including its first j elements by $\mathbf{t}^{\downarrow}_{(j)}$, $\mathbf{r}^{\downarrow}_{(j)}$, $\mathbf{t}^{\uparrow}_{(j)}$, and $\mathbf{r}^{\uparrow}_{(j)}$ ($j = 1,2, \ldots, N-1$), we may represent the procedure of involving the $j+1$th element (see Section 7.3) as follows:

(i) if the $j+1$th element is an interface,

$$\mathbf{t}^{\downarrow}_{(j+1)} = \mathbf{t}^{\downarrow}_{j+1}(\mathbf{U} - \mathbf{r}^{\uparrow}_{(j)}\mathbf{r}^{\downarrow}_{j+1})^{-1}\mathbf{t}^{\downarrow}_{(j)}, \tag{8.239a}$$

$$\mathbf{r}^{\downarrow}_{(j+1)} = \mathbf{r}^{\downarrow}_{(j)} + \mathbf{t}^{\uparrow}_{(j)}\mathbf{r}^{\downarrow}_{j+1}(\mathbf{U} - \mathbf{r}^{\uparrow}_{(j)}\mathbf{r}^{\downarrow}_{j+1})^{-1}\mathbf{t}^{\downarrow}_{(j)}, \tag{8.239b}$$

$$\mathbf{t}^{\uparrow}_{(j+1)} = \mathbf{t}^{\uparrow}_{(j)}(\mathbf{U} - \mathbf{r}^{\downarrow}_{j+1}\mathbf{r}^{\uparrow}_{(j)})^{-1}\mathbf{t}^{\uparrow}_{j+1}, \tag{8.239c}$$

$$\mathbf{r}^{\uparrow}_{(j+1)} = \mathbf{r}^{\uparrow}_{j+1} + \mathbf{t}^{\downarrow}_{j+1}\mathbf{r}^{\uparrow}_{(j)}(\mathbf{U} - \mathbf{r}^{\downarrow}_{j+1}\mathbf{r}^{\uparrow}_{(j)})^{-1}\mathbf{t}^{\uparrow}_{j+1}; \tag{8.239d}$$

(ii) if the $j+1$th element is a bulk,

$$\mathbf{t}^{\downarrow}_{(j+1)} = \mathbf{t}^{\downarrow}_{j+1}\mathbf{t}^{\downarrow}_{(j)}, \tag{8.240a}$$

$$\mathbf{r}^{\downarrow}_{(j+1)} = \mathbf{r}^{\downarrow}_{(j)}, \tag{8.240b}$$

$$\mathbf{t}^{\uparrow}_{(j+1)} = \mathbf{t}^{\uparrow}_{(j)}\mathbf{t}^{\uparrow}_{j+1}, \tag{8.240c}$$

$$\mathbf{r}^{\uparrow}_{(j+1)} = \mathbf{t}^{\downarrow}_{j+1}\mathbf{r}^{\uparrow}_{(j)}\mathbf{t}^{\uparrow}_{j+1}. \tag{8.240d}$$

[2] See also remarks at the end of Section 7.2.1.

The computation starts from

$$\mathbf{t}^{\downarrow}_{(1)} = \mathbf{t}^{\downarrow}_{1}, \quad \mathbf{r}^{\downarrow}_{(1)} = \mathbf{r}^{\downarrow}_{1}, \quad \mathbf{t}^{\uparrow}_{(1)} = \mathbf{t}^{\uparrow}_{1}, \quad \mathbf{r}^{\uparrow}_{(1)} = \mathbf{r}^{\uparrow}_{1}$$

if the first element is an interface, or from

$$\mathbf{t}^{\downarrow}_{(1)} = \mathbf{t}^{\downarrow}_{1}, \quad \mathbf{r}^{\downarrow}_{(1)} = \mathbf{0}, \quad \mathbf{t}^{\uparrow}_{(1)} = \mathbf{t}^{\uparrow}_{1}, \quad \mathbf{r}^{\uparrow}_{(1)} = \mathbf{0}$$

if the first element is a bulk. Successive calculations by (8.239)–(8.240) for $j = 2, \ldots, N\text{–}1$ give the transmission and reflection operators for the whole layered system, which are $\mathbf{t}^{\downarrow}_{(N)}$, $\mathbf{r}^{\downarrow}_{(N)}$, $\mathbf{t}^{\uparrow}_{(N)}$, and $\mathbf{r}^{\uparrow}_{(N)}$.

An Algorithm Using the Transfer Matrices of Elements to be Added

The step (i) of the above algorithm, in principle, can be performed without calculating the matrices $\mathbf{t}^{\downarrow}_{j}$, $\mathbf{r}^{\downarrow}_{j}$, $\mathbf{t}^{\uparrow}_{j}$, and $\mathbf{r}^{\uparrow}_{j}$, but using directly the EW transfer matrix of the interface to be added. In particular, the matrices $\mathbf{t}^{\downarrow}_{(j+1)}$, $\mathbf{r}^{\downarrow}_{(j+1)}$, $\mathbf{t}^{\uparrow}_{(j+1)}$, and $\mathbf{r}^{\uparrow}_{(j+1)}$ can be calculated as follows:

$$\begin{aligned}
\mathbf{t}^{\downarrow}_{(j+1)} &= (\mathbf{I}_{11k} - \mathbf{r}^{\uparrow}_{(j)}\mathbf{I}_{21k})^{-1}\mathbf{t}^{\downarrow}_{(j)},\\
\mathbf{r}^{\uparrow}_{(j+1)} &= (\mathbf{I}_{11k} - \mathbf{r}^{\uparrow}_{(j)}\mathbf{I}_{21k})^{-1}\,(\mathbf{r}^{\uparrow}_{(j)}\mathbf{I}_{22k} - \mathbf{I}_{12k}),\\
\mathbf{r}^{\downarrow}_{(j+1)} &= \mathbf{r}^{\downarrow}_{(j)} + \mathbf{t}^{\uparrow}_{(j)}\mathbf{I}_{21k}\,\mathbf{t}^{\downarrow}_{(j+1)},\\
\mathbf{t}^{\uparrow}_{(j+1)} &= \mathbf{t}^{\uparrow}_{(j)}(\mathbf{I}_{21k}\,\mathbf{r}^{\uparrow}_{(j+1)} + \mathbf{I}_{22k}),
\end{aligned}$$

$\mathbf{I}_{11k}$, $\mathbf{I}_{12k}$, $\mathbf{I}_{21k}$, and $\mathbf{I}_{22k}$ being the 2×2 blocks of the matrix

$$\mathbf{I}_k = \begin{pmatrix} \mathbf{I}_{11k} & \mathbf{I}_{12k} \\ \mathbf{I}_{21k} & \mathbf{I}_{22k} \end{pmatrix} = \boldsymbol{\Psi}_k^{-1}\boldsymbol{\Psi}_{k+1},$$

where $\boldsymbol{\Psi}_k$ and $\boldsymbol{\Psi}_{k+1}$ are respectively the $\boldsymbol{\Psi}$-matrix for the last layer of the subsystem of j elements and that for the next layer of the system ($\mathbf{I}_k$ is the inverse of the EW transfer matrix for the interface to be added). With step (i) performed in this way, the method under consideration becomes practically equivalent to the method of Ko and Sambles [40].

Being applied in ordinary situations (without TIR mode and strongly absorbing layers), both presented variants of the method are somewhat more expensive in computational cost as compared with the transfer matrix method; however, in view of their higher numerical stability, it is often reasonable to use them in ordinary situations as well.

8.5 Transmissivities and Reflectivities

In this section, we consider how the transmissivities and reflectivities of OTR units (see Section 7.1) and the transmissivities of transfer channels as well as the extreme values of these transmissivities and reflectivities can be calculated from corresponding transmission and reflection EW Jones matrices.

Normal Flux

The time-averaged Poynting vector $\langle\mathbf{S}\rangle$ of an arbitrary field $\{\mathbf{E}, \mathbf{H}\}$ of the form (8.57) can be expressed as follows:

$$\langle\mathbf{S}(z,t)\rangle = \frac{c}{8\pi}\mathrm{Re}\left(\tilde{\mathbf{E}}(z) \times \tilde{\mathbf{H}}(z)^*\right). \tag{8.241}$$

The time-averaged energy flux density of this field through the plane z = ξ, where ξ is a given value of z, can be represented as

$$S_z(\xi) = \mathbf{z}\langle \mathbf{S}(\xi, t)\rangle = \frac{c}{8\pi}\mathrm{Re}\left[\mathbf{z}\left(\tilde{\mathbf{E}}(\xi)\times\tilde{\mathbf{H}}(\xi)^*\right)\right]. \tag{8.242}$$

Such flux densities will be called the *normal fluxes*. The notion of the normal flux is important for us here since it is customary in optics of stratified media to define transmissivities and reflectivities for fields of the form (8.57) as ratios of the absolute values of corresponding normal fluxes. In terms of the Berreman vector $\mathbf{\Psi}$ the normal flux can be expressed as follows:

$$S_z(\xi) = \frac{c}{16\pi}\Psi(\xi)^\dagger \mathbf{I}_0 \Psi(\xi). \tag{8.243}$$

In terms of the vector **A** (8.68)

$$S_z(\xi) = \frac{c}{16\pi}\mathbf{A}(\xi)^\dagger \tilde{\mathbf{N}}_0(\xi)\mathbf{A}(\xi) = \frac{c}{16\pi}\sum_{j=1}^{4}\sum_{k=1}^{4} N_{0jk}(\xi)A_j^*(\xi)A_k(\xi), \tag{8.244}$$

where

$$\tilde{\mathbf{N}}_0 \equiv \begin{pmatrix} N_{011} & N_{012} & N_{013} & N_{014} \\ N_{021} & N_{022} & N_{023} & N_{024} \\ N_{031} & N_{032} & N_{033} & N_{034} \\ N_{041} & N_{042} & N_{043} & N_{044} \end{pmatrix} \equiv \mathbf{\Psi}^\dagger \mathbf{I}_0 \mathbf{\Psi}. \tag{8.245}$$

The elements of the matrix $\tilde{\mathbf{N}}_0$ can be represented as

$$N_{0jk} = \mathbf{z}(\mathbf{e}_j^* \times \mathbf{h}_k) + \mathbf{z}(\mathbf{e}_k \times \mathbf{h}_j^*) = \boldsymbol{\psi}_j^\dagger \mathbf{I}_0 \boldsymbol{\psi}_k. \tag{8.246}$$

If the plane z = ξ lies within a nonabsorbing layer out of TIR mode, in the absence of polarization degeneracy or otherwise with an optimal eigenwave basis (see Section 8.4.1), due to the orthogonality (8.160) of the eigenwave basis, the matrix $\tilde{\mathbf{N}}_0$ is diagonal:

$$\tilde{\mathbf{N}}_0(\xi) = \begin{pmatrix} N_{011}(\xi) & 0 & 0 & 0 \\ 0 & N_{022}(\xi) & 0 & 0 \\ 0 & 0 & N_{033}(\xi) & 0 \\ 0 & 0 & 0 & N_{044}(\xi) \end{pmatrix}. \tag{8.247}$$

In this case, expression (8.244) takes the form

$$S_z(\xi) = \frac{c}{16\pi}\sum_{j=1}^{4} N_{0jj}(\xi)A_j^*(\xi)A_j(\xi). \tag{8.248}$$

If we denote the normal flux of a wave field $\mathbf{X}$ through the plane z = ξ as $S_z\{\mathbf{X}(\xi)\}$, we may write, according to (8.248), the following relation:

$$S_z\left\{\sum_{j=1}^{4}\mathbf{X}_j(\xi)\right\} = \sum_{j=1}^{4} S_z\left\{\mathbf{X}_j(\xi)\right\}, \tag{8.249}$$

where $\mathbf{X}_j$ (j = 1, 2, 3, 4) are the eigenwave constituents of the field $\{\mathbf{E}, \mathbf{H}\}$; that is, the contribution of each eigenwave constituent to the normal flux of the field $\{\mathbf{E}, \mathbf{H}\}$ is unaffected by the other eigenwave constituents. If relations (8.160) are violated, the matrix $\tilde{\mathbf{N}}_0(\xi)$ is not diagonal and the sum in (8.244) may contain nonzero interference terms $N_{0jk}(\xi)A_j^*(\xi)A_k(\xi)$ $(k \neq j)$, and consequently the relation (8.249) may also be violated. Estimates of the effect of off-diagonal elements of $\tilde{\mathbf{N}}_0$ in the cases of an absorbing medium and TIR mode may be found in Section 9.2.

According to (8.244), the normal flux of a forward propagating field $\mathbf{X}^{\downarrow}$ through the plane $z = \xi$ may be represented as follows:

$$S_z\{\mathbf{X}^{\downarrow}(\xi)\} = \frac{c}{16\pi}\mathbf{a}^{\downarrow}(\xi)^{\dagger}\mathbf{n}_0^{\downarrow}(\xi)\mathbf{a}^{\downarrow}(\xi), \tag{8.250}$$

where $\mathbf{a}^{\downarrow}$ is the $\mathbf{a}$-vector of the field $\mathbf{X}^{\downarrow}$ and

$$\mathbf{n}_0^{\downarrow} = \begin{pmatrix} N_{011} & N_{012} \\ N_{021} & N_{022} \end{pmatrix}. \tag{8.251}$$

Analogously, for a backward propagating field $\mathbf{X}^{\uparrow}$, characterized by an $\mathbf{a}$-vector $\mathbf{a}^{\uparrow}$, we may write

$$S_z\{\mathbf{X}^{\uparrow}\} = \frac{c}{16\pi}\mathbf{a}^{\uparrow\dagger}\mathbf{n}_0^{\uparrow}\mathbf{a}^{\uparrow}, \tag{8.252}$$

where

$$\mathbf{n}_0^{\uparrow} = \begin{pmatrix} N_{033} & N_{034} \\ N_{043} & N_{044} \end{pmatrix}. \tag{8.253}$$

Since

$$\left|S_z\{\mathbf{X}^{\downarrow}\}\right| = S_z\{\mathbf{X}^{\downarrow}\}, \ \left|S_z\{\mathbf{X}^{\uparrow}\}\right| = -S_z\{\mathbf{X}^{\uparrow}\}, \tag{8.254}$$

the corresponding irradiances, the absolute values of $S_z\left\{\mathbf{X}^{\downarrow}(z)\right\}$ and $S_z\left\{\mathbf{X}^{\uparrow}(z)\right\}$, can be expressed as follows:

$$E\{\mathbf{X}^{\downarrow}\} = \frac{c}{16\pi}\mathbf{a}^{\downarrow\dagger}\mathbf{n}_0^{\downarrow}\mathbf{a}^{\downarrow}, \quad E\{\mathbf{X}^{\uparrow}\} = \frac{c}{16\pi}\mathbf{a}^{\uparrow\dagger}\bar{\mathbf{n}}_0^{\uparrow}\mathbf{a}^{\uparrow}, \tag{8.255}$$

where

$$\bar{\mathbf{n}}_0^{\uparrow} \equiv -\mathbf{n}_0^{\uparrow}. \tag{8.256}$$

In most cases of interest, the matrices $\mathbf{n}_0^{\downarrow}$ and $\bar{\mathbf{n}}_0^{\uparrow}$ can be represented as

$$\mathbf{n}_0^{\downarrow} = \mathbf{q}^{\downarrow\dagger}\mathbf{q}^{\downarrow}, \quad \bar{\mathbf{n}}_0^{\uparrow} = \mathbf{q}^{\uparrow\dagger}\mathbf{q}^{\uparrow}. \tag{8.257}$$

Using (8.257), we may rewrite expressions (8.255) as follows:

$$E\{\mathbf{X}^{\downarrow}\} = \frac{c}{16\pi}\underline{\mathbf{a}}^{\downarrow\dagger}\underline{\mathbf{a}}^{\downarrow}, \quad E\{\mathbf{X}^{\uparrow}\} = \frac{c}{16\pi}\underline{\mathbf{a}}^{\uparrow\dagger}\underline{\mathbf{a}}^{\uparrow}, \tag{8.258}$$

where

$$\underline{\mathbf{a}}^{\downarrow} = \mathbf{q}^{\downarrow}\mathbf{a}^{\downarrow}, \quad \underline{\mathbf{a}}^{\uparrow} = \mathbf{q}^{\uparrow}\mathbf{a}^{\uparrow}, \tag{8.259}$$

$\mathbf{q}^{\downarrow}$ and $\mathbf{q}^{\uparrow}$ are matrices satisfying (8.257)—these matrices are discussed at the end of this section.

Transmissivities and Reflectivities of an OTR unit

Returning to the general situation illustrated by Figure 8.12, we may define the transmissivities, $t^{\downarrow}(z', z'')$ and $t^{\uparrow}(z', z'')$, and reflectivities, $r^{\downarrow}(z', z'')$ and $r^{\uparrow}(z', z'')$, of the fragment (z', z'') as

$$\begin{aligned} t^{\downarrow}(z', z'') &\equiv \frac{E\{\mathbf{X}^{\downarrow}_{\mathrm{tr}}(z'')\}}{E\{\mathbf{X}^{\downarrow}_{\mathrm{inc}}(z')\}}, \quad r^{\downarrow}(z', z'') \equiv \frac{E\{\mathbf{X}^{\uparrow}_{\mathrm{ref}}(z')\}}{E\{\mathbf{X}^{\downarrow}_{\mathrm{inc}}(z')\}}, \\ t^{\uparrow}(z', z'') &\equiv \frac{E\{\mathbf{X}^{\uparrow}_{\mathrm{tr}}(z')\}}{E\{\mathbf{X}^{\uparrow}_{\mathrm{inc}}(z'')\}}, \quad r^{\uparrow}(z', z'') \equiv \frac{E\{\mathbf{X}^{\downarrow}_{\mathrm{ref}}(z'')\}}{E\{\mathbf{X}^{\uparrow}_{\mathrm{inc}}(z'')\}}. \end{aligned} \tag{8.260}$$

According to (8.255) and (8.107)–(8.110), given **a**-vectors of the incident fields, these transmissivities and reflectivities can be calculated as

$$\begin{aligned} t^{\downarrow}(z', z'') &= \frac{\mathbf{a}^{\downarrow}_{\mathrm{tr}}(z'')^{\dagger}\mathbf{n}^{\downarrow}_{0}(z'')\mathbf{a}^{\downarrow}_{\mathrm{tr}}(z'')}{\mathbf{a}^{\downarrow}_{\mathrm{inc}}(z')^{\dagger}\mathbf{n}^{\downarrow}_{0}(z')\mathbf{a}^{\downarrow}_{\mathrm{inc}}(z')}, \quad r^{\downarrow}(z', z'') = \frac{\mathbf{a}^{\uparrow}_{\mathrm{ref}}(z')^{\dagger}\bar{\mathbf{n}}^{\uparrow}_{0}(z')\mathbf{a}^{\uparrow}_{\mathrm{ref}}(z')}{\mathbf{a}^{\downarrow}_{\mathrm{inc}}(z')^{\dagger}\mathbf{n}^{\downarrow}_{0}(z')\mathbf{a}^{\downarrow}_{\mathrm{inc}}(z')}, \\ t^{\uparrow}(z', z'') &= \frac{\mathbf{a}^{\uparrow}_{\mathrm{tr}}(z')^{\dagger}\bar{\mathbf{n}}^{\uparrow}_{0}(z')\mathbf{a}^{\uparrow}_{\mathrm{tr}}(z')}{\mathbf{a}^{\uparrow}_{\mathrm{inc}}(z'')^{\dagger}\bar{\mathbf{n}}^{\uparrow}_{0}(z'')\mathbf{a}^{\uparrow}_{\mathrm{inc}}(z'')}, \quad r^{\uparrow}(z', z'') = \frac{\mathbf{a}^{\downarrow}_{\mathrm{ref}}(z'')^{\dagger}\mathbf{n}^{\downarrow}_{0}(z'')\mathbf{a}^{\downarrow}_{\mathrm{ref}}(z'')}{\mathbf{a}^{\uparrow}_{\mathrm{inc}}(z'')^{\dagger}\bar{\mathbf{n}}^{\uparrow}_{0}(z'')\mathbf{a}^{\uparrow}_{\mathrm{inc}}(z'')}, \end{aligned} \tag{8.261}$$

where

$$\mathbf{a}^{\downarrow}_{\mathrm{tr}}(z'') = \mathbf{t}^{\downarrow}(z', z'')\mathbf{a}^{\downarrow}_{\mathrm{inc}}(z'), \tag{8.262}$$

$$\mathbf{a}^{\uparrow}_{\mathrm{ref}}(z') = \mathbf{r}^{\downarrow}(z', z'')\mathbf{a}^{\downarrow}_{\mathrm{inc}}(z'), \tag{8.263}$$

$$\mathbf{a}^{\uparrow}_{\mathrm{tr}}(z') = \mathbf{t}^{\uparrow}(z', z'')\mathbf{a}^{\uparrow}_{\mathrm{inc}}(z''), \tag{8.264}$$

$$\mathbf{a}^{\downarrow}_{\mathrm{ref}}(z'') = \mathbf{r}^{\uparrow}(z', z'')\mathbf{a}^{\uparrow}_{\mathrm{inc}}(z''). \tag{8.265}$$

Evaluation of Extreme Values of Transmissivities and Reflectivities for an OTR unit

With a transmission (reflection) EW Jones matrix known, one may easily determine the maximum and minimum values of the corresponding transmissivity (reflectivity) over the set of all possible states of the incident field at the given ω and **b**. We denote such extreme values of the transmissivities and reflectivities for the fragment (z', z'') by $\underline{\max}[t^{\downarrow}(z', z'')]$, $\underline{\min}[t^{\downarrow}(z', z'')]$, $\underline{\max}[r^{\downarrow}(z', z'')]$, $\underline{\min}[r^{\downarrow}(z', z'')]$, and so on. Let us find, for instance, $\underline{\max}[t^{\downarrow}(z', z'')]$ and $\underline{\min}[t^{\downarrow}(z', z'')]$. From (8.261) and (8.262) we have

$$t^{\downarrow}(z', z'') = \frac{\mathbf{a}^{\downarrow}_{\mathrm{inc}}(z')^{\dagger}\mathbf{t}^{\downarrow}(z', z'')^{\dagger}\mathbf{n}^{\downarrow}_{0}(z'')\mathbf{t}^{\downarrow}(z', z'')\mathbf{a}^{\downarrow}_{\mathrm{inc}}(z')}{\mathbf{a}^{\downarrow}_{\mathrm{inc}}(z')^{\dagger}\mathbf{n}^{\downarrow}_{0}(z')\mathbf{a}^{\downarrow}_{\mathrm{inc}}(z')}. \tag{8.266}$$

To solve the problem, we must find the maximum and minimum values of $t^{\downarrow}(z', z'')$ as a function of $\mathbf{a}^{\downarrow}_{\mathrm{inc}}(z')$. Substitution of $\mathbf{a}^{\downarrow}_{\mathrm{inc}}(z') = \mathbf{q}^{\downarrow}(z')^{-1}\mathbf{x}$ into (8.266) gives the following equation:

$$t^{\downarrow}(z', z'') = \frac{(\mathbf{x}^{\dagger}\bar{\mathbf{t}}^{\downarrow}(z', z'')\mathbf{x})}{(\mathbf{x}^{\dagger}\mathbf{x})}, \tag{8.267}$$

where

$$\bar{\mathbf{t}}^{\downarrow}(z', z'') = \left(\mathbf{q}^{\downarrow}(z')^{-1}\right)^{\dagger}\mathbf{t}^{\downarrow}(z', z'')^{\dagger}\mathbf{n}^{\downarrow}_{0}(z'')\mathbf{t}^{\downarrow}(z', z'')\mathbf{q}^{\downarrow}(z')^{-1}. \tag{8.268}$$

The problem now is to determine the extrema of $t^{\downarrow}(z', z'')$ as a function of $\mathbf{x}$. The matrix $\bar{\mathbf{t}}^{\downarrow}(z', z'')$ is Hermitian. According to the Rayleigh–Ritz theorem [41], for any Hermitian matrix $\mathbf{t}$

$$\max_{\mathbf{x}\neq 0}\left(\frac{\mathbf{x}^{\dagger}\mathbf{t}\mathbf{x}}{\mathbf{x}^{\dagger}\mathbf{x}}\right) = \lambda_{\max}[\mathbf{t}], \quad \min_{\mathbf{x}\neq 0}\left(\frac{\mathbf{x}^{\dagger}\mathbf{t}\mathbf{x}}{\mathbf{x}^{\dagger}\mathbf{x}}\right) = \lambda_{\min}[\mathbf{t}], \tag{8.269}$$

where $\lambda_{\max}[\mathbf{t}]$ and $\lambda_{\min}[\mathbf{t}]$ are respectively the maximum and minimum eigenvalues of the matrix $\mathbf{t}$. Appliying this theorem, we find that

$$\underline{\max}[t^{\downarrow}(z', z'')] = \lambda_{\max}[\bar{\mathbf{t}}^{\downarrow}(z', z'')], \quad \underline{\min}[t^{\downarrow}(z', z'')] = \lambda_{\min}[\bar{\mathbf{t}}^{\downarrow}(z', z'')]. \tag{8.270}$$

This solution may be also represented in the following form (this form is often more convenient for calculations):

$$\left.\begin{array}{l}\underline{\max}[t^{\downarrow}(z', z'')] \\ \underline{\min}[t^{\downarrow}(z', z'')]\end{array}\right\} = \frac{1}{2}\left[\left\|\widehat{\mathbf{t}}^{\downarrow}(z', z'')\right\|_{\mathrm{E}}^{2} \pm \sqrt{\left\|\widehat{\mathbf{t}}^{\downarrow}(z', z'')\right\|_{\mathrm{E}}^{4} - 4\left|\det \widehat{\mathbf{t}}^{\downarrow}(z', z'')\right|^{2}}\right], \tag{8.271}$$

where

$$\widehat{\mathbf{t}}^{\downarrow}(z', z'') = \mathbf{q}^{\downarrow}(z'')\mathbf{t}^{\downarrow}(z', z'')\mathbf{q}^{\downarrow}(z')^{-1},$$

$\|\bullet\|_{\mathrm{E}}$ stands for the Euclidean norm [see Section 5.1.4 and Eq. (5.44) therein]. The matrices $\bar{\mathbf{t}}^{\downarrow}(z', z'')$ and $\widehat{\mathbf{t}}^{\downarrow}(z', z'')$ are related by

$$\bar{\mathbf{t}}^{\downarrow}(z', z'') = \widehat{\mathbf{t}}^{\downarrow}(z', z'')^{\dagger}\widehat{\mathbf{t}}^{\downarrow}(z', z''). \tag{8.272}$$

Similarly, we may find that

$$\begin{aligned}
&\left.\begin{array}{l}\underline{\max}[t^{\uparrow}(z', z'')] \\ \underline{\min}[t^{\uparrow}(z', z'')]\end{array}\right\} = \frac{1}{2}\left[\left\|\widehat{\mathbf{t}}^{\uparrow}(z', z'')\right\|_{\mathrm{E}}^{2} \pm \sqrt{\left\|\widehat{\mathbf{t}}^{\uparrow}(z', z'')\right\|_{\mathrm{E}}^{4} - 4\left|\det \widehat{\mathbf{t}}^{\uparrow}(z', z'')\right|^{2}}\right], \\
&\left.\begin{array}{l}\underline{\max}[r^{\downarrow}(z', z'')] \\ \underline{\min}[r^{\downarrow}(z', z'')]\end{array}\right\} = \frac{1}{2}\left[\left\|\widehat{\mathbf{r}}^{\downarrow}(z', z'')\right\|_{\mathrm{E}}^{2} \pm \sqrt{\left\|\widehat{\mathbf{r}}^{\downarrow}(z', z'')\right\|_{\mathrm{E}}^{4} - 4\left|\det \widehat{\mathbf{r}}^{\downarrow}(z', z'')\right|^{2}}\right], \\
&\left.\begin{array}{l}\underline{\max}[r^{\uparrow}(z', z'')] \\ \underline{\min}[r^{\uparrow}(z', z'')]\end{array}\right\} = \frac{1}{2}\left[\left\|\widehat{\mathbf{r}}^{\uparrow}(z', z'')\right\|_{\mathrm{E}}^{2} \pm \sqrt{\left\|\widehat{\mathbf{r}}^{\uparrow}(z', z'')\right\|_{\mathrm{E}}^{4} - 4\left|\det \widehat{\mathbf{r}}^{\uparrow}(z', z'')\right|^{2}}\right],
\end{aligned} \tag{8.273}$$

where

$$\begin{aligned}
\widehat{\mathbf{t}}^{\uparrow}(z', z'') &\equiv \mathbf{q}^{\uparrow}(z')\mathbf{t}^{\uparrow}(z', z'')\mathbf{q}^{\uparrow}(z'')^{-1}, \\
\widehat{\mathbf{r}}^{\downarrow}(z', z'') &\equiv \mathbf{q}^{\uparrow}(z')\mathbf{r}^{\downarrow}(z', z'')\mathbf{q}^{\downarrow}(z')^{-1}, \\
\widehat{\mathbf{r}}^{\uparrow}(z', z'') &\equiv \mathbf{q}^{\downarrow}(z'')\mathbf{r}^{\uparrow}(z', z'')\mathbf{q}^{\uparrow}(z'')^{-1}.
\end{aligned}$$

Of course, when using these results, one should remember that here we consider the fragment (z', z'') in isolation from the other elements of the system and neglect the interference effects connected with the presence of the adjacent layers. Moreover, considering light incidence from an absorbing medium, let it be, for instance, the medium $\mathbf{M}_A$ (see Figure 8.12), in the analysis of energy transfer, one should take into account that in this case, the relation

$$S_z\left\{\mathbf{X}_{\text{inc}}^{\downarrow}(z') + \mathbf{X}_{\text{ref}}^{\uparrow}(z')\right\} = S_z\left\{\mathbf{X}_{\text{inc}}^{\downarrow}(z')\right\} + S_z\left\{\mathbf{X}_{\text{ref}}^{\uparrow}(z')\right\}$$

is in general violated.

Transmissivity of a Transfer Channel

The transmissivity of a transfer channel $\mathbf{C}$ for an input field $\mathbf{X}_{\text{inp}}$ can be defined as

$$t\{\mathbf{C}\} \equiv \frac{E\left\{\mathbf{X}_{\text{out}}(z_{\text{co}})\right\}}{E\left\{\mathbf{X}_{\text{inp}}(z_{\text{ci}})\right\}}, \tag{8.274}$$

where $\mathbf{X}_{\text{out}} = \mathbf{C}\mathbf{X}_{\text{inp}}$ is the output field; $z = z_{\text{ci}}$ and $z = z_{\text{co}}$ are respectively the entrance and exit planes of this channel. Denoting the $\mathbf{a}$-vector of a wave field $\mathbf{X}$ at $z = \xi$ by $\tilde{\mathbf{a}}\{\mathbf{X}(\xi)\}$, we may express the transmissivity $t\{\mathbf{C}\}$ in terms of the $\mathbf{a}$-vectors of the fields $\mathbf{X}_{\text{inp}}$ and $\mathbf{X}_{\text{out}}$ as follows:

$$t\{\mathbf{C}\} = \frac{\tilde{\mathbf{a}}\left\{\mathbf{X}_{\text{out}}(z_{\text{co}})\right\}^{\dagger}\bar{\mathbf{n}}\left\{\mathbf{X}_{\text{out}}, z_{\text{co}}\right\}\tilde{\mathbf{a}}\left\{\mathbf{X}_{\text{out}}(z_{\text{co}})\right\}}{\tilde{\mathbf{a}}\left\{\mathbf{X}_{\text{inp}}(z_{\text{ci}})\right\}^{\dagger}\bar{\mathbf{n}}\left\{\mathbf{X}_{\text{inp}}, z_{\text{ci}}\right\}\tilde{\mathbf{a}}\left\{\mathbf{X}_{\text{inp}}(z_{\text{ci}})\right\}}, \tag{8.275}$$

where

$$\bar{\mathbf{n}}\{\mathbf{X}, z\} = \begin{cases} \mathbf{n}_0^{\downarrow}(z) \text{ if } \mathbf{X} \text{ is forward propagating} \\ \bar{\mathbf{n}}_0^{\uparrow}(z) \text{ if } \mathbf{X} \text{ is backward propagating.} \end{cases} \tag{8.276}$$

The vectors $\tilde{\mathbf{a}}\left\{\mathbf{X}_{\text{inp}}(z_{\text{ci}})\right\}$ and $\tilde{\mathbf{a}}\left\{\mathbf{X}_{\text{out}}(z_{\text{co}})\right\}$ are related by

$$\tilde{\mathbf{a}}\left\{\mathbf{X}_{\text{out}}(z_{\text{co}})\right\} = \mathbf{t}\{\mathbf{C}\}\,\tilde{\mathbf{a}}\left\{\mathbf{X}_{\text{inp}}(z_{\text{ci}})\right\},$$

where $\mathbf{t}\{\mathbf{C}\}$ is the EW Jones matrix of the channel $\mathbf{C}$.

The extreme values of the transmissivity $t\{\mathbf{C}\}$ as a function of $\tilde{\mathbf{a}}\{\mathbf{X}_{\text{inp}}(z_{\text{ci}})\}$ may be calculated from the matrix $\mathbf{t}\{\mathbf{C}\}$ by the following formulas:

$$\left.\begin{matrix} \underline{\max}[t\{\mathbf{C}\}] \\ \underline{\min}[t\{\mathbf{C}\}] \end{matrix}\right\} = \frac{1}{2}\left[\left\|\widehat{\mathbf{t}}\{\mathbf{C}\}\right\|_{\text{E}}^{2} \pm \sqrt{\left\|\widehat{\mathbf{t}}\{\mathbf{C}\}\right\|_{\text{E}}^{4} - 4\left|\det\widehat{\mathbf{t}}\{\mathbf{C}\}\right|^{2}}\right], \tag{8.277}$$

where

$$\widehat{\mathbf{t}}\{\mathbf{C}\} \equiv \bar{\mathbf{q}}\left\{\mathbf{X}_{\text{out}}, z_{\text{co}}\right\}\cdot\mathbf{t}\{\mathbf{C}\}\cdot\bar{\mathbf{q}}\left\{\mathbf{X}_{\text{inp}}, z_{\text{ci}}\right\}^{-1},$$

$$\bar{\mathbf{q}}\{\mathbf{X}, z\} = \begin{cases} \mathbf{q}^{\downarrow}(z) \text{ if } \mathbf{X} \text{ is forward propagating} \\ \mathbf{q}^{\uparrow}(z) \text{ if } \mathbf{X} \text{ is backward propagating.} \end{cases}$$

The average transmissivity of the channel $\mathbf{C}$ may be expressed as

$$\underline{\text{avr}}[t\{\mathbf{C}\}] = \frac{1}{2}\left(\underline{\max}[t\{\mathbf{C}\}] + \underline{\min}[t\{\mathbf{C}\}]\right) = \frac{1}{2}\left\|\widehat{\mathbf{t}}\{\mathbf{C}\}\right\|_{\text{E}}^{2}. \tag{8.278}$$

The quantity $\underline{\text{avr}}[t\{\mathbf{C}\}]$ may be treated as the transmissivity of the channel $\mathbf{C}$ for unpolarized quasi-monochromatic incident light with infinitely narrow bandwidth.

How to Simplify the Calculations

The matrices $\mathbf{n}_0^{\downarrow}$, $\bar{\mathbf{n}}_0^{\uparrow}$, $\mathbf{q}^{\downarrow}$, and $\mathbf{q}^{\uparrow}$ entering into the above formulas depend on the parameters of the eigenwave basis, and a suitable choice of the eigenwave basis may simplify calculations. In the case of an isotropic medium, whether nonabsorbing or absorbing, the standard variant of the eigenwave basis presented in Section 9.2 with any normalization gives the matrices $\mathbf{n}_0^{\downarrow}$ and $\bar{\mathbf{n}}_0^{\uparrow}$ of the form

$$\mathbf{n}_0^{\downarrow} = \begin{pmatrix} N_{011} & 0 \\ 0 & N_{022} \end{pmatrix}, \quad \bar{\mathbf{n}}_0^{\uparrow} = \begin{pmatrix} -N_{033} & 0 \\ 0 & -N_{044} \end{pmatrix}. \tag{8.279}$$

For nonabsorbing anisotropic media out of TIR mode in the absence of polarization degeneracy as well as in the presence of polarization degeneracy provided that an optimal eigenwave basis is used (see Sections 8.4.1, 9.1, 9.3, and 9.4), these matrices also have the form (8.279) under any normalization. In all these cases, we can use the following matrices $\mathbf{q}^{\downarrow}$ and $\mathbf{q}^{\uparrow}$:

$$\mathbf{q}^{\downarrow} = \begin{pmatrix} \sqrt{N_{011}} & 0 \\ 0 & \sqrt{N_{022}} \end{pmatrix}, \quad \mathbf{q}^{\uparrow} = \begin{pmatrix} \sqrt{-N_{033}} & 0 \\ 0 & \sqrt{-N_{044}} \end{pmatrix}. \tag{8.280}$$

In any case where the flux normalization (8.164) is applicable (see Section 8.4.1), the use of this normalization gives

$$\mathbf{n}_0^{\downarrow} = \begin{pmatrix} 1 & N_{012} \\ N_{012}^* & 1 \end{pmatrix}, \quad \bar{\mathbf{n}}_0^{\uparrow} = \begin{pmatrix} 1 & -N_{034} \\ -N_{034}^* & 1 \end{pmatrix}. \tag{8.281}$$

If the eigenwave basis is orthogonal in the sense of (8.160), with the flux normalization

$$\mathbf{n}_0^{\downarrow} = \mathbf{U}, \quad \bar{\mathbf{n}}_0^{\uparrow} = \mathbf{U}. \tag{8.282}$$

With such entrance and exit bases, we may omit the matrices $\mathbf{n}_0^{\downarrow}$, $\bar{\mathbf{n}}_0^{\uparrow}$, $\mathbf{q}^{\downarrow}$, and $\mathbf{q}^{\uparrow}$ in the above formulas for transmissivities and reflectivities. In the case of a nonabsorbing isotropic medium, with the standard choice of the eigenwave basis and the standard "electrical" (E-) normalization

$$\mathbf{e}_j^*\mathbf{e}_j = 1 \qquad j = 1, 2, 3, 4, \tag{8.283}$$

each of the matrices $\mathbf{n}_0^{\downarrow}$ and $\bar{\mathbf{n}}_0^{\uparrow}$ is a product of a scalar and the unit matrix [see (9.27)]. With such entrance and exit bases, we may replace the matrices $\mathbf{n}_0^{\downarrow}$, $\bar{\mathbf{n}}_0^{\uparrow}$, $\mathbf{q}^{\downarrow}$, $\mathbf{q}^{\uparrow}$, $(\mathbf{q}^{\downarrow})^{-1}$, and $(\mathbf{q}^{\uparrow})^{-1}$ in the above formulas for transmissivities and reflectivities by the corresponding scalar factors. In the case of an absorbing anisotropic medium, using the flux normalization to have (8.281), one can use the following variant of factorization to obtain $\mathbf{q}^{\downarrow}$ and $\mathbf{q}^{\uparrow}$: any matrix $\boldsymbol{C}$ of the form

$$\boldsymbol{C} = \begin{pmatrix} 1 & C \\ C^* & 1 \end{pmatrix}$$

with $|C| < 1$ can be represented as $C = \mathbf{q}^{\dagger}\mathbf{q}$ with

$$\mathbf{q} = \begin{pmatrix} (a_1 + a_2)/2 & b_1 (a_1 - a_2)/2 \\ b_1^* (a_1^* - a_2^*)/2 & (a_1 + a_2)/2 \end{pmatrix},$$

where

$$a_1 = \sqrt{1 + b_2 b_2^*}, \quad a_2 = \sqrt{1 - b_2 b_2^*}, \quad b_1 = \frac{b_2}{b_2^*}, \quad b_2 = \sqrt{C}.$$

8.6 Mathematical Properties of Transfer Matrices and Transmission and Reflection EW Jones Matrices of Lossless Media and Reciprocal Media

In this section, we will consider properties of Berreman transfer matrices, EW transfer matrices, and EW Jones matrices peculiar to reciprocal media (see Section 8.1.1) and nonabsorbing media. Relations presented here play the same role in the rigorous theory as relation (1.225) for the Jones matrices of lossless systems and reciprocity relation (1.254) do in the classical Jones calculus.

8.6.1 Properties of Matrix Operators for Nonabsorbing Regions

Let us assume that the medium in Figure 8.12 is nonabsorbing within the region (z′, z″). Any field satisfying the Maxwell equation throughout this region will also satisfy (8.18) for any volume domain situated between the planes z = z′ and z = z″. The time-averaged Poynting vector of a field of the form (8.57) is independent of x and y, that is,

$$\langle \mathbf{S}(\mathbf{r}, t) \rangle = \langle \mathbf{S}(z, t) \rangle . \tag{8.284}$$

In view of this, taking as S in (8.18) the surface of a rectangular parallelepiped two sides of which lie on the planes z = z′ and z = z″ and two sides are parallel to the plane of incidence (see Figure 8.20) and substituting (8.57) into (8.18), we may see that the normal flux [see Section 8.5, Eq. (8.242)] through the plane z = z″ is equal to that through the plane z = z′, that is,

$$S_z(z'') = S_z(z'). \tag{8.285}$$

Let us consider how property (8.285) is manifested in properties of matrix operators characterizing this region.

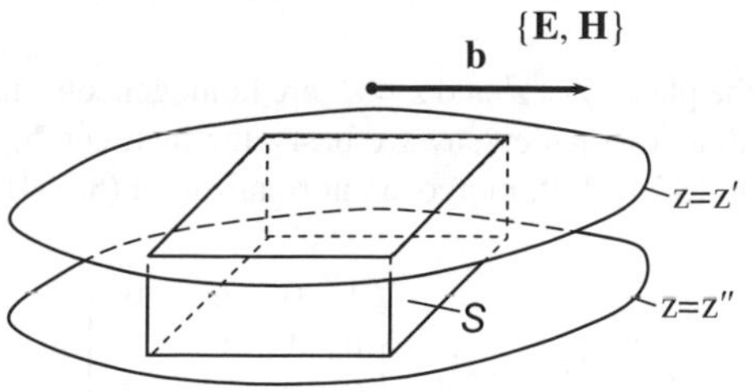

Figure 8.20 Nonabsorbing fragment. Geometry of the problem

Transfer Matrices

According to (8.285) and (8.57), we have the following relation for $\tilde{\mathbf{E}}$ and $\tilde{\mathbf{H}}$

$$\mathbf{z}\left(\tilde{\mathbf{E}}(z'')\times\tilde{\mathbf{H}}(z'')^*\right)+\mathbf{z}\left(\tilde{\mathbf{E}}(z'')^*\times\tilde{\mathbf{H}}(z'')\right)=\mathbf{z}\left(\tilde{\mathbf{E}}(z')\times\tilde{\mathbf{H}}(z')^*\right)+\mathbf{z}\left(\tilde{\mathbf{E}}(z')^*\times\tilde{\mathbf{H}}(z')\right). \quad (8.286)$$

Relation (8.286) is equivalent to the following ones:

$$\boldsymbol{\Psi}(z'')^{\dagger}\mathbf{I}_0\boldsymbol{\Psi}(z'')=\boldsymbol{\Psi}(z')^{\dagger}\mathbf{I}_0\boldsymbol{\Psi}(z'), \quad (8.287)$$

$$\mathbf{A}(z'')^{\dagger}\tilde{\mathbf{N}}_0(z'')\mathbf{A}(z'')=\mathbf{A}(z')^{\dagger}\tilde{\mathbf{N}}_0(z')\mathbf{A}(z'), \quad (8.288)$$

where

$$\tilde{\mathbf{N}}_0(z)=\boldsymbol{\Psi}(z)^{\dagger}\mathbf{I}_0\boldsymbol{\Psi}(z) \quad (8.289)$$

[see (8.243)–(8.245)]. Using (8.287) and (8.288), we can easily find restrictions that are imposed by the absence of absorption on the transfer matrices $\mathbf{P}(z'', z')$ and $\mathbf{T}(z'', z')$. Substitution of the expression

$$\boldsymbol{\Psi}(z'')=\mathbf{P}(z'',z')\boldsymbol{\Psi}(z')$$

into (8.287) leads to the relation

$$\boldsymbol{\Psi}(z')^{\dagger}\mathbf{P}(z'',z')^{\dagger}\mathbf{I}_0\mathbf{P}(z'',z')\boldsymbol{\Psi}(z')=\boldsymbol{\Psi}(z')^{\dagger}\mathbf{I}_0\boldsymbol{\Psi}(z'). \quad (8.290)$$

Since this relation is an identity [i.e., it is satisfied at any $\boldsymbol{\Psi}(z')$], it determines the following property of the matrix $\mathbf{P}(z'', z')$:

$$\mathbf{P}(z'',z')^{\dagger}\mathbf{I}_0\mathbf{P}(z'',z')=\mathbf{I}_0. \quad (8.291)$$

This is a general property of the Berreman matrices for nonabsorbing regions. A more convenient form of relation (8.291) is

$$\mathbf{P}(z'',z')^{-1}=\mathbf{I}_0\mathbf{P}(z'',z')^{\dagger}\mathbf{I}_0, \quad (8.292)$$

where we have used the fact that $\mathbf{I}_0^{-1}=\mathbf{I}_0$. From (8.288), we may derive an analogous relation for the matrix $\mathbf{T}(z'', z')$:

$$\mathbf{T}(z'',z')^{-1}=\tilde{\mathbf{N}}_0(z')^{-1}\mathbf{T}(z'',z')^{\dagger}\tilde{\mathbf{N}}_0(z''). \quad (8.293)$$

If the waves passing across the planes $z = z'$ and $z = z''$ are homogeneous, in the absence of polarization degeneracy or otherwise with an optimal eigenwave basis, the matrices $\tilde{\mathbf{N}}_0$ [see (8.289)] in (8.293) are diagonal [see Section 8.5, Eq. (8.247)]. If, moreover, normalization (8.164) is used,

$$\tilde{\mathbf{N}}_0=\tilde{\mathbf{N}}_0^{-1}=\tilde{\mathbf{U}}_0\equiv\begin{pmatrix}1&0&0&0\\0&1&0&0\\0&0&-1&0\\0&0&0&-1\end{pmatrix}. \quad (8.294)$$

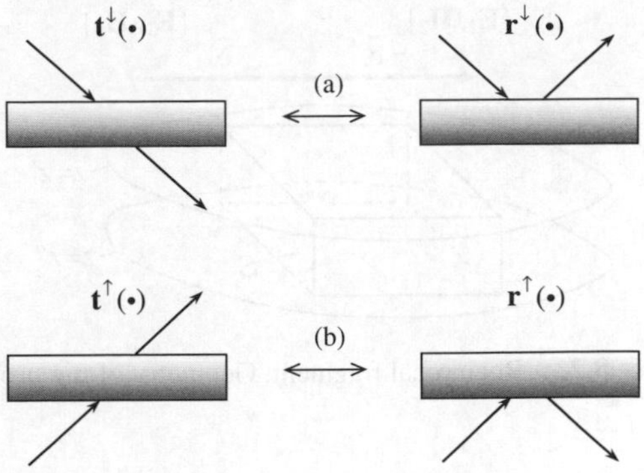

Figure 8.21 Transmission and reflection matrices related by equations (8.295) and (8.296)

Transmission and Reflection Matrices

Let the nonabsorbing fragment (z′, z″) be considered as an OTR unit (see Section 7.1), and let all the eigenwaves in the planes z = z′ and z = z″ be homogeneous. One can show [starting, say, from relation (8.288); we will give a similar derivation in Section 8.6.2] that in this case, the matrices $\mathbf{t}^{\downarrow}(z', z'')$, $\mathbf{t}^{\uparrow}(z', z'')$, $\mathbf{r}^{\downarrow}(z', z'')$, and $\mathbf{r}^{\uparrow}(z', z'')$ satisfy the following relations (see Figure 8.21):

$$\mathbf{t}^{\downarrow}(z', z'')^{\dagger}\mathbf{n}_0^{\downarrow}(z'')\mathbf{t}^{\downarrow}(z', z'') - \mathbf{r}^{\downarrow}(z', z'')^{\dagger}\mathbf{n}_0^{\uparrow}(z')\mathbf{r}^{\downarrow}(z', z'') = \mathbf{n}_0^{\downarrow}(z'), \tag{8.295a}$$

$$\mathbf{t}^{\uparrow}(z', z'')^{\dagger}\mathbf{n}_0^{\uparrow}(z')\mathbf{t}^{\uparrow}(z', z'') - \mathbf{r}^{\uparrow}(z', z'')^{\dagger}\mathbf{n}_0^{\downarrow}(z'')\mathbf{r}^{\uparrow}(z', z'') = \mathbf{n}_0^{\uparrow}(z''), \tag{8.295b}$$

the matrices $\mathbf{n}_0^{\downarrow}$ and $\mathbf{n}_0^{\uparrow}$ have been defined and examined in Section 8.5. With normalization (8.164), the relations (8.295) take the form

$$\mathbf{t}^{\downarrow}(z', z'')^{\dagger}\mathbf{t}^{\downarrow}(z', z'') + \mathbf{r}^{\downarrow}(z', z'')^{\dagger}\mathbf{r}^{\downarrow}(z', z'') = \mathbf{U}, \tag{8.296a}$$

$$\mathbf{t}^{\uparrow}(z', z'')^{\dagger}\mathbf{t}^{\uparrow}(z', z'') + \mathbf{r}^{\uparrow}(z', z'')^{\dagger}\mathbf{r}^{\uparrow}(z', z'') = \mathbf{U}. \tag{8.296b}$$

Relevant sources: References 19–21 and 42.

8.6.2 Properties of Matrix Operators for Reciprocal Regions

Transfer Matrices

Let us assume that within the region (z′, z″) the medium is reciprocal, being possibly absorbing. Using the Lorentz lemma (8.19), one can show that the transfer matrices for this region, considered as functions of **b**, satisfy the conditions

$$\mathbf{P}(z'', z', -\tilde{\mathbf{b}}) = -\mathbf{I}_r \left(\mathbf{P}(z'', z', \tilde{\mathbf{b}})^{-1}\right)^{\mathrm{T}} \mathbf{I}_r, \tag{8.297}$$

$$\mathbf{T}(z'', z', -\tilde{\mathbf{b}}) = \mathbf{N}_r(z'', \tilde{\mathbf{b}})^{-1} \left(\mathbf{T}(z'', z', \tilde{\mathbf{b}})^{-1}\right)^{\mathrm{T}} \mathbf{N}_r(z', \tilde{\mathbf{b}}), \tag{8.298}$$

where

$$\mathbf{N}_r(z, \tilde{\mathbf{b}}) \equiv \boldsymbol{\Psi}(z, \tilde{\mathbf{b}})^{\mathrm{T}}\mathbf{I}_r\boldsymbol{\Psi}(z, -\tilde{\mathbf{b}}) \tag{8.299}$$

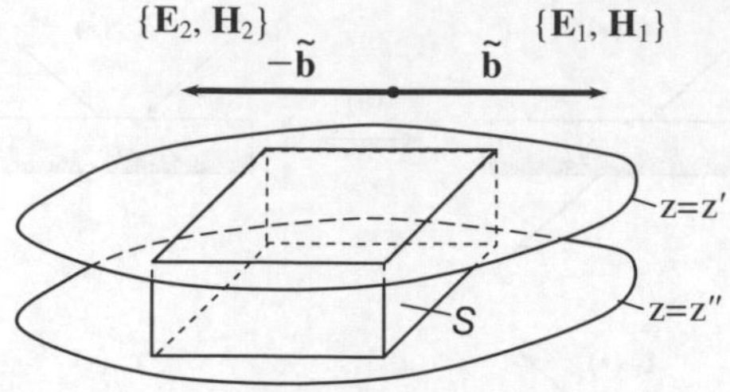

Figure 8.22 Reciprocal fragment. Geometry of the problem

and $\tilde{\mathbf{b}}$ is any given value of the vector $\mathbf{b}$; in these and all other formulas of this section where there are $\mathbf{P}$, $\boldsymbol{\Psi}$, and $\boldsymbol{\psi}$ corresponding to different values of $\mathbf{b}$, these quantities are assumed to be referred to the same reference system (see Section 8.3.4). Let $\{\mathbf{E}_1, \mathbf{H}_1\}$ and $\{\mathbf{E}_2, \mathbf{H}_2\}$ be arbitrary fields of the form

$$\begin{pmatrix} \mathbf{E}_1(\mathbf{r},t) \\ \mathbf{H}_1(\mathbf{r},t) \end{pmatrix} = \begin{pmatrix} \tilde{\mathbf{E}}_1(\mathrm{z}) \\ \tilde{\mathbf{H}}_1(\mathrm{z}) \end{pmatrix} \exp\left[i\left(k_0\tilde{\mathbf{b}}\mathbf{r} - \omega t\right)\right], \tag{8.300a}$$

$$\begin{pmatrix} \mathbf{E}_2(\mathbf{r},t) \\ \mathbf{H}_2(\mathbf{r},t) \end{pmatrix} = \begin{pmatrix} \tilde{\mathbf{E}}_2(\mathrm{z}) \\ \tilde{\mathbf{H}}_2(\mathrm{z}) \end{pmatrix} \exp\left[i\left(-k_0\tilde{\mathbf{b}}\mathbf{r} - \omega t\right)\right], \tag{8.300b}$$

each satisfying the Maxwell equations throughout the space between the planes $\mathrm{z} = \mathrm{z}'$ and $\mathrm{z} = \mathrm{z}''$. Substituting (8.300) into (8.19) and choosing the surface of integration in the same manner as in the previous section (see Figure 8.22), we may see that, according to the Lorentz lemma,

$$\begin{aligned} &\mathbf{z}\left(\tilde{\mathbf{E}}_1(\mathrm{z}') \times \tilde{\mathbf{H}}_2(\mathrm{z}')\right) - \mathbf{z}\left(\tilde{\mathbf{E}}_2(\mathrm{z}') \times \tilde{\mathbf{H}}_1(\mathrm{z}')\right) \\ &= \mathbf{z}\left(\tilde{\mathbf{E}}_1(\mathrm{z}'') \times \tilde{\mathbf{H}}_2(\mathrm{z}'')\right) - \mathbf{z}\left(\tilde{\mathbf{E}}_2(\mathrm{z}'') \times \tilde{\mathbf{H}}_1(\mathrm{z}'')\right). \end{aligned} \tag{8.301}$$

We may rewrite this relation in terms of the $\boldsymbol{\Psi}$- and $\mathbf{A}$-vectors: labeling quantities pertaining to the fields $\{\mathbf{E}_1, \mathbf{H}_1\}$ and $\{\mathbf{E}_2, \mathbf{H}_2\}$ with the subscripts 1 and 2, respectively, we have

$$\Psi_1(\mathrm{z}')^{\mathrm{T}}\mathbf{I}_r\Psi_2(\mathrm{z}') = \Psi_1(\mathrm{z}'')^{\mathrm{T}}\mathbf{I}_r\Psi_2(\mathrm{z}''), \tag{8.302}$$

$$\mathbf{A}_1(\mathrm{z}')^{\mathrm{T}}\mathbf{N}_r(\mathrm{z}', \tilde{\mathbf{b}})\mathbf{A}_2(\mathrm{z}') = \mathbf{A}_1(\mathrm{z}'')^{\mathrm{T}}\mathbf{N}_r(\mathrm{z}'', \tilde{\mathbf{b}})\mathbf{A}_2(\mathrm{z}''); \tag{8.303}$$

the matrix $\mathbf{I}_r$ has been defined in Section 8.4.1 [see (8.213)]. Substitution of the expressions

$$\Psi_1(\mathrm{z}'') = \mathbf{P}(\mathrm{z}'', \mathrm{z}', \tilde{\mathbf{b}})\Psi_1(\mathrm{z}'), \quad \Psi_2(\mathrm{z}'') = \mathbf{P}(\mathrm{z}'', \mathrm{z}', -\tilde{\mathbf{b}})\Psi_2(\mathrm{z}')$$

into (8.302) gives

$$\Psi_1(\mathrm{z}')^{\mathrm{T}}\mathbf{I}_r\Psi_2(\mathrm{z}') = \Psi_1(\mathrm{z}')^{\mathrm{T}}\mathbf{P}(\mathrm{z}'', \mathrm{z}', \tilde{\mathbf{b}})^{\mathrm{T}}\mathbf{I}_r\mathbf{P}(\mathrm{z}'', \mathrm{z}', -\tilde{\mathbf{b}})\Psi_2(\mathrm{z}').$$

This relation must be satisfied at any $\Psi_1(\mathrm{z}')$ and $\Psi_2(\mathrm{z}')$, which is possible only if

$$\mathbf{I}_r = \mathbf{P}(\mathrm{z}'', \mathrm{z}', \tilde{\mathbf{b}})^{\mathrm{T}}\mathbf{I}_r\mathbf{P}(\mathrm{z}'', \mathrm{z}', -\tilde{\mathbf{b}}). \tag{8.304}$$

Condition (8.304) is equivalent to condition (8.297). Similarly, starting from (8.303), one can prove the validity of (8.298).

Transmission and Reflection Matrices

For any z from the interval (z′, z″), parameters of the eigenwave bases satisfy (8.210), which determines the following form of the matrix $\mathbf{N}_r(z, \tilde{\mathbf{b}})$ [see (8.299)]:

$$\mathbf{N}_r(z, \tilde{\mathbf{b}}) = \begin{pmatrix} \mathbf{O} & \mathbf{n}_{r12}(z, \tilde{\mathbf{b}}) \\ \mathbf{n}_{r21}(z, \tilde{\mathbf{b}}) & \mathbf{O} \end{pmatrix}, \tag{8.305}$$

where

$$\mathbf{n}_{r12} = \begin{pmatrix} n_{r13} & n_{r14} \\ n_{r23} & n_{r24} \end{pmatrix}, \quad \mathbf{n}_{r21} = \begin{pmatrix} n_{r31} & n_{r32} \\ n_{r41} & n_{r42} \end{pmatrix}, \quad \mathbf{O} = \begin{pmatrix} 0 & 0 \\ 0 & 0 \end{pmatrix}, \tag{8.306a}$$

$$n_{rjk}(z, \tilde{\mathbf{b}}) = \boldsymbol{\psi}_j(z, \tilde{\mathbf{b}})^{\mathrm{T}} \mathbf{I}_r \boldsymbol{\psi}_k(z, -\tilde{\mathbf{b}}) = \mathbf{z} \cdot \left[\mathbf{e}_j(z, \tilde{\mathbf{b}}) \times \mathbf{h}_k(z, -\tilde{\mathbf{b}}) - \mathbf{e}_k(z, -\tilde{\mathbf{b}}) \times \mathbf{h}_j(z, \tilde{\mathbf{b}})\right]. \tag{8.306b}$$

Representing the vectors $\mathbf{A}_1$ and $\mathbf{A}_2$ as

$$\mathbf{A}_1 = \begin{pmatrix} \mathbf{a}_1^{\downarrow} \\ \mathbf{a}_1^{\uparrow} \end{pmatrix}, \quad \mathbf{A}_2 = \begin{pmatrix} \mathbf{a}_2^{\downarrow} \\ \mathbf{a}_2^{\uparrow} \end{pmatrix}$$

[see (8.93)] and using (8.305), we may rewrite (8.303) as follows:

$$\begin{aligned} &\mathbf{a}_1^{\downarrow}(z')^{\mathrm{T}} \mathbf{n}_{r12}(z', \tilde{\mathbf{b}}) \mathbf{a}_2^{\uparrow}(z') + \mathbf{a}_1^{\uparrow}(z')^{\mathrm{T}} \mathbf{n}_{r21}(z', \tilde{\mathbf{b}}) \mathbf{a}_2^{\downarrow}(z') \\ &\quad = \mathbf{a}_1^{\downarrow}(z'')^{\mathrm{T}} \mathbf{n}_{r12}(z'', \tilde{\mathbf{b}}) \mathbf{a}_2^{\uparrow}(z'') + \mathbf{a}_1^{\uparrow}(z'')^{\mathrm{T}} \mathbf{n}_{r21}(z'', \tilde{\mathbf{b}}) \mathbf{a}_2^{\downarrow}(z''). \end{aligned} \tag{8.307}$$

This relation must be satisfied at any values of the **a**-vectors characterizing the incident fields, which implies the validity of the following equations:

$$\mathbf{t}^{\uparrow}(z', z'', -\tilde{\mathbf{b}}) = \mathbf{n}_{r12}(z', \tilde{\mathbf{b}})^{-1} \mathbf{t}^{\downarrow}(z', z'', \tilde{\mathbf{b}})^{\mathrm{T}} \mathbf{n}_{r12}(z'', \tilde{\mathbf{b}}), \tag{8.308a}$$

$$\mathbf{t}^{\downarrow}(z', z'', -\tilde{\mathbf{b}}) = \mathbf{n}_{r21}(z'', \tilde{\mathbf{b}})^{-1} \mathbf{t}^{\uparrow}(z', z'', \tilde{\mathbf{b}})^{\mathrm{T}} \mathbf{n}_{r21}(z', \tilde{\mathbf{b}}), \tag{8.308b}$$

$$\mathbf{r}^{\downarrow}(z', z'', -\tilde{\mathbf{b}}) = -\mathbf{n}_{r12}(z', \tilde{\mathbf{b}})^{-1} \mathbf{r}^{\downarrow}(z', z'', \tilde{\mathbf{b}})^{\mathrm{T}} \mathbf{n}_{r21}(z', \tilde{\mathbf{b}}), \tag{8.308c}$$

$$\mathbf{r}^{\uparrow}(z', z'', -\tilde{\mathbf{b}}) = -\mathbf{n}_{r21}(z'', \tilde{\mathbf{b}})^{-1} \mathbf{r}^{\uparrow}(z', z'', \tilde{\mathbf{b}})^{\mathrm{T}} \mathbf{n}_{r12}(z'', \tilde{\mathbf{b}}) \tag{8.308d}$$

(see Figure 8.23). Thus, at $\mathbf{a}_1^{\uparrow}(z'') = \widehat{\mathbf{0}}$ and $\mathbf{a}_2^{\downarrow}(z') = \widehat{\mathbf{0}}$, the following relations for the vectors $\mathbf{a}_1^{\downarrow}(z')$, $\mathbf{a}_1^{\downarrow}(z'')$, $\mathbf{a}_2^{\uparrow}(z')$, and $\mathbf{a}_2^{\uparrow}(z'')$ are satisfied:

$$\mathbf{a}_1^{\downarrow}(z'') = \mathbf{t}^{\downarrow}(z', z'', \tilde{\mathbf{b}}) \mathbf{a}_1^{\downarrow}(z'), \quad \mathbf{a}_2^{\uparrow}(z') = \mathbf{t}^{\uparrow}(z', z'', -\tilde{\mathbf{b}}) \mathbf{a}_2^{\uparrow}(z''), \tag{8.309}$$

$$\mathbf{a}_1^{\downarrow}(z')^{\mathrm{T}} \mathbf{n}_{r12}(z', \tilde{\mathbf{b}}) \mathbf{a}_2^{\uparrow}(z') = \mathbf{a}_1^{\downarrow}(z'')^{\mathrm{T}} \mathbf{n}_{r12}(z'', \tilde{\mathbf{b}}) \mathbf{a}_2^{\uparrow}(z''). \tag{8.310}$$

On substituting from (8.309) into (8.310), we see that these relations cannot hold simultaneously at any $\mathbf{a}_1^{\downarrow}(z')$ and $\mathbf{a}_2^{\uparrow}(z'')$ unless the transmission matrices meet (8.308a).

The operation characterized by the matrix $\mathbf{t}^{\uparrow}(z', z'', -\tilde{\mathbf{b}})$ may be considered as the reversed operation in relation to the operation characterized by the matrix $\mathbf{t}^{\downarrow}(z', z'', \tilde{\mathbf{b}})$, the operation characterized by the matrix $\mathbf{r}^{\downarrow}(z', z'', -\tilde{\mathbf{b}})$ may be considered as the reversed operation relatively to that characterized by the matrix $\mathbf{r}^{\downarrow}(z', z'', \tilde{\mathbf{b}})$ (see Figure 8.23) and so on. Using the term "reversed operation" in this sense, we may write any of relations (8.308) as follows:

$$\mathbf{t}\left\{\mathsf{O}^R\right\} = \mathbf{n}_R\left\{\mathbf{X}_{\mathrm{out}}(\mathsf{O}^R), z_{\mathrm{out}}(\mathsf{O}^R)\right\}^{-1} \mathbf{t}\left\{\mathsf{O}\right\}^{\mathrm{T}} \mathbf{n}_R\left\{\mathbf{X}_{\mathrm{inp}}(\mathsf{O}^R), z_{\mathrm{inp}}(\mathsf{O}^R)\right\}, \tag{8.311}$$

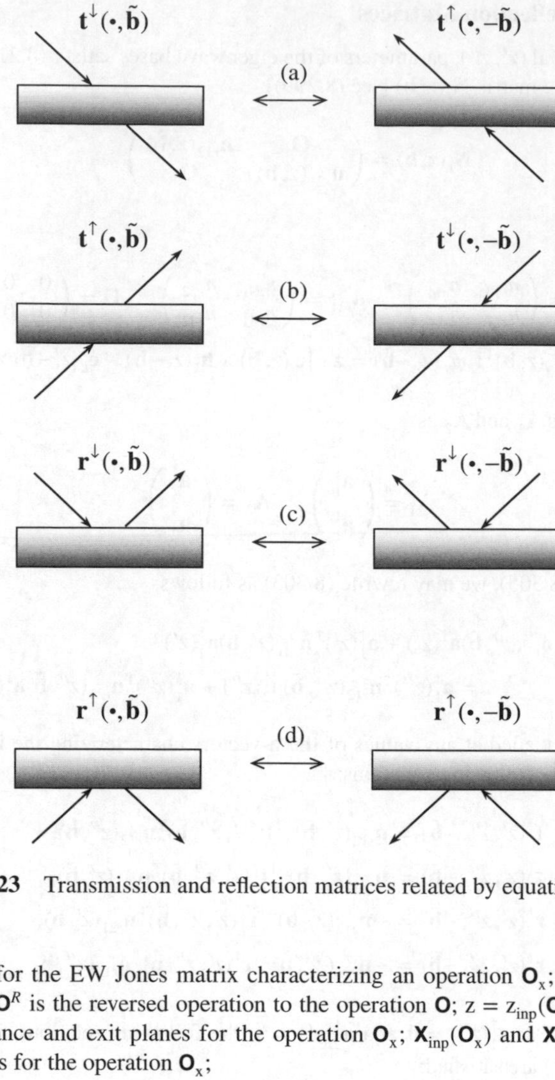

Figure 8.23 Transmission and reflection matrices related by equations (8.308)

where $\mathbf{t}\{\mathsf{O}_x\}$ stands for the EW Jones matrix characterizing an operation O_x; O is a transmission or reflection operation, O^R is the reversed operation to the operation O; $z = z_{inp}(\mathsf{O}_x)$ and $z = z_{out}(\mathsf{O}_x)$ are respectively the entrance and exit planes for the operation O_x; $\mathsf{X}_{inp}(\mathsf{O}_x)$ and $\mathsf{X}_{out}(\mathsf{O}_x)$ are respectively input and output fields for the operation O_x;

$$\mathbf{n}_R\{\mathsf{X}, z\} = \begin{cases} \mathbf{n}_{r21}(z,\tilde{\mathbf{b}}) \text{ if } \mathsf{X} \text{ is forward propagating} \\ -\mathbf{n}_{r12}(z,\tilde{\mathbf{b}}) \text{ if } \mathsf{X} \text{ is backward propagating,} \end{cases} \tag{8.312}$$

where $\tilde{\mathbf{b}}$ is the value of $\mathbf{b}$ for the operation O.

General Reciprocity Relation for Elementary Transfer Channels

It is important that reciprocity relations similar to (8.311) are fulfilled not only for the operators of the overall transmission and reflection but also for the operators of nonbranched transfer channels like those shown in Figures 8.13–8.15. Let C be a chain of N elementary operations $\mathsf{O}_1, \mathsf{O}_2, \ldots, \mathsf{O}_N$,

$$\mathsf{C} = \mathsf{O}_N \mathsf{O}_{N-1} \cdot \ldots \cdot \mathsf{O}_3 \mathsf{O}_2 \mathsf{O}_1. \tag{8.313}$$

The reversed channel $\mathbf{C}^R$ to the channel $\mathbf{C}$ may be defined as

$$\mathbf{C}^R = \mathbf{O}_1^R\mathbf{O}_2^R\mathbf{O}_3^R \cdot \ldots \cdot \mathbf{O}_{N-1}^R\,\mathbf{O}_N^R, \tag{8.314}$$

where $\mathbf{O}_j^R$ is the reversed operation to the operation $\mathbf{O}_j$ ($j = 1,2, \ldots, N$); the operators of the operations $\mathbf{O}_j$ and $\mathbf{O}_j^R$ are assumed to satisfy (8.311). One can show that the transmission EW Jones matrices of the channels $\mathbf{C}$ and $\mathbf{C}^R$ are related by

$$\mathbf{t}\left\{\mathbf{C}^R\right\} = \mathbf{n}_R\left\{\mathbf{X}_{\text{out}}(\mathbf{C}^R), z_{\text{out}}(\mathbf{C}^R)\right\}^{-1}\mathbf{t}\left\{\mathbf{C}\right\}^{\mathrm{T}}\mathbf{n}_R\left\{\mathbf{X}_{\text{inp}}(\mathbf{C}^R), z_{\text{inp}}(\mathbf{C}^R)\right\}. \tag{8.315}$$

Now we give a proof of (8.315) for the case $N = 2$. Generalization to the case of a channel being a chain of an arbitrary number of elementary operations can trivially be done by induction. With $N = 2$, we may represent the EW Jones matrices of the channels $\mathbf{C}$ and $\mathbf{C}^R$ as follows:

$$\mathbf{t}\{\mathbf{C}\} = \mathbf{t}\left\{\mathbf{O}_2\mathbf{O}_1\right\} = \mathbf{t}\left\{\mathbf{O}_2\right\}\mathbf{t}\left\{\mathbf{O}_1\right\}, \tag{8.316}$$

$$\mathbf{t}\left\{\mathbf{C}^R\right\} = \mathbf{t}\left\{\mathbf{O}_1^R\mathbf{O}_2^R\right\} = \mathbf{t}\left\{\mathbf{O}_1^R\right\}\mathbf{t}\left\{\mathbf{O}_2^R\right\}. \tag{8.317}$$

From (8.311) we have

$$\begin{aligned}\mathbf{t}\left\{\mathbf{O}_1^R\right\} &= \mathbf{n}_R\left\{\mathbf{X}_{\text{out}}\left(\mathbf{O}_1^R\right), z_{\text{out}}\left(\mathbf{O}_1^R\right)\right\}^{-1}\mathbf{t}\{\mathbf{O}_1\}^{\mathrm{T}}\mathbf{n}_R\left\{\mathbf{X}_{\text{inp}}\left(\mathbf{O}_1^R\right), z_{\text{inp}}\left(\mathbf{O}_1^R\right)\right\},\\ \mathbf{t}\left\{\mathbf{O}_2^R\right\} &= \mathbf{n}_R\left\{\mathbf{X}_{\text{out}}\left(\mathbf{O}_2^R\right), z_{\text{out}}\left(\mathbf{O}_2^R\right)\right\}^{-1}\mathbf{t}\{\mathbf{O}_2\}^{\mathrm{T}}\mathbf{n}_R\left\{\mathbf{X}_{\text{inp}}\left(\mathbf{O}_2^R\right), z_{\text{inp}}\left(\mathbf{O}_2^R\right)\right\}.\end{aligned} \tag{8.318}$$

Substitution from (8.318) into (8.317) gives

$$\begin{aligned}\mathbf{t}\left\{\mathbf{C}^R\right\} = {}&\mathbf{n}_R\left\{\mathbf{X}_{\text{out}}\left(\mathbf{O}_1^R\right), z_{\text{out}}\left(\mathbf{O}_1^R\right)\right\}^{-1}\mathbf{t}\{\mathbf{O}_1\}^{\mathrm{T}}\mathbf{n}_R\left\{\mathbf{X}_{\text{inp}}\left(\mathbf{O}_1^R\right), z_{\text{inp}}\left(\mathbf{O}_1^R\right)\right\}\cdot\\ &\mathbf{n}_R\left\{\mathbf{X}_{\text{out}}\left(\mathbf{O}_2^R\right), z_{\text{out}}\left(\mathbf{O}_2^R\right)\right\}^{-1}\mathbf{t}\{\mathbf{O}_2\}^{\mathrm{T}}\mathbf{n}_R\left\{\mathbf{X}_{\text{inp}}\left(\mathbf{O}_2^R\right), z_{\text{inp}}\left(\mathbf{O}_2^R\right)\right\}.\end{aligned} \tag{8.319}$$

Since $\mathbf{X}_{\text{inp}}(\mathbf{O}_1^R) = \mathbf{X}_{\text{out}}(\mathbf{O}_2^R)$ and $z_{\text{inp}}(\mathbf{O}_1^R) = z_{\text{out}}(\mathbf{O}_2^R)$, we may reduce (8.319) as follows:

$$\begin{aligned}\mathbf{t}\left\{\mathbf{C}^R\right\} &= \mathbf{n}_R\left\{\mathbf{X}_{\text{out}}\left(\mathbf{O}_1^R\right), z_{\text{out}}\left(\mathbf{O}_1^R\right)\right\}^{-1}\mathbf{t}\{\mathbf{O}_1\}^{\mathrm{T}}\mathbf{t}\{\mathbf{O}_2\}^{\mathrm{T}}\mathbf{n}_R\left\{\mathbf{X}_{\text{inp}}\left(\mathbf{O}_2^R\right), z_{\text{inp}}\left(\mathbf{O}_2^R\right)\right\}\\ &= \mathbf{n}_R\left\{\mathbf{X}_{\text{out}}\left(\mathbf{O}_1^R\right), z_{\text{out}}\left(\mathbf{O}_1^R\right)\right\}^{-1}\left(\mathbf{t}\{\mathbf{O}_2\}\mathbf{t}\{\mathbf{O}_1\}\right)^{\mathrm{T}}\mathbf{n}_R\left\{\mathbf{X}_{\text{inp}}\left(\mathbf{O}_2^R\right), z_{\text{inp}}\left(\mathbf{O}_2^R\right)\right\}\\ &= \mathbf{n}_R\left\{\mathbf{X}_{\text{out}}\left(\mathbf{O}_1^R\right), z_{\text{out}}\left(\mathbf{O}_1^R\right)\right\}^{-1}\mathbf{t}\{\mathbf{O}_2\mathbf{O}_1\}^{\mathrm{T}}\mathbf{n}_R\left\{\mathbf{X}_{\text{inp}}\left(\mathbf{O}_2^R\right), z_{\text{inp}}\left(\mathbf{O}_2^R\right)\right\}.\end{aligned} \tag{8.320}$$

Taking into account that

$$\begin{aligned}\mathbf{X}_{\text{out}}\left(\mathbf{O}_1^R\right) &= \mathbf{X}_{\text{out}}(\mathbf{C}^R), \quad z_{\text{out}}\left(\mathbf{O}_1^R\right) = z_{\text{out}}(\mathbf{C}^R),\\ \mathbf{X}_{\text{inp}}\left(\mathbf{O}_2^R\right) &= \mathbf{X}_{\text{inp}}(\mathbf{C}^R), \quad z_{\text{inp}}\left(\mathbf{O}_2^R\right) = z_{\text{inp}}(\mathbf{C}^R)\end{aligned}$$

and (8.316), we see that relation (8.320) is equivalent to (8.315).

Reciprocity Matrices and Particular Forms of the Reciprocity Relations

We will call the matrices $\mathbf{n}_{r12}$ and $\mathbf{n}_{r21}$ entering into the above equations the *reciprocity matrices*. In the absence of polarization degeneracy, as well as in the presence of polarization degeneracy provided that the optimal reciprocal bases [see Section 8.4.1, Eq. (8.214)] are used, the matrices $\mathbf{n}_{r12}$ and $\mathbf{n}_{r21}$ are diagonal and, according to (8.306) and (8.214), can be represented as

$$\mathbf{n}_{r12}(z,\tilde{\mathbf{b}}) = \begin{pmatrix} -2\mathbf{z}\left[\mathbf{e}_1(z,\tilde{\mathbf{b}})\times\mathbf{h}_1(z,\tilde{\mathbf{b}})\right]b_{31}(z) & 0 \\ 0 & -2\mathbf{z}\left[\mathbf{e}_2(z,\tilde{\mathbf{b}})\times\mathbf{h}_2(z,\tilde{\mathbf{b}})\right]b_{42}(z) \end{pmatrix}, \tag{8.321a}$$

$$\mathbf{n}_{r21}(z,\tilde{\mathbf{b}}) = \begin{pmatrix} -2\mathbf{z}\left[\mathbf{e}_3(z,\tilde{\mathbf{b}})\times\mathbf{h}_3(z,\tilde{\mathbf{b}})\right]b_{13}(z) & 0 \\ 0 & -2\mathbf{z}\left[\mathbf{e}_4(z,\tilde{\mathbf{b}})\times\mathbf{h}_4(z,\tilde{\mathbf{b}})\right]b_{24}(z) \end{pmatrix}. \tag{8.321b}$$

It is seen from (8.321) that if the symmetrical normalization (8.167) is used,

$$\mathbf{n}_{r12}(z,\tilde{\mathbf{b}}) = \begin{pmatrix} -b_{31}(z) & 0 \\ 0 & -b_{42}(z) \end{pmatrix}, \quad \mathbf{n}_{r21}(z,\tilde{\mathbf{b}}) = \begin{pmatrix} b_{13}(z) & 0 \\ 0 & b_{24}(z) \end{pmatrix}. \tag{8.322}$$

If the eigenwave bases are chosen so that

$$\mathbf{e}_1(-\tilde{\mathbf{b}}) = \mathbf{e}_3(\tilde{\mathbf{b}}), \quad \mathbf{e}_2(-\tilde{\mathbf{b}}) = \mathbf{e}_4(\tilde{\mathbf{b}}), \quad \mathbf{e}_3(-\tilde{\mathbf{b}}) = \mathbf{e}_1(\tilde{\mathbf{b}}), \quad \mathbf{e}_4(-\tilde{\mathbf{b}}) = \mathbf{e}_2(\tilde{\mathbf{b}}) \tag{8.323}$$

(the argument z is omitted for brevity), according to (8.214) and (8.322)

$$\mathbf{n}_{r12} = -\mathbf{U}, \quad \mathbf{n}_{r21} = \mathbf{U}. \tag{8.324}$$

If the reciprocity matrices meet (8.324) in the entrance and exit planes of a channel C, the relation (8.315) for this channel is reduced to

$$\mathbf{t}\left\{\mathsf{C}^R\right\} = \mathbf{t}\left\{\mathsf{C}\right\}^{\mathrm{T}}, \tag{8.325}$$

that is, the matrix $\mathbf{t}\left\{\mathsf{C}^R\right\}$ may be obtained by transposing the matrix $\mathbf{t}\left\{\mathsf{C}\right\}$ [cf. (1.257)]. If the eigenwave bases are such that

$$\mathbf{e}_1(-\tilde{\mathbf{b}}) = \mathbf{e}_3(\tilde{\mathbf{b}}), \quad \mathbf{e}_2(-\tilde{\mathbf{b}}) = -\mathbf{e}_4(\tilde{\mathbf{b}}), \quad \mathbf{e}_3(-\tilde{\mathbf{b}}) = \mathbf{e}_1(\tilde{\mathbf{b}}), \quad \mathbf{e}_4(-\tilde{\mathbf{b}}) = -\mathbf{e}_2(\tilde{\mathbf{b}}), \tag{8.326}$$

according to (8.214) and (8.322)

$$\mathbf{n}_{r21} = -\mathbf{n}_{r12} = \mathbf{I}_1 \equiv \begin{pmatrix} 1 & 0 \\ 0 & -1 \end{pmatrix}. \tag{8.327}$$

If the matrices $\mathbf{n}_{r12}$ and $\mathbf{n}_{r21}$ in the entrance and exit planes of a channel C satisfy (8.327), the relation (8.315) for this channel takes the form

$$\mathbf{t}\{\mathsf{C}^R\} = \mathbf{I}_1\mathbf{t}\{\mathsf{C}\}^{\mathrm{T}}\mathbf{I}_1, \tag{8.328}$$

or, in components,

$$\begin{pmatrix} t_{11}\{\mathsf{C}^R\} & t_{12}\{\mathsf{C}^R\} \\ t_{21}\{\mathsf{C}^R\} & t_{22}\{\mathsf{C}^R\} \end{pmatrix} = \begin{pmatrix} t_{11}\{\mathsf{C}\} & -t_{21}\{\mathsf{C}\} \\ -t_{12}\{\mathsf{C}\} & t_{22}\{\mathsf{C}\} \end{pmatrix}, \tag{8.329}$$

where $t_{kj}\{\mathsf{C}\}$ and $t_{kj}\{\mathsf{C}^R\}$ are components of the matrices $\mathbf{t}\{\mathsf{C}\}$ and $\mathbf{t}\{\mathsf{C}^R\}$, respectively [cf. (1.256)]. The relation connecting $\mathbf{t}\{\mathsf{C}\}$ and $\mathbf{t}\{\mathsf{C}^R\}$ may have a simple form, (8.325) or (8.328), not only with normalization (8.167), but also with the electrical normalization (8.283) provided that the entrance and exit media of the channel **C** are isotropic and identical (see Section 9.2).

Berreman Transfer Matrices for a Nonabsorbing Reciprocal Region

Suppose that the medium within the region (z′, z″) is not only reciprocal but also nonabsorbing. In this case, $\mathbf{P}(z'', z', \mathbf{b})$ satisfies both (8.297) and (8.292). The conditions (8.297) and (8.292) together determine the following property of $\mathbf{P}(z'', z', \mathbf{b})$

$$\mathbf{P}(z'', z', -\tilde{\mathbf{b}}) = \mathbf{U}_0\mathbf{P}(z'', z', \tilde{\mathbf{b}})^*\mathbf{U}_0, \tag{8.330}$$

where

$$\mathbf{U}_0 = \begin{pmatrix} 1 & 0 & 0 & 0 \\ 0 & -1 & 0 & 0 \\ 0 & 0 & 1 & 0 \\ 0 & 0 & 0 & -1 \end{pmatrix}.$$

The relations presented in this section are useful in analysis of situations like those discussed in Sections 1.4.2 and 1.4.3, allowing consideration of oblique incidence and much more complicated optical systems. In many cases, with the help of these relations one can significantly reduce the computational cost in calculating the viewing angle characteristics of LCDs.

Relevant sources: References 43–47 and 31.

8.7 Calculation of EW 4 × 4 Transfer Matrices for LC Layers

We should make some remarks concerning the calculation of EW 4 × 4 transfer matrices for LC layers with the use of the staircase approximation.

As an example, we consider a 1D-inhomogeneous layer of a locally uniaxial LC with LC director field $\mathbf{n}(z)$. Let $z = z_1$ and $z = z_2$ ($z_1 < z_2$) be the planes of the interfaces of this LC layer with the adjacent layers. As usual, the principal refractive indices of the liquid crystal are assumed to be spatially invariant.

Another Variant of the Staircase Model

In Section 8.3.3, we considered a staircase model (approximating multilayer) consisting of homogeneous slices of equal thickness. When dealing with LC layers, in many cases, it is more convenient to use staircase models where the two outermost slices are half as thick as the other slices. The optical parameters of these outermost slices are taken to be equal to the corresponding parameters of the LC layer just at its boundaries. In the example under consideration, the internal interfaces of such an approximating multilayer will coincide with the planes

$$z = z_{1j} = z_1 + h\left(j - \frac{1}{2}\right) \qquad j = 1, \ldots N-1, \qquad h = d/(N-1), \tag{8.331a}$$

where N is the number of the slices and d is the thickness of the LC layer. Denote the thickness of the jth slice and a unit vector directed along the optic axis of this slice by d_j and $\mathbf{c}_j$, respectively. In our case,

$$d_j = \begin{cases} h/2 & j = 1, N \\ h & j = 2, \ldots, N-1. \end{cases} \tag{8.331b}$$

The optic axes of the slices are taken as

$$\mathbf{c}_j = \begin{cases} \mathbf{n}(\mathrm{z}_1 + 0) & j = 1 \\ \mathbf{n}\left(\mathrm{z}_1 + (j-1)h\right) & j = 2, \ldots, N-1 \\ \mathbf{n}(\mathrm{z}_2 - 0) & j = N. \end{cases} \tag{8.331b}$$

In contrast to the staircase model described in Section 8.3.3, for this model (i) all parameters of the outermost slices except their thicknesses are independent of N, and hence the eigenwave bases for these slices can be chosen independent of N, (ii) at any N, the EW transfer matrices of the fragments adjacent to the bulk of the LC layer (the fragments including the interfaces $\mathrm{z} = \mathrm{z}_1$ and $\mathrm{z} = \mathrm{z}_2$) are exact. It is these points that make this variant of staircase model preferable. We employed this variant of staircase model in all numerical examples given in this book where the EW 4 × 4 transfer matrix method is applied to LC layers.

EW Transfer Matrices for Different Variants of OTR Units Including the LC Layer

Recall that the 4 × 4 transfer matrix methods are used for calculating characteristics of the overall transmission and overall reflection of layered systems. In the modeling of LCDs, they are employed to calculate transmission and reflection operators for fragments of LCD panels that are regarded as OTR units (see Section 7.1). In different kinds of calculations, the OTR unit including the LC layer is chosen differently (see, e.g., Figures 8.13b, 8.13c, 8.13e, and 8.13f and examples in Section 10.2). It may be the bulk of the LC layer only, or the bulk of the LC layer + the interfaces of this layer with the adjacent layers or one of them, or the bulk of the LC layer + the adjacent thin-layer systems or one of them, and so on. Let us give expressions for EW transfer matrices for three of the listed variants of the OTR unit, taken as examples, adopting the above staircase model for the LC layer.

Bulk of LC layer. The EW transfer matrix for the bulk of the LC layer, $\mathbf{T}(\mathrm{z}_2 - 0, \mathrm{z}_1 + 0)$, may be expressed as follows:

$$\mathbf{T}(\mathrm{z}_2 - 0, \mathrm{z}_1 + 0) = \mathbf{C}_N \mathbf{\Psi}_N^{-1} \mathbf{P}_{(N-1,2)} \mathbf{\Psi}_1 \mathbf{C}_1, \tag{8.332}$$

where $\mathbf{C}_j$ is the transfer matrix for the bulk of the jth slice of the staircase model of the LC layer, $\mathbf{\Psi}_j$ is the $\mathbf{\Psi}$-matrix for the jth slice, and $\mathbf{P}_{(k,j)}$ is the Berreman matrix of the pile consisting of the slices with numbers from j to k. The diagonal matrix $\mathbf{C}_j$ can be represented as

$$\mathbf{C}_j = \tilde{\mathbf{T}}\left(\sigma_l^{(j)}, d_j\right) \tag{8.333}$$

[see (8.85)], where $\sigma_l^{(j)}$ $(l = 1, 2, 3, 4)$ are values of σ_l for the jth slice. The matrix $\mathbf{P}_{(k,j)}$ can be expressed by the recurrent formula

$$\mathbf{P}_{(k,j)} = \mathbf{P}_k \mathbf{P}_{(k-1,j)}, \tag{8.334}$$

where $\mathbf{P}_k$ is the Berreman matrix of the kth slice; $\mathbf{P}_{(j,j)} = \mathbf{P}_j$.

Another representation of the matrix $\mathbf{T}(z_2 - 0, z_1 + 0)$ is

$$\mathbf{T}(z_2 - 0, z_1 + 0) = \mathbf{T}_{(N,2)}\mathbf{C}_1, \quad (8.335)$$

where $\mathbf{T}_{(k,j)}$ is the EW transfer matrix of the domain consisting of the fragments {interface with the previous slice + bulk of the slice} for the slices with numbers from j to k. It is convenient to express the matrix $\mathbf{T}_{(k,j)}$ by the following recurrent formulas

$$\mathbf{T}_{(k,j)} = \mathbf{C}_k\mathbf{B}_{(k,j)}, \quad (8.336a)$$

$$\mathbf{B}_{(k,j)} = \mathbf{T}_{|k,k-1}\mathbf{T}_{(k-1,j)}, \quad (8.336b)$$

$$\mathbf{T}_{|k,k-1} = \mathbf{\Psi}_k^{-1}\mathbf{\Psi}_{k-1}, \quad (8.336c)$$

$$\mathbf{B}_{(j,j)} = \mathbf{T}_{|j,j-1}. \quad (8.336d)$$

Here the matrix $\mathbf{T}_{|k,k-1}$ is the EW transfer matrix of the interface between the kth and k–1th layers.

From the matrix $\mathbf{T}(z_2 - 0, z_1 + 0)$ determined to sufficient accuracy, one can find accurate values of transmission and reflection EW Jones matrices of the LC layer bulk. This was used, for example, in numerical experiments whose results are presented in Section 11.2

LC layer between thin-layer systems. Suppose that the LC layer is situated between two thin-layer systems, a system A with external interfaces at $z = z_{S1}$ and $z = z_1$ and a system B with external interfaces at $z = z_2$ and $z = z_{S2}$, characterized by the Berreman matrices $\mathbf{P}_A$ and $\mathbf{P}_B$, respectively. Let the whole system be surrounded by media S_1 ($z < z_{S1}$) and S_2 ($z > z_{S2}$) (usually S_1 and S_2 represent glass substrates) characterized by their $\mathbf{\Psi}$-matrices $\mathbf{\Psi}_{S1}$ and $\mathbf{\Psi}_{S2}$, respectively. Then the EW transfer matrix characterizing the whole system, $\mathbf{T}(z_{S2} + 0, z_{S1} - 0)$, can be expressed as follows:

$$\mathbf{T}(z_{S2} + 0, z_{S1} - 0) = \mathbf{\Psi}_{S2}^{-1}\mathbf{P}_B\mathbf{P}_{(N,1)}\mathbf{P}_A\mathbf{\Psi}_{S1} \quad (8.337)$$

and

$$\mathbf{T}(z_{S2} + 0, z_{S1} - 0) = \mathbf{\Psi}_{S2}^{-1}\mathbf{P}_B\mathbf{\Psi}_N\mathbf{T}_{(N,2)}\mathbf{C}_1\mathbf{\Psi}_1^{-1}\mathbf{P}_A\mathbf{\Psi}_{S1}. \quad (8.338)$$

Thin-layer system + bulk of LC layer. For the same layered system as in the previous example, the EW transfer matrix of the fragment consisting of system A and the bulk of the LC layer can be expressed as

$$\mathbf{T}(z_2 - 0, z_{S1} - 0) = \mathbf{C}_N\mathbf{\Psi}_N^{-1}\mathbf{P}_{(N-1,1)}\mathbf{P}_A\mathbf{\Psi}_{S1} \quad (8.339)$$

and

$$\mathbf{T}(z_2 - 0, z_{S1} - 0) = \mathbf{T}_{(N,2)}\mathbf{C}_1\mathbf{\Psi}_1^{-1}\mathbf{P}_A\mathbf{\Psi}_{S1}. \quad (8.340)$$

The last two variants of the OTR unit including the LC layer are used in examples of Section 10.2.

Efficient Techniques of Involving Slices of the Approximating Multilayer

It is seen from the above formulas that in numerical calculations of the transfer matrices, involving the slices can be performed using the recursion (8.334) or the recursion (8.336). Numerical tests show that optimized (in speed) computational techniques employing the recursion (8.334) with calculation of the matrices $\mathbf{P}_j$ for slices by using the polynomial representations (see Section 8.3.2) and that employing the recursion (8.336) with the use of the fast method of inverting $\mathbf{\Psi}$-matrices described in Section 8.4.1 and formulas for the parameters of the basis eigenwaves presented in Section 9.3 are alike in computational cost. Our tests have shown that in the most interesting case of a nonabsorbing LC, the latter technique is a little more efficient than the former ones. We note here some calculation details increasing the efficiency of the technique based on the recursion (8.336).

In the case of a nonabsorbing LC, the recursion (8.336) allows performing the calculations without matrix multiplications of general complex matrices, since the matrices $\mathbf{\Psi}_j$ are real and the matrices $\mathbf{C}_j$ are diagonal: stage (8.336c) includes multiplying real matrices; stage (8.336b) is a multiplication of a real matrix by a complex matrix; in stage (8.336a), a general complex matrix is pre-multiplied by a diagonal complex matrix. For any diagonal $n \times n$ matrix $\mathbf{C} = [c_{jk}]$ and any $n \times n$ matrix $\mathbf{A} = [a_{jk}]$, the elements of the matrix $\mathbf{B} = [b_{jk}] = \mathbf{CA}$ can be expressed as follows:

$$b_{jk} = c_{jj} a_{jk} \qquad j, k = 1, \ldots, n. \tag{8.341}$$

This formula is used in stage (8.336a) of the optimized technique. In our tests, when calculations by this technique were performed without normalization of eigenwave bases (except for the outermost slices), the computation times were ~15% less than those for the techniques using (8.334). When basis normalization was used for all slices (for various reasons, we prefer this variant in practical calculations), the computation times for this technique were only a few percent smaller than for the techniques using (8.334). Possibly, this difference can be compensated by further optimization of computational procedures for the algorithm (8.334). In view of this, we regard the computational efficiencies of the compared techniques as nearly the same.

8.8 Transformation of the Elements of EW Jones Vectors and EW Jones Matrices Under Changes of Eigenwave Bases

As in the classical Jones matrix method, when dealing with EW Jones vectors and matrices, it is useful to know how the elements of EW Jones vectors change with altering the eigenwave basis and how to calculate from a given EW Jones matrix corresponding to one pair of input and output bases the EW Jones matrix representing the same operator but corresponding to another pair of bases. In this section, we give a set of conversion relations for EW Jones vectors and EW Jones matrices.

8.8.1 Coordinates of the EW Jones Vector of a Wave Field in Different Eigenwave Bases

Let $\mathbf{e}_a^{(1)}$ and $\mathbf{e}_b^{(1)}$ be electric vibration vectors of a pair of basis waves of an eigenwave basis (basis 1) used to represent a forward or backward propagating wave field, and let $\mathbf{e}_a^{(2)}$ and $\mathbf{e}_b^{(2)}$ be electric vibration vectors of basis waves of another eigenwave basis (basis 2) that can be used to represent the same wave field. Let columns

$$\mathbf{a}^{(1)} = \begin{pmatrix} A_a^{(1)} \\ A_b^{(1)} \end{pmatrix} \text{ and } \mathbf{a}^{(2)} = \begin{pmatrix} A_a^{(2)} \\ A_b^{(2)} \end{pmatrix}$$

represent the EW Jones vector of the field in basis 1 and basis 2, respectively. The problem is to express $\mathbf{a}^{(2)}$ in terms of $\mathbf{a}^{(1)}$.

Since the columns $\mathbf{a}^{(1)}$ and $\mathbf{a}^{(2)}$ describe the same field, according to (8.105) or (8.106) we may write

$$\mathbf{e}_a^{(1)} A_a^{(1)} + \mathbf{e}_b^{(1)} A_b^{(1)} = \mathbf{e}_a^{(2)} A_a^{(2)} + \mathbf{e}_b^{(2)} A_b^{(2)}. \tag{8.342}$$

Multiplying (8.342) scalarly by $\mathbf{e}_a^{(2)*}$ and $\mathbf{e}_b^{(2)*}$, we have

$$\begin{aligned}(\mathbf{e}_a^{(2)*}\cdot\mathbf{e}_a^{(1)})A_a^{(1)}+(\mathbf{e}_a^{(2)*}\cdot\mathbf{e}_b^{(1)})A_b^{(1)}&=(\mathbf{e}_a^{(2)*}\cdot\mathbf{e}_a^{(2)})A_a^{(2)}+(\mathbf{e}_a^{(2)*}\cdot\mathbf{e}_b^{(2)})A_b^{(2)},\\(\mathbf{e}_b^{(2)*}\cdot\mathbf{e}_a^{(1)})A_a^{(1)}+(\mathbf{e}_b^{(2)*}\cdot\mathbf{e}_b^{(1)})A_b^{(1)}&=(\mathbf{e}_b^{(2)*}\cdot\mathbf{e}_a^{(2)})A_a^{(2)}+(\mathbf{e}_b^{(2)*}\cdot\mathbf{e}_b^{(2)})A_b^{(2)}.\end{aligned}\tag{8.343}$$

Casting (8.343) in matrix form, we get

$$\begin{pmatrix}\mathbf{e}_a^{(2)*}\cdot\mathbf{e}_a^{(1)} & \mathbf{e}_a^{(2)*}\cdot\mathbf{e}_b^{(1)}\\ \mathbf{e}_b^{(2)*}\cdot\mathbf{e}_a^{(1)} & \mathbf{e}_b^{(2)*}\cdot\mathbf{e}_b^{(1)}\end{pmatrix}\begin{pmatrix}A_a^{(1)}\\ A_b^{(1)}\end{pmatrix}=\begin{pmatrix}\mathbf{e}_a^{(2)*}\cdot\mathbf{e}_a^{(2)} & \mathbf{e}_a^{(2)*}\cdot\mathbf{e}_b^{(2)}\\ \mathbf{e}_b^{(2)*}\cdot\mathbf{e}_a^{(2)} & \mathbf{e}_b^{(2)*}\cdot\mathbf{e}_b^{(2)}\end{pmatrix}\begin{pmatrix}A_a^{(2)}\\ A_b^{(2)}\end{pmatrix},\tag{8.344}$$

or, more concisely,

$$\boldsymbol{F}_{(2*1)}\mathbf{a}^{(1)}=\boldsymbol{F}_{(2*2)}\mathbf{a}^{(2)},\tag{8.345}$$

where

$$\boldsymbol{F}_{(j*l)}=\begin{pmatrix}\mathbf{e}_a^{(j)*}\cdot\mathbf{e}_a^{(l)} & \mathbf{e}_a^{(j)*}\cdot\mathbf{e}_b^{(l)}\\ \mathbf{e}_b^{(j)*}\cdot\mathbf{e}_a^{(l)} & \mathbf{e}_b^{(j)*}\cdot\mathbf{e}_b^{(l)}\end{pmatrix}.\tag{8.346}$$

From (8.345) we find the desired expression:

$$\mathbf{a}^{(2)}=\boldsymbol{G}_{1\to 2}\mathbf{a}^{(1)},\tag{8.347}$$

where

$$\boldsymbol{G}_{1\to 2}=\boldsymbol{F}_{(2*2)}^{-1}\boldsymbol{F}_{(2*1)}.\tag{8.348}$$

The transformation matrix $\boldsymbol{G}_{1\to 2}$ may also be expressed as follows:

$$\boldsymbol{G}_{1\to 2}=\boldsymbol{F}_{(1*2)}^{-1}\boldsymbol{F}_{(1*1)}.\tag{8.349}$$

We would have arrived at (8.349), had we multiplied (8.342) by $\mathbf{e}_a^{(1)*}$ and $\mathbf{e}_b^{(1)*}$ rather than by $\mathbf{e}_a^{(2)*}$ and $\mathbf{e}_b^{(2)*}$. Equation (8.347) may be regarded as the general law of coordinate transformation for EW Jones vectors.

Of special interest are the following two cases of basis transformation.

Term-by-Term Transformation

Let basis 2 be such that

$$\mathbf{e}_a^{(2)}=c_a^{(2/1)}\mathbf{e}_a^{(1)},\quad \mathbf{e}_b^{(2)}=c_b^{(2/1)}\mathbf{e}_b^{(1)},\tag{8.350}$$

where $c_a^{(2/1)}$ and $c_b^{(2/1)}$ are scalar (possibly complex) numbers; that is, the vectors $\mathbf{e}_k^{(1)}$ and $\mathbf{e}_k^{(2)}$ ($k = a, b$) describe the same vibration mode but may have different lengths and phases. It is easy to find that in this case, the transformation matrix $\boldsymbol{G}_{1\to2}$ can be written as follows:

$$\boldsymbol{G}_{1\to2} = \begin{pmatrix} \dfrac{1}{c_a^{(2/1)}} & 0 \\ 0 & \dfrac{1}{c_b^{(2/1)}} \end{pmatrix}. \tag{8.351}$$

The coefficients $c_k^{(j/l)}$ may be defined as

$$c_k^{(j/l)} = \frac{\mathbf{e}_k^{(l)*} \cdot \mathbf{e}_k^{(j)}}{\mathbf{e}_k^{(l)*} \cdot \mathbf{e}_k^{(l)}} \qquad k = a,\ b. \tag{8.352}$$

Using this definition, we may also represent the matrix $\boldsymbol{G}_{1\to2}$ as follows:

$$\boldsymbol{G}_{1\to2} = \begin{pmatrix} c_a^{(1/2)} & 0 \\ 0 & c_b^{(1/2)} \end{pmatrix}. \tag{8.353}$$

The matrix of the inverse transformation, $\mathbf{a}^{(2)} \to \mathbf{a}^{(1)}$, defined by the relation

$$\mathbf{a}^{(1)} = \boldsymbol{G}_{2\to1}\mathbf{a}^{(2)}, \tag{8.354}$$

may be expressed as

$$\boldsymbol{G}_{2\to1} = \boldsymbol{G}_{1\to2}^{-1} = \begin{pmatrix} \dfrac{1}{c_a^{(1/2)}} & 0 \\ 0 & \dfrac{1}{c_b^{(1/2)}} \end{pmatrix} = \begin{pmatrix} c_a^{(2/1)} & 0 \\ 0 & c_b^{(2/1)} \end{pmatrix}. \tag{8.355}$$

We may use (8.351), (8.353), and (8.355), for example, when *changing the normalization* of an EW basis or when changing the *direction* of a basis **e**-vector *to opposite*.

A typical situation is that in which one of the bases is normalized using the electrical normalization (8.283), and the other using the flux normalization (8.164), the vectors $\mathbf{e}_k^{(1)}$ and $\mathbf{e}_k^{(2)}$ being codirectional ($k = a, b$). In this situation, if basis 1 has the electrical normalization, we may write

$$\mathbf{e}_a^{(1)} = \underline{\mathbf{e}}_a, \quad \mathbf{e}_b^{(1)} = \underline{\mathbf{e}}_b, \tag{8.356}$$

where $\underline{\mathbf{e}}_a$ and $\underline{\mathbf{e}}_b$ are unit vectors ($\underline{\mathbf{e}}_k^*\underline{\mathbf{e}}_k = 1$), for basis 1 and

$$\mathbf{e}_a^{(2)} = \frac{1}{\sqrt{2\left|\mathrm{Re}\left(\mathbf{z}(\underline{\mathbf{e}}_a^* \times \underline{\mathbf{h}}_a)\right)\right|}}\underline{\mathbf{e}}_a, \quad \mathbf{e}_b^{(2)} = \frac{1}{\sqrt{2\left|\mathrm{Re}\left(\mathbf{z}(\underline{\mathbf{e}}_b^* \times \underline{\mathbf{h}}_b)\right)\right|}}\underline{\mathbf{e}}_b, \tag{8.357}$$

where $\underline{\mathbf{h}}_k = \mathbf{m}_k \times \underline{\mathbf{e}}_k$ with $\mathbf{m}_k$ being the refraction vector of the kth basis wave ($k = a, b$), for basis 2, which has the flux normalization. From (8.356) and (8.357) we see that in this situation

$$\begin{aligned} c_k^{(1/2)} &= \frac{1}{c_k^{(2/1)}} = \sqrt{2\left|\mathrm{Re}\left(\mathbf{z}(\underline{\mathbf{e}}_k^* \times \underline{\mathbf{h}}_k)\right)\right|} = \sqrt{2\left|\mathrm{Re}\left(\mathbf{z}\left(\underline{\mathbf{e}}_k^* \times (\mathbf{m}_k \times \underline{\mathbf{e}}_k)\right)\right)\right|} \\ &= \sqrt{2\left|\mathrm{Re}\left(\mathbf{z}\left(\mathbf{m}_k\left(\underline{\mathbf{e}}_k^* \cdot \underline{\mathbf{e}}_k\right) - \underline{\mathbf{e}}_k\left(\underline{\mathbf{e}}_k^* \cdot \mathbf{m}_k\right)\right)\right)\right|} = \sqrt{2\left|\mathrm{Re}\left(\sigma_k - \left(\mathbf{z} \cdot \underline{\mathbf{e}}_k\right)\left(\mathbf{m}_k \cdot \underline{\mathbf{e}}_k^*\right)\right)\right|}, \end{aligned} \tag{8.358}$$

$k = a, b$. Here are some particularly simple cases where

$$\left(\mathbf{z} \cdot \underline{\mathbf{e}}_k\right)\left(\mathbf{m}_k \cdot \underline{\mathbf{e}}_k^*\right) = 0. \tag{8.359}$$

(i) *Nonabsorbing isotropic medium, normal incidence*. In this case, $|\sigma_k| = n$, where n is the refractive index of the medium. Hence we may write

$$c_k^{(1/2)} = \frac{1}{c_k^{(2/1)}} = \sqrt{2n} \qquad k = a,\ b. \tag{8.360}$$

(ii) *Nonabsorbing uniaxial medium whose optic axis is perpendicular to* **z**, *normal incidence*. In this case (see Section 9.3), if the wave a is extraordinary, $|\sigma_a| = n_{\|}$ and $|\sigma_b| = n_{\perp}$, where $n_{\|}$ and $n_{\perp}$ are the principal refractive indices. Therefore,

$$c_a^{(1/2)} = \frac{1}{c_a^{(2/1)}} = \sqrt{2n_{\|}}, \quad c_b^{(1/2)} = \frac{1}{c_b^{(2/1)}} = \sqrt{2n_{\perp}}. \tag{8.361}$$

(iii) *Nonabsorbing uniaxial medium whose optic axis is parallel to* **z**, *normal incidence*. In this case,

$$c_k^{(1/2)} = \frac{1}{c_k^{(2/1)}} = \sqrt{2n_{\perp}} \qquad k = a,\ b. \tag{8.362}$$

Unitary Transformations With Changing Basis Polarizations

As in the classical Jones matrix method, often useful are basis transformations with a change of basis polarizations. The eigenwave representation permits such basis transformations only in the case of polarization degeneracy, with some additional conditions. For example, for an absorbing isotropic medium, such transformations are allowed only at normal incidence. In the case of a nonabsorbing isotropic medium out of TIR mode, such transformations are possible at both normal and oblique incidence. Here we restrict ourselves to considering the most typical situation when the vectors $\mathbf{e}_a^{(1)}$, $\mathbf{e}_b^{(1)}$, $\mathbf{e}_a^{(2)}$, and $\mathbf{e}_b^{(2)}$ have identical length (L), that is,

$$\mathbf{e}_a^{(1)*} \cdot \mathbf{e}_a^{(1)} = \mathbf{e}_b^{(1)*} \cdot \mathbf{e}_b^{(1)} = \mathbf{e}_a^{(2)*} \cdot \mathbf{e}_a^{(2)} = \mathbf{e}_b^{(2)*} \cdot \mathbf{e}_b^{(2)} = L^2, \tag{8.363}$$

and are orthogonal to the wave normal $\mathbf{l}_a$ ($\mathbf{l}_b = \mathbf{l}_a$), that is,

$$\mathbf{l}_a \cdot \mathbf{e}_a^{(1)} = \mathbf{l}_a \cdot \mathbf{e}_b^{(1)} = \mathbf{l}_a \cdot \mathbf{e}_a^{(2)} = \mathbf{l}_a \cdot \mathbf{e}_b^{(2)} = 0, \tag{8.364}$$

and, moreover, the vectors $\mathbf{e}_a^{(j)}$ and $\mathbf{e}_b^{(j)}$ ($j = 1, 2$) are mutually orthogonal in the sense that

$$\mathbf{e}_a^{(j)*} \cdot \mathbf{e}_b^{(j)} = 0. \tag{8.365}$$

According to (8.363) and (8.365), the matrices $\boldsymbol{F}_{(l*l)}$ $(l = 1, 2)$ entering into (8.348) and (8.349) may be expressed as

$$\boldsymbol{F}_{(l*l)} = \begin{pmatrix} L^2 & 0 \\ 0 & L^2 \end{pmatrix} = L^2 \begin{pmatrix} 1 & 0 \\ 0 & 1 \end{pmatrix}. \tag{8.366}$$

In view of this, writing the matrices $\boldsymbol{F}_{(j*l)}$ with $l \neq j$ in (8.348) and (8.349) as

$$\boldsymbol{F}_{(j*l)} = L^2 \begin{pmatrix} \dfrac{\mathbf{e}_a^{(j)*} \cdot \mathbf{e}_a^{(l)}}{L^2} & \dfrac{\mathbf{e}_a^{(j)*} \cdot \mathbf{e}_b^{(l)}}{L^2} \\ \dfrac{\mathbf{e}_b^{(j)*} \cdot \mathbf{e}_a^{(l)}}{L^2} & \dfrac{\mathbf{e}_b^{(j)*} \cdot \mathbf{e}_b^{(l)}}{L^2} \end{pmatrix}, \tag{8.367}$$

we obtain

$$\boldsymbol{G}_{1\to 2} = \begin{pmatrix} \dfrac{\mathbf{e}_a^{(2)*} \cdot \mathbf{e}_a^{(1)}}{L^2} & \dfrac{\mathbf{e}_a^{(2)*} \cdot \mathbf{e}_b^{(1)}}{L^2} \\ \dfrac{\mathbf{e}_b^{(2)*} \cdot \mathbf{e}_a^{(1)}}{L^2} & \dfrac{\mathbf{e}_b^{(2)*} \cdot \mathbf{e}_b^{(1)}}{L^2} \end{pmatrix} = \begin{pmatrix} \dfrac{\mathbf{e}_a^{(1)*} \cdot \mathbf{e}_a^{(2)}}{L^2} & \dfrac{\mathbf{e}_a^{(1)*} \cdot \mathbf{e}_b^{(2)}}{L^2} \\ \dfrac{\mathbf{e}_b^{(1)*} \cdot \mathbf{e}_a^{(2)}}{L^2} & \dfrac{\mathbf{e}_b^{(1)*} \cdot \mathbf{e}_b^{(2)}}{L^2} \end{pmatrix}^{-1}. \tag{8.368}$$

In the case of the electrical normalization ($L^2 = 1$),

$$\boldsymbol{G}_{1\to 2} = \begin{pmatrix} \mathbf{e}_a^{(2)*} \cdot \mathbf{e}_a^{(1)} & \mathbf{e}_a^{(2)*} \cdot \mathbf{e}_b^{(1)} \\ \mathbf{e}_b^{(2)*} \cdot \mathbf{e}_a^{(1)} & \mathbf{e}_b^{(2)*} \cdot \mathbf{e}_b^{(1)} \end{pmatrix} = \begin{pmatrix} \mathbf{e}_a^{(1)*} \cdot \mathbf{e}_a^{(2)} & \mathbf{e}_a^{(1)*} \cdot \mathbf{e}_b^{(2)} \\ \mathbf{e}_b^{(1)*} \cdot \mathbf{e}_a^{(2)} & \mathbf{e}_b^{(1)*} \cdot \mathbf{e}_b^{(2)} \end{pmatrix}^{-1} \tag{8.369}$$

[cf. (1.67)].

8.8.2 EW Jones Operators in Different Eigenwave Bases

Let $\mathbf{t}_{1-1'}$ be a given EW Jones matrix relating the input EW Jones vector represented in basis 1, $\mathbf{a}_{\text{inp}}^{(1)}$, and the output EW Jones vector represented in basis 1′, $\mathbf{a}_{\text{out}}^{(1')}$:

$$\mathbf{a}_{\text{out}}^{(1')} = \mathbf{t}_{1-1'}\,\mathbf{a}_{\text{inp}}^{(1)}. \tag{8.370}$$

The problem is to find the matrix $\mathbf{t}_{2-2'}$ representing the same operator as $\mathbf{t}_{1-1'}$ but using basis 2 for the input vector and basis 2′ for the output vector, that is, linking the vectors $\mathbf{a}_{\text{out}}^{(2')}$ and $\mathbf{a}_{\text{inp}}^{(2)}$ as

$$\mathbf{a}_{\text{out}}^{(2')} = \mathbf{t}_{2-2'}\,\mathbf{a}_{\text{inp}}^{(2)}. \tag{8.371}$$

By using the transformation matrices $\boldsymbol{G}_{2\to 1} = (\boldsymbol{G}_{1\to 2})^{-1}$ and $\boldsymbol{G}_{1'\to 2'} = (\boldsymbol{G}_{2'\to 1'})^{-1}$, we can express the vectors $\mathbf{a}_{\text{inp}}^{(1)}$ and $\mathbf{a}_{\text{out}}^{(2')}$ in terms of $\mathbf{a}_{\text{inp}}^{(2)}$ and $\mathbf{a}_{\text{out}}^{(1')}$ as follows:

$$\mathbf{a}_{\text{inp}}^{(1)} = \boldsymbol{G}_{2\to 1}\,\mathbf{a}_{\text{inp}}^{(2)}, \tag{8.372}$$

$$\mathbf{a}_{\text{out}}^{(2')} = \boldsymbol{G}_{1'\to 2'}\,\mathbf{a}_{\text{out}}^{(1')}. \tag{8.373}$$

Combining (8.370), (8.372), and (8.373) we obtain

$$\mathbf{a}_{\text{out}}^{(2')} = \boldsymbol{G}_{1'\to 2'}\,\mathbf{t}_{1-1'}\,\boldsymbol{G}_{2\to 1}\,\mathbf{a}_{\text{inp}}^{(2)}. \tag{8.374}$$

Consequently,

$$\mathbf{t}_{2-2'} = \boldsymbol{G}_{1'\to 2'}\,\mathbf{t}_{1-1'}\,\boldsymbol{G}_{2\to 1}. \tag{8.375}$$

Relation (8.375) represents the law of transformation of EW Jones matrices under changes of the eigenwave bases.

Cases of Identity of EW Jones Matrices Corresponding to Different Variants of Normalization of the Eigenwave Bases

In many cases of practical interest, EW Jones matrices, of an operation, corresponding to different variants of normalization of the input and output EW bases are identical. Let us consider a layered system S, with the external interfaces coincident with the planes $z = z_1$ and $z = z_2$, confined between a homogeneous isotropic nonabsorbing medium A ($z < z_1$) and a homogeneous medium B ($z > z_2$). Let a homogeneous plane monochromatic wave X_{inc} fall on S from medium A. Let the wave field $\mathsf{X}_{\text{ref}} = \mathsf{R}\mathsf{X}_{\text{inc}}$ be the total or a partial reflected field propagating in A and let $\mathsf{X}_{\text{tr}} = \mathsf{T}\mathsf{X}_{\text{inc}}$ be the total or a partial transmitted field propagating in B. Define operators $\mathbf{r}$ and $\mathbf{t}$ of the operations R and T by the relations

$$\mathbf{a}_{\text{ref}} = \mathbf{r}\,\mathbf{a}_{\text{inc}}, \qquad \mathbf{a}_{\text{tr}} = \mathbf{t}\,\mathbf{a}_{\text{inc}},$$

where $\mathbf{a}_{\text{inc}}$ is the EW Jones vector of the wave X_{inc} at $z = z_1 - 0$, $\mathbf{a}_{\text{ref}}$ is the EW Jones vector of the wave X_{ref} at $z = z_1 - 0$, and $\mathbf{a}_{\text{tr}}$ is the EW Jones vector of the wave X_{tr} at $z = z_2 + 0$. With the standard choice of the EW basis (Section 9.2), columns representing the vector $\mathbf{a}_{\text{inc}}$ under E-, S-, and F-normalizations, respectively $\mathbf{a}_{\text{inc(E)}}$, $\mathbf{a}_{\text{inc(S)}}$, and $\mathbf{a}_{\text{inc(F)}}$, are related as follows:

$$\mathbf{a}_{\text{inc(S)}} = \mathbf{a}_{\text{inc(F)}} = c_{\text{A}}\mathbf{a}_{\text{inc(E)}}, \tag{8.376}$$

where $c_{\text{A}} = \sqrt{2\sigma_{\text{A}}}$ with σ_{A} being the value of σ_1 in the medium A. Analogously, for $\mathbf{a}_{\text{ref(E)}}$, $\mathbf{a}_{\text{ref(S)}}$, and $\mathbf{a}_{\text{ref(F)}}$,

$$\mathbf{a}_{\text{ref(S)}} = \mathbf{a}_{\text{ref(F)}} = c_{\text{A}}\mathbf{a}_{\text{ref(E)}}. \tag{8.377}$$

Relations (8.376) and (8.377) can easily be derived by using (8.350)–(8.359) or (9.23). Let the EW Jones matrices $\mathbf{r}_{\text{(E)}}$, $\mathbf{r}_{\text{(S)}}$, and $\mathbf{r}_{\text{(F)}}$ represent the operator $\mathbf{r}$ under E-, S-, and F-normalizations of the EW basis in A, that is,

$$\mathbf{a}_{\text{ref(E)}} = \mathbf{r}_{\text{(E)}}\mathbf{a}_{\text{inc(E)}}, \quad \mathbf{a}_{\text{ref(S)}} = \mathbf{r}_{\text{(S)}}\mathbf{a}_{\text{inc(S)}}, \quad \mathbf{a}_{\text{ref(F)}} = \mathbf{r}_{\text{(F)}}\mathbf{a}_{\text{inc(F)}}. \tag{8.378}$$

As can be seen from (8.376)–(8.378),

$$\mathbf{r}_{\text{(E)}} = \mathbf{r}_{\text{(S)}} = \mathbf{r}_{\text{(F)}}. \tag{8.379}$$

E-normalization of the input and output bases imparts to the matrix $\mathbf{r}_{\text{(E)}}$ the sense of a *canonical* Jones matrix, a matrix relating Jones vectors of the kind (1.21) or (1.63). It follows from (8.379) that in the case under consideration, we may directly obtain the canonical Jones matrix of $\mathbf{r}$ not only with E-normalization but also with S- and F-normalizations.

In a similar way, one can find that in situations when medium B is isotropic, nonabsorbing, and out of TIR mode, the matrices $\mathbf{t}_{(E)}$, $\mathbf{t}_{(S)}$, and $\mathbf{t}_{(F)}$ representing the operator $\mathbf{t}$ under respectively E-, S-, and F-normalizations of the input and output EW bases will satisfy the relations

$$\mathbf{t}_{(E)} = \mathbf{t}_{(S)}, \quad \mathbf{t}_{(E)} = \mathbf{t}_{(F)} \tag{8.380}$$

if the refractive indices of media A and B are equal.

As an example, we refer to Figures 8.13 and 8.14. Relation (8.379) is valid for the transfer channels shown in Figure 8.14, and relations (8.380) hold for the transfer channels shown in Figure 8.13.

References

[1] L. D. Landau, E. M. Lifshitz, and L. P. Pitaevskii, *Electrodynamics of Continuous Media*, 2nd ed. (Pergamon Press, Oxford, 1984).
[2] D. W. Berreman, "Optics in stratified and anisotropic media: 4 × 4 matrix formulation," *J. Opt. Soc. Am.* **62**, 502 (1972).
[3] E. J. Rothwell and M. J. Cloud, *Electromagnetics* (CRC Press, Boca Raton, FL, 2001).
[4] M. Born and E. Wolf, *Principles of Optics*, 7th ed. (Pergamon Press, New York, 1999).
[5] J. D. Jackson, *Classical Electrodynamics*, 3rd ed. (John Wiley & Sons, Ltd, New York, 1998).
[6] E. Hecht, *Optics*, 4th ed. (Addison Wesley, San Francisco, 2002).
[7] J. A. Stratton, *Electromagnetic Theory*, 1st ed. (McGraw-Hill, New York, 1941).
[8] A. Yariv and P. Yeh, *Optical Waves in Crystals* (John Wiley & Sons, Inc., New York, 1984).
[9] P. Yeh, "Electromagnetic propagation in birefringent layered media," *J. Opt. Soc. Am.* **69**, 742 (1979).
[10] F. I. Fedorov, *Optics of Anisotropic Media* (Academy of Sciences of Belarus, Minsk, 1958) (in Russian).
[11] F. I. Fedorov, *The Theory of Gyrotropy* (Nauka i tekhnika, Minsk, 1976).
[12] G. N. Borzdov, "Waves with linear, quadratic and cubic coordinate dependence of amplitude in crystals," *Pramana J. Phys.* **46**, 245 (1996).
[13] S. Pancharatnam, "The propagation of light in absorbing biaxial crystals - I. Theoretical," *Proc. Indian Acad. Sci. A* **42**, 86 (1955).
[14] G. S. Ranganath, "Optics of absorbing anisotropic media," *Curr. Sci.* **67**, 231 (1994).
[15] S. Teitler and B. Henvis, "Refraction in stratified and anisotropic media," *J. Opt. Soc. Am.* **60**, 830 (1970).
[16] D. W. Berreman, "Optics in smoothly varying anisotropic planar structures: application to liquid-crystal twist cells," *J. Opt. Soc. Am.* **63**, 1374 (1973).
[17] P. J. Lin-Chung and S. Teitler, "4 × 4 matrix formalisms for optics in stratified anisotropic media," *J. Opt. Soc. Am. A* **1**, 703 (1984).
[18] P. Allia, C. Oldano, and L. Trossi, "Polarization transfer matrix for the transmission of light through liquid-crystal slabs," *J. Opt. Soc. Am. B* **5**, 2452 (1988).
[19] C. Oldano, "Electromagnetic-wave propagation in stratified media," *Phys. Rev. A* **40**, 6014 (1989).
[20] K. Eidner, G. Mayer, M. Schmidt, and H. Schmiedel, "Optics in stratified media—the use of optical eigenmodes of uniaxial crystals in the 4 × 4-matrix formalism," *Mol. Cryst. Liq. Cryst.* **172**, 191 (1989).
[21] K. Eidner, "Light propagation in stratified anisotropic media: orthogonality and symmetry properties of the 4 × 4 matrix formalisms," *J. Opt. Soc. Am. A* **6**, 1657 (1989).
[22] D. A. Yakovlev, "Orthogonality relations for polarization characteristics of plane waves induced in a uniform anisotropic layer by an arbitrary incident plane wave and their applications," *Opt. Spectrosc.* **84**, 1017 (1998) (in Russian).
[23] H. Yuan, W. E, T. Kosa, and P. Palffy-Muhoray, "Analytic 4 × 4 propagation matrices for linear optical media," *Mol. Cryst. Liq. Cryst.* **331**, 491 (1999).
[24] P. Yeh, "Extended Jones matrix method," *J. Opt. Soc. Am.* **72**, 507 (1982).
[25] C. Gu and P. Yeh, "Extended Jones matrix method. II," *J. Opt. Soc. Am. A* **10**, 966 (1993).
[26] P. Yeh and C. Gu, *Optics of Liquid Crystal Displays* (John Wiley & Sons, Ltd, New York, 1999).
[27] A. Lien, "Extended Jones matrix representation for the twisted nematic liquid crystal display at oblique incidence," *Appl. Phys. Lett.* **57**, 2767 (1990).
[28] A. Lien, "A detailed derivation of extended Jones matrix representation for twisted nematic liquid crystal displays," *Liq. Cryst.* **22**, 171 (1997).

[29] H. L. Ong, "2 × 2 propagation matrix for electromagnetic waves propagating obliquely in layered inhomogeneous uniaxial media," *J. Opt. Soc. Am. A* **10**, 283 (1993).
[30] C.-J. Chen, A. Lien, and M. I. Nathan, "4 × 4 and 2 × 2 matrix formulations for the optics in stratified and biaxial media," *J. Opt. Soc. Am. A* **14**, 3125 (1997).
[31] Z. Ge, T. X. Wu, X. Zhu, and S.-T. Wu, "Reflective liquid-crystal displays with asymmetric incident and exit angles," *J. Opt. Soc. Am. A* **22**, 966 (2005).
[32] C. Moler and C. Van Loan, "Nineteen dubious ways to compute the exponential of a matrix, twenty-five years later," *SIAM Rev*. **45**, 1 (2003).
[33] H. Wöhler, G. Haas, M. Fritsch, and D. A. Mlynski, "Faster 4 × 4 matrix method for inhomogeneous uniaxial media," *J. Opt. Soc. Am. A* **5**, 1554 (1988).
[34] S. P. Palto, "An algorithm for solving the optical problem for stratified anisotropic media," *JETP*. **92**, 552 (2001).
[35] G. F. Barrick, "An interpolating polynomial method for liquid crystal display optics," *Jpn. J. Appl. Phys.* **41**, 5480 (2002).
[36] G. A. Korn and T. M. Korn, *Mathematical Handbook for Scientists and Engineers*, 2nd ed. (Dover, New York, 2000).
[37] H.-W. Cheng and S. S.-T. Yau, "More explicit formulas for the matrix exponential," *Linear Algebra Appl.* **262**, 131 (1997).
[38] R. J. Gagnon, "Liquid-crystal twist-cell optics," *J. Opt. Soc. Am.* **71**, 348 (1981).
[39] D. A. Yakovlev, "Simple formulas for the amplitude transmission and reflection coefficients for the interface of anisotropic media," *Opt. Spectrosc*. **84**, 829 (1998) (in Russian).
[40] D. Y. K. Ko and J. R. Sambles, "Scattering matrix method for propagation of radiation in stratified media: attenuated total reflection studies of liquid crystals," *J. Opt. Soc. Am. A* **5**, 1863 (1988).
[41] R. A. Horn and C. R. Johnson, *Matrix Analysis* (Cambridge University Press, Cambridge, 1986).
[42] M. Schmidt and H. Schmiedel, "The conservation of energy in Berreman's 4 × 4-formalism and conditions for the validity of numerical approximations," *Mol. Cryst. Liq. Cryst.* **172**, 223 (1989).
[43] P. Yeh and C. Gu, *Optics of Liquid Crystal Displays*, 2nd ed. (John Wiley & Sons, Ltd, New Jersey, 2010).
[44] C. Gu and P. Yeh, "Reciprocity in photorefractive wave mixing," *Opt. Lett.* **16**, 455 (1991).
[45] C. Altman and S. G. Lipson, "Reciprocity relations in light propagation through a multilayer birefringent system," *J. Opt. Soc. Am.* **61**, 1373 (1971).
[46] S. Stallinga, "Berreman 4 × 4 matrix method for reflective liquid crystal displays," *J. Appl. Phys.* **85**, 3023 (1999).
[47] G. P. Montgomery, "Efficient calculation of liquid-crystal-display viewing characteristics," *J. Opt. Soc. Am. A* **1**, 1163–1165 (1984).

9

Choice of Eigenwave Bases for Isotropic, Uniaxial, and Biaxial Media

As we saw in Chapter 8 and will see in the next chapters, the eigenwave representation underlies several highly flexible methods for accurate modeling of the optical characteristics of multilayer systems, such as LCDs. All these methods require the specification of eigenwave bases (EWBs). In this chapter, we discuss how to choose eigenwave bases for different types of optical media that we deal with in modeling LCDs. Moreover, in this chapter, we describe the routines of the modeling library LMOPTICS (Fortran 90; see the companion website) that are intended for generating EWBs. In Section 9.1, we give some common information concerning the specification of EWBs. In Sections 9.2, 9.3, and 9.4, the problem of EWB specification is considered for isotropic, uniaxial, and biaxial media, respectively.

9.1 General Aspects of EWB Specification. EWB-generating routines

Coordinate Systems

Coordinate systems used in this chapter and in the EWB-generating routines of LMOPTICS are shown in Figure 9.1. The (X, Y, Z) system is a reference system for specifying the orientation of the principal axes of anisotropic layers and the plane of incidence. The coordinate system (x, y, z), as before, is attached to the plane of incidence. This coordinate system is used in most computational formulas for determining the EWB parameters. Writing a vector in coordinates, where necessary, we will indicate the corresponding coordinate system by subscripts: $(\cdot)_{xyz}$ or $(\cdot)_{XYZ}$. For example, for the wave normal $\mathbf{l}_{\text{inc}}$ of the incident wave in Figure 9.1 we can write

$$\mathbf{l}_{\text{inc}} = \begin{pmatrix} \sin\beta_{\text{inc}}\cos\alpha_{\text{inc}} \\ \sin\beta_{\text{inc}}\sin\alpha_{\text{inc}} \\ \cos\beta_{\text{inc}} \end{pmatrix}_{XYZ} = \begin{pmatrix} \sin\beta_{\text{inc}} \\ 0 \\ \cos\beta_{\text{inc}} \end{pmatrix}_{xyz},$$

where β_{inc} and α_{inc} are the polar and azimuthal angles of incidence.

Modeling and Optimization of LCD Optical Performance, First Edition.
Dmitry A. Yakovlev, Vladimir G. Chigrinov and Hoi-Sing Kwok.

Website Companion: www.wiley.com/go/yakovlev/modelinglcd

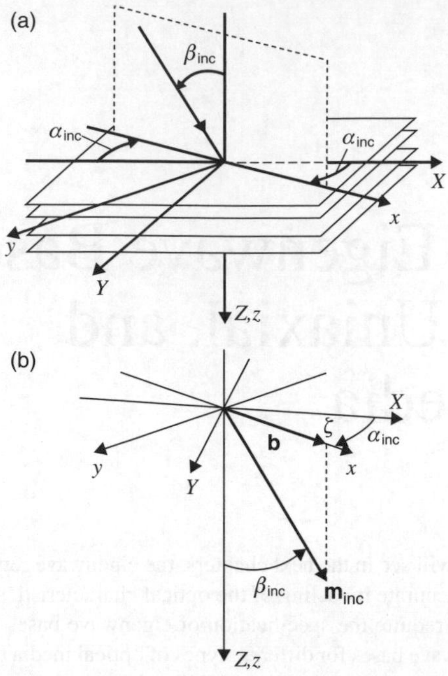

Figure 9.1 Geometry of the problem. Here and in Figure 9.2, sketches are given in two variants: one variant is more visual (a); the other is more geometrically correct (b)

Propagation Constants

Under the conditions of the basic model problem (Section 8.1.3, Fig. 8.4), the refraction vector of any basis eigenwave for any homogeneous layer of the layered system under consideration must have the form

$$\mathbf{m} = \mathbf{b} + \mathbf{z}\sigma, \tag{9.1}$$

where $\mathbf{z}$ is the unit vector directed along the axes Z and z, and $\mathbf{b}$ is the tangential component of the refraction vector of the incident wave $\mathbf{m}_{\text{inc}} = n_{\text{inc}}\mathbf{l}_{\text{inc}}$ with n_{inc} being the refractive index of the medium from which the light falls. In coordinates,

$$\mathbf{b} = \begin{pmatrix} \zeta \\ 0 \\ 0 \end{pmatrix}_{xyz} = \begin{pmatrix} \zeta \cos\alpha_{\text{inc}} \\ \zeta \sin\alpha_{\text{inc}} \\ 0 \end{pmatrix}_{XYZ}, \tag{9.2}$$

where

$$\zeta = n_{\text{inc}} \sin\beta_{\text{inc}}. \tag{9.3}$$

The spatial frequency parameter ζ and the angle α_{inc} describing the orientation of the plane of incidence are the only parameters of the incident field that are directly involved in calculating EWBs. The frequency of the incident field, commonly specified in programs in terms of the free-space wavelength $\lambda = c/\nu = 2\pi c/\omega$, at this stage of calculation is only used in specifying the material optical constants of the layers (real refractive indices, absorption coefficients). Because the parameters ζ and α_{inc} are constants in the model problem under consideration and are used in many routines, we prefer to represent them in programs by global variables. In LMOPTICS, these are variables MT (ζ) and FIR (α_{inc} in radians), which are declared in an auxiliary module MTWFIR of LMOPTICS. The EWB-generating routines are contained in the main module OPTSM_1 of LMOPTICS.

Parameters to be Determined

In standard calculations, the specification of the jth basis wave for a homogeneous layer consists in the specification of the normal component of the refraction vector, σ_j, and the four-component column vector

$$\boldsymbol{\psi}_j = \begin{pmatrix} \mathrm{e}_{jx} \\ \mathrm{h}_{jy} \\ \mathrm{e}_{jy} \\ -\mathrm{h}_{jx} \end{pmatrix}_{xyz}$$

composed of Cartesian components of the electric vibration vector $\mathbf{e}_j$ (e_{jx}, e_{jy}, e_{jz}) and magnetic vibration vector $\mathbf{h}_j$ (h_{jx}, h_{jy}, h_{jz}) of this wave. Different algorithms of evaluating the optical characteristics of layered media employ the parameters of whether the four basis waves or—for example, when the transmission characteristics of a layered system are calculated ignoring the multiple reflections—only two basis waves (usually, the forward propagating ones). In the first case, the calculation is performed using the matrices

$$\boldsymbol{\Psi} \equiv (\boldsymbol{\psi}_1 \quad \boldsymbol{\psi}_2 \quad \boldsymbol{\psi}_3 \quad \boldsymbol{\psi}_4)$$

and

$$\boldsymbol{\Psi}^{-1} \equiv \begin{pmatrix} \overline{\boldsymbol{\psi}}_1 \\ \overline{\boldsymbol{\psi}}_2 \\ \overline{\boldsymbol{\psi}}_3 \\ \overline{\boldsymbol{\psi}}_4 \end{pmatrix}$$

[see (8.162)]. An array of σ_j and the matrices $\boldsymbol{\Psi}$ and $\boldsymbol{\Psi}^{-1}$ constitute a standard set of output parameters of most EWB-generating routines of LMOPTICS; we will call the routines with this set of output parameters *standard-output* routines. In the second case, if forward propagation is considered, only columns $\boldsymbol{\psi}_1$ and $\boldsymbol{\psi}_2$ and rows $\overline{\boldsymbol{\psi}}_1$ and $\overline{\boldsymbol{\psi}}_2$ are involved in the calculations (see Section 8.4.2). Due to orthogonality (8.161), calculating the row $\overline{\boldsymbol{\psi}}_j$ of the matrix $\boldsymbol{\Psi}^{-1}$ requires knowledge of only the column $\boldsymbol{\psi}_j$. Therefore, one can deal with the parameters of only the first two basis waves. The EWB-generating routines can operate in a mode when the parameters of only forward propagating basis waves are computed and only the first two columns of $\boldsymbol{\Psi}$ and the first two rows of $\boldsymbol{\Psi}^{-1}$ are specified. These routines have an input parameter NWAVE (NWAVE may be set to 2 or 4) indicating the number of basis waves the parameters of which will be computed.

Some routines of LMOPTICS return arrays containing all components of the vectors $\mathbf{e}_j$ and $\mathbf{h}_j$. We will call such routines *full-output*. We used these routines when debugging the standard-output routines and preparing demonstration examples for this chapter. These routines are included in LMOPTICS because they give the most complete and clear information on the eigenwaves. Furthermore, these routines may be useful in modeling optical elements having nonparallel interfaces—in such calculations all the components of $\mathbf{e}_j$ and $\mathbf{h}_j$ may be required.

Orthogonality of the EWB

As we saw in Chapter 8, the orthogonality of the EWB expressed by the relations

$$\boldsymbol{\psi}_j^{\mathrm{T}} \mathbf{I}_0 \boldsymbol{\psi}_k \equiv \mathbf{z}(\mathbf{e}_j \times \mathbf{h}_k + \mathbf{e}_k \times \mathbf{h}_j) = 0 \quad j \neq k \tag{9.4}$$

allows one to greatly simplify calculations (see Section 8.4). In the absence of polarization degeneracy, for the optical media considered in this book, orthogonality condition (9.4) is satisfied for any particular choice of the basis waves. When polarization degeneracy occurs, the situation is different, but the EWB can be chosen such that condition (9.4) is satisfied. Therefore, it is reasonable to consider the orthogonality in the sense of (9.4) in the case of polarization degeneracy as a requirement to EWBs. In the next three sections, we show how to construct EWBs satisfying this requirement in different cases. The routines of LMOPTICS in the case of polarization degeneracy always give EWBs satisfying (9.4).

Normalization of the EWB

The normalization of EWBs is an optional (in many cases) but a very useful operation. The normalization (i) defines exactly the physical meaning of the scalar complex amplitudes of the basis waves and matrix operators describing the light propagation, (ii) simplifies calculations and analysis in most cases (Sections 8.4, 8.5, 11.2; Chapter 12), (iii) protects against computational errors caused by incommensurability of the leading (largest in magnitude) elements in the columns of the matrix $\boldsymbol{\Psi}$, and so on. In our programs, we use the following three types of normalization, each of which provides advantages in certain situations:

1. *Electrical (E-) normalization*:

$$\mathbf{e}_j^* \mathbf{e}_j = 1 \quad j = 1,2,3,4. \tag{9.5}$$

   When applied to homogeneous basis waves in an isotropic medium, this normalization defines the EW Jones vectors as *canonical* Jones vectors (Jones vectors of the kind (1.21) or (1.63)) [1–4]. This makes the normalization (9.5) appropriate when the transmission and reflection matrices to be calculated are canonical Jones matrices. For instance, if in the example in Figure 8.12 both layer A and layer B are nonabsorbing, isotropic, and out of TIR mode and the normalization (9.5) is applied to these layers, any of the matrices $\mathbf{r}^{\downarrow}$, $\mathbf{t}^{\downarrow}$, $\mathbf{r}^{\uparrow}$, and $\mathbf{t}^{\uparrow}$ of fragment F is a canonical Jones matrix. A disadvantage of this normalization is that the corresponding metric matrices, which are used in the energy transfer analysis (see Section 8.5), are dependent on the optical constants of the medium.
2. *Flux (F-) normalization*:

$$\boldsymbol{\psi}_j^{\dagger} \mathbf{I}_0 \boldsymbol{\psi}_j \equiv 2\mathrm{Re}(\mathbf{z}(\mathbf{e}_j^* \times \mathbf{h}_j)) = \begin{cases} 1 & j = 1,2 \\ -1 & j = 3,4. \end{cases} \tag{9.6}$$

   With this normalization, the mentioned metric matrices have their simplest form and are independent of the optical constants of the medium, which greatly simplifies the energy transfer analysis [5] and makes it very similar mathematically to that performed in the framework of the classical Jones calculus. F-normalization should not be used in TIR mode, because in this case we, as a rule, deal with the situation when

$$\boldsymbol{\psi}_j^{\dagger} \mathbf{I}_0 \boldsymbol{\psi}_j \equiv 2\mathrm{Re}(\mathbf{z}(\mathbf{e}_j^* \times \mathbf{h}_j)) = 0,$$

   that is, when an eigenwave, taken alone, gives a zero normal flux [see (8.242)].

3. *Symmetrical (S-) normalization*:

$$\boldsymbol{\psi}_j^{\mathrm{T}}\mathbf{I}_0\boldsymbol{\psi}_j \equiv 2\mathbf{z}(\mathbf{e}_j \times \mathbf{h}_j) = \begin{cases} 1 & j = 1{,}2 \\ -1 & j = 3{,}4. \end{cases} \tag{9.7}$$

This normalization simplifies many key formulas (Sections 8.4, 8.6.2, 11.2) [6,7]. In contrast to the flux normalization, S-normalization is applicable in the case of TIR mode. Out of TIR mode, for any nonabsorbing medium of interest to us here (see Section 8.1.1), one can choose the vectors $\mathbf{e}_j$ real. With this choice, the vectors $\mathbf{h}_j$ are also real, and hence basis waves satisfying (9.7) also satisfy (9.6), so that normalization (9.7) gives the same benefits as (9.6). In some practically important situations, S-normalization, being applied to the entrance and exit media, ensures a direct evaluation of the canonical Jones matrices (see Section 8.8.2).

Assume that we know the normal components of the refraction vectors, σ_j, and nonnormalized or arbitrarily normalized electric vibration vectors $\boldsymbol{e}_j$ (e_{jx}, e_{jy}, e_{jz}) for all four basis waves (j = 1,2,3,4). Here is an example of a general algorithm for calculating the matrices $\boldsymbol{\Psi}$ and $\boldsymbol{\Psi}^{-1}$ from σ_j and $\boldsymbol{e}_j$ that allows the choice of any of the three normalization conditions:

1. Calculating the electric vibration vectors $\underline{\mathbf{e}}_j$ ($\underline{\mathrm{e}}_{jx}$, $\underline{\mathrm{e}}_{jy}$, $\underline{\mathrm{e}}_{jz}$) normalized by the condition $\underline{\mathbf{e}}_j^*\underline{\mathbf{e}}_j = 1$. These vectors are calculated from the vectors $\boldsymbol{e}_j$ as follows:

$$\underline{\mathbf{e}}_j = \boldsymbol{e}_j/\sqrt{\boldsymbol{e}_j^*\boldsymbol{e}_j}.$$

2. Calculating components of the magnetic vibration vectors $\underline{\mathbf{h}}_j$ ($\underline{\mathrm{h}}_{jx}$, $\underline{\mathrm{h}}_{jy}$, $\underline{\mathrm{h}}_{jz}$) corresponding to $\underline{\mathbf{e}}_j$:

$$\underline{\mathbf{h}}_j = \mathbf{m}_j \times \underline{\mathbf{e}}_j.$$

   In the usual calculations, the z-components of the magnetic vectors are not used.
3. Determining the matrices $\boldsymbol{\Psi}$ and $\boldsymbol{\Psi}^{-1}$. Calculations in this stage depend on which normalization is chosen for the EWB.
   A. *Electrical normalization*. Since the vectors $\underline{\mathbf{e}}_j$ satisfy (9.5), we set $\mathbf{e}_j = \underline{\mathbf{e}}_j$ and $\mathbf{h}_j = \underline{\mathbf{h}}_j$ and assign to elements of the matrix $\boldsymbol{\Psi}$ (ψ_{jk}) the following values:

$$\psi_{1j} = \underline{\mathrm{e}}_{jx}, \quad \psi_{2j} = \underline{\mathrm{h}}_{jy}, \quad \psi_{3j} = \underline{\mathrm{e}}_{jy}, \quad \psi_{4j} = -\underline{\mathrm{h}}_{jx} \qquad j = 1{,}2{,}3{,}4.$$

   Calculating $\boldsymbol{\Psi}^{-1}$ is performed by the following general formula, which is valid owing to the orthogonality (9.4):

$$\boldsymbol{\Psi}^{-1} = \begin{pmatrix} c_1\psi_{21} & c_1\psi_{11} & c_1\psi_{41} & c_1\psi_{31} \\ c_2\psi_{22} & c_2\psi_{12} & c_2\psi_{42} & c_2\psi_{32} \\ c_3\psi_{23} & c_3\psi_{13} & c_3\psi_{43} & c_3\psi_{33} \\ c_4\psi_{24} & c_4\psi_{14} & c_4\psi_{44} & c_4\psi_{34} \end{pmatrix}, \tag{9.8}$$

   where

$$c_j = \frac{1}{2(\psi_{1j}\psi_{2j} + \psi_{3j}\psi_{4j})} \qquad j = 1{,}2{,}3{,}4.$$

B. *Flux normalization*. In this case, the elements of $\mathbf{\Psi}$ are calculated as follows:

$$\psi_{ij} = \underline{\psi}_{ij}/N_j \quad i,j = 1,2,3,4, \tag{9.9}$$

$$N_j = \sqrt{2|\mathrm{Re}(\underline{\psi}^*_{2j}\underline{\psi}_{1j} + \underline{\psi}^*_{4j}\underline{\psi}_{3j})|} \quad j = 1,2,3,4, \tag{9.10}$$

where

$$\underline{\psi}_{1j} = \underline{e}_{jx}, \quad \underline{\psi}_{2j} = \underline{h}_{jy}, \quad \underline{\psi}_{3j} = \underline{e}_{jy}, \quad \underline{\psi}_{4j} = -\underline{h}_{jx} \quad j = 1,2,3,4. \tag{9.11}$$

The matrix $\mathbf{\Psi}^{-1}$ is computed using the general formula (9.8).

C. *Symmetrical normalization*. In this case, the elements of the matrix $\mathbf{\Psi}$ are calculated by formula (9.9) with

$$N_j = \begin{cases} \sqrt{2(\underline{\psi}_{2j}\underline{\psi}_{1j} + \underline{\psi}_{4j}\underline{\psi}_{3j})} & j = 1,2 \\ \sqrt{-2(\underline{\psi}_{2j}\underline{\psi}_{1j} + \underline{\psi}_{4j}\underline{\psi}_{3j})} & j = 3,4, \end{cases} \tag{9.12}$$

where, as in the previous case, $\underline{\psi}_{ij}$ are given by (9.11). The matrix $\mathbf{\Psi}^{-1}$ in this case is composed of the elements of the matrix $\mathbf{\Psi}$:

$$\mathbf{\Psi}^{-1} = \begin{pmatrix} \psi_{21} & \psi_{11} & \psi_{41} & \psi_{31} \\ \psi_{22} & \psi_{12} & \psi_{42} & \psi_{32} \\ -\psi_{23} & -\psi_{13} & -\psi_{43} & -\psi_{33} \\ -\psi_{24} & -\psi_{14} & -\psi_{44} & -\psi_{34} \end{pmatrix} \tag{9.13}$$

[see (8.168)].

If the electrical normalization is not required (as an option), step 1 is unnecessary, and $\underline{\psi}_{ij}$ in (9.9), (9.10), and (9.12) can be taken as

$$\underline{\psi}_{1j} = e_{jx}, \quad \underline{\psi}_{2j} = h_{jy}, \quad \underline{\psi}_{3j} = e_{jy}, \quad \underline{\psi}_{4j} = -h_{jx} \quad j = 1,2,3,4, \tag{9.14}$$

where components of the nonnormalized (or arbitrarily normalized) electric vibration vectors $\boldsymbol{e}_j$ and vectors $\boldsymbol{h}_j$ (h_{jx}, h_{jy}, h_{jz}) defined as

$$\boldsymbol{h}_j = \mathbf{m}_j \times \boldsymbol{e}_j \tag{9.15}$$

are used. We will call the vectors $\boldsymbol{e}_j$ and $\boldsymbol{h}_j$ as well as the vectors

$$\boldsymbol{d}_j = \varepsilon \boldsymbol{e}_j \tag{9.16}$$

"*raw*" *vibration vectors*. The determination of the normal components of the refraction vectors, σ_j, and the "raw" electric vibration vectors $\boldsymbol{e}_j$ is the main subject matter of the next three sections.

If the electrical normalization must be provided for but the vectors $\underline{\mathbf{e}}_j$ are not required as output parameters, step 1 is also unnecessary, because in this case the elements of the matrix $\mathbf{\Psi}$ can be calculated by formula (9.9) with $\underline{\psi}_{ij}$ given by (9.14) and N_j calculated as

$$N_j = \sqrt{\boldsymbol{e}_j^* \boldsymbol{e}_j}.$$

In standard-output EWB-generating routines of LMOPTICS, the type of normalization to be used is indicated with the aid of an input parameter NRM (1, 2, or 3): NRM = 1 corresponds to the electrical (E-) normalization, NRM = 2 to the flux (F-) normalization, and NRM = 3 to the symmetrical (S-) normalization. In the full-output routines, only E-normalization is used.

Testing the EWB-Generating Routines. Orthogonality Test

Obvious ways to test an EWB-generating routine are (i) to consider situations for which the correct answer is known and (ii) to examine how accurately the calculated EW bases satisfy corresponding orthogonality relations: (8.160) in the absence of absorption and TIR mode and (9.4) [(8.161)] in the general case. When calculation of the matrix $\mathbf{\Psi}^{-1}$ is carried out by using the orthogonality relations (9.4), as in our case, one can estimate the error of determining the EWB by evaluating an orthogonality parameter

$$\delta_{\text{ort}} = \max_{j \neq k} \left| [\mathbf{\Psi}_{\text{c-ortinv}} \mathbf{\Psi}_{\text{c}}]_{jk} \right|, \tag{9.17}$$

where $\mathbf{\Psi}_{\text{c}}$ is a computed value of $\mathbf{\Psi}$, $\mathbf{\Psi}_{\text{c-ortinv}}$ is the value of $\mathbf{\Psi}^{-1}$ obtained from $\mathbf{\Psi}_{\text{c}}$ by formula (9.8) or (9.13). Errors in the computed matrix $\mathbf{\Psi}$ violate the orthogonality and lead to deviation of off-diagonal elements of the matrix $\mathbf{\Psi}_{\text{c-ortinv}} \cdot \mathbf{\Psi}_{\text{c}}$ from zero, so we can use the values of these elements to estimate the error in the computed $\mathbf{\Psi}$. We will call this way of error estimation *the orthogonality test*. In Sections 9.3 and 9.4, devoted to uniaxial and biaxial media, we give many numerical examples with tabulated results, demonstrating EWBs generated by routines of LMOPTICS. In the tables, we present rounded values of components of the refraction and vibration vectors and the δ_{ort} values calculated using double-precision values of the matrices $\mathbf{\Psi}$ and $\mathbf{\Psi}^{-1}$. Since in all the examples, the orthogonality test indicates very good accuracy of the results, these tables may be used as a collection of "correct answers" [see item (i)].

Routines for Isotropic, Uniaxial, and Biaxial Media. "Real-Arithmetic" Routines

The library LMOPTICS contains three sets of EWB-generating routines. The first set of routines is for calculating EWBs for isotropic media, the second for uniaxial media, and the third for biaxial media. In the case of a nonabsorbing medium out of TIR mode, calculating the EWB can be performed without using complex arithmetic. In this case, the use of real-type variables instead of complex-type ones (used in the general case) reduces the computational cost. For this reason, we include in the library, along with universal routines, "real-arithmetic" routines for nonabsorbing media out of TIR mode.

Data Types

Table 9.1 shows the names and types of some variables that are input or output parameters of the standard-output EWB-generating routines of LMOPTICS. The other input parameters of these routines are of `real(8)` type.

Table 9.1 Data types of some input and output parameters of the EWB-generating routines

Parameter	Variable	
	General routines	Real-arithmetic routines
$\mathbf{\Psi}$	`complex(8) YY(4,4)`	`real(8) YY(4,4)`
$\mathbf{\Psi}^{-1}$	`complex(8) YI(4,4)`	`real(8) YI(4,4)`
$\sigma_j, j = 1,2,3,4$	`complex(8) X(4)`	`complex(8) X(4)`
NRM	`integer(4) NRM`	`integer(4) NRM`
NWAVE	`integer(4) NWAVE`	`integer(4) NWAVE`

9.2 Isotropic Media

In the case of an isotropic medium, the roots of (8.73) are

$$\sigma_1 = \sigma_2 = \sqrt{n^2 - \zeta^2}, \quad \sigma_3 = \sigma_4 = -\sqrt{n^2 - \zeta^2}, \tag{9.18}$$

where n is the complex refractive index of the medium. Since polarization degeneracy takes place, there is much freedom in choosing the vibration vectors of the basis waves. The only restriction on the vectors $\boldsymbol{e}_j$ is

$$\mathbf{m}_j \boldsymbol{e}_j = 0 \quad j = 1,2,3,4 \tag{9.19}$$

[see (1.10)]. We use the following set of $\boldsymbol{e}_j$:

$$\boldsymbol{e}_1 = \begin{pmatrix} C_{\beta n} \\ 0 \\ -S_{\beta n} \end{pmatrix}_{xyz}, \quad \boldsymbol{e}_2 = \begin{pmatrix} 0 \\ 1 \\ 0 \end{pmatrix}_{xyz}, \quad \boldsymbol{e}_3 = \begin{pmatrix} -C_{\beta n} \\ 0 \\ -S_{\beta n} \end{pmatrix}_{xyz}, \quad \boldsymbol{e}_4 = \begin{pmatrix} 0 \\ 1 \\ 0 \end{pmatrix}_{xyz}, \tag{9.20}$$

where

$$S_{\beta n} = \zeta / n, \quad C_{\beta n} = \sigma_1 / n = \sqrt{1 - S_{\beta n}^2}. \tag{9.21}$$

Figure 9.2 shows the orientation of the vectors $\mathbf{e}_j$ corresponding to this choice in a situation when the basis waves are homogeneous. For a nonabsorbing medium, at $\zeta < n$, $S_{\beta n}$ and $C_{\beta n}$ equal respectively

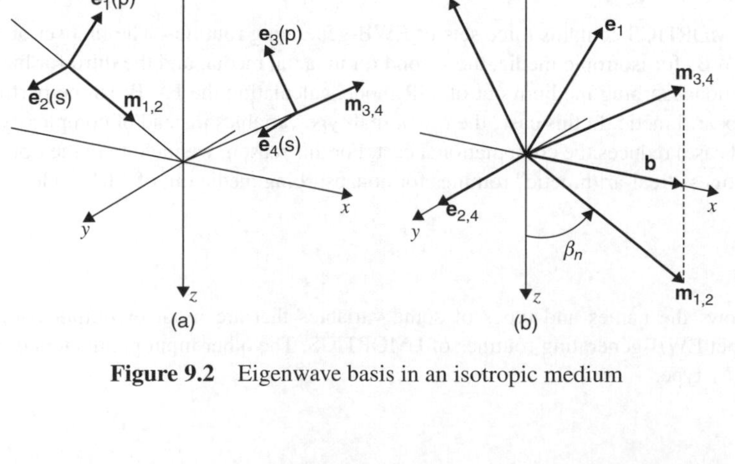

Figure 9.2 Eigenwave basis in an isotropic medium

Table 9.2 Expressions for the normalizing factors $a_{\psi j}$

Normalization type	$a_{\psi j}$ General case	$a_{\psi j}$ Nonabsorbing medium out of TIR mode
Electrical (E-) normalization	$\begin{cases} \sqrt{C^*_{\beta n} C_{\beta n} + S^*_{\beta n} S_{\beta n}} & j = 1,3 \\ 1 & j = 2,4 \end{cases}$	1
Flux (F-) normalization	$\begin{cases} \sqrt{2\mathrm{Re}\left(n^* C_{\beta n}\right)} & j = 1,3 \\ \sqrt{2\mathrm{Re}\left(n C_{\beta n}\right)} & j = 2,4 \end{cases}$	$\sqrt{2nC_{\beta n}}$
Symmetrical (S-) normalization	$\sqrt{2nC_{\beta n}}$	$\sqrt{2nC_{\beta n}}$

the sine and cosine of the angle β_n between the axis z and $\mathbf{m}_1$ (see Figure 9.2b). This choice of the electric vibration vectors of the basis waves is consistent with the usual choice of the basis orthogonal polarizations for incident, transmitted, and reflected waves in the problem on reflection and transmission of a plane wave obliquely incident on a plane boundary between two homogeneous isotropic media [8–10, 4] (see Section 1.2): basis waves 1 and 3 have electric vectors parallel to the plane of incidence, that is, are p-polarized, and waves 2 and 4 have electric vectors perpendicular to the plane of incidence, that is, are s-polarized (see Figure 9.2a). The magnetic vibration vectors $\boldsymbol{h}_j = \mathbf{m}_j \times \boldsymbol{e}_j$ for $\boldsymbol{e}_j$ expressed by (9.20) are

$$\boldsymbol{h}_1 = n\begin{pmatrix} 0 \\ 1 \\ 0 \end{pmatrix}_{xyz}, \quad \boldsymbol{h}_2 = n\begin{pmatrix} -C_{\beta n} \\ 0 \\ S_{\beta n} \end{pmatrix}_{xyz}, \quad \boldsymbol{h}_3 = n\begin{pmatrix} 0 \\ 1 \\ 0 \end{pmatrix}_{xyz}, \quad \boldsymbol{h}_2 = n\begin{pmatrix} C_{\beta n} \\ 0 \\ S_{\beta n} \end{pmatrix}_{xyz}. \tag{9.22}$$

Using (9.20) and (9.22), we obtain the following expression for the columns of the matrix $\mathbf{\Psi}$:

$$\boldsymbol{\psi}_1 = \frac{1}{a_{\psi 1}}\begin{pmatrix} C_{\beta n} \\ n \\ 0 \\ 0 \end{pmatrix}_{xyz}, \quad \boldsymbol{\psi}_2 = \frac{1}{a_{\psi 2}}\begin{pmatrix} 0 \\ 0 \\ 1 \\ nC_{\beta n} \end{pmatrix}_{xyz}, \quad \boldsymbol{\psi}_3 = \frac{1}{a_{\psi 3}}\begin{pmatrix} -C_{\beta n} \\ n \\ 0 \\ 0 \end{pmatrix}_{xyz}, \quad \boldsymbol{\psi}_4 = \frac{1}{a_{\psi 4}}\begin{pmatrix} 0 \\ 0 \\ 1 \\ -nC_{\beta n} \end{pmatrix}_{xyz}, \tag{9.23}$$

where $a_{\psi j}$ are normalizing factors determined by the normalization conditions used (see Table 9.2). The rows of the corresponding matrix $\mathbf{\Psi}^{-1}$ can be expressed as follows:

$$\begin{aligned} \overline{\boldsymbol{\psi}}_1 &= \frac{a_{\psi 1}}{2nC_{\beta n}}(n \quad C_{\beta n} \quad 0 \quad 0), \quad \overline{\boldsymbol{\psi}}_2 = \frac{a_{\psi 2}}{2nC_{\beta n}}(0 \quad 0 \quad nC_{\beta n} \quad 1), \\ \overline{\boldsymbol{\psi}}_3 &= -\frac{a_{\psi 3}}{2nC_{\beta n}}(n \quad -C_{\beta n} \quad 0 \quad 0), \quad \overline{\boldsymbol{\psi}}_4 = -\frac{a_{\psi 4}}{2nC_{\beta n}}(0 \quad 0 \quad -nC_{\beta n} \quad 1). \end{aligned} \tag{9.24}$$

For a nonabsorbing medium out of TIR mode, all quantities in (9.18)–(9.24) are real, and the EWB is orthogonal both in the sense of (9.4) and in the sense of (8.160). In the presence of absorption and in TIR mode, the EWB satisfies (9.4) but in general does not satisfy (8.160).

Metric Matrices

Substitution from (9.23) into (8.245) leads to a rather simple general expression for the metric matrix $\tilde{\mathbf{N}}_0$, the matrix whose elements are involved in the calculation of the power characteristics of wave fields (see Section 8.5):

$$\tilde{\mathbf{N}}_0 \equiv [N_{0ij}] = \begin{pmatrix} \frac{2\,\mathrm{Re}\,(n^* C_{\beta n})}{a^*_{\psi 1} a_{\psi 1}} & 0 & -i\frac{2\,\mathrm{Im}\,(n^* C_{\beta n})}{a^*_{\psi 1} a_{\psi 3}} & 0 \\ 0 & \frac{2\,\mathrm{Re}(n C_{\beta n})}{a^*_{\psi 2} a_{\psi 2}} & 0 & -i\frac{2\,\mathrm{Im}(n C_{\beta n})}{a^*_{\psi 2} a_{\psi 4}} \\ i\frac{2\,\mathrm{Im}\,(n^* C_{\beta n})}{a^*_{\psi 3} a_{\psi 1}} & 0 & -\frac{2\,\mathrm{Re}\,(n^* C_{\beta n})}{a^*_{\psi 3} a_{\psi 3}} & 0 \\ 0 & i\frac{2\,\mathrm{Im}(n C_{\beta n})}{a^*_{\psi 4} a_{\psi 2}} & 0 & -\frac{2\,\mathrm{Re}(n C_{\beta n})}{a^*_{\psi 4} a_{\psi 4}} \end{pmatrix} \tag{9.25}$$

It is interesting to examine the structure of this matrix in the cases when the orthogonality relations (8.160) are violated, that is, when the medium is absorbing or/and TIR mode occurs. Considering that $nC_{\beta n} = \sigma_1$ [see (9.21)], we see that in the usual situation when $\mathrm{Im}(n) \ll \mathrm{Re}(n)$, the ratio of any of the elements N_{013}, N_{031}, N_{024}, and N_{042}—these elements describe the interference contributions to the energy flux—to any of the diagonal elements of $\tilde{\mathbf{N}}_0$ is approximately equal in absolute magnitude to $\mathrm{Im}(\sigma_1)/\mathrm{Re}(\sigma_1)$. If the medium is nonabsorbing and TIR mode is realized, all diagonal elements of the matrix $\tilde{\mathbf{N}}_0$ are equal to zero and only combinations of forward and backward propagating waves can give nonzero normal fluxes [see (8.242)].

As seen from (9.25), in the absence of absorption and TIR mode, the electrical normalization gives

$$\tilde{\mathbf{N}}_0 = 2nC_{\beta n} \begin{pmatrix} 1 & 0 & 0 & 0 \\ 0 & 1 & 0 & 0 \\ 0 & 0 & -1 & 0 \\ 0 & 0 & 0 & -1 \end{pmatrix}, \tag{9.26}$$

and, consequently, for the metric 2×2 matrices (Section 8.5),

$$\mathbf{n}_0^{\downarrow} = -\mathbf{n}_0^{\uparrow} = 2nC_{\beta n} \begin{pmatrix} 1 & 0 \\ 0 & 1 \end{pmatrix}. \tag{9.27}$$

From (9.25) it is seen that in the absence of absorption and TIR mode, the flux and symmetrical normalizations put the matrices $\tilde{\mathbf{N}}_0$, $\mathbf{n}_0^{\downarrow}$, and $\mathbf{n}_0^{\uparrow}$ in their simplest forms:

$$\tilde{\mathbf{N}}_0 = \begin{pmatrix} 1 & 0 & 0 & 0 \\ 0 & 1 & 0 & 0 \\ 0 & 0 & -1 & 0 \\ 0 & 0 & 0 & -1 \end{pmatrix}, \tag{9.28}$$

$$\mathbf{n}_0^{\downarrow} = -\mathbf{n}_0^{\uparrow} = \begin{pmatrix} 1 & 0 \\ 0 & 1 \end{pmatrix}. \tag{9.29}$$

Reciprocity Matrices

The reciprocity matrices $\mathbf{n}_{r12}(\tilde{\mathbf{b}})$ and $\mathbf{n}_{r21}(\tilde{\mathbf{b}})$ enter into the reciprocity relations for EW Jones matrices (Section 8.6.2). Here we consider what forms these matrices have with the above (our) choice of the EWBs. The general expression for the elements of $\mathbf{n}_{r12}(\tilde{\mathbf{b}})$ and $\mathbf{n}_{r21}(\tilde{\mathbf{b}})$ is

$$n_{rjk}(\tilde{\mathbf{b}}) = \boldsymbol{\psi}_j(\tilde{\mathbf{b}})^{\mathrm{T}}\mathbf{I}_r\boldsymbol{\psi}_k(-\tilde{\mathbf{b}}) = \mathbf{z}\cdot\left[\mathbf{e}_j(\tilde{\mathbf{b}})\times\mathbf{h}_k(-\tilde{\mathbf{b}}) - \mathbf{e}_k(-\tilde{\mathbf{b}})\times\mathbf{h}_j(\tilde{\mathbf{b}})\right] \tag{9.30}$$

[see (8.306)]. With our choice of the EWBs, the vectors $\mathbf{e}_j(-\tilde{\mathbf{b}})$ and $\mathbf{h}_j(-\tilde{\mathbf{b}})$ ($j = 1,2,3,4$) are obtained by the rotation of the vectors $\mathbf{e}_j(\tilde{\mathbf{b}})$ and $\mathbf{h}_j(\tilde{\mathbf{b}})$ through an angle of 180° about the z-axis, and the following relations hold:

$$\begin{aligned}
\mathbf{e}_1(-\tilde{\mathbf{b}}) &= \mathbf{e}_3(\tilde{\mathbf{b}}), \quad \mathbf{h}_1(-\tilde{\mathbf{b}}) = -\mathbf{h}_3(\tilde{\mathbf{b}}),\\
\mathbf{e}_2(-\tilde{\mathbf{b}}) &= -\mathbf{e}_4(\tilde{\mathbf{b}}), \quad \mathbf{h}_2(-\tilde{\mathbf{b}}) = \mathbf{h}_4(\tilde{\mathbf{b}}),\\
\mathbf{e}_3(-\tilde{\mathbf{b}}) &= \mathbf{e}_1(\tilde{\mathbf{b}}), \quad \mathbf{h}_3(-\tilde{\mathbf{b}}) = -\mathbf{h}_1(\tilde{\mathbf{b}}),\\
\mathbf{e}_4(-\tilde{\mathbf{b}}) &= -\mathbf{e}_2(\tilde{\mathbf{b}}), \quad \mathbf{h}_4(-\tilde{\mathbf{b}}) = \mathbf{h}_2(\tilde{\mathbf{b}}).
\end{aligned} \tag{9.31}$$

Since both bases are orthogonal in the sense of (8.161) and relations (8.214) are valid, these bases are optimal reciprocal (see Section 8.4.1) and hence only diagonal elements of the matrices $\mathbf{n}_{r12}(\tilde{\mathbf{b}})$ and $\mathbf{n}_{r21}(\tilde{\mathbf{b}})$ are different from zero. As is seen from (9.30) and (9.31), the nonzero elements of these matrices [see (8.306a)] may be expressed as follows:

$$\begin{aligned}
n_{r13}(\tilde{\mathbf{b}}) &= -2\mathbf{z}\cdot\left[\mathbf{e}_1(\tilde{\mathbf{b}})\times\mathbf{h}_1(\tilde{\mathbf{b}})\right] = -\boldsymbol{\psi}_1(\tilde{\mathbf{b}})^{\mathrm{T}}\mathbf{I}_0\boldsymbol{\psi}_1(\tilde{\mathbf{b}}),\\
n_{r24}(\tilde{\mathbf{b}}) &= 2\mathbf{z}\cdot\left[\mathbf{e}_2(\tilde{\mathbf{b}})\times\mathbf{h}_2(\tilde{\mathbf{b}})\right] = \boldsymbol{\psi}_2(\tilde{\mathbf{b}})^{\mathrm{T}}\mathbf{I}_0\boldsymbol{\psi}_2(\tilde{\mathbf{b}}),\\
n_{r31}(\tilde{\mathbf{b}}) &= -2\mathbf{z}\cdot\left[\mathbf{e}_3(\tilde{\mathbf{b}})\times\mathbf{h}_3(\tilde{\mathbf{b}})\right] = -\boldsymbol{\psi}_3(\tilde{\mathbf{b}})^{\mathrm{T}}\mathbf{I}_0\boldsymbol{\psi}_3(\tilde{\mathbf{b}}),\\
n_{r42}(\tilde{\mathbf{b}}) &= 2\mathbf{z}\cdot\left[\mathbf{e}_4(\tilde{\mathbf{b}})\times\mathbf{h}_4(\tilde{\mathbf{b}})\right] = \boldsymbol{\psi}_4(\tilde{\mathbf{b}})^{\mathrm{T}}\mathbf{I}_0\boldsymbol{\psi}_4(\tilde{\mathbf{b}}).
\end{aligned} \tag{9.32}$$

From these expressions we see that in the case of S-normalization the matrices $\mathbf{n}_{r12}(\tilde{\mathbf{b}})$ and $\mathbf{n}_{r21}(\tilde{\mathbf{b}})$ are as in (8.327). In the general case, according to (9.23) and (9.32),

$$\mathbf{n}_{r12}(\tilde{\mathbf{b}}) = 2nC_{\beta n}\begin{pmatrix} -a_{\psi 1}^{-2} & 0\\ 0 & a_{\psi 2}^{-2}\end{pmatrix}, \quad \mathbf{n}_{r21}(\tilde{\mathbf{b}}) = 2nC_{\beta n}\begin{pmatrix} a_{\psi 3}^{-2} & 0\\ 0 & -a_{\psi 4}^{-2}\end{pmatrix}, \tag{9.33}$$

where $C_{\beta n}$ corresponds to $\mathbf{b} = \tilde{\mathbf{b}}$. As can be seen from (9.33) and Table 9.2, for nonabsorbing isotropic media out of TIR mode the electrical normalization gives

$$\mathbf{n}_{r21}(\tilde{\mathbf{b}}) = -\mathbf{n}_{r12}(\tilde{\mathbf{b}}) = 2nC_{\beta n}\begin{pmatrix} 1 & 0\\ 0 & -1\end{pmatrix} = 2nC_{\beta n}\mathbf{I}_1 \tag{9.34}$$

[cf. (8.327)]. From (9.34), (8.312), and (8.315) it is seen that when the entrance and exit media of the transfer channel **C** (8.313) are identical, being isotropic, nonabsorbing, and out of TIR mode, with our choice of the EWBs in these media the reciprocity relation (8.315) is reduced to (8.328).

Program Implementation

The library LMOPTICS includes two EWB-generating routines for isotropic media: ISOTRR and ISOTRC. The routine ISOTRR is real-arithmetic. It can be used for nonabsorbing media out of TIR mode. ISOTRC is applicable in any case. The interfaces of these routines are

```
CALL ISOTRR(N, X, YY, YI, NWAVE, NRM)
CALL ISOTRC(N, A, X, YY, YI, NWAVE, NRM)
```

In ISOTRR, N is the refractive index of the medium. In ISOTRC, N and A are respectively the real and imaginary parts of the complex refractive index of the medium. Both ISOTRR and ISOTRC are standard-output routines. The output parameters of these routines are the array of σ_j values (X(j)), matrix $\boldsymbol{\Psi}$ (YY), and matrix $\boldsymbol{\Psi}^{-1}$ (YI). Input parameters NRM and NWAVE have been defined in Section 9.1.

9.3 Uniaxial Media

Most optically anisotropic elements of LCD panels are considered in modeling as uniaxial or locally uniaxial layers. Therefore, computing the parameters of basis waves for a uniaxial medium is one of the most frequently executed operations when the optical characteristics of an LCD are calculated by using methods based on eigenwave representation. Parameters of EWBs are also used in theoretical analyses of light propagation through LC layers (see Chapter 11). It is clear that an accurate and yet the simplest possible representation of the EWB parameters for uniaxial media is wanted. Here we describe probably the simplest exact representation. This representation is employed in EWB-generating routines of LMOPTICS. Some important formulas of this book were derived using this representation (see Sections 8.4.2 and 11.2).

For a uniaxial medium, in the absence of polarization degeneracy, the pair of forward propagating basis waves as well as the pair of backward propagating basis waves consist of an extraordinary wave and an ordinary wave. We assign number 1 to the extraordinary forward propagating wave, number 2 to the ordinary forward propagating wave, number 3 to the extraordinary backward propagating wave, and number 4 to the ordinary backward propagating wave. Let

$$\mathbf{c} = \begin{pmatrix} c_x \\ c_y \\ c_z \end{pmatrix}_{xyz}$$

be a unit vector directed along the optic axis of the medium, and let $\varepsilon_{\parallel}$ and $\varepsilon_{\perp}$ be the principal values of the complex permittivity tensor $\boldsymbol{\varepsilon}$ of the medium. In the optical region,

$$\varepsilon_{\parallel} = n_{\parallel}^2, \quad \varepsilon_{\perp} = n_{\perp}^2, \tag{9.35}$$

where $n_{\parallel}$ and $n_{\perp}$ are the principal refractive indices of the medium. Let, as usual,

$$\mathbf{b} = \begin{pmatrix} \zeta \\ 0 \\ 0 \end{pmatrix}_{xyz}, \quad \mathbf{m}_j = \mathbf{b} + \mathbf{z}\sigma_j = \begin{pmatrix} \zeta \\ 0 \\ \sigma_j \end{pmatrix}_{xyz}.$$

Then the basic formulas of the mentioned simple representation can be written as follows:

(a) *the normal components of the refraction vectors*:

$$\sigma_1 = \frac{-b_1 + b_4}{b_2}, \quad \sigma_3 = \frac{-b_1 - b_4}{b_2}, \tag{9.36}$$

$$\sigma_2 = \sqrt{\varepsilon_{\perp} - \zeta^2}, \quad \sigma_4 = -\sigma_2, \tag{9.37}$$

where

$$b_4 = \sqrt{b_1^2 - b_2(b_3 - \varepsilon_\perp \varepsilon_\parallel)},$$
$$b_1 = \Delta\varepsilon(\mathbf{bc})(\mathbf{zc}), \quad b_2 = \varepsilon_\perp + \Delta\varepsilon(\mathbf{zc})^2, \quad b_3 = \varepsilon_\perp \zeta^2 + \Delta\varepsilon(\mathbf{bc})^2,$$
$$\Delta\varepsilon = \varepsilon_\parallel - \varepsilon_\perp;$$

in components,

$$b_1 = \Delta\varepsilon\zeta c_x c_z, \quad b_2 = \varepsilon_\perp + \Delta\varepsilon c_z^2, \quad b_3 = \zeta^2\left(\varepsilon_\perp + \Delta\varepsilon c_x^2\right); \tag{9.38}$$

(b) *the electric vibration vectors*:

$$\boldsymbol{e}_j = \mathbf{c} - \frac{(\mathbf{m}_j\mathbf{c})}{\varepsilon_\perp}\mathbf{m}_j \qquad j = 1,3, \tag{9.39}$$

$$\boldsymbol{e}_j = \mathbf{m}_j \times \mathbf{c} \qquad j = 2,4, \tag{9.40}$$

or, in components, for the extraordinary waves ($j = 1,3$)

$$\boldsymbol{e}_j = \begin{pmatrix} e_{jx} \\ e_{jy} \\ e_{jz} \end{pmatrix}_{xyz} = \begin{pmatrix} c_x - m_{jc}\zeta \\ c_y \\ c_z - m_{jc}\sigma_j \end{pmatrix}_{xyz}, \tag{9.41}$$
$$m_{jc} = (\zeta c_x + \sigma_j c_z)/\varepsilon_\perp,$$

for the ordinary waves ($j = 2,4$)

$$\boldsymbol{e}_j = \begin{pmatrix} e_{jx} \\ e_{jy} \\ e_{jz} \end{pmatrix}_{xyz} = \begin{pmatrix} -\sigma_j c_y \\ \sigma_j c_x - \zeta c_z \\ \zeta c_y \end{pmatrix}_{xyz}; \tag{9.42}$$

(c) *the magnetic vibration vectors*:

$$\boldsymbol{h}_j = \mathbf{m}_j \times \boldsymbol{e}_j \quad j = 1, 2, 3, 4, \tag{9.43}$$

or, in components,

$$\boldsymbol{h}_j = \begin{pmatrix} h_{jx} \\ h_{jy} \\ h_{jz} \end{pmatrix}_{xyz} = \begin{pmatrix} -\sigma_j e_{jy} \\ \sigma_j e_{jx} - \zeta e_{jz} \\ \zeta e_{jy} \end{pmatrix}_{xyz}.$$

The chief advantage of this representation is the simplicity of the expression for the electric vibration vectors $\boldsymbol{e}_j$ of the extraordinary waves (9.39). This expression—we found it in a monograph [11] (a similar solution for the $\boldsymbol{\psi}$ vectors of the basis extraordinary waves was obtained in Reference 12 in the framework of the Berreman formalism)—looks very simple in comparison with the commonly used expressions for such vectors [13–17]. We will derive this expression to remove all doubts about its correctness. In passing, we will show the background of the other expressions used in this representation.

The permittivity tensor $\boldsymbol{\varepsilon}$ of a uniaxial medium and its inverse $\boldsymbol{\varepsilon}^{-1}$ can be expressed as

$$\boldsymbol{\varepsilon} = \varepsilon_\perp \mathbf{U} + \Delta\varepsilon \mathbf{c} \otimes \mathbf{c}, \tag{9.44}$$

$$\boldsymbol{\varepsilon}^{-1} = \frac{1}{\varepsilon_\perp}\mathbf{U} + \left(\frac{1}{\varepsilon_\parallel} - \frac{1}{\varepsilon_\perp}\right)\mathbf{c} \otimes \mathbf{c} \tag{9.45}$$

with $\otimes$ denoting the dyadic product [see (8.26) and (5.66)]. If the tensor $\boldsymbol{\varepsilon}$ has the form (9.44), equation (8.36) splits into two equations:

$$\mathbf{m}\boldsymbol{\varepsilon}\mathbf{m} = \varepsilon_\parallel \varepsilon_\perp \tag{9.46}$$

and

$$\mathbf{m}^2 = \varepsilon_\perp. \tag{9.47}$$

The former equation gives solutions for extraordinary waves, and the latter for ordinary waves. Using (9.44), one can rewrite (9.46) as follows:

$$\varepsilon_\perp \mathbf{m}^2 + \Delta\varepsilon(\mathbf{mc})^2 = \varepsilon_\parallel \varepsilon_\perp. \tag{9.48}$$

Substituting $\mathbf{m} = \mathbf{b} + \mathbf{z}\sigma$ into (9.48) and (9.47) gives the following equations for σ:

$$(\varepsilon_\perp + \Delta\varepsilon(\mathbf{zc})^2)\sigma^2 + (2\Delta\varepsilon(\mathbf{bc})(\mathbf{zc}))\sigma + \varepsilon_\perp \mathbf{b}^2 + \Delta\varepsilon\,(\mathbf{bc})^2 - \varepsilon_\parallel \varepsilon_\perp = 0, \tag{9.49}$$

$$\sigma^2 = \varepsilon_\perp - \mathbf{b}^2. \tag{9.50}$$

The roots of (9.49) are σ_1 and σ_3 expressed by (9.36). The roots of (9.50) are σ_2 and σ_4 expressed by (9.37). If the tensor $\boldsymbol{\varepsilon}^{-1}$ has the form (9.45), the following relations hold with any $\mathbf{m}$:

$$\boldsymbol{\varepsilon}^{-1}\,(\mathbf{m} \times \mathbf{c}) = \frac{1}{\varepsilon_\perp}\,(\mathbf{m} \times \mathbf{c}), \tag{9.51}$$

$$\boldsymbol{\varepsilon}^{-1}\,(\mathbf{m} \times (\mathbf{m} \times \mathbf{c})) = \frac{1}{\varepsilon_\perp}\,(\mathbf{m}\,(\mathbf{mc})) - \frac{1}{\varepsilon_\perp \varepsilon_\parallel}(\varepsilon_\perp \mathbf{m}^2 + \Delta\varepsilon\,(\mathbf{mc})^2)\mathbf{c}. \tag{9.52}$$

These relations can easily be obtained by using the formulas

$$\mathbf{c} \otimes \mathbf{c}\,(\mathbf{m} \times \mathbf{c}) = \mathbf{c}\,(\mathbf{c}\,(\mathbf{m} \times \mathbf{c})) = 0,$$

$$\mathbf{c} \otimes \mathbf{c}(\mathbf{m} \times (\mathbf{m} \times \mathbf{c})) = \mathbf{c} \otimes \mathbf{c}(\mathbf{m}(\mathbf{mc}) - \mathbf{cm}^2) = \mathbf{c}(\mathbf{mc})^2 - \mathbf{cm}^2,$$

where we have used the identity

$$(\mathbf{A} \otimes \mathbf{B})\,\mathbf{C} = \mathbf{A}(\mathbf{BC}), \tag{9.53}$$

which is valid for any vectors $\mathbf{A}$, $\mathbf{B}$, and $\mathbf{C}$. Substituting

$$\boldsymbol{e} = a\boldsymbol{\varepsilon}^{-1}\,(\mathbf{m} \times \mathbf{c}) \tag{9.54}$$

(where a is an arbitrary scalar coefficient) into the wave equation for the electric vector

$$\mathbf{m}(\mathbf{m}\boldsymbol{e}) - \mathbf{m}^2\boldsymbol{e} + \varepsilon\boldsymbol{e} = 0, \tag{9.55}$$

on account of (9.51), leads to the following condition:

$$\left(1 - \frac{\mathbf{m}^2}{\varepsilon_\perp}\right)(\mathbf{m}\times\mathbf{c}) = 0. \tag{9.56}$$

If $\mathbf{m}$ satisfies (9.47), condition (9.56) is satisfied. This means that expression (9.54) is suitable for representation of the vectors $\boldsymbol{e}_j$ of the ordinary basis waves [see (9.40) and (9.51)]. Then, one can notice that for $\mathbf{m}$ satisfying condition (9.48), according to (9.52),

$$\varepsilon^{-1}(\mathbf{m}\times(\mathbf{m}\times\mathbf{c})) = \mathbf{m}\frac{(\mathbf{mc})}{\varepsilon_\perp} - \mathbf{c}. \tag{9.57}$$

Substitution of

$$\boldsymbol{e} = a\varepsilon^{-1}(\mathbf{m}\times(\mathbf{m}\times\mathbf{c})) \tag{9.58}$$

into (9.55) shows that, subject to (9.57), the vector $\boldsymbol{e}$ expressed by (9.58) satisfies (9.55). This means that the expression

$$\boldsymbol{e} = a\left(\mathbf{m}\frac{(\mathbf{mc})}{\varepsilon_\perp} - \mathbf{c}\right) \tag{9.59}$$

can be used for calculating the vectors $\boldsymbol{e}_j$ of the extraordinary basis waves. This fact is employed in the representation under consideration [see (9.39)].

Polarization Degeneracy

In the case of a uniaxial medium, polarization degeneracy occurs for waves whose refraction vectors $\mathbf{m}$ are parallel to the optic axis $\mathbf{c}$ [$\mathbf{m} = n_\perp\mathbf{c}$ and $\mathbf{m} = -n_\perp\mathbf{c}$ satisfy both (9.47) and (9.48)]. With $\mathbf{m} = \pm n_\perp\mathbf{c}$, any $\boldsymbol{e}$ satisfying the condition $\mathbf{m}\boldsymbol{e} = 0$ satisfies (9.55), just as in the case of an isotropic medium. From (8.202) it follows that the choice of the electric vibration vectors ($\boldsymbol{e}_j$ and $\boldsymbol{e}_{j+1}$) of the basis waves having identical refraction vectors ($\mathbf{m}_j = \mathbf{m}_{j+1}$) in accordance with the condition

$$\boldsymbol{e}_j\boldsymbol{e}_{j+1} = 0 \tag{9.60}$$

ensures the fulfillment of orthogonality relations (9.4). Since here we deal with the same situation as in the case of an isotropic medium, the electric vibration vectors of the basis waves with $\mathbf{m}_j = \pm n_\perp\mathbf{c}$ can be chosen in a similar manner. We can take

$$\boldsymbol{e}_1 = \begin{pmatrix}\sigma_1\\ 0\\ -\zeta\end{pmatrix}_{xyz}, \quad \boldsymbol{e}_2 = \begin{pmatrix}0\\ 1\\ 0\end{pmatrix}_{xyz} \tag{9.61}$$

if polarization degeneracy takes place for the forward propagating waves ($\mathbf{m}_1 = \mathbf{m}_2 = \pm n_\perp \mathbf{c}$), and

$$e_3 = \begin{pmatrix} \sigma_3 \\ 0 \\ -\zeta \end{pmatrix}_{xyz}, \quad e_4 = \begin{pmatrix} 0 \\ 1 \\ 0 \end{pmatrix}_{xyz} \tag{9.62}$$

when polarization degeneracy occurs for the backward propagating waves ($\mathbf{m}_3 = \mathbf{m}_4 = \pm n_\perp \mathbf{c}$). Such a choice of the electric vibration vectors in the case of polarization degeneracy is used in the routines of LMOPTICS.

Small-Birefringence Approximation

The subsection *Numerical Tests* at the end of this section contains some numerical examples which provide, in particular, estimates of the error of determining the EWB parameters in the small-birefringence approximation, which uses the condition $|n_\parallel - n_\perp| \ll n_\parallel, n_\perp$. These estimates are for the most popular variant of the approximate calculations, the variant used by Gu and Yeh [18, 19] in their version of the extended Jones matrix method. In our notation, the basic formulas for calculating the electric vibration vectors of the basis waves in this approximation can be written as follows:

(i) for the ordinary waves,

$$\mathbf{e}_j = \frac{\mathbf{c} \times \mathbf{m}_j}{|\mathbf{c} \times \mathbf{m}_j|} \quad j = 2,4, \tag{9.63}$$

(ii) for the extraordinary waves,

$$\mathbf{e}_j \approx \underline{\mathbf{e}}_{jSBA} \equiv \frac{\mathbf{m}_{j+1} \times \mathbf{e}_{j+1}}{|\mathbf{m}_{j+1} \times \mathbf{e}_{j+1}|} \quad j = 1,3, \tag{9.64}$$

that is, the electric vibration vectors of the ordinary waves are calculated here exactly [cf. (9.40) and (9.63)], and the electric vibration vector of each extraordinary wave is taken to be perpendicular to the refraction vector and electric vibration vector of the corresponding ordinary wave.

Program Implementation

The library LMOPTICS contains three routines for generating EWBs for uniaxial media: UNAXLR, UNAXLC, and UNAXLFC. UNAXLR is a real-arithmetic standard-output routine. UNAXLC is a universal standard-output routine. UNAXLFC is a universal full-output routine. In all these routines, the representation (9.36)–(9.43) is used. The routines have the following interfaces:

```
CALL UNAXLR(NE, NO, TE, FI, X, YY, YI, NWAVE, NRM)
CALL UNAXLC(NE, AE, NO, AO, TE, FI, X, YY, YI, NWAVE, NRM)
CALL UNAXLFC(NE, AE, NO, AO, TE, FI, X, EE, HH, NWAVE)
```

In the case of UNAXLR, the principal refractive indices are assumed to be real: NE = $n_\parallel$ and NO = $n_\perp$. For UNAXLC and UNAXLFC, NE = $\mathrm{Re}(n_\parallel)$, AE = $\mathrm{Im}(n_\parallel)$, NO = $\mathrm{Re}(n_\perp)$, and AO = $\mathrm{Im}(n_\perp)$. The orientation of the optic axis of the medium is specified by the arguments TE and FI: TE = θ and FI = φ, where θ is the angle between **c** and the *XY*-plane, and φ is the angle between the *XZ*-plane and the plane containing **c** and the *Z*-axis (the **c–z** plane) (Figure 9.3). The components of **c** in the coordinate system

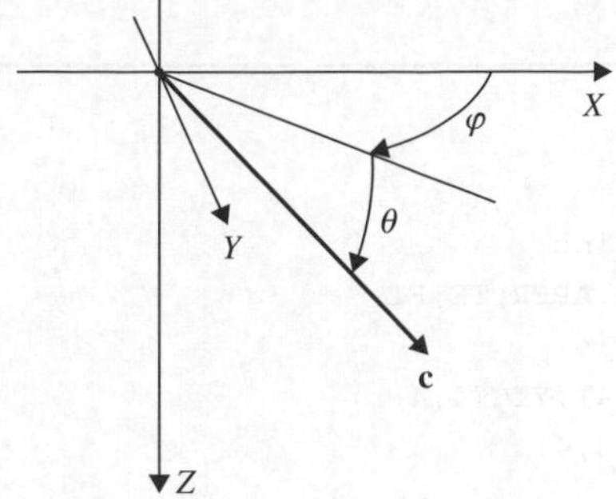

Figure 9.3 Specification of the orientation of the optic axis (**c**) of a uniaxial medium

(x, y, z) attached to the plane of incidence can be expressed in terms of θ and φ as follows:

$$\mathbf{c} = \begin{pmatrix} c_x \\ c_y \\ c_z \end{pmatrix}_{xyz} = \begin{pmatrix} \cos\theta\cos(\varphi - \alpha_{\rm inc}) \\ \cos\theta\sin(\varphi - \alpha_{\rm inc}) \\ \sin\theta \end{pmatrix}_{xyz}. \tag{9.65}$$

As in the routines for isotropic media described in the previous section, X in UNAXLR and UNAXLC is the output array of the σ_j values, YY = $\mathbf{\Psi}$, and YI = $\mathbf{\Psi}^{-1}$. The parameters NWAVE and NRM were defined in Section 9.1. The routine UNAXLFC returns the arrays of the coordinates of the vectors $\underline{\mathbf{e}}_j$ (EE(:, j)) and $\underline{\mathbf{h}}_j$ (HH(:, j)) in the coordinate system (x, y, z). Recall that $\underline{\mathbf{e}}_j$ is the electric vibration vector of the jth basis wave normalized by the condition $\underline{\mathbf{e}}_j{}^*\underline{\mathbf{e}}_j = 1$, and $\underline{\mathbf{h}}_j = \mathbf{m}_j \times \underline{\mathbf{e}}_j$.

Table 9.3 presents fragments of the code of a program **test_uniax**, which exemplify usage of UNAXLR, UNAXLC, and UNAXLFC. As can be seen from this code, before calling any of these routines, one should specify the global variables MT (ζ) and FIR ($\alpha_{\rm inc}$ in radians), which characterize the incident field (see Section 9.1). In the program **test_uniax**, the medium from which the light falls is taken to be of refractive index ($n_{\rm inc}$) 1, therefore $\zeta = \sin\beta_{\rm inc}$ in the code. If $n_{\rm inc} \neq 1$, MT should be calculated using the general formula $\zeta = n_{\rm inc}\sin\beta_{\rm inc}$.

Numerical Tests

Here we present the results of several numerical tests. In these tests, we consider nonabsorbing and absorbing uniaxial layers with a relatively large difference of the principal refractive indices, typical for practical LC materials, under different illumination conditions, including those under which TIR mode is realized. The tables show parameters of EWBs generated by routines of LMOPTICS and the results of the orthogonality test (see Section 9.1). In the tables, the calculated values of σ_j, $\underline{\mathbf{e}}_j$, and $n_{wj} \equiv (\mathbf{mm})^{1/2}$ are presented. The parameter n_{wj} can be considered as the refractive index for the jth basis wave if this wave is homogeneous [see (8.47)–(8.49)]. The orthogonality parameters $\delta_{\rm ort}$ computed for the matrices $\mathbf{\Psi}$ and $\mathbf{\Psi}^{-1}$ generated by UNAXLR and UNAXLC are denoted by $[\delta_{\rm ort}]_{\rm UNAXLR}$ and $[\delta_{\rm ort}]_{\rm UNAXLC}$, respectively.

In the first four tests, we consider nonabsorbing media out of TIR mode. In the tables, the results of the orthogonality tests for matrices $\mathbf{\Psi}$ calculated by the exact formulas ($[\delta_{\rm ort}]_{\rm UNAXLR}$) and those calculated by using the approximation (9.64) ($[\delta_{\rm ort}]_{\rm SBA}$) are compared.

Table 9.3 Usage of routines UNAXLR, UNAXLC, and UNAXLFC

Program **test_uniax**	Comments
`program test_uniax`	
`  use OPTSM_1`	
`  use MTWFIR`	
`  real(8) BETA_inc,ALFA_inc`	
`  real(8) NPAR,NPER,APAR,APER,TE,FI`	
`  real(8) YYR(4,4),YIR(4,4)`	
`  complex(8) X(4),YYC(4,4),YIC(4,4)`	
`  complex(8) EE(3,4),HH(3,4)`	
`  integer(4) NWAVE,NRM`	
`  ...`	
`! Specifying the incident light parameters and the`	
`! global variables`	
`  BETA_inc=45.0_8*PI3 !polar angle of incidence`	β_{inc} (rad)
`  ALFA_inc=30.0_8*PI3 !azimuthal angle of incidence`	α_{inc} (rad)
`                     ! PI3=PI/180`	
`  MT=sin(BETA_inc)    ! PI3, MT and FIR are global`	$\texttt{MT} = \zeta$
`  FIR=ALFA_inc        ! variables declared in MTWFIR`	$\texttt{FIR} = \beta_{\text{inc}}$
`  NWAVE=4       ! calculation for 4 basis waves`	
`  NRM=3         ! normalization type`	
`! Parameters of the layer`	
`! Real parts of the principal refractive indices`	$\texttt{NPAR} = \text{Re}(n_\parallel)$
`  NPAR=1.7_8; NPER=1.5_8`	$\texttt{NPER} = \text{Re}(n_\perp)$
`! Orientation of the principal axis`	$\texttt{TE} = \theta$ (rad)
`  TE=45.0_8*PI3; FI=90.0_8*PI3`	$\texttt{FI} = \varphi$ (rad)
`! Nonabsorbing layer`	
`  call UNAXLR(NPAR,NPER,TE,FI,X,YYR,YIR,NWAVE,NRM)`	$\sigma_j = \texttt{X}(j)$
`  ...`	$\mathbf{\Psi} = \texttt{YYR}$, $\mathbf{\Psi}^{-1} = \texttt{YIR}$
`! Absorbing layer`	
`! Imaginary parts of the principal refractive indices`	
`  APAR=0.01_8; APER=0.0001_8`	$\texttt{APAR} = \text{Im}(n_\parallel)$
`  ...`	$\texttt{APER} = \text{Im}(n_\perp)$
`  call UNAXLC(NPAR,APAR,NPER,APER,TE,FI,X,YYC,YIC, &`	
`               NWAVE,NRM)`	
`  ...`	$\mathbf{\Psi} = \texttt{YYC}$, $\mathbf{\Psi}^{-1} = \texttt{YIC}$
`  call UNAXLFC(NPAR,APAR,NPER,APER,TE,FI,X,EE,HH, &`	
`               NWAVE)`	
`  ...`	$\mathbf{e}_j = \texttt{EE}(:,j)$
`end program test_uniax`	$\mathbf{h}_j = \texttt{HH}(:,j)$

Test 1. Parameters: $n_{\|} = 1.7$, $n_{\perp} = 1.5$, $\theta = 45°$, $\varphi = 0°$, $\beta_{inc} = 45°$, $\alpha_{inc} = 90°$, $n_{inc} = 1$. The plane of incidence is perpendicular to the plane of symmetry of the layer (the **c–z** plane).

j	1	2	3	4
σ_j	1.446517	1.322876	−1.446517	−1.322876
n_{wj}	1.610097	1.500000	1.610097	1.500000
$\underline{e}_{jx}$	−0.413003	0.797724	0.413003	−0.797724
$\underline{e}_{jy}$	−0.908504	−0.426401	−0.908504	−0.426401
$\underline{e}_{jz}$	0.063631	−0.426401	0.063631	−0.426401
$[\delta_{ort}]_{UNAXLR}$		5.6×10^{-17}		
$[\delta_{ort}]_{SBA}$		3.1×10^{-2}		

Test 2. Parameters: $n_{\|} = 1.7$, $n_{\perp} = 1.5$, $\theta = 45°$, $\varphi = 0°$, $\beta_{inc} = 45°$, $\alpha_{inc} = 45°$, $n_{inc} = 1$. The same layer as in the previous test is considered, but the plane of incidence is rotated by 45°.

j	1	2	3	4
σ_j	1.374810	1.322876	−1.499323	−1.322876
n_{wj}	1.545995	1.500000	1.657700	1.500000
$\underline{e}_{jx}$	0.161191	0.862170	0.793920	−0.478420
$\underline{e}_{jy}$	−0.966655	0.210431	−0.549751	−0.840071
$\underline{e}_{jz}$	−0.198987	−0.460849	0.259740	−0.255726
$[\delta_{ort}]_{UNAXLR}$		1.2×10^{-16}		
$[\delta_{ort}]_{SBA}$		3.3×10^{-2}		

Test 3. Parameters: $n_{\|} = 1.7$, $n_{\perp} = 1.5$, $\theta = 70°$, $\varphi = 0°$, $\beta_{inc} = 30.865882473°$, $\alpha_{inc} = 0°$, $n_{inc} = 1$. In this case, the orientation of the optic axis **c** and the propagation direction of the incident light are chosen so that polarization degeneracy takes place for the forward propagating basis waves.

j	1	2	3	4
σ_j	1.409539	1.409539	−1.484510	−1.409539
n_{wj}	1.500000	1.500000	1.570659	1.500000
$\underline{e}_{jx}$	0.939693	0.000000	0.977086	0.000000
$\underline{e}_{jy}$	0.000000	1.000000	0.000000	−1.000000
$\underline{e}_{jz}$	−0.342020	0.000000	0.212846	0.000000
$[\delta_{ort}]_{UNAXLR}$		5.6×10^{-17}		
$[\delta_{ort}]_{SBA}$		2.3×10^{-2}		

From the results of the orthogonality test it is seen that the routine UNAXLR gives very accurate values of $\boldsymbol{\Psi}$ and $\boldsymbol{\Psi}^{-1}$ both in the absence (tests 1 and 2) and in the presence (test 3) of polarization degeneracy. The tiny deviations of $[\delta_{ort}]_{UNAXLR}$ from zero are caused by round-off errors. At the same time, the orthogonality test reveals a significant error in the values of $\boldsymbol{\Psi}$ obtained using the small-birefringence approximation.

Test 4. This example allows us to assess the degree of discrepancy between the true values of the electric vibration vector of an extraordinary wave, $\underline{\mathbf{e}}_j$, and corresponding values of $\underline{\mathbf{e}}_{jSBA}$ (9.64) for media with Δn (here $\Delta n \equiv n_{\|} - n_{\perp}$) typical of LC materials used in LCDs. In this test, as in the previous three tests,

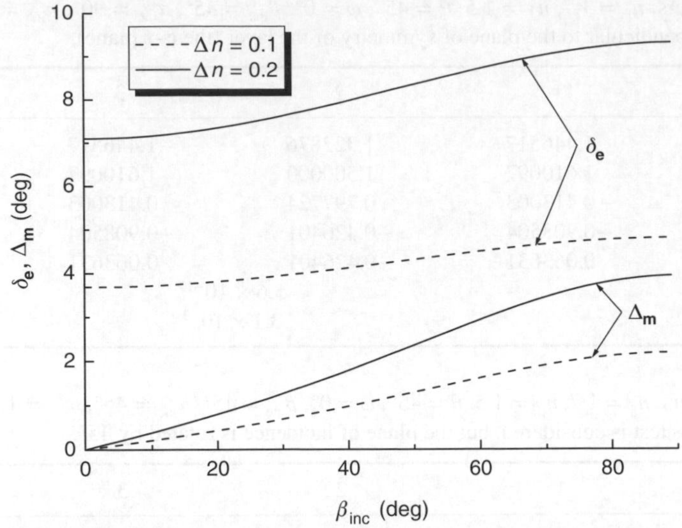

Figure 9.4 The angle of deviation of the electric vibration vector of the extraordinary wave calculated in the small-birefringence approximation from its true value ($\delta_{\mathbf{e}}$) and the angle between the refraction vector of the extraordinary wave and that of the ordinary wave ($\Delta_{\mathbf{m}}$)

the light falls on a nonabsorbing uniaxial layer from an isotropic medium with $n_{inc} = 1$. The uniaxial layer has a tilted optic axis with $\theta = 45°$ and $\varphi = 0°$; $n_{\perp} = 1.5$. The plane of incidence is oriented so that $\alpha_{inc} = 90°$. Figure 9.4 shows dependences of the angle between the true $\underline{\mathbf{e}}_1$ and $\underline{\mathbf{e}}_{1SBA}$—this angle is denoted by $\delta_{\mathbf{e}}$—on the polar angle of incidence (β_{inc}) for $n_{\parallel} = 1.6$ ($\Delta n = 0.1$) and $n_{\parallel} = 1.7$ ($\Delta n = 0.2$). Analogous curves for the angle between the refraction vectors of the extraordinary ($\mathbf{m}_1$) and ordinary ($\mathbf{m}_2$) waves ($\Delta_{\mathbf{m}}$) are shown in this figure for comparison. As is seen from Figure 9.4, the deviation of approximate values from the true ones may be large even for media with moderate (for LC materials) birefringence, so that we cannot be sure that we will obtain good accuracy using the approximation (9.64) for LC layers in common situations.

In the next two tests, absorbing media out of TIR mode are considered.

Test 5. Parameters: $n_{\parallel} = 1.7 + i0.03$, $n_{\perp} = 1.5 + i0.003$, $\theta = 45°$, $\varphi = 0°$, $\beta_{inc} = 45°$, $\alpha_{inc} = 45°$, $n_{inc} = 1$. The situation is the same as in Test 2 except that the medium is absorbing.

j	1	2	3	4
σ_j	$1.374912 + i0.009550$	$1.322877 + i0.003402$	$-1.499607 - i0.024953$	$-1.322877 - i0.003402$
$\underline{e}_{jx}$	$0.161122 - i0.000881$	$0.862166 + i0.002217$	$0.793815 + i0.005119$	$-0.478419 - i0.001230$
$\underline{e}_{jy}$	$-0.966547 + i0.000000$	$0.210430 + i0.002217$	$-0.549616 + i0.000000$	$-0.840069 - i0.001230$
$\underline{e}_{jz}$	$-0.199167 - i0.012591$	$-0.460847 + i0.000000$	$0.259566 - i0.019473$	$-0.255726 + i0.000000$
$[\delta_{ort}]_{UNAXLC}$		8.3×10^{-17}		

Test 6. Parameters: $n_{\|} = 1.7 + i0.03$, $n_{\perp} = 1.5 + i0.003$, $\theta = 0°$, $\varphi = 0°$, $\beta_{inc} = 45°$, $\alpha_{inc} = 45°$, $n_{inc} = 1$. In this case, the direction of maximum absorbance is along the optic axis and parallel to the layer boundaries, so that the layer behaves as a nonideal *O*-type polarizer.

j	1	2	3	4
σ_j	1.522813 + *i*0.030191	1.322877 + *i*0.003402	−1.522813 − *i*0.030191	−1.322877 − *i*0.003402
$\underline{e}_{jx}$	0.574314 + *i*0.000656	0.661436 + *i*0.001701	0.574314 + *i*0.000656	−0.661436 − *i*0.001701
$\underline{e}_{jy}$	−0.738402 + *i*0.000000	0.661436 + *i*0.001701	−0.738402 + *i*0.000000	−0.661436 − *i*0.001701
$\underline{e}_{jz}$	−0.353404 − *i*0.005592	−0.353552 + *i*0.000000	0.353404 + *i*0.005592	−0.353552 + *i*0.000000
$[\delta_{ort}]_{UNAXLC}$		8.3×10^{-17}		

Tests 7 and 8 demonstrate the capability of UNAXLC and UNAXLFC to properly work in the case of TIR mode. In Test 7, the layer is absorbing. In Test 8, a nonabsorbing medium is considered. In these two tests, we specify directly the parameter ζ rather than β_{inc}. The value 1.72 assigned to ζ corresponds to $\beta_{inc} \approx 73°$ if the light falls on the layer from an isotropic medium whose refractive index is equal to 1.8.

Test 7. Parameters: $n_{\|} = 1.7 + i0.03$, $n_{\perp} = 1.5 + i0.003$, $\theta = 45°$, $\varphi = 0°$, $\alpha_{inc} = 45°$, $\zeta = 1.72$.

j	1	2	3	4
σ_j	−0.074756 + *i*0.457503	0.005346 + *i*0.841687	−0.228558 − *i*0.494970	−0.005346 − *i*0.841687
$\underline{e}_{jx}$	−0.118827 − *i*0.246556	0.001669 + *i*0.262692	−0.031716 + *i*0.260774	−0.113607 − *i*0.262029
$\underline{e}_{jy}$	−0.503526 + *i*0.000000	−0.757503 + *i*0.262692	−0.483478 + *i*0.000000	−0.758920 − *i*0.262029
$\underline{e}_{jz}$	0.804724 − *i*0.154824	−0.536816 + *i*0.000000	0.827245 + *i*0.113607	−0.535461 + *i*0.000000
$[\delta_{ort}]_{UNAXLC}$		1.1×10^{-16}		

Test 8. Parameters: $n_{\|} = 1.7$, $n_{\perp} = 1.5$, $\theta = 45°$, $\varphi = 0°$, $\alpha_{inc} = 45°$, $\zeta = 1.72$.

j	1	2	3	4
σ_j	−0.151436 + *i*0.470246	0.000000 + *i*0.841665	−0.151436 − *i*0.470246	0.000000 − *i*0.841665
$\underline{e}_{jx}$	−0.074713 − *i*0.251324	0.000000 + *i*0.262355	−0.074713 + *i*0.251324	0.000000 − *i*0.262355
$\underline{e}_{jy}$	−0.494365 + *i*0.000000	−0.758217 + *i*0.262355	−0.494365 + *i*0.000000	−0.758217 − *i*0.262355
$\underline{e}_{jz}$	0.817953 − *i*0.133457	−0.536140 + *i*0.000000	0.817953 + *i*0.133457	−0.536140 + *i*0.000000
$[\delta_{ort}]_{UNAXLC}$		1.2×10^{-16}		

The orthogonality test in the last four examples demonstrates that the routines UNAXLC and UNAXLFC give very accurate eigenwave bases both for absorbing media and in the case of TIR mode.

9.4 Biaxial Media

Calculating the parameters of eigenwaves in a biaxial medium generally requires knowledge of the matrices of the permittivity tensor $\boldsymbol{\varepsilon}$ and the inverse tensor $\boldsymbol{\varepsilon}^{-1}$. Therefore, we will first say a few words about specification of these matrices. The permittivity tensor of a nonabsorbing biaxial medium can always be represented as follows:

$$\boldsymbol{\varepsilon} = \varepsilon_1 \mathbf{u}_1 \otimes \mathbf{u}_1 + \varepsilon_2 \mathbf{u}_2 \otimes \mathbf{u}_2 + \varepsilon_3 \mathbf{u}_3 \otimes \mathbf{u}_3, \tag{9.66}$$

where $\mathbf{u}_1$, $\mathbf{u}_2$, and $\mathbf{u}_3$ are the unit basis vectors of a Cartesian coordinate system in which the matrix of $\boldsymbol{\varepsilon}$ is diagonal:

$$\begin{pmatrix} \varepsilon_1 & 0 & 0 \\ 0 & \varepsilon_2 & 0 \\ 0 & 0 & \varepsilon_3 \end{pmatrix}_{\mathbf{u}_1\mathbf{u}_2\mathbf{u}_3} \tag{9.67}$$

The vectors $\mathbf{u}_1$, $\mathbf{u}_2$, and $\mathbf{u}_3$ show the directions of the principal axes of the tensor $\boldsymbol{\varepsilon}$; ε_1, ε_2, and ε_3 are the principal dielectric constants, related to the principal refractive indices of the medium, n_1, n_2, and n_3, by

$$\varepsilon_j = n_j^2 \quad j = 1,2,3. \tag{9.68}$$

In this case (a nonabsorbing medium), the tensor $\boldsymbol{\varepsilon}^{-1}$ has the same system of principal axes as the tensor $\boldsymbol{\varepsilon}$. It can therefore be expressed as follows:

$$\boldsymbol{\varepsilon}^{-1} = \frac{1}{\varepsilon_1}\mathbf{u}_1 \otimes \mathbf{u}_1 + \frac{1}{\varepsilon_2}\mathbf{u}_2 \otimes \mathbf{u}_2 + \frac{1}{\varepsilon_3}\mathbf{u}_3 \otimes \mathbf{u}_3. \tag{9.69}$$

Using the following general relation

$$\mathbf{u}_1 \otimes \mathbf{u}_1 + \mathbf{u}_2 \otimes \mathbf{u}_2 + \mathbf{u}_3 \otimes \mathbf{u}_3 = \mathbf{U},$$

where $\mathbf{U}$ is the unit tensor, we can eliminate one of the vectors $\mathbf{u}_j$ from the above expressions for $\boldsymbol{\varepsilon}$ and $\boldsymbol{\varepsilon}^{-1}$ to obtain another representation of these tensors, more convenient for their specification. For instance, these tensors can be expressed as

$$\boldsymbol{\varepsilon} = \varepsilon_2 \mathbf{U} + (\varepsilon_1 - \varepsilon_2)\mathbf{u}_1 \otimes \mathbf{u}_1 + (\varepsilon_3 - \varepsilon_2)\mathbf{u}_3 \otimes \mathbf{u}_3, \tag{9.70}$$

$$\boldsymbol{\varepsilon}^{-1} = \frac{1}{\varepsilon_2}\mathbf{U} + \left(\frac{1}{\varepsilon_1} - \frac{1}{\varepsilon_2}\right)\mathbf{u}_1 \otimes \mathbf{u}_1 + \left(\frac{1}{\varepsilon_3} - \frac{1}{\varepsilon_2}\right)\mathbf{u}_3 \otimes \mathbf{u}_3. \tag{9.71}$$

Thus, to calculate the matrices of the tensors $\boldsymbol{\varepsilon}$ and $\boldsymbol{\varepsilon}^{-1}$ referred to any particular coordinate system, it suffices to substitute the coordinates of the vectors $\mathbf{u}_1$ and $\mathbf{u}_3$ in this coordinate system into (9.70) and (9.71). The representation (9.66) and consequently expressions (9.68)–(9.71) are valid not only for

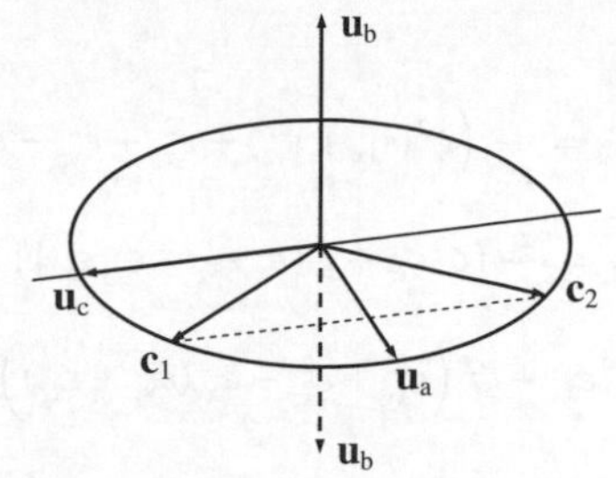

Figure 9.5 Orientation of the principal axes of the permittivity tensor ($\mathbf{u}_a$, $\mathbf{u}_b$, $\mathbf{u}_c$) and the optic axes ($\mathbf{c}_1$, $\mathbf{c}_2$) of a biaxial medium

nonabsorbing media, but also for absorbing media of the orthorhombic system. In the latter case, n_1, n_2, and n_3 in (9.68) are the principal complex refractive indices.

We should mention yet another representation of the tensor ε^{-1}. This representation is very important because it explicitly reflects the main peculiarities of the optical properties of biaxial media. Let n_a, n_b, and n_c be the principal refractive indices of an arbitrary nonabsorbing biaxial medium ordered so that $n_a < n_b < n_c$ and let $\mathbf{u}_a$, $\mathbf{u}_b$, and $\mathbf{u}_c$ be unit vectors directed along the corresponding principal axes. Then the tensor ε^{-1} can be expressed as

$$\varepsilon^{-1} = \frac{1}{\varepsilon_b}\mathbf{U} + \frac{1}{2}\left(\frac{1}{\varepsilon_a} - \frac{1}{\varepsilon_c}\right)(\mathbf{c}_1 \otimes \mathbf{c}_2 + \mathbf{c}_2 \otimes \mathbf{c}_1), \tag{9.72}$$

where

$$\begin{gathered}\varepsilon_a = n_a^2, \quad \varepsilon_b = n_b^2, \quad \varepsilon_c = n_c^2, \\ \mathbf{c}_1 = k_a\mathbf{u}_a + k_c\mathbf{u}_c, \quad \mathbf{c}_2 = k_a\mathbf{u}_a - k_c\mathbf{u}_c, \\ k_a = \sqrt{\frac{\varepsilon_c(\varepsilon_b - \varepsilon_a)}{\varepsilon_b(\varepsilon_c - \varepsilon_a)}}, \quad k_c = \sqrt{\frac{\varepsilon_a(\varepsilon_c - \varepsilon_b)}{\varepsilon_b(\varepsilon_c - \varepsilon_a)}}\end{gathered} \tag{9.73}$$

[11]. The unit vectors $\mathbf{c}_1$ and $\mathbf{c}_2$ defined by (9.73) show the directions of two optic axes of the medium (the optic axes should not be confused with the principal axes of the permittivity tensor, see Figure 9.5). As is seen from (9.73), both optic axes are perpendicular to the axis $\mathbf{u}_b$ corresponding to the middle principal refractive index. Homogeneous eigenwaves of this medium are linearly polarized. For waves whose wave normals are parallel to any of the optic axes, any orientation of the polarization plane is allowed, that is, the optic axes show the propagation directions for which polarization degeneracy takes place. Waves (of equal frequency) with different polarization states, propagating along an optic axis, have the same phase velocity. The refractive index n_w [see (8.47)] for such waves is equal to n_b. For any other propagation direction, only two polarization modes are allowed, and equinormal waves of different polarization modes have different phase velocities: for waves of one of the allowed modes, $n_a \le n_w \le n_b$, and for waves of the other allowed mode, $n_b \le n_w \le n_c$.

The normal components of the refraction vectors of the basis eigenwaves for a biaxial medium can be found from the following equation [it is equation (8.73) written in a convenient form]:

$$\sigma^4 + C_3\sigma^3 + C_2\sigma^2 + C_1\sigma + C_0 = 0, \tag{9.74}$$

where

$$C_3 = \frac{2\varepsilon_{xz}}{\varepsilon_{zz}}\zeta, \quad C_2 = \frac{1}{\varepsilon_{zz}}\left(\zeta^2(\varepsilon_{xx}+\varepsilon_{zz})+\varepsilon_{xz}^2+\varepsilon_{yz}^2-\varepsilon_{zz}(\varepsilon_{xx}+\varepsilon_{yy})\right),$$
$$C_1 = \frac{2}{\varepsilon_{zz}}\left(\zeta^3\varepsilon_{xz}+\zeta\left(\varepsilon_{xy}\varepsilon_{yz}-\varepsilon_{yy}\varepsilon_{xz}\right)\right), \tag{9.75}$$
$$C_0 = \frac{1}{\varepsilon_{zz}}\left(\zeta^4\varepsilon_{xx}+\zeta^2\left(\varepsilon_{xy}^2+\varepsilon_{xz}^2-\varepsilon_{xx}(\varepsilon_{yy}+\varepsilon_{zz})\right)+\det\varepsilon\right),$$

where ε_{kj} are components of ε in the system (x, y, z) with the xz-plane coincident with the plane of incidence (see Figure 9.1); ζ, as usual, is the tangential component of the refraction vectors ($\zeta = \mathbf{b}\boldsymbol{x}$ with $\boldsymbol{x}$ being the unit vector along the x-axis); $\det\varepsilon$ is the determinant of ε. When representation (9.66) is applicable, $\det\varepsilon$ can be calculated as $\det\varepsilon = \varepsilon_1\varepsilon_2\varepsilon_3$; a more general formula is

$$\det\varepsilon = \varepsilon_{xx}\left(\varepsilon_{yy}\varepsilon_{zz}-\varepsilon_{xz}^2\right)+2\varepsilon_{xy}\varepsilon_{xz}\varepsilon_{yz}-\varepsilon_{xy}^2\varepsilon_{zz}-\varepsilon_{xz}^2\varepsilon_{yy}.$$

In modeling LCDs, we are faced with the need to consider biaxial media when dealing with biaxial compensators (retarders). For a usual biaxial compensation film, one of the vectors $\mathbf{u}_j$ is perpendicular to the film surfaces. It greatly simplifies the problem. If one of the vectors $\mathbf{u}_j$ is parallel to the z-axis, then

$$\varepsilon_{xz} = 0, \quad \varepsilon_{yz} = 0. \tag{9.76}$$

As can be seen from (9.75), subject to (9.76), equation (9.74) is biquadratic ($C_3 = 0$, $C_1 = 0$), so that its roots can easily be found:

$$\sigma_1 = \sqrt{\frac{-C_2+D_C}{2}}, \quad \sigma_2 = \sqrt{\frac{-C_2-D_C}{2}}, \quad D_C = \sqrt{C_2^2-4C_0}, \tag{9.77}$$
$$\sigma_3 = -\sigma_1, \quad \sigma_4 = -\sigma_2.$$

Formulas (9.77) can also be used in the case of normal incidence ($\zeta = 0$) for any orientation of the axes $\mathbf{u}_j$.

Considering the case of oblique incidence and a biaxial medium with arbitrary orientation of the principal axes (we may deal with such a situation, e.g., when modeling the optical properties of smectic C^* layers with allowance for the optical biaxiality of the liquid crystal), we need to solve (9.74) as a general quartic equation. In principle, roots of this equation may be found by using the classical explicit formulas for roots of quartic and cubic equations (Ferrari's and Cardano's solutions). This approach is used, for example, in References 20–22. However, the computational algorithms based on using these exact formulas are numerically unstable (see, e.g., Reference 20) and often give inaccurate results, even when double-precision arithmetic is used (we observed this in numerical experiments). For this reason, and also because of the complexity of the exact expressions for the roots, it is considered to be preferable to solve (9.74) by using iterative algorithms. Some efficient iterative algorithms for solving (8.73) may be found in old books on optics of anisotropic media (e.g., [11]). Descriptions of other iterative techniques for finding σ_j, which are currently used in modeling LCDs, are given in References 23 and 24. Of particular interest, in our opinion, is the algorithm proposed in Reference 24. In this algorithm, finding the roots of (9.74) is carried out with the help of Laguerre's method [25, 26]. In our programs, we use a slightly modified version of this algorithm: First, we find the root σ_1 by using Laguerre's method with the following initial approximation:

$$\sigma = \sqrt{\varepsilon_{\mathrm{m}}-\zeta^2}, \tag{9.78}$$

where ε_m is the principal dielectric constant whose real part is middle in magnitude among ε_1, ε_2, and ε_3. Then, by the formulas

$$B_2 = C_3 + \sigma_1, \quad B_1 = C_2 + B_2\sigma_1, \quad B_0 = C_1 + B_1\sigma_1,$$

we calculate the coefficients of the cubic equation

$$\sigma^3 + B_2\sigma^2 + B_1\sigma + B_0 = 0, \tag{9.79}$$

which is the result of division of (9.74) by $(\sigma - \sigma_1)$. From (9.79), using Laguerre's method and initial approximation (9.78) again, we find the root σ_2. Then, by the formulas

$$D_1 = B_2 + \sigma_2, \quad D_0 = B_1 + D_1\sigma_2,$$

the coefficients of the quadratic equation $\sigma^2 + D_1\sigma + D_0 = 0$ are calculated. Solving this equation in the usual way, we find σ_3 and σ_4. Numerous numerical experiments have shown that this algorithm is very robust and finds the roots with high speed and accuracy. Note that to attain satisfactory accuracy of the final results, σ_j should be determined with an absolute error less than 10^{-8}. Therefore, all the computations should be performed using double-precision arithmetic. In some special cases, renumbering of the roots thus obtained is required.

Once σ_j are found, the vectors $\mathbf{e}_j$ and $\mathbf{h}_j$ can be calculated. There are several ways of calculating these vectors [11, 17, 21]. We prefer to use the following technique. It is based on the approach described in Reference 21. This approach uses the fact that the displacement vibration vector **d** of an eigenwave with refraction vector **m** satisfies the relation **md** = 0 [see (8.50)]. This allows one to represent the corresponding "raw" vector $\boldsymbol{d}$ as a linear combination of two mutually orthogonal vectors each of which is orthogonal to **m**:

$$\boldsymbol{d} = \boldsymbol{y}d_y + \boldsymbol{g}d_g, \tag{9.80}$$

where, as before, $\boldsymbol{y}$ is the unit vector along the y-axis, perpendicular to the plane of incidence; $\boldsymbol{g} = \mathbf{m}\times\boldsymbol{y}$. According to (8.40)–(8.42), $\boldsymbol{d}$ obeys the equation

$$\mathbf{m}\times(\mathbf{m}\times(\varepsilon^{-1}\boldsymbol{d})) + \boldsymbol{d} = 0.$$

We rewrite this equation in the form

$$\mathbf{m}(\mathbf{m}(\varepsilon^{-1}\boldsymbol{d})) - (\varepsilon^{-1}\boldsymbol{d})m^2 + \boldsymbol{d} = 0, \tag{9.81}$$

where $m^2 = \mathbf{mm}$. Multiplying (9.81) scalarly by $\boldsymbol{y}$, we have

$$-\boldsymbol{y}(\varepsilon^{-1}\boldsymbol{d})m^2 + \boldsymbol{yd} = 0. \tag{9.82}$$

Substitution from (9.80) into (9.82) leads to the equation

$$A_{yy}d_y - A_{yg}d_g = 0, \tag{9.83}$$

where

$$A_{yy} = 1 - m^2\boldsymbol{y}(\varepsilon^{-1}\boldsymbol{y}), \quad A_{yg} = m^2\boldsymbol{y}(\varepsilon^{-1}\boldsymbol{g}).$$

Similarly, multiplying (9.81) scalarly by $\boldsymbol{g}$, we obtain

$$-\boldsymbol{g}(\varepsilon^{-1}\boldsymbol{d})m^2 + \boldsymbol{gd} = 0$$

and, further,

$$-A_{yg}d_y + A_{gg}d_g = 0, \tag{9.84}$$

where

$$A_{gg} = m^2(1 - \boldsymbol{g}(\varepsilon^{-1}\boldsymbol{g})).$$

Here we have used the relations $\boldsymbol{gg} = m^2$ and

$$\boldsymbol{g}(\varepsilon^{-1}\boldsymbol{y}) = \boldsymbol{y}(\varepsilon^{-1}\boldsymbol{g}).$$

The latter relation holds due to the symmetry of the tensor ε^{-1}. If at least one of the coefficients A_{yy}, A_{yg}, A_{gg} differs from zero, equations (9.83) and (9.84) uniquely determine the relation between d_y and d_g. If all three coefficients are nonzero, they are such that $A_{yy}/A_{yg} = A_{yg}/A_{gg}$. The case $A_{yy} = A_{yg} = A_{gg} = 0$ corresponds to polarization degeneracy.

From (9.83) and (9.84) it is easy to obtain the following working formulas for calculating the coordinates of $\boldsymbol{d}_j$ in the system (x, y, z):

$$\boldsymbol{d}_j = \begin{pmatrix} -\sigma_j d_g^{(j)} \\ d_y^{(j)} \\ \zeta d_g^{(j)} \end{pmatrix}_{xyz}, \tag{9.85}$$

where

$$\begin{cases} d_y^{(j)} = A_{yg}^{(j)},\ d_g^{(j)} = A_{yy}^{(j)} & \left|A_{yy}^{(j)}\right| > \left|A_{gg}^{(j)}\right| \\ d_y^{(j)} = A_{gg}^{(j)},\ d_g^{(j)} = A_{yg}^{(j)} & \left|A_{yy}^{(j)}\right| < \left|A_{gg}^{(j)}\right|, \end{cases}$$
$$A_{yy}^{(j)} = 1 - m_j^2\tilde{\varepsilon}_{yy}, \quad A_{yg}^{(j)} = m_j^2(\tilde{\varepsilon}_{yz}\zeta - \tilde{\varepsilon}_{xy}\sigma_j), \tag{9.86}$$
$$A_{gg}^{(j)} = m_j^2\left(1 - \tilde{\varepsilon}_{xx}\sigma_j^2 + 2\tilde{\varepsilon}_{xz}\zeta\sigma_j - \tilde{\varepsilon}_{zz}\zeta^2\right), \quad m_j^2 = \zeta^2 + \sigma_j^2;$$

$\tilde{\varepsilon}_{kl}$ are components of the tensor ε^{-1} in the system (x, y, z). These formulas are used when polarization degeneracy is absent. If polarization degeneracy takes place, it is convenient to choose vectors $\boldsymbol{d}_j$ as follows:

$$\boldsymbol{d}_j = \begin{cases} \boldsymbol{y} + \mathbf{u}_b & \mathbf{u}_b\boldsymbol{y} \geq 0 \\ \boldsymbol{y} - \mathbf{u}_b & \mathbf{u}_b\boldsymbol{y} < 0, \end{cases} \qquad \boldsymbol{d}_{j+1} = \mathbf{m}_j \times \boldsymbol{d}_j, \tag{9.87}$$

where $j = 1$ if the calculations are being carried out for the forward propagating waves (when $\sigma_1 = \sigma_2$), and $j = 3$ if for the backward propagating waves (when $\sigma_3 = \sigma_4$). This choice of $\boldsymbol{d}_j$ ensures the orthogonality of the EWB in the sense of (9.4) at both normal and oblique incidence.

The vectors $\boldsymbol{e}_j$ and $\boldsymbol{h}_j$ are calculated in terms of $\boldsymbol{d}_j$ by the formulas

$$\boldsymbol{e}_j = \varepsilon^{-1}\boldsymbol{d}_j, \quad \boldsymbol{h}_j = \mathbf{m}_j \times \boldsymbol{e}_j \quad j = 1,2,3,4. \tag{9.88}$$

To complete the picture, let us mention one old method of calculating the vectors $\boldsymbol{e}_j$ and $\boldsymbol{h}_j$. This method is described in detail in Reference 11, where it is applied to nonabsorbing media. In the absence of polarization degeneracy, refraction vectors of two basis waves, one forward propagating wave and one backward propagating wave, satisfy the following equation:

$$\frac{1}{\varepsilon_b}\mathbf{m}_j^2 + \frac{1}{2}\left(\frac{1}{\varepsilon_a} - \frac{1}{\varepsilon_c}\right)\left\{(\mathbf{m}_j \times \mathbf{c}_1)(\mathbf{m}_j \times \mathbf{c}_2) + \sqrt{(\mathbf{m}_j \times \mathbf{c}_1)^2(\mathbf{m}_j \times \mathbf{c}_2)^2}\right\} = 1. \quad (9.89)$$

The refraction vectors of the other two basis waves satisfy another equation:

$$\frac{1}{\varepsilon_b}\mathbf{m}_j^2 + \frac{1}{2}\left(\frac{1}{\varepsilon_a} - \frac{1}{\varepsilon_c}\right)\left\{(\mathbf{m}_j \times \mathbf{c}_1)(\mathbf{m}_j \times \mathbf{c}_2) - \sqrt{(\mathbf{m}_j \times \mathbf{c}_1)^2(\mathbf{m}_j \times \mathbf{c}_2)^2}\right\} = 1. \quad (9.90)$$

Equations (9.89) and (9.90) are consistent with (8.73). The "raw" vectors $\boldsymbol{h}_j$ for the waves satisfying (9.89) can be calculated as

$$\boldsymbol{h}_j = \sqrt{(\mathbf{m}_j \times \mathbf{c}_2)^2}(\mathbf{m}_j \times \mathbf{c}_1) + \sqrt{(\mathbf{m}_j \times \mathbf{c}_1)^2}(\mathbf{m}_j \times \mathbf{c}_2), \quad (9.91)$$

and for the waves satisfying (9.90) as

$$\boldsymbol{h}_j = \sqrt{(\mathbf{m}_j \times \mathbf{c}_2)^2}(\mathbf{m}_j \times \mathbf{c}_1) - \sqrt{(\mathbf{m}_j \times \mathbf{c}_1)^2}(\mathbf{m}_j \times \mathbf{c}_2). \quad (9.92)$$

The vectors $\boldsymbol{e}_j$ in this case can be calculated by the formula:

$$\boldsymbol{e}_j = -\varepsilon^{-1}(\mathbf{m}_j \times \boldsymbol{h}_j) \quad j = 1,2,3,4.$$

Note that expressions (9.91) and (9.92) cannot be used when $\mathbf{m}_j\mathbf{u}_b = 0$.

Program Implementation

In the library LMOPTICS, there are six subroutines intended for calculating the parameters of basis waves in biaxial media. Here are their interfaces:

```
CALL BIAXLR (N1, N2, N3, PMIN, PMAX, X, YY, YI, NWAVE, NRM)
CALL BIAXLC (N1, A1, N2, A2, N3, A3, PMIN, PMAX, X, YY, YI, NWAVE, NRM)
CALL BIAXLFR (N1, N2, N3, PMIN, PMAX, X, EE, HH, NWAVE)
CALL BIAXLFC (N1, A1, N2, A2, N3, A3, PMIN, PMAX, X, EE, HH, NWAVE)
CALL BIAXSORT1R (N01, N02, N03, AA, BB, GG, N1, N2, N3, PMIN, PMAX)
CALL BIAXSORT1C (N01, A01, N02, A02, N03, A03, AA, BB, GG, N1, A1, N2, A2, N3, A3, PMIN,
PMAX)
```

The main routines for practical calculations are the standard-output routines BIAXLR and BIAXLC. The output parameters of these routines are, as usual, the array of σ_j values (X(j)), matrix $\boldsymbol{\Psi}$ (YY), and matrix $\boldsymbol{\Psi}^{-1}$ (YI). The input parameters describing the medium in these routines are the principal refractive indices n_a, n_b, and n_c, ordered so that $\mathrm{Re}(n_a) \le \mathrm{Re}(n_b) \le \mathrm{Re}(n_c)$ [see the paragraph with equation (9.72)], and arrays of the coordinates of the vectors $\mathbf{u}_a$ (PMIN) and $\mathbf{u}_c$ (PMAX) in the system (x, y, z). BIAXLR is a real-arithmetic routine, BIAXLC is a complex-arithmetic one (see Section 9.1). The routine BIAXLR is applicable only to nonabsorbing media out of TIR mode. The routine BIAXLC is free of these restrictions. BIAXLFR and BIAXLFC are full-output analogs of BIAXLR and BIAXLC,

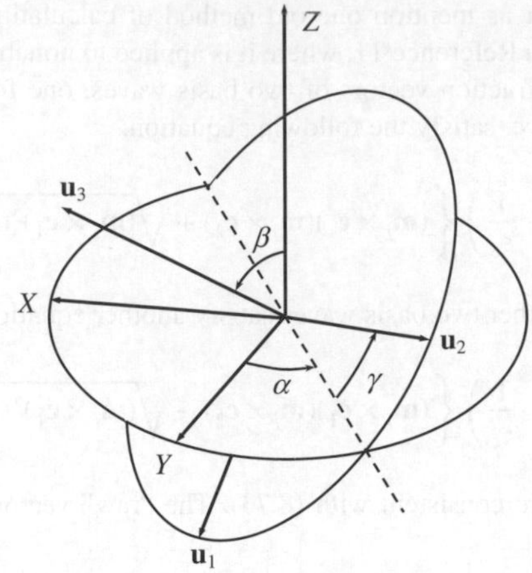

Figure 9.6 Specification of the orientation of the principal axes ($\mathbf{u}_1$, $\mathbf{u}_2$, $\mathbf{u}_3$) of the permittivity tensor of a biaxial medium by using the Euler angles α, β, and γ. The *Z–Y–Z* convention is used

respectively. These routines return the arrays of the coordinates of the vectors $\mathbf{e}_j$ (EE(:, *j*)) and $\mathbf{h}_j$ (HH(:, *j*)) in the system (*x*, *y*, *z*). In BIAXLR and BIAXLFR, the input refractive indices are real: N1 = n_a, N2 = n_b, N3 = n_c. In BIAXLC and BIAXLFC, the complex principal refractive indices are used: N1 = Re(n_a), A1 = Im(n_a), N2 = Re(n_b), A2 = Im(n_b), N3 = Re(n_c), A3 = Im(n_c). Calculating Ψ^{-1} from Ψ in BIAXLR and BIAXLC is carried out by using formulas (9.8) and (9.13). The input parameters NWAVE and NRM perform their usual functions (see Section 9.1). The subroutines BIAXSORT1R/BIAXSORT1C and BIAXSORT2R/BIAXSORT2C facilitate preparing input data for routines BIAXLR/BIAXLC and BIAXLFR/BIAXLFC. Let α, β, and γ be the Euler angles specifying the orientation of the principal axes $\mathbf{u}_1$, $\mathbf{u}_2$, and $\mathbf{u}_3$ (the corresponding principal refractive indices, as before, are n_1, n_2, and n_3) with respect to the reference system (*X*, *Y*, *Z*) as shown in Figure 9.6. The subroutine BIAXSORT1R/BIAXSORT1C calculates the coordinates of the vectors $\mathbf{u}_1$, $\mathbf{u}_2$, and $\mathbf{u}_3$ in the system (*x*, *y*, *z*) by the formulas

$$\mathbf{u}_1 = \begin{pmatrix} \cos\alpha' \cos\beta \cos\gamma - \sin\alpha' \sin\gamma \\ \sin\alpha' \cos\beta \cos\gamma + \cos\alpha' \sin\gamma \\ -\sin\beta \cos\gamma \end{pmatrix}_{xyz},$$

$$\mathbf{u}_2 = \begin{pmatrix} -\cos\alpha' \cos\beta \sin\gamma - \sin\alpha' \cos\gamma \\ -\sin\alpha' \cos\beta \sin\gamma + \cos\alpha' \cos\gamma \\ \sin\beta \sin\gamma \end{pmatrix}_{xyz}, \quad \mathbf{u}_3 = \begin{pmatrix} \cos\alpha' \sin\beta \\ \sin\alpha' \sin\beta \\ \cos\beta \end{pmatrix}_{xyz},$$

$$\alpha' = \alpha - \alpha_{inc},$$

where, as before, α_{inc} is the angle between *XZ*-plane and the plane of incidence (the *xz*-plane). Then, n_a, n_b, and n_c are selected from the input set of the principal refractive indices (n_1, n_2, n_3) (nonordered in magnitude). The results are the ordered set of the principal refractive indices (n_a, n_b, n_c) and the arrays of the coordinates of the vectors $\mathbf{u}_a$ (PMIN) and $\mathbf{u}_c$ (PMAX) in the system (*x*, *y*, *z*)—the parameters required for the routines BIAXLR/BIAXLC and BIAXLFR/BIAXLFC. For BIAXSORT1R and BIAXSORT1C: AA = α, BB = β, and GG = γ (input). For BIAXSORT1R: N01 = n_1, N02 = n_2, N03 = n_3 (input), N1 = n_a, N2 = n_b, N3 = n_c (output). For BIAXSORT1C: N01 = Re(n_1), A01 = Im(n_1), N02 = Re(n_2), A02 = Im(n_2), N03 = Re(n_3), A03 = Im(n_3) (input), N1 = Re(n_a), A1 = Im(n_a), N2 = Re(n_b), A2 = Im(n_b), N3 = Re(n_c), A3 = Im(n_c) (output).

Examples and Numerical Tests

Tables 9.4 and 9.5 present two sample programs, **test_biax1** and **test_biax2** (the output parts of these programs are omitted), for the routines described in this section. Either program calculates the EWB parameters for a biaxial layer at oblique incidence; the light is assumed to fall on the layer from an isotropic medium of refractive index 1. In **test_biax1**, the biaxial medium is nonabsorbing. In this program, after specification of the input data, BIAXSORT1R is called to prepare the input data for BIAXLR and BIAXLFR. Then, BIAXLR is called to calculate σ_j, $\mathbf{\Psi}$, and $\mathbf{\Psi}^{-1}$. Further, the matrix T $= \mathbf{\Psi}^{-1}\mathbf{\Psi}$ is computed to evaluate the calculation accuracy (the orthogonality test). Then, BIAXLFR calculates σ_j (once again), $\mathbf{e}_j$, and $\mathbf{h}_j$. The program **test_biax2** carries out the same calculations for an absorbing biaxial layer using BIAXSORT1C, BIAXLC, and BIAXLFC.

Now we present results of several numerical tests performed with the use of these programs and their slightly modified versions. In the tables we give the values of the same set of parameters as in the tests of the preceding section. As in the previous examples, we estimate accuracy of calculated EWBs by using the orthogonality test.

Tests 1–6 are for the routines BIAXLR and BIAXLFR. These examples illustrate some statements made above concerning nonabsorbing biaxial media.

Test 1. Parameters: $n_1 = 1.7$, $n_2 = 1.5$, $n_3 = 1.6$, $\alpha = -10°$, $\beta = 75°$, $\gamma = 0°$, $\beta_{inc} = 45°$, $\alpha_{inc} = -10°$. In this case, the plane of incidence is perpendicular to the principal axis $\mathbf{u}_2$. In such a situation, the electric vibration vectors of two basis waves, a forward propagating one and a backward propagating one, must be parallel to $\mathbf{u}_2$. The parameter n_{wj} for these waves must be equal to n_2. The calculations gave the following results.

j	1	2	3	4
σ_j	1.479783	1.322876	−1.322876	−1.439101
n_{wj}	1.640048	1.500000	1.500000	1.603438
$\underline{e}_{jx}$	0.926206	0.000000	0.000000	−0.906904
$\underline{e}_{jy}$	0.000000	1.000000	1.000000	0.000000
$\underline{e}_{jz}$	−0.377018	0.000000	0.000000	−0.421338
$[\delta_{ort}]_{BIAXLR}$		1.1×10^{-15}		

We see that two basis waves, namely, the second and third ones, meet the mentioned requirements: their vectors $\underline{\mathbf{e}}_j$ are perpendicular to the plane of incidence (i.e., parallel to $\boldsymbol{y}$) and consequently parallel to $\mathbf{u}_2$ and $n_{w2} = n_{w3} = n_2$.

Test 2. Parameters: $n_1 = 1.7$, $n_2 = 1.5$, $n_3 = 1.6$, $\alpha = -10°$, $\beta = 75°$, $\gamma = 90°$, $\beta_{inc} = 45°$, $\alpha_{inc} = -10°$. This situation is similar to the previous one, but the permittivity tensor is rotated about the $\mathbf{u}_3$ axis by 90° relative to its orientation in the previous test. In this case, the plane of incidence is perpendicular to the principal axis $\mathbf{u}_1$, and hence two basis waves must have $\underline{\mathbf{e}}_j \parallel \boldsymbol{y}$ and $n_{wj} = n_1$. The calculation results are as follows.

j	1	2	3	4
σ_j	1.382301	1.545962	−1.430567	−1.545962
n_{wj}	1.552661	1.700000	1.595783	1.700000
$\underline{e}_{jx}$	−0.859099	0.000000	0.884489	0.000000
$\underline{e}_{jy}$	0.000000	−1.000000	0.000000	−1.000000
$\underline{e}_{jz}$	0.511809	0.000000	0.466560	0.000000
$[\delta_{ort}]_{BIAXLR}$		7.8×10^{-16}		

We see that $\underline{\mathbf{e}}_2 \parallel \boldsymbol{y}$, $\underline{\mathbf{e}}_4 \parallel \boldsymbol{y}$, and $n_{w2} = n_{w4} = n_1$.

Table 9.4 Usage of routines BIAXLR, BIAXLFR, and BIAXSORT1R

Program **test_biax1**

```
program test_biax1
  use OPTSM_1
  use MTWFIR
  real(8) BETA_inc,ALFA_inc
  real(8) N01,N02,N03,AA,BB,GG
  real(8) N1,N2,N3,PMIN(3),PMAX(3)
  complex(8) X(4)
  real(8) YY(4,4),YI(4,4),T(4,4),EE(3,4),HH(3,4)
  integer(4) NWAVE,NRM

  NWAVE=4        ! calculation for 4 basis waves
  NRM=3          ! normalization type

! Incident light parameters
  BETA_inc=45.0_8*PI3 !polar angle of incidence
  ALFA_inc=30.0_8*PI3 !azimuthal angle of incidence
                      ! PI3=PI/180
  MT=sin(BETA_inc)    ! PI3, MT and FIR are global
  FIR=ALFA_inc        ! variables declared in MTWFIR

! Parameters of the layer
! The principal refractive indices
  N01=1.7_8; N02=1.5_8; N03=1.6_8

! Orientation of the principal axes (Euler angles)
  AA=-10.0_8*PI3; BB=75.0_8*PI3; GG=20.0_8*PI3

! Preparing the input data for BIAXLR/BIAXLFR
  call BIAXSORT1R(N01,N02,N03,AA,  &
                            BB,GG,N1,N2,N3,PMIN,PMAX)
! Calling BIAXLR
  call BIAXLR(N1,N2,N3,PMIN,PMAX,X,YY,YI,NWAVE,NRM)

! Accuracy check
  T=matmul(YI,YY)

! Calling BIAXLFR
  call BIAXLFR(N1,N2,N3,PMIN,PMAX,X,EE,HH,NWAVE)

  ...
end program test_biax1
```

Comments

- BETA_inc line: β_{inc} (rad)
- ALFA_inc line: α_{inc} (rad)
- MT line: MT $= \zeta$
- FIR line: FIR $= \beta_{inc}$
- N01, N02, N03 line: n_1, n_2, n_3
- Euler angles: α, β, γ (rad)
- BIAXLR call: $\sigma_j =$ X(j); $\mathbf{\Psi} =$ YY, $\mathbf{\Psi}^{-1} =$ YI
- Accuracy check: T $= \mathbf{\Psi}^{-1}\mathbf{\Psi}$
- BIAXLFR call: $\mathbf{e}_j =$ EE(:, j); $\mathbf{h}_j =$ HH(:, j)

Table 9.5 Usage of routines BIAXLC, BIAXLFC, and BIAXSORT1C

Program **test_biax2**	Comments
program test_biax2	
use OPTSM_1	
use MTWFIR	
real(8) BETA_inc,ALFA_inc	
real(8) N01,N02,N03,A01,A02,A03,AA,BB,GG	
real(8) N1,N2,N3,A1,A2,A3,PMIN(3),PMAX(3)	
complex(8) X(4),YY(4,4),YI(4,4),T(4,4)	
complex(8) EE(3,4),HH(3,4)	
integer(4) NWAVE,NRM	
NWAVE=4 ! calculation for 4 basis waves	
NRM=3 ! normalization type	
! Incident light parameters	
BETA_inc=45.0_8*PI3 !polar angle of incidence	β_{inc} (rad)
ALFA_inc=30.0_8*PI3 !azimuthal angle of incidence	α_{inc} (rad)
! PI3=PI/180	
MT=sin(BETA_inc) ! PI3, MT and FIR are global	MT $= \zeta$
FIR=ALFA_inc ! variables declared in MTWFIR	FIR $= \beta_{\text{inc}}$
! Parameters of the layer	
! Real parts of the principal refractive indices	$\text{Re}(n_1)$, $\text{Re}(n_2)$,
N01=1.7_8; N02=1.5_8; N03=1.6_8	$\text{Re}(n_3)$
! Imaginary parts of the principal refractive indices	$\text{Im}(n_1)$, $\text{Im}(n_2)$,
A01=0.01_8; A02=0.02_8; A03=0.03_8	$\text{Im}(n_3)$
! Orientation of the principal axes (Euler angles)	
AA=-10.0_8*PI3; BB=75.0_8*PI3; GG=20.0_8*PI3	α, β, γ (rad)
! Preparing the input data for BIAXLC/BIAXLFC	
call BIAXSORT1C(N01,A01,N02,A02,N03,A03,AA,BB,GG, &	
N1,A1,N2,A2,N3,A3,PMIN,PMAX)	
! Calling BIAXLC	
call BIAXLC(N1,A1,N2,A2,N3,A3,PMIN,PMAX,X,YY,YI, &	$\sigma_j =$ X(j)
NWAVE, NRM)	$\mathbf{\Psi} =$ YY, $\mathbf{\Psi}^{-1} =$ YI
! Accuracy check	
T = matmul(YI,YY)	T $= \mathbf{\Psi}^{-1}\,\mathbf{\Psi}$
! Calling BIAXLFC	
call BIAXLFC(N1,A1,N2,A2,N3,A3,PMIN,PMAX,X,EE,HH, &	
NWAVE)	$\mathbf{e}_j =$ EE(:,j)
	$\mathbf{h}_j =$ HH(:,j)
...	
end program test_biax2	

Test 3. Parameters: $n_1 = 1.7$, $n_2 = 1.5$, $n_3 = 1.6$, $\alpha = -10°$, $\beta = 75°$, $\gamma = 0°$, $\beta_{inc} = 45°$, $\alpha_{inc} = 30°$. In this case, the layer has the same parameters as in Test 1, but the plane of incidence is nonparallel to the $\mathbf{u}_3$-$\mathbf{z}$ plane, which is the plane of symmetry of the layer.

j	1	2	3	4
σ_j	1.471356	1.335977	−1.339830	−1.436338
n_{wj}	1.632449	1.511567	1.514973	1.600958
$\underline{e}_{jx}$	0.790483	0.474030	0.530222	0.736471
$\underline{e}_{jy}$	−0.525600	0.858488	0.821978	−0.579772
$\underline{e}_{jz}$	−0.314454	−0.195690	0.207887	0.348530
$[\delta_{ort}]_{BIAXLR}$		4.3×10^{-14}		

Test 4. Parameters: $n_1 = 1.7$, $n_2 = 1.5$, $n_3 = 1.6$, $\alpha = -10°$, $\beta = 75°$, $\gamma = 20°$, $\beta_{inc} = 45°$, $\alpha_{inc} = 30°$. In contrast to the previous tests, in this case, the permittivity tensor is oriented so that the layer has no symmetry plane.

j	1	2	3	4
σ_j	1.486063	1.362524	−1.323153	−1.436189
n_{wj}	1.645716	1.535081	1.500245	1.600824
$\underline{e}_{jx}$	0.913173	0.188982	0.553433	0.700218
$\underline{e}_{jy}$	−0.196791	0.981830	0.774905	−0.631683
$\underline{e}_{jz}$	−0.356916	−0.017200	0.305342	0.332672
$[\delta_{ort}]_{BIAXLR}$		2.2×10^{-14}		

As can be seen from the results of Tests 3 and 4, the n_{wj} value for one forward (backward) propagating basis wave lies between n_a (in this case $n_a = n_2$) and n_b (n_3), while that for the other forward (backward) propagating basis wave lies between n_b and n_c (n_1), although these waves are not equinormal.

Test 5. Parameters: $n_1 = 1.7$, $n_2 = 1.5$, $n_3 = 1.6$, $\alpha = -10°$, $\beta = 0°$, $\gamma = 0°$, $\beta_{inc} = 45°$, $\alpha_{inc} = 30°$. In this case, equation (9.74) is biquadratic (because $\mathbf{u}_3 \parallel \mathbf{z}$).

j	1	2	3	4
σ_j	1.533145	1.332914	−1.533145	−1.332914
n_{wj}	1.688352	1.508861	1.688352	1.508861
$\underline{e}_{jx}$	0.681029	0.580653	0.681029	0.580653
$\underline{e}_{jy}$	−0.638554	0.769586	−0.638554	0.769586
$\underline{e}_{jz}$	−0.358399	−0.265666	0.358399	0.265666
$[\delta_{ort}]_{BIAXLR}$		3.1×10^{-15}		

Test 6. Parameters: $n_1 = 1.7$, $n_2 = 1.5$, $n_3 = 1.55$, $\alpha = 0°$, $\beta = 90°$, $\gamma = 0°$, $\beta_{inc} = 56.082206°$, $\alpha_{inc} = 90°$. Here the orientation of the principal axes and the propagation direction of the incident light are chosen so that polarization degeneracy takes place for both forward and backward propagating basis waves. For the given orientation of the principal axes, the value of β_{inc} for which the polarization degeneracy occurs can be calculated by the formula

$$\sin \beta_{inc} = \zeta = \sqrt{\frac{\varepsilon_c(\varepsilon_b - \varepsilon_a)}{\varepsilon_c - \varepsilon_a}}.$$

This formula can easily be derived from (9.73).

j	1	2	3	4
σ_j	1.309147	1.309147	−1.309147	1.309147
n_{wj}	1.550000	1.550000	1.550000	1.550000
$\underline{e}_{jx}$	0.000000	−0.896745	0.000000	0.896745
$\underline{e}_{jy}$	1.000000	0.000000	1.000000	0.000000
$\underline{e}_{jz}$	0.000000	0.442547	0.000000	0.442547
$[\delta_{ort}]_{BIAXLR}$		1.4×10^{-17}		

One can see that in this case, as in the above examples, the orthogonality test shows a very good accuracy of the results. Considering this example, one can demonstrate that the electric vibration vector of a wave propagating along an optic axis of a biaxial medium, in contrast to the case of a uniaxial medium, may be nonperpendicular to the refraction vector of this wave (this fact is closely related to the conical refraction phenomenon [3, 8]). In this example, $|\mathbf{m}_2\underline{\mathbf{e}}_2| = |\mathbf{m}_4\underline{\mathbf{e}}_4| \approx 0.165$ (while $|\mathbf{m}_1\underline{\mathbf{e}}_1| = |\mathbf{m}_3\underline{\mathbf{e}}_3| = 0$).

The next five tests demonstrate capabilities of the routines BIAXLC and BIAXLFC. Tests 7, 8, and 9 are similar to Tests 1, 4, and 5, respectively. But, in Tests 7–9, the media are absorbing.

Test 7. Parameters: $n_1 = 1.7 + i0.01$, $n_2 = 1.5 + i0.02$, $n_3 = 1.6 + i0.03$, $\alpha = -10°$, $\beta = 75°$, $\gamma = 0°$, $\beta_{inc} = 45°$, $\alpha_{inc} = -10°$

j	1	2	3	4
σ_j	1.479783 + i0.024199	1.322919 + i0.022677	−1.322919 − i0.022677	−1.439109 − i0.032384
$\underline{e}_{jx}$	0.926036 − i0.015570	0.000000 + i0.000000	0.000000 + i0.000000	−0.905246 − i0.054420
$\underline{e}_{jy}$	0.000000 + i0.000000	0.999980 − i0.006321	0.997945 + i0.064074	0.000000 + i0.000000
$\underline{e}_{jz}$	−0.377114 − i0.000341	0.000000 + i0.000000	0.000000 + i0.000000	−0.420847 − i0.021337
$[\delta_{ort}]_{BIAXLC}$		8.9×10^{-16}		

Test 8. Parameters: $n_1 = 1.7 + i0.01$, $n_2 = 1.5 + i0.02$, $n_3 = 1.6 + i0.03$, $\alpha = -10°$, $\beta = 75°$, $\gamma = 20°$, $\beta_{inc} = 45°$, $\alpha_{inc} = 30°$

j	1	2	3	4
σ_j	1.485810 + i0.021638	1.362834 + i0.024591	−1.323197 − i0.022731	−1.436234 − i0.033134
$\underline{e}_{jx}$	0.910234 − i0.062778	0.189632 + i0.032690	0.551827 + i0.042362	0.697578 + i0.060792
$\underline{e}_{jy}$	−0.197986 − i0.034072	0.979577 − i0.053664	0.772688 + i0.058277	−0.629571 − i0.051164
$\underline{e}_{jz}$	−0.356267 + i0.015709	−0.017262 − i0.014874	0.304783 + i0.018899	0.331842 + i0.024358
$[\delta_{ort}]_{BIAXLC}$		3.6×10^{-14}		

Test 9. Parameters: $n_1 = 1.7 + i0.01$, $n_2 = 1.5 + i0.02$, $n_3 = 1.6 + i0.03$, $\alpha = -10^\circ$, $\beta = 0^\circ$, $\gamma = 0^\circ$, $\beta_{inc} = 45^\circ$, $\alpha_{inc} = 30^\circ$

j	1	2	3	4
σ_j	1.533237 + i0.013927	1.332992 + i0.023173	−1.533237 − i0.013927	−1.332992 − i0.023173
$\underline{e}_{jx}$	0.680042 − i0.037399	0.580250 − i0.024842	0.680042 − i0.037399	0.580250 − i0.024842
$\underline{e}_{jy}$	−0.636761 + i0.048522	0.769146 − i0.023671	−0.636761 + i0.048522	0.769146 − i0.023671
$\underline{e}_{jz}$	−0.356695 + i0.033069	−0.264920 + i0.019105	0.356695 − i0.033069	0.264920 − i0.019105
$[\delta_{ort}]_{BIAXLC}$		1.8×10^{-16}		

Tests 10 and 11 are for TIR mode. These tests are analogous to Tests 7 and 8 of the previous section.

Test 10. Parameters: $n_1 = 1.7 + i0.01$, $n_2 = 1.5 + i0.02$, $n_3 = 1.6 + i0.03$, $\beta = 75^\circ$, $\gamma = 20^\circ$, $\zeta = 1.72$, $\alpha_{inc} = 30^\circ$

j	1	2	3	4
σ_j	0.066110 + i0.782840	0.140982 + i0.339907	0.033223 − i0.372330	−0.023307 − i0.773859
$\underline{e}_{jx}$	−0.008543 − i0.102671	0.033813 + i0.216910	−0.092710 − i0.197237	0.007271 + i0.111808
$\underline{e}_{jy}$	0.937125 + i0.180130	0.214285 − i0.057628	0.234671 + i0.055841	0.865560 − i0.391104
$\underline{e}_{jz}$	0.258040 − i0.110231	−0.949999 + i0.008330	−0.917392 + i0.229576	0.269966 + i0.111393
$[\delta_{ort}]_{BIAXLC}$		4.8×10^{-10}		

Test 11. Parameters: $n_1 = 1.7$, $n_2 = 1.5$, $n_3 = 1.6$, $\beta = 75^\circ$, $\gamma = 20^\circ$, $\zeta = 1.72$, $\alpha_{inc} = 30^\circ$

j	1	2	3	4
σ_j	0.021740 + i0.777081	0.086801 + i0.350181	0.086801 − i0.350181	0.021740 − i0.777081
$\underline{e}_{jx}$	−0.000594 − i0.107201	−0.029884 + i0.210857	−0.029884 − i0.210857	−0.000594 + i0.107201
$\underline{e}_{jy}$	0.908503 + i0.285266	0.224866 − i0.055035	0.224866 + i0.055035	0.908503 − i0.285266
$\underline{e}_{jz}$	0.264031 − i0.109730	−0.942849 − i0.109945	−0.942849 + i0.109945	0.264031 + i0.109730
$[\delta_{ort}]_{BIAXLC}$		7.2×10^{-10}		

As is seen from the results of the orthogonality test, the accuracy of the calculated values of the EWB parameters in these examples is also very good.

References

[1] R. C. Jones, "A new calculus for the treatment of optical systems I. Description and discussion of the calculus," *J. Opt. Soc. Am.* **31**, 488 (1941).
[2] R. C. Jones, "A new calculus for the treatment of optical systems V. A more general formulation, and description of another calculus," *J. Opt. Soc. Am.* **38**, 671 (1948).
[3] A. Yariv and P. Yeh, *Optical Waves in Crystals* (John Wiley & Sons, Inc., New York, 2003).
[4] R. M. A. Azzam and N. M. Bashara, *Ellipsometry and Polarized Light* (North-Holland, Amsterdam, 1977).
[5] C. Oldano, "Electromagnetic-wave propagation in stratified media," *Phys. Rev. A* **40**, 6014 (1989).
[6] D. A. Yakovlev, "Simple formulas for the amplitude transmission and reflection coefficients at the interface of anisotropic media," *Opt. Spectr.* **84**, 748 (1998).
[7] D. A. Yakovlev, "Calculation of transmission characteristics of smoothly inhomogeneous anisotropic media in the approximation of negligible bulk reflection: I. Basic equation," *Opt. Spectr.* **87**, 903 (1999).
[8] M. Born and E. Wolf, *Principles of Optics*, 7th ed. (Pergamon Press, New York, 1999).
[9] E. Hecht, *Optics*, 4th ed. (Addison Wesley, San Francisco, 2002).
[10] P. Yeh, "Extended Jones matrix method," *J. Opt. Soc. Am.* **72**, 507 (1982).
[11] F. I. Fedorov, *Optics of Anisotropic Media* (Academy of Sciences of Belarus, Minsk, 1958) (in Russian).
[12] K. Eidner, G. Mayer, M. Schmidt, and H. Schmiedel, "Optics in stratified media—the use of optical eigenmodes of uniaxial crystals in the 4×4-matrix formalism," *Mol. Cryst. Liq. Cryst.* **172**, 191 (1989).
[13] H. L. Ong, "Electro-optics of a twisted nematic liquid crystal display by 2×2 propagation matrix at oblique incidence," *Jpn. J. Appl. Phys.* **30**, L1028 (1991).
[14] A. Lien, "Extended Jones matrix representation for the twisted nematic liquid crystal display at oblique incidence," *Appl. Phys. Lett.* **57**, 2767 (1990).
[15] A. Lien, "A detailed derivation of extended Jones matrix representation for twisted nematic liquid crystal displays," *Liq. Cryst.* **22**, 171 (1997).
[16] T. Scharf, *Polarized Light in Liquid Crystals and Polymers* (Wiley Interscience, 2006).
[17] P. Yeh, "Electromagnetic propagation in birefringent layered media," *J. Opt. Soc. Am.* **69**, 742 (1979).
[18] C. Gu and P. Yeh, "Extended Jones matrix method. II," *J. Opt. Soc. Am. A* **10**, 966 (1993).
[19] P. Yeh and C. Gu, *Optics of Liquid Crystal Displays* (John Wiley & Sons, Inc., New York, 1999).
[20] M. Mansuripur, "Analysis of multilayer thin-film structures containing magneto-optic and anisotropic media at oblique incidence using 2×2 matrices," *J. Appl. Phys.* **67**, 6466 (1990).
[21] H. Yuan, W. E, T. Kosa, and P. Palffy-Muhoray, "Analytic 4×4 propagation matrices for linear optical media," *Mol. Cryst. Liq. Cryst.* **331**, 491 (1999).
[22] S. Stallinga, "Berreman 4×4 matrix method for reflective liquid crystal displays," *J. Appl. Phys.* **85**, 3023 (1999).
[23] C.-J. Chen, A. Lien, M. I. Nathan, "4×4 and 2×2 matrix formulations for the optics in stratified and biaxial media," *J. Opt. Soc. Am. A* **14**, 3125 (1997).
[24] S. P. Palto, "An algorithm for solving the optical problem for stratified anisotropic media," *JETP* **92**, 552 (2001).
[25] F. S. Acton, *Numerical Methods that Work*, 2nd ed. (Mathematical Association of America, Washington, 1990).
[26] W. H. Press, S. A. Teukolsky, W. T. Vetterling, and B. P. Flannery, *Numerical Recipes in Fortran 77: The Art of Scientific Computing*, 2nd ed. (Cambridge University Press, Cambridge, 1997).

10

Efficient Methods for Calculating Optical Characteristics of Layered Systems for Quasimonochromatic Incident Light. Main Routines of LMOPTICS Library

If we wish to simulate with the highest possible accuracy a spectral experiment in which the characteristics of the overall transmission or overall reflection of a layered system including one or more "thick" layers[1] are measured (such as the experiment with LC cells that was described in Section 6.2 or the experiment with ITO-coated and uncoated glass substrates whose results are shown in Figure 7.5), we are faced with the necessity of taking account of multiple reflections from interfaces of "thick" layers and "thin" layered systems adjacent to them. Such multiple reflections give rise to dense interference oscillations in the spectra of the monochromatic overall transmittance and reflectance (and corresponding Mueller matrices) of the approximating medium for the layered system (which can be calculated by using the methods described in Chapter 8) but cannot give any noticeable interference pattern in experimental spectra of the overall transmittance and reflectance of the system because the measured quantities in fact characterize the interaction with the system not monochromatic but quasimonochromatic incident light with a coherence length less than thicknesses of "thick" layers (see Section 7.1). One of the ways to make the model spectra realistic is the spectral averaging of the monochromatic transmittance/reflectance (see Section 7.1). However, as we have noted, for such complex layered systems as LCDs the spectral averaging method is rather expensive computationally. In this section, we discuss an approach leading to much faster methods of calculating realistic transmission and reflection spectra of layered systems consisting of "thin" and "thick" layers. This approach exploits features of quasimonochromatic light propagation in layered systems and allows treating coherent and incoherent interactions between different fractions of light propagating in the layered system with equal ease. The underlying theory is presented in Section 10.1. There are many practical situations connected with optimization of LCDs and characterization of elements

[1] see Section 7.1.

Modeling and Optimization of LCD Optical Performance, First Edition.
Dmitry A. Yakovlev, Vladimir G. Chigrinov and Hoi-Sing Kwok.

Website Companion: www.wiley.com/go/yakovlev/modelinglcd

of LCDs where this theory can be useful. In Section 10.2, we consider two efficient computational methods derived from this theory and a few examples of their application to LC devices.

In Section 10.3, we proceed with description of the main routines of LMOPTICS library which helps to implement optical methods described in Chapter 8 and this chapter in modeling programs.

10.1 EW Stokes Vectors and EW Mueller Matrices

EW Stokes Vectors

To begin with we consider, as in Section 8.1.3, the layered structure shown in Figure 8.3, that is, a 1D-inhomogeneous medium M_{SM} sandwiched between two semi-infinite homogeneous media $\mathsf{M}_{\mathrm{inc}}$ and M_{tr}, the medium $\mathsf{M}_{\mathrm{inc}}$ being nonabsorbing and isotropic. Here, for the sake of simplicity, we assume that the refractive index of $\mathsf{M}_{\mathrm{inc}}$ is equal to 1. Let a quasimonochromatic plane wave $\mathbf{X}_{\mathrm{inc}}$ with electric field

$$\mathbf{E}_{\mathrm{inc}}(\mathbf{r}, t) = \mathbf{E}_{\mathrm{inc}}\left(\frac{\mathbf{m}_{\mathrm{inc}}\mathbf{r}}{c} - t\right), \tag{10.1}$$

where the refraction vector $\mathbf{m}_{\mathrm{inc}}$ is real, and the length of coherence l_{coh} fall on M_{SM} from $\mathsf{M}_{\mathrm{inc}}$. Since the refractive index of $\mathsf{M}_{\mathrm{inc}}$ is assumed to be equal to unity, $\mathbf{m}_{\mathrm{inc}} = \mathbf{l}_{\mathrm{inc}}$, where $\mathbf{l}_{\mathrm{inc}}$ is the wave normal of $\mathbf{X}_{\mathrm{inc}}$. Let $\tilde{\lambda}$ be the mean wavelength of $\mathbf{X}_{\mathrm{inc}}$ and let $\Omega_\lambda = [\tilde{\lambda} - \Delta\lambda/2,\ \tilde{\lambda} + \Delta\lambda/2]$ ($\Delta\lambda \ll \tilde{\lambda}$, $l_{\mathrm{coh}} \approx \tilde{\lambda}^2/\Delta\lambda$) be the effective spectral range of $\mathbf{X}_{\mathrm{inc}}$. Let $\{\mathbf{m}_j(z; \tilde{\lambda}),\ \mathbf{e}_j(z; \tilde{\lambda}),\ \mathbf{h}_j(z; \tilde{\lambda})\}$ ($j = 1,2,3,4$) be an eigenwave basis for harmonic wave fields of the form (8.57) with

$$\mathbf{b} = \mathbf{l}_{\mathrm{inc}} - \mathbf{z}(\mathbf{z} \cdot \mathbf{l}_{\mathrm{inc}})$$

and $\omega = 2\pi c/\tilde{\lambda}$, that is, for fields that can be generated in M_{SM}, $\mathsf{M}_{\mathrm{inc}}$, and M_{tr} by incident plane monochromatic waves with wave normal $\mathbf{l}_{\mathrm{inc}}$ and wavelength $\tilde{\lambda}$. Neglecting the variation of the optical constants of the media within the range Ω_λ, we may use this eigenwave basis for all harmonic components of $\mathbf{X}_{\mathrm{inc}}$. By doing so, we can represent any forward propagating field $\mathbf{X}^{\downarrow}$ generated by $\mathbf{X}_{\mathrm{inc}}$, in any plane $z = \xi$ within the domain where $\mathbf{X}^{\downarrow}$ propagates, as follows:

$$\begin{pmatrix} \mathbf{E}(\mathbf{r}_\xi, t) \\ \mathbf{H}(\mathbf{r}_\xi, t) \end{pmatrix} = \sum_{j=1}^{2} \begin{pmatrix} \mathbf{e}_j(\xi; \tilde{\lambda}) \\ \mathbf{h}_j(\xi; \tilde{\lambda}) \end{pmatrix} a_j^{\downarrow}(\mathbf{r}_\xi, t), \tag{10.2}$$

$\mathbf{r}_\xi = (x, y, \xi)$. The analogous representation of a backward propagating field $\mathbf{X}^{\uparrow}$ generated by $\mathbf{X}_{\mathrm{inc}}$ and passing through the plane $z = \xi$ is

$$\begin{pmatrix} \mathbf{E}(\mathbf{r}_\xi, t) \\ \mathbf{H}(\mathbf{r}_\xi, t) \end{pmatrix} = \sum_{j=1}^{2} \begin{pmatrix} \mathbf{e}_{j+2}(\xi; \tilde{\lambda}) \\ \mathbf{h}_{j+2}(\xi; \tilde{\lambda}) \end{pmatrix} a_j^{\uparrow}(\mathbf{r}_\xi, t). \tag{10.3}$$

If the plane $z = \xi$ lies within a homogeneous layer, each term of these decompositions characterizes a corresponding plane quasimonochromatic wave. These representations allow us to introduce, by analogy with usual Stokes vectors for quasimonochromatic fields, the following characteristics of the fields $\mathbf{X}^{\downarrow}$ and $\mathbf{X}^{\uparrow}$:

$$\begin{aligned} S^{\downarrow}(\xi) = S^{\downarrow}(\mathbf{r}_\xi) \equiv \mathbf{L}\left\langle \mathbf{a}^{\downarrow}(\mathbf{r}_\xi, t) \otimes \mathbf{a}^{\downarrow}(\mathbf{r}_\xi, t)^* \right\rangle, \\ S^{\uparrow}(\xi) = S^{\uparrow}(\mathbf{r}_\xi) \equiv \mathbf{L}\left\langle \mathbf{a}^{\uparrow}(\mathbf{r}_\xi, t) \otimes \mathbf{a}^{\uparrow}(\mathbf{r}_\xi, t)^* \right\rangle, \end{aligned} \tag{10.4}$$

where

$$\mathbf{a}^{\downarrow}(\mathbf{r}_\xi, t) \equiv \begin{pmatrix} a_1^{\downarrow}(\mathbf{r}_\xi, t) \\ a_2^{\downarrow}(\mathbf{r}_\xi, t) \end{pmatrix}, \quad \mathbf{a}^{\uparrow}(\mathbf{r}_\xi, t) \equiv \begin{pmatrix} a_1^{\uparrow}(\mathbf{r}_\xi, t) \\ a_2^{\uparrow}(\mathbf{r}_\xi, t) \end{pmatrix}, \tag{10.5}$$

$$\mathbf{L} = \begin{pmatrix} 1 & 0 & 0 & 1 \\ 1 & 0 & 0 & -1 \\ 0 & 1 & 1 & 0 \\ 0 & -i & i & 0 \end{pmatrix}; \tag{10.6}$$

the brackets $\langle\ \rangle$ denote time averaging, and $\otimes$ denotes the Kronecker matrix multiplication (see Section 5.1.5). In an explicit form,

$$S^{\downarrow}(\xi) = \begin{pmatrix} \langle a_1^{\downarrow}(\mathbf{r}_\xi, t) a_1^{\downarrow}(\mathbf{r}_\xi, t)^* \rangle + \langle a_2^{\downarrow}(\mathbf{r}_\xi, t) a_2^{\downarrow}(\mathbf{r}_\xi, t)^* \rangle \\ \langle a_1^{\downarrow}(\mathbf{r}_\xi, t) a_1^{\downarrow}(\mathbf{r}_\xi, t)^* \rangle - \langle a_2^{\downarrow}(\mathbf{r}_\xi, t) a_2^{\downarrow}(\mathbf{r}_\xi, t)^* \rangle \\ 2\,\mathrm{Re}\,\langle a_1^{\downarrow}(\mathbf{r}_\xi, t) a_2^{\downarrow}(\mathbf{r}_\xi, t)^* \rangle \\ 2\,\mathrm{Im}\,\langle a_1^{\downarrow}(\mathbf{r}_\xi, t) a_2^{\downarrow}(\mathbf{r}_\xi, t)^* \rangle \end{pmatrix},$$

$$S^{\uparrow}(\xi) = \begin{pmatrix} \langle a_1^{\uparrow}(\mathbf{r}_\xi, t) a_1^{\uparrow}(\mathbf{r}_\xi, t)^* \rangle + \langle a_2^{\uparrow}(\mathbf{r}_\xi, t) a_2^{\uparrow}(\mathbf{r}_\xi, t)^* \rangle \\ \langle a_1^{\uparrow}(\mathbf{r}_\xi, t) a_1^{\uparrow}(\mathbf{r}_\xi, t)^* \rangle - \langle a_2^{\uparrow}(\mathbf{r}_\xi, t) a_2^{\uparrow}(\mathbf{r}_\xi, t)^* \rangle \\ 2\,\mathrm{Re}\,\langle a_1^{\uparrow}(\mathbf{r}_\xi, t) a_2^{\uparrow}(\mathbf{r}_\xi, t)^* \rangle \\ 2\,\mathrm{Im}\,\langle a_1^{\uparrow}(\mathbf{r}_\xi, t) a_2^{\uparrow}(\mathbf{r}_\xi, t)^* \rangle \end{pmatrix}$$

[cf. (5.83)]. We will call the column-vectors $S^{\downarrow}$ and $S^{\uparrow}$ the *EW Stokes vectors* (*eigenwave Stokes vectors*) of the fields $\mathbf{X}^{\downarrow}$ and $\mathbf{X}^{\uparrow}$. In (10.4) we used and expressed the fact that the vector $S^{\downarrow}$, as well as $S^{\uparrow}$, has a constant value in all points of the plane $z = \xi$, which is due to spatial invariance of time-averaged local characteristics of the incident wave $\mathbf{X}_{\mathrm{inc}}$. Like usual Stokes vector, EW Stokes vector is real and is an additive characteristic for mutually incoherent wave fields. The use of EW Stokes vector as traced characteristic (see Sections 7.1 and 7.2) is a key feature of the methods described in Section 10.2.

Spectral Representation of EW Stokes Vectors

Let S_{inc} and $S_{\lambda\,\mathrm{inc}}(\lambda)$ be respectively the EW Stokes vector of the wave $\mathbf{X}_{\mathrm{inc}}$ and the spectral density of this Stokes vector. By definition,

$$S_{\mathrm{inc}} = \int S_{\lambda\,\mathrm{inc}}(\lambda)\mathrm{d}\lambda. \tag{10.7}$$

Let $\mathbf{X}$ be a forward or backward propagating wave field produced by $\mathbf{X}_{\mathrm{inc}}$ in M_{SM}, $\mathsf{M}_{\mathrm{inc}}$, or M_{tr}. Using the symbolic notation introduced in Section 7.1, we may express the relationship between $\mathbf{X}_{\mathrm{inc}}$ and $\mathbf{X}$ as $\mathbf{X} = \mathbf{O}\mathbf{X}_{\mathrm{inc}}$, where $\mathbf{O}$ is the corresponding operation. Denote the EW Stokes vector of the wave field $\mathbf{X}$ by $S\{\mathbf{X}\}$ and the spectral density of this vector by $S_\lambda\{\mathbf{X}; \lambda\}$. Let $\tilde{\mathbf{t}}_{\mathbf{O}}(\lambda)$ be the EW Jones matrix of the operation $\mathbf{O}$ for incident monochromatic plane waves with wave normal $\mathbf{l}_{\mathrm{inc}}$ and wavelength $\lambda \in \Omega_\lambda$, that is, for an incident wave $\mathbf{X}_{\lambda\,\mathrm{inc}}$ of this kind

$$\mathbf{a}_{\lambda\,\mathrm{out}} = \tilde{\mathbf{t}}_{\mathbf{O}}(\lambda)\mathbf{a}_{\lambda\,\mathrm{inc}}, \tag{10.8}$$

where $\mathbf{a}_{\lambda\,\mathrm{inc}}$ and $\mathbf{a}_{\lambda\,\mathrm{out}}$ are the EW Jones vectors of the fields $\mathbf{X}_{\lambda\,\mathrm{inc}}$ and $\mathbf{X}_{\lambda\,\mathrm{out}} = \mathbf{O}\mathbf{X}_{\lambda\,\mathrm{inc}}$, respectively. The matrix $\tilde{\mathbf{t}}_{\mathbf{O}}(\lambda)$ is assumed to be calculated using the chosen EW basis (for $\lambda = \tilde{\lambda}$) at least in the domains where $\mathbf{X}_{\lambda\,\mathrm{inc}}$ and $\mathbf{X}_{\lambda\,\mathrm{out}}$ travel. The EW Stokes vectors of the fields $\mathbf{X}_{\lambda\,\mathrm{inc}}$ and $\mathbf{X}_{\lambda\,\mathrm{out}}$ may be expressed as

$$S\{\mathbf{X}_{\lambda\,\mathrm{inc}}\} = \mathbf{L}\left(\mathbf{a}_{\lambda\,\mathrm{inc}} \otimes \mathbf{a}^*_{\lambda\,\mathrm{inc}}\right), \tag{10.9}$$

$$S\{\mathbf{X}_{\lambda\,\mathrm{out}}\} = \mathbf{L}\left(\mathbf{a}_{\lambda\,\mathrm{out}} \otimes \mathbf{a}^*_{\lambda\,\mathrm{out}}\right). \tag{10.10}$$

Substituting (10.8) into (10.10) gives

$$\begin{aligned} S\{\mathbf{X}_{\lambda\,\mathrm{out}}\} &= \mathbf{L}\left[(\tilde{\mathbf{t}}_{\mathbf{O}}(\lambda)\mathbf{a}_{\lambda\,\mathrm{inc}}) \otimes \left(\tilde{\mathbf{t}}_{\mathbf{O}}(\lambda)^*\mathbf{a}^*_{\lambda\,\mathrm{inc}}\right)\right] = \mathbf{L}\left(\tilde{\mathbf{t}}_{\mathbf{O}}(\lambda) \otimes \tilde{\mathbf{t}}_{\mathbf{O}}(\lambda)^*\right)\left(\mathbf{a}_{\lambda\,\mathrm{inc}} \otimes \mathbf{a}^*_{\lambda\,\mathrm{inc}}\right) \\ &= \mathbf{L}\left(\tilde{\mathbf{t}}_{\mathbf{O}}(\lambda) \otimes \tilde{\mathbf{t}}_{\mathbf{O}}(\lambda)^*\right)\mathbf{L}^{-1}\mathbf{L}\left(\mathbf{a}_{\lambda\,\mathrm{inc}} \otimes \mathbf{a}^*_{\lambda\,\mathrm{inc}}\right) = \mathbf{L}\left(\tilde{\mathbf{t}}_{\mathbf{O}}(\lambda) \otimes \tilde{\mathbf{t}}_{\mathbf{O}}(\lambda)^*\right)\mathbf{L}^{-1}S\{\mathbf{X}_{\lambda\,\mathrm{inc}}\}, \end{aligned} \tag{10.11}$$

where matrix identity (5.70) and expression (10.9) have been used. As is seen from (10.11), the EW Mueller matrix $\tilde{\mathbf{M}}_{\mathbf{O}}(\lambda)$ relating the EW Stokes vectors of the waves $\mathbf{X}_{\lambda\,\mathrm{inc}}$ and $\mathbf{X}_{\lambda\,\mathrm{out}}$ as

$$S\{\mathbf{X}_{\lambda\,\mathrm{out}}\} = \tilde{\mathbf{M}}_{\mathbf{O}}(\lambda)S\{\mathbf{X}_{\lambda\,\mathrm{inc}}\} \tag{10.12}$$

can be expressed as follows:

$$\tilde{\mathbf{M}}_{\mathbf{O}}(\lambda) = \mathbf{L}\left(\tilde{\mathbf{t}}_{\mathbf{O}}(\lambda) \otimes \tilde{\mathbf{t}}_{\mathbf{O}}(\lambda)^*\right)\mathbf{L}^{-1}. \tag{10.13}$$

Notation. Let $\mathbf{t}$ be a 2×2 matrix. By $\tilde{\mathbf{T}}\{\mathbf{t}\}$ we will denote the 4×4 matrix calculated as

$$\tilde{\mathbf{T}}\{\mathbf{t}\} = \mathbf{L}(\mathbf{t} \otimes \mathbf{t}^*)\mathbf{L}^{-1}. \tag{10.14}$$

It is customary to call a Mueller matrix that can be represented as $\tilde{\mathbf{T}}\{\mathbf{t}\}$, where $\mathbf{t}$ is a Jones matrix, a *Mueller–Jones matrix*. Note the following property of the transformation $\mathbf{t} \to \tilde{\mathbf{T}}\{\mathbf{t}\}$:

$$\tilde{\mathbf{T}}\{\mathbf{t}_N\} \ldots \tilde{\mathbf{T}}\{\mathbf{t}_2\}\tilde{\mathbf{T}}\{\mathbf{t}_1\} = \tilde{\mathbf{T}}\{\mathbf{t}_N \ldots \mathbf{t}_2\mathbf{t}_1\}, \tag{10.15}$$

where $\mathbf{t}_1, \mathbf{t}_2, \ldots, \mathbf{t}_N$ are arbitrary 2×2 matrices.

Using notation (10.14), we may rewrite expression (10.13) as follows:

$$\tilde{\mathbf{M}}_{\mathbf{O}}(\lambda) = \tilde{\mathbf{T}}\{\tilde{\mathbf{t}}_{\mathbf{O}}(\lambda)\}. \tag{10.16}$$

Relation (10.12) applies to corresponding monochromatic components of the fields $\mathbf{X}_{\mathrm{inc}}$ and $\mathbf{X}$. Therefore, the spectral density of the vector $S\{\mathbf{X}\}$ can be expressed in terms of the spectral density of S_{inc} as

$$S_\lambda\{\mathbf{X}; \lambda\} = \tilde{\mathbf{M}}_{\mathbf{O}}(\lambda)S_{\lambda\,\mathrm{inc}}(\lambda). \tag{10.17}$$

Using (10.17) and the equation

$$S\{\mathbf{X}\} = \int S_\lambda\{\mathbf{X}; \lambda\}\mathrm{d}\lambda, \tag{10.18}$$

we may express the vector $S\{\mathbf{X}\}$ as

$$S\{\mathbf{X}\} = \int \tilde{\mathbf{M}}_{\mathbf{O}}(\lambda)S_{\lambda\,\mathrm{inc}}(\lambda)\mathrm{d}\lambda. \tag{10.19}$$

If the spectral density $S_{\lambda\,\mathrm{inc}}(\lambda)$ can be represented in the form

$$S_{\lambda\,\mathrm{inc}}(\lambda) = g(\lambda)S_{\mathrm{inc}}, \tag{10.20}$$

where $g(\lambda)$ is the spectral form-factor of the wave $\mathbf{X}_{\mathrm{inc}}$ [$g(\lambda)$ is a scalar function such that $\int g(\lambda)\mathrm{d}\lambda = 1$], which is a common situation, equation (10.19) can be rewritten as follows:

$$S\{\mathbf{X}\} = \mathbf{M}_{\mathbf{O}}S_{\mathrm{inc}}, \tag{10.21}$$

where

$$\mathbf{M}_{\mathbf{O}} = \int \tilde{\mathbf{M}}_{\mathbf{O}}(\lambda)g(\lambda)\mathrm{d}\lambda = \int \tilde{\mathbf{T}}\{\tilde{\mathbf{t}}_{\mathbf{O}}(\lambda)\}g(\lambda)\mathrm{d}\lambda \tag{10.22}$$

is the EW Mueller matrix of the operation $\mathbf{O}$ for incident quasimonochromatic waves with the given spectral form-factor.

Similarly, if the wave field $\mathbf{X}$ propagates within $\mathrm{M_{SM}}$ and a wave field $\mathbf{X}'$ is the result of an operation $\mathbf{O}'$ on $\mathbf{X}$, the spectral densities of the EW Stokes vectors $S\{\mathbf{X}\}$ and $S\{\mathbf{X}'\}$ are related by

$$S_{\lambda}\{\mathbf{X}'; \lambda\} = \tilde{\mathbf{M}}_{\mathbf{O}'}(\lambda)S_{\lambda}\{\mathbf{X}; \lambda\}, \tag{10.23}$$

where

$$\tilde{\mathbf{M}}_{\mathbf{O}'}(\lambda) = \tilde{\mathbf{T}}\{\tilde{\mathbf{t}}_{\mathbf{O}'}(\lambda)\} \tag{10.24}$$

with $\tilde{\mathbf{t}}_{\mathbf{O}'}(\lambda)$ being the corresponding EW Jones matrix of the operation $\mathbf{O}'$. Note also the following representations of the spectral density $S_{\lambda}\{\mathbf{X}'; \lambda\}$:

$$S_{\lambda}\{\mathbf{X}'; \lambda\} = \tilde{\mathbf{M}}_{\mathbf{O}'}(\lambda)\tilde{\mathbf{M}}_{\mathbf{O}}(\lambda)S_{\lambda\,\mathrm{inc}}(\lambda), \tag{10.25}$$

$$S_{\lambda}\{\mathbf{X}'; \lambda\} = \tilde{\mathbf{T}}\{\tilde{\mathbf{t}}_{\mathbf{O}'}(\lambda)\tilde{\mathbf{t}}_{\mathbf{O}}(\lambda)\}S_{\lambda\,\mathrm{inc}}(\lambda). \tag{10.26}$$

From (10.23) and the equation

$$S\{\mathbf{X}'\} = \int S_{\lambda}\{\mathbf{X}'; \lambda\}\mathrm{d}\lambda, \tag{10.27}$$

we have the following expression for $S\{\mathbf{X}'\}$:

$$S\{\mathbf{X}'\} = \int \tilde{\mathbf{M}}_{\mathbf{O}'}(\lambda)S_{\lambda}\{\mathbf{X}; \lambda\}\mathrm{d}\lambda. \tag{10.28}$$

Monochromatic Approximation for EW Mueller Matrices

If the variations of the matrix $\tilde{\mathbf{M}}_{\mathbf{O}'}(\lambda)$ within the region Ω_{λ} are negligible or absent, we may replace $\tilde{\mathbf{M}}_{\mathbf{O}'}(\lambda)$ in (10.28), in the former case to some approximation, by $\tilde{\mathbf{M}}_{\mathbf{O}'}(\tilde{\lambda})$ to obtain

$$S\{\mathbf{X}'\} = \tilde{\mathbf{M}}_{\mathbf{O}'}(\tilde{\lambda}) \int S_{\lambda}\{\mathbf{X}; \lambda\}\mathrm{d}\lambda = \tilde{\mathbf{M}}_{\mathbf{O}'}(\tilde{\lambda})S\{\mathbf{X}\} \tag{10.29}$$

[see (10.18)]. This means that the EW Mueller matrix $\mathbf{M}_{\mathrm{O}'}$ linking the vectors $S\{\mathbf{X}\}$ and $S\{\mathbf{X}'\}$,

$$S\{\mathbf{X}'\} = \mathbf{M}_{\mathrm{O}'} S\{\mathbf{X}\}, \tag{10.30}$$

can be estimated as

$$\mathbf{M}_{\mathrm{O}'} = \tilde{\mathbf{M}}_{\mathrm{O}'}(\tilde{\lambda}). \tag{10.31}$$

In Section 7.1, such estimation of operators for quasimonochromatic light has been called the monochromatic approximation.

Let us consider some situations where the monochromatic approximation gives acceptable results. The quasimonochromatic EW Mueller matrices corresponding to operations $\mathbf{T}^{\downarrow}$, $\mathbf{R}^{\downarrow}$, $\mathbf{T}^{\uparrow}$, and $\mathbf{R}^{\uparrow}$ will be denoted by $\mathcal{T}^{\downarrow}$, $\mathcal{R}^{\downarrow}$, $\mathcal{T}^{\uparrow}$, and $\mathcal{R}^{\uparrow}$. Recall that the variations of the optical constants of the media within the region Ω_λ are assumed to be negligible.

1. *An interface*. Let the plane $z = \underline{z}$ coincide with an interface. Since the chosen EW basis is common for all $\lambda \in \Omega_\lambda$, according to (8.217a) and (8.121), for all $\lambda \in \Omega_\lambda$

$$\mathbf{t}^{\downarrow}(\underline{z}-0, \underline{z}+0; \lambda) = \mathbf{t}^{\downarrow}(\underline{z}-0, \underline{z}+0; \tilde{\lambda}),$$

and similarly for $\mathbf{r}^{\downarrow}(\underline{z}-0, \underline{z}+0; \lambda)$, $\mathbf{t}^{\uparrow}(\underline{z}-0, \underline{z}+0; \lambda)$, and $\mathbf{r}^{\uparrow}(\underline{z}-0, \underline{z}+0; \lambda)$. Therefore, according to (10.28) and (10.29),

$$\mathcal{T}^{\downarrow}(\underline{z}-0, \underline{z}+0) = \tilde{\mathbf{T}}\left\{\mathbf{t}^{\downarrow}(\underline{z}-0, \underline{z}+0; \tilde{\lambda})\right\}, \tag{10.32}$$

and similarly for the pairs $(\mathcal{R}^{\downarrow}, \mathbf{r}^{\downarrow})$, $(\mathcal{T}^{\uparrow}, \mathbf{t}^{\uparrow})$, and $(\mathcal{R}^{\uparrow}, \mathbf{r}^{\uparrow})$.

2. *Bulk of a homogeneous layer*. Let a domain (z', z'') be the bulk of a homogeneous layer. The transmission EW Jones matrices of this domain can be expressed as

$$\mathbf{t}^{\downarrow}(z', z'') = \begin{pmatrix} e^{i\frac{2\pi\bar{\sigma}_1 d}{\lambda}} & 0 \\ 0 & e^{i\frac{2\pi\bar{\sigma}_2 d}{\lambda}} \end{pmatrix}, \quad \mathbf{t}^{\uparrow}(z', z'') = \begin{pmatrix} e^{-i\frac{2\pi\bar{\sigma}_3 d}{\lambda}} & 0 \\ 0 & e^{-i\frac{2\pi\bar{\sigma}_4 d}{\lambda}} \end{pmatrix}, \tag{10.33}$$

where $\bar{\sigma}_j \equiv \sigma_j(z') = \sigma_j(z'')$ and d is the thickness of the layer. If the layer is isotropic ($\bar{\sigma}_1 = \bar{\sigma}_2 = -\bar{\sigma}_3 = -\bar{\sigma}_4$), $\mathbf{t}^{\uparrow}(z', z'') = \mathbf{t}^{\downarrow}(z', z'')$ and hence $\mathcal{T}^{\uparrow}(z', z'') = \mathcal{T}^{\downarrow}(z', z'')$. According to (10.28), (10.24), and (10.33),

(a) if the layer is isotropic and nonabsorbing ($\mathrm{Im}(\bar{\sigma}_j) = 0$),[2]

$$\mathcal{T}^{\uparrow}(z', z'') = \mathcal{T}^{\downarrow}(z', z'') = \tilde{\mathbf{T}}\left\{\mathbf{t}^{\downarrow}(z', z''; \tilde{\lambda})\right\} = \mathbf{U}; \tag{10.34}$$

(b) if the layer is isotropic and absorbing, to a good approximation,

$$\mathcal{T}^{\uparrow}(z', z'') = \mathcal{T}^{\downarrow}(z', z'') \approx \tilde{\mathbf{T}}\left\{\mathbf{t}^{\downarrow}(z', z''; \tilde{\lambda})\right\} = \mathbf{U}\exp\left[-4\pi \mathrm{Im}(\bar{\sigma}_1) d/\tilde{\lambda}\right]; \tag{10.35}$$

(c) if the layer is anisotropic and

$$l_{\mathrm{coh}} \gg |\mathrm{Re}(\sigma_2 - \sigma_1)|d, \ |\mathrm{Re}(\sigma_4 - \sigma_3)|d, \tag{10.36}$$

---

[2] The conditions of the problem under consideration do not allow the realization of TIR mode in any layer.

to a good approximation,

$$\mathcal{T}^{\downarrow}(z',z'') \approx \tilde{\mathbf{T}}\left\{\mathbf{t}^{\downarrow}(z',z'';\tilde{\lambda})\right\}, \quad \mathcal{T}^{\uparrow}(z',z'') \approx \tilde{\mathbf{T}}\left\{\mathbf{t}^{\uparrow}(z',z'';\tilde{\lambda})\right\}. \tag{10.37}$$

3. *Bulk of a smoothly inhomogeneous layer.* Let a domain (z', z'') be the bulk of a smoothly inhomogeneous layer with negligible bulk reflection. An analysis of expressions of Chapter 11 for transmission EW Jones matrices of such layers shows that in this case approximation (10.37) is accurate when

$$l_{\text{coh}} \gg \int_{z'}^{z''} |\text{Re}(\sigma_2(z) - \sigma_1(z))|\text{d}z, \quad l_{\text{coh}} \gg \int_{z'}^{z''} |\text{Re}(\sigma_4(z) - \sigma_3(z))|\text{d}z. \tag{10.38}$$

To our knowledge, anisotropic layers for which conditions (10.36) and (10.38) are violated are not used in LCD panels. So, in what follows, we exclude such layers from consideration.

4. *A "thin" layered system.* We have noted in Section 7.1 that the monochromatic approximation can be used for estimating the quasimonochromatic Mueller matrices of "thin" layered systems. This applies to the EW Mueller matrices as well. A more accurate criterion of applicability of the monochromatic approximation to a layered system than $d_{\text{sys}} \ll l_{\text{coh}}$, where d_{sys} is the thickness of the system, is

$$l_{\text{coh}} \gg \int_{\underline{z}}^{\bar{z}} \sigma_{\max}(z)\text{d}z, \tag{10.39}$$

where $\sigma_{\max}(z) \equiv \max_j\{|\text{Re}\sigma_j(z)|\}$, and $\underline{z}$ and $\bar{z}$ ($\bar{z} = \underline{z} + d_{\text{sys}}$) are the z-coordinates of the external interfaces of the system. It can be seen from (10.39) that for situations typical of LCDs the simpler criterion $d_{\text{sys}} \ll l_{\text{coh}}$ is quite good. The monochromatic approximation in this case gives the following computational formulas for the quasimonochromatic EW Mueller matrices:

$$\begin{aligned}
\mathcal{T}^{\downarrow}(\underline{z}-0,\bar{z}+0) &= \tilde{\mathbf{T}}\left\{\mathbf{t}^{\downarrow}(\underline{z}-0,\bar{z}+0;\tilde{\lambda})\right\}, \quad \mathcal{R}^{\downarrow}(\underline{z}-0,\bar{z}+0) = \tilde{\mathbf{T}}\left\{\mathbf{r}^{\downarrow}(\underline{z}-0,\bar{z}+0;\tilde{\lambda})\right\}, \\
\mathcal{T}^{\uparrow}(\underline{z}-0,\bar{z}+0) &= \tilde{\mathbf{T}}\left\{\mathbf{t}^{\uparrow}(\underline{z}-0,\bar{z}+0;\tilde{\lambda})\right\}, \quad \mathcal{R}^{\uparrow}(\underline{z}-0,\bar{z}+0) = \tilde{\mathbf{T}}\left\{\mathbf{r}^{\uparrow}(\underline{z}-0,\bar{z}+0;\tilde{\lambda})\right\}.
\end{aligned} \tag{10.40}$$

A Simple Example of Using the Additivity of EW Stokes Vectors for Mutually Incoherent Wave Fields

If a forward or backward propagating wave field X produced by the wave X_{inc} in M_{SM}, M_{inc}, or M_{tr} is the superposition of mutually incoherent fields $\mathsf{X}_1 = \mathsf{O}_1\mathsf{X}_{\text{inc}}$ and $\mathsf{X}_2 = \mathsf{O}_2\mathsf{X}_{\text{inc}}$,

$$S\{\mathsf{X}\} \equiv S\{\mathsf{X}_1 + \mathsf{X}_2\} = S\{\mathsf{X}_1\} + S\{\mathsf{X}_2\} = \mathbf{M}_{\mathsf{O}_1}S_{\text{inc}} + \mathbf{M}_{\mathsf{O}_2}S_{\text{inc}} = (\mathbf{M}_{\mathsf{O}_1} + \mathbf{M}_{\mathsf{O}_2})S_{\text{inc}}, \tag{10.41}$$

where $\mathbf{M}_{\mathsf{O}_1}$ and $\mathbf{M}_{\mathsf{O}_2}$ are the quasimonochromatic EW Mueller matrices of the operations of O_1 and O_2, respectively. This means that the matrix $\mathbf{M}_{\mathsf{O}_1+\mathsf{O}_2}$ linking the vectors S_{inc} and $S\{\mathsf{X}\}$ as

$$S\{\mathsf{X}\} = \mathbf{M}_{\mathsf{O}_1+\mathsf{O}_2}S_{\text{inc}} \tag{10.42}$$

can be calculated as follows:

$$\mathbf{M}_{\mathsf{O}_1+\mathsf{O}_2} = \mathbf{M}_{\mathsf{O}_1} + \mathbf{M}_{\mathsf{O}_2}. \tag{10.43}$$

This property is useful when the matrices $\mathbf{M}_{\mathsf{O}_1}$ and $\mathbf{M}_{\mathsf{O}_2}$ can be calculated using the monochromatic approximation. For example, estimating the partial transmittance being measured in the situation shown in Figure 7.6b, we can calculate the approximating EW Mueller matrix of the combined channel M_{t1} + M_{t2} (see the discussion of this situation in Section 7.1) by the formulas

$$\begin{aligned}
\mathbf{M}_{\mathsf{M}_{t1}+\mathsf{M}_{t2}} &= \mathbf{M}_{\mathsf{M}_{t1}} + \mathbf{M}_{\mathsf{M}_{t2}}, \\
\mathbf{M}_{\mathsf{M}_{t1}} &= \tilde{\mathbf{T}}\{\mathbf{t}^{\downarrow}(z_1-0, z_1+0; \tilde{\lambda})\mathbf{t}^{\downarrow}(z_0+0, z_1-0; \tilde{\lambda})\mathbf{t}^{\downarrow}(z_0-0, z_0+0; \tilde{\lambda})\}, \\
\mathbf{M}_{\mathsf{M}_{t2}} &= \tilde{\mathbf{T}}\{\mathbf{t}^{\downarrow}(z_1-0, z_1+0; \tilde{\lambda})\mathbf{t}^{\downarrow}(z_0+0, z_1-0; \tilde{\lambda})\mathbf{r}^{\uparrow}(z_0-0, z_0+0; \tilde{\lambda}) \\
&\quad \cdot\mathbf{t}^{\uparrow}(z_0+0, z_1-0; \tilde{\lambda})\mathbf{r}^{\downarrow}(z_1-0, z_1+0; \tilde{\lambda}) \\
&\quad \cdot\mathbf{t}^{\downarrow}(z_0+0, z_1-0; \tilde{\lambda})\mathbf{t}^{\downarrow}(z_0-0, z_0+0; \tilde{\lambda})\}
\end{aligned} \tag{10.44}$$

in all cases where the quasimonochromatic EW Mueller matrices for the bulk of the layer, $\mathcal{T}^{\downarrow}(z_0+0, z_1-0)$ and $\mathcal{T}^{\uparrow}(z_0+0, z_1-0)$, can be estimated as

$$\begin{aligned}
\mathcal{T}^{\downarrow}(z_0+0, z_1-0) &= \tilde{\mathbf{T}}\left\{\mathbf{t}^{\downarrow}(z_0+0, z_1-0; \tilde{\lambda})\right\}, \\
\mathcal{T}^{\uparrow}(z_0+0, z_1-0) &= \tilde{\mathbf{T}}\left\{\mathbf{t}^{\uparrow}(z_0+0, z_1-0; \tilde{\lambda})\right\}.
\end{aligned} \tag{10.45}$$

Relationship Between EW Stokes Vectors and Other Types of Stokes Vectors for Quasimonochromatic Plane Waves in Nonabsorbing Isotropic Media

In order to use the plane wave approximation (7.36) we need to find the Mueller matrix relating irradiance-based Stokes vectors (see Section 5.3). It is sometimes necessary to find the intensity-based Stokes vector of an emergent wave. It is often convenient to specify the incident wave by its intensity-based Stokes vector. To use EW Mueller matrices and EW Stokes vectors in such cases we need to know the relationship between EW Stokes vectors and the mentioned types of Stokes vectors for waves in isotropic media. Let $\mathbf{X}$ be a wave propagating in an isotropic nonabsorbing domain M of the M_{inc}–M_{SM}–M_{tr} system, for example, it may be the incident wave $\mathbf{X}_{inc}$, or a reflected wave propagating in the medium M_{inc}, or a transmitted wave propagating in M_{tr} if M_{tr} is isotropic and nonabsorbing. Let the eigenwave basis in M be chosen in the standard manner (as in Section 9.2) and let the polarization reference axis (see Section 5.3) for Stokes vectors $S_{(E)}\{\mathbf{X}\}$and $S_{(I)}\{\mathbf{X}\}$ be chosen parallel to $\mathbf{e}_1$ if $\mathbf{X}$ is forward propagating or to $\mathbf{e}_3$ otherwise. Then the following relations hold:

(a) in the case of the flux normalization (8.164) or symmetrical normalization (8.167) of the EW basis in M

$$S_{(E)}\{\mathbf{X}\} = \frac{c}{16\pi}S\{\mathbf{X}\}, \tag{10.46}$$

$$S_{(I)}\{\mathbf{X}\} = \frac{c}{16\pi\,|\mathbf{lz}|}S\{\mathbf{X}\}, \tag{10.47}$$

(b) in the case of the electrical normalization (8.164)

$$S_{(E)}\{\mathbf{X}\} = \frac{cn\,|\mathbf{lz}|}{8\pi}S\{\mathbf{X}\}, \tag{10.48}$$

$$S_{(I)}\{\mathbf{X}\} = \frac{cn}{8\pi}S\{\mathbf{X}\}, \tag{10.49}$$

where $\mathbf{l}$ is the wave normal of $\mathbf{X}$ and n is the refractive index of M.

If $\mathbf{X}_{\text{ref}} = \mathbf{R}\mathbf{X}_{\text{inc}}$ is a reflected wave propagating in M_{inc}, and $\mathbf{M}_{\mathsf{R}(E)}$ and $\mathbf{M}_{\mathsf{R}}$ are Mueller matrices such that

$$S_{(E)}\{\mathbf{X}_{\text{ref}}\} = \mathbf{M}_{\mathsf{R}(E)}S_{(E)}\{\mathbf{X}_{\text{inc}}\}, \tag{10.50}$$

$$S\{\mathbf{X}_{\text{ref}}\} = \mathbf{M}_{\mathsf{R}}S\{\mathbf{X}_{\text{inc}}\}, \tag{10.51}$$

according to (10.46) and (10.48), with any of the three types of normalization in M_{inc}

$$\mathbf{M}_{\mathsf{R}(E)} = \mathbf{M}_{\mathsf{R}}. \tag{10.52}$$

Assume that the medium M_{tr} is isotropic and nonabsorbing, $\mathbf{X}_{\text{tr}} = \mathbf{T}\mathbf{X}_{\text{inc}}$ is a transmitted wave propagating in M_{tr}, and $\mathbf{M}_{\mathsf{T}(E)}$ and $\mathbf{M}_{\mathsf{T}}$ are the Mueller matrices such that

$$S_{(E)}\{\mathbf{X}_{\text{tr}}\} = \mathbf{M}_{\mathsf{T}(E)}S_{(E)}\{\mathbf{X}_{\text{inc}}\}, \tag{10.53}$$

$$S\{\mathbf{X}_{\text{tr}}\} = \mathbf{M}_{\mathsf{T}}S\{\mathbf{X}_{\text{inc}}\}. \tag{10.54}$$

From (10.46) and (10.48) it is seen that the relation

$$\mathbf{M}_{\mathsf{T}(E)} = \mathbf{M}_{\mathsf{T}} \tag{10.55}$$

will be valid whatever the refractive index of M_{tr} is if the flux normalization or symmetrical normalization is used in M_{inc} and M_{tr}. Moreover, equation (10.55) will be valid in the case of the electrical normalization if the refractive indices of M_{inc} and M_{tr} are equal.

10.2 Calculation of the EW Mueller Matrices of the Overall Transmission and Reflection of a System Consisting of "Thin" and "Thick" Layers

Suppose that the system M_{SM} that we dealt with in the previous section consists only of elements of the following types:

- "thick" homogeneous layer satisfying (10.36);
- "thick" smoothly inhomogeneous layer with negligible bulk reflection satisfying (10.38); and
- "thin" layered system (situated between "thick" layers or between a "thick" layer and M_{inc} or M_{tr}).

Then we can arbitrarily divide this system into fragments of the following kinds:

- interface between "thick" layers;
- interface between a "thick" layer and M_{inc} or M_{tr};
- bulk of a "thick" layer;
- interface between "thick" layers plus the bulk of one of these layers;
- interface between "thick" layers plus the bulks of these layers;
- "thin" layered system (including its interfaces with adjacent "thick" layers, M_{inc}, or M_{tr});
- "thin" layered system plus the bulk of an adjacent "thick" layer; and
- "thin" layered system plus the bulks of adjacent "thick" layers.

The fragments of these kinds—we will call them **D**-*fragments*—have two important features. First, the transmission and reflection EW Mueller matrices for any of such fragments may be estimated with good accuracy by using the monochromatic approximation. Second, for any of such fragments, EW Stokes vector can be considered, at least to some approximation, a TR-additive characteristic (see Section 7.2.1). The assumption of TR-additivity of EW Stokes vector for a fragment (z', z'') is equivalent to the assumption that the wave fields $\mathbf{X}_{tr}{}^{\downarrow}(z'')$ and $\mathbf{X}_{ref}{}^{\downarrow}(z'')$ as well as the fields $\mathbf{X}_{tr}{}^{\uparrow}(z')$ and $\mathbf{X}_{ref}{}^{\uparrow}(z')$ (see Figure 7.10) are mutually incoherent. Such initial assumptions are often used in calculations of transmission and reflection characteristics of "thick" layers [in particular, they lead to the well-known formulas (7.38), (7.39), (7.42), and (7.43)], systems of "thick" layers (see, e.g., [1]), and systems including both "thick" and "thin" layers (see, e.g., [2–4]).

We consider two fast methods for calculating characteristics of transmission and reflection for systems of "thick" and "thin" layers that exploit division into **D**-fragments. One of them, namely, the 8 × 8 transfer matrix method [4,5], uses the transfer matrix technique (Section 7.2.1). The second method employs the adding technique (Section 7.2.2).

8 × 8 Transfer Matrix Method

By definition, the transmission and reflection EW Mueller matrices of a **D**-fragment (z', z'') can be evaluated by using the formulas

$$\begin{aligned} \mathcal{T}^{\downarrow}(z',z'') &= \tilde{\mathbf{T}}\{\mathbf{t}^{\downarrow}(z',z'';\tilde{\lambda})\}, \quad \mathcal{R}^{\downarrow}(z',z'') = \tilde{\mathbf{T}}\{\mathbf{r}^{\downarrow}(z',z'';\tilde{\lambda})\}, \\ \mathcal{T}^{\uparrow}(z',z'') &= \tilde{\mathbf{T}}\{\mathbf{t}^{\uparrow}(z',z'';\tilde{\lambda})\}, \quad \mathcal{R}^{\uparrow}(z',z'') = \tilde{\mathbf{T}}\{\mathbf{r}^{\uparrow}(z',z'';\tilde{\lambda})\}. \end{aligned} \tag{10.56}$$

Following the transfer matrix technique (Section 7.2.1), we may use this fact to calculate the EW Mueller matrices describing the overall transmission and reflection of the layered system under consideration with the help of transfer matrices for the state vector

$$\vec{\mathcal{S}} \equiv \begin{pmatrix} S^{\downarrow} \\ S^{\uparrow} \end{pmatrix} \tag{10.57}$$

[see (10.4)]. According to (7.72), the transfer matrix $\mathbf{D}(z'', z')$ linking the vectors $\vec{\mathcal{S}}(z')$ and $\vec{\mathcal{S}}(z'')$,

$$\vec{\mathcal{S}}(z'') = \mathbf{D}(z'', z')\vec{\mathcal{S}}(z'), \tag{10.58}$$

can be represented as

$$\mathbf{D}(z'', z') \equiv \begin{pmatrix} \underline{\mathcal{T}}^{\downarrow} - \underline{\mathcal{R}}^{\uparrow}\left(\underline{\mathcal{T}}^{\uparrow}\right)^{-1}\underline{\mathcal{R}}^{\downarrow} & \underline{\mathcal{R}}^{\uparrow}\left(\underline{\mathcal{T}}^{\uparrow}\right)^{-1} \\ -\left(\underline{\mathcal{T}}^{\uparrow}\right)^{-1}\underline{\mathcal{R}}^{\downarrow} & \left(\underline{\mathcal{T}}^{\uparrow}\right)^{-1} \end{pmatrix},$$
$$\underline{\mathcal{T}}^{\downarrow} \equiv \mathcal{T}^{\downarrow}(z',z''),\ \underline{\mathcal{R}}^{\downarrow} \equiv \mathcal{R}^{\downarrow}(z',z''),\ \underline{\mathcal{T}}^{\uparrow} \equiv \mathcal{T}^{\uparrow}(z',z''),\ \underline{\mathcal{R}}^{\uparrow} \equiv \mathcal{R}^{\uparrow}(z',z''). \tag{10.59}$$

Using (10.59), (10.56), (8.121), (8.122), (10.15), and (5.70), one can express the matrix $\mathbf{D}(z'', z')$ in terms of the EW 4 × 4 transfer matrix (see Section 8.2.2) of the fragment (z', z'') for $\lambda = \tilde{\lambda}$, $\mathbf{T}(z'', z'; \tilde{\lambda})$. We write the corresponding expression for $\mathbf{D}(z'', z')$ as follows:

$$\mathbf{D}(z'', z') = \widehat{\mathbf{D}}\{\mathbf{T}(z'', z'; \tilde{\lambda})\}, \tag{10.60}$$

where $\widehat{\mathbf{D}}\{\bar{\mathbf{T}}\}$ denotes the real 8×8 matrix calculated from a 4×4 matrix $\bar{\mathbf{T}}$ as

$$\widehat{\mathbf{D}}\{\bar{\mathbf{T}}\} = \begin{pmatrix} \mathbf{L}[\mathbf{t}_{11} \otimes (\mathbf{t}_{11} - \mathbf{t}_1)^* - \mathbf{t}_1 \otimes \mathbf{t}_{11}^*]\mathbf{L}^{-1} & \mathbf{L}(\mathbf{t}_{12} \otimes \mathbf{t}_{12}^*)\mathbf{L}^{-1} \\ -\mathbf{L}(\mathbf{t}_{21} \otimes \mathbf{t}_{21}^*)\mathbf{L}^{-1} & \mathbf{L}(\mathbf{t}_{22} \otimes \mathbf{t}_{22}^*)\mathbf{L}^{-1} \end{pmatrix}, \quad \mathbf{t}_1 = \mathbf{t}_{12}\mathbf{t}_{22}^{-1}\mathbf{t}_{21}, \tag{10.61}$$

with $\mathbf{t}_{kl}$ $(k,l = 1,2)$ being 2×2 blocks of the matrix $\bar{\mathbf{T}}$,

$$\bar{\mathbf{T}} \equiv \begin{pmatrix} \mathbf{t}_{11} & \mathbf{t}_{12} \\ \mathbf{t}_{21} & \mathbf{t}_{22} \end{pmatrix}.$$

The calculation of $\mathbf{D}(z'', z')$ with use of (10.60) is generally faster than that by using (10.59).

Further, according to the transfer matrix technique, we can calculate the 8×8 transfer matrix $\mathbf{D}_S$ that characterizes the whole layered system by multiplying the 8×8 transfer matrices of its **D**-fragments. Suppose that the layered system is divided into N **D**-fragments with boundaries at $z = \underline{z}_0, \underline{z}_1, \underline{z}_2, \ldots, \underline{z}_N$ ($\underline{z}_0 < \underline{z}_1 < \underline{z}_2 < \ldots < \underline{z}_N$). Then

$$\mathbf{D}_S \equiv \mathbf{D}(\underline{z}_N, \underline{z}_0) = \mathbf{D}_N \ldots \mathbf{D}_2\mathbf{D}_1, \tag{10.62}$$

where $\mathbf{D}_j \equiv \mathbf{D}(\underline{z}_j, \underline{z}_{j-1})$ is the 8×8 transfer matrix of the jth **D**-fragment ($j = 1, 2, \ldots, N$). The EW Mueller matrices characterizing the overall transmission and overall reflection of the system can be calculated as

$$\begin{aligned} \mathcal{T}^{\downarrow}(\underline{z}_0, \underline{z}_N) &= \mathrm{t}^{\downarrow}\{\mathbf{D}_S\}, \quad \mathcal{R}^{\downarrow}(\underline{z}_0, \underline{z}_N) = \mathrm{r}^{\downarrow}\{\mathbf{D}_S\}, \\ \mathcal{T}^{\uparrow}(\underline{z}_0, \underline{z}_N) &= \mathrm{t}^{\uparrow}\{\mathbf{D}_S\}, \quad \mathcal{R}^{\uparrow}(\underline{z}_0, \underline{z}_N) = \mathrm{r}^{\uparrow}\{\mathbf{D}_S\} \end{aligned} \tag{10.63}$$

[see (7.84)]. It is clear that this approach can be employed for calculating the transmission and reflection matrices not only for a layered system as a whole but also for its OTR units including two or more **D**-fragments (this may be needed, e.g., when a partial transmission or a partial reflection of the system must be estimated). For example, the operators $\mathcal{T}^{\downarrow}$ and $\mathcal{R}^{\downarrow}$ for the domain $(\underline{z}_1, \underline{z}_3)$ of the above system can be calculated as

$$\mathcal{T}^{\downarrow}(\underline{z}_1, \underline{z}_3) = \mathrm{t}^{\downarrow}\{\mathbf{D}_3\mathbf{D}_2\}, \quad \mathcal{R}^{\downarrow}(\underline{z}_1, \underline{z}_3) = \mathrm{r}^{\downarrow}\{\mathbf{D}_3\mathbf{D}_2\}.$$

Let us consider some examples showing capabilities and features of the 8×8 transfer matrix method.

Example 1 *Two-layer system consisting of a "thin" layer and a "thick" layer.* Assume that the boundary planes of the "thin" layer are $z = z_0$ and $z = z_1$, and those of the "thick" layer are $z = z_1$ and $z = z_2$ (see the inset in Figure 10.1a). There are three possible divisions of this system into **D**-fragments:

(i) $(z_0{-}0, z_1{+}0)$, $(z_1{+}0, z_2{+}0)$;
(ii) $(z_0{-}0, z_2{-}0)$, $(z_2{-}0, z_2{+}0)$; and
(iii) $(z_0{-}0, z_1{+}0)$, $(z_1{+}0, z_2{-}0)$, $(z_2{-}0, z_2{+}0)$.

The corresponding expressions for the matrix $\mathbf{D}_S \equiv \mathbf{D}(z_2{+}0, z_0{-}0)$ in terms of EW 4×4 transfer matrices of **D**-fragments are

$$\mathbf{D}_S = \widehat{\mathbf{D}}\{\mathbf{T}(z_2 + 0, z_1 + 0; \tilde{\lambda})\}\widehat{\mathbf{D}}\{\mathbf{T}(z_1 + 0, z_0 - 0; \tilde{\lambda})\}, \tag{10.64}$$

$$\mathbf{D}_S = \widehat{\mathbf{D}}\{\mathbf{T}(z_2 + 0, z_2 - 0; \tilde{\lambda})\}\widehat{\mathbf{D}}\{\mathbf{T}(z_2 - 0, z_0 - 0; \tilde{\lambda})\}, \tag{10.65}$$

$$\mathbf{D}_S = \widehat{\mathbf{D}}\{\mathbf{T}(z_2 + 0, z_2 - 0; \tilde{\lambda})\}\widehat{\mathbf{D}}\{\mathbf{T}(z_2 - 0, z_1 + 0; \tilde{\lambda})\}\widehat{\mathbf{D}}\{\mathbf{T}(z_1 + 0, z_0 - 0; \tilde{\lambda})\}. \tag{10.66}$$

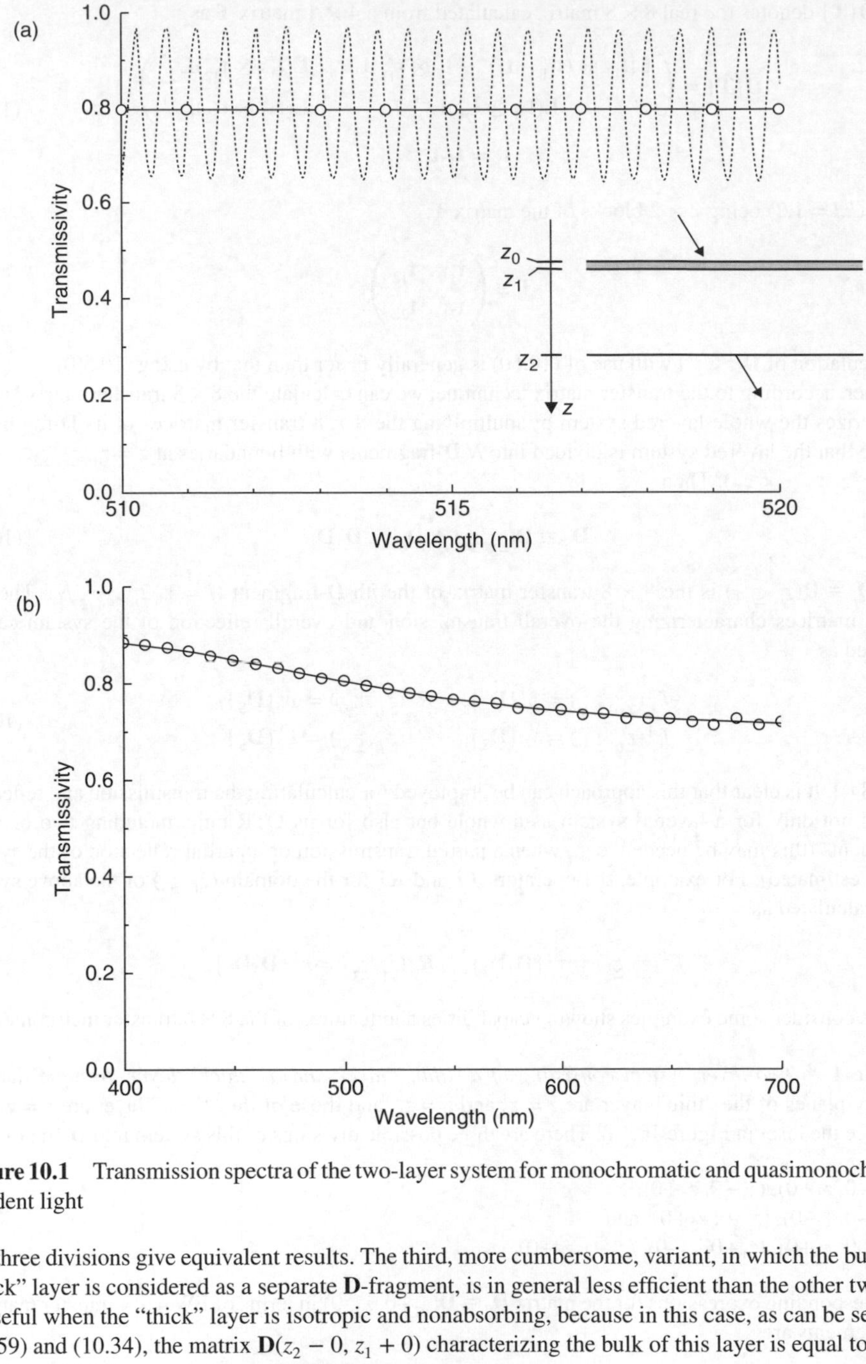

Figure 10.1 Transmission spectra of the two-layer system for monochromatic and quasimonochromatic incident light

All three divisions give equivalent results. The third, more cumbersome, variant, in which the bulk of the "thick" layer is considered as a separate **D**-fragment, is in general less efficient than the other two, but it is useful when the "thick" layer is isotropic and nonabsorbing, because in this case, as can be seen from (10.59) and (10.34), the matrix $\mathbf{D}(z_2 - 0, z_1 + 0)$ characterizing the bulk of this layer is equal to the unit matrix, which allows reduction of (10.66) to

$$\mathbf{D}_S = \widehat{\mathbf{D}}\{\mathbf{T}(z_2 + 0, z_2 - 0; \tilde{\lambda})\}\widehat{\mathbf{D}}\{\mathbf{T}(z_1 + 0, z_0 - 0; \tilde{\lambda})\}, \tag{10.67}$$

where the bulk of the "thick" layer is excluded from the calculations.

Figure 10.1 shows calculated spectra of transmissivities of such a system. In the calculations both layers were taken to be isotropic and nonabsorbing. The "thin" layer: $n = 2$, $d = 0.1$ μm (n is the refractive index, d is the thickness). The "thick" layer: $n = 1.52$, $d = 200$ μm. The surrounding media: $n = 1$. The angle of incidence was taken to be 20°. The incident light was assumed to be s-polarized. According to the criterion $d > l_{coh}$ and the estimate $l_{coh} \approx \tilde{\lambda}^2/\Delta\lambda$, we can regard the second layer as "thick" for all wavelengths of the visible region if $\Delta\lambda > 3$ nm. Figure 10.1 demonstrates the monochromatic transmissivity spectrum (dashed line in Figure 10.1a), points (shown by circles) of the quasimonochromatic transmissivity spectrum calculated as the convolution of the monochromatic transmissivity spectrum with a trapezoidal spectral window of width 4 nm (i.e., by the spectral averaging method), and the quasimonochromatic transmissivity spectrum calculated by the 8×8 transfer matrix method (solid lines). The results presented in Figure 10.1a well illustrate what was said about spectra of thin and thick layers and layered systems including thin and thick layers in Section 7.1. The monochromatic transmissivity spectrum of the system has fast oscillations (the width of the spectral zone shown in Figure 10.1a is only 10 nm) due to a large thickness of the thick layer. Convolving the monochromatic spectrum with a relatively narrow spectral window removes these fast oscillations but retains the pattern of FP interference in the thin layer. As is seen from Figure 10.1, the 8×8 transfer matrix method gives almost the same results as the spectral averaging.

We should note that dealing with a layered system including two or more "thick" layers one may observe somewhat worse agreement between results obtained by the spectral averaging and those obtained by the 8×8 transfer matrix method. This is connected with transformation of the correlation properties of quasimonochromatic light as it propagates through a layered system. Light leaving a "thick" layer, because of multiple reflections, contains many derivatives of the same wavetrain of the incident wave, which emerge from the layer one after another with a temporal delay exceeding the coherence time of the incident wave. Derivatives of these related wavetrains, due to multiple reflections in a following "thick" layer, may overlap and hence interfere (a situation of this kind is discussed in Section 7.6 of [6]). The spectral averaging method allows for such "secondary" interference, while the 8×8 transfer matrix method does not. Using theoretical estimates given in [6] and considering real geometries of LCD panels, it is easy to see that the "secondary" interference plays no role in LCDs, at least in most practical situations. In this sense, the 8×8 transfer matrix method seems more suitable for modeling LCDs than the spectral averaging method.

Example 2 *A model TN LCD.* We included in this model, along with the LC layer, glass substrates, and polarizers, the electrode–alignment layer systems (EASs) (Figure 10.2). The polarizers are regarded

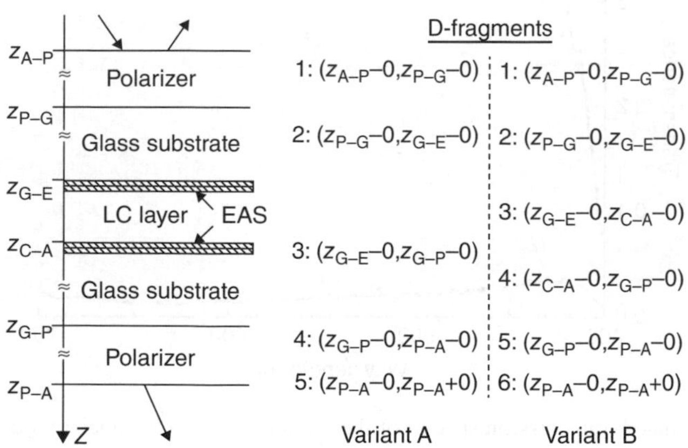

Figure 10.2 Structure of the model TN LCD and two variants of dividing it into **D**-fragments

as homogeneous uniaxial dichroic layers. Film polarizers in standard spectral measurements do not give any Fabry–Perot interference (FPI) pattern and can be considered as "thick" layers, as well as glass substrates. LC layers in LCDs are usually sufficiently thin to give FPI patterns in experimental spectra. For a real LC device having structure shown in Figure 10.2, with an LC layer thickness less than 10 μm, in experimental spectra obtained with $\Delta\lambda < 2$ nm one may observe an almost perfect FPI pattern produced by the thin-layer system including the LC layer, that is, the LC layer can be considered as a "thin" layer. With greater $\Delta\lambda$ this FPI pattern is blurred to a greater or lesser extent. A possible division of the model LCD into **D**-fragments for the case when the LC layer is considered as an element of the "thin" layered system including, along with the LC layer, EASs is shown in Figure 10.2 (variant A). If necessary (see, e.g., [7] or Section 12.5), blurred spectra corresponding to measurements with relatively large $\Delta\lambda$ can be calculated from those obtained with treating the LC layer as "thin" by formula (7.31). It is often desirable to remove the FPI ripples related to the LC layer from the modeled spectra at all: smoothed spectra are more convenient for comparison, for use in optimization procedures, for graphic and tabular representations, for calculating the color characteristics, and so on. To obtain such a smoothed spectrum directly, without spectral averaging, it suffices to treat the LC layer as a "thick" layer. A possible division of the LCD into **D**-fragments in this case is shown in Figure 10.2 as variant B.

Figure 10.3 demonstrates calculated transmittance spectra of the LCD for unpolarized incident light at normal incidence, corresponding to variants A and B. The calculations were performed with the following parameters of the LCD elements. For the glass substrates, d = 100 μm, n = 1.52. For the electrodes, d = 0.03 μm, n = 2. For the alignment layers: d = 0.1 μm, n = 1.6. Parameters of the LC layer: $K_{11} = 1.3 \times 10^{-6}$ dyn, $K_{22} = 7.1 \times 10^{-7}$ dyn, $K_{33} = 1.95 \times 10^{-6}$ dyn, $\varepsilon_{\|} = 15.1$, $\varepsilon_{\perp} = 3.8$ at

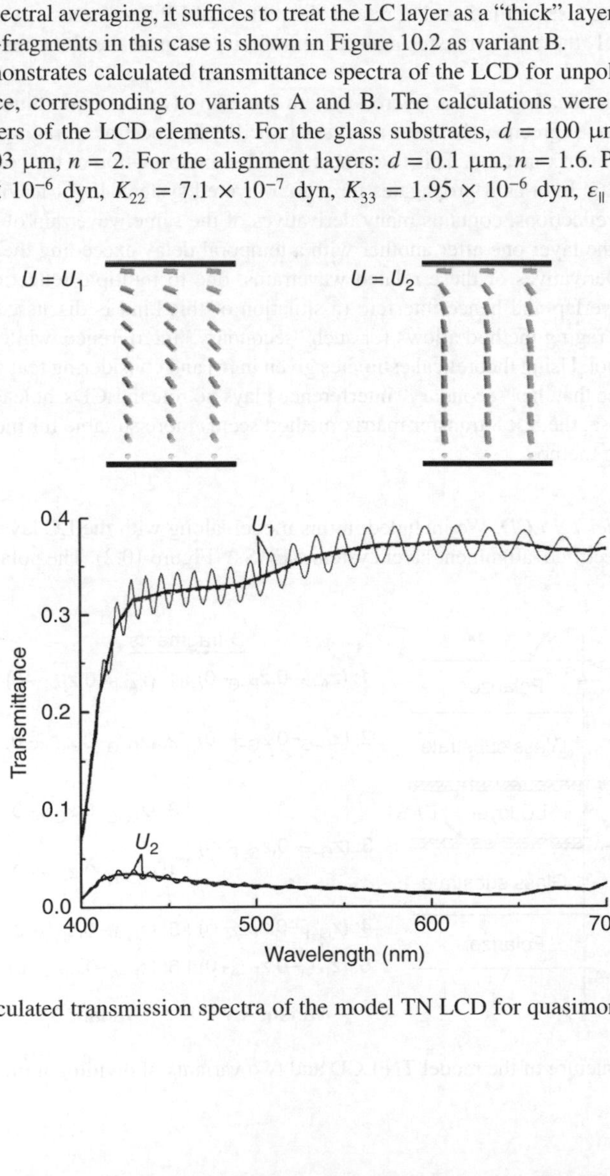

Figure 10.3 Calculated transmission spectra of the model TN LCD for quasimonochromatic incident light

the frequency of the applied voltage, the natural helical pitch $p_0 = 17.1$ μm, $d = 6$ μm. The principal refractive indices of the liquid crystal:

Wavelength (nm)	$n_\perp$	$n_\parallel$
437	1.518	1.681
546	1.503	1.650
644	1.497	1.637

The surface anchoring is assumed to be infinitely strong. The twist angle $\Phi = 90°$, the pretilt angles are equal to 4°. The polarizers have $d = 200$ μm, $\mathrm{Re}(n_\perp) = 1.5$, $\mathrm{Re}(n_\parallel) = 1.501$, and absorption properties identical to those of one of the commercial neutral polarizers of O-type for LCDs. The polarizers are crossed. The transmission axis of each of the polarizers is parallel to the projection of the easy axis at the nearest LC layer boundary onto the boundary plane. In other words, according to the standard classification, this model device is a normally white TN LCD in E-mode. The spectra were calculated for two values of the voltage applied to the LC layer: $U_1 = 1.7$ V (bright state) and $U_2 = 3.5$ V (dark state). The spectra corresponding to variant A are shown by thin solid lines, and those corresponding to variant B by thick solid lines.

For comparison, in Figure 10.3, we also plotted the corresponding transmittance spectra for the "useful" transfer channel of this LCD that is defined by the chain of transmission operations successively performed by the air–frontal polarizer interface, the bulk of the frontal polarizer, the frontal polarizer–glass interface, the bulk of the frontal glass substrate, the frontal EAS, the bulk of the LC layer, the rear EAS, the bulk of the rear glass substrate, the glass–rear polarizer interface, the bulk of the rear polarizer, and the rear polarizer–air interface. The quasimonochromatic transmittance of this channel can be evaluated by using the monochromatic approximation (10.31). According to (10.31), the transmittance of this channel (t_{UC}) for unpolarized incident light with mean wavelength $\tilde{\lambda}$ can be calculated from the EW Jones matrix of this channel ($\mathbf{t}_{\mathrm{UC}}$) as

$$t_{\mathrm{UC}} = \frac{1}{2}\|\mathbf{t}_{\mathrm{UC}}(\tilde{\lambda})\|_{\mathrm{E}}^2 \tag{10.68}$$

provided that the normalization of the EW basis in air is the same at both sides of the LCD. The corresponding curves are shown in Figure 10.3 by dashed lines. One of the two dashed curves, namely, that for the voltage U_2, almost coincides with the corresponding curve of the overall transmittance calculated by variant B. For the voltage U_1, the curves of these kinds are also close to each other. This testifies that for this LCD, the light circulation between **D**-fragments gives a small contribution to the overall transmittance. In high-contrast devices with phase compensation, the light circulation between and within **D**-fragments may strongly affect the contrast ratio. The 8 × 8 transfer matrix method is an appropriate instrument for simulating such situations.

Example 3 *A model single-polarizer RLCD.* The structure of this model device is shown in Figure 10.4. The last element of the device is a metal reflector. In Figure 10.4, we show two variants of division of the LCD into **D**-fragments: the LC layer is treated as "thin" in variant A and as "thick" in variant B, as in the previous example. Figure 10.5 shows the calculated reflectance spectra of this LCD for unpolarized normally incident light. In the calculations, the parameters of the polarizer, glass substrate, frontal electrode, and alignment layers were taken to be the same as in the previous example. The parameters of the LC layer are the following. For the LC material, $K_{11} = 1.32 \times 10^{-6}$ dyn, $K_{22} = 6.5 \times 10^{-7}$ dyn,

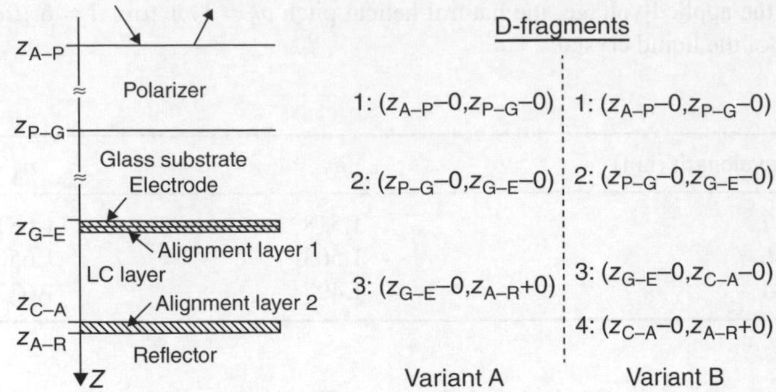

Figure 10.4 Structure of the model single-polarizer RLCD and two variants of dividing it into **D**-fragments

$K_{33} = 1.38 \times 10^{-6}$ dyn, $\varepsilon_{\parallel} = 8.3$, $\varepsilon_{\perp} = 3.1$ at the frequency of the applied voltage, $p_0 = \infty$ (a pure nematic), and the principal refractive indices:

Wavelength (nm)	$n_{\perp}$	$n_{\parallel}$
436	1.4939	1.5987
546	1.4819	1.5809
633	1.4774	1.5734

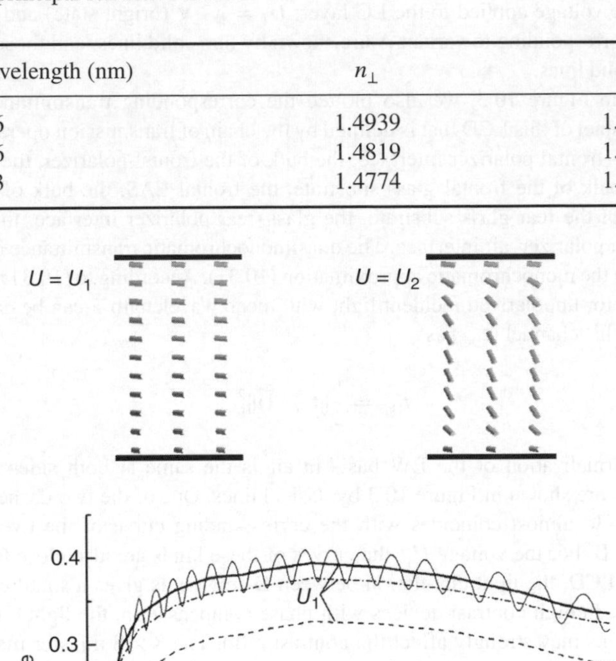

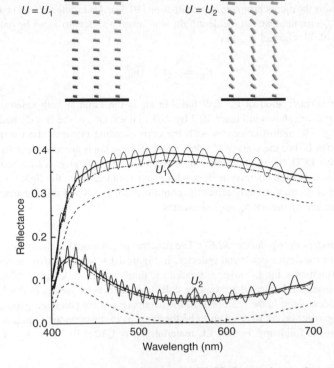

Figure 10.5 Calculated reflection spectra of the model RLCD for quasimonochromatic incident light

The thickness of the LC layer is 5.36 μm. The anchoring is infinitely strong, the twist angle $\Phi = 52°$, and the pretilt angles are equal to 4°. For the metal reflector, we used spectral data for the complex refractive index of aluminum. The transmission axis of the polarizer is parallel to the projection of the easy axis at the frontal LC layer boundary onto the boundary plane. The spectra of the device were calculated for the bright-state voltage $U_1 = 0$ V and the dark-state voltage $U_2 = 2.65$ V. As in Figure 10.3, the spectra for variant A are shown by thin solid lines, and those for variant B by thick solid lines.

As in the previous example, for comparison, we give in Figure 10.5 the transmittance spectra for the "useful" transfer channel of the device (dashed lines). In this case, the "useful" channel is defined by the chain: transmission of the air–polarizer interface, transmission of the polarizer bulk, transmission of the polarizer–glass interface, transmission of the glass substrate bulk, transmission of the frontal EAS, transmission of the LC layer bulk, reflection from the rear alignment layer–reflector system, and back, transmission of the LC layer bulk, transmission of the frontal EAS, and so forth.

As is seen from Figure 10.5, the contrast ratio of the RLCD defined in terms of its overall reflectances is low, much lower than that for the "useful" channel. This is mainly because of reflection from the external boundary of the device and that from the frontal EAS. In the next example, illustrated by Figure 10.6, we significantly weakened these reflections by incorporating in the device antireflective (AR) layers. The structure and parameters of the AR-system for the external boundary of the RLCD in these calculations are shown in Figure 10.6a. The position and parameters of the AR-layers in the EAS are shown in Figure 10.6b. Figure 10.6c shows reflection spectra of the device corresponding to different variants of calculation. As in Figure 10.5, the spectra of the overall reflectance of the RLCD are shown by solid lines: thin and thick solid lines correspond to the variants with treating the LC layer as "thin" and "thick," respectively. The dashed lines show the transmittance spectra for the "useful" transfer channel of the RLCD. We see that in this case the parasitic reflections play much smaller role than in the case

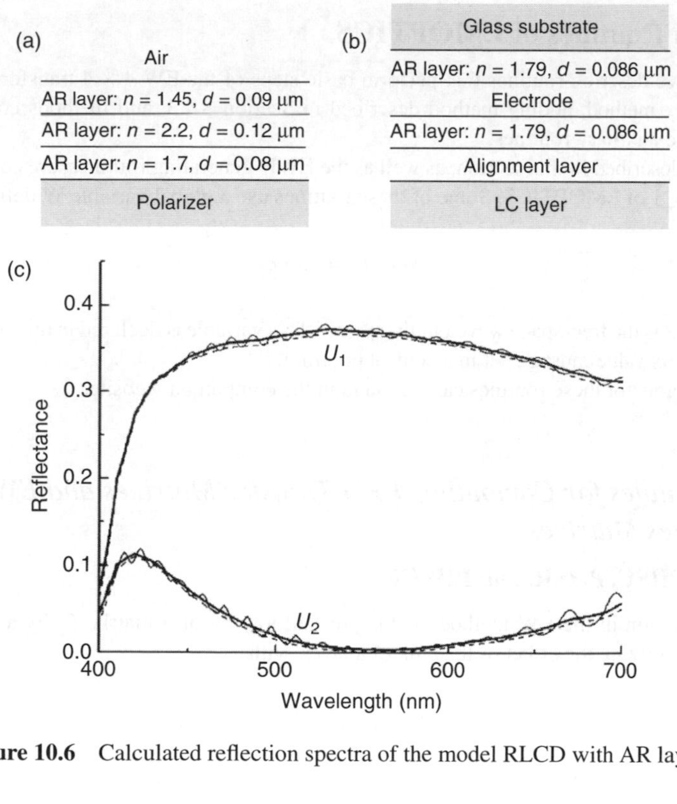

Figure 10.6 Calculated reflection spectra of the model RLCD with AR layers

of the RLCD without AR-layers, and the overall reflectance is close to the transmittance of the "useful" channel of the device.

EW-Mueller-Matrix Adding Method

The 8×8 transfer matrix method has the same disadvantage as any other method using the transfer matrix technique: it is numerically unstable in the presence of layers with strong attenuation (see Section 7.2.1). For instance, we could not perform the above calculations for LCDs because of overflows if the minimum transmittance of the polarizers were less than 10^{-6}. The EW-Mueller-matrix adding method works well in such situations. This method operates with the matrices $\mathcal{T}^{\downarrow}$, $\mathcal{R}^{\downarrow}$, $\mathcal{T}^{\uparrow}$, and $\mathcal{R}^{\uparrow}$ of **D**-fragments. These matrices are calculated by formulas (10.56) from the corresponding EW Jones matrices of **D**-fragments. The EW Jones matrices for a **D**-fragment are calculated by any appropriate method when there is no danger of obtaining the overflow or a large error and by the adding method described in Section 8.4.3 otherwise. The adding of **D**-fragments can be carried out by using the flexible adding technique described in Section 7.2.2. Recall that the flexible adding technique allows calculating the characteristics of both overall and partial transmission and reflection. This is another advantage of the EW-Mueller-matrix adding method. This method allows one to calculate any of the curves shown in Figures 10.1, 10.3, 10.5, and 10.6. As an individual example demonstrating the versatility of this method, we give in Figure 10.5 the curves for the reflectance calculated with allowance for reflections from each of **D**-fragments (for variant B) but ignoring re-reflections between them (dash-dot lines). A chosen variant of the adding may be assigned through the interface of a modeling program. This enables users to easily evaluatc the effect of different reflections on the performance of modeled devices and find the optimal options of calculations.

10.3 Main Routines of LMOPTICS

In this section, we describe routines that perform basic steps of the EW 4×4 transfer matrix method, EW Jones matrix method, adding method described in Section 8.4.3, and methods considered in this chapter and some auxiliary routines.

The routines described in this section, as well as the EWB-generating routines, are components of the module OPTSM_1 of LMOPTICS. Some of these routines use a global variable W defined as

$$\mathrm{W} = k_0 = 2\pi/\lambda, \tag{10.69}$$

where, as usual, λ is the free-space wavelength (in μm). This variable is declared in the module MTWFIR of LMOPTICS. Its value must be set in a control program.

Sample programs for these routines can be found in the companion website.

10.3.1 Routines for Computing 4 × 4 Transfer Matrices and EW Jones Matrices

Subroutines PHSC, PHSR, and PHSCR

A standard operation in the EW methods is the pre-multiplying of a matrix $\mathbf{T}_1$ by a diagonal matrix $\mathbf{C} = [c_{jk}]$, characterizing the effect of the bulk of a layer, with

$$c_{jj} = \mathrm{e}^{ik_0\sigma_j h},$$

where h is the thickness of the layer and σ_j is the normal component of the refraction vector of the jth basis eigenwave [see, e.g., (8.336a)]. In LMOPTICS, this operation is carried out with the help of routines PHSC, PHSR, and PHSCR. These routines can perform this operation for both 4×4 matrices and 2×2 matrices. In all three routines, the multiplication $\mathbf{CT}_1$ is performed by formula (8.341).

Interfaces:
CALL PHSC(X, H, T1C, T2, NW)
CALL PHSR(X, H, T1R, T2, NW)
CALL PHSCR(X, H, T1C, T2, NW)

Data types:
NW—integer(4)
X—(4) array, complex(8)
H—real(8)
T1C, T2—(NW,NW) arrays, complex(8)
T1R—(NW,NW) array, real(8)

Input parameters:
The parameter NW must be set to 2 if the calculations are performed for 2×2 matrices, or to 4 in the case of 4×4 matrices. H is the thickness (in μm) of the layer (H=h). X is the array of σ_j values (see Table 9.1). When NW=2, only X(1) and X(2) are used in the calculations. Arrays X can be calculated using EWB-generating routines presented in Chapter 9 with NWAVE = 2 (if NW = 2) or NWAVE = 4 (in any case). The routines PHSC and PHSCR are used when $\mathbf{T}_1$ is complex; in this case T1C = $\mathbf{T}_1$. The routine PHSR can be used when $\mathbf{T}_1$ and σ_j (j=1,… ,NW) are real; in this case T1R = $\mathbf{T}_1$. The routine PHSCR is applicable only in the case of real σ_j.

This routines use the global variable W [see (10.69)].

Output parameters:
On exit, T2 = $\mathbf{T}_2 = \mathbf{CT}_1$.

Comments. The routine PHSR can be used in computing the Berreman matrices for homogeneous nonabsorbing layers out of TIR mode by formula (8.132). PHSCR can be used for involving homogeneous nonabsorbing layers out of TIR mode when computing the EW 4×4 transfer matrix of a layered system (see the algorithm described in Section 8.7) or the EW Jones matrix of a transfer channel. The most universal routine PHSC can be used in such calculations for involving absorbing layers and layers in TIR mode.

Function BT2

The function subprogram BT2 calculates the transmission EW Jones matrix $t = \mathbf{t}^{\downarrow}$ for an interface {medium 1 → medium 2} according to formulas (8.219) and (8.215).

Interface:
t = BT2(YI1, YY2)

Data types:
YI1—(4,4) array, real(8) or complex(8)
YY2—(4,4) array, real(8) or complex(8)

Input parameters:
The first two rows of the array YI1 must contain the first two rows of the matrix $\mathbf{\Psi}^{-1}$ for medium 1; the first two columns of YY2 must contain those of the matrix $\mathbf{\Psi}$ for medium 2. Only these rows of YI1 and columns of YY2 are used in the calculations. Arrays YI1 and YY2 can be calculated using EWB-generating routines described in Chapter 9 with NWAVE = 2 or NWAVE = 4.

On exit:
The function BT2 returns the (2,2) array representing the matrix $\mathbf{t}^{\downarrow}$. This array is of the real(8) type if both YI1 and YY2 are real, and of the complex(8) type otherwise.

Comments. The routine BT2 is useful, in particular, in computing transmission EW Jones matrices for LC layers (see Section 11.3.1) and EW Jones matrices for "useful" channels of transmissive devices.

Subroutines TRYJ1, TRYJ2, and REFTR1DJones

The routines TRYJ1, TRYJ2, and REFTR1DJones are intended for calculating the transmission and reflection EW Jones matrices, $\mathbf{t}^{\downarrow}$, $\mathbf{r}^{\downarrow}$, $\mathbf{t}^{\uparrow}$, and $\mathbf{r}^{\uparrow}$, of a layered system or a fragment (an OTR unit) of a layered system from the EW 4×4 transfer matrix of this system or fragment, **T**. The calculations are performed according to formulas (8.121) and (7.84).

Interfaces:
CALL TRYJ1(T, TJF, RJF)
CALL TRYJ2(T, TJB, RJB)
CALL REFTR1DJones(T, TJF, RJF, TJB, RJB)

Data types:
T—(4,4) array, complex(8)
TJF, RJF, TJB, RJB—(2,2) arrays, complex(8)

Input and output parameters:
The array T represents the transfer matrix **T**. On exit, TJF = $\mathbf{t}^{\downarrow} = \mathtt{t}^{\downarrow}\{\mathbf{T}\}$, RJF $=\mathbf{r}^{\downarrow} = \mathtt{r}^{\downarrow}\{\mathbf{T}\}$, TJB = $\mathbf{t}^{\uparrow} = \mathtt{t}^{\uparrow}\{\mathbf{T}\}$, RJB = $\mathbf{r}^{\uparrow} = \mathtt{r}^{\uparrow}\{\mathbf{T}\}$.

Subroutine ISOTRBR

This routine calculates the Berreman matrix of a homogeneous nonabsorbing isotropic layer according to expression (8.140)

Interface:
CALL ISOTRBR(N, H, P)

Data types:
N, H—real(8)
P—(4,4) array, complex(8)

Input parameters:
N and H are respectively the refractive index and thickness (in μm) of the layer. This routine uses the global variable W [see (10.69)].

Output parameters:
P is the Berreman matrix of the layer.

Subroutine BRMNC

This routine is intended for calculating the Berreman matrix of a homogeneous layer according to formula (8.132).

Interface:
CALL BRMNC(X, YY, YI, P, H)

Data types:
X—(4) array, complex(8)
YY,YI, P—(4,4) arrays, complex(8)
H—real(8)

Input parameters:
X, YY, YI, and H are respectively the array of σ_j values, matrix $\mathbf{\Psi}$, matrix $\mathbf{\Psi}^{-1}$ (see Table 9.1), and thickness (in μm) of the layer. The input parameters of this routine (X, YY, YI) are consistent in data type with the output parameters of the routines ISOTRC (isotropic medium; Section 9.2), UNAXLC (uniaxial medium; Section 9.3), and BIAXLC (biaxial medium; Section 9.4).
This routine uses the global variable W [see (10.69)].

Output parameters:
P is the Berreman matrix of the layer.

Subroutine BRMNR

This routine is for calculating the Berreman matrix of a homogeneous nonabsorbing layer out of TIR mode according to formula (8.132).

Interface:
CALL BRMNR(X, YY, YI, P, H)

Data types:
X—(4) array, complex(8)
YY,YI—(4,4) arrays, real (8)
P—(4,4) array, complex(8)
H—real(8)

Input parameters:
As for BRMNC, X, YY, YI, and H are respectively the array of σ_j values, matrix $\mathbf{\Psi}$, matrix $\mathbf{\Psi}^{-1}$ (see Table 9.1), and thickness (in μm) of the layer. The input parameters of BRMNR are consistent in data type with the output parameters of the routines ISOTRR (Section 9.2), UNAXLR (Section 9.3), and BIAXLR (Section 9.4).
This routine uses the global variable W [see (10.69)].

Output parameters:
P is the Berreman matrix of the layer.

Routines for Calculating the EW Jones Matrices of Layered Systems by the Adding Technique

The following two routines are intended for calculating the transmission and reflection EW Jones matrices of layered systems according to the flexible adding technique (see Sections 7.2.2 and 8.4.3)

Subroutine AddElementJones

This routine is for adding a new element E to a system S. This routine uses, as input parameters, the EW Jones matrices of the system S, $\mathbf{t}^{\downarrow}_{\mathsf{S}}$, $\mathbf{r}^{\downarrow}_{\mathsf{S}}$, $\mathbf{t}^{\uparrow}_{\mathsf{S}}$, and $\mathbf{r}^{\uparrow}_{\mathsf{S}}$, and the EW Jones matrices of the element E, $\mathbf{t}^{\downarrow}_{\mathsf{E}}$, $\mathbf{r}^{\downarrow}_{\mathsf{E}}$, $\mathbf{t}^{\uparrow}_{\mathsf{E}}$, and $\mathbf{r}^{\uparrow}_{\mathsf{E}}$, and returns the EW Jones matrices of the system S + E, $\mathbf{t}^{\downarrow}_{\mathsf{S+E}}$, $\mathbf{r}^{\downarrow}_{\mathsf{S+E}}$, $\mathbf{t}^{\uparrow}_{\mathsf{S+E}}$, and $\mathbf{r}^{\uparrow}_{\mathsf{S+E}}$.

Interface:
CALL AddElementJones(TJF, RJF, TJB, RJB, TJFE, RJFE, TJBE, RJBE, RAO)

Table 10.1 Values of the parameter RAO for different modes of adding

RAO	Mode of adding
0	Ignoring all re-reflections between S and E [see (7.111)]
1	With allowance for only one-fold re-reflections between S and E [see (7.113)]
2	With allowance for all re-reflections between S and E [see (7.110)]

Data types:
TJF, RJF, TJB, RJB, TJFE, RJFE, TJBE, RJBE—(2,2) arrays, complex(8)
RAO—integer(4)

Input and output parameters:
On entry: TJF = $\mathbf{t}_\mathsf{S}^\downarrow$, RJF = $\mathbf{r}_\mathsf{S}^\downarrow$, TJB = $\mathbf{t}_\mathsf{S}^\uparrow$, RJB = $\mathbf{r}_\mathsf{S}^\uparrow$, TJFE = $\mathbf{t}_\mathsf{E}^\downarrow$, RJFE = $\mathbf{r}_\mathsf{E}^\downarrow$, TJBE = $\mathbf{t}_\mathsf{E}^\uparrow$, RJBE = $\mathbf{r}_\mathsf{E}^\uparrow$. The flag RAO (0, 1, or 2) specifies the mode of adding (see Table 10.1).
On exit: TJF = $\mathbf{t}_\mathsf{S+E}^\downarrow$, RJF = $\mathbf{r}_\mathsf{S+E}^\downarrow$, TJB = $\mathbf{t}_\mathsf{S+E}^\uparrow$, RJB = $\mathbf{r}_\mathsf{S+E}^\uparrow$.

Comments. This routine can be used to perform step (8.239) of the algorithm described in Section 8.4.3; in this case, RAO = 2. The input matrices $\mathbf{t}_\mathsf{E}^\downarrow$, $\mathbf{r}_\mathsf{E}^\downarrow$, $\mathbf{t}_\mathsf{E}^\uparrow$, and $\mathbf{r}_\mathsf{E}^\uparrow$ can be calculated with the help of the routine REFTR1DJones.

Subroutine BULK1DJones

This routine is for adding the bulk E of a homogeneous layer to a system S. This routine uses, as input parameters, the EW Jones matrices of the system S, $\mathbf{t}_\mathsf{S}^\downarrow$, $\mathbf{t}_\mathsf{S}^\uparrow$, and $\mathbf{r}_\mathsf{S}^\uparrow$, the array of σ_j values for E, and the thickness h of E and returns the EW Jones matrices of the system S + E, $\mathbf{t}_\mathsf{S+E}^\downarrow$, $\mathbf{t}_\mathsf{S+E}^\uparrow$, and $\mathbf{r}_\mathsf{S+E}^\uparrow$ (the matrix $\mathbf{r}_\mathsf{S+E}^\downarrow$ is not calculated because here $\mathbf{r}_\mathsf{S+E}^\downarrow = \mathbf{r}_\mathsf{S}^\downarrow$). This routine is mainly intended for performing step (8.240) of the algorithm described in Section 8.4.3.

Interface:
CALL BULK1DJones(TJF, TJB, RJB, X, H)

Data types:
TJF, TJB, RJB—(2,2) arrays, complex(8)
X—(4) array, complex(8)
H—real(8)

Input and output parameters:
On entry: TJF = $\mathbf{t}_\mathsf{S}^\downarrow$, TJB = $\mathbf{t}_\mathsf{S}^\uparrow$, RJB = $\mathbf{r}_\mathsf{S}^\uparrow$; X is the array of σ_j values (see Table 9.1); H = h (in μm). On exit: TJF = $\mathbf{t}_\mathsf{S+E}^\downarrow$, TJB = $\mathbf{t}_\mathsf{S+E}^\uparrow$, RJB = $\mathbf{r}_\mathsf{S+E}^\uparrow$. This routine uses the global variable W [see (10.69)].

10.3.2 Routines for Computing EW Mueller Matrices

Subroutine TRJMC

The routine TRJMC calculates from a Jones matrix **t** the corresponding Mueller–Jones matrix $\mathbf{M} = \tilde{\mathbf{T}}\{\mathbf{t}\}$ [see (10.14)].

Interface:
CALL TRJMC(TJ, TM)

Data types:
TJ—(2,2) array, complex(8)
TM—(4,4) array, real(8)

Input and output parameters:
On entry, TJ = **t**. On exit, TM = **M**.

Subroutines TRYM1, TRYM2, and REFTR1DMueller

This set of routines is similar to the set (TRYJ1, TRYJ2, REFTR1DJones), but in this case the EW Mueller matrices rather than EW Jones matrices are computed. These routines are intended for calculating the transmission and reflection EW Mueller matrices, $\mathcal{T}^{\downarrow}$, $\mathcal{R}^{\downarrow}$, $\mathcal{T}^{\uparrow}$, and $\mathcal{R}^{\uparrow}$, of a layered system or a fragment (an OTR unit) of a layered system from the EW 4×4 transfer matrix of this system or fragment, **T**, in the monochromatic approximation.

Interfaces:
CALL TRYM1(T, TF, RF)
CALL TRYM2(T, TB, RB)
CALL REFTR1DMueller(T, TF, RF, TB, RB)

Data types:
T—(4,4) array, complex(8)
TF, RF, TB, RB—(4,4) array, real(8)

Input and output parameters:
On entry, T = **T**. On exit, TF = $\mathcal{T}^{\downarrow}$, RF = $\mathcal{R}^{\downarrow}$, TB = $\mathcal{T}^{\uparrow}$, RB = $\mathcal{R}^{\uparrow}$.

Routines of the 8 × 8 Transfer Matrix Method

Subroutine TR48C

This routine calculates from the EW 4×4 transfer matrix of a **D**-fragment, **T**, the 8×8 transfer matrix $\mathbf{D} = \widehat{\mathbf{D}}\{\mathbf{T}\}$ of this fragment [see (10.61)].

Interface:
CALL TR48C(T, D)

Data types:
T—(4,4) array, complex(8)
D—(8,8) array, real(8)

Input and output parameters:
On entry, T=**T**. On exit, D=**D**.

Subroutines REFTR1 and REFTR2

These routines calculate from the 8×8 transfer matrix of a layered system, **D**, the transmission and reflection EW Mueller matrices, $\mathcal{T}^{\downarrow}$, $\mathcal{R}^{\downarrow}$, $\mathcal{T}^{\uparrow}$, and $\mathcal{R}^{\uparrow}$, of this system [see (10.63)].

Interfaces:
CALL REFTR1(D, TF, RF)
CALL REFTR2(D, TB, RB)

Data types:
D—(8,8) array, real(8)
TF, RF, TB, RB—(4,4) arrays, real(8)

Input and output parameters:
On entry, D=**D**. On exit, TF=$\mathcal{T}^{\downarrow}$, RF=$\mathcal{R}^{\downarrow}$, TB=$\mathcal{T}^{\uparrow}$, RB=$\mathcal{R}^{\uparrow}$.

Subroutine REF1

This routine calculates the reflection EW Mueller matrix $\mathcal{R}^{\downarrow}_{\mathsf{A+R}}$ for the system consisting of a layered system A and a reflecting system R from the 8 × 8 transfer matrix $\mathbf{D}_{\mathsf{A}}$ of the system A and the reflection EW Mueller matrix $\mathcal{R}^{\downarrow}_{\mathsf{R}}$ of the system R. The calculations are carried out according to formula (7.86).

Interface:
CALL REF1(DA, RFR, RFAR)

Data types:
DA—(8,8) array, real(8)
RFR, RFAR—(4,4) arrays, real(8)

Input and output parameters:
On entry, D = $\mathbf{D}_{\mathsf{A}}$, RFR = $\mathcal{R}^{\downarrow}_{\mathsf{R}}$. On exit, RFAR=$\mathcal{R}^{\downarrow}_{\mathsf{A+R}}$.

Routines for Calculating the EW Mueller Matrices of Layered Systems by the Adding Technique (EW-Mueller-Matrix Adding Method)

Subroutine AddElementMueller

This routine is for adding a new element E to a system S. This routine uses the EW Mueller matrices of the system S, $\mathcal{T}^{\downarrow}_{\mathsf{S}}$, $\mathcal{R}^{\downarrow}_{\mathsf{S}}$, $\mathcal{T}^{\uparrow}_{\mathsf{S}}$, and $\mathcal{R}^{\uparrow}_{\mathsf{S}}$, and the EW Mueller matrices of the element E, $\mathcal{T}^{\downarrow}_{\mathsf{E}}$, $\mathcal{R}^{\downarrow}_{\mathsf{E}}$, $\mathcal{T}^{\uparrow}_{\mathsf{E}}$, and $\mathcal{R}^{\uparrow}_{\mathsf{E}}$, to calculate the EW Mueller matrices of the system S+E, $\mathcal{T}^{\downarrow}_{\mathsf{S+E}}$, $\mathcal{R}^{\downarrow}_{\mathsf{S+E}}$, $\mathcal{T}^{\uparrow}_{\mathsf{S+E}}$, and $\mathcal{R}^{\uparrow}_{\mathsf{S+E}}$.

Interface:
CALL AddElementMueller(TF, RF, TB, RB, TFE, RFE, TBE, RBE, RAO)

Data types:
TF, RF, TB, RB, TFE, RFE, TBE, RBE—(4,4) arrays, real(8)
RAO—integer(4)

Input and output parameters:
On entry: TF=$\mathcal{T}^{\downarrow}_{\mathsf{S}}$, RF=$\mathcal{R}^{\downarrow}_{\mathsf{S}}$, TB=$\mathcal{T}^{\uparrow}_{\mathsf{S}}$, RB=$\mathcal{R}^{\uparrow}_{\mathsf{S}}$, TFE=$\mathcal{T}^{\downarrow}_{\mathsf{E}}$, RFE=$\mathcal{R}^{\downarrow}_{\mathsf{E}}$, TBE=$\mathcal{T}^{\uparrow}_{\mathsf{E}}$, RBE=$\mathcal{R}^{\uparrow}_{\mathsf{E}}$. The input parameter RAO (0, 1, or 2) specifies the mode of adding (see Table 10.1).
On exit: TF=$\mathcal{T}^{\downarrow}_{\mathsf{S+E}}$, RF=$\mathcal{R}^{\downarrow}_{\mathsf{S+E}}$, TB=$\mathcal{T}^{\uparrow}_{\mathsf{S+E}}$, RB=$\mathcal{R}^{\uparrow}_{\mathsf{S+E}}$.

Subroutine AddNRElementMueller

This routine is for adding a new element E with negligible reflection to a system S. This routine uses, as input parameters, the EW Mueller matrices of the system S, $\mathcal{T}^{\downarrow}_{\mathsf{S}}$, $\mathcal{T}^{\uparrow}_{\mathsf{S}}$, and $\mathcal{R}^{\uparrow}_{\mathsf{S}}$, and the transmission EW Mueller matrices of the element E, $\mathcal{T}^{\downarrow}_{\mathsf{E}}$ and $\mathcal{T}^{\uparrow}_{\mathsf{E}}$, and calculates the EW Mueller matrices of the system S+E, $\mathcal{T}^{\downarrow}_{\mathsf{S+E}}$, $\mathcal{T}^{\uparrow}_{\mathsf{S+E}}$, and $\mathcal{R}^{\uparrow}_{\mathsf{S+E}}$. The matrices $\mathcal{R}^{\downarrow}_{\mathsf{S}}$ and $\mathcal{R}^{\downarrow}_{\mathsf{S+E}}$ are not involved in the calculations because in the case under consideration $\mathcal{R}^{\downarrow}_{\mathsf{S+E}} = \mathcal{R}^{\downarrow}_{\mathsf{S}}$.

Interface:
CALL AddNRElementMueller(TF, TB, RB, TFE, TBE)

Data types:
TF, TB, RB, TFE, TBE—(4,4) arrays, real(8)

Input and output parameters:
On entry: TF = $\mathcal{T}^{\downarrow}_{\mathsf{S}}$, TB = $\mathcal{T}^{\uparrow}_{\mathsf{S}}$, RB = $\mathcal{R}^{\downarrow}_{\mathsf{S}}$, TFE = $\mathcal{T}^{\downarrow}_{\mathsf{E}}$, TBE = $\mathcal{T}^{\uparrow}_{\mathsf{E}}$.
On exit: TF = $\mathcal{T}^{\downarrow}_{\mathsf{S+E}}$, TB = $\mathcal{T}^{\uparrow}_{\mathsf{S+E}}$, RB = $\mathcal{R}^{\uparrow}_{\mathsf{S+E}}$.

Subroutine AddMirrorMueller

This routine calculates the reflection EW Mueller matrix $\mathcal{R}^{\downarrow}_{\mathsf{A+R}}$ for the system consisting of a layered system A and a reflecting system R, using the EW Mueller matrices of the system A, $\mathcal{T}^{\downarrow}_{\mathsf{A}}$, $\mathcal{R}^{\downarrow}_{\mathsf{A}}$, $\mathcal{T}^{\uparrow}_{\mathsf{A}}$, and $\mathcal{R}^{\uparrow}_{\mathsf{A}}$, and the reflection EW Mueller matrix $\mathcal{R}^{\downarrow}_{\mathsf{R}}$ of the system R.

Interface:
CALL AddMirrorMueller(TFA, RFA, TBA, RBA, RFR, RFAR, RAO)

Data types:
TFA, RFA, TBA, RBA, RFR, RFAR—(4,4) arrays, real(8)
RAO—integer(4)

Input and output parameters:
On entry: TFA = $\mathcal{T}^{\downarrow}_{\mathsf{A}}$, RFA = $\mathcal{R}^{\downarrow}_{\mathsf{A}}$, TBA = $\mathcal{T}^{\uparrow}_{\mathsf{A}}$, RBA = $\mathcal{R}^{\uparrow}_{\mathsf{A}}$, RFR = $\mathcal{R}^{\downarrow}_{\mathsf{R}}$. For this routine, the flag RAO specifying the mode of adding can be set to 0, 1, 2, or 3. The modes corresponding RAO = 0, 1, 2 are indicated in Table 10.1. If RAO = 3, the matrix $\mathcal{R}^{\downarrow}_{\mathsf{A+R}}$ is calculated by the formula

$$\mathcal{R}^{\downarrow}_{\mathsf{A+R}} = \mathcal{T}^{\uparrow}_{\mathsf{A}}\mathcal{R}^{\downarrow}_{\mathsf{R}}\mathcal{T}^{\downarrow}_{\mathsf{A}}, \tag{10.70}$$

that is, ignoring all reflections except for the reflection from R.
On exit: RFAR=$\mathcal{R}^{\downarrow}_{\mathsf{A+R}}$.

Comments. The mode RAO = 3 can be used, for instance, when it is desired to estimate the EW Mueller matrix of the "useful" channel of a reflective LCD (see the example for an RLCD in Section 10.2).

10.3.3 Other Useful Routines

Subroutine TRANS

For a given 2 × 2 matrix $\mathbf{t} = [t_{ij}]$, this routine calculates the quantity

$$t = \frac{1}{2}\,\|\mathbf{t}\|^2_{\mathrm{E}} = \frac{1}{2}\left(t_{11}t^*_{11} + t_{12}t^*_{12} + t_{21}t^*_{21} + t_{22}t^*_{22}\right). \tag{10.71}$$

It can be used to calculate the transmissivity or reflectivity of a system or the transmissivity of a transfer channel for unpolarized incident light in the monochromatic approximation from the corresponding Jones matrix [see (8.278)].

Interface:
CALL TRANS(TJ, T)

Data types:
TJ—(2,2) array, complex(8)
T—real(8)

Input and output parameters:
On entry, TJ=**t**. On exit, T=t.

Subroutine MMID

This routine uses the EW Mueller matrix $\mathbf{M}\{\mathbf{O}\}$ of an operation $\mathbf{O}$ to calculate the EW Mueller matrix $\mathbf{M}\{\mathbf{O}^R\}$ for the reversed operation $\mathbf{O}^R$ (see Section 8.6.2) provided that the eigenwave bases are chosen so that the EW Jones matrices $\mathbf{t}\{\mathbf{O}\}$ and $\mathbf{t}\{\mathbf{O}^R\}$ satisfy the relation

$$\begin{pmatrix} t_{11}\{\mathbf{O}^R\} & t_{12}\{\mathbf{O}^R\} \\ t_{21}\{\mathbf{O}^R\} & t_{22}\{\mathbf{O}^R\} \end{pmatrix} = \begin{pmatrix} t_{11}\{\mathbf{O}\} & -t_{21}\{\mathbf{O}\} \\ -t_{12}\{\mathbf{O}\} & t_{22}\{\mathbf{O}\} \end{pmatrix} \tag{10.72}$$

(see the discussion of the reciprocity matrices in Section 8.6.2).

Interface:
CALL MMID(M, MR)

Data types:
M, MR—(4,4) arrays, real(8)

Input and output parameters:
On entry, M = $\mathbf{M}\{\mathbf{O}\}$. On exit, MR = $\mathbf{M}\{\mathbf{O}^R\}$.

References

[1] Š. Višňovský and R. Krishnan, "Complex Faraday effect in multilayer structures," *J. Opt. Soc. Am.* **71**, 315 (1981).
[2] L. Harris, J. K. Beasley, and A. L. Loeb, "Reflection and transmission of radiation by metal films and the influence of nonabsorbing backings," *J. Opt. Soc. Am.* **41**, 604 (1951).
[3] B. Harbecke, "Coherent and incoherent reflection and transmission of multilayer structures," *Appl. Phys. B* **39**, 165 (1986).
[4] D. A. Yakovlev, "Method of calculating Mueller matrices of reflection and transmission of quasi-monochromatic light by a planar structure consisting of thin and thick layers," *Opt. Spectr.* **64**, 604 (1988) (in Russian).
[5] D. A. Yakovlev, V. G. Chigrinov, and H. S. Kwok, "The novel 8 × 8 transfer matrix method of the LC optics simulations: application to reflective LCDs", *Mol. Cryst. Liq. Cryst.* **366**, 327 (2001).
[6] M. Born and E. Wolf, *Principles of Optics*, 7th ed. (Pergamon Press, New York, 1999).
[7] D. A. Yakovlev and V. G. Chigrinov, "A robust polarization-spectral method for determination of twisted liquid crystal layer parameters," *J. Appl. Phys.* **102**, 023510 (2007).

11

Calculation of Transmission Characteristics of Inhomogeneous Liquid Crystal Layers with Negligible Bulk Reflection

This chapter is devoted to instruments developed within the frameworks of the classical Jones matrix method and the EW Jones matrix method for modeling and analysis of the transmission properties of smoothly inhomogeneous anisotropic media and, in particular, inhomogeneous liquid crystal layers. In the classical Jones's method, the basic instrument for such purposes is the differential calculus [1]. In Section 11.1.1, we present basic equations, general computational techniques, and analytical solutions that are used when modeling and analysis of the optical properties of inhomogeneous LC layers are performed within the framework of this method. As has been noted, the Jones calculus (JC) often gives quite satisfactory results; however, it is applicable only in the case of normal incidence. Furthermore, the differential JC is in good agreement with electromagnetic theory only for media with very weak birefringence. Application of this approach in other situations implies an artificial interpretation of Jones vectors (see Section 1.4.5). More general approaches are offered by the extended Jones matrix method (EJMM) variants [2–6]. Two most popular variants of EJMM [2,3] are discussed in Section 11.1.2. The methods [2,3] are more closely related to electromagnetic theory than JC and cover the case of oblique incidence. Nevertheless, the basic approaches used in these methods for treatment of inhomogeneous LC layers also have serious limitations. In particular, they do not guarantee attaining good accuracy for media with a relatively large difference of the principal refractive indices, typical of practical LC materials. As will be shown, even in the case of normal incidence, the absolute error in calculated values of transmittances of a nematic layer may be of the order of $|n_{\|} - n_{\perp}|/n_{\|}$, where $n_{\|}$ and $n_{\perp}$ are the principal refractive indices of the LC. For many LC materials used in LCDs, the ratio $|n_{\|} - n_{\perp}|/n_{\|}$ is of the order of 0.1 and larger.

Significantly more accurate estimates of the transmission characteristics of LC layers, at nearly the same computational cost as with [2,3], may be obtained using the methods [7–9] which are based on the use of the negligible-bulk-reflection approximation (NBR approximation, NBRA) [10–13]. It should be noted that NBRA is used explicitly or implicitly in all variants of the Jones matrix method considered in this chapter, including the methods [2,3], but in [2,3] this approximation is used in combination with other

Modeling and Optimization of LCD Optical Performance, First Edition.
Dmitry A. Yakovlev, Vladimir G. Chigrinov and Hoi-Sing Kwok.

Website Companion: www.wiley.com/go/yakovlev/modelinglcd

approximations which imply low birefringence of the medium (the small-birefringence approximation, SBA). The methods [7–9] use NBRA only.

Our special interest in NBRA is determined by the fact that application of this approximation is fully justified in considering most types of LCDs. One of the principal advantages of the theory of NBRA [7–10] is that it successfully combines high accuracy with mathematical simplicity. As we will see, this theory is very similar mathematically to the differential Jones calculus and gives a simple solution in any situation where a simple solution can be obtained using the classical JC. In Section 11.2, we present the basic differential equations for state vectors and transmission operators in the NBR approximation. Section 11.3 is devoted to numerical methods of calculating the transmission operators of inhomogeneous layers in this approximation. In Section 11.4, we consider analytical solutions for the transmission operators of inhomogeneous layers provided by the NBRA theory.

11.1 Application of Jones Matrix Methods to Inhomogeneous LC Layers

11.1.1 Calculation of Transmission Jones Matrices of LC Layers Using the Classical Jones Calculus

We begin by considering several representations used for calculating transmission Jones matrices of inhomogeneous LC layers within the framework of the classical Jones matrix method.

Let the LC layer be an element of a layered system like that shown in Figure 11.1. Let z be a coordinate along a normal to the interfaces, and let the boundaries of the LC layer with the adjacent layers (the glass plates in the system shown in Figure 11.1) coincide with the planes $z = z_1$ and $z = z_2$ ($z_2 > z_1$) (see Figure 11.1). The LC is assumed to be optically uniaxial and to have spatially independent principal refractive indices. The LC director field $\mathbf{n}(\mathbf{r})$ is assumed to be uniform over any plane perpendicular to the z-axis (i.e., 2D-homogeneous). The dependence of $\mathbf{n}$—the local optic axis vector $\mathbf{c}$ (see Section 9.3) is assumed to coincide with $\mathbf{n}$—on z will be represented as

$$\mathbf{n}(z) = \begin{pmatrix} \cos\theta(z)\cos\varphi(z) \\ \cos\theta(z)\sin\varphi(z) \\ \sin\theta(z) \end{pmatrix}_{XYZ}, \tag{11.1}$$

where θ and φ are respectively the tilt angle and azimuthal angle of the director in the reference frame (X, Y, Z), attached to the layered system [see Figures 9.1 and 9.3 ($\mathbf{c} = \mathbf{n}$)]. The following symbols will be used: $\theta_1 = \theta(z')$, $\theta_2 = \theta(z'')$, $\varphi_1 = \varphi(z')$, and $\varphi_2 = \varphi(z'')$, where $z' = z_1 + 0$ and $z'' = z_2 - 0$ are the z-coordinates of the outer planes of the bulk of the LC layer. The twist angle is denoted by Φ ($\Phi = \varphi_2 - \varphi_1$),

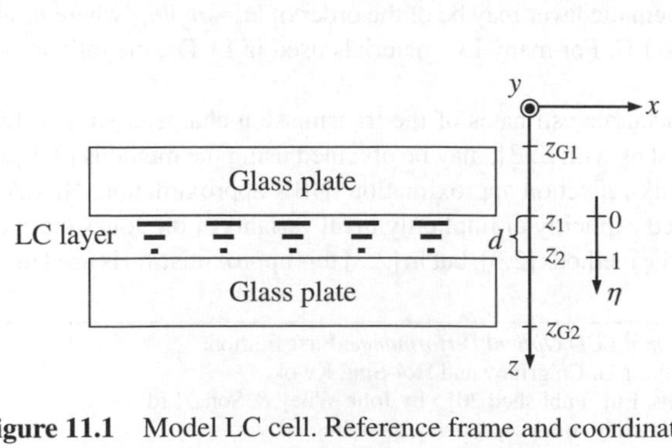

Figure 11.1 Model LC cell. Reference frame and coordinates

and the thickness of the LC layer by d ($d = z_2 - z_1$). Along with the coordinate z, we will use a normalized spatial coordinate

$$\eta = (z - z')/(z'' - z') \tag{11.2}$$

(see Figure 11.1).

Let a plane monochromatic wave fall on the system normally, in the positive z-direction. Consider the problem of finding the Jones matrix $\mathbf{t}_{\text{JL}}$ of the LC layer defined by the relation

$$\mathbf{J}_{x-y}(z'') = \mathbf{t}_{\text{JL}}\mathbf{J}_{x-y}(z'), \tag{11.3}$$

where $\mathbf{J}_{x-y}$ is a Cartesian Jones vector referred to the basis (x, y) with the axes x and y directed along the axes X and Y, respectively. The vector $\mathbf{J}_{x-y}(z')$ describes the light entering the LC layer. $\mathbf{J}_{x-y}(z'')$ is the Jones vector of the transmitted light at the exit plane of the LC layer.

One of the general representations of the matrix $\mathbf{t}_{\text{JL}}$ offered by the Jones calculus is

$$\mathbf{t}_{\text{JL}} = \lim_{N\to\infty} \mathbf{t}_{\text{JL}}^{(N)}, \tag{11.4a}$$

$$\mathbf{t}_{\text{JL}}^{(N)} = \mathbf{t}_{\text{JS}}(z_{SN}, h)\mathbf{t}_{\text{JS}}(z_{SN-1}, h)\ldots\mathbf{t}_{\text{JS}}(z_{S2}, h)\mathbf{t}_{\text{JS}}(z_{S1}, h), \tag{11.4b}$$

where $\mathbf{t}_{\text{JS}}(z_{Sj}, h)$ is the transmission Jones matrix of the jth sublayer of a staircase model of the LC layer (z_{Sj} is the z-coordinate of the midplane of this sublayer and h is its thickness). The matrix $\mathbf{t}_{\text{JS}}(z_{Sj}, h)$ is expressed as

$$\mathbf{t}_{\text{JS}}(z_{Sj}, h) = \widehat{R}_{\text{C}}(-\varphi(z_{Sj}))\mathbf{t}_{\text{JD}}(z_{Sj}, h)\widehat{R}_{\text{C}}(\varphi(z_{Sj})), \tag{11.5}$$

where

$$\mathbf{t}_{\text{JD}}(z_{Sj}, h) = \begin{pmatrix} \exp(ik_0\sigma_1(z_{Sj})h) & 0 \\ 0 & \exp(ik_0\sigma_2(z_{Sj})h) \end{pmatrix}, \tag{11.6a}$$

$$\widehat{R}_{\text{C}}(\alpha) = \begin{pmatrix} \cos\alpha & \sin\alpha \\ -\sin\alpha & \cos\alpha \end{pmatrix}, \tag{11.6b}$$

$$\sigma_1(z) = \frac{n_{\|}n_{\perp}}{\sqrt{n_{\perp}^2\cos^2\theta(z) + n_{\|}^2\sin^2\theta(z)}}, \tag{11.6c}$$

$$\sigma_2(z) = n_{\perp}, \tag{11.6d}$$

$$h = d/N, \quad z_{Sj} = z_1 + (j - 0.5)h. \tag{11.6e}$$

The representation (11.4)–(11.6) is applicable to both nonabsorbing and absorbing media. In the latter case, the principal refractive indices $n_{\|}$ and $n_{\perp}$ are complex.

Another convenient representation of the matrix $\mathbf{t}_{\text{JL}}^{(N)}$, mathematically equivalent to (11.4b), is

$$\begin{aligned}\mathbf{t}_{\text{JL}}^{(N)} = {}& \widehat{R}_{\text{C}}(-\varphi(z_{SN}))\mathbf{t}_{\text{JD}}(z_{SN}, h)\widehat{R}_{\text{C}}(\Delta\varphi_N)\mathbf{t}_{\text{JD}}(z_{SN-1}, h) \\ & \ldots\widehat{R}_{\text{C}}(\Delta\varphi_3)\mathbf{t}_{\text{JD}}(z_{S2}, h)\widehat{R}_{\text{C}}(\Delta\varphi_2)\mathbf{t}_{\text{JD}}(z_{S1}, h)\widehat{R}_{\text{C}}(\varphi(z_{S1})),\end{aligned} \tag{11.7}$$

where

$$\Delta\varphi_j = \varphi(z_{Sj}) - \varphi(z_{Sj-1}).$$

Unimodular Representation

In considering nonabsorbing LC layers, the following representation is often used:

$$\mathbf{t}_{\mathrm{JL}} = C_{\mathrm{AP}} \lim_{N\to\infty} \mathbf{t}_{\mathrm{JUL}}^{(N)}, \tag{11.8a}$$

$$\mathbf{t}_{\mathrm{JUL}}^{(N)} = \mathbf{t}_{\mathrm{JU}}(z_{SN}, h)\mathbf{t}_{\mathrm{JU}}(z_{SN-1}, h)\ldots\mathbf{t}_{\mathrm{JU}}(z_{S2}, h)\mathbf{t}_{\mathrm{JU}}(z_{S1}, h), \tag{11.8b}$$

where

$$\mathbf{t}_{\mathrm{JU}}(z_{Sj}, h) = \widehat{R}_{\mathrm{C}}(-\varphi(z_{Sj}))\mathbf{t}_{\mathrm{JDU}}(z_{Sj}, h)\widehat{R}_{\mathrm{C}}(\varphi(z_{Sj})), \tag{11.9a}$$

$$\mathbf{t}_{\mathrm{JDU}}(z_{Sj}, h) = \begin{pmatrix} \exp\left(i\frac{\pi}{\lambda}\Delta\sigma(z_{Sj})h\right) & 0 \\ 0 & \exp\left(-i\frac{\pi}{\lambda}\Delta\sigma(z_{Sj})h\right) \end{pmatrix}, \quad \Delta\sigma = \sigma_1 - \sigma_2, \tag{11.9b}$$

$$C_{\mathrm{AP}} = \exp\left(i\frac{\pi d}{\lambda}\int_0^1 [\sigma_1(\eta) + \sigma_2(\eta)]\mathrm{d}\eta\right). \tag{11.9c}$$

If the medium is nonabsorbing, the factor C_{AP} is usually omitted (see Sections 1.4.5 and 5.4.3). The matrix

$$\mathbf{t}_{\mathrm{JUL}} = \lim_{N\to\infty} \mathbf{t}_{\mathrm{JUL}}^{(N)} = \frac{1}{C_{\mathrm{AP}}}\mathbf{t}_{\mathrm{JL}}, \tag{11.10}$$

which is considered as the Jones matrix of the LC layer in this case, is unitary and unimodular and has the form

$$\begin{pmatrix} a & b \\ -b^* & a^* \end{pmatrix}, \tag{11.11}$$

where a and b are complex numbers such that $a^*a + b^*b = 1$, since all matrices whose product is $\mathbf{t}_{\mathrm{JUL}}$ are unitary and unimodular and have the form (11.11) (see Section 5.1.3).

Differential Equations for the Jones Vectors and Matrices in a Fixed Reference System

In the differential Jones calculus, the spatial evolution of a Jones vector $\mathbf{J}$ in a layer with continuously varying parameters is described by a differential equation of the form

$$\frac{\mathrm{d}\mathbf{J}}{\mathrm{d}z} = \mathbf{N}_{\mathrm{J}}\mathbf{J} \tag{11.12}$$

[1]. The matrix $\mathbf{N}_\mathrm{J}(\xi)$, where ξ is an arbitrary value of the variable z inside the layer, may be expressed as follows:

$$\mathbf{N}_\mathrm{J}(\xi) = \lim_{h\to 0} \frac{\mathbf{t}_\mathrm{J}\left(\xi - \frac{h}{2}, \xi + \frac{h}{2}\right) - \mathbf{U}}{h}, \tag{11.13}$$

where $\mathbf{t}_\mathrm{J}\,(\xi - h/2, \xi + h/2)$ is the Jones matrix of the region $(\xi - h/2, \xi + h/2)$, such that

$$\mathbf{J}\left(\xi + \frac{h}{2}\right) = \mathbf{t}_\mathrm{J}\left(\xi - \frac{h}{2}, \xi + \frac{h}{2}\right)\mathbf{J}\left(\xi - \frac{h}{2}\right).$$

The general solution of (11.12) may be written as

$$\mathbf{J}(z) = \mathbf{t}_\mathrm{J}(z', z)\mathbf{J}(z'). \tag{11.14}$$

Substituting (11.14) into (11.12) leads to the following operator equation for the function $\mathbf{t}_\mathrm{J}(z', z)$:

$$\frac{\mathrm{d}\mathbf{t}_\mathrm{J}(z', z)}{\mathrm{d}z} = \mathbf{N}_\mathrm{J}\mathbf{t}_\mathrm{J}(z', z) \quad [\mathbf{t}_\mathrm{J}(z', z') = \mathbf{U}]. \tag{11.15}$$

Thus, one may associate a representation of the Jones matrix of a layer with continuously varying parameters in terms of Jones matrices of its infinitesimal sublayers with a certain differential equation, integration of which also gives the Jones matrix of the layer. For instance, the differential equation corresponding to representation (11.4) is

$$\frac{\mathrm{d}\mathbf{t}_{x-y}(z', z)}{\mathrm{d}z} = \mathbf{N}_{x-y}\mathbf{t}_{x-y}(z', z) \quad [\mathbf{t}_{x-y}(z', z') = \mathbf{U}], \tag{11.16}$$

where

$$\mathbf{N}_{x-y} = ik_0\bar{\sigma}\mathbf{U} + ik_0\frac{\Delta\sigma}{2}\begin{pmatrix} \cos 2\varphi & \sin 2\varphi \\ \sin 2\varphi & -\cos 2\varphi \end{pmatrix}, \quad \bar{\sigma} = \frac{\sigma_1 + \sigma_2}{2}; \tag{11.17}$$

the function $\mathbf{t}_{x-y}(z', z)$ is defined by the relation $\mathbf{J}_{x-y}(z) = \mathbf{t}_{x-y}(z', z)\mathbf{J}_{x-y}(z')$. The matrix $\mathbf{t}_\mathrm{JL}$ is equal to $\mathbf{t}_{x-y}(z', z'')$. The corresponding equation for $\mathbf{J}_{x-y}(z)$ is

$$\frac{\mathrm{d}\mathbf{J}_{x-y}}{\mathrm{d}z} = \mathbf{N}_{x-y}\mathbf{J}_{x-y}. \tag{11.18}$$

Expression (11.17) may be obtained in the following way. According to (11.4) and (11.5), as $h \to 0$,

$$\mathbf{t}_{x-y}\left(\xi - \frac{h}{2}, \xi + \frac{h}{2}\right) \simeq \widehat{R}_\mathrm{C}\left(-\varphi(\xi)\right)\mathbf{t}_\mathrm{JD}(\xi, h)\widehat{R}_\mathrm{C}\left(\varphi(\xi)\right)$$

and hence

$$\begin{aligned}\mathbf{t}_{x-y}(\xi - h/2, \xi + h/2) - \mathbf{U} &\simeq \widehat{R}_\mathrm{C}\left(-\varphi(\xi)\right)[\mathbf{t}_\mathrm{JD}(\xi, h) - \mathbf{U}]\widehat{R}_\mathrm{C}\left(\varphi(\xi)\right) \\ &= \widehat{R}_\mathrm{C}(-\varphi(\xi))\begin{pmatrix} \exp(ik_0\sigma_1(\xi)h) - 1 & 0 \\ 0 & \exp(ik_0\sigma_2(\xi)h) - 1 \end{pmatrix}\widehat{R}_\mathrm{C}(\varphi(\xi)).\end{aligned}$$

Substituting this expression into (11.13) yields

$$\mathbf{N}_{x-y} = ik_0 \widehat{R}_{\mathrm{C}}(-\varphi) \begin{pmatrix} \sigma_1 & 0 \\ 0 & \sigma_2 \end{pmatrix} \widehat{R}_{\mathrm{C}}(\varphi). \tag{11.19}$$

Performing matrix multiplications in (11.19), we arrive at (11.17). Due to the specific form of the matrix $\mathbf{N}_{x-y}$ [see (11.17)], the function $\mathbf{t}_{x-y}(z', z)$ can be factorized as follows:

$$\mathbf{t}_{x-y}(z', z) = c_{\mathrm{AP}}(z', z)\, \mathbf{t}_{\mathrm{U}x-y}(z', z), \tag{11.20}$$

where

$$c_{\mathrm{AP}}(z', z) = \exp\left(ik_0 \int_{z'}^{z} \bar{\sigma}(\xi) \mathrm{d}\xi \right) \tag{11.21}$$

is a function satisfying the equation

$$\frac{\mathrm{d}c_{\mathrm{AP}}(z', z)}{\mathrm{d}z} = ik_0 \bar{\sigma} c_{\mathrm{AP}}(z', z),$$

and $\mathbf{t}_{\mathrm{U}x-y}(z', z)$ is the solution of the equation

$$\frac{\mathrm{d}\mathbf{t}_{\mathrm{U}x-y}(z', z)}{\mathrm{d}z} = ik_0 \frac{\Delta\sigma}{2} \begin{pmatrix} \cos 2\varphi & \sin 2\varphi \\ \sin 2\varphi & -\cos 2\varphi \end{pmatrix} \mathbf{t}_{\mathrm{U}x-y}(z', z) \tag{11.22}$$

[$\mathbf{t}_{\mathrm{U}x-y}(z', z') = \mathbf{U}$]. It is obvious that

$$C_{\mathrm{AP}} = c_{\mathrm{AP}}(z', z''), \quad \mathbf{t}_{\mathrm{JUL}} = \mathbf{t}_{\mathrm{U}x-y}(z', z'') \tag{11.23}$$

[see (11.9) and (11.10)].

Differential Equations for the Jones Vectors and Matrices in the Local Proper Reference System

In calculations of transmission Jones matrices for LC layers, one often uses appropriately chosen local reference frames for Jones vectors (see, e.g., Section 2.1). Let us introduce a z-dependent proper reference system $(x'(z), y'(z))$ such that for any fixed z the angle between the axes x and $x'(z)$ is equal to $\varphi(z)$. We denote the column representing a Jones vector expressed in the system (x', y') by $\mathbf{J}_{x'-y'}$. The columns $\mathbf{J}_{x'-y'}$ and $\mathbf{J}_{x-y}$ representing the same Jones vector are related by

$$\mathbf{J}_{x'-y'}(z) = \widehat{R}_{\mathrm{C}}(\varphi(z))\, \mathbf{J}_{x-y}(z) \tag{11.24a}$$

and, conversely,

$$\mathbf{J}_{x-y}(z) = \widehat{R}_{\mathrm{C}}(-\varphi(z))\, \mathbf{J}_{x'-y'}(z). \tag{11.24b}$$

By substituting (11.24b) into (11.18), we obtain

$$\frac{\mathrm{d}\widehat{R}_{\mathrm{C}}(-\varphi)\,\mathbf{J}_{x'-y'}}{\mathrm{d}z} = \frac{\mathrm{d}\widehat{R}_{\mathrm{C}}(-\varphi)}{\mathrm{d}z}\mathbf{J}_{x'-y'} + \widehat{R}_{\mathrm{C}}(-\varphi)\,\frac{\mathrm{d}\mathbf{J}_{x'-y'}}{\mathrm{d}z} = \mathbf{N}_{x-y}\widehat{R}_{\mathrm{C}}(-\varphi)\,\mathbf{J}_{x'-y'}, \tag{11.25}$$

and then

$$\begin{aligned}\frac{\mathrm{d}\mathbf{J}_{x'-y'}}{\mathrm{d}z} &= \left[\widehat{R}_{\mathrm{C}}(\varphi)\,\mathbf{N}_{x-y}\widehat{R}_{\mathrm{C}}(-\varphi) - \widehat{R}_{\mathrm{C}}(\varphi)\,\frac{\mathrm{d}\widehat{R}_{\mathrm{C}}(-\varphi)}{\mathrm{d}z}\right]\mathbf{J}_{x'-y'} \\ &= \left[ik_0\begin{pmatrix}\sigma_1 & 0\\ 0 & \sigma_2\end{pmatrix} + \frac{\mathrm{d}\varphi}{\mathrm{d}z}\begin{pmatrix}0 & 1\\ -1 & 0\end{pmatrix}\right]\mathbf{J}_{x'-y'}.\end{aligned} \tag{11.26}$$

Thus, we have arrived at the following differential equation for the function $\mathbf{J}_{x'-y'}(z)$:

$$\frac{\mathrm{d}\mathbf{J}_{x'-y'}}{\mathrm{d}z} = \mathbf{N}_{x'-y'}\mathbf{J}_{x'-y'}, \tag{11.27}$$

where

$$\mathbf{N}_{x'-y'} = \begin{pmatrix} ik_0\sigma_1 & \varphi_z \\ -\varphi_z & ik_0\sigma_2 \end{pmatrix}, \tag{11.28}$$

$$\varphi_z \equiv \frac{\mathrm{d}\varphi}{\mathrm{d}z}. \tag{11.29}$$

The corresponding equation for the function $\mathbf{t}_{x'-y'}(z', z)$, such that

$$\mathbf{J}_{x'-y'}(z) = \mathbf{t}_{x'-y'}(z', z)\mathbf{J}_{x'-y'}(z') \tag{11.30}$$

[see (11.14)], is

$$\frac{\mathrm{d}\mathbf{t}_{x'-y'}(z', z)}{\mathrm{d}z} = \mathbf{N}_{x'-y'}\mathbf{t}_{x'-y'}(z', z) \quad [\mathbf{t}_{x'-y'}(z', z') = \mathbf{U}]. \tag{11.31}$$

As in the above case, the matrix $\mathbf{N}_{x'-y'}$ may be split into two components [see (11.17)],

$$\mathbf{N}_{x'-y'} = ik_0\bar{\sigma}\mathbf{U} + \mathbf{N}_{\mathrm{U}x'-y'}, \tag{11.32}$$

$$\mathbf{N}_{\mathrm{U}x'-y'} = \begin{pmatrix} i\dfrac{\pi}{\lambda}\Delta\sigma & \varphi_z \\ -\varphi_z & -i\dfrac{\pi}{\lambda}\Delta\sigma \end{pmatrix}, \tag{11.33}$$

and hence the function $\mathbf{t}_{x'-y'}(z', z)$ may also be factorized as follows:

$$\mathbf{t}_{x'-y'}(z', z) = c_{\mathrm{AP}}(z', z)\mathbf{t}_{\mathrm{U}x'-y'}(z', z), \tag{11.34}$$

where $\mathbf{t}_{\mathrm{U}x'-y'}(z', z)$ satisfies the equation

$$\frac{\mathrm{d}\mathbf{t}_{\mathrm{U}x'-y'}(z', z)}{\mathrm{d}z} = \mathbf{N}_{\mathrm{U}x'-y'}\mathbf{t}_{\mathrm{U}x'-y'}(z', z) \quad [\mathbf{t}_{\mathrm{U}x'-y'}(z', z') = \mathbf{U}]; \tag{11.35}$$

the function $c_{\text{AP}}(z', z)$ is defined by (11.21). In the case of a nonabsorbing medium, the matrices $\mathbf{t}_{\text{U}x-y}(z', z)$ [see (11.22)] and $\mathbf{t}_{\text{U}x'-y'}(z', z)$ at any z are unitary and unimodular and have the form (11.11).

In the general case, the matrix $\mathbf{t}_{x'-y'}(z', z'')$ may be represented as follows:

$$\mathbf{t}_{x'-y'}(z', z'') = \lim_{N\to\infty} \mathbf{t}^{(N)}_{\text{JRL1}}, \tag{11.36a}$$

$$\mathbf{t}^{(N)}_{\text{JRL1}} = \mathbf{t}_{\text{JD}}(z_{\text{S}N}, h)\widehat{R}_{\text{C}}(\Delta\varphi_N)\mathbf{t}_{\text{JD}}(z_{\text{S}N-1}, h)$$

$$\ldots \mathbf{t}_{\text{JD}}(z_{\text{S}3}, h)\widehat{R}_{\text{C}}(\Delta\varphi_3)\mathbf{t}_{\text{JD}}(z_{\text{S}2}, h)\widehat{R}_{\text{C}}(\Delta\varphi_2)\mathbf{t}_{\text{JD}}(z_{\text{S}1}, h) \tag{11.36b}$$

[cf. (11.7)]. In this expression, the same staircase model as in representation (11.4) is used. This expression is exact, but a more convenient expression for $\mathbf{t}_{x'-y'}(z', z'')$ in this case is

$$\mathbf{t}_{x'-y'}(z', z'') = \lim_{N\to\infty} \mathbf{t}^{(N)}_{\text{JRL2}}, \tag{11.37a}$$

$$\mathbf{t}^{(N)}_{\text{JRL2}} = \mathbf{t}_{\text{JD}}(z_{\text{C}N}, h/2)\widehat{R}_{\text{C}}(\Delta\bar{\varphi}_N)\mathbf{t}_{\text{JD}}(z_{\text{C}N-1}, h)$$

$$\ldots \mathbf{t}_{\text{JD}}(z_{\text{C}2}, h)\widehat{R}_{\text{C}}(\Delta\bar{\varphi}_2)\mathbf{t}_{\text{JD}}(z_{\text{C}1}, h)\widehat{R}_{\text{C}}(\Delta\bar{\varphi}_1)\mathbf{t}_{\text{JD}}(z_{\text{C}0}, h/2), \tag{11.37b}$$

where

$$\Delta\bar{\varphi}_j = \varphi(z_{\text{C}j}) - \varphi(z_{\text{C}j-1}),$$
$$z_{\text{C}j} = z' + j(z'' - z')/N.$$

Here the outermost nodes of the grid are just at the layer boundaries ($z_{\text{C}0} = z'$ and $z_{\text{C}N} = z''$), and the first and last sublayers are half as thick as the other sublayers [see also (8.331)]. In most cases of practical interest, at large N $\mathbf{t}^{(N)}_{\text{JRL2}}$ approximates $\mathbf{t}_{x'-y'}(z', z'')$ better than $\mathbf{t}^{(N)}_{\text{JRL1}}$.

According to (11.24), the matrix $\mathbf{t}_{x-y}(z', z'')$ may be calculated from $\mathbf{t}_{x'-y'}(z', z'')$ as follows:

$$\mathbf{t}_{x-y}(z', z'') = \widehat{R}_{\text{C}}(-\varphi(z''))\mathbf{t}_{x'-y'}(z', z'')\widehat{R}_{\text{C}}(\varphi(z')). \tag{11.38}$$

Now we recall some well-known exact (within the accuracy of the Jones approach) and approximate analytical expressions for the matrices $\mathbf{t}_{x'-y'}(z', z'')$ of inhomogeneous layers.

Nontwisted Layer

If $\varphi(z) = \varphi_1 = const$, with any $\theta(z)$

$$\mathbf{t}_{x'-y'}(z', z'') = \begin{pmatrix} \exp\left(ik_0 d \int_0^1 \sigma_1(\eta)\mathrm{d}\eta\right) & 0 \\ 0 & \exp(ik_0\sigma_2 d) \end{pmatrix}. \tag{11.39}$$

This expression can easily be obtained from (11.27), (11.36), or (11.37).

Ideal Twisted Layer (General Case)

In the case of an ideal twisted layer, for which, by definition,

$$\theta(z) = \theta_c = const, \tag{11.40a}$$

$$\varphi(z) = \varphi_1 + \Phi(z - z')/d, \tag{11.40b}$$

the matrix $\mathbf{N}_{x'-y'}$ in (11.31) is independent of z and, in a standard way (see Section 2.1 and Appendix B.5), one can obtain an analytical expression for $\mathbf{t}_{x'-y'}(z', z'')$, such as the following one:

$$\mathbf{t}_{x'-y'}(z', z'') = C_{\mathrm{AP}} \begin{pmatrix} \cos Q + i\dfrac{G}{Q}\sin Q & \dfrac{\Phi}{Q}\sin Q \\ -\dfrac{\Phi}{Q}\sin Q & \cos Q - i\dfrac{G}{Q}\sin Q \end{pmatrix}, \tag{11.41}$$

where

$$Q = \sqrt{G^2 + \Phi^2}, \tag{11.42}$$

$$G = \frac{\pi(n_{ec} - n_{\perp})d}{\lambda}, \quad n_{ec} = \frac{n_{\parallel} n_{\perp}}{\sqrt{n_{\perp}^2 \cos^2\theta_c + n_{\parallel}^2 \sin^2\theta_c}}$$

[see also (11.9c), (11.6c), and (11.6d)]. As far as we know, for the first time such an expression for the transmission matrix of a uniformly twisted crystal was obtained by Jones [1] (see Appendix B.5). It should be noted that expression (11.41) is valid not only for nonabsorbing media but also for absorbing ones. In the latter case, Q is complex. Recall that in any case,

$$\cos Q = \frac{\mathrm{e}^{iQ} + \mathrm{e}^{-iQ}}{2}, \quad \sin Q = \frac{\mathrm{e}^{iQ} - \mathrm{e}^{-iQ}}{2i}. \tag{11.43}$$

To the best of our knowledge, there are no other cases where an exact (within the accuracy of the Jones approach) analytical expression for the transmission Jones matrix of a smoothly inhomogeneous layer can be obtained. Now we consider two approximate expressions for Jones matrices of inhomogeneous layers, used in LCD optics.

A Quasi-Planar Twisted Nonabsorbing LC Layer

An ideal twisted layer is a very good model of a nematic or cholesteric layer with zero pretilt angles ($\theta_1 = \theta_2 = 0°$). In practical TN and STN displays, a tilted orientation of the LC director at the boundaries ($\theta_1 \neq 0°, \theta_2 \neq 0°$) is commonly used. When the angles θ_1 and θ_2 are nonzero, the equilibrium configuration of the LC director field in the field-off state generally differs from an ideal one. Usually, when θ_1 and θ_2 are relatively small and equal, the LC director at the center of the layer has a smaller tilt than at the boundaries (see an example in Figure 11.2), and the difference in director tilt at the center of the layer and at its boundaries increases as θ_1 and θ_2 increase, while the dependence of φ on z remains almost linear (Figure 11.2). LC structures with small θ_1 and θ_2 that just slightly differ from ideal ones will be called *quasi-planar*. The transmission characteristics of quasi-planar twisted layers are close to those of ideal twisted layers. Lien [14], by some numerical examples, has shown that the Jones matrix of a quasi-planar twisted LC layer is well approximated by the Jones matrix of an ideal twisted LC layer having the same twist angle and the same principal refractive indices as the quasi-planar layer and the tilt angle θ_c equal to the average tilt angle in the quasi-planar layer. The same or a little better correspondence is achieved when approximate values of $\mathbf{t}_{x'-y'}(z', z'')$ are calculated by formulas (11.41) and (11.42) with

$$G = \frac{\pi(\bar{n}_{\mathrm{e}} - n_{\perp})d}{\lambda}, \tag{11.44}$$

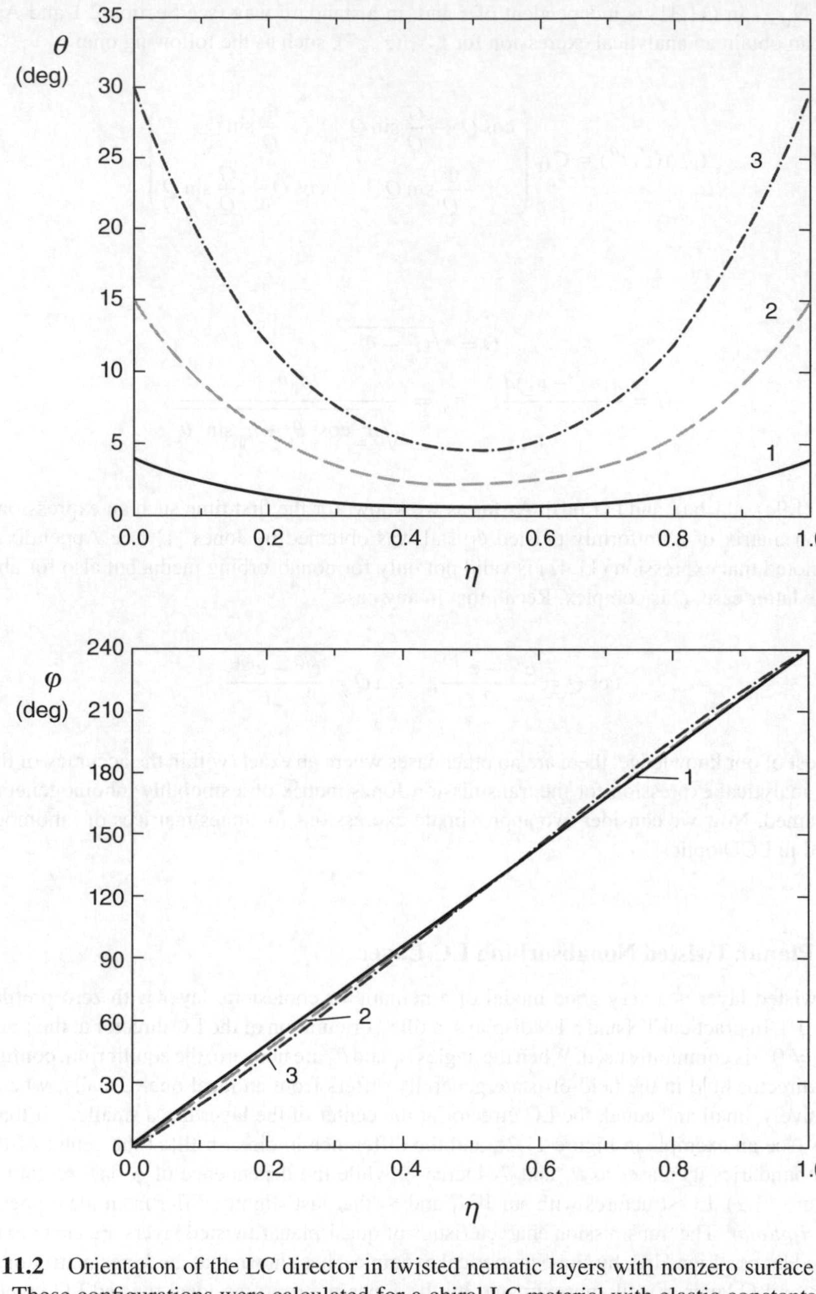

Figure 11.2 Orientation of the LC director in twisted nematic layers with nonzero surface tilt angles (θ_1, θ_2). These configurations were calculated for a chiral LC material with elastic constants $K_{11} = 1.3 \times 10^{-6}$ dyn, $K_{22} = 7.1 \times 10^{-7}$ dyn, and $K_{33} = 1.95 \times 10^{-6}$ dyn (HR-8596) at $d/p_0 = 0.7$, where p_0 is the natural helix pitch, for $\theta_1 = \theta_2 = 4°$ (1), 15° (2), and 30° (3) (infinitely strong anchoring is assumed). The calculations were performed with the MOUSE-LCD program

where

$$\bar{n}_{\mathrm{e}} = \int_{0}^{1} \frac{n_{\|} n_{\perp}}{\sqrt{n_{\perp}^{2} \cos^{2} \theta(\eta) + n_{\|}^{2} \sin^{2} \theta(\eta)}} \mathrm{d}\eta. \quad (11.45)$$

Practically the same accuracy of approximation for quasi-planar layers with $\theta_1, \theta_2 < 15°$ is provided by the following representation:

$$G \approx \frac{\pi \Delta n L_{\theta}}{\lambda}, \quad (11.46)$$

where

$$\Delta n = n_{\|} - n_{\perp}, \quad L_{\theta} = d\langle \cos^2 \theta \rangle, \quad \langle \cos^2 \theta \rangle = \int_{0}^{1} \cos^2 \theta(\eta) \mathrm{d}\eta. \quad (11.47)$$

Here use is made of the fact that for small θ and usual values of the principal refractive indices of LC

$$n_{\mathrm{e}} - n_{\perp} \approx \Delta n \left(\cos^2 \theta - \frac{\Delta n}{2 n_{\perp}} \left(3 + \frac{\Delta n}{n_{\perp}} \right) \sin^2 \theta \right) \approx \Delta n \cos^2 \theta. \quad (11.48)$$

Representation (11.46) is very convenient in solving inverse problems for twisted layers [15] (see Section 12.5).

Adiabatic Approximation (Mauguin Mode)

When the condition

$$\left| \frac{\pi \Delta \sigma}{\lambda} \right| \gg |\varphi_z| \quad (11.49)$$

[see (11.33)] is satisfied at any z in the range z' to z'', the following approximation may be good:

$$\mathbf{t}_{x'-y'}(z', z'') \approx \begin{pmatrix} \exp\left(ik_0 d \int_{0}^{1} \sigma_1(\eta) \mathrm{d}\eta \right) & 0 \\ 0 & \exp(ik_0 \sigma_2 d) \end{pmatrix}. \quad (11.50)$$

This solution describes the so-called Mauguin mode. In this case, the forward (backward) propagating wave field in the layer may be considered as a superposition of two separate linearly polarized waves having spatially dependent vibration directions that are locally coincident with vibration directions of the corresponding basis eigenmodes of the infinitesimal sublayers. This approximation is closely related to the geometrical optics approximation (GOA). GOA and allied, more accurate, approximations are considered in Section 11.4.

Numerical Calculation of Jones Matrices of Inhomogeneous Layers

The most-used approach to calculating Jones matrices of inhomogeneous LC layers in situations where the mentioned simple solutions are inapplicable is based on the use of any of representations (11.4),

(11.7), (11.8), (11.10), (11.36), or (11.37) with approximation of the matrix to be estimated by the matrix of the approximating system of homogeneous sublayers,

$$\mathbf{t}_{\mathrm{JL}} \approx \mathbf{t}_{\mathrm{JL}}^{(N)}, \text{ or } \mathbf{t}_{\mathrm{JUL}} \approx \mathbf{t}_{\mathrm{JUL}}^{(N)}, \text{ or } \mathbf{t}_{x'-y'}(z', z) \approx \mathbf{t}_{\mathrm{JRL1}}^{(N)} \tag{11.51}$$

and so on, with sufficiently large N. We will refer to this approach as the *discretization method* (DM). In a way analogous to that used in Section 8.3.4 when the accuracy of the staircase approximation in the Berreman method was discussed, it is easy to find that the error of approximations of the kind (11.51) may also be estimated as $O(1/N^2)$. The validity of this estimate is quite evident in view of the formal analogy between DM and the staircase approximation as used within the framework of the Berreman method. Some tricks used to reduce computational effort when DM is applied to nonabsorbing layers will be presented in Section 11.3. There are methods for calculation of Jones matrices of smoothly inhomogeneous layers that do not use approximation of the inhomogeneous layer by a system of homogeneous sublayers [14, 16]. For example, in the method [14] the approximating layered system is composed of sublayers with ideal twisted structure. Approximating systems including both ideal twisted and nontwisted sublayers were also used. But, in practical situations, it is difficult to estimate the range of applicability and predict the accuracy of such approximations because of their artificial character.

In the 1980–1990s, in modeling TN and STN LCDs [14] and solving some diffraction problems for LC objects, in particular, in modeling the scattering properties of nematic droplets in PDLC films [17, 18] (in this case, JC is used within the framework of the anomalous diffraction approximation [19, 20]), as an alternative to the relatively time-consuming DM, a technique based on the representation of the matrix $\mathbf{t}_{\mathrm{JUL}}$ [see (11.10)] by a power series in a parameter proportional to d/λ was successfully used. This technique is much more efficient than DM when it is desired to calculate the dependence $\mathbf{t}_{\mathrm{JUL}}$ on λ or on d, assuming that the functions $\theta(\eta)$ and $\varphi(\eta)$ are independent of d. The *power series method* (PSM) is a special technique for solving differential equations of the form (11.22) with a real-valued $\Delta\sigma(z)$ (nonabsorbing media). The underlying mathematics of this method is presented in Appendix B.4. Computational aspects of PSM are discussed in Section 11.3.3. Note that PSM does not require introducing any intermediate model of the inhomogeneous medium under consideration at all.

11.1.2 Extended Jones Matrix Methods

Many attempts were made to create a matrix formalism similar to the Jones calculus but applicable in the case of oblique incidence and more consistent with electromagnetic theory. As a result, several variants of the extended Jones matrix method (EJMM) were developed [2–6]. As may be seen from the literature, the most commonly used variants of EJMM in LCD modeling are that developed by Gu and Yeh [2] and that proposed by Lien [3]. Both these methods use the rigorous eigenwave representation (see Chapter 8) and the staircase approximation when treating inhomogeneous LC layers. The transmission operators for interfaces between anisotropic layers (in particular, the interfaces between sublayers of the layered system approximating the LC layer) are calculated approximately. Gu and Yeh [2, 21] used SBA (see the subsection about SBA in Section 9.3) and introduced an intermediate zero-thickness fictitious isotropic layer to obtain their approximate expression for transmission operators of such interfaces. We have estimated the accuracy of this expression at the end of Section 8.4.2. Lien's expression for interface transmission operators that is used in the method [3, 22] for all interfaces in the model LCD except for the external interfaces with air was derived from the boundary condition for the electric field on the assumption that reflected waves are absent. In many cases these two methods give rather good results (see Section 12.2). In many other cases the accuracy of these methods turns out far from satisfactory. The following examples allow one to see some features of these methods and estimate the level of their accuracy from the standpoint of electromagnetic theory.

Let us consider a nonabsorbing LC layer with nontwisted ($\varphi(z)$ = const) homeoplanar structure (Figure 11.3a). Let $\theta_1 = \theta(z') = 90°$ and $\theta_2 = \theta(z'') = 0°$, that is, the tilt angle θ decreases monotonically

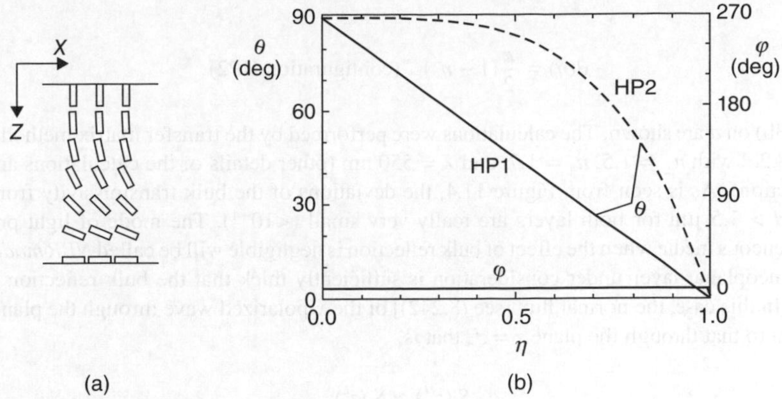

Figure 11.3 (a) Homeoplanar layer; (b) homeoplanar configurations HP1 and HP2 used in numerical examples of this chapter

with z from 90° to 0°. Also we assume that the layer is oriented so that $\varphi = 0°$. As before, the light is assumed to be incident on the layer in the normal direction. In the chosen geometry, light linearly polarized along the y-axis will pass through the layer bulk without any losses, propagating with a constant refractive index, equal to $n_\perp$. We consider an alternative situation when the light entering the layer is polarized along the x-axis. In this case, the light, traveling through the layer, suffers losses due to bulk reflection, since the refractive index for it changes with z from $n_\perp$ to $n_\parallel$. For usual LC materials at typical values of d, these losses are very small and the bulk transmissivity of the layer is very close to unity. As an illustration, in Figure 11.4, accurately calculated dependences of the bulk transmissivity for layers with

$$\theta(\eta) = \frac{\pi}{2}(1 - \eta) \quad \text{(configuration HP1)} \tag{11.52}$$

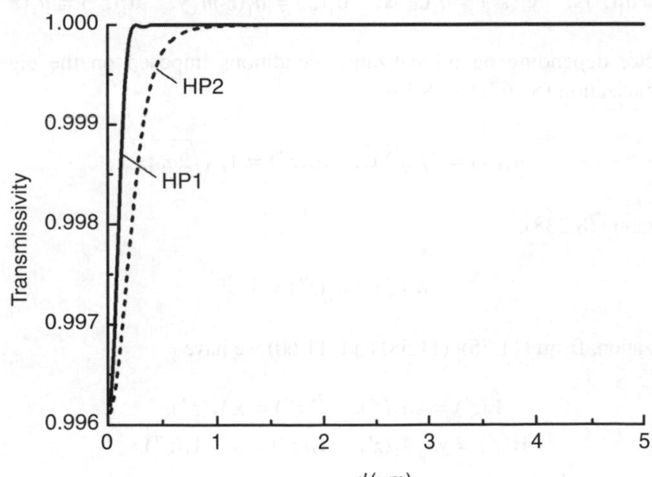

Figure 11.4 Transmissivities of the bulks of homeoplanar layers with configurations HP1 and HP2 for a normally propagating extraordinary wave

and

$$\theta(\eta) = \frac{\pi}{2}(1 - \eta^4) \quad \text{(configuration HP2)} \tag{11.53}$$

(Figure 11.3b) on d are shown. The calculations were performed by the transfer matrix method described in Section 8.2.2 with $n_\perp = 1.5$, $n_{||} = 1.7$, and $\lambda = 550$ nm (other details of the calculations are given in the next section). As is seen from Figure 11.4, the deviations of the bulk transmissivity from unity in the region $d > 1.5$ μm for both layers are really very small ($<10^{-4}$). The mode of light propagation in inhomogeneous media when the effect of bulk reflection is negligible will be called *NBR mode*. Assume that the homeoplanar layer under consideration is sufficiently thick that the bulk reflection losses are very small. In this case, the normal flux [see (8.242)] of the x-polarized wave through the plane $z = z''$ is almost equal to that through the plane $z = z'$, that is,

$$S_z(z'') \approx S_z(z'). \tag{11.54}$$

From (11.54) and (8.242), we obtain the following relation for the fields of the wave at $z = z'$ and $z = z''$:

$$\mathrm{Re}(\mathbf{z}(\tilde{\mathbf{E}}(z'') \times \tilde{\mathbf{H}}(z'')^*)) \approx \mathrm{Re}(\mathbf{z}(\tilde{\mathbf{E}}(z') \times \tilde{\mathbf{H}}(z')^*)). \tag{11.55}$$

Using EWB parameters (see Section 9.3), we may express $\tilde{\mathbf{E}}(z)$ and $\tilde{\mathbf{H}}(z)$ as follows:

$$\tilde{\mathbf{E}}(z) = \mathbf{e}_1(z)A_1(z), \quad \tilde{\mathbf{H}}(z) = \mathbf{h}_1(z)A_1(z), \tag{11.56}$$

where

$$\mathbf{h}_1(z) = \sigma_1(z)[\mathbf{z} \times \mathbf{e}_1(z)] \tag{11.57}$$

with $\sigma_1(z)$ given by (11.6c). Since $\mathbf{n}(z')\|z$ and $\mathbf{n}(z'')\|x$ (Figure 11.3), the parameters of the eigenwave basis at the layer boundaries may be represented as

$$\mathbf{e}_1(z') = a_1(z')\mathbf{x}, \quad \mathbf{e}_1(z'') = a_1(z'')\mathbf{x}, \quad \mathbf{h}_1(z') = a_1(z')n_\perp\mathbf{y}, \quad \mathbf{h}_1(z'') = a_1(z'')n_{||}\mathbf{y}, \tag{11.58}$$

where a_1 is a factor depending on normalization conditions imposed on the eigenwave basis. In particular, for normalization (8.167) [or (8.164)],

$$a_1(z') = 1/\sqrt{2n_\perp}, \quad a_1(z'') = 1/\sqrt{2n_{||}}; \tag{11.59}$$

for Yeh's normalization (8.238),

$$a_1(z') = a_1(z'') = 1. \tag{11.60}$$

For Yeh's normalization, from (11.56)–(11.58) and (11.60) we have

$$\begin{aligned} \tilde{\mathbf{E}}(z') &= \mathbf{x}A_1(z'), \quad \tilde{\mathbf{E}}(z'') = \mathbf{x}A_1(z''), \\ \tilde{\mathbf{H}}(z') &= \mathbf{y}n_\perp A_1(z'), \quad \tilde{\mathbf{H}}(z'') = \mathbf{y}n_{||}A_1(z''). \end{aligned} \tag{11.61}$$

Substituting (11.61) into (11.55) leads to the following relation:

$$n_\perp |A_1(z')|^2 \approx n_{||}|A_1(z'')|^2. \tag{11.62}$$

According to (11.61), this relation may be rewritten as

$$n_{\perp}|E_x(z')|^2 \approx n_{\parallel}|E_x(z'')|^2, \tag{11.63}$$

where E_x is the x-component of the electric field vector $\tilde{\mathbf{E}}$. Hence for NBR mode, to a very good approximation,

$$\frac{|A_1(z'')|^2}{|A_1(z')|^2} \approx \frac{n_{\perp}}{n_{\parallel}}, \tag{11.64}$$

$$\frac{|E_x(z'')|^2}{|E_x(z')|^2} \approx \frac{n_{\perp}}{n_{\parallel}}. \tag{11.65}$$

We see that in this situation, on passing through the layer, the amplitudes A_1 (if Yeh's normalization is used) and E_x of the wave significantly change in magnitude (by a factor $\sqrt{n_{\perp}/n_{\parallel}}$), while the normal flux remains almost constant. Figure 11.5 shows how $|E_x|$ changes as the wave in NBR mode propagates through the layer with configuration HP1. It is easy to verify that neither Gu and Yeh's method [2] nor Lien's method [3, 22] takes these changes in $|E_x|$ and $|A_1|$ into account. Gu and Yeh's method in this situation yields the following relation for the state vectors:

$$\begin{pmatrix} A_1(z'') \\ A_2(z'') \end{pmatrix} = \begin{pmatrix} \exp\left(ik_0 d \int_0^1 \sigma_1(\eta)\mathrm{d}\eta\right) & 0 \\ 0 & \exp\left(ik_0\sigma_2 d\right) \end{pmatrix} \begin{pmatrix} A_1(z') \\ A_2(z') \end{pmatrix} \tag{11.66}$$

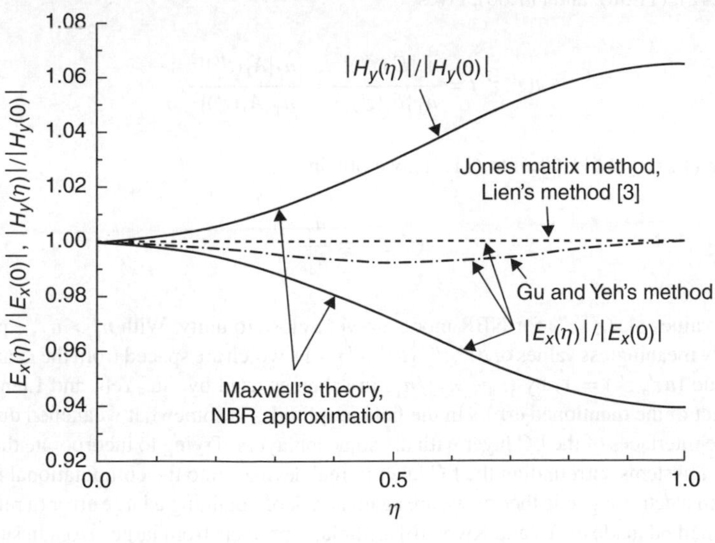

Figure 11.5 Spatial evolution of the amplitudes of field components of an extraordinary wave propagating through the homeoplanar layer with configuration HP1 in NBR mode. Comparison of results obtained by different methods

for Yeh's normalization. Lien's method [3, 22] gives the relation

$$\begin{pmatrix} E_x(z'') \\ E_y(z'') \end{pmatrix} = \begin{pmatrix} \exp\left(ik_0 d \int_0^1 \sigma_1(\eta) \mathrm{d}\eta \right) & 0 \\ 0 & \exp\left(ik_0 \sigma_2 d \right) \end{pmatrix} \begin{pmatrix} E_x(z') \\ E_y(z') \end{pmatrix}. \tag{11.67}$$

In our case $A_2(z) = 0$ and $E_y(z) = 0$ and, according to these relations,

$$\frac{|A_1(z'')|^2}{|A_1(z')|^2} = 1 \tag{11.68}$$

and

$$\frac{|E_x(z'')|^2}{|E_x(z')|^2} = 1. \tag{11.69}$$

Comparing these relations with (11.64) and (11.65), we see that the level of accuracy of the methods under consideration in the context of electromagnetic theory is not very good and becomes worse with increasing the deviation of $n_\perp / n_\parallel$ from unity. Significant difference between estimates for the fields given by these methods and the rigorous theory explains the fact that use of exact electromagnetic formulas within the frameworks of these methods sometimes leads to large errors and even nonphysical results. Thus, in the above example, the exact formula for the bulk transmissivity of the LC layer,

$$t(z', z'') \equiv S_z(z'')/S_z(z'), \tag{11.70}$$

in view of (8.242), (11.58), and (11.60), gives

$$t(z', z'') = \frac{n_\parallel |E_x(z'')|^2}{n_\perp |E_x(z')|^2} = \frac{n_\parallel |A_1(z'')|^2}{n_\perp |A_1(z')|^2}. \tag{11.71}$$

By substituting (11.68) or (11.69) into (11.71), we obtain

$$t(z', z'') = \frac{n_\parallel}{n_\perp}, \tag{11.72}$$

while the true values of $t(z', z'')$ for NBR mode are very close to unity. With $n_\parallel > n_\perp$, equation (11.72) gives physically meaningless values of $t(z', z'')$ ($t(z', z'') > 1$), which are spaced from the nearest physically acceptable value ($t(z', z'') = 1$) by $(n_\parallel - n_\perp)/n_\perp$. In schemes used by Gu, Yeh, and Lien for modeling LCDs, the effect of the mentioned errors in the field amplitudes is somewhat weakened due to a specific inclusion of the interfaces of the LC layer with the adjacent layers. Trying to incorporate these interfaces, or the thin-layer systems surrounding the LC layer in real devices, into the computational model in strict accordance with electromagnetic theory, we are again at risk of obtaining a large error (a reformulation of Gu and Yeh's method made by Yu and Kwok [6] artificially protects from large errors in such situations). To illustrate, suppose that the LC layer is situated between glass plates as in Figure 11.1, and estimate the transmissivity of the domain $(z_1 - 0, z_2 + 0)$ for an incident wave linearly polarized along the x-axis. The domain $(z_1 - 0, z_2 + 0)$ includes the frontal glass plate–LC layer interface $(z_1 - 0, z_1 + 0)$, the bulk of the LC layer $(z_1 + 0, z_2 - 0)$, and the LC layer–rear glass plate interface $(z_2 - 0, z_2 + 0)$. Neglecting multiple reflections, we may connect the amplitudes of the forward propagating light at the interfaces as follows:

$$E_x(z_1 + 0) = \frac{2n_g}{n_g + n_\perp} E_x(z_1 - 0), \quad E_x(z_2 + 0) = \frac{2n_{||}}{n_{||} + n_g} E_x(z_2 - 0), \tag{11.73}$$

where n_g is the refractive index of the glass plates. These are accurate relations given by electromagnetic theory; they take account of single reflections from the interfaces. Let us try to use these relations in combination with the following one, given by Lien's method [3, 22] [see (11.67)]:

$$E_x(z_2 - 0) = E_x(z_1 + 0) \exp\left(ik_0 d \int_0^1 \sigma_1(\eta) d\eta \right). \tag{11.74}$$

Upon substituting (11.74) into (11.73), we have

$$|E_x(z_2 + 0)|^2 = \left(\frac{2n_{||}}{n_{||} + n_g} \right)^2 \left(\frac{2n_g}{n_g + n_\perp} \right)^2 |E_x(z_1 - 0)|^2. \tag{11.75}$$

The transmissivity

$$t(z_1 - 0, z_2 + 0) \equiv S_z(z_2 + 0)/S_z(z_1 - 0) \tag{11.76}$$

may be expressed as follows:

$$t(z_1 - 0, z_2 + 0) = \frac{n_g |E_x(z_2 + 0)|^2}{n_g |E_x(z_1 - 0)|^2} = \frac{|E_x(z_2 + 0)|^2}{|E_x(z_1 - 0)|^2}. \tag{11.77}$$

From (11.77) and (11.75), we obtain

$$t(z_1 - 0, z_2 + 0) = \left(\frac{2n_{||}}{n_{||} + n_g} \right)^2 \left(\frac{2n_g}{n_g + n_\perp} \right)^2. \tag{11.78}$$

It is easy to see that this expression is erroneous. For example, taking $n_g = n_\perp$, we see that with $n_{||} > n_g$ expression (11.78) gives $t(z_1 - 0, z_2 + 0) > 1$. Thus, we again face the situation when an attempt to use the variants [2, 3] of EJMM as truly electromagnetic methods leads to bad results.

Recognizing the disadvantages of the method [3], Lien with coauthors developed later a more accurate method (see expression (72) in [5]) which is fully consistent with electromagnetic theory, for media with arbitrary birefringence. But this method is significantly slower than [3]. Then, more efficient and simple methods of the same level of accuracy were proposed in [7–9]. We proceed to describe these methods.

11.2 NBRA. Basic Differential Equations

Differential Equations for State Vectors

According to the rigorous electromagnetic theory of light propagation in stratified media, which has been set forth in detail in Chapter 8, spatial evolution of a harmonic wave field of the form (8.57) in a layer

(z', z'') with smoothly varying parameters may be described by equation (8.86) which, in view of (8.93), may be rewritten as follows:

$$\frac{\mathbf{da}^{\downarrow}}{\mathrm{d}z} = \mathbf{N}_{11}\mathbf{a}^{\downarrow} + \mathbf{N}_{12}\mathbf{a}^{\uparrow}, \tag{11.79a}$$

$$\frac{\mathbf{da}^{\uparrow}}{\mathrm{d}z} = \mathbf{N}_{21}\mathbf{a}^{\downarrow} + \mathbf{N}_{22}\mathbf{a}^{\uparrow}, \tag{11.79b}$$

where $\mathbf{a}^{\uparrow}$ and $\mathbf{a}^{\downarrow}$ are, as usual, the EW Jones vectors characterizing the forward and backward propagating fields, respectively; $\mathbf{N}_{ij}$ $(i, j = 1, 2)$ are 2×2 blocks of the matrix

$$\widehat{\boldsymbol{\Delta}}_{\mathrm{A}} = \begin{pmatrix} \mathbf{N}_{11} & \mathbf{N}_{12} \\ \mathbf{N}_{21} & \mathbf{N}_{22} \end{pmatrix} \tag{11.80}$$

expressed, in terms of the parameters of the eigenwave basis, by (8.87). Since the layer is inhomogeneous, the forward and backward propagating fields within it are coupled, that is, bulk reflection takes place. The coupling between forward and backward propagating waves is described by nonzero elements of off-diagonal 2×2 blocks of the matrix $\widehat{\boldsymbol{\Delta}}_{\mathrm{A}}$, $\mathbf{N}_{12}$, and $\mathbf{N}_{21}$. Considering practical LCDs, we mostly deal with LC layers whose bulk reflection under illumination conditions of interest is very small. In such cases, we may expect to obtain accurate estimates of the transmission characteristics of the LC layer even if we neglect the difference of elements of the blocks $\mathbf{N}_{12}$ and $\mathbf{N}_{21}$ from zero. Calculating the transmission characteristics of layers with smoothly varying parameters on the assumption that all elements of the blocks $\mathbf{N}_{12}$ and $\mathbf{N}_{21}$ are zero is the essence of the negligible-bulk-reflection approximation (NBRA) in the context of the rigorous theory. Under this assumption, the simultaneous equations (11.79) split into two independent equations: one equation,

$$\frac{\mathbf{da}^{\downarrow}}{\mathrm{d}z} = \mathbf{N}^{\downarrow}\mathbf{a}^{\downarrow} \tag{11.81}$$

with $\mathbf{N}^{\downarrow} \equiv \mathbf{N}_{11}$, is for the forward propagating fields, and the other,

$$\frac{\mathbf{da}^{\uparrow}}{\mathrm{d}z} = \mathbf{N}^{\uparrow}\mathbf{a}^{\uparrow} \tag{11.82}$$

with $\mathbf{N}^{\uparrow} \equiv \mathbf{N}_{22}$, for the backward propagating fields. Using (11.81), one can estimate the operator $\mathbf{t}^{\downarrow}(z', z'')$. From (11.82) one can find an approximate value of the matrix $\mathbf{t}^{\uparrow}(z', z'')$. Since the ways of the use of (11.81) and (11.82) for estimating the corresponding transmission operators are identical, we confine our further discussion to the methods of estimation of the matrix $\mathbf{t}^{\downarrow}(z', z'')$, which stem from (11.81).

Differential Equation for the Transmission Operator for Forward Propagating Fields

Substitution of the expression $\mathbf{a}^{\downarrow}(z) = \widehat{\mathbf{t}}^{\downarrow}(z', z)\mathbf{a}^{\downarrow}(z')$ into (11.81) gives the following operator equation:

$$\frac{\mathrm{d}\widehat{\mathbf{t}}^{\downarrow}(z', z)}{\mathrm{d}z} = \mathbf{N}^{\downarrow}(z)\widehat{\mathbf{t}}^{\downarrow}(z', z) \quad [\widehat{\mathbf{t}}^{\downarrow}(z', z') = \mathbf{U}]. \tag{11.83}$$

The solution of this equation, $\widehat{\mathbf{t}}^{\downarrow}(z', z'')$, is the value of the operator $\mathbf{t}^{\downarrow}(z', z'')$ yielded by NBRA. According to (8.87), the general expression for the matrix $\mathbf{N}^{\downarrow}$ may be written as follows:

$$\mathbf{N}^{\downarrow} = \begin{pmatrix} ik_0\sigma_1 - \bar{\boldsymbol{\psi}}_1\boldsymbol{\psi}_{z1} & -\bar{\boldsymbol{\psi}}_1\boldsymbol{\psi}_{z2} \\ -\bar{\boldsymbol{\psi}}_2\boldsymbol{\psi}_{z1} & ik_0\sigma_2 - \bar{\boldsymbol{\psi}}_2\boldsymbol{\psi}_{z2} \end{pmatrix}, \tag{11.84}$$

where

$$\boldsymbol{\psi}_{zj} \equiv \frac{\mathrm{d}\boldsymbol{\psi}_j}{\mathrm{d}z} \quad j = 1, 2.$$

This general expression for the matrix $\mathbf{N}^{\downarrow}$ is too complicated to see any advantage of NBRA. In this section, we derive much simpler expressions for this matrix that make NBRA a really useful tool in both numerical calculations and analysis. But, before proceeding to these derivations, we provide some numerical examples demonstrating the validity of NBRA and accuracy of the relation

$$\mathbf{t}^{\downarrow}(z', z'') \approx \widehat{\mathbf{t}}^{\downarrow}(z', z'') \tag{11.85}$$

in situations typical of LCDs.

Numerical Estimates of the Accuracy of NBRA

No simple exact general criteria that allow evaluating the accuracy of the approximation (11.85) *a priori* are known. The simplest way to evaluate the accuracy of this approximation in solving a particular problem is to compare results obtained using NBRA with the corresponding exact solutions under conditions typical of this problem or under somewhat more severe conditions to estimate the upper limit of the errors in typical situations. Here (see Figures 11.6–11.9) we demonstrate several examples of such estimates for locally uniaxial layers with spatially invariant principal refractive indices. In these examples we consider five model configurations of the optic axis field. The first two are the homeoplanar configurations HP1 (11.52) and HP2 (11.53), which have been considered in the previous section (Figure 11.3). The third is an ideal twisted configuration with $\Phi = 270°$ and tilt angle $\theta = 20°$ (Figure 11.7):

$$\theta(\eta) = \pi/9, \quad \varphi(\eta) = 3\pi\eta/2 \tag{11.86}$$

[see (11.2)]. The fourth configuration (Figure 11.8) is

$$\theta(\eta) = \pi/9 + (3\pi/9)\sin(\pi\eta), \quad \varphi(\eta) = 3\pi\eta/2 - (2\pi/9)\sin(2\pi\eta). \tag{11.87}$$

This is a distorted twisted configuration with $\Phi = 270°$, $\theta_1 = \theta_2 = 20°$, and $\theta(\eta = 0.5) = 80°$. The fifth configuration (Figure 11.9) is an ideal twisted one with $\Phi = 90°$ and $\theta = 4°$:

$$\theta(\eta) = \pi/45, \quad \varphi(\eta) = \pi\eta/2. \tag{11.88}$$

For these layers, the transmission matrices $\mathbf{t}^{\downarrow}(z', z'')$ and $\widehat{\mathbf{t}}^{\downarrow}(z', z'')$ as well as the reflection matrix $\mathbf{r}^{\downarrow}(z', z'')$ were calculated. The matrices $\mathbf{t}^{\downarrow}(z', z'')$ and $\mathbf{r}^{\downarrow}(z', z'')$ were computed by transfer matrix method (the eigenwave variant, see Section 8.2.2) with the use of the staircase approximation. The staircase model of the kind (8.331) with 801 sublayers was used. The maximum errors in the calculated $\mathbf{t}^{\downarrow}(z', z'')$ and $\mathbf{r}^{\downarrow}(z', z'')$ were of the order of 3×10^{-5}. The matrix $\widehat{\mathbf{t}}^{\downarrow}(z', z'')$ was calculated with the help of an approximating multilayer method (AMM) that will be described in Section 11.3.1, using the same

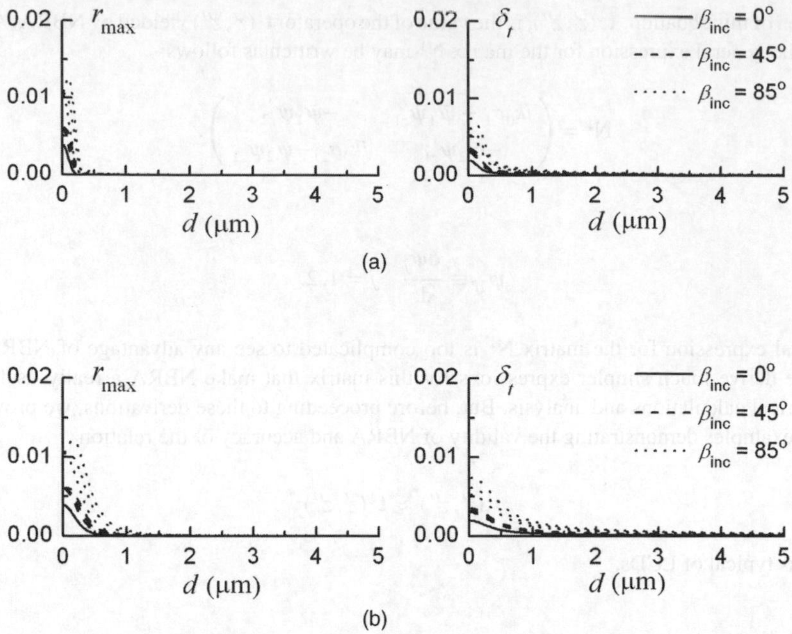

Figure 11.6 The maximum bulk reflectivity r_{max} and the error of the NBR approximation for the layers with homeoplanar configurations HP1 (a) and HP2 (b) (see Figure 11.3)

staircase model, which gave approximately the same accuracy as in the calculations of $\mathbf{t}^{\downarrow}(z', z'')$ and $\mathbf{r}^{\downarrow}(z', z'')$. From the obtained $\mathbf{r}^{\downarrow}(z', z'')$, the maximum (over the set of all possible values of the EW Jones vector of the incident light) reflectivity $r_{max} = \underline{\max}[r^{\downarrow}(z', z'')]$ was calculated (see Section 8.5). The quantity

$$\delta_t \equiv \|\mathbf{t}^{\downarrow}(z', z'') - \widehat{\mathbf{t}}^{\downarrow}(z', z'')\|_E, \tag{11.89}$$

where $\|\cdot\|_E$, as usual, denotes the Euclidean norm (see Sections 5.1.4 and 11.5), was considered as a measure of accuracy of approximation (11.85). In the calculations, we assumed that the media were nonabsorbing and had $n_{\|} = 1.7$ and $n_{\perp} = 1.5$. All the calculations were performed with $\lambda = 550$ nm for different propagation directions of the incident light which was assumed to fall on the layered system containing the layer under consideration from air. Figures 11.6–11.9 present the calculated r_{max} and δ_t as functions of the layer thickness d for the case of normal incidence ($\beta_{inc} = 0°$) and for the cases $\beta_{inc} = 45°$ and $\beta_{inc} = 85°$. For both cases of oblique incidence, curves corresponding to three different orientations of the plane of incidence ($\alpha_{inc} = 0°$, 45°, and 90°) are shown. As is seen from the plots for r_{max}, in all the cases, for layers with $d > 1$ μm the reflectivities are very small. The results for the homeoplanar configurations (Figure 11.6) in the region $d < 2$ μm allow one to estimate the reflectivity of fragments of LC layers where the derivative $d\theta/dz$ is large (for example, the near-surface regions at high voltages). The results for the layers with twisted configurations are in fair agreement with the following simple criterion of a weak bulk reflection:

$$\frac{\lambda|\Phi|}{2\pi d} \ll 1, \tag{11.90}$$

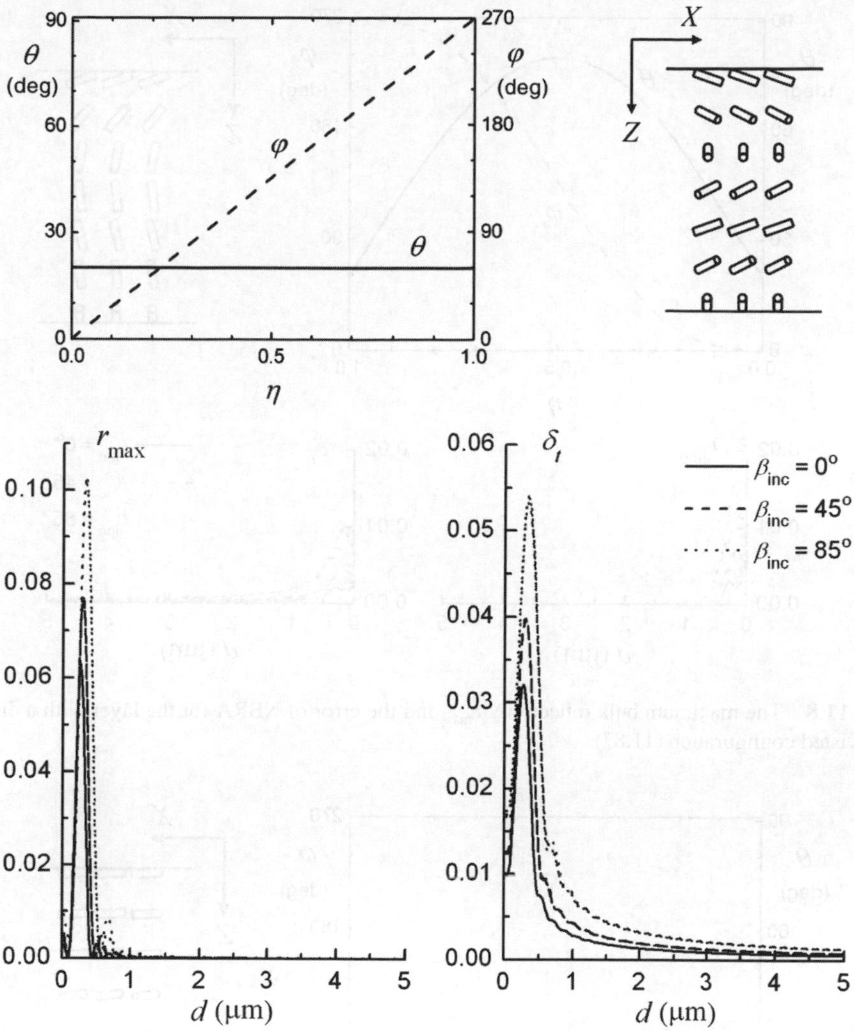

Figure 11.7 The maximum bulk reflectivity r_{max} and the error of NBRA for the layer with a supertwisted configuration (11.86)

which is well known in optics of cholesteric LCs. For all the twisted configurations, including the distorted one (11.87), the maxima of reflectivity correspond to thicknesses for which

$$\frac{\lambda|\Phi|}{2\pi d} \sim 1, \tag{11.91}$$

and for thicknesses satisfying (11.90) the reflectivity is very small. As might be expected, the largest values of δ_t were obtained for the ideal supertwisted structure (11.86). But for $d > 0.4\ \mu\text{m}/\Delta n$ (this condition is satisfied for the practical STN modes) δ_t is relatively small. For instance, at normal incidence and at $\beta_{inc} = 45°$, for layers with $d > 2\ \mu\text{m}$ δ_t does not exceed 0.002 [for comparison, SBA for these conditions gives errors (determined analogously to δ_t) of the order of 0.06]. Note that in practical STN devices,

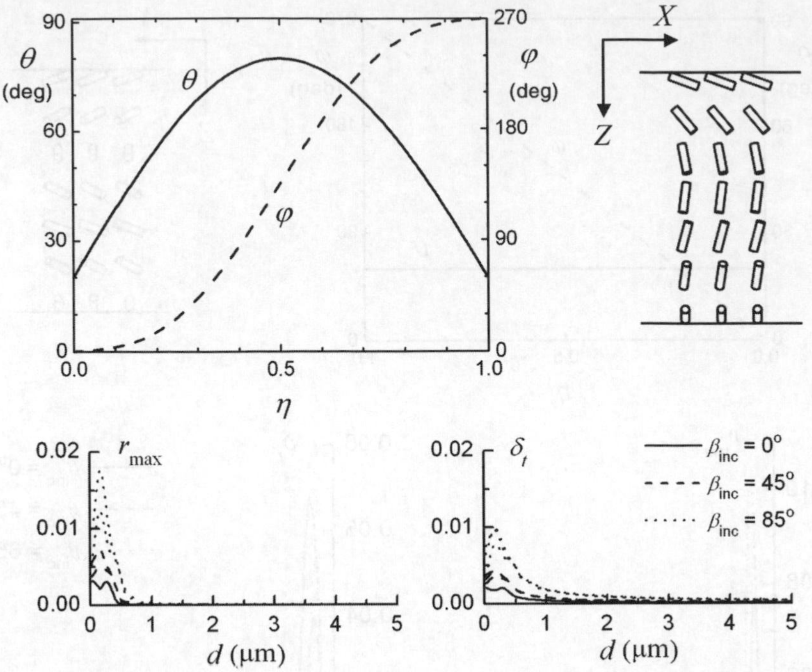

Figure 11.8 The maximum bulk reflectivity r_{max} and the error of NBRA for the layer with a distorted supertwisted configuration (11.87)

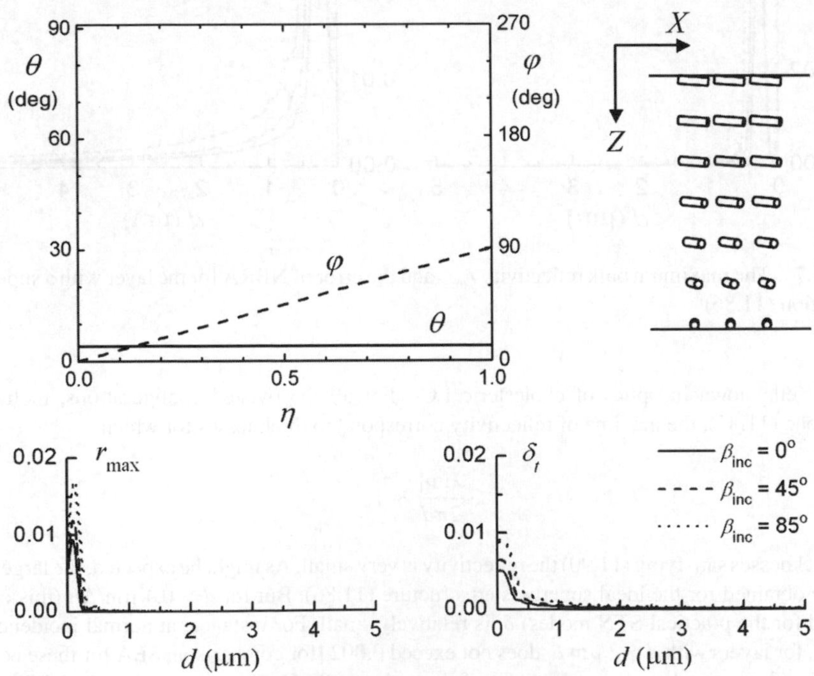

Figure 11.9 The maximum bulk reflectivity r_{max} and the error of NBRA for the layer with a 90°-twisted configuration (11.88)

as a rule, conditions are more favorable for NBRA (in particular, smaller Φ) than in this example. The values of δ_t for the layers with configurations HP1, HP2, (11.87), and (11.88) are much smaller than for configuration (11.86), and high accuracy of NBRA, at least for $d > 2$ μm, is quite evident. These examples show that NBRA really has a very wide field of application in LCD optics. Now we will obtain the simpler expressions for the matrix $\mathbf{N}^{\downarrow}$, applicable in considering LC layers. We note that the simplifications will be made without invoking any additional physical approximations, and the resulting expressions for $\mathbf{N}^{\downarrow}$ are exact.

General Case of a Reciprocal Optically Locally Centrosymmetric Medium

A standard model of an LC layer is a medium characterized by constitutive relations (8.6) with a symmetric permittivity tensor, that is, the LC is regarded as a reciprocal optically locally centrosymmetric medium. As we saw in Section 8.4.1, in the absence of polarization degeneracy, the basis waves for such media always satisfy the condition

$$\boldsymbol{\psi}_k^{\mathrm{T}}\mathbf{I}_0\boldsymbol{\psi}_j \equiv \mathbf{z}\cdot(\mathbf{e}_k \times \mathbf{h}_j + \mathbf{e}_j \times \mathbf{h}_k) = 0 \qquad k,j = 1,2,3,4, \quad k \neq j \tag{11.92}$$

[see (8.161) and (8.97)]. In the presence of polarization degeneracy, the eigenwave basis can always be chosen so that condition (11.92) is satisfied. Assuming that the EW basis satisfies (11.92) and has been normalized by condition (8.167) (S-normalization), we may represent the first two rows of the matrix $\boldsymbol{\Psi}^{-1}$ $(\bar{\boldsymbol{\psi}}_k)$ as

$$\bar{\boldsymbol{\psi}}_k = \boldsymbol{\psi}_k^{\mathrm{T}}\mathbf{I}_0 = (\,\mathbf{h}_k\mathbf{y} \quad \mathbf{e}_k\mathbf{x} \quad -\mathbf{h}_k\mathbf{x} \quad \mathbf{e}_k\mathbf{y}\,) \quad k = 1,2 \tag{11.93}$$

(see Section 8.4.1). By making use of (11.93), we may rewrite expression (11.84) as follows:

$$\mathbf{N}^{\downarrow} = \begin{pmatrix} ik_0\sigma_1 - \boldsymbol{\psi}_1^{\mathrm{T}}\mathbf{I}_0\boldsymbol{\psi}_{z1} & -\boldsymbol{\psi}_1^{\mathrm{T}}\mathbf{I}_0\boldsymbol{\psi}_{z2} \\ -\boldsymbol{\psi}_2^{\mathrm{T}}\mathbf{I}_0\boldsymbol{\psi}_{z1} & ik_0\sigma_2 - \boldsymbol{\psi}_2^{\mathrm{T}}\mathbf{I}_0\boldsymbol{\psi}_{z2} \end{pmatrix}. \tag{11.94}$$

Differentiating (11.92),

$$\frac{\mathrm{d}\left(\boldsymbol{\psi}_k^{\mathrm{T}}\mathbf{I}_0\boldsymbol{\psi}_j\right)}{\mathrm{d}z} = \left(\frac{\mathrm{d}\boldsymbol{\psi}_k}{\mathrm{d}z}\right)^{\mathrm{T}}\mathbf{I}_0\boldsymbol{\psi}_j + \boldsymbol{\psi}_k^{\mathrm{T}}\mathbf{I}_0\left(\frac{\mathrm{d}\boldsymbol{\psi}_j}{\mathrm{d}z}\right) = \boldsymbol{\psi}_j^{\mathrm{T}}\mathbf{I}_0\left(\frac{\mathrm{d}\boldsymbol{\psi}_k}{\mathrm{d}z}\right) + \boldsymbol{\psi}_k^{\mathrm{T}}\mathbf{I}_0\left(\frac{\mathrm{d}\boldsymbol{\psi}_j}{\mathrm{d}z}\right) = 0$$

(here the property $\mathbf{I}_0{}^{\mathrm{T}} = \mathbf{I}_0$ has been used), we find that

$$\boldsymbol{\psi}_1^{\mathrm{T}}\mathbf{I}_0\boldsymbol{\psi}_{z2} = -\boldsymbol{\psi}_2^{\mathrm{T}}\mathbf{I}_0\boldsymbol{\psi}_{z1}. \tag{11.95a}$$

Similarly, differentiating (8.167), we obtain

$$\boldsymbol{\psi}_1^{\mathrm{T}}\mathbf{I}_0\boldsymbol{\psi}_{z1} = \boldsymbol{\psi}_2^{\mathrm{T}}\mathbf{I}_0\boldsymbol{\psi}_{z2} = 0. \tag{11.95b}$$

In view of (11.95), expression (11.94) may be rewritten as

$$\mathbf{N}^{\downarrow}(z) = \begin{pmatrix} ik_0\sigma_1(z) & \vartheta_z(z) \\ -\vartheta_z(z) & ik_0\sigma_2(z) \end{pmatrix}, \tag{11.96}$$

where

$$\vartheta_z = -\boldsymbol{\psi}_1^{\mathrm{T}} \mathbf{I}_0 \boldsymbol{\psi}_{z2} = \boldsymbol{\psi}_2^{\mathrm{T}} \mathbf{I}_0 \boldsymbol{\psi}_{z1}. \tag{11.97}$$

Thus, we may specify the function $\mathbf{N}^{\downarrow}(z)$ by specifying three scalar functions $\sigma_1(z)$, $\sigma_2(z)$, and $\vartheta_z(z)$. If the medium is nonabsorbing, the vectors $\mathbf{e}_1(z)$ and $\mathbf{e}_2(z)$ can be chosen real (in what follows, when considering nonabsorbing media, we assume that the EWB is chosen in that way). The function $\vartheta_z(z)$ will then take only real values, as will $\sigma_1(z)$ and $\sigma_2(z)$. We must emphasize that expression (11.96) corresponds to S-normalization of the eigenwave basis.

One cannot but notice that expression (11.96) is very similar to expression (11.28) for the matrix $\mathbf{N}_{x'-y'}$ entering into equations (11.27) and (11.31) of the Jones method. The parameter φ_z determining the off-diagonal elements of the matrix $\mathbf{N}_{x'-y'}$ is a purely configurational parameter ($\varphi_z = \mathrm{d}\varphi/\mathrm{d}z$). As we will see below, considering the case of an optically uniaxial medium, the parameter ϑ_z is also rather simply connected with the geometry of the medium, and, furthermore, in the case of normal incidence, is close to φ_z.

Optically Uniaxial Medium

Let the medium be locally uniaxial and have spatially invariant principal refractive indices ($n_\parallel$ and $n_\perp$). The permittivity tensor $\boldsymbol{\varepsilon}$ of this medium as a function of z can be represented as

$$\boldsymbol{\varepsilon}(z) = \varepsilon_\perp \mathbf{U} + \Delta\varepsilon \mathbf{c}(z) \otimes \mathbf{c}(z),$$

where

$$\Delta\varepsilon = \varepsilon_\parallel - \varepsilon_\perp, \quad \varepsilon_\parallel = n_\parallel^2, \quad \varepsilon_\perp = n_\perp^2,$$

and $\mathbf{c}$ is, as usual, the unit vector parallel to the local optic axis. For the sake of generality, we assume that, for the forward propagating waves, there is polarization degeneracy in the planes $z = z_{Dk}$ ($k = 1, 2, \ldots, N_D$) (i.e., $\sigma_1(z_{Dk}) = \sigma_2(z_{Dk})$). Using the template for EW bases in uniaxial media given in Section 9.3, we choose the eigenwave basis (S-normalized) as follows:

$$\sigma_1 = \frac{-b_1 + \sqrt{b_1^2 - b_2(b_3 - \varepsilon_\perp \varepsilon_\parallel)}}{b_2}, \quad \sigma_2 = \sqrt{\varepsilon_\perp - \mathbf{b}^2},$$
$$b_1 = \Delta\varepsilon(\mathbf{bc})(\mathbf{zc}), \quad b_2 = \varepsilon_\perp + \Delta\varepsilon(\mathbf{zc})^2, \quad b_3 = \varepsilon_\perp \mathbf{b}^2 + \Delta\varepsilon(\mathbf{bc})^2;$$
$$\mathbf{e}_j = \frac{\boldsymbol{e}_j}{2\mathbf{z}\cdot(\boldsymbol{e}_j \times \boldsymbol{h}_j)}, \quad \mathbf{h}_j = \frac{\boldsymbol{h}_j}{2\mathbf{z}\cdot(\boldsymbol{e}_j \times \boldsymbol{h}_j)}, \tag{11.98}$$
$$\boldsymbol{h}_j = \mathbf{m}_j \times \boldsymbol{e}_j, \quad \mathbf{m}_j = \mathbf{b} + \mathbf{z}\sigma_j, \quad j = 1, 2;$$
$$\boldsymbol{e}_1(z) = \frac{a(z)}{b(z)}\widehat{\mathbf{k}}(z), \quad \boldsymbol{e}_2(z) = \frac{a(z)}{b(z)}\widehat{\mathbf{j}}(z) \ \text{ for } z \neq z_{Dk}$$
$$\boldsymbol{e}_1(z_{Dk}) = \frac{a(z_{Dk}-0)}{b(z_{Dk}-0)}\widehat{\mathbf{k}}(z_{Dk}-0), \quad \boldsymbol{e}_2(z_{Dk}) = \frac{a(z_{Dk}-0)}{b(z_{Dk}-0)}\widehat{\mathbf{j}}(z_{Dk}-0),$$

where

$$\widehat{\mathbf{k}} \equiv -b\varepsilon^{-1}(\mathbf{m}_1 \times (\mathbf{m}_1 \times \mathbf{c})) = b\left(\mathbf{c} - \frac{\mathbf{m}_1(\mathbf{m}_1\mathbf{c})}{\varepsilon_\perp}\right),$$
$$\widehat{\mathbf{j}} \equiv b(\mathbf{m}_2 \times \mathbf{c}), \quad b \equiv \frac{1}{\sqrt{(\mathbf{m}_2 \times \mathbf{c})^2}}, \tag{11.99}$$

$a(z)$ is a step function taking the values 1 and -1, remaining constant in the regions that do not contain points $z = z_{Dk}$, and satisfying the condition

$$a(z_{Dk}+0) = a(z_{Dk}-0)\operatorname{sign}\left(\widehat{\mathbf{j}}(z_{Dk}-0)\cdot\widehat{\mathbf{j}}(z_{Dk}+0)\right) \quad k = 1, 2, \ldots, N_D.$$

Recall that $\mathbf{m}_1$, σ_1, $\mathbf{e}_1$, and $\mathbf{h}_1$ are parameters of the extraordinary basis wave, and $\mathbf{m}_2$, σ_2, $\mathbf{e}_2$, and $\mathbf{h}_2$ of the ordinary basis wave. The factors a and b have been introduced to ensure the continuity of the functions $\mathbf{e}_j(z)$ and $\mathbf{h}_j(z)$ ($j = 1, 2$) at $z = z_{Dk}$. These factors may be omitted if polarization degeneracy is absent.

Let us introduce the vector $\mathbf{i} = \mathbf{m}_2/M_2$, where $M_2 \equiv \sqrt{\mathbf{m}_2^2} = n_\perp$ ($\mathbf{m}_2$ and σ_2 are independent of z within the medium) and the vector functions

$$\mathbf{j}(z) = \begin{cases} a(z)\widehat{\mathbf{j}}(z) & z \neq z_{Dk} \\ a(z-0)\widehat{\mathbf{j}}(z-0) & z = z_{Dk} \end{cases} \quad k = 1, 2, \ldots N_D \tag{11.100}$$

and $\mathbf{k}(z) = \mathbf{i} \times \mathbf{j}(z)$. Note that

$$\widehat{\mathbf{j}}(z) = \frac{\mathbf{i} \times \mathbf{c}(z)}{p(z)}$$

[see (11.99)], where $p = \sqrt{(\mathbf{i} \times \mathbf{c})^2}$. The vectors $\mathbf{i}$, $\mathbf{j}$, and $\mathbf{k}$ are unit and orthogonal in the sense that

$$\mathbf{i}^2 = \mathbf{j}^2 = \mathbf{k}^2 = 1, \quad \mathbf{i}\cdot\mathbf{j} = \mathbf{j}\cdot\mathbf{k} = \mathbf{i}\cdot\mathbf{k} = 0. \tag{11.101}$$

In the absence of absorption, the vectors $\mathbf{i}$, $\mathbf{j}$, and $\mathbf{k}$ are real; otherwise, they may be complex. Due to (11.101), the functions $\mathbf{j}(z)$ and $\mathbf{k}(z)$ can be represented as

$$\mathbf{j}(z) = \mathbf{j}(z')\cos \upsilon(z) + \mathbf{k}(z')\sin \upsilon(z),$$
$$\mathbf{k}(z) = -\mathbf{j}(z')\sin \upsilon(z) + \mathbf{k}(z')\cos \upsilon(z), \tag{11.102}$$

where $\upsilon(z)$ is a continuous function satisfying the conditions

$$\cos \upsilon(z) = \mathbf{j}(z)\cdot\mathbf{j}(z'),$$
$$\sin \upsilon(z) = \mathbf{j}(z)\cdot\mathbf{k}(z'), \text{ and } \upsilon(z') = 0. \tag{11.103}$$

If the vectors $\mathbf{i}$, $\mathbf{j}(\xi)$, and $\mathbf{k}(\xi)$, where ξ is any given value of z, are real, they as well as the parameter $\upsilon(\xi)$ have a simple geometrical interpretation (see Figure 11.10). The vector $\mathbf{i}$ coincides with the wave normal of the ordinary basis wave; $\mathbf{j}(\xi)$ is a normal to the plane subtending the vectors $\mathbf{c}(\xi)$ and $\mathbf{i}$, $\mathbf{k}(\xi)$ is

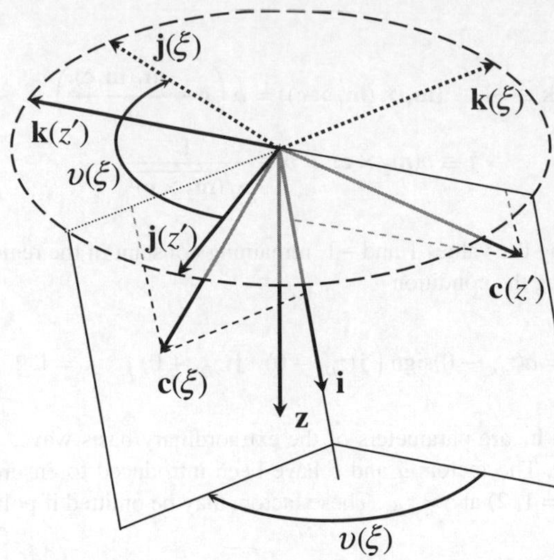

Figure 11.10 To the problem on light propagation in a locally uniaxial medium. Basic vectors

a vector perpendicular to **i** and $\mathbf{j}(\xi)$, and $\upsilon(\xi)$ is the angle between the vectors $\mathbf{j}(z')$ and $\mathbf{j}(\xi)$ [and between $\mathbf{k}(z')$ and $\mathbf{k}(\xi)$]. According to (11.97) and (11.93), the function $\vartheta_z(z)$ may be represented as

$$\vartheta_z = -\mathbf{z} \cdot \left(\mathbf{e}_1 \times \frac{\mathrm{d}\mathbf{h}_2}{\mathrm{d}z} + \frac{\mathrm{d}\mathbf{e}_2}{\mathrm{d}z} \times \mathbf{h}_1 \right). \tag{11.104}$$

Using (11.98), (11.99), and (11.101), we find that

$$\mathbf{e}_2(z) = g\mathbf{j}(z), \quad \mathbf{h}_2(z) = M_2\mathbf{k}(z),$$

where

$$g = \frac{1}{\sqrt{2M_2(\mathbf{i} \cdot \mathbf{z})}}.$$

In view of (11.102), this gives

$$\frac{\mathrm{d}\mathbf{e}_2}{\mathrm{d}z} = \upsilon_z(z)g\mathbf{k}(z), \qquad \frac{\mathrm{d}\mathbf{h}_2}{\mathrm{d}z} = -\upsilon_z(z)gM_2\mathbf{j}(z), \tag{11.105}$$

where $\upsilon_z \equiv \mathrm{d}\upsilon/\mathrm{d}z$. Substitution of the expression

$$\mathbf{m}_1 = M_2\left(\mathbf{i} + \gamma\mathbf{z}\right),$$

where

$$\gamma \equiv \frac{\sigma_1 - \sigma_2}{M_2},$$

into (11.98) and (11.99) followed by substitution of the resulting formulas for $\mathbf{e}_1$ and $\mathbf{h}_1$ and (11.105) into (11.104) gives

$$\vartheta_z(z) = \upsilon_z(z)\left(\frac{1+\dfrac{h'}{2}\gamma(z)}{\sqrt{1+h'\gamma(z)-h''(z)\gamma(z)^2}}\right), \tag{11.106}$$

where

$$h' = (\mathbf{i}\cdot\mathbf{z})^{-1}, \quad h'' = h'(\mathbf{z}\cdot\mathbf{c})(\mathbf{z}\times\mathbf{c})\cdot(\mathbf{z}\times\mathbf{i})p^{-2}.$$

Note that the function $\gamma(z)$ can be expressed in terms of $\mathbf{i}$ and $\mathbf{c}(z)$ as

$$\gamma = \frac{w}{v}\left(\sqrt{1+\bar{\delta}}-1\right), \tag{11.107}$$

where

$$\bar{\delta} = v\delta_\varepsilon\left(\frac{p}{w}\right)^2, \quad w = \mathbf{i}\cdot\mathbf{z} + \delta_\varepsilon(\mathbf{i}\cdot\mathbf{c})(\mathbf{z}\cdot\mathbf{c}), \quad v = +\delta_\varepsilon(\mathbf{z}\cdot\mathbf{c})^2, \quad \delta_\varepsilon = \frac{\Delta\varepsilon}{\varepsilon_\perp}.$$

From (11.107) it follows that $\gamma = O(p^2)$ as $p \to 0$. Using this estimate, one can easily find that for $p = 0$ $\vartheta_z = \upsilon_z$ $[p(z_{Dk}) = 0]$.

Thus, in the case of a uniaxial medium with spatially invariant principal refractive indices, all elements of the matrix $\mathbf{N}^{\downarrow}(z)$ are rather simply expressed in terms of parameters of the medium. It is easy to see that in the typical situation when

$$\delta_\varepsilon \ll 1, \quad |\mathbf{i}\cdot\mathbf{z}| \sim 1, \tag{11.108}$$

the difference between $\vartheta_z(z)$ and $\upsilon_z(z)$ is small. Let us introduce the parameter

$$\gamma' = \frac{\gamma}{p^2\,(\mathbf{i}\cdot\mathbf{z})}.$$

It is seen from (11.106) that, as $\gamma' \to 0$,

$$\begin{aligned} \vartheta_z &= \upsilon_z(1 + h\gamma'^2 + O(\gamma'^3)), \\ h &= \frac{p^4}{8} + \frac{p^2}{2}(\mathbf{i}\cdot\mathbf{z})(\mathbf{z}\cdot\mathbf{c})(\mathbf{z}\times\mathbf{c})\cdot(\mathbf{z}\times\mathbf{i}) \end{aligned} \tag{11.109}$$

(in the absence of absorption, the factor h satisfies the condition $|h| \leq 1/4$). When conditions (11.108) are met, we have $|\gamma'| \ll 1$ because, according to (11.107),

$$|\gamma'| \leq \left|\frac{\delta_\varepsilon}{2w\,(\mathbf{i}\cdot\mathbf{z})}\right|.$$

In the case of normal incidence, if the field of the local optic axis is given by (11.1), according to (11.102), we have $\upsilon(z) = \varphi(z)$ and, consequently, $\upsilon_z = \varphi_z \equiv \mathrm{d}\varphi/\mathrm{d}z$, so that the relation $\vartheta_z(z) \approx \upsilon_z(z)$ is equivalent to the following one:

$$\vartheta_z(z) \approx \varphi_z(z). \tag{11.110}$$

We see that in the case of normal incidence, the matrix of coefficients of equation (11.81), describing the spatial evolution of the vector $\mathbf{a}^{\downarrow}$ with S-normalization, is almost identical to that of equation (11.27) for the Jones vector $\mathbf{J}_{x'-y'}$. But one should remember that the mentioned state vectors are of absolutely different nature (see Section 1.4.5).

Equation (11.81) successfully passes the test with the homeoplanar layer from the previous section. For this layer, ϑ_z is equal to zero and integration of (11.81) with $\mathbf{N}^{\downarrow}$ of the form (11.96) leads to relations (11.66) but for input and output state vectors corresponding to S-normalization. The medium in the test is assumed to be nonabsorbing, and consequently $\sigma_1(z)$ and $\sigma_2(z)$ are real. Therefore, the matrix

$$\widehat{\mathbf{t}}^{\downarrow}(z', z'') = \begin{pmatrix} \exp\left(ik_0 d \int_0^1 \sigma_1(\eta) \mathrm{d}\eta \right) & 0 \\ 0 & \exp\left(ik_0 \sigma_2 d \right) \end{pmatrix}$$

is unitary. S-normalization in this situation is equivalent to the flux (F-) normalization (8.164), and hence we can use (8.282). According to (8.266) and (8.282), the transmittance of a system characterized by a unitary transmission EW Jones matrix under F-normalization is equal to unity at any value of the state vector of the input light. Thus, in this test, equation (11.81) gives the expected result: $t(z', z'') = 1$. This result may also be verified directly, by using (11.56), (11.58), (11.59), (11.68), and (8.242).

11.3 NBRA. Numerical Methods

In this section, we present three numerical methods for calculating EW Jones matrices of smoothly inhomogeneous layers in the NBR approximation.

For the sake of convenience, in subsequent formulas we often use the normalized coordinate η in place of z [see (11.2)]. In terms of η, we may write the general equation for finding the transmission matrix of an arbitrary part (η', η'') $(0 \le \eta' < \eta'' \le 1)$ of an inhomogeneous layer in the NBR approximation as follows:

$$\frac{\mathrm{d}\widehat{\mathbf{t}}^{\downarrow}(\eta', \eta)}{\mathrm{d}\eta} = \left(\mathbf{N}_{\sigma}^{\downarrow}(\eta) - \mathbf{N}_{\psi}^{\downarrow}(\eta) \right) \widehat{\mathbf{t}}^{\downarrow}(\eta', \eta) \quad [\widehat{\mathbf{t}}^{\downarrow}(\eta', \eta') = \mathbf{U}], \tag{11.111}$$

where

$$\mathbf{N}_{\sigma}^{\downarrow}(\eta) \equiv \begin{pmatrix} ik_0\sigma_1(\eta)d & 0 \\ 0 & ik_0\sigma_2(\eta)d \end{pmatrix}, \quad \mathbf{N}_{\psi}^{\downarrow}(\eta) \equiv \begin{pmatrix} \bar{\boldsymbol{\psi}}_1(\eta)\boldsymbol{\psi}_{\eta 1}(\eta) & \bar{\boldsymbol{\psi}}_1(\eta)\boldsymbol{\psi}_{\eta 2}(\eta) \\ \bar{\boldsymbol{\psi}}_2(\eta)\boldsymbol{\psi}_{\eta 1}(\eta) & \bar{\boldsymbol{\psi}}_2(\eta)\boldsymbol{\psi}_{\eta 2}(\eta) \end{pmatrix}, \tag{11.112}$$

$$\boldsymbol{\psi}_{\eta k} \equiv d\boldsymbol{\psi}_k / d\eta \tag{11.113}$$

[see (11.83) and (11.84)]. The matrix of the layer, $\widehat{\mathbf{t}}_{\mathrm{L}}^{\downarrow} \equiv \widehat{\mathbf{t}}^{\downarrow}(0,1)$, may be calculated by integrating (11.111) with $\eta' = 0$. The first of the three methods, which is described in Section 11.3.1, is a general method of integrating (11.111). This method is applicable whatever be $\sigma_k(\eta)$ and $\boldsymbol{\psi}_k(\eta)$.

The simplified equation for the case of an optically locally centrosymmetric medium in terms of η may be written as follows:

$$\frac{\mathrm{d}\widehat{\mathbf{t}}^{\downarrow}(0, \eta)}{\mathrm{d}\eta} = \begin{pmatrix} ik_0\sigma_1(\eta)d & \vartheta_\eta(\eta) \\ -\vartheta_\eta(\eta) & ik_0\sigma_2(\eta)d \end{pmatrix} \widehat{\mathbf{t}}^{\downarrow}(0, \eta), \quad \widehat{\mathbf{t}}^{\downarrow}(0, 0) = \mathbf{U}, \tag{11.114}$$

$$\vartheta_\eta = -\boldsymbol{\psi}_1^{\mathrm{T}}\mathbf{I}_0\boldsymbol{\psi}_{\eta 2} = \boldsymbol{\psi}_2^{\mathrm{T}}\mathbf{I}_0\boldsymbol{\psi}_{\eta 1}$$

[see (11.96)]. If the medium is nonabsorbing, with real-valued $\sigma_k(\eta)$ and $\vartheta_\eta(\eta)$, the matrix $\widehat{\mathbf{t}}^{\downarrow}_{\mathrm{L}}$ may be represented as

$$\widehat{\mathbf{t}}^{\downarrow}_{\mathrm{L}} = e_{\mathrm{p}}\widehat{R}(\vartheta(1))\tilde{\mathbf{t}}, \quad \tilde{\mathbf{t}} \equiv \begin{pmatrix} c_1 & c_2 \\ -c_2^* & c_1^* \end{pmatrix} \equiv \breve{\mathbf{t}}(1), \tag{11.115}$$

$$e_{\mathrm{p}} \equiv \exp\left\{\frac{i\pi d}{\lambda}\int_0^1 (\sigma_1(\eta)+\sigma_2(\eta))\,d\eta\right\}, \quad \widehat{R}(\varphi) \equiv \begin{pmatrix} \cos\varphi & \sin\varphi \\ -\sin\varphi & \cos\varphi \end{pmatrix}, \tag{11.116}$$

$$\vartheta(\eta) = \int_0^\eta \vartheta_\eta(\bar{\eta})d\bar{\eta}, \tag{11.117}$$

where $\breve{\mathbf{t}}(\eta)$ is the solution of the equation

$$\frac{d\breve{\mathbf{t}}(\eta)}{d\eta} = \frac{i\pi d}{\lambda}\Delta\sigma(\eta)\begin{pmatrix} \cos 2\vartheta(\eta) & \sin 2\vartheta(\eta) \\ \sin 2\vartheta(\eta) & -\cos 2\vartheta(\eta) \end{pmatrix}\breve{\mathbf{t}}(\eta) \quad [\breve{\mathbf{t}}(0) = \mathbf{U}]; \quad \Delta\sigma(\eta) = \sigma_1(\eta) - \sigma_2(\eta). \tag{11.118}$$

How this representation may be derived from (11.114) is evident from the discussion of differential equations of the Jones method in Section 11.1.1 [see (11.16), (11.27), and (11.22)]. Equation (11.118) is mathematically equivalent to (11.22). The two computational methods that will be presented in Sections 11.3.2 and 11.3.3 use the same mathematical approaches that are used to solve (11.22). In both methods, great efficiency is attained due to the fact that the matrix $\breve{\mathbf{t}}(\eta)$ for any $\eta \in [0, 1]$ is unitary and has the form (5.31).

11.3.1 Approximating Multilayer Method

In this section, we will consider the most universal method of the three numerical methods being presented. Unlike the other two methods, this method is applicable to absorbing media and, generally speaking, to any media characterized by constitutive relations of the form (8.5) [8]. We have briefly described it in Section 8.2.1 but without mentioning NBRA. The method is based on the approximation of the transmission matrix of the smoothly inhomogeneous layer by the transmission matrix of the elementary transmission channel of a staircase model (approximating multilayer) of this layer, the channel defined by the chain of transmission operations performed by the bulks of sublayers of the staircase model and interfaces between them (see Figure 8.15). We call it the *approximating multilayer method* (AMM).

Here is a convenient AMM-based algorithm for the case of a medium characterized by constitutive relations (8.6). Let the values of the permittivity tensor of the medium, we denote it by $\tilde{\varepsilon}$, be given at the nodes $\eta_j = j/N, j = 0, 1, \ldots, N$. We assign the dielectric tensor to the staircase model of this medium as follows:

$$\varepsilon(\eta) = \tilde{\varepsilon}(\eta_j) \quad \text{for } \bar{\eta}_j < \eta < \bar{\eta}_{j+1}, \quad j = 0, 1, \ldots, N;$$
$$\bar{\eta}_l = \begin{cases} 0 & l = 0 \\ (\eta_l + \eta_{l+1})/2 & l = 1, 2, \ldots, N \\ 1 & l = N+1; \end{cases} \tag{11.119}$$

$N+1$ is the number of sublayers in the staircase model. The transmission matrix approximating the matrix $\widehat{\mathbf{t}}_{\mathrm{L}}^{\downarrow}$ in this case is calculated as

$$\mathbf{t}_{\mathrm{a}N}^{\downarrow} = \left\{ {}^{N}\prod_{j=1} \left(\mathbf{t}_j^{\downarrow(\mathrm{B})} \mathbf{t}_j^{\downarrow(\mathrm{I})} \right) \right\} \mathbf{t}_0^{\downarrow(\mathrm{B})}, \tag{11.120}$$

where $\mathbf{t}_j^{\downarrow(\mathrm{B})}$ is the transmission matrix of the bulk of the jth sublayer (this is the layer with the boundaries $\eta = \bar{\eta}_j$ and $\eta = \bar{\eta}_{j+1}$), expressed by

$$\mathbf{t}_j^{\downarrow(\mathrm{B})} = \widehat{P}(\eta_j, \Delta\eta_j),$$

$$\widehat{P}(\bar{\eta}, \Delta\bar{\eta}) \equiv \begin{pmatrix} \exp(ik_0\sigma_1(\bar{\eta})\Delta\bar{\eta}d) & 0 \\ 0 & \exp(ik_0\sigma_2(\bar{\eta})\Delta\bar{\eta}d) \end{pmatrix}, \tag{11.121}$$

$$\Delta\eta_j = \begin{cases} 1/N & j = 1, 2, \ldots, N-1 \\ 1/(2N) & j = 0, N; \end{cases}$$

$\mathbf{t}_j^{\downarrow(\mathrm{I})}$ is the transmission matrix of the interface between the $j-1$th and jth layers of the approximating multilayer. The interface matrix $\mathbf{t}_j^{\downarrow(\mathrm{I})}$ in the most general case may be expressed as

$$\mathbf{t}_j^{\downarrow(\mathrm{I})} = \mathbf{w}_j^{-1} = \frac{1}{\tilde{t}_{j11}\tilde{t}_{j22} - \tilde{t}_{j12}\tilde{t}_{j21}} \begin{pmatrix} \tilde{t}_{j22} & -\tilde{t}_{j12} \\ -\tilde{t}_{j21} & \tilde{t}_{j11} \end{pmatrix},$$
$$\mathbf{w}_j = \begin{pmatrix} \tilde{t}_{j11} & \tilde{t}_{j12} \\ \tilde{t}_{j21} & \tilde{t}_{j22} \end{pmatrix} \equiv \begin{pmatrix} \bar{\boldsymbol{\psi}}_1(\eta_{j-1})\boldsymbol{\psi}_1(\eta_j) & \bar{\boldsymbol{\psi}}_1(\eta_{j-1})\boldsymbol{\psi}_2(\eta_j) \\ \bar{\boldsymbol{\psi}}_2(\eta_{j-1})\boldsymbol{\psi}_1(\eta_j) & \bar{\boldsymbol{\psi}}_2(\eta_{j-1})\boldsymbol{\psi}_2(\eta_j) \end{pmatrix}, \tag{11.122}$$

$\tilde{t}_{jlk}$ being elements of the matrix $\tilde{T}_j \equiv \boldsymbol{\Psi}(\eta_{j-1})^{-1}\boldsymbol{\Psi}(\eta_j)$ (see Section 8.4.2). One can show that $\widehat{\mathbf{t}}_{\mathrm{L}}^{\downarrow} = \lim_{N\to\infty} \mathbf{t}_{\mathrm{a}N}^{\downarrow}$ and that the following asymptotic estimate is valid:

$$\left\| \widehat{\mathbf{t}}_{\mathrm{L}}^{\downarrow} - \mathbf{t}_{\mathrm{a}N}^{\downarrow} \right\|_{\mathrm{E}} = O(1/N) \tag{11.123}$$

as $N \to \infty$.The fact of convergence of the transmission matrix of the approximating multilayer as $N\to\infty$ to the transmission matrix of the original smoothly inhomogeneous layer in the NBR approximation is not so very obvious. The transmission matrix of the approximating multilayer, $\mathbf{t}_{\mathrm{a}N}^{\downarrow}$, is calculated taking account of single reflections from interfaces, and the illusion may arise that bulk reflection is taken into account. To destroy this illusion, we will derive relation (11.123) here.

Derivation of Relation (11.123)

Let us rewrite expression (11.120) for the matrix $\mathbf{t}_{\mathrm{a}N}^{\downarrow}$ in the form

$$\mathbf{t}_{\mathrm{a}N}^{\downarrow} = {}^{N}\prod_{j=1} \mathbf{t}_{\mathrm{a}j}, \tag{11.124}$$

where $\mathbf{t}_{aj}$ is the transmission matrix of the region (η_{j-1}, η_j) of the approximating multilayer, which can be expressed as

$$\mathbf{t}_{aj} = \widehat{P}(\eta_j, \Delta\eta/2)\mathbf{t}_j^{\downarrow(\mathrm{I})}\widehat{P}(\eta_{j-1}, \Delta\eta/2), \tag{11.125}$$

where $\Delta\eta = \eta_j - \eta_{j-1} = 1/N$. Let us obtain an asymptotic estimate for the matrix $\mathbf{t}_{aj}$ with $\Delta\eta \to 0$ $(N \to \infty)$. Representing the matrix $\mathbf{\Psi}(\eta_j)$ in the form

$$\mathbf{\Psi}(\eta_j) = \mathbf{\Psi}(\eta_{j-1}) + \mathbf{\Psi}_\eta(\eta_{j-1})\Delta\eta + O(\Delta\eta^2) \quad (\Delta\eta \to 0),$$

we find that

$$\tilde{T}_j = \mathbf{U} + \mathbf{\Psi}(\eta_{j-1})^{-1}\mathbf{\Psi}_\eta(\eta_{j-1})\Delta\eta + O(\Delta\eta^2) \quad (\Delta\eta \to 0)$$

[see (11.122)]. From this equation it follows that, as $\Delta\eta \to 0$,

$$\mathbf{w}_j = \mathbf{U} + \mathbf{N}_\psi^\downarrow(\eta_{j-1})\Delta\eta + O(\Delta\eta^2). \tag{11.126}$$

Using (11.126) and the algebraic expression

$$(\mathbf{U} + \mathbf{H})^{-1} = \mathbf{U} + \sum_{k=1}^{\infty}(-\mathbf{H})^k,$$

which is valid for any matrix $\mathbf{H}$ with $\|\mathbf{H}\| < 1$ if the matrix $\mathbf{U} + \mathbf{H}$ is invertible (see Section 5.1.6), we obtain the following estimate for the interface matrix $\mathbf{t}_j^{\downarrow(\mathrm{I})}$:

$$\mathbf{t}_j^{\downarrow(\mathrm{I})} = \mathbf{U} - \mathbf{N}_\psi^\downarrow(\eta_{j-1})\Delta\eta + O(\Delta\eta^2) \quad (\Delta\eta \to 0). \tag{11.127}$$

Proceeding from (11.121), we can obtain similar estimates for the matrices $\widehat{P}(\eta_{j-1}, \Delta\eta/2)$ and $\widehat{P}(\eta_j, \Delta\eta/2)$ for $\Delta\eta \to 0$:

$$\widehat{P}(\eta_{j-1}, \Delta\eta/2) = \mathbf{U} + \mathbf{N}_\sigma^\downarrow(\eta_{j-1})\Delta\eta/2 + O(\Delta\eta^2), \tag{11.128}$$

$$\widehat{P}(\eta_j, \Delta\eta/2) = \mathbf{U} + \mathbf{N}_\sigma^\downarrow(\eta_{j-1})\Delta\eta/2 + O(\Delta\eta^2). \tag{11.129}$$

To obtain (11.129), one can use the relation

$$\mathbf{N}_\sigma^\downarrow(\eta_j) = \mathbf{N}_\sigma^\downarrow(\eta_{j-1}) + \mathbf{N}_{\sigma\eta}^\downarrow(\eta_{j-1})\Delta\eta + O(\Delta\eta^2) \quad (\Delta\eta \to 0),$$

where $\mathbf{N}_{\sigma\eta}^\downarrow \equiv d\mathbf{N}_\sigma^\downarrow/d\eta$. Substitution of (11.127)–(11.129) into (11.125) leads to the following expression for the matrix $\mathbf{t}_{aj}$:

$$\mathbf{t}_{aj} = \mathbf{U} + \left(\mathbf{N}_\sigma^\downarrow(\eta_{j-1}) - \mathbf{N}_\psi^\downarrow(\eta_{j-1})\right)\Delta\eta + O(\Delta\eta^2) \quad (\Delta\eta \to 0). \tag{11.130}$$

According to (11.111), the matrix $\widehat{\mathbf{t}}^\downarrow(\eta_{j-1}, \eta_j)$, characterizing the region (η_{j-1}, η_j) of the original medium in the NBR approximation, can also be represented as

$$\widehat{\mathbf{t}}^\downarrow(\eta_{j-1}, \eta_j) = \mathbf{U} + \left(\mathbf{N}_\sigma^\downarrow(\eta_{j-1}) - \mathbf{N}_\psi^\downarrow(\eta_{j-1})\right)\Delta\eta + O(\Delta\eta^2) \quad (\Delta\eta \to 0). \tag{11.131}$$

It follows from (11.130) and (11.131) that

$$\mathbf{t}_{aj} = \widehat{\mathbf{t}}^{\downarrow}(\eta_{j-1}, \eta_j) + O(1/N^2) \quad (N \to \infty) \tag{11.132}$$

(recall that $\Delta\eta = 1/N$). Substitution of (11.132) into (11.124) leads to the relation

$$\mathbf{t}^{\downarrow}_{aN} = \widehat{\mathbf{t}}^{\downarrow}(0, 1) + O(1/N) \quad (N \to \infty),$$

from which we see that estimate (11.123) is correct.

Application of AMM to Liquid Crystal Layers

When applying AMM to LC layers, we may calculate transmission matrices for interfaces between sublayers of the approximating multilayer by the simple formulas of Section 8.2.2. In this case, there is no need to compute parameters of the backward propagating basis waves and deal with 4 × 4 matrices. Use of these formulas and the simplified expressions for EWB parameters of Section 9.3 makes AMM close in computational cost to the variants [2, 3] of EJMM.

Although, according to (11.123), AMM can be considered as a method of integrating (11.111), it does not require knowledge of the matrix $\mathbf{N}^{\downarrow}_{\psi}$. The derivatives of functions $\boldsymbol{\psi}_k(\eta)$ are not used in AMM at all. Except in some special cases, this allows one to choose the eigenwave basis ignoring the requirement that the functions $\boldsymbol{\psi}_k(\eta)$ be continuous. In the methods presented in the following two sections, this requirement must be met.

Numerical Tests

The results of numerical experiments that are presented in Figure 11.11 allow one to estimate the accuracy of AMM and the other methods presented in Section 11.3 and to see how the accuracy of the computed matrix $\widehat{\mathbf{t}}^{\downarrow}_{\mathrm{L}}$ depends on computational parameters of these methods (see also Section 11.5). The computational error Σ, whose values are shown in Figure 11.11, was determined as the Euclidean norm of the difference between the exact and computed values of the matrix $\widehat{\mathbf{t}}^{\downarrow}_{\mathrm{L}}$. The "exact" values of this matrix were calculated with an error less than 1.5×10^{-4}. The calculations were carried out for nonabsorbing locally uniaxial layers with supertwisted configurations (11.86) (configuration A) and (11.87) (configuration B). The data presented in Figures 11.11–11.13 correspond to the case of oblique incidence from air at an angle of 45° (β_{inc}), the angle between the plane of incidence and the XZ plane (α_{inc}) being equal to 45° (Figure 9.1). Since the layers are nonabsorbing, their exact matrices $\widehat{\mathbf{t}}^{\downarrow}_{\mathrm{L}}$ are unitary. Figure 11.12 shows, for comparison, the dependences of the characteristic parameters of the matrix $\widehat{\mathbf{t}}^{\downarrow}_{\mathrm{L}}$

$$t_e \equiv \left| \left[\widehat{\mathbf{t}}^{\downarrow}_{\mathrm{L}} \right]_{11} \right|^2$$

and

$$t_{eo} \equiv \mathrm{Re}\left(\left[\widehat{\mathbf{t}}^{\downarrow}_{\mathrm{L}} \right]^{*}_{11} \left[\widehat{\mathbf{t}}^{\downarrow}_{\mathrm{L}} \right]_{21} \right)$$

for layers with configurations A and B on d.

In practical calculations for LC layers, the following rough estimate of the error of AMM for usual values of $N(\geq 100)$ may be useful: $\Sigma(N, d) \approx A_{\Sigma}/N + dB_{\Sigma}/N^2$. As a rule, with $N \geq 100$, the term A_{Σ}/N

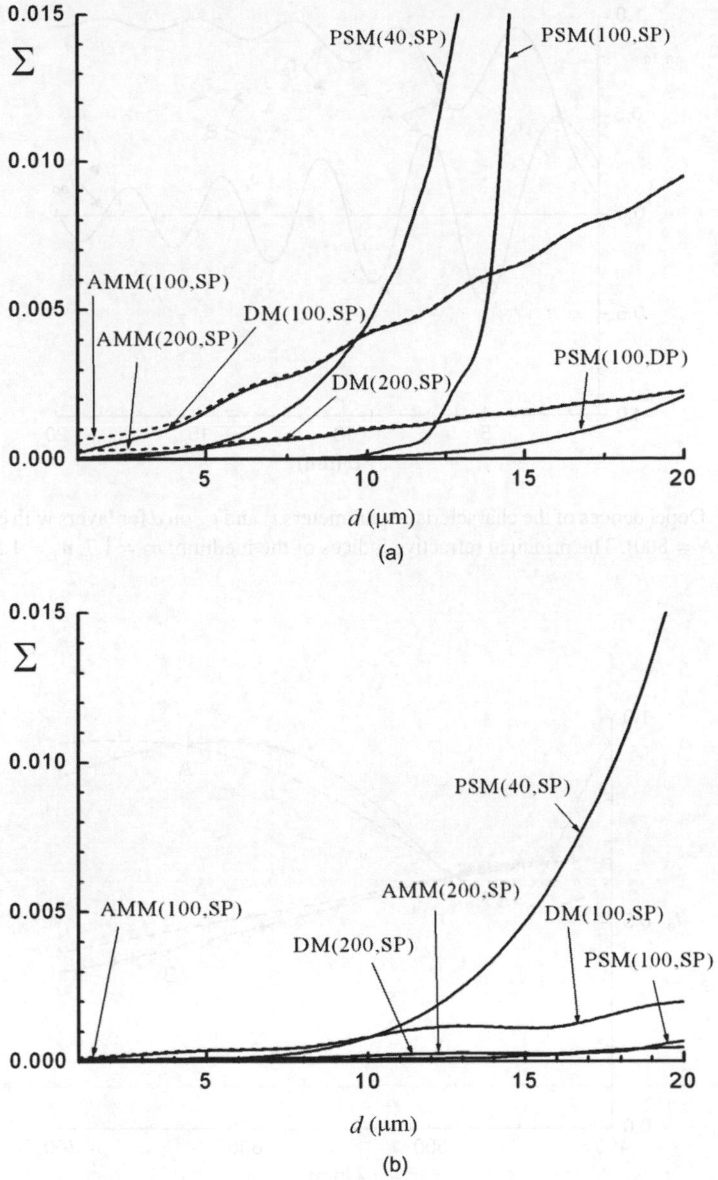

Figure 11.11 Dependences of the accuracy of calculated values of the matrix $\widehat{\mathbf{t}}_{\mathrm{L}}^{\downarrow}$ for layers with configurations A (a) and B (b) on the layer thickness d for AMM, DM, and PSM [8]. The number in parentheses after the name of the method is the value of N (for AMM and DM) or M (for PSM); SP and DP denote the use of single- and double-precision arithmetic, respectively. Principal refractive indices: $n_{\parallel} = 1.7$, $n_{\perp} = 1.5$. Wavelength $\lambda = 550$ nm

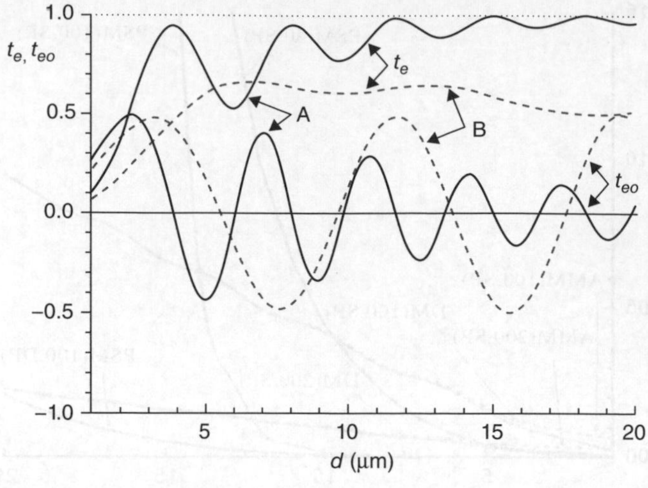

Figure 11.12 Dependences of the characteristic parameters t_e and t_{eo} on d for layers with configurations A and B (DM, $N = 800$). The principal refractive indices of the medium: $n_{\|} = 1.7, n_{\perp} = 1.5$; wavelength $\lambda = 550$ nm

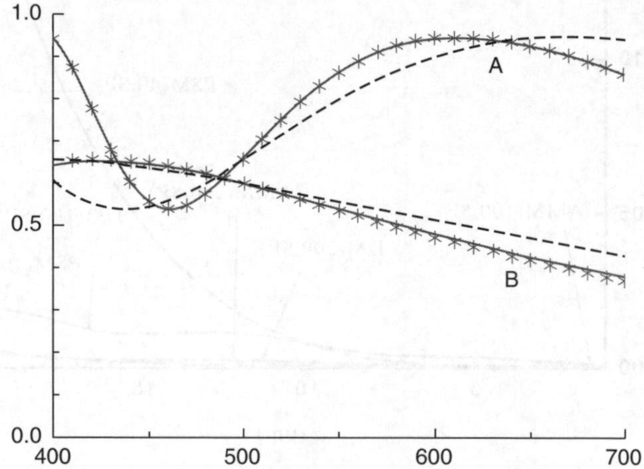

Figure 11.13 Effect of the approximate correction for the dispersion of refractive indices in PSM. The solid curves show the exactly calculated spectra of the parameter t_e for layers with configurations A and B; $d = 5$ μm; the wavelength dependence of the principal refractive indices is identical to that of nematic 5CB at 27°C. The dashed curves and asterisks show the same spectra but calculated approximately, using PSM with $\lambda_c = 500$ nm. The spectra shown by dashed lines were calculated ignoring the dispersion, and those shown by asterisks with the approximate correction for the dispersion

is very small. For large values of d, if N is not very large, doubling N decreases the error by approximately a factor 4 (see Figure 11.11).

11.3.2 Discretization Method

Let us assume that the medium is nonabsorbing and the EWB parameters are specified for the same nodes $\eta_j = j/N$ ($j = 0, 1, \ldots, N$) as in the above realization of AMM. Application of DM (see Section 11.1.1) to integrate (11.118) yields the following computational formula for the matrix $\tilde{\mathbf{t}}$ [see (11.115)]:

$$\tilde{\mathbf{t}} \cong \tilde{\mathbf{t}}_N \tilde{\mathbf{t}}_{N-1} \ldots \tilde{\mathbf{t}}_1 \tilde{\mathbf{t}}_0, \tag{11.133}$$

where

$$\tilde{\mathbf{t}}_j = \widehat{R}_{\mathrm{C}}(-\vartheta_j) \begin{pmatrix} e^{i\delta_j} & 0 \\ 0 & e^{-i\delta_j} \end{pmatrix} \widehat{R}_{\mathrm{C}}(\vartheta_j) = \begin{pmatrix} a_1^{(j)} + i a_2^{(j)} & i a_4^{(j)} \\ i a_4^{(j)} & a_1^{(j)} - i a_2^{(j)} \end{pmatrix} \tag{11.134}$$

with

$$a_1^{(j)} = \cos\delta_j, \quad a_2^{(j)} = \sin\delta_j \cos 2\vartheta_j, \quad a_4^{(j)} = \sin\delta_j \sin 2\vartheta_j,$$

$$\delta_j = \frac{\pi \Delta\sigma(\eta_j) \Delta\eta_j d}{\lambda}, \quad \vartheta_j \equiv \vartheta(\eta_j)$$

[see (11.8) and (11.9)]. The form of the matrices $\tilde{\mathbf{t}}_j$ allows calculation of the matrix $\tilde{\mathbf{t}}$ [see (11.115)] by the following recurrence formulas:

$$\begin{gathered} c_1^{(0)} = a_1^{(0)}, \quad c_2^{(0)} = a_2^{(0)}, \quad c_3^{(0)} = 0, \quad c_4^{(0)} = a_4^{(0)}, \\ c_1^{(j)} = a_1^{(j)} c_1^{(j-1)} - a_2^{(j)} c_2^{(j-1)} - a_4^{(j)} c_4^{(j-1)}, \quad c_2^{(j)} = a_1^{(j)} c_2^{(j-1)} + a_2^{(j)} c_1^{(j-1)} - a_4^{(j)} c_3^{(j-1)}, \\ c_3^{(j)} = a_1^{(j)} c_3^{(j-1)} - a_2^{(j)} c_4^{(j-1)} + a_4^{(j)} c_2^{(j-1)}, \quad c_4^{(j)} = a_1^{(j)} c_4^{(j-1)} + a_2^{(j)} c_3^{(j-1)} + a_4^{(j)} c_1^{(j-1)}, \\ j = 1, 2, \ldots, N; \\ \mathrm{Re}c_1 = c_1^{(N)}, \quad \mathrm{Im}c_1 = c_2^{(N)}, \quad \mathrm{Re}c_2 = c_3^{(N)}, \quad \mathrm{Im}c_2 = c_4^{(N)}. \end{gathered} \tag{11.135}$$

The values of $\vartheta(\eta)$ at the nodes $\eta = \eta_j$ [see (11.115), (11.117), and (11.134)] can be calculated as

$$\begin{gathered} \vartheta(\eta_j) = \vartheta(\eta_{j-1}) + \Delta\vartheta_j \qquad j = 1, 2, \ldots, N; \\ \vartheta(\eta_0) = 0, \\ \Delta\vartheta_j \cong \arcsin\left[\frac{1}{2}\left(\boldsymbol{\psi}_2(\eta_{j-1})^{\mathrm{T}} \mathbf{I}_0 \boldsymbol{\psi}_1(\eta_j) - \boldsymbol{\psi}_1(\eta_{j-1})^{\mathrm{T}} \mathbf{I}_0 \boldsymbol{\psi}_2(\eta_j)\right)\right]. \end{gathered} \tag{11.136}$$

For locally uniaxial media with spatially invariant principal refractive indices it is more convenient to calculate the increments $\Delta\vartheta_j$ by the formulas

$$\begin{gathered} \Delta\vartheta_j \cong \frac{f_{\mathrm{C}}(\eta_j) + f_{\mathrm{C}}(\eta_{j-1})}{2} \arctan\left(\frac{\mathbf{j}(\eta_j)\mathbf{k}(\eta_{j-1})}{\mathbf{j}(\eta_j)\mathbf{j}(\eta_{j-1})}\right), \\ f_{\mathrm{C}}(\eta) = \frac{1 + (h'/2)\gamma(\eta)}{\sqrt{1 + h'\gamma(\eta) - h''(\eta)\gamma(\eta)^2}}, \end{gathered} \tag{11.137}$$

in terms of quantities defined in subsection *Optically Uniaxial Medium* of Section 11.2.

We have noted that the error of DM decreases as $O(1/N^2)$ as $N \to \infty$. This estimate is supported by the results of the numerical experiments (see Figure 11.11). For LC layers of LCDs, sufficiently good accuracy of evaluation of the matrix $\widehat{\mathbf{t}}_{\mathrm{L}}^{\downarrow}$ is usually achieved with $N = 100$–300 for both DM and AMM, the accuracy of DM and AMM at the same N being almost the same (Figure 11.11). With the same N, DM is more efficient than AMM, being approximately 1.5 times faster (when the medium is locally uniaxial and the recurrence formulas (11.135) are used). In DM and the method presented in the next section, the number of operations with complex numbers is relatively small, which is convenient if a programming language without complex arithmetic is used.

11.3.3 Power Series Method

The PSM [8] enables one to rapidly calculate dependences of the matrix $\widehat{\mathbf{t}}_{\mathrm{L}}^{\downarrow}$ of a nonabsorbing layer on the layer thickness and the wavelength with approximate correction for the dispersion of the refractive indices, tens of times faster than DM. The basic theory of PSM is presented in Appendix B.4. Using the approach described in Appendix B.4 to integrate (11.118), one can obtain the following expressions for the elements of the matrix $\tilde{\mathbf{t}}$:

$$c_1 = 1 + \sum_{j=1}^{\infty} q_j' h^j, \quad c_2 = \sum_{j=1}^{\infty} q_j'' h^j, \quad h = \frac{id}{\lambda w}, \tag{11.138a}$$

$$q_j' \equiv \mathrm{Re}(f_j(1)), \quad q_j'' \equiv \mathrm{Im}(f_j(1)), \tag{11.138b}$$

$$f_1(\eta) = \int_0^{\eta} F(\bar{\eta}) d\bar{\eta}, \quad f_j(\eta) = \begin{cases} \displaystyle\int_0^{\eta} F(\bar{\eta})^* f_{j-1}(\bar{\eta}) d\bar{\eta} & j = 2, 4, 6, \ldots \\ \displaystyle\int_0^{\eta} F(\bar{\eta}) f_{j-1}(\bar{\eta}) d\bar{\eta} & j = 3, 5, 7, \ldots \end{cases}, \tag{11.138c}$$

$$F(\eta) = w\pi\Delta\sigma(\eta)\exp(2i\vartheta(\eta)), \tag{11.138d}$$

where w is a real number (see below). Representation (11.138) may be directly used to evaluate c_1 and c_2. Let $d_{\max}$ be the maximum value of d for which we want to calculate the matrix $\tilde{\mathbf{t}}$. The minimum number of terms in the series (11.138a) that must be taken into account for evaluating the matrix $\tilde{\mathbf{t}}$ with a prescribed tolerance ε_t for $d < d_{\max}$ is equal to the least value of J satisfying the inequality

$$|(\pi\bar{\delta})^J / J!| < \varepsilon_t, \tag{11.139}$$

where

$$\bar{\delta} = \Delta\widehat{\sigma} d_{\max}/\lambda, \quad \Delta\widehat{\sigma} = \int_0^1 \Delta\sigma(\eta) d\eta. \tag{11.140}$$

Integral parameters $f_j(1)$ which enter into formulas for coefficients of the series may be calculated recursively. When calculating $f_1(1)$, one obtains the values of $F(\eta_m)$ and $f_1(\eta_m)$, where $\eta_m = m/M$ ($m = 0,1,\ldots,M$) are the integration nodes. The calculated $F(\eta_m)$ and $f_1(\eta_m)$ are then used in calculating $f_2(1)$. The values of $f_2(\eta_m)$, obtained during the calculation of $f_2(1)$, are used to calculate $f_3(\eta_m)$, and so on. The integrals in (11.138c) may be calculated with the help of quadrature formulas based on quadratic approximation (like Simpson's formula). The recurrent formulas for computing $f_j(\eta_m)$ ($j = 1, 2, \ldots, J$) in this case are

$$\begin{gathered} f_j(\eta_0) = \bar{f}_0, \\ f_j(\eta_{m+1}) = f_j(\eta_m) + \left[5\bar{f}_m + 8\bar{f}_{m+1} - \bar{f}_{m+2}\right]/12, \\ f_j(\eta_{m+2}) = f_j(\eta_m) + \left[\bar{f}_m + 4\bar{f}_{m+1} + \bar{f}_{m+2}\right]/3, \\ m = 0, 2, 4, \ldots, M-2, \end{gathered}$$

where

$$\begin{gathered} \bar{f}_m = \begin{cases} F_m & j = 1 \\ F^*_m f_{j-1}(\eta_m) & j = 2, 4, 6, \ldots \\ F_m f_{j-1}(\eta_m) & j = 3, 5, 7, \ldots \end{cases} \\ F_m = F(\eta_m)/M, \quad m = 0, 1, 2, \ldots, M; \end{gathered}$$

M is assumed to be an even number. As a rule, sufficient accuracy for LC layers of LCDs is achieved with $M = 40$. The factor w in (11.138) is a computational parameter introduced to decrease the effect of round-off errors. Numerical experiments showed that in usual calculations, $w = 20$ is a very good choice.

In calculating spectra for a wavelength region $[\lambda_{\min}, \lambda_{\max}]$ the dispersion of the refractive indices can be approximately taken into account by using as the expansion parameter the quantity

$$h = \frac{i\Delta\widehat{\sigma}(\lambda)d}{\Delta\widehat{\sigma}(\lambda_c)w\lambda},$$

where λ_c is the wavelength for which coefficients of the series are computed; $\lambda_c \in [\lambda_{\min}, \lambda_{\max}]$. The value of λ_c should be chosen so that the value of $\Delta\widehat{\sigma}(\lambda_c)$ will be close to the mean value of $\Delta\widehat{\sigma}(\lambda)$ in the region $[\lambda_{\min}, \lambda_{\max}]$. For common nematic and cholesteric materials, without significant loss in accuracy, the values of $\Delta\widehat{\sigma}(\lambda)$ may be calculated by the formula

$$\Delta\widehat{\sigma}(\lambda) \simeq \Delta\widehat{\sigma}_P(\lambda) = A + B/\lambda^2 + C/\lambda^4$$

with coefficients A, B, and C determined from the conditions

$$\Delta\widehat{\sigma}_P(\lambda_c) = \Delta\widehat{\sigma}(\lambda_c), \quad \Delta\widehat{\sigma}_P(\lambda_{\max}) = \Delta\widehat{\sigma}(\lambda_{\max}), \quad \Delta\widehat{\sigma}_P(\lambda_{\min}) = \Delta\widehat{\sigma}(\lambda_{\min}),$$

where the right-hand-side quantities are calculated by the exact formula (11.140). In estimating the number of necessary terms of the series in the case of the approximate correction for dispersion of refractive indices [see (11.139)], one should use the following expression for $\bar{\delta}$:

$$\bar{\delta} = \Delta\widehat{\sigma}(\lambda_{\min})d_{\max}/\lambda_{\min}.$$

An example demonstrating the effect of the approximate correction for dispersion of refractive indices in calculating spectra for the visible region in a typical situation is presented in Figure 11.13. As a rule, the accuracy of calculating the transmittance spectra of an LCD panel with the use of this correction is sufficient for accurate estimation of its colorimetric characteristics.

11.4 NBRA. Analytical Solutions

As in the case of the classical Jones approach (see Section 11.1.1), we may distinguish two basic cases when exact analytical expressions for the matrix $\widehat{\mathbf{t}}_{\mathrm{L}}^{\downarrow}$ of an inhomogeneous layer can be obtained:

(i) the case where the matrix

$$\mathbf{N}_{\eta}^{\downarrow} \equiv \mathbf{N}_{\sigma}^{\downarrow} - \mathbf{N}_{\psi}^{\downarrow} \tag{11.141}$$

[see (11.111)] is independent of η; and

(ii) the case where the matrix $\mathbf{N}_{\eta}^{\downarrow}$ is diagonal.

The first case is considered in Section 11.4.1, the second in Section 11.4.2. Along with the exact solutions, we will consider some allied approximate solutions expressed by analytical formulas. In particular, in Section 11.4.3 the adiabatic and quasiadiabatic approximations are discussed.

11.4.1 Twisted Structures

Ideal Twisted Layer

Fulfillment of the condition $\mathbf{N}_{\eta}^{\downarrow} = const$ can be secured by proper choice of the eigenwave basis in considering the case of normal incidence on an ideal twisted layer. The spatial dependence of the permittivity tensor in a layer with an ideal twisted structure can be represented as

$$\varepsilon(\eta) = \mathbf{R}_{3Z}(\Phi\eta)\varepsilon(0)\mathbf{R}_{3Z}(-\Phi\eta),$$

where $\mathbf{R}_{3Z}(\varphi)$ is a rotation matrix given by

$$\mathbf{R}_{3Z}(\varphi) = \begin{pmatrix} \cos\varphi & -\sin\varphi & 0 \\ \sin\varphi & \cos\varphi & 0 \\ 0 & 0 & 1 \end{pmatrix},$$

and Φ is the twist angle of the structure. In the case of normal incidence, the eigenwave basis in this layer can be chosen such that

$$\sigma_1(\eta) = \bar{\sigma}_1 \equiv \sigma_1(0), \quad \sigma_2(\eta) = \bar{\sigma}_2 \equiv \sigma_2(0), \tag{11.142}$$

$$\mathbf{e}_k(\eta) = \mathbf{R}_{3Z}(\Phi\eta)\mathbf{e}_k(0), \quad \mathbf{h}_k(\eta) = \mathbf{R}_{3Z}(\Phi\eta)\mathbf{h}_k(0). \tag{11.143}$$

Having chosen such a basis, we may represent the spatial dependence of the vectors $\boldsymbol{\psi}_k$ as follows:

$$\boldsymbol{\psi}_k(\eta) = \mathbf{R}_{\mathrm{B}}(\Phi\eta)\boldsymbol{\psi}_k(0), \tag{11.144}$$

where

$$\mathbf{R}_{\mathrm{B}}(\varphi) = \begin{pmatrix} \cos\varphi & 0 & -\sin\varphi & 0 \\ 0 & \cos\varphi & 0 & -\sin\varphi \\ \sin\varphi & 0 & \cos\varphi & 0 \\ 0 & \sin\varphi & 0 & \cos\varphi \end{pmatrix}$$

$[\mathbf{R}_B(\varphi) = \mathbf{R}_{BR}(-\varphi)$, see (8.152)]. According to (11.144), the derivative of $\boldsymbol{\psi}_k$ with respect to η may be expressed as

$$\boldsymbol{\psi}_{\eta k}(\eta) = \Phi \mathbf{R}_B\left(\Phi\eta + \frac{\pi}{2}\right)\boldsymbol{\psi}_k(0). \tag{11.145}$$

Using (11.145), we can obtain the following expressions for the elements of $\mathbf{N}^{\downarrow}_{\psi}(\eta)$:

$$\begin{aligned}\left[\mathbf{N}^{\downarrow}_{\psi}(\eta)\right]_{jk} &= \bar{\boldsymbol{\psi}}_j(\eta)\boldsymbol{\psi}_{\eta k}(\eta)\\ &= \Phi\bar{\boldsymbol{\psi}}_j(0)\mathbf{R}_B(-\Phi\eta)\mathbf{R}_B\left(\Phi\eta+\frac{\pi}{2}\right)\boldsymbol{\psi}_k(0) = \Phi\left(\bar{\boldsymbol{\psi}}_j(0)\mathbf{R}_B\left(\frac{\pi}{2}\right)\boldsymbol{\psi}_k(0)\right).\end{aligned} \tag{11.146}$$

From (11.112), (11.142), and (11.146) we see that with the chosen eigenwave basis, the elements of the matrix $\mathbf{N}^{\downarrow}_{\eta}$ do not depend on η and can be expressed in terms of parameters of this basis in the plane $\eta = 0$.

If the medium is optically *locally centrosymmetric*, under S-normalization, the matrix $\mathbf{N}^{\downarrow}_{\psi}$ has the form

$$\mathbf{N}^{\downarrow}_{\psi} = \begin{pmatrix} 0 & -\vartheta_\eta \\ \vartheta_\eta & 0 \end{pmatrix}, \tag{11.147}$$

where

$$\vartheta_\eta = -\boldsymbol{\psi}_1^{\mathrm{T}}\mathbf{I}_0\boldsymbol{\psi}_{\eta 2} = \boldsymbol{\psi}_2^{\mathrm{T}}\mathbf{I}_0\boldsymbol{\psi}_{\eta 1} \tag{11.148}$$

[see (11.114) and (11.112)]. The parameter ϑ_η in the case in question is independent of η. Using (11.148) and (11.145), we may express this parameter as follows:

$$\vartheta_\eta = \bar{\Phi} \equiv -\Phi\left(\boldsymbol{\psi}_1(0)^{\mathrm{T}}\mathbf{I}_4\boldsymbol{\psi}_2(0)\right), \tag{11.149}$$

where

$$\mathbf{I}_4 = \begin{pmatrix} 0 & 0 & 0 & -1 \\ 0 & 0 & -1 & 0 \\ 0 & 1 & 0 & 0 \\ 1 & 0 & 0 & 0 \end{pmatrix}. \tag{11.150}$$

From (11.114), on the assumption that σ_1, σ_2, and ϑ_η are independent of η, by analogy with (11.41), we may obtain the following expression for the matrix $\mathbf{t}^{\downarrow}_{\mathrm{L}}$:

$$\widehat{\mathbf{t}}^{\downarrow}_{\mathrm{L}} = C_{AP}\begin{pmatrix} \cos Q + i\dfrac{G}{Q}\sin Q & \dfrac{\bar{\Phi}}{Q}\sin Q \\ -\dfrac{\bar{\Phi}}{Q}\sin Q & \cos Q - i\dfrac{G}{Q}\sin Q \end{pmatrix}, \tag{11.151}$$
$$G = \frac{\pi(\bar{\sigma}_1 - \bar{\sigma}_2)d}{\lambda}, \quad Q = \sqrt{G^2 + \bar{\Phi}^2}$$

[see also (11.9c)]. Here, in order for this expression for $\widehat{\mathbf{t}}^{\downarrow}_{\mathrm{L}}$ to be more similar to expression (11.41) for the Jones matrix of a twisted layer, we use $\bar{\Phi}$ in place of ϑ_η [see (11.149)]. Expression (11.151) is very general, being valid for both locally uniaxial and locally biaxial media, both nonabsorbing and absorbing [see (11.43)].

In the case of a *locally uniaxial medium* [see (11.40)], the basic parameters of formula (11.151) can be explicitly expressed in terms of parameters of the medium:

$$\bar{\sigma}_1 = \frac{n_\perp n_\parallel}{\sqrt{n_\perp^2 + \left(n_\parallel^2 - n_\perp^2\right)\sin^2\theta_c}}, \quad \bar{\sigma}_2 = n_\perp, \tag{11.152}$$

$$\bar{\Phi} = \Phi\left(\frac{1+\gamma/2}{\sqrt{1+\gamma}}\right), \tag{11.153}$$

where

$$\gamma = \frac{\bar{\sigma}_1 - n_\perp}{n_\perp}. \tag{11.154}$$

Equation (11.153) may easily be obtained from (11.106) by using the relations $\vartheta_\eta = \vartheta_z d$ and $\upsilon_z = \Phi/d$. Comparing expressions (11.41)–(11.42) with (11.151)–(11.152), we see that these expressions are alike except that expression (11.151) is for the matrix $\widehat{\mathbf{t}}^{\downarrow}_{\mathrm{L}} \equiv \widehat{\mathbf{t}}^{\downarrow}(z', z'')$ rather than $\mathbf{t}_{x'-y'}(z', z'')$ and contains $\bar{\Phi}$ rather than Φ. As can be seen from (11.153), in the case of low birefringence, $\widehat{\mathbf{t}}^{\downarrow}_{\mathrm{L}} \approx \mathbf{t}_{x'-y'}(z', z'')$ since $\bar{\Phi}$ differs little from Φ because of the smallness of γ. One may also notice that with decreasing γ the difference between $\bar{\Phi}$ and Φ decreases proportionally to γ^2. But again [see the paragraph under equation (11.110)] we must say that the closeness of the matrices $\widehat{\mathbf{t}}^{\downarrow}(z', z'')$ and $\mathbf{t}_{x'-y'}(z', z'')$ gives no grounds to conclude that, physically, a high degree of correspondence exists between the solutions considered, as the matrices $\widehat{\mathbf{t}}^{\downarrow}(z', z'')$ and $\mathbf{t}_{x'-y'}(z', z'')$ are associated with state vectors of different nature.

A Quasi-Planar Twisted Nonabsorbing LC Layer

As in the case of the classical Jones method (see Section 11.1), for quasi-planar twisted layers, as a rule, a rather accurate estimate for the matrix $\widehat{\mathbf{t}}^{\downarrow}(z', z'')$ is obtained when approximate values of this matrix are calculated by formulas (11.151) with G calculated by (11.44) or (11.46). In this case γ is approximately calculated as $\gamma \approx (n_{||} - n_\perp)/n_\perp$.

11.4.2 Nontwisted Structures

In the cases where the matrix $\mathbf{N}^{\downarrow}_{\eta}(\eta)$ is diagonal for all $\eta \in [0,1]$, the resulting expression for the matrix $\widehat{\mathbf{t}}^{\downarrow}_{\mathrm{L}}$ is the simplest, namely

$$\widehat{\mathbf{t}}^{\downarrow}_{\mathrm{L}} = \begin{pmatrix} \exp\left\{ik_0 d \int_0^1 \sigma_1(\eta)\mathrm{d}\eta\right\} & 0 \\ 0 & \exp\left\{ik_0 d \int_0^1 \sigma_2(\eta)\mathrm{d}\eta\right\} \end{pmatrix}, \tag{11.155}$$

when the diagonal elements of the matrix $\mathbf{N}_{\psi}^{\downarrow}(\eta)$ are zero. As we know, for an arbitrary optically locally centrosymmetric medium, the diagonal elements of the matrix $\mathbf{N}_{\psi}^{\downarrow}(\eta)$ can be made equal to zero by using S-normalization, which leads to (11.114). For simplicity, in this section and Section 11.4.3, we assume that (11.114) holds. From (11.114), we may write the validity condition for (11.155) as follows:

$$\vartheta_{\eta}(\eta) = 0 \text{ for all } \eta \in [0, 1]. \tag{11.156}$$

Here is a typical situation where solution (11.155) may be applied. Let us consider a nematic layer with a nontwisted structure ($\varphi(\eta) = \varphi_1$, $\theta(\eta)$ is arbitrary). As may be seen from (11.106), (11.103), and (11.98), condition (11.156) is satisfied if the vibration vectors ($\mathbf{e}_2$, $\mathbf{h}_2$) of the ordinary basis wave do not depend on η throughout the layer. Thus it is clear that for the layer under consideration condition (11.156) will be satisfied at normal incidence as well as at oblique incidence if the plane of incidence is oriented so that $\alpha_{\text{inc}} = \varphi_1$ or $\alpha_{\text{inc}} = \varphi_1 + \pi$ (see Figures 11.14 and 9.1). The matrix $\widehat{\mathbf{t}}_{\text{L}}^{\downarrow}$ for these cases may be written as

$$\widehat{\mathbf{t}}_{\text{L}}^{\downarrow} = \begin{pmatrix} \exp\left\{ ik_0 d \int\limits_0^1 \sigma_1(\eta)\mathrm{d}\eta \right\} & 0 \\ 0 & \exp(ik_0 d\bar{\sigma}_2) \end{pmatrix}, \tag{11.157}$$

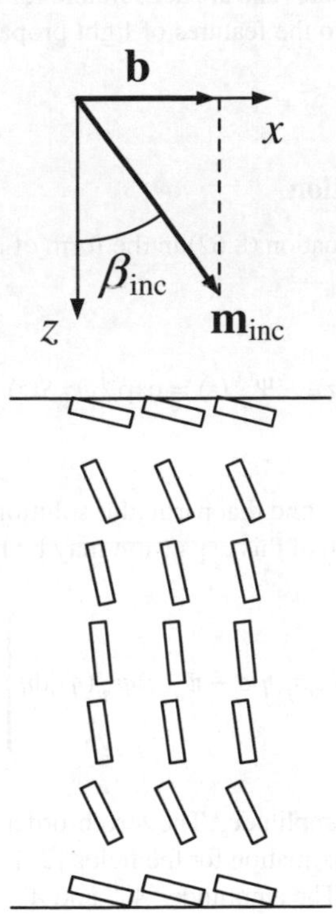

Figure 11.14 A nontwisted layer. The orientation of the incident plane (x–z) for which the transmission matrix $\widehat{\mathbf{t}}_{\text{L}}^{\downarrow}$ has a simple form

with

$$\sigma_1 = \frac{n_\parallel \sqrt{1 + \delta_\varepsilon \sin^2\theta - (\zeta/n_\perp)^2} - \delta_\varepsilon p_\varphi \zeta \sin\theta\cos\theta}{1 + \delta_\varepsilon \sin^2\theta},$$

$$\bar{\sigma}_2 = \sqrt{n_\perp^2 - \zeta^2}, \quad \zeta = |\mathbf{b}|, \quad \delta_\varepsilon \equiv \frac{n_\parallel^2}{n_\perp^2} - 1,$$

$$p_\varphi = \begin{cases} 1 & \alpha_{\text{inc}} = \varphi_1 \\ -1 & \alpha_{\text{inc}} = \varphi_1 + \pi. \end{cases}$$

It should be noted that this solution agrees completely with expressions for the wave fields obtained for this case in [23] by using GOA.

11.4.3 NBRA and GOA. Adiabatic and Quasiadiabatic Approximations

In this section, we consider two approximations, namely, the geometrical optics approximation (GOA) and quasiadiabatic approximation (QAA), which give relatively simple expressions for the matrix $\widehat{\mathbf{t}}_{\text{L}}^{\downarrow}$, being applicable to a much wider variety of layer configurations than the exact analytical solutions discussed above without strict restrictions on the light incidence direction. Although in many cases these approximations yield only rough estimates and are not suitable for accurate numerical modeling, they are very useful, giving a better insight into the features of light propagation in inhomogencous anisotropic media.

Geometrical Optics Approximation

Seeking solutions of the Berreman equation (8.62) in the form of an asymptotic power series expansion in $1/(ik_0)$

$$\Psi(z) = \sum_{n=0}^{\infty} \Psi_{\text{B}}^{(n)}(z), \quad \Psi_{\text{B}}^{(n)}(z) = \exp\left\{ik_0\widehat{S}(z)\right\}(ik_0)^{-n}\Psi^{(n)}(z), \tag{11.158}$$

where $\widehat{S}(z)$ is a scalar function, one can find that particular solutions, solutions that describe the forward propagating fields, for the leading term of this expansion may be represented (with η in place of z) as

$$\Psi_{\text{B}j}^{(0)}(\eta) = \boldsymbol{\psi}_j(\eta)A_j^{(0)}(\eta) = \exp\left\{\int_0^{\eta} [ik_0\sigma_j(\eta)d - \bar{\boldsymbol{\psi}}_j(\eta)\boldsymbol{\psi}_{\eta j}(\eta)]d\eta\right\}\boldsymbol{\psi}_j(\eta)A_j^{(0)}(0) \quad (j = 1, 2), \tag{11.159}$$

where $A_1^{(0)}$ and $A_2^{(0)}$ are scalar complex amplitudes. The zeroth-order terms in field expansions like (11.158) represent the geometrical optics approximation for the fields [24]. Therefore, functions (11.159) may be regarded as GOA solutions of (8.62). The amplitudes $A_1^{(0)}$ and $A_2^{(0)}$ satisfy the equation

$$\frac{d}{d\eta}\begin{pmatrix} A_1^{(0)} \\ A_2^{(0)} \end{pmatrix} = \begin{pmatrix} ik_0\sigma_1 d - \bar{\boldsymbol{\psi}}_1\boldsymbol{\psi}_{\eta 1} & 0 \\ 0 & ik_0\sigma_2 - \bar{\boldsymbol{\psi}}_2\boldsymbol{\psi}_{\eta 2} \end{pmatrix}\begin{pmatrix} A_1^{(0)} \\ A_2^{(0)} \end{pmatrix}. \tag{11.160}$$

Comparing this equation with (11.81) and (11.84), we see that GOA gives the same results as NBRA in situations when the matrix $\mathbf{N}_\eta^\downarrow$ is diagonal. We have pointed to this correspondence considering the example with a nontwisted layer in the previous section. If the matrix $\mathbf{N}_\eta^\downarrow$ is not diagonal, GOA implies neglecting the off-diagonal elements of this matrix. Since we have assumed that the medium is locally centrosymmetric and that the EW basis is S-normalized, in subsequent formulas of this section, we omit the terms $\bar{\boldsymbol{\Psi}}_j \boldsymbol{\Psi}_{\eta j}$ [see (11.95b)]. The GOA expression for the transmission matrix $\mathbf{t}^\downarrow(z', z'')$ in any case is

$$\mathbf{t}^\downarrow(z', z'') \approx \begin{pmatrix} \exp\left\{ ik_0 d \int_0^1 \sigma_1(\eta)\mathrm{d}\eta \right\} & 0 \\ 0 & \exp\left\{ ik_0 d \int_0^1 \sigma_2(\eta)\mathrm{d}\eta \right\} \end{pmatrix}. \tag{11.161}$$

GOA in this application is equivalent to the adiabatic approximation (see Section 11.1.1): the fields $\{\mathbf{E}_j^{(0)}(\eta), \mathbf{H}_j^{(0)}(\eta)\}$ characterized by the functions $\Psi_{\mathrm{B}j}^{(0)}(\eta)$ have the form

$$\begin{pmatrix} \mathbf{E}_j^{(0)}(\eta) \\ \mathbf{H}_j^{(0)}(\eta) \end{pmatrix} = \begin{pmatrix} \mathbf{e}_j(\eta) \\ \mathbf{h}_j(\eta) \end{pmatrix} A_j^{(0)}(\eta), \tag{11.162}$$

meeting the description of the fields of the Mauguin mode. The condition for the validity of the adiabatic approximation is

$$|\vartheta_z(z)| \ll k_0 |\sigma_2(z) - \sigma_1(z)| \quad \text{for all } z \in [z', z''] \tag{11.163}$$

[see (11.96)] ($\vartheta_z = \vartheta_\eta / d$). For a layer with a given dependence of the dielectric tensor $\boldsymbol{\varepsilon}$ on the normalized coordinate η, in the absence of polarization degeneracy, the GOA solution may be considered as strictly valid in the limit $d \to \infty$. For LC layers, GOA, as a rule, gives sufficiently accurate results only in situations when $\vartheta_z = 0$, as in the example illustrated by Figure 11.14. For twisted LC layers, including LC layers of TN LCDs, which are commonly associated with the adiabatic mode, GOA in most cases gives only a rough estimate. Such an estimate is good for a qualitative description of the optical properties of the layer but insufficient for solving optimization and inverse problems. More accurate but still relatively simple analytical expressions for the matrix $\widehat{\mathbf{t}}_{\mathrm{L}}^\downarrow$ for cases where ϑ_z is nonzero but satisfies (11.163) were obtained with the aid of the so-called quasiadiabatic approximation [10].

Quasiadiabatic Approximation

This approximation is derived from the following representation of the solution of (11.111):

$$\widehat{\mathbf{t}}^\downarrow(0, \eta) = \mathbf{W}(\eta)\mathbf{V}(\eta), \tag{11.164}$$

where $\mathbf{W}(\eta)$ is the solution of the equation

$$\frac{\mathrm{d}\mathbf{W}}{\mathrm{d}\eta} = \mathbf{N}_\sigma^\downarrow \mathbf{W} \quad [\mathbf{W}(0) = \mathbf{U}], \tag{11.165}$$

in other words, $\mathbf{W}(\eta)$ is the GOA estimate for $\widehat{\mathbf{t}}^{\downarrow}(0,\eta)$, and $\mathbf{V}(\eta)$ is the solution of the equation

$$\frac{\mathrm{d}\mathbf{V}}{\mathrm{d}\eta} = \mathbf{N}_{\mathrm{V}}\mathbf{V} \quad [\mathbf{V}(0) = \mathbf{U}], \tag{11.166}$$

where $\mathbf{N}_{\mathrm{V}} = -\mathbf{W}^{-1}\mathbf{N}_{\psi}^{\downarrow}\mathbf{W}$. In QAA, $\mathbf{V}(1)$ is approximated by the sum of leading terms of the expansion

$$\mathbf{V}(1) = \mathbf{U} + \sum_{j=1}^{\infty}\mathbf{V}_j(1), \tag{11.167}$$

where

$$\mathbf{V}_1(\eta) = \int_0^{\eta}\mathbf{N}_{\mathrm{V}}(\bar{\eta})\mathrm{d}\bar{\eta}, \quad \mathbf{V}_j(\eta) = \int_0^{\eta}\mathbf{N}_{\mathrm{V}}(\bar{\eta})\mathbf{V}_{j-1}(\bar{\eta})\mathrm{d}\bar{\eta} \quad (j = 2, 3 \ldots).$$

The zero-, first-, and second-order expressions [where respectively one, two, and three terms of the series (11.167) are taken into account] of QAA for the matrix $\widehat{\mathbf{t}}_{\mathrm{L}}^{\downarrow}$ are

(zeroth order)

$$\widehat{\mathbf{t}}_{\mathrm{L}}^{\downarrow} \approx \begin{pmatrix} b_1 & 0 \\ 0 & b_2 \end{pmatrix}, \tag{11.168}$$

(first order)

$$\widehat{\mathbf{t}}_{\mathrm{L}}^{\downarrow} \approx \begin{pmatrix} b_1 & b_1 a_- \\ -b_2 a_+ & b_2 \end{pmatrix}, \tag{11.169}$$

(second order)

$$\widehat{\mathbf{t}}_{\mathrm{L}}^{\downarrow} \approx \begin{pmatrix} b_1(1-c_-) & b_1 a_- \\ -b_2 a_+ & b_2(1-c_+) \end{pmatrix}, \tag{11.170}$$

where

$$b_j \equiv \exp\left\{ik_0 d\int_0^1 \sigma_j(\eta)\mathrm{d}\eta\right\} \quad j = 1, 2;$$

$$a_{\pm} \equiv \int_0^1 f_{\pm}(\eta)\mathrm{d}\eta, \quad c_{\pm} \equiv \int_0^1 f_{\pm}(\eta)\int_0^{\eta} f_{\mp}(\bar{\eta})\mathrm{d}\bar{\eta}\mathrm{d}\eta,$$

$$f_{\pm}(\eta) = \vartheta_{\eta}(\eta)\exp\left\{\pm ik_0 d\int_0^{\eta}\Delta\sigma(\bar{\eta})\mathrm{d}\bar{\eta}\right\}, \quad \Delta\sigma = \sigma_1 - \sigma_2.$$

The zeroth-order expression (11.168) corresponds to GOA [cf. (11.161)]. The first-order expression (11.169) gives estimates for the off-diagonal elements of the matrix $\widehat{\mathbf{t}}_{\mathrm{L}}^{\downarrow}$, which characterize the deviation from the adiabatic regime. However, one may notice that the approximating matrix in this case is nonunitary (here we assume that the medium is nonabsorbing) and yields values greater than unity for the layer transmissivity. The second-order expression (11.170) provides more accurate estimates for the diagonal elements of $\mathbf{t}_{\mathrm{L}}^{\downarrow}$. The approximating matrix in this case is generally closer to a unitary one than that given by the first-order expression.

11.5 Effect of Errors in Values of the Transmission Matrix of the LC Layer on the Accuracy of Modeling the Transmittance of the LCD Panel

In this section, we give a few useful formulas for estimating the level of possible errors in calculated values of the transmittance of an LCD panel caused by an approximate calculation of the transmission matrix of the LC layer.

We consider the standard situation when the transmittance of an LCD panel for unpolarized incident light is calculated as

$$t = \frac{1}{2}\,\|\mathbf{t}\|_{\mathrm{E}}^2\,,$$

where $\mathbf{t}$ is the transmission (EW) Jones matrix of the LCD panel, and the matrix $\mathbf{t}$ can be represented as

$$\mathbf{t} = \mathbf{t}_3\mathbf{t}_2\mathbf{t}_1,$$

where $\mathbf{t}_2$ is the transmission matrix of the LC layer, and $\mathbf{t}_1$ and $\mathbf{t}_3$ are the transmission matrices of the layered systems preceding and following the LC layer, respectively. We also assume that the flux normalization of EW bases in the entrance and exit planes for the operators $\mathbf{t}_1$, $\mathbf{t}_2$, and $\mathbf{t}_3$ is used.

We denote the matrix approximating the matrix $\mathbf{t}_2$ by $\mathbf{t}_{2\mathrm{C}}$. By t_{C} we denote the approximate value of t calculated with the matrix $\mathbf{t}_{2\mathrm{C}}$, that is,

$$t_{\mathrm{C}} \equiv \frac{1}{2}\|\mathbf{t}_3\mathbf{t}_{2\mathrm{C}}\mathbf{t}_1\|_{\mathrm{E}}^2.$$

We represent $\mathbf{t}_{2\mathrm{C}}$ in the form

$$\mathbf{t}_{2\mathrm{C}} = (\mathbf{U} + \widehat{\boldsymbol{\alpha}})\mathbf{t}_2, \tag{11.171}$$

where $\mathbf{U}$ is the unit matrix. Then

$$t_{\mathrm{C}} = \frac{1}{2}\|\mathbf{t}_3(\mathbf{U} + \widehat{\boldsymbol{\alpha}})\mathbf{t}_2\mathbf{t}_1\|_{\mathrm{E}}^2 = \frac{1}{2}\|\mathbf{t}_3\mathbf{t}_2\mathbf{t}_1 + \mathbf{t}_3\widehat{\boldsymbol{\alpha}}\mathbf{t}_2\mathbf{t}_1\|_{\mathrm{E}}^2. \tag{11.172}$$

By using property (5.58) of matrix norms, one can find from (11.172) that

$$|\sqrt{t} - \delta_{\mathrm{A}}| \le \sqrt{t_{\mathrm{C}}} \le \sqrt{t} + \delta_{\mathrm{A}}, \tag{11.173}$$

where

$$\delta_{\mathrm{A}} = \frac{1}{\sqrt{2}}\|\mathbf{t}_3\widehat{\boldsymbol{\alpha}}\mathbf{t}_2\mathbf{t}_1\|_{\mathrm{E}}.$$

According to (11.173),

$$|\sqrt{t} - \sqrt{t_\mathrm{C}}| \leq \delta_\mathrm{A},$$

and, consequently, the absolute error in the transmittance, $\delta \equiv |t - t_\mathrm{C}|$, may be estimated as

$$\delta \leq \left(2\sqrt{t} + \delta_\mathrm{A}\right)\delta_\mathrm{A}. \tag{11.174}$$

By using properties (5.54) and (5.59) of matrix norms, it is easy to find that

$$\delta_\mathrm{A} \leq \frac{1}{\sqrt{2}} \|\mathbf{t}_3\|_\mathrm{S} \|\mathbf{t}_2\mathbf{t}_1\|_\mathrm{S} \|\widehat{\boldsymbol{\alpha}}\|_\mathrm{E}.$$

Physically, $\|\mathbf{t}_j\|_S^2$ ($j = 1,2,3$), with the flux normalization of the EW-bases, is the maximum transmissivity of the system characterized by the matrix $\mathbf{t}_j$ with respect to variations of the state vector of the wave field incident on this system [see (8.277) and (5.48)]. From this it is obvious that $\|\mathbf{t}_3\|_\mathrm{S}\|\mathbf{t}_2\mathbf{t}_1\|_\mathrm{S} \leq 1$. Therefore, the following inequality is valid:

$$\delta_\mathrm{A} \leq \frac{1}{\sqrt{2}} \|\widehat{\boldsymbol{\alpha}}\|_\mathrm{S}. \tag{11.175}$$

According to (11.171), we may write

$$\widehat{\boldsymbol{\alpha}}\mathbf{t}_2 = \mathbf{t}_{2\mathrm{C}} - \mathbf{t}_2.$$

If $\mathbf{t}_2$ is a unitary matrix, at any $\mathbf{t}_{2\mathrm{C}}$, according to (5.60),

$$\|\widehat{\boldsymbol{\alpha}}\|_\mathrm{E} = \|\mathbf{t}_{2\mathrm{C}} - \mathbf{t}_2\|_\mathrm{E}. \tag{11.176}$$

In this case, using (11.174), (11.175), and (11.176), we may estimate the error δ as

$$\delta \leq \left(\sqrt{2t} + \frac{1}{2}\delta_\mathbf{t}\right)\delta_\mathbf{t}, \tag{11.177}$$

where

$$\delta_\mathbf{t} = \|\mathbf{t}_{2\mathrm{C}} - \mathbf{t}_2\|_\mathrm{E}. \tag{11.178}$$

In the alternative case where $\mathbf{t}_{2\mathrm{C}}$ is unitary while $\mathbf{t}_2$ is general, formula (11.177) is applicable as well (strictly, on replacing t by t_C).

Estimates of accuracy parameters, defined in the same manner as $\delta_\mathbf{t}$, for different approximations used in calculations of transmission matrices of LC layers can be found in preceding sections of this chapter.

References

[1] R. C. Jones, "A new calculus for the treatment of optical systems. VII. Properties of the N-matrices," *J. Opt. Soc. Am.* **38**, 671 (1948).

[2] C. Gu and P. Yeh, "Extended Jones matrix method. II," *J. Opt. Soc. Am. A.* **10**, 966 (1993).

[3] A. Lien, "Extended Jones matrix representation for the twisted nematic liquid crystal display at oblique incidence," *Appl. Phys. Lett.* **57**, 2767 (1990).

[4] H. L. Ong, "Electro-optics of a twisted nematic liquid crystal display by 2 × 2 propagation matrix at oblique incidence," *Jpn. J. Appl. Phys.* **30**, L1028 (1991).

[5] C.-J. Chen, A. Lien, and M. I. Nathan, "4 × 4 and 2 × 2 matrix formulations for the optics in stratified and biaxial media," *J. Opt. Soc. Am. A.* **14**, 3125 (1997).

[6] F. H. Yu and H. S. Kwok, "Comparison of extended Jones matrices for twisted nematic liquid-crystal displays at oblique angles of incidence," *J. Opt. Soc. Am. A.* **16**, 2772 (1999).

[7] D. A. Yakovlev, "Calculation of transmission characteristics of smoothly inhomogeneous anisotropic media in the approximation of negligible bulk reflection: I. Basic equation," *Opt. Spectr.* **87**, 990 (1999) (in Russian).

[8] D. A. Yakovlev, "Calculation of transmission characteristics of smoothly inhomogeneous anisotropic media in the approximation of negligible bulk reflection: II. Numerical methods," *Opt. Spectr.* **94**, 655 (2003) (in Russian).

[9] D. A. Yakovlev, "Calculation of transmission characteristics of smoothly inhomogeneous anisotropic media in the approximation of negligible bulk reflection: III. Analytical solutions," *Opt. Spectr.* **95**, 1010 (2003) (in Russian).

[10] P. Allia, C. Oldano, and L. Trossi, "Polarization transfer matrix for the transmission of light through liquid-crystal slabs," *J. Opt. Soc. Am. B.* **5**, 2452 (1988).

[11] P. Allia, C. Oldano, and L. Trossi, "Light propagation in anisotropic stratified media in the quasi adiabatic limit," *Mol. Cryst. Liq. Cryst.* **143**, 17 (1987).

[12] P. Allia, C. Oldano, and L. Trossi, "Jones matrix treatment of electromagnetic wave propagation in anisotropic stratified media," *Phys. Scripta* **37**, 755 (1988).

[13] K. Lu and B. E. A. Saleh, "Reducing Berreman's 4 × 4 formulation of liquid crystal display optics to 2 × 2 Jones vector equations," *Opt. Lett.* **17**, 1557 (1993).

[14] A. Lien, "The general and simplified Jones matrix representations for the high pretilt twisted nematic cell," *J. Appl. Phys.* **67**, 2853 (1990).

[15] D. A. Yakovlev and V. G. Chigrinov, "A robust polarization-spectral method for determination of twisted liquid crystal layer parameters," *J. Appl. Phys.* **102**, 023510 (2007).

[16] D. A. Yakovlev, "Efficient method for calculating the polarization-optical characteristics of nematic layers with one-dimensional deformation of the director field," *Opt. Spectr.* **71**, 788 (1991).

[17] O. A. Aphonin, Yu. V. Panina, A. B. Pravdin, and D. A. Yakovlev, "Optical properties of stretched polymer dispersed liquid crystal films," *Liq. Cryst.* **15**, 395 (1993).

[18] D. A. Yakovlev and O.A. Afonin, "Method for calculating the amplitude scattering matrix for nonuniform anisotropic particles in the approximation of anomalous diffraction," *Opt. Spectr.* **82**, 86 (1997) (in Russian).

[19] H. C. van de Hulst, *Light Scattering by Small Particles* (Wiley, New York, 1957).

[20] S. Zumer, "Light scattering from nematic droplets: anomalous-diffraction approach," *Phys. Rev. A.* **37**, 4006 (1988).

[21] C. Gu and P. Yeh, "Extended Jones matrix method and its application in the analysis of compensators for liquid crystal displays," *Displays* **20**, 237 (1999).

[22] A. Lien, "A detailed derivation of extended Jones matrix representation for twisted nematic liquid crystal displays," *Liq. Cryst.* **22**, 171 (1997).

[23] H. L. Ong and R. B. Meyer, "Geometrical-optics approximation for the electromagnetic fields in layered-inhomogeneous liquid-crystalline structures," *J. Opt. Soc. Am. A.* **2**, 198 (1985).

[24] M. Born and E. Wolf, *Principles of Optics*, 7th ed. (Pergamon Press, New York, 1999).

12

Some Approximate Representations in EW Jones Matrix Method and Their Application in Solving Optimization and Inverse Problems for LCDs

As has been noted, many optical optimization problems for LCDs are stated or may be stated as a problem of optimization of transfer characteristics of a "useful" channel of light propagation in the device under consideration (see Section 6.4). The EW Jones matrix method presented in the previous chapters enables accurate evaluation of the transfer characteristics of "useful" channels for realistic optical models of LCDs with all important optical effects taken into account. The realistic optical model of an LCD panel, a model sufficiently full to ensure an adequate estimation of the transmittance of the "useful" channel of the panel, as a rule, includes, along with the basic polarization elements of the LCD panel (LC layer, polarizers, compensators), a number of isotropic layers. These are such layers as glass substrates, electrodes, alignment layers, color filters, and so on. On the other hand, many problems, including the search for optimal parameters of the polarization elements, can be solved, at least to a first approximation, neglecting the real effect of the isotropic layers and surface effects, employing simplified optical models of the LC device at hand, in terms of polarization Jones matrices and reduced transmittance (see Sections 6.4 and 6.5). Commonly, solutions of this kind are sought using the classical Jones calculus (JC). The EW Jones matrix method supplemented with a set of approximate representations, which will be discussed in this section, can also be used for finding such solutions, and guarantees higher accuracy than JC. The approximate representations considered here are also useful in solving inverse problems for LC layers (see Section 12.5). It is important that some of these representations are applicable in the case of oblique incidence and can be used in optimization calculations aimed at improvement of the viewing angle performance of LCDs (Section 12.6).

Modeling and Optimization of LCD Optical Performance, First Edition.
Dmitry A. Yakovlev, Vladimir G. Chigrinov and Hoi-Sing Kwok.

Website Companion: www.wiley.com/go/yakovlev/modelinglcd

Most approximate representations considered in this chapter involve approximating true EW Jones matrices by STU matrices (see Sections 5.1.3). In general, an FI–EW Jones matrix [FI indicates that the flux (F-) normalization is used for the input and output EW bases of this matrix] is an STU matrix if the operation described by this Jones matrix is an operation without diattenuation. Examples of operations without diattenuation are: transmissions and reflections by an interface between isotropic media at normal incidence, transmissions and reflections by a system of isotropic layers sandwiched between isotropic media at normal incidence, transmissions by the bulk of a homogeneous (nonabsorbing or absorbing) isotropic layer, transmissions by the bulk of a nonabsorbing homogeneous anisotropic layer out of TIR mode. Transmissions by the bulk of a nonabsorbing inhomogeneous LC layer in the NBR approximation (see Chapter 11) are also considered as operations without diattenuation. Operations with a weak diattenuation are characterized by FI–EW Jones matrices close to STU matrices. The weaker the diattenuation, the closer the corresponding FI–EW Jones matrix to an STU matrix.

One of the mathematical instruments used in this section is STUM approximation. STUM approximation is the approximation of a general 2×2 matrix $\mathbf{t}$ by an STU matrix $\mathbf{t}_{STU}$ closest to the matrix $\mathbf{t}$. The theory of this approximation, which is presented in Section 12.1, shows how to calculate the matrix $\mathbf{t}_{STU}$ and how to estimate the degree of closeness of the matrices $\mathbf{t}$ and $\mathbf{t}_{STU}$.

12.1 Theory of STUM Approximation

STUM Approximation

Let $\mathbf{t} = [t_{jk}]$ be an arbitrary 2×2 matrix with $\det \mathbf{t} \neq 0$. It is reasonable to regard as the best approximating STU matrix for the matrix $\mathbf{t}$ an STU matrix $\mathbf{t}_{STU}$ such that at $\mathbf{B}_{STU} = \mathbf{t}_{STU}$ the global minimum of the function

$$f_{\mathbf{t}}(\mathbf{B}_{STU}) = \|\mathbf{t} - \mathbf{B}_{STU}\|_E \quad \mathbf{B}_{STU} \in M_{STU}, \tag{12.1}$$

where M_{STU} is the set of all possible STU matrices, is attained, that is,

$$f_{\mathbf{t}}(\mathbf{t}_{STU}) \leq f_{\mathbf{t}}(\mathbf{B}_{STU}) \tag{12.2}$$

for any STU matrix $\mathbf{B}_{STU}$. The approximation $\mathbf{t} \approx \mathbf{t}_{STU}$ will be called *STUM* (STU Matrix) *approximation*. One can show that

$$f_{\mathbf{t}}(\mathbf{t}_{STU}) \equiv \|\mathbf{t} - \mathbf{t}_{STU}\|_E = \sqrt{\frac{1}{2}\|\mathbf{t}\|_E^2 - |\det \mathbf{t}|} \tag{12.3}$$

and that the matrix $\mathbf{t}_{STU}$ may be expressed as follows:

$$\mathbf{t}_{STU} = w\mathbf{A}_{UM}, \tag{12.4}$$

where

$$w = \sqrt{\frac{\det \mathbf{t}}{|\det \mathbf{t}|}} \cdot \frac{\sqrt{\|\mathbf{t}\|_E^2 + 2|\det \mathbf{t}|}}{2} \tag{12.5}$$

and

$$\mathbf{A}_{UM} = \begin{pmatrix} a & b \\ -b^* & a^* \end{pmatrix} \tag{12.6}$$

with

$$a = \frac{l_1}{\sqrt{l_1 l_1^* + l_2 l_2^*}}, \quad b = \frac{l_2}{\sqrt{l_1 l_1^* + l_2 l_2^*}}, \tag{12.7}$$

$$l_1 = t_{11} w^* + t_{22}^* w, \quad l_2 = t_{12} w^* - t_{21}^* w. \tag{12.8}$$

Note that the matrix $\mathbf{A}_{\mathrm{UM}}$ is unitary and unimodular and has determinant 1 and that

$$|w| = \frac{\sqrt{\|\mathbf{t}\|_{\mathrm{E}}^2 + 2\,|\det \mathbf{t}|}}{2}. \tag{12.9}$$

A derivation of equations (12.3)–(12.8) is given at the end of this section.

If the matrix $\mathbf{t}$ is diagonal, the best approximating STU matrix for it may be calculated by the following formula:

$$\mathbf{t}_{\mathrm{STU}} = \frac{|t_{11}| + |t_{22}|}{2} \begin{pmatrix} t_{11}/|t_{11}| & 0 \\ 0 & t_{22}/|t_{22}| \end{pmatrix}. \tag{12.10}$$

In this case, the error of STUM approximation may be estimated using the expression

$$\|\mathbf{t} - \mathbf{t}_{\mathrm{STU}}\|_{\mathrm{E}} = \frac{||t_{11}| - |t_{22}||}{\sqrt{2}}. \tag{12.11}$$

Let $\mathbf{t}_1$ be an arbitrary nonsingular 2×2 matrix. Let $\mathbf{A}_{\mathrm{U}}$ and $\mathbf{B}_{\mathrm{U}}$ be any unitary 2×2 matrices. Let $\mathbf{t}_{1\mathrm{STU}}$ be the best approximating STU matrix for $\mathbf{t}_1$. It is easy to show that the best approximating STU matrix for the matrix

$$\mathbf{t}_2 = \mathbf{A}_{\mathrm{U}} \mathbf{t}_1 \mathbf{B}_{\mathrm{U}} \tag{12.12}$$

can be represented as

$$\mathbf{t}_{2\mathrm{STU}} = \mathbf{A}_{\mathrm{U}}\, \mathbf{t}_{1\mathrm{STU}}\, \mathbf{B}_{\mathrm{U}}. \tag{12.13}$$

This expression can easily be derived from (12.3). According to (12.3), the matrix $\mathbf{t}_{2\mathrm{STU}}$ must satisfy the relation

$$\|\mathbf{t}_2 - \mathbf{t}_{2\mathrm{STU}}\|_{\mathrm{E}} = \sqrt{\frac{1}{2}\|\mathbf{t}_2\|_{\mathrm{E}}^2 - |\det \mathbf{t}_2|}. \tag{12.14}$$

Substitution of (12.12) and (12.13) into (12.14) gives

$$\|\mathbf{A}_{\mathrm{U}}\mathbf{t}_1\mathbf{B}_{\mathrm{U}} - \mathbf{A}_{\mathrm{U}}\mathbf{t}_{1\mathrm{STU}}\mathbf{B}_{\mathrm{U}}\|_{\mathrm{E}} = \sqrt{\frac{1}{2}\|\mathbf{A}_{\mathrm{U}}\mathbf{t}_1\mathbf{B}_{\mathrm{U}}\|_{\mathrm{E}}^2 - |\det(\mathbf{A}_{\mathrm{U}}\mathbf{t}_1\mathbf{B}_{\mathrm{U}})|}. \tag{12.15}$$

For the left-hand side of (12.15), we have

$$\|\mathbf{A}_{\mathrm{U}}\mathbf{t}_1\mathbf{B}_{\mathrm{U}} - \mathbf{A}_{\mathrm{U}}\mathbf{t}_{1\mathrm{STU}}\mathbf{B}_{\mathrm{U}}\|_{\mathrm{E}} = \|\mathbf{A}_{\mathrm{U}}(\mathbf{t}_1 - \mathbf{t}_{1\mathrm{STU}})\mathbf{B}_{\mathrm{U}}\|_{\mathrm{E}} = \|\mathbf{t}_1 - \mathbf{t}_{1\mathrm{STU}}\|_{\mathrm{E}}, \tag{12.16}$$

where property (5.60) of Euclidian norms has been used. In view of the same property of Euclidian norms, $\|\mathbf{A}_U\mathbf{t}_1\mathbf{B}_U\|_E^2 = \|\mathbf{t}_1\|_E^2$. Then, using property (5.17) of determinants and the fact that the modulus of the determinant of any unitary matrix is equal to unity, we find that

$$|\det(\mathbf{A}_U\mathbf{t}_1\mathbf{B}_U)| = |\det(\mathbf{A}_U)\det(\mathbf{t}_1)\det(\mathbf{B}_U)| = |\det(\mathbf{t}_1)|. \tag{12.17}$$

Therefore, for the right-hand side of (12.15), we have

$$\sqrt{\frac{1}{2}\|\mathbf{A}_U\mathbf{t}_1\mathbf{B}_U\|_E^2 - |\det(\mathbf{A}_U\mathbf{t}_1\mathbf{B}_U)|} = \sqrt{\frac{1}{2}\|\mathbf{t}_1\|_E^2 - |\det \mathbf{t}_1|}. \tag{12.18}$$

According to (12.3), the matrix $\mathbf{t}_{1STU}$ meets the relations

$$\|\mathbf{t}_1 - \mathbf{t}_{1STU}\|_E = \sqrt{\frac{1}{2}\|\mathbf{t}_1\|_E^2 - |\det \mathbf{t}_1|}. \tag{12.19}$$

Equations (12.16), (12.18), and (12.19) show that relation (12.15) is valid, which means that the matrix $\mathbf{t}_{2STU}$ expressed by (12.13) does indeed satisfy (12.14) and hence is the best approximating STU matrix for $\mathbf{t}_2$.

The singular values of an arbitrary 2×2 matrix $\mathbf{t}$ can be expressed as

$$\sigma_{1,2}(\mathbf{t}) = \sqrt{\frac{\|\mathbf{t}\|_E^2 \pm \sqrt{\|\mathbf{t}\|_E^4 - 4\,|\det \mathbf{t}|^2}}{2}} \tag{12.20}$$

[see (5.12)]. According to (12.20),

$$\sigma_1(\mathbf{t})^2 + \sigma_2(\mathbf{t})^2 = \|\mathbf{t}\|_E^2, \tag{12.21}$$

$$\sigma_1(\mathbf{t})\sigma_2(\mathbf{t}) = |\det \mathbf{t}|. \tag{12.22}$$

Using these relations, we may express the error of STUM approximation in terms of the singular values of the matrix to be approximated:

$$\|\mathbf{t} - \mathbf{t}_{STU}\|_E = \frac{|\sigma_1(\mathbf{t}) - \sigma_2(\mathbf{t})|}{\sqrt{2}}. \tag{12.23}$$

Physical Estimation of the STUM Approximation Accuracy

Let $\mathbf{t}$ be a transmission matrix that relates the FI Jones vectors of the incident ($\mathbf{J}_{inc}$) and transmitted ($\mathbf{J}_{tr}$) wave fields, $\mathbf{J}_{tr} = \mathbf{t}\mathbf{J}_{inc}$. In this case, the transmissivity as a function of $\mathbf{J}_{inc}$ may be represented as

$$t(\mathbf{J}_{inc}) = \frac{\mathbf{J}_{tr}^\dagger \mathbf{J}_{tr}}{\mathbf{J}_{inc}^\dagger \mathbf{J}_{inc}} = \frac{\mathbf{J}_{inc}^\dagger \mathbf{t}^\dagger \mathbf{t} \mathbf{J}_{inc}}{\mathbf{J}_{inc}^\dagger \mathbf{J}_{inc}}. \tag{12.24}$$

The maximum and minimum values of $t(\mathbf{J}_{inc})$ at the given $\mathbf{t}$, t_{max} and t_{min}, are equal to eigenvalues of the matrix $\mathbf{t}^\dagger\mathbf{t}$, $\lambda_{max}[\mathbf{t}^\dagger\mathbf{t}]$ and $\lambda_{min}[\mathbf{t}^\dagger\mathbf{t}]$, respectively (see Section 8.5). The singular values of the matrix $\mathbf{t}$, by definition, are equal to the square roots of $\lambda_{max}[\mathbf{t}^\dagger \mathbf{t}]$ and $\lambda_{min}[\mathbf{t}^\dagger \mathbf{t}]$. We may therefore rewrite (12.23) as follows:

$$\|\mathbf{t} - \mathbf{t}_{STU}\|_E = \frac{\sqrt{t_{max}} - \sqrt{t_{min}}}{\sqrt{2}}. \qquad (12.25)$$

When the extreme transmittances t_{max} and t_{min} for an element or a system are known a priori, equation (12.25) allows one to estimate the accuracy of STUM approximation for this element or system without any additional calculations.

STUM Approximation of Real Matrices

If the matrix $\mathbf{t} = [t_{jk}]$ to be approximated is real, the best approximating STU matrix for it may be calculated by the formula

$$\mathbf{t}_{STU} = \frac{1}{2}\begin{pmatrix} t_{11} + s_t t_{22} & t_{12} - s_t t_{21} \\ -s_t t_{12} + t_{21} & s_t t_{11} + t_{22} \end{pmatrix}, \qquad (12.26)$$

where

$$s_t = \begin{cases} 1 & \text{if } \det \mathbf{t} > 0 \\ -1 & \text{if } \det \mathbf{t} < 0. \end{cases}$$

The base matrix $\mathbf{A}_B$ and loss factor w_B of the matrix $\mathbf{t}_{STU}$ ($\mathbf{t}_{STU} = w_B \mathbf{A}_B$, see Section 5.1.3) can be taken to be

$$\mathbf{A}_B = \frac{1}{\sqrt{(t_{11} + s_t t_{22})^2 + (t_{12} - s_t t_{21})^2}}\begin{pmatrix} t_{11} + s_t t_{22} & t_{12} - s_t t_{21} \\ -s_t t_{12} + t_{21} & s_t t_{11} + t_{22} \end{pmatrix}, \qquad (12.27)$$

$$w_B = \frac{1}{2}\sqrt{(t_{11} + s_t t_{22})^2 + (t_{12} - s_t t_{21})^2}. \qquad (12.28)$$

The accuracy of STUM approximation in this case may be estimated using the following expression:

$$\|\mathbf{t} - \mathbf{t}_{STU}\|_E = \sqrt{\frac{(t_{11} - s_t t_{22})^2 + (t_{12} + s_t t_{21})^2}{2}}. \qquad (12.29)$$

In the general case, the *relative error of STUM approximation* may be defined as

$$\delta_{STU} = \frac{\sqrt{2}\|\mathbf{t} - \mathbf{t}_{STU}\|_E}{\|\mathbf{t}\|_E}. \qquad (12.30)$$

In the case of a real $\mathbf{t}$, this quantity can be calculated by the formula

$$\delta_{STU} = \sqrt{\frac{(t_{11} - s_t t_{22})^2 + (t_{12} + s_t t_{21})^2}{t_{11}^2 + t_{12}^2 + t_{21}^2 + t_{22}^2}}. \qquad (12.31)$$

Derivation of Equations (12.3)–(12.8)

According to (5.46),

$$\|\mathbf{t}-\mathbf{t}_{\mathrm{STU}}\|_{\mathrm{E}}^2 = \|\mathbf{t}\|_{\mathrm{E}}^2 + \|\mathbf{t}_{\mathrm{STU}}\|_{\mathrm{E}}^2 - \mathrm{Tr}(\mathbf{t}^{\dagger}\mathbf{t}_{\mathrm{STU}} + \mathbf{t}_{\mathrm{STU}}^{\dagger}\mathbf{t}). \tag{12.32}$$

Since the matrix $\mathbf{t}_{\mathrm{STU}}$ is an STU matrix, we may always represent it in the form (12.4) with $\mathbf{A}_{\mathrm{UM}}$ of the form (12.6) with a and b satisfying the condition

$$a^*a + b^*b = 1. \tag{12.33}$$

On substituting (12.4) into (12.32) and making use of the fact that $\|\mathbf{A}_{\mathrm{UM}}\|_{\mathrm{E}}^2 = 2$, we have

$$\|\mathbf{t}-\mathbf{t}_{\mathrm{STU}}\|_{\mathrm{E}}^2 = \|\mathbf{t}\|_{\mathrm{E}}^2 + 2|w|^2 - k_w, \tag{12.34}$$

where

$$k_w = \mathrm{Tr}\left(\mathbf{t}^{\dagger} w\mathbf{A}_{\mathrm{UM}} + w^*\mathbf{A}_{\mathrm{UM}}^{\dagger}\mathbf{t}\right) \tag{12.35}$$

or, explicitly,

$$k_w = t_{11}^* wa - t_{21}^* wb^* + t_{12}^* wb + t_{22}^* wa^* + w^* t_{11} a^* - w^* t_{21} b + w^* t_{12} b^* + w^* t_{22} a. \tag{12.36}$$

The first two terms on the right-hand side of (12.34) are independent of a, b, and $\arg w$ ($w = |w|\mathrm{e}^{i\arg w}$). We can therefore find these parameters from the requirement that k_w be maximum. Equation (12.36) may be rewritten as

$$k_w = 2[\mathrm{Re}(l_1^* a) + \mathrm{Re}(l_2^* b)], \tag{12.37}$$

where l_1 and l_2 are expressed by (12.8). Using the representation $a = \mathrm{e}^{i\varphi}\cos\rho$, $b = \mathrm{e}^{i\psi}\sin\rho$ [see (12.33)], where ρ, φ, and ψ are real, it is easy to find that k_w is maximum at a and b expressed by (12.7). With these a and b

$$k_w = 2[\mathrm{Re}(l_1^* a) + \mathrm{Re}(l_2^* b)] = 2\sqrt{l_1 l_1^* + l_2 l_2^*} = 2|w|\sqrt{\|\mathbf{t}\|_{\mathrm{E}}^2 + 2\mathrm{Re}[(\det\mathbf{t})^* \mathrm{e}^{2i\arg w}]}.$$

From this equation we see that k_w assumes its largest value, namely,

$$k_w = 2\,|w|\,\sqrt{\|\mathbf{t}\|_{\mathrm{E}}^2 + 2\,|\det\mathbf{t}|}, \tag{12.38}$$

when

$$\mathrm{e}^{i\arg w} = \sqrt{\frac{\det\mathbf{t}}{|\det\mathbf{t}|}}. \tag{12.39}$$

On substituting (12.38) into (12.34), we obtain the expression

$$\|\mathbf{t}-\mathbf{t}_{\mathrm{STU}}\|_{\mathrm{E}}^2 = \|\mathbf{t}\|_{\mathrm{E}}^2 + 2|w|^2 - 2|w|\sqrt{\|\mathbf{t}\|_{\mathrm{E}}^2 + 2|\det\mathbf{t}|}. \tag{12.40}$$

Minimizing the right-hand side of (12.40) with respect to $|w|$, we arrive at (12.9) and (12.3).

12.2 Exact and Approximate Expressions for Transmission Operators of Interfaces at Normal Incidence

In calculations of transmission characteristics of birefringent layered systems at normal incidence, the interfaces between layers are often represented by transmission operators having the form of a 2×2 rotation matrix, a 2×2 rotation matrix multiplied by a scalar, or simply a scalar (as amplitude transmission coefficients). The applicability of such representations is connected, in one way or another, with the possibility to represent exactly or approximately, with good accuracy, exact transmission EW Jones matrices of interfaces by STU matrices. In most cases of interest, EW Jones matrices of interfaces between isotropic and anisotropic media as well as interfaces between anisotropic media are not STU matrices. In this section, we will discuss the accuracy of approximations of the EW Jones matrices of interfaces by STU matrices and other important points concerning the application of such approximations in LCD optics. In particular, we will show that the accuracy of STUM approximation of a transmission EW Jones matrix of the interface between an isotropic or anisotropic medium and an anisotropic medium depends not only on the material parameters of the media but also on the normalization of the eigenwave basis, that is, in other words, on the kind of Jones vectors related by this matrix. As will be shown, the highest accuracy is attained, in general, with S- (or F-) normalization. In many situations typical of LCD optics, the accuracy of the STUM approximation of the interface transmission matrices with S- or F-normalization is sufficient for solving practical problems. With E- and ET-normalizations (the latter is defined below), which are frequently used in LCD optics, the error of STUM approximation for the interface matrices in the same situations may be unacceptably large.

We will deal with various kinds of normalization of EW bases and, correspondingly, various kinds of EW Jones vectors. We will refer to EW Jones vectors corresponding to the symmetrical (S-) normalization (8.167) as *SN Jones vectors*. With the flux (F-) normalization (8.164), EW Jones vectors are FI Jones vectors. If an EW basis is such that both conditions (8.167) and (8.164) are satisfied (this is possible when a nonabsorbing medium out of TIR mode is considered), to mark this fact we will use the term *S–F-normalization*. Corresponding EW Jones vectors will be simultaneously SN Jones vectors and FI Jones vectors (see Section 5.4.2). We will call vectors of this class *SN–FI Jones vectors*.

The best approximating STU matrix for a matrix $\mathbf{t}$ will be denoted as $(\mathbf{t})_{\text{STU}}$.

Simplified Exact Formulas for the Transmission EW Jones Matrix of an Interface

In Section 8.4.2, we have shown that the transmission matrix $\mathbf{t}^{\downarrow}_{\text{AB}}$ of the interface between arbitrary media A and B with symmetric permittivity tensors can be expressed as follows:

$$\mathbf{t}^{\downarrow}_{\text{AB}} = \widehat{\mathbf{t}}_{\text{BA}}^{-1} = \frac{1}{t_{11}^{(\text{BA})}t_{22}^{(\text{BA})} - t_{12}^{(\text{BA})}t_{21}^{(\text{BA})}} \begin{pmatrix} t_{22}^{(\text{BA})} & -t_{12}^{(\text{BA})} \\ -t_{21}^{(\text{BA})} & t_{11}^{(\text{BA})} \end{pmatrix}, \tag{12.41}$$

where

$$\widehat{\mathbf{t}}_{\text{BA}} = \begin{pmatrix} t_{11}^{(\text{BA})} & t_{12}^{(\text{BA})} \\ t_{21}^{(\text{BA})} & t_{22}^{(\text{BA})} \end{pmatrix}, \tag{12.42}$$

$$t_{jk}^{(\text{BA})} = \frac{\mathbf{z}(\mathbf{e}_{j\text{A}} \times \mathbf{h}_{k\text{B}}) + \mathbf{z}(\mathbf{e}_{k\text{B}} \times \mathbf{h}_{j\text{A}})}{2\mathbf{z}(\mathbf{e}_{j\text{A}} \times \mathbf{h}_{j\text{A}})} \qquad j, k = 1, 2. \tag{12.43}$$

This expression is always valid in the absence of polarization degeneracy. It also holds in the presence of polarization degeneracy if the vibration vectors of the basis eigenwaves are chosen so that condition (8.161) is satisfied. Recall that in the absence of polarization degeneracy, condition (8.161) is satisfied whatever the EW basis. In what follows, condition (8.161) is assumed to be satisfied for both media. In the case of normal incidence ($\mathbf{b} = 0$), the refraction vectors of the basis eigenwaves can be represented as

$$\mathbf{m}_j = \sigma_j \mathbf{z}, \quad j = 1, 2. \tag{12.44}$$

By making use of (12.44) and (8.197), it is easy to find that at normal incidence the orthogonality relation (8.161) implies that

$$\mathbf{e}_{t1}\mathbf{e}_{t2} = 0, \tag{12.45}$$

where $\mathbf{e}_{tj} \equiv \mathbf{e}_j - \mathbf{z}(\mathbf{e}_j\mathbf{z})$ is the tangential constituent of the vector $\mathbf{e}_j$ ($j = 1,2$), and that

$$\mathbf{z}(\mathbf{e}_j \times \mathbf{h}_j) = \sigma_j(\mathbf{e}_{tj}\mathbf{e}_{tj}) \tag{12.46}$$

and

$$\mathbf{z}(\mathbf{e}_{j\mathrm{A}} \times \mathbf{h}_{k\mathrm{B}}) + \mathbf{z}(\mathbf{e}_{k\mathrm{B}} \times \mathbf{h}_{j\mathrm{A}}) = (\sigma_{k\mathrm{B}} + \sigma_{j\mathrm{A}})(\mathbf{e}_{tj\mathrm{A}}\mathbf{e}_{tk\mathrm{B}}) \tag{12.47}$$

($j, k = 1, 2$). Using (12.43), (12.46), and (12.47), we can express the elements of the matrix $\widehat{\mathbf{t}}_{\mathrm{BA}}$ [see (12.42)] as follows:

$$t_{jk}^{(\mathrm{BA})} = \frac{(\sigma_{k\mathrm{B}} + \sigma_{j\mathrm{A}})(\mathbf{e}_{tk\mathrm{B}}\mathbf{e}_{tj\mathrm{A}})}{2\sigma_{j\mathrm{A}}(\mathbf{e}_{tj\mathrm{A}}\mathbf{e}_{tj\mathrm{A}})}. \tag{12.48}$$

Dealing with any isotropic medium, or any uniaxial medium, or any nonabsorbing biaxial medium at normal incidence, we can choose the vibration vectors of the basis eigenwaves so that the vectors $\mathbf{e}_{t1}$ and $\mathbf{e}_{t2}$ are real and the triple ($\mathbf{e}_{t1}$, $\mathbf{e}_{t2}$, $\mathbf{z}$) is right-handed. Suppose that the EW bases for the media A and B satisfy these requirements. Then we can represent the vectors $\mathbf{e}_{t1\mathrm{A}}$, $\mathbf{e}_{t2\mathrm{A}}$, $\mathbf{e}_{t1\mathrm{B}}$, and $\mathbf{e}_{t2\mathrm{B}}$ as

$$\mathbf{e}_{t1\mathrm{A}} = c_{1\mathrm{A}}\mathbf{x}_\mathrm{A}, \ \mathbf{e}_{t2\mathrm{A}} = c_{2\mathrm{A}}\mathbf{y}_\mathrm{A}, \ \mathbf{e}_{t1\mathrm{B}} = c_{1\mathrm{B}}\mathbf{x}_\mathrm{B}, \ \mathbf{e}_{t2\mathrm{B}} = c_{2\mathrm{B}}\mathbf{y}_\mathrm{B}, \tag{12.49}$$

where $\mathbf{x}_\mathrm{A}$, $\mathbf{y}_\mathrm{A}$, $\mathbf{x}_\mathrm{B}$, and $\mathbf{y}_\mathrm{B}$ are real unit vectors directed along $\mathbf{e}_{t1\mathrm{A}}$, $\mathbf{e}_{t2\mathrm{A}}$, $\mathbf{e}_{t1\mathrm{B}}$, and $\mathbf{e}_{t2\mathrm{B}}$, respectively; $c_{1\mathrm{A}}$, $c_{2\mathrm{A}}$, $c_{1\mathrm{B}}$, and $c_{2\mathrm{B}}$ are positive constants determined by the normalization conditions used. The triples ($\mathbf{x}_\mathrm{A}$, $\mathbf{y}_\mathrm{A}$, $\mathbf{z}$) and ($\mathbf{x}_\mathrm{B}$, $\mathbf{y}_\mathrm{B}$, $\mathbf{z}$) are right-handed; $\mathbf{y}_\mathrm{A}$ is perpendicular to $\mathbf{x}_\mathrm{A}$; $\mathbf{y}_\mathrm{B}$ is perpendicular to $\mathbf{x}_\mathrm{B}$ (Figure 12.1). We consider two variants of normalization. In the first variant, the symmetrical (S-) normalization (8.167) is used for both media. In this case,

$$c_{j\mathrm{A}} = \frac{1}{\sqrt{2\sigma_{j\mathrm{A}}}}, \ c_{j\mathrm{B}} = \frac{1}{\sqrt{2\sigma_{j\mathrm{B}}}}, \quad j = 1, 2. \tag{12.50}$$

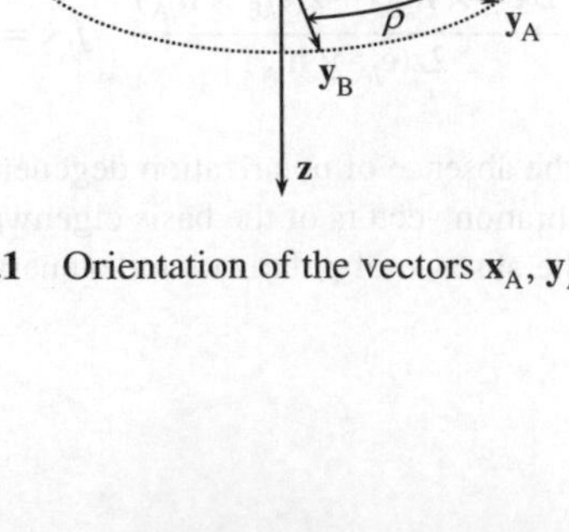

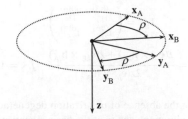

Figure 12.1 Orientation of the vectors $\mathbf{x}_\mathrm{A}$, $\mathbf{y}_\mathrm{A}$, $\mathbf{x}_\mathrm{B}$, and $\mathbf{y}_\mathrm{B}$

In the second variant, for both media the following normalization is used:

$$\mathbf{e}_{tj}\mathbf{e}_{tj} = 1, \quad j = 1, 2. \tag{12.51}$$

We refer to this kind of normalization as *ET-normalization*. With this normalization,

$$c_{jA} = 1, \quad c_{jB} = 1, \quad j = 1, 2, \tag{12.52}$$

and both state vectors related by the matrix $\mathbf{t}^{\downarrow}_{AB}$ are FTCEF Jones vectors (see Section 5.4.1). With this choice of the bases, the SN Jones vector $\mathbf{J}_{(S)}$ and the FTCEF Jones vector $\mathbf{J}_{(ET)}$ of a wave field are related by

$$\mathbf{J}_{(ET)} = \boldsymbol{G}_{(S\rightarrow ET)}\mathbf{J}_{(S)}, \tag{12.53}$$

where

$$\boldsymbol{G}_{(S\rightarrow ET)} = \begin{pmatrix} \sigma_1^{-1/2} & 0 \\ 0 & \sigma_2^{-1/2} \end{pmatrix}. \tag{12.54}$$

Where necessary, quantities corresponding to S-normalization will be labeled by the subscript (S), and those corresponding to ET-normalization by the subscript (ET).

From (12.48), (12.49), (12.50), and (12.52) we obtain the following expressions for the matrix $\widehat{\mathbf{t}}_{BA}$:

(i) for S-normalization

$$\widehat{\mathbf{t}}_{BA(S)} = \begin{pmatrix} \dfrac{\sigma_{1B}+\sigma_{1A}}{2\sqrt{\sigma_{1A}\sigma_{1B}}}(\mathbf{x}_B\mathbf{x}_A) & \dfrac{\sigma_{2B}+\sigma_{1A}}{2\sqrt{\sigma_{1A}\sigma_{2B}}}(\mathbf{y}_B\mathbf{x}_A) \\ \dfrac{\sigma_{1B}+\sigma_{2A}}{2\sqrt{\sigma_{2A}\sigma_{1B}}}(\mathbf{x}_B\mathbf{y}_A) & \dfrac{\sigma_{2B}+\sigma_{2A}}{2\sqrt{\sigma_{2A}\sigma_{2B}}}(\mathbf{y}_B\mathbf{y}_A) \end{pmatrix}, \tag{12.55}$$

(ii) for ET-normalization

$$\widehat{\mathbf{t}}_{BA(ET)} = \begin{pmatrix} \dfrac{\sigma_{1B}+\sigma_{1A}}{2\sigma_{1A}}(\mathbf{x}_B\mathbf{x}_A) & \dfrac{\sigma_{2B}+\sigma_{1A}}{2\sigma_{1A}}(\mathbf{y}_B\mathbf{x}_A) \\ \dfrac{\sigma_{1B}+\sigma_{2A}}{2\sigma_{2A}}(\mathbf{x}_B\mathbf{y}_A) & \dfrac{\sigma_{2B}+\sigma_{2A}}{2\sigma_{2A}}(\mathbf{y}_B\mathbf{y}_A) \end{pmatrix}. \tag{12.56}$$

It is convenient to write these expressions in the form

$$\widehat{\mathbf{t}}_{BA} = \begin{pmatrix} g_{11}\cos\rho & -g_{12}\sin\rho \\ g_{21}\sin\rho & g_{22}\cos\rho \end{pmatrix}, \tag{12.57}$$

where ρ is the angle between the vectors $\mathbf{x}_A$ and $\mathbf{x}_B$ (see Figure 12.1) and (i) for S-normalization

$$g_{jk} = \frac{\sigma_{jA}+\sigma_{kB}}{2\sqrt{\sigma_{jA}\sigma_{kB}}} = \frac{1}{\sqrt{1-\left(\dfrac{\Delta\sigma_{jkAB}}{2\bar{\sigma}_{jkAB}}\right)^2}} \tag{12.58}$$

and (ii) for ET-normalization

$$g_{jk} = \frac{\sigma_{kB} + \sigma_{jA}}{2\sigma_{jA}} = 1 - \frac{\Delta\sigma_{jkAB}}{2\sigma_{jA}} \tag{12.59}$$

with

$$\Delta\sigma_{jkAB} = \sigma_{jA} - \sigma_{kB}, \quad \bar{\sigma}_{jkAB} = \frac{\sigma_{jA} + \sigma_{kB}}{2}, \quad j,k = 1,2. \tag{12.60}$$

Using (12.57) and (12.41), we may express the matrix $\mathbf{t}^{\dagger}_{AB}$ (from now on, for convenience sake, we denote this matrix simply as $\mathbf{t}_{AB}$) as follows:

$$\mathbf{t}_{AB} = \frac{1}{g_\rho}\begin{pmatrix} g_{22}\cos\rho & g_{12}\sin\rho \\ -g_{21}\sin\rho & g_{11}\cos\rho \end{pmatrix}, \tag{12.61}$$

where

$$g_\rho = g_{11}g_{22}\cos^2\rho + g_{12}g_{21}\sin^2\rho. \tag{12.62}$$

In the presence of polarization degeneracy in the medium A (i.e., when $\sigma_{1A} = \sigma_{2A}$)

$$g_{11} = g_{21}, \quad g_{12} = g_{22}, \tag{12.63}$$

which permits the following representation of the matrix $\mathbf{t}_{AB}$:

$$\mathbf{t}_{AB} = \mathbf{t}'_{AB}\widehat{R}_C(\rho), \tag{12.64}$$

where

$$\mathbf{t}'_{AB} = \begin{pmatrix} g_{11}^{-1} & 0 \\ 0 & g_{22}^{-1} \end{pmatrix} \tag{12.65}$$

and

$$\widehat{R}_C(\rho) \equiv \begin{pmatrix} \cos\rho & \sin\rho \\ -\sin\rho & \cos\rho \end{pmatrix}. \tag{12.66}$$

If polarization degeneracy takes place in the medium B [$\sigma_{1B} = \sigma_{2B}$],

$$g_{11} = g_{12}, \quad g_{21} = g_{22}. \tag{12.67}$$

In this case, the matrix $\mathbf{t}_{AB}$ can be represented as

$$\mathbf{t}_{AB} = \widehat{R}_C(\rho)\mathbf{t}'_{AB}. \tag{12.68}$$

The status of the matrix $\mathbf{t}'_{AB}$ is evident. This is the transmission matrix of the interface in the situation where the EW basis in the medium with polarization degeneracy is chosen so that $\mathbf{x}_A = \mathbf{x}_B$ and $\mathbf{y}_A = \mathbf{y}_B$.

STUM Approximation of Transmission Matrices for Interfaces Between Nonabsorbing Media

Now we consider what the best approximating STU matrices for $\mathbf{t}_{AB}$ are and estimate the error of STUM approximation for the two specified variants of EWB normalization in the case when both A and B are nonabsorbing. S-normalization in this case is tantamount to F-normalization and hence can be called S–F-normalization.

If the media A and B are nonabsorbing, the matrix $\mathbf{t}_{AB}$ expressed by (12.61) is real and has a positive determinant. According to (12.61) and (12.26), the best approximating STU matrix for it is

$$(\mathbf{t}_{AB})_{STU} = \frac{1}{2g_\rho}\begin{pmatrix} (g_{11}+g_{22})\cos\rho & (g_{12}+g_{21})\sin\rho \\ -(g_{12}+g_{21})\sin\rho & (g_{11}+g_{22})\cos\rho \end{pmatrix}. \tag{12.69}$$

This matrix can be represented as

$$(\mathbf{t}_{AB})_{STU} = w_{AB}\mathbf{t}_{ABU}, \tag{12.70}$$

where

$$\mathbf{t}_{ABU} = \frac{1}{\sqrt{(g_{11}+g_{22})^2\cos^2\rho + (g_{12}+g_{21})^2\sin^2\rho}}\begin{pmatrix} (g_{11}+g_{22})\cos\rho & (g_{12}+g_{21})\sin\rho \\ -(g_{12}+g_{21})\sin\rho & (g_{11}+g_{22})\cos\rho \end{pmatrix} \tag{12.71}$$

is the base matrix (det $\mathbf{t}_{ABU} = 1$) and

$$w_{AB} = \frac{\sqrt{(g_{11}+g_{22})^2\cos^2\rho + (g_{12}+g_{21})^2\sin^2\rho}}{2g_\rho} \tag{12.72}$$

is the loss factor of $(\mathbf{t}_{AB})_{STU}$ [see (12.26)–(12.28)]. The relative error of STUM approximation δ_{STU} [see (12.30) and (12.31)] in this case can be estimated as

$$\delta_{STU} = \sqrt{\frac{(g_{11}-g_{22})^2\cos^2\rho + (g_{12}-g_{21})^2\sin^2\rho}{(g_{11}^2+g_{22}^2)\cos^2\rho + (g_{12}^2+g_{21}^2)\sin^2\rho}}. \tag{12.73}$$

Inspection of this equation shows that

$$\delta_{STU} \le \delta_g \equiv \max\{\delta_{g11}, \delta_{g12}\}, \tag{12.74}$$

where

$$\delta_{g11} = \frac{|g_{11}-g_{22}|}{\sqrt{g_{11}^2+g_{22}^2}}, \quad \delta_{g12} = \frac{|g_{12}-g_{21}|}{\sqrt{g_{12}^2+g_{21}^2}}, \tag{12.75}$$

and $\max\{\delta_{g11}, \delta_{g12}\}$ is the larger number in the pair $\{\delta_{g11}, \delta_{g12}\}$.

S–F-normalization. Usually

$$2\bar{\sigma}_{jkAB} \gg |\Delta\sigma_{jkAB}| \tag{12.76}$$

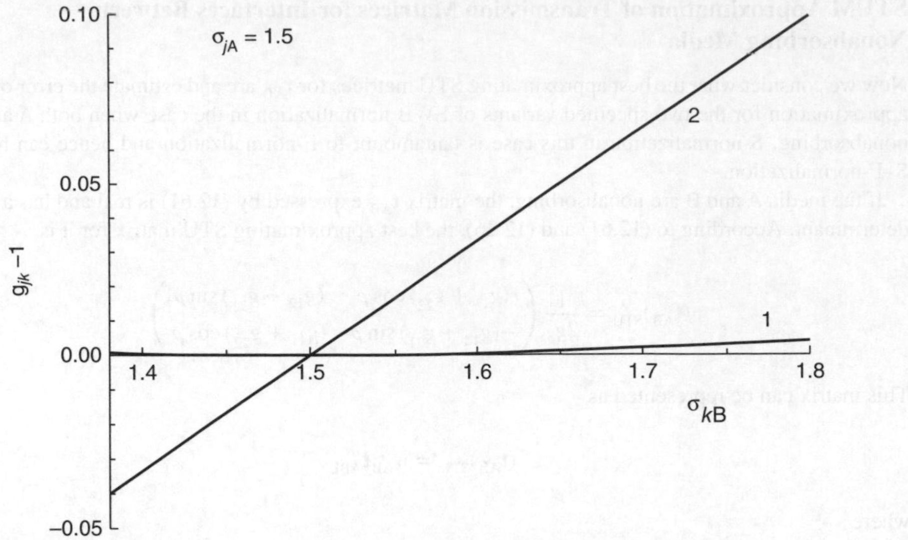

Figure 12.2 Dependence of $g_{jk}-1$ on σ_{kB} at $\sigma_{jA} = 1.5$ in the cases of S-normalization (1) and ET-normalization (2) of the EW bases

[see (12.60)], which, according to (12.58), allows one to approximately express the coefficients g_{jk} in the case of S-normalization as follows:

$$g_{jk} = \frac{1}{\sqrt{1-\delta\sigma_{jk}^2}} \approx 1 + \frac{1}{2}\delta\sigma_{jk}^2, \tag{12.77}$$

where

$$\delta\sigma_{jk} = \frac{\Delta\sigma_{jk\mathrm{AB}}}{2\bar{\sigma}_{jk\mathrm{AB}}}. \tag{12.78}$$

Because of a quadratic dependence of g_{jk} on $\delta\sigma_{jk}$ at small values of $\delta\sigma_{jk}$, for a wide range of values of $\sigma_{1\mathrm{A}}$, $\sigma_{2\mathrm{A}}$, $\sigma_{1\mathrm{B}}$, and $\sigma_{2\mathrm{B}}$ the coefficients g_{jk} are very close to 1. As an illustration, Figure 12.2 shows the exact dependence of $g_{jk} - 1$ on $\sigma_{k\mathrm{B}}$ at $\sigma_{j\mathrm{A}} = 1.5$. In this example, $|g_{jk}-1| < 2 \times 10^{-3}$ in the range $1.33 < \sigma_{k\mathrm{B}} < 1.7$, and $|g_{jk} - 1| < 10^{-3}$ in the range $1.37 < \sigma_{k\mathrm{B}} < 1.64$. As is seen from (12.71), the closer g_{jk} to 1, the more accurate the relation

$$\mathbf{t}_{\mathrm{ABU}} \approx \widehat{R}_{\mathrm{C}}(\rho) \tag{12.79}$$

[see (12.66)].

According to (12.77), at small values of $\delta\sigma_{jk}$,

$$\delta_{g11} \approx \frac{|\delta\sigma_{11}^2 - \delta\sigma_{22}^2|}{2\sqrt{2}}, \quad \delta_{g12} \approx \frac{|\delta\sigma_{12}^2 - \delta\sigma_{21}^2|}{2\sqrt{2}} \tag{12.80}$$

[see (12.74) and (12.75)], whence it is evident that the relative error of STUM approximation for a wide range of values of $\sigma_{1\mathrm{A}}$, $\sigma_{2\mathrm{A}}$, $\sigma_{1\mathrm{B}}$, and $\sigma_{2\mathrm{B}}$ in the case of S-normalization is very small. With

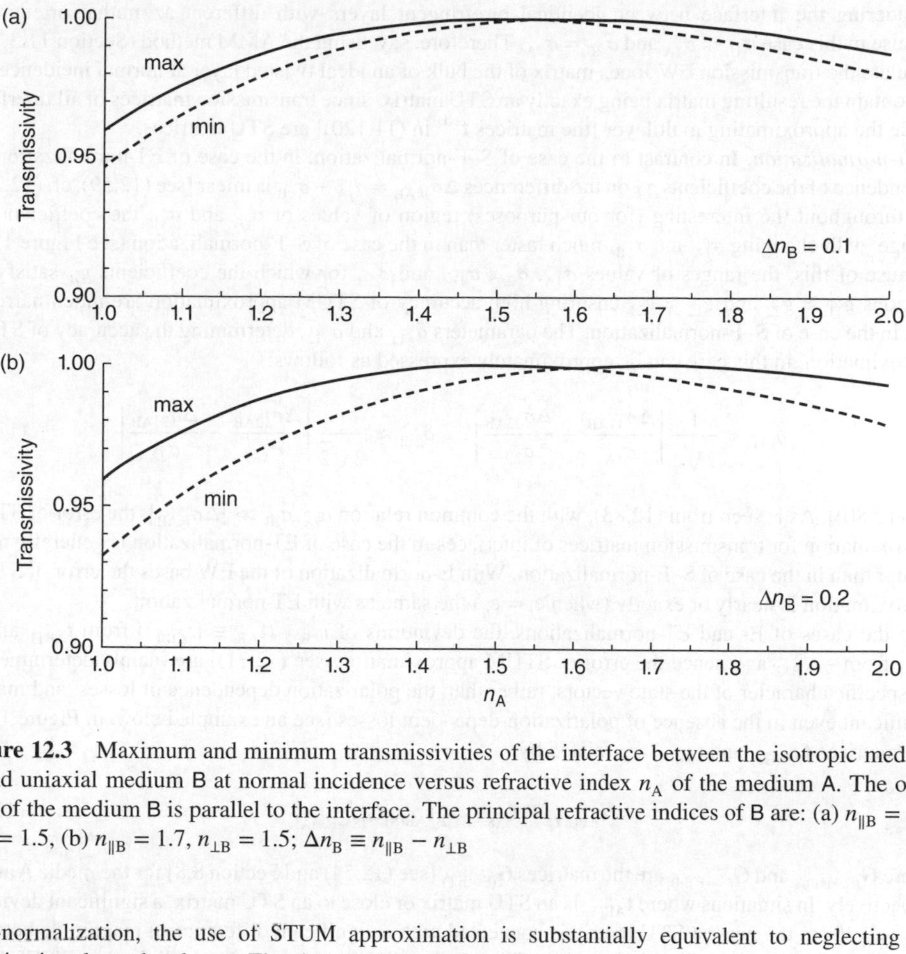

Figure 12.3 Maximum and minimum transmissivities of the interface between the isotropic medium A and uniaxial medium B at normal incidence versus refractive index n_A of the medium A. The optic axis of the medium B is parallel to the interface. The principal refractive indices of B are: (a) $n_{\|B} = 1.6$, $n_{\perp B} = 1.5$, (b) $n_{\|B} = 1.7$, $n_{\perp B} = 1.5$; $\Delta n_B \equiv n_{\|B} - n_{\perp B}$

S–F-normalization, the use of STUM approximation is substantially equivalent to neglecting the polarization-dependent losses. The above estimates of the accuracy of STUM approximation, in essence, reflect the fact that in usual situations, the transmissivities of interfaces of anisotropic media at normal incidence depend very weakly on the polarization state of incident light. As an illustration, Figure 12.3 shows the extreme (over all possible values of the EW Jones vector of the incident light; see Section 8.5) values of the transmissivity of the interface between an isotropic medium and a uniaxial medium as functions of the refractive index n_A of the isotropic medium. As can be seen from this figure, in a wide range of values of n_A, the maximum and minimum values of the transmissivity are close to each other even when the difference of the principal refractive indices of the uniaxial medium, $\Delta n_B \equiv n_{\|B} - n_{\perp B}$, is relatively large ($\Delta n_B = 0.2$).

Note that with σ_{1A}, σ_{2A}, σ_{1B}, and σ_{2B} satisfying the condition

$$\sigma_{1A}\sigma_{2A} = \sigma_{1B}\sigma_{2B}, \tag{12.81}$$

the coefficients g_{jk} satisfy the relations

$$g_{11} = g_{22},\ g_{12} = g_{21}, \tag{12.82}$$

and the matrix $\mathbf{t}_{AB}$ is exactly an STU matrix, that is, the amount of the reflection losses does not depend on the polarization state of the incident light at all. We might deal with such a situation, for instance, considering the interface between identical birefringent layers with different azimuthal orientation, because in this case $\sigma_{1B} = \sigma_{1A}$ and $\sigma_{2B} = \sigma_{2A}$. Therefore, say, using the AMM method (Section 11.3.1) to calculate the transmission EW Jones matrix of the bulk of an ideal twisted layer at normal incidence, we will obtain the resulting matrix being exactly an STU matrix, since transmission matrices of all interfaces inside the approximating multilayer [the matrices $\mathbf{t}_j^{\downarrow(\mathrm{I})}$ in (11.120)] are STU matrices.

ET-normalization. In contrast to the case of S–F-normalization, in the case of ET-normalization the dependence of the coefficients g_{jk} on the differences $\Delta\sigma_{jk\mathrm{AB}} = \sigma_{j\mathrm{A}} - \sigma_{k\mathrm{B}}$ is linear [see (12.59), cf. (12.77)], and throughout the interesting (for our purposes) region of values of $\sigma_{j\mathrm{A}}$ and $\sigma_{k\mathrm{B}}$ the coefficients g_{jk} change, with changing $\sigma_{j\mathrm{A}}$ and $\sigma_{k\mathrm{B}}$, much faster than in the case of S–F-normalization (see Figure 12.2). Because of this, the ranges of values $\sigma_{1\mathrm{A}}$, $\sigma_{2\mathrm{A}}$, $\sigma_{1\mathrm{B}}$, and $\sigma_{2\mathrm{B}}$ for which the coefficients g_{jk} satisfy the relations $g_{11} \approx g_{22}$ and $g_{12} \approx g_{21}$ ensuring high accuracy of STUM approximation are much narrower than in the case of S–F-normalization. The parameters δ_{g11} and δ_{g12}, determining the accuracy of STUM approximation, in this case can be approximately expressed as follows:

$$\delta_{g11} \approx \frac{1}{2\sqrt{2}}\left|\frac{\Delta\sigma_{11\mathrm{AB}}}{\sigma_{1\mathrm{A}}} - \frac{\Delta\sigma_{22\mathrm{AB}}}{\sigma_{2\mathrm{A}}}\right|, \quad \delta_{g12} \approx \frac{1}{2\sqrt{2}}\left|\frac{\Delta\sigma_{12\mathrm{AB}}}{\sigma_{1\mathrm{A}}} - \frac{\Delta\sigma_{21\mathrm{AB}}}{\sigma_{2\mathrm{A}}}\right| \tag{12.83}$$

[cf. (12.80)]. As is seen from (12.83), with the common relation $\sigma_{j\mathrm{A}}, \sigma_{j\mathrm{B}} \gg |\Delta\sigma_{jk\mathrm{AB}}|$, the error of STUM approximation for transmission matrices of interfaces in the case of ET-normalization is generally much greater than in the case of S–F-normalization. With E-normalization of the EW bases the error of STUM approximation is nearly or exactly (when $\mathbf{e}_j = \mathbf{e}_{tj}$) the same as with ET-normalization.

In the cases of E- and ET-normalizations, the deviations of $t_{\mathrm{AB}22}$ ($\mathbf{t}_{\mathrm{AB}} \equiv [t_{\mathrm{AB}jk}]$) from $t_{\mathrm{AB}11}$ and of $t_{\mathrm{AB}21}$ from $-t_{\mathrm{AB}12}$ and hence the error of STUM approximation [see (12.31)] are mainly determined by the specific character of the state vectors, rather than the polarization dependence of losses, and may be significant even in the absence of polarization-dependent losses (see an example below, in Figure 12.4). The matrices $\mathbf{t}_{\mathrm{AB(ET)}}$ and $\mathbf{t}_{\mathrm{AB(S)}}$ are related by

$$\mathbf{t}_{\mathrm{AB(ET)}} = \boldsymbol{G}_{(\mathrm{S}\to\mathrm{ET})\mathrm{B}}\mathbf{t}_{\mathrm{AB(S)}}\boldsymbol{G}^{-1}_{(\mathrm{S}\to\mathrm{ET})\mathrm{A}}, \tag{12.84}$$

where $\boldsymbol{G}_{(\mathrm{S}\to\mathrm{ET})\mathrm{A}}$ and $\boldsymbol{G}_{(\mathrm{S}\to\mathrm{ET})\mathrm{B}}$ are the matrices $\boldsymbol{G}_{(\mathrm{S}\to\mathrm{ET})}$ [see (12.54) and Section 8.8] for the media A and B, respectively. In situations where $\mathbf{t}_{\mathrm{AB(S)}}$ is an STU matrix or close to an STU matrix, a significant deviation of $\mathbf{t}_{\mathrm{AB(ET)}}$ from the nearest STU matrix is connected with a significant difference in magnitude between the elements (1,1) and (2,2) of the matrices $\boldsymbol{G}_{(\mathrm{S}\to\mathrm{ET})}$. Recall that the matrix $\mathbf{t}_{\mathrm{AB(ET)}}$ links the FTCEF Jones vectors of the incident and transmitted light (see Section 5.4.1). In the absence of polarization degeneracy, the modules of the complex FTCEF (and FEF) amplitudes of differently polarized eigenwaves of equal irradiance are different (see Section 5.4.2). It is this disproportion that results in the difference of the diagonal elements of the matrices $\boldsymbol{G}_{(\mathrm{S}\to\mathrm{ET})}$ and is the main cause of a poor accuracy of STUM approximation of transmission matrices of interfaces under E- and ET-normalization.

Base matrix and geometry. In the presence of polarization degeneracy at least in one of the media (i.e., if $\sigma_{1\mathrm{A}} = \sigma_{2\mathrm{A}}$ or $\sigma_{1\mathrm{B}} = \sigma_{2\mathrm{B}}$),

$$g_{11} + g_{22} = g_{12} + g_{21} \tag{12.85}$$

for both variants of normalization. As is seen from (12.71), subject to (12.85),

$$\mathbf{t}_{\mathrm{ABU}} = \widehat{R}_{\mathrm{C}}(\rho) \tag{12.86}$$

[see (12.66)]. Here the base matrix $\mathbf{t}_{\mathrm{ABU}}$ depends only on the angle between the vectors $\mathbf{x}_{\mathrm{A}}$ and $\mathbf{x}_{\mathrm{B}}$, being independent of the optical constants of the media, that is, it has a purely geometrical nature. This is the case for interfaces between isotropic media as well as for interfaces between isotropic and anisotropic media. For interfaces between anisotropic media, relation (12.86) is in general violated, though, as a rule, not strongly [see (12.79)]. We can see the effect of the approximation

$$\mathbf{t}_{\mathrm{AB}} \approx w_{\mathrm{AB}}\widehat{R}_{\mathrm{C}}(\rho) \tag{12.87}$$

for interfaces between anisotropic layers, for example, comparing expressions (11.41) and (11.151) for transmission matrices of ideal twisted layers at normal incidence. Expression (11.151) is obtained from (11.120) in the limit $N \to \infty$ if the exact expressions for the interface matrices $\mathbf{t}_j^{\downarrow(\mathrm{I})}$ are used. If we use approximation (12.87) for the interface matrices, the resulting matrix will be exactly the same as the matrix on the right of (11.41).

Polarization Jones Matrices of Interfaces

High accuracy of STUM approximation of the transmission EW Jones matrix of an interface under S–F-normalization allows one to use a polarization transmission Jones matrix for approximate description of the optical action of this interface. With this normalization, the EW Jones matrix links the FI Jones vectors of the incident and transmitted fields. In Section 5.4.3 we defined the polarization Jones vector of a wave field as a unit vector collinear to the FI Jones vector of this field. A polarization transmission Jones matrix is a matrix linking polarization Jones vectors of the incident and transmitted fields. In accordance with these definitions, in the above example we can choose the polarization Jones matrix of the interface A–B, $\mathbf{t}_{\mathrm{AB}}^{(p)}$, equal to the base matrix $\mathbf{t}_{\mathrm{ABU}}$ of the matrix $(\mathbf{t}_{\mathrm{AB(S)}})_{\mathrm{STU}}$. With this choice, $\mathbf{t}_{\mathrm{AB}}^{(p)}$ will be equal to the rotation matrix $\widehat{R}_{\mathrm{C}}(\rho)$ if one of the media, A or B, is isotropic or if both media are anisotropic but in one of them polarization degeneracy takes place. In the absence of polarization degeneracy in both media, the relation $\mathbf{t}_{\mathrm{AB}}^{(p)} \approx \widehat{R}_{\mathrm{C}}(\rho)$ will be as accurate as relation (12.87).

A Homogeneous Layer Between Isotropic Media. Exact and Approximate Formulas

Let us consider a homogeneous layer B surrounded by isotropic media A and C. Let a plane wave fall normally on the layer B from the medium A. We consider the standard situation where the basis vectors $\mathbf{x}_{\mathrm{C}}$ and $\mathbf{y}_{\mathrm{C}}$ in the medium C are codirectional with the vectors $\mathbf{x}_{\mathrm{A}}$ and $\mathbf{y}_{\mathrm{A}}$, respectively, that is, $\mathbf{x}_{\mathrm{C}} = \mathbf{x}_{\mathrm{A}}$ and $\mathbf{y}_{\mathrm{C}} = \mathbf{y}_{\mathrm{A}}$. Neglecting multiple reflections, we represent the transmission matrix of the layer $\mathbf{t}_{\mathrm{layer}}$ as the product of the transmission matrix of the interface A–B ($\mathbf{t}_{\mathrm{AB}}$), the transmission matrix of the bulk of B ($\mathbf{t}_{\mathrm{B}}$), and that of the interface B–C ($\mathbf{t}_{\mathrm{BC}}$):

$$\mathbf{t}_{\mathrm{layer}} = \mathbf{t}_{\mathrm{BC}}\mathbf{t}_{\mathrm{B}}\mathbf{t}_{\mathrm{AB}}. \tag{12.88}$$

The transmission matrix of the bulk of the layer B is given by

$$\mathbf{t}_{\mathrm{B}} = \begin{pmatrix} \exp(ik_0\sigma_{1\mathrm{B}}d_{\mathrm{B}}) & 0 \\ 0 & \exp(ik_0\sigma_{2\mathrm{B}}d_{\mathrm{B}}) \end{pmatrix}, \tag{12.89}$$

where d_{B} is the thickness of this layer. Let α be the angle between $\mathbf{x}_{\mathrm{A}}$ ($\mathbf{x}_{\mathrm{C}} = \mathbf{x}_{\mathrm{A}}$) and $\mathbf{x}_{\mathrm{B}}$ and let $\mathbf{t}'_{\mathrm{AB}}$ and $\mathbf{t}'_{\mathrm{BC}}$ be the matrices $\mathbf{t}_{\mathrm{AB}}$ and $\mathbf{t}_{\mathrm{BC}}$ at $\mathbf{x}_{\mathrm{C}} = \mathbf{x}_{\mathrm{A}} = \mathbf{x}_{\mathrm{B}}$ and $\mathbf{y}_{\mathrm{C}} = \mathbf{y}_{\mathrm{A}} = \mathbf{y}_{\mathrm{B}}$ [see (12.64) and (12.68)], that is, at $\alpha = 0$. Using (12.64) and (12.68), we may express the matrix $\mathbf{t}_{\mathrm{layer}}$ in terms of the diagonal matrices $\mathbf{t}'_{\mathrm{AB}}$, $\mathbf{t}'_{\mathrm{BC}}$, and $\mathbf{t}_{\mathrm{B}}$ as follows:

$$\mathbf{t}_{\mathrm{layer}} = \widehat{R}_{\mathrm{C}}(-\alpha)\mathbf{t}'_{\mathrm{BC}}\mathbf{t}_{\mathrm{B}}\mathbf{t}'_{\mathrm{AB}}\widehat{R}_{\mathrm{C}}(\alpha). \tag{12.90}$$

Denoting

$$\mathbf{t}'_{\text{layer}} \equiv \mathbf{t}'_{\text{BC}}\mathbf{t}_{\text{B}}\mathbf{t}'_{\text{AB}}, \tag{12.91}$$

we may rewrite expression (12.90) as

$$\mathbf{t}_{\text{layer}} = \widehat{R}_{\text{C}}(-\alpha)\mathbf{t}'_{\text{layer}}\widehat{R}_{\text{C}}(\alpha). \tag{12.92}$$

The diagonal matrix $\mathbf{t}'_{\text{layer}}$ is the transmission matrix of the layer B in the EW basis tied to the principal axes (see Section 1.3.3) of this layer: $\mathbf{t}_{\text{layer}} = \mathbf{t}'_{\text{layer}}$ when $\mathbf{x}_{\text{C}} = \mathbf{x}_{\text{A}} = \mathbf{x}_{\text{B}}$ and $\mathbf{y}_{\text{C}} = \mathbf{y}_{\text{A}} = \mathbf{y}_{\text{B}}$. Since the matrices $\mathbf{t}'_{\text{AB}}$, $\mathbf{t}_{\text{B}}$, and $\mathbf{t}'_{\text{BC}}$ are diagonal, the matrix $\mathbf{t}'_{\text{layer}}$ can be represented as

$$\mathbf{t}'_{\text{layer}} = \mathbf{t}_{\text{int}}\mathbf{t}_{\text{B}}, \tag{12.93}$$

where the matrix

$$\mathbf{t}_{\text{int}} \equiv \mathbf{t}'_{\text{BC}}\mathbf{t}'_{\text{AB}} \tag{12.94}$$

describes the effect of both interfaces simultaneously. Solving the same problem with the help of the classical Jones matrix method, we might use the following representation of the transmission matrix of the layer:

$$\mathbf{t}_{\text{layer}} \approx k\widehat{R}_{\text{C}}(-\alpha)\mathbf{t}_{\text{B}}\widehat{R}_{\text{C}}(\alpha), \tag{12.95}$$

where k is a scalar factor introduced to take into account reflection losses. It is obvious that the transition from the exact representation (12.92)–(12.94) to the approximate one (12.95) is equivalent to making use of the approximation

$$\mathbf{t}_{\text{int}} \approx k\mathbf{U} \tag{12.96}$$

which involves ignoring the polarization dependence of the reflection losses.

The normal components of the refraction vectors of the basis eigenwaves in the media A ($\sigma_{1\text{A}}$, $\sigma_{2\text{A}}$) and C ($\sigma_{1\text{C}}$, $\sigma_{2\text{C}}$) in this example are

$$\sigma_{1\text{A}} = \sigma_{2\text{A}} = n_{\text{A}}, \quad \sigma_{1\text{C}} = \sigma_{2\text{C}} = n_{\text{C}}, \tag{12.97}$$

where n_{A} and n_{C} are the refractive indices of the media A and C, respectively. Assume that the medium B is uniaxial. Denote the principal refractive indices of this medium by $n_{\|\text{B}}$ and $n_{\perp\text{B}}$ and a unit vector directed along its optic axis by $\mathbf{c}_{\text{B}}$. Let the optic axis of B be not perpendicular to interfaces and let basis eigenwave 1 in this layer be extraordinary. In this case,

$$\sigma_{1\text{B}} = n_{e\text{B}} = \frac{n_{\|\text{B}}n_{\perp\text{B}}}{\sqrt{n_{\perp\text{B}}^2 + \left(n_{\|\text{B}}^2 - n_{\perp\text{B}}^2\right)(\mathbf{c}_{\text{B}}\mathbf{z})^2}}, \quad \sigma_{2\text{B}} = n_{\perp\text{B}}. \tag{12.98}$$

At $\text{Re}(n_{\|\text{B}}) > \text{Re}(n_{\perp\text{B}})$, the vectors $\mathbf{x}_{\text{B}}$ and $\mathbf{y}_{\text{B}}$ are oriented along the slow axis and fast axis of the layer, respectively, and vice versa at $\text{Re}(n_{\|\text{B}}) < \text{Re}(n_{\perp\text{B}})$. Table 12.1 presents expressions for the transmission matrices of the interfaces A–B and B–C in terms of material parameters of the media for the cases of S-normalization and ET-normalization of the EW basis (for all three media). One may notice that in the case of S-normalization, the matrix $\mathbf{t}_{\text{AB}}$ at $n_{\text{A}} = \sqrt{n_{e\text{B}}n_{\perp\text{B}}}$ and the matrix $\mathbf{t}_{\text{BC}}$ at $n_{\text{C}} = \sqrt{n_{e\text{B}}n_{\perp\text{B}}}$ are STU

Table 12.1 Exact and approximate expressions for the transmission matrices of the interfaces A–B and B–C (nonabsorbing media)

	$\mathbf{t}'^{\downarrow}_{AB}$, $\mathbf{t}^{\downarrow}_{AB}$	$\mathbf{t}'^{\downarrow}_{BC}$, $\mathbf{t}^{\downarrow}_{BC}$
S-normalization	$\mathbf{t}'_{AB(S)} = \begin{pmatrix} 2\dfrac{\sqrt{n_A n_{eB}}}{n_A + n_{eB}} & 0 \\ 0 & 2\dfrac{\sqrt{n_A n_{\perp B}}}{n_A + n_{\perp B}} \end{pmatrix}$ $\mathbf{t}_{AB(S)} = \mathbf{t}'_{AB(S)}\widehat{R}_C(\alpha)$	$\mathbf{t}'_{BC(S)} = \begin{pmatrix} 2\dfrac{\sqrt{n_C n_{eB}}}{n_C + n_{eB}} & 0 \\ 0 & 2\dfrac{\sqrt{n_C n_{\perp B}}}{n_C + n_{\perp B}} \end{pmatrix}$ $\mathbf{t}_{BC(S)} = \widehat{R}_C(-\alpha)\mathbf{t}'_{BC(S)}$
STUM approximation at S-normalization*	$(\mathbf{t}'_{AB(S)})_{STU} = w_{AB(S)}\mathbf{U}$ $(\mathbf{t}_{AB(S)})_{STU} = w_{AB(S)}\widehat{R}_C(\alpha)$ $w_{AB(S)} = \dfrac{\sqrt{n_A n_{eB}}}{n_A + n_{eB}} + \dfrac{\sqrt{n_A n_{\perp B}}}{n_A + n_{\perp B}}$	$(\mathbf{t}'_{BC(S)})_{STU} = w_{BC(S)}\mathbf{U}$ $(\mathbf{t}_{BC(S)})_{STU} = w_{BC(S)}\widehat{R}_C(-\alpha)$ $w_{BC(S)} = \dfrac{\sqrt{n_C n_{eB}}}{n_C + n_{eB}} + \dfrac{\sqrt{n_C n_{\perp B}}}{n_C + n_{\perp B}}$
ET-normalization	$\mathbf{t}'_{AB(ET)} = \begin{pmatrix} \dfrac{2n_A}{n_A + n_{eB}} & 0 \\ 0 & \dfrac{2n_A}{n_A + n_{\perp B}} \end{pmatrix}$ $\mathbf{t}_{AB(ET)} = \mathbf{t}'_{AB(ET)}\widehat{R}_C(\alpha)$	$\mathbf{t}'_{BC(ET)} = \begin{pmatrix} \dfrac{2n_{eB}}{n_C + n_{eB}} & 0 \\ 0 & \dfrac{2n_{\perp B}}{n_C + n_{\perp B}} \end{pmatrix}$ $\mathbf{t}_{BC(ET)} = \widehat{R}_C(-\alpha)\mathbf{t}'_{BC(ET)}$
STUM approximation at ET-normalization	$(\mathbf{t}'_{AB(ET)})_{STU} = w_{AB(ET)}\mathbf{U}$ $(\mathbf{t}_{AB(ET)})_{STU} = w_{AB(ET)}\widehat{R}_C(\alpha)$ $w_{AB(ET)} = \dfrac{n_A}{n_A + n_{eB}} + \dfrac{n_A}{n_A + n_{\perp B}}$	$(\mathbf{t}'_{BC(ET)})_{STU} = w_{BC(ET)}\mathbf{U}$ $(\mathbf{t}_{BC(ET)})_{STU} = w_{BC(ET)}\widehat{R}_C(-\alpha)$ $w_{BC(ET)} = \dfrac{n_{eB}}{n_C + n_{eB}} + \dfrac{n_{\perp B}}{n_C + n_{\perp B}}$

*U is the unit matrix.

matrices [see (12.81)]. In the case of ET-normalization, the matrices $\mathbf{t}_{AB}$ and $\mathbf{t}_{BC}$ at any $n_{eB} \neq n_{\perp B}$ are not STU matrices. One may also notice that $\mathbf{t}'_{BC(S)} = \mathbf{t}'_{AB(S)}$ when $n_C = n_A$. Note that the expressions for exact transmission matrices in Table 12.1 are valid in the presence of absorption as well. In the subsequent discussion of this example we assume that the media A, B, and C are nonabsorbing.

As an example, Figure 12.4 shows the dependences of the nonzero elements of the matrices $\mathbf{t}'_{AB(S)}$, $\mathbf{t}'_{BC(S)}$, $\mathbf{t}'_{AB(ET)}$, and $\mathbf{t}'_{BC(ET)}$ on n_A in the situation when $n_C = n_A$, at $n_{\parallel B} = 1.7$, $n_{\perp B} = 1.5$, and $\mathbf{c}_B \perp \mathbf{z}$ ($n_{eB} = n_{\parallel B}$). As can be seen from this figure, the elements (1,1) and (2,2) for both interface matrices in the case of ET-normalization are much more different from each other than in the case of S-normalization, and this difference is significant even in the absence of polarization-dependent losses (when $t'_{AB(S)11} = t'_{AB(S)22}$) (see above). It is clear that in this situation, the matrices $\mathbf{t}'_{AB(ET)}$ and $\mathbf{t}'_{BC(ET)}$ are approximated by STU matrices much worse than the matrices $\mathbf{t}'_{AB(S)}$ and $\mathbf{t}'_{BC(S)}$. Explicit expressions for the best approximating STU matrices for the transmission matrices of the interfaces A–B and B–C in terms of the material parameters are given in Table 12.1. The relative errors of STUM approximation for the matrices $\mathbf{t}'_{AB(S)}$ and $\mathbf{t}'_{AB(ET)}$ at Δn_B equal to 0.1 ($n_{\perp B} = 1.5$, $n_{\parallel B} = 1.6$) and 0.2 ($n_{\perp B} = 1.5$, $n_{\parallel B} = 1.7$) are compared in Figure 12.5 ($n_C = n_A$; $\mathbf{c}_B$ is perpendicular to $\mathbf{z}$). We see that in both cases, at any n_A, the error of STUM approximation for the matrix $\mathbf{t}'_{AB(S)}$ is many times smaller than for $\mathbf{t}'_{AB(ET)}$.

In the case of nonabsorbing media, the most correct transition from representation (12.92) to representation (12.95) is carried out not by replacement of the matrices $\mathbf{t}_{AB}$ and $\mathbf{t}_{BC}$ by the matrices $(\mathbf{t}_{AB})_{STU}$

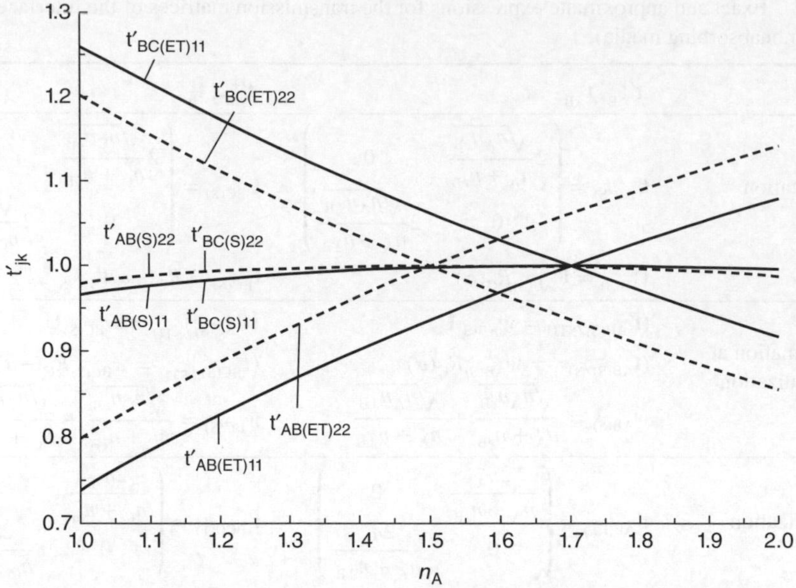

Figure 12.4 Elements of the interface matrices $\mathbf{t}'_{AB(S)}$, $\mathbf{t}'_{BC(S)}$, $\mathbf{t}'_{AB(ET)}$, and $\mathbf{t}'_{BC(ET)}$ versus n_A

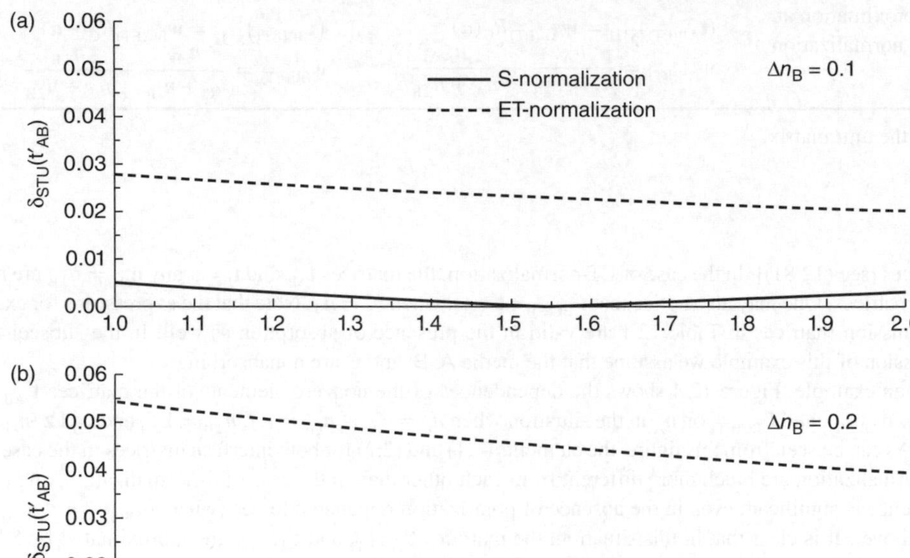

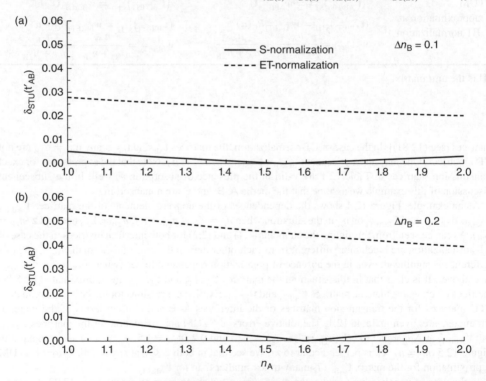

Figure 12.5 Relative error of STUM approximation for the interface matrix $\mathbf{t}'_{AB}$ under different normalizations of the EW basis

and $(\mathbf{t}_{BC})_{STU}$ but by replacing $\mathbf{t}_{int}$ by $(\mathbf{t}_{int})_{STU}$. According to (12.13), we have from (12.92)

$$(\mathbf{t}_{layer})_{STU} = \widehat{R}_C(-\alpha)(\mathbf{t}'_{layer})_{STU}\widehat{R}_C(\alpha) \quad (12.99)$$

and, from (12.93),

$$(\mathbf{t}'_{layer})_{STU} = (\mathbf{t}_{int})_{STU}\mathbf{t}_B. \quad (12.100)$$

Since in general

$$(\mathbf{t}_{int})_{STU} = (\mathbf{t}_{BC}\mathbf{t}_{AB})_{STU} \neq (\mathbf{t}_{BC})_{STU}(\mathbf{t}_{AB})_{STU}, \quad (12.101)$$

the approximation

$$\begin{aligned}\mathbf{t}_{layer} \approx (\mathbf{t}_{layer})_{STU} = \widehat{R}_C(-\alpha)(\mathbf{t}'_{layer})_{STU}\widehat{R}_C(\alpha) &= \widehat{R}_C(-\alpha)(\mathbf{t}_{int})_{STU}\mathbf{t}_B\widehat{R}_C(\alpha) \\ &= w_{ABC}\widehat{R}_C(-\alpha)\mathbf{t}_B\widehat{R}_C(\alpha)\end{aligned} \quad (12.102)$$

with

$$w_{ABC} = \frac{\mathrm{Tr}\,\mathbf{t}_{int}}{2} \quad (12.103)$$

is more accurate than the approximation

$$\mathbf{t}_{layer} \approx (\mathbf{t}_{BC})_{STU}\mathbf{t}_B(\mathbf{t}_{AB})_{STU} = w_{BC}w_{AB}\widehat{R}_C(-\alpha)\mathbf{t}_B\widehat{R}_C(\alpha). \quad (12.104)$$

At usual values of the optical constants, to a good approximation,

$$w_{ABC(S)} \approx w_{BC(S)}w_{AB(S)}, \quad (12.105)$$

$$w^2_{ABC(S)} \approx t_{layer-unpol}, \quad (12.106)$$

where $t_{layer\text{-}unpol} = \underline{\mathrm{avr}}[\mathbf{t}_{layer(S)}]$ is the average transmissivity of the layer [see (8.278)]. To illustrate, in Table 12.2 we compare the values of $w_{ABC(S)}$, $w_{BC(S)}w_{AB(S)}$, and $(t_{layer\text{-}unpol})^{1/2}$ at $n_{\|B} = 1.7$, $n_{\perp B} = 1.5$, $\mathbf{c}_B \perp \mathbf{z}$, and $n_C = n_A$. In addition, in this table, the values of $w_{ABC(ET)}$ and $w_{BC(ET)}w_{AB(ET)}$ are shown. We see that for these quantities, also to a good approximation,

$$w_{ABC(ET)} \approx w_{BC(ET)}w_{AB(ET)}. \quad (12.107)$$

Table 12.2 Numerical estimates of quantities appearing in relations (12.105)–(12.107)

	$n_A = 1$	$n_A = 1.52$
$w_{ABC(S)}$	0.946392	0.998416
$w_{BC(S)}w_{AB(S)}$	0.946343	0.998415
$(t_{layer\text{-}unpol})^{1/2}$	0.946490	0.998417
$w_{ABC(ET)}$	0.946392	0.998416
$w_{BC(ET)}w_{AB(ET)}$	0.947270	0.999393

The discrepancies between $w_{ABC(ET)}$ and $w_{BC(ET)}w_{AB(ET)}$ are just a little larger than the discrepancies between $w_{ABC(S)}$ and $w_{BC(S)}w_{AB(S)}$. Note that $w_{ABC(ET)}$ equals $w_{ABC(S)}$ (see Table 12.2) in general only at $n_C = n_A$. It is clear that with such small differences in the values of w_{ABC} and $w_{BC}w_{AB}$ as in Table 12.2, the accuracies of approximations (12.102) and (12.104) both in the case of S-normalization and in the case of ET-normalization are almost the same. Table 12.3 illustrates this statement. In this table, we show the values of the relative error of STUM approximation for the matrices $\mathbf{t}_{AB(S)}$, $\mathbf{t}_{BC(S)}$, $\mathbf{t}_{AB(ET)}$, and $\mathbf{t}_{BC(ET)}$ and the relative errors of approximation (12.102) (approximation 1), approximation (12.104) (approximation 2), and the following one

$$\mathbf{t}_{layer} \approx \mathbf{t}_{BC}\mathbf{t}_{B}\left(\mathbf{t}_{AB}\right)_{STU} \tag{12.108}$$

(approximation 3) for four sets of values of the material parameters. In all four cases, $n_{\perp B} = 1.5$, $\mathbf{c}_B \perp \mathbf{z}$, and $n_C = n_A$. The values of n_A and Δn_B for each case are shown in the table. In Table 12.3 and in what follows, $\delta_{STU}(\mathbf{t})$ denotes the relative error of STUM approximation of a matrix $\mathbf{t}$ [see (12.30)]. The relative errors of the approximations were calculated by the formula

$$\delta_{appr} = \frac{\sqrt{2}\|\mathbf{t}_{exact} - \mathbf{t}_{appr}\|_E}{\|\mathbf{t}_{exact}\|_E}, \tag{12.109}$$

where $\mathbf{t}_{exact}$ and $\mathbf{t}_{appr}$ are respectively the exact value of the matrix being approximated and its approximate value. The relative errors of approximations (12.102), (12.104), and (12.108) are denoted by δ_{appr1}, δ_{appr2}, and δ_{appr3}, respectively.

From Table 12.3 we see that in all the cases

$$\delta_{appr1(S)} = \delta_{appr1(ET)}. \tag{12.110}$$

This relation is valid when the media A and C are isotropic, even if they are different. At $n_C \neq n_A$, the matrices $\mathbf{t}_{int(S)}$ and $\mathbf{t}_{int(ET)}$ [see (12.94)] are different and related by

$$\mathbf{t}_{int(ET)} = \sqrt{\frac{n_A}{n_C}}\mathbf{t}_{int(S)}. \tag{12.111}$$

Table 12.3 The relative errors of STUM approximation for the interface transmission matrices and approximations (12.102), (12.104), and (12.108)

	$n_A = 1$ $\Delta n_B = 0.1$	$n_A = 1$ $\Delta n_B = 0.2$	$n_A = 1.52$ $\Delta n_B = 0.1$	$n_A = 1.52$ $\Delta n_B = 0.2$
$\delta_{STU}(\mathbf{t}_{AB(S)})$	0.0049	0.0102	0.0002	0.0011
$\delta_{STU}(\mathbf{t}_{BC(S)})$	0.0049	0.0102	0.0002	0.0011
$\delta_{STU}(\mathbf{t}_{AB(ET)})$	0.0277	0.0544	0.0230	0.0453
$\delta_{STU}(\mathbf{t}_{BC(ET)})$	0.0179	0.0341	0.0226	0.0431
$\delta_{appr1(S)}$	0.0098	0.0203	0.0004	0.0022
$\delta_{appr1(ET)}$	0.0098	0.0203	0.0004	0.0022
$\delta_{appr2(S)}$	0.0098	0.0203	0.0004	0.0022
$\delta_{appr2(ET)}$	0.0098	0.0204	0.0006	0.0026
$\delta_{appr3(S)}$	0.0049	0.0102	0.0002	0.0011
$\delta_{appr3(ET)}$	0.0277	0.0545	0.0230	0.0454

The matrices $\mathbf{t}_{\text{layer(S)}}$ and $\mathbf{t}_{\text{layer(ET)}}$ are related similarly:

$$\mathbf{t}_{\text{layer(ET)}} = \sqrt{\frac{n_{\text{A}}}{n_{\text{C}}}}\mathbf{t}_{\text{layer(S)}} \tag{12.112}$$

[see (12.90)–(12.93)]. It is obvious that subject to (12.112), the relative errors of STUM approximation of the matrices $\mathbf{t}_{\text{layer(S)}}$ and $\mathbf{t}_{\text{layer(ET)}}$ [see (12.102)] are equivalent, which explains (12.110). If the media A and C were anisotropic, in the absence of polarization degeneracy at least in one of these media, relation (12.110) would be violated.

In approximations 2 and 3, in contrast to approximation 1, STUM approximation is applied to interfaces. In approximation 2, STUM approximation is applied to both interfaces (separately). In approximation 3, only the matrix of the interface A–B is replaced by the corresponding STU matrix, while the matrix of the interface B–C is retained exact. As is seen from Table 12.3, the error of approximation 2 is close to the error of approximation 1 for both variants of normalization. We have explained why this takes place [see (12.105) and (12.107)]. Note that the error of approximation 2 under ET-normalization, $\delta_{\text{appr2(ET)}}$, is significantly less than the errors of the interface matrices $\delta_{\text{STU}}(\mathbf{t}_{\text{AB(ET)}})$ and $\delta_{\text{STU}}(\mathbf{t}_{\text{BC(ET)}})$. At $n_{\text{A}} = 1.52$, $\delta_{\text{appr2(ET)}}$ is almost 40 times less than $\delta_{\text{STU}}(\mathbf{t}_{\text{AB(ET)}})$ and $\delta_{\text{STU}}(\mathbf{t}_{\text{BC(ET)}})$! In this case, the error introduced in calculations of the matrix $\mathbf{t}_{\text{layer(ET)}}$ by the inaccuracy of the matrix approximating $\mathbf{t}_{\text{AB(ET)}}$ is almost completely compensated by the error in the matrix approximating $\mathbf{t}_{\text{BC(ET)}}$. In the case of approximation 3, only one of the interface matrices is approximate, and it would seem that the accuracy of the matrix $\mathbf{t}_{\text{layer(ET)}}$ must be higher than in the case of approximation 2. But actually we have exactly the opposite of this. The error in $\mathbf{t}_{\text{layer(ET)}}$ for approximation 3 is much larger than that for approximation 2, because in approximation 3 the error cancellation is absent and hence $\delta_{\text{appr3(ET)}} \approx \delta_{\text{STU}}(\mathbf{t}_{\text{AB(ET)}})$.

The points noted have a direct bearing on many popular variants of Jones matrix method used for modeling LCDs. The discussed error cancellation explains the fact that the methods [1–3] and some other methods using FTCEF or FEF Jones vectors and relatively rough approximations for interface matrices, allied to the approximation $\mathbf{t}_{\text{AB(ET)}} \approx (\mathbf{t}_{\text{AB(ET)}})_{\text{STU}}$, in many cases provide sufficiently accurate results. But a combined use of approximate and exact operators for different elements in calculations, as in the case of approximation 3, often leads to large errors. We dealt with a situation of this kind in Section 11.1.2.

According to (12.102), the polarization matrix of the layer, $\mathbf{t}^{(p)}_{\text{layer}}$, may be taken to be

$$\mathbf{t}^{(p)}_{\text{layer}} = \widehat{R}_{\text{C}}(-\alpha)\mathbf{t}_{\text{B}}\widehat{R}_{\text{C}}(\alpha). \tag{12.113}$$

Since we regard the polarization vectors as prescribed-phase (see Section 5.4.3), we can replace the matrix $\mathbf{t}_{\text{B}}$ by the unimodular matrix

$$\mathbf{t}_{\text{BUM}} = \begin{pmatrix} \exp(i\pi\Delta\sigma_{\text{B}}d_{\text{B}}/\lambda) & 0 \\ 0 & \exp(-i\pi\Delta\sigma_{\text{B}}d_{\text{B}}/\lambda) \end{pmatrix}, \tag{12.114}$$

where $\Delta\sigma_{\text{B}} = \sigma_{1\text{B}} - \sigma_{2\text{B}}$, to deal with the unimodular polarization matrix

$$\mathbf{t}^{(p)}_{\text{layer}} = \widehat{R}_{\text{C}}(-\alpha)\mathbf{t}_{\text{BUM}}\widehat{R}_{\text{C}}(\alpha), \tag{12.115}$$

which has the standard form (5.31). It is needless to say that these representations of the polarization Jones matrix are equivalent to those used in the classical JC. Note that approximation (12.104) leads to the same variants of the polarization matrix of the layer.

Absorbing Media

In the presence of absorption at least in one of the neighboring media (A and B), the exact transmission EW Jones matrix of the interface A–B, $\mathbf{t}_{AB}$, is in general complex. If absorption is weak (the term "weakly absorbing" is commonly used for media whose extinction coefficients are less than 0.03), the imaginary parts of the elements of the matrix $\mathbf{t}_{AB(S)}$ are very small and, to a good approximation,

$$\mathbf{t}_{AB(S)} \approx \mathbf{t}_{A'B'(S)}, \tag{12.116}$$

where $\mathbf{t}_{A'B'}$ is the transmission EW Jones matrix of the interface between media A′ and B′ whose refractive indices are equal to the real parts of the corresponding complex refractive indices of the media A and B. Table 12.4 demonstrates three numerical examples showing how accurate approximation (12.116) is. We see that in all these examples, the imaginary parts of the elements of $\mathbf{t}_{AB}$ and the differences of the corresponding elements of $\mathrm{Re}(\mathbf{t}_{AB})$ and $\mathbf{t}_{A'B'}$ are many times less in magnitude than the largest of the extinction coefficients. According to (12.57)–(12.58), such is always the case for weakly absorbing media under S-normalization. The relation $\mathbf{t}_{AB(ET)} \approx \mathbf{t}_{A'B'(ET)}$ is usually much less accurate than relation (12.116). It is clear that in situations when the relations (12.116) and $\mathbf{t}_{A'B'(S)} \approx (\mathbf{t}_{A'B'(S)})_{STU}$ are sufficiently

Table 12.4 Numerical examples to the approximation (12.116)*

Media A and B	$\mathbf{t}_{AB(S)}$	$\mathbf{t}_{A'B'(S)}$
Medium A: isotropic, nonabsorbing, $n_A = 1.52$. Medium B: uniaxial, absorbing, $n_{\parallel B} = 1.7+i0.02$, $n_{\perp B} = 1.5+i0.002$, $\mathbf{c}_B \perp \mathbf{z}$. $\mathbf{x}_A \parallel \mathbf{x}_B$	Real part: $\begin{pmatrix} 0.998452 & 0 \\ 0 & 0.999978 \end{pmatrix}$ Imaginary part: $\begin{pmatrix} -0.000329 & 0 \\ 0 & 0.000004 \end{pmatrix}$	$\begin{pmatrix} 0.998436 & 0 \\ 0 & 0.999978 \end{pmatrix}$
Medium A: isotropic, nonabsorbing, $n_A = 1$. Medium B: as above.	Real part: $\begin{pmatrix} 0.965814 & 0 \\ 0 & 0.979796 \end{pmatrix}$ Imaginary part: $\begin{pmatrix} -0.001473 & 0 \\ 0 & -0.000131 \end{pmatrix}$	$\begin{pmatrix} 0.965808 & 0 \\ 0 & 0.979796 \end{pmatrix}$
Medium A: isotropic, nonabsorbing, $n_A = 1$. Medium B: uniaxial, absorbing, $n_{\parallel B} = 1.51+i0.02$, $n_{\perp B} = 1.5+i0.002$, $\mathbf{c}_B \perp \mathbf{z}$. $\mathbf{x}_A \parallel \mathbf{x}_B$	Real part: $\begin{pmatrix} 0.979151 & 0 \\ 0 & 0.979796 \end{pmatrix}$ Imaginary part: $\begin{pmatrix} -0.001318 & 0 \\ 0 & -0.000131 \end{pmatrix}$	$\begin{pmatrix} 0.979140 & 0 \\ 0 & 0.979796 \end{pmatrix}$

*The first basis wave in the medium B is extraordinary.

accurate, the matrix $\mathbf{t}_{AB(S)}$ can be approximated, with good accuracy, by a real STU matrix, such as the matrix $(\mathbf{t}_{A'B'(S)})_{STU}$.

12.3 Polarization Jones Matrix of an Inhomogeneous Nonabsorbing Anisotropic Layer with Negligible Bulk Reflection at Normal Incidence. Simple Representations of Polarization Matrices of LC Layers at Normal Incidence

Let us consider an inhomogeneous nonabsorbing anisotropic layer B surrounded by isotropic nonabsorbing media A and C. Assume that the transmission properties of the bulk of B can be adequately treated using the NBR approximation and that the action of the interfaces A–B and B–C can be, with sufficient accuracy, represented by polarization transmission matrices. Neglecting multiple reflections, we can represent the transmission EW Jones matrix of the layer as

$$\mathbf{t}_{\text{layer}} = \mathbf{t}_{BC}\mathbf{t}_{B(NBR)}\mathbf{t}_{AB}, \tag{12.117}$$

where $\mathbf{t}_{B(NBR)}$ is the transmission matrix of the bulk of B in the NBR approximation, and the matrices $\mathbf{t}_{AB}$ and $\mathbf{t}_{BC}$ have the same meaning as in (12.88). Let S–F-normalization be used for all the media. In this case, the matrix $\mathbf{t}_{B(NBR)}$ is unitary and may in principle be considered as a polarization matrix of the bulk of the layer. The most accurate variant of the polarization matrix of the layer, $\mathbf{t}^{(p)}_{\text{layer}}$, is

$$\mathbf{t}^{(p)}_{\text{layer}} = [\mathbf{t}_{\text{layer}}]_{STU}, \tag{12.118}$$

where $[\mathbf{t}]_{STU}$ denotes the base matrix of the best approximating STU matrix for a given matrix $\mathbf{t}$. Another useful variant of the polarization matrix is

$$\mathbf{t}^{(p)}_{\text{layer}} = \mathbf{t}^{(p)}_{BC}\mathbf{t}^{(p)}_{B}\mathbf{t}^{(p)}_{AB}, \tag{12.119}$$

where $\mathbf{t}^{(p)}_{AB}$, $\mathbf{t}^{(p)}_{B}$, and $\mathbf{t}^{(p)}_{BC}$ are the polarization Jones matrices characterizing the interface A–B, the bulk of the layer B, and the interface B–C, respectively. The matrix $\mathbf{t}^{(p)}_{B}$ can be chosen unimodular and of determinant 1:

$$\mathbf{t}^{(p)}_{B} = \mathbf{t}_{B(NBR)UM} \equiv \frac{1}{\sqrt{\det \mathbf{t}_{B(NBR)}}}\mathbf{t}_{B(NBR)}. \tag{12.120}$$

With the assumptions made, variants (12.118) and (12.119) are almost equivalent in accuracy. In principle, these representations can be used at both normal and oblique incidence. Representation (12.119) is convenient when normal incidence is considered and an analytical expression for the matrix $\mathbf{t}^{(p)}_{B}$ is known (see Section 11.4). In this case, from (12.119), one can obtain an analytical expression for the matrix $\mathbf{t}^{(p)}_{\text{layer}}$, using the fact that the interface matrices $\mathbf{t}^{(p)}_{AB}$ and $\mathbf{t}^{(p)}_{BC}$ at normal incidence are explicitly and simply expressed in terms of geometrical parameters of the system. We give here some useful expressions and examples for this case.

Let the EW basis in A and C be chosen so that $\mathbf{x}_C = \mathbf{x}_A$ and $\mathbf{y}_C = \mathbf{y}_A$, as in the above example for a homogeneous layer. The parameters of the EW basis within the layer B are functions of z. Suppose that the EW basis is chosen so that the vectors $\mathbf{e}_{t1}$, $\mathbf{e}_{t2}$, and $\mathbf{z}$ form a right-handed triple in the surrounding media as well as in B at least just at the interfaces, that is, at $z = z_1 + 0$ and $z = z_2 - 0$, where z_1 and z_2 ($z_1 < z_2$) are the z-coordinates of the interfaces A–B and B–C, respectively. Then the interface matrices $\mathbf{t}^{(p)}_{\mathrm{AB}}$ and $\mathbf{t}^{(p)}_{\mathrm{BC}}$ can be expressed as follows (see the previous section):

$$\mathbf{t}^{(p)}_{\mathrm{AB}} = \widehat{R}_{\mathrm{C}}(\alpha_1), \quad \mathbf{t}^{(p)}_{\mathrm{BC}} = \widehat{R}_{\mathrm{C}}(-\alpha_2), \tag{12.121}$$

where α_1 is the angle between $\mathbf{x}_{\mathrm{A}}$ and $\mathbf{e}_{t1}(z_1 + 0)$, and α_2 is the angle between $\mathbf{x}_{\mathrm{A}}$ and $\mathbf{e}_{t1}(z_2 - 0)$ (both α_1 and α_2 are measured from $\mathbf{x}_{\mathrm{A}}$). Using (12.120) and (12.121), we obtain the following expression for the polarization matrix $\mathbf{t}^{(p)}_{\mathrm{layer}}$:

$$\mathbf{t}^{(p)}_{\mathrm{layer}} = \widehat{R}_{\mathrm{C}}(-\alpha_2)\mathbf{t}_{\mathrm{B(NBR)UM}}\widehat{R}_{\mathrm{C}}(\alpha_1). \tag{12.122}$$

In this case, the matrix $\mathbf{t}^{(p)}_{\mathrm{layer}}$ is unimodular and has the standard form (5.31).

Locally Uniaxial Layer

Let the layer B be locally uniaxial, and suppose that the orientation of its local optic axis is given by (11.1) where $\theta(z)$ and $\varphi(z)$ are continuous functions. Let the vector $\mathbf{x}_{\mathrm{A}}$ be chosen codirectional with the X-axis and let the EW basis in B be chosen so that

$$\mathbf{e}_{t1}(z) = c_{1\mathrm{B}}(z)\mathbf{x}_{\mathrm{B}}(z), \quad \mathbf{e}_{t2}(z) = c_{2\mathrm{B}}(z)\mathbf{y}_{\mathrm{B}}(z) \tag{12.123}$$

with

$$\mathbf{x}_{\mathrm{B}}(z) = \begin{pmatrix} \cos\varphi(z) \\ \sin\varphi(z) \\ 0 \end{pmatrix}_{XYZ}, \quad \mathbf{y}_{\mathrm{B}}(z) = \begin{pmatrix} -\sin\varphi(z) \\ \cos\varphi(z) \\ 0 \end{pmatrix}_{XYZ}, \tag{12.124}$$

and $c_{1\mathrm{B}}$ and $c_{2\mathrm{B}}$ being positive scalar quantities. In this case, we may rewrite expression (12.122) as follows:

$$\mathbf{t}^{(p)}_{\mathrm{layer}} = \widehat{R}_{\mathrm{C}}(-\varphi_2)\mathbf{t}_{\mathrm{B(NBR)UM}}\widehat{R}_{\mathrm{C}}(\varphi_1), \tag{12.125}$$

or equivalently

$$\mathbf{t}^{(p)}_{\mathrm{layer}} = \widehat{R}_{\mathrm{C}}(-\varphi_1)[\widehat{R}_{\mathrm{C}}(-\Phi)\mathbf{t}_{\mathrm{B(NBR)UM}}]\widehat{R}_{\mathrm{C}}(\varphi_1), \tag{12.126}$$

where $\varphi_1 \equiv \varphi(z_1 + 0)$, $\varphi_2 \equiv \varphi(z_2 - 0)$, and $\Phi = \varphi_2 - \varphi_1$. Note that the EW basis adopted for the layer B here is a variant of the basis used for the arbitrary locally uniaxial medium in Section 11.2. The final formulas of Section 11.2 for uniaxial media are adapted to the case under consideration by taking $\mathbf{k} = -\mathbf{x}_{\mathrm{B}}$ and $\mathbf{j} = \mathbf{y}_{\mathrm{B}}$ in those formulas.

Ideal Twisted Layer. NBRA and SBA Representations of the Polarization Matrix

For an ideal twisted layer, the functions $\theta(z)$ and $\varphi(z)$ are given by (11.40). According to (11.151), the matrix $\mathbf{t}_{\mathrm{B(NBR)UM}}$ of this layer can be expressed as

$$\mathbf{t}_{\mathrm{B(NBR)UM}} = \begin{pmatrix} \cos Q + i\dfrac{G}{Q}\sin Q & \dfrac{\tilde{\Phi}}{Q}\sin Q \\ -\dfrac{\tilde{\Phi}}{Q}\sin Q & \cos Q - i\dfrac{G}{Q}\sin Q \end{pmatrix}, \tag{12.127}$$

$$Q = \sqrt{G^2 + \tilde{\Phi}^2},$$

where

$$G = \frac{\pi\left(n_{\mathrm{e}} - n_{\perp}\right)d}{\lambda}, \quad n_{\mathrm{e}} = \frac{n_{\parallel}n_{\perp}}{\sqrt{n_{\perp}^2\cos^2\theta_c + n_{\parallel}^2\sin^2\theta_c}},$$
$$\tilde{\Phi} = \bar{\Phi} = \Phi\Gamma, \quad \Gamma = \frac{1+\gamma/2}{\sqrt{1+\gamma}}, \quad \gamma = \frac{n_{\mathrm{e}} - n_{\perp}}{n_{\perp}}; \tag{12.128}$$

$n_{\parallel}$ and $n_{\perp}$ are the principal refractive indices of the layer. Since all the matrices on the right-hand side of (12.126) are unimodular and have determinant 1, the matrix $\mathbf{t}_{\mathrm{layer}}^{(p)}$ can be written as

$$\mathbf{t}_{\mathrm{layer}}^{(p)} = \begin{pmatrix} a' + ia'' & b' + ib'' \\ -b' + ib'' & a' - ia'' \end{pmatrix}, \tag{12.129}$$

where a', a'', b', and b'' are real. On substituting (12.127) into (12.126), we find that

$$a' = a'_{\Phi}\cos\Phi + b'_{\Phi}\sin\Phi, \quad a'' = a''_{\Phi}\cos\left(\Phi + 2\varphi_1\right), \tag{12.130a}$$

$$b' = -a'_{\Phi}\sin\Phi + b'_{\Phi}\cos\Phi, \quad b'' = a''_{\Phi}\sin\left(\Phi + 2\varphi_1\right), \tag{12.130b}$$

where

$$a'_{\Phi} = \cos Q, \quad a''_{\Phi} = \frac{G}{Q}\sin Q, \quad b'_{\Phi} = \frac{\tilde{\Phi}}{Q}\sin Q. \tag{12.131}$$

Representation (12.129)–(12.131) with G and $\tilde{\Phi}$ expressed by (12.128) will be referred to as *the NBRA representation* of the polarization Jones matrix of an ideal twisted layer. The same representation with the same G but with $\tilde{\Phi} = \Phi$ will be called *the SBA* (small-birefringence approximation) *representation* of this matrix [see discussion under formula (11.154)]. The SBA representation gives the same polarization matrix as the classical JC.

Nonideal Quasi-Planar Twisted Layer. PITL Approximation

An approximate value of the matrix $\mathbf{t}_{\mathrm{B(NBR)UM}}$ for a quasi-planar twisted layer (see Sections 11.1.1 and 11.4.1) can be calculated by formulas (12.127) with G given by (11.46) and

$$\tilde{\Phi} = \Phi\frac{1+\gamma_0/2}{\sqrt{1+\gamma_0}}, \tag{12.132}$$

where

$$\gamma_0 = \frac{n_{\|} - n_{\perp}}{n_{\perp}}. \tag{12.133}$$

The corresponding approximate value of the matrix $\mathbf{t}^{(p)}_{\mathrm{layer}}$ can be calculated by formulas (12.129)–(12.131) with these G and $\tilde{\Phi}$. This approximation will be called *the PITL* (*planar ideal twisted layer*) *approximation* (the approximating matrix in this case is an exact polarization matrix of a planar ideal twisted layer of thickness L_θ [see (11.46)]). Some estimates of the accuracy of the PITL approximation will be given in Section 12.4, where we use this approximation and other representations presented here in an experimental method for determining configurational and optical parameters of LC layers with a twisted structure.

12.4 Immersion Model of the Polarization-Converting System of an LCD

In our modeling program MOUSE-LCD (see Chapter 4), calculations in terms of polarization matrices are used in some optimization procedures, in particular in those where techniques described in Sections 6.4 and 6.5 are employed. In these calculations, polarization matrices are utilized as characteristics of the polarization-converting system (PCS) and its polarization elements. In most practical cases, diattenuation in PCSs of LCDs is relatively weak at both normal and oblique incidence. The latter allows calculations in terms of polarization matrices for PCSs in the case of oblique incidence. Use of the EW Jones matrix method and STUM approximation makes the calculation of polarization matrices most accurate and consistent. In MOUSE-LCD, polarization matrices for the elements of the PCS and the PCS as a whole are calculated using a model that will be referred to as *the immersion model*.

Immersion Model

In the immersion model, the model PCS is a layered system consisting of anisotropic layers modeling the polarization elements of the PCS being modeled (LC layer, compensation films, reflector in the case of a reflective PCS), which are surrounded and separated from each other by an isotropic nonabsorbing immersion medium whose refractive index n_{IM} is between 1.5 and 1.55 (in MOUSE-LCD, we set $n_{\mathrm{IM}} =$ 1.52). The isotropic strata are regarded as input and output media for polarization matrices of polarization elements, so that the polarization matrix of any polarization element links the polarization Jones vectors of waves propagating in isotropic layers of the same refractive index. The polarization matrix $\mathbf{t}^{(p)}$ of a transmissive polarization element, the LC layer or a compensation film, a homogeneous or a smoothly inhomogeneous layer, is calculated as

$$\mathbf{t}^{(p)} = [\mathbf{t}_{\mathrm{L}\to\mathrm{I}}\mathbf{t}_{\mathrm{Lbulk}}\mathbf{t}_{\mathrm{I}\to\mathrm{L}}]_{\mathrm{STU}} \tag{12.134}$$

[cf. (12.118)], where $\mathbf{t}_{I\to L}$ and $\mathbf{t}_{L\to I}$ are the transmission EW Jones matrices of the frontal and rear interfaces of the layer, respectively, and $\mathbf{t}_{Lbulk}$ is the transmission EW Jones matrix for the bulk of the layer. If the layer is inhomogeneous, the matrix $\mathbf{t}_{Lbulk}$ can be calculated in the NBR approximation. The polarization matrix of the PCS is calculated as the product of polarization matrices of its polarization elements. In the case of a reflective PCS with a specular metal reflector, the polarization matrix of the reflector can be calculated by the formula

$$\mathbf{r}_R^{(p)} = [\mathbf{r}_R^{\downarrow}]_{STU}, \tag{12.135}$$

where $\mathbf{r}_R^{\downarrow}$ is the EW Jones matrix of the interface between the immersion medium and the metal.

In MOUSE-LCD, polarization matrices of the PCS, calculated in this way, are used in particular to find the unitary polarization transport coefficients for this system (see Sections 6.1 and 6.4). Polarization matrices of the LC layer are used in calculations of the modulation efficiency (see Sections 6.5 and 12.6).

Now we will give some numerical examples allowing one to estimate the effect of factors that are neglected in the immersion model, and in particular diattenuation, in typical situations. In some of these examples we use a standard characteristic of diattenuation that can be defined as follows.

Consider a transfer channel (possibly defined by a single transmission or reflection operation performed by an element) with nonabsorbing input and output media. In accordance with the common conventions [4], the diattenuation of this channel can be characterized by the following parameter:

$$D = \frac{t_{max} - t_{min}}{t_{max} + t_{min}}, \tag{12.136}$$

where t_{max} and t_{min} are respectively the maximum and minimum transmissivity (or transmittance) of the channel over the set of all possible polarization states of the incident light. Note the following useful relation. Let $\mathbf{t}$ be the EW Jones matrix of this channel with F-normalization of the input and output EW bases. Using (12.25), (12.30), and (8.278), one can find that at

$$t_{max} - t_{min} \ll 2(t_{max} + t_{min}), \tag{12.137}$$

to a good approximation,

$$\delta_{STU} \approx \frac{1}{\sqrt{2}} D, \tag{12.138}$$

where δ_{STU} is the relative error of STUM approximation for the matrix $\mathbf{t}$ [see (12.30)].

Transmission Diattenuation at "Soft" Interfaces

Most interfaces in standard PCSs of LCDs are those between layers with relatively close refractive indices. Such interfaces will be referred to as "soft" interfaces. The reflectivity of "soft" interfaces is small for a wide cone of incident directions as is the transmission diattenuation. As an example, Figure 12.6 shows the relative errors of STUM approximation δ_{STU} [see (12.138)] for the transmission matrix of a "soft" interface between an isotropic layer and a uniaxial layer at different angles of incidence for different sets of values of the refractive indices of the layers. The layers are assumed to be elements of a layered structure on which a plane monochromatic wave falls from an isotropic medium of refractive index 1 (this system is considered in all examples that follow in this section). The propagation direction of this wave is specified by the angles β_{inc} (the polar angle of incidence) and α_{inc} (the azimuthal angle

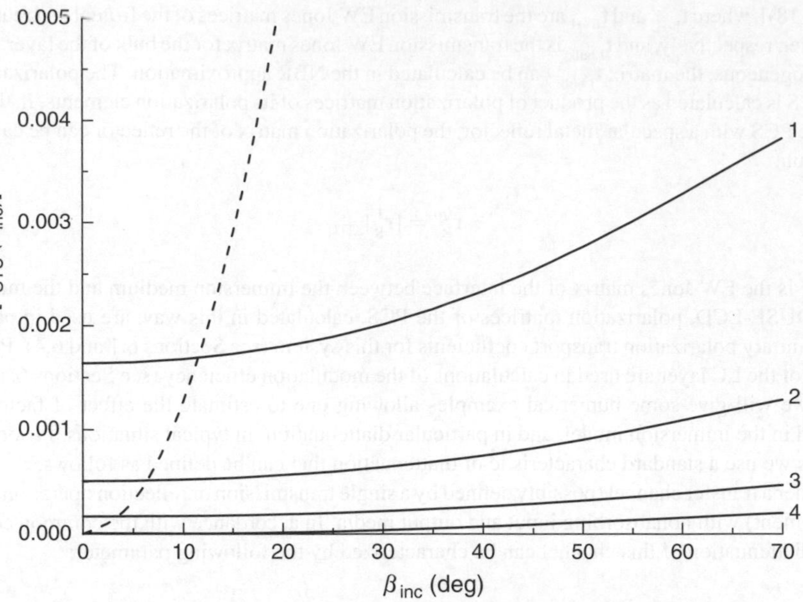

Figure 12.6 Relative error of STUM approximation for the transmission EW Jones matrix of the interface between an isotropic layer (refractive index n) and a uniaxial layer (principal refractive indices $n_\perp$ and $n_\parallel$, the optic axis is parallel to the interface): (1) $n = 1.5$, $n_\perp = 1.52$, $n_\parallel = 1.72$; (2) $n = 1.5$, $n_\perp = 1.52$, $n_\parallel = 1.62$; (3) $n = 1.5$, $n_\perp = 1.52$, $n_\parallel = 1.57$; (4) $n = 1.55$, $n_\perp = 1.5$, $n_\parallel = 1.6$. For comparison, the dashed curve represents the data for the interface between isotropic media of refractive indices 1 and 1.52

specifying the orientation of the plane of incidence) shown in Figure 9.1. The quantity max $[\delta_{\mathrm{STU}}(\alpha_{\mathrm{inc}})]$, whose values are shown in Figure 12.6, represents the maximum value of the function $\delta_{\mathrm{STU}}(\alpha_{\mathrm{inc}})$ at a given β_{inc}. As is seen from this figure, for "soft" interfaces the values of δ_{STU}, being small at normal incidence, remain small at oblique incidence for a wide range of incident angles, even at relatively strong birefringence. With such small δ_{STU} for the interface matrices, the transmission matrix $\mathbf{t}_{\mathrm{L}\to\mathrm{I}}\mathbf{t}_{\mathrm{Lbulk}}\mathbf{t}_{\mathrm{I}\to\mathrm{L}}$ in (12.134) will be accurately approximated by the closest STU matrix, which justifies the use of (12.134). Note also the following property of the transmission matrices of "soft" interfaces. Let $\mathbf{t}_|$ be a transmission FI–EW Jones matrix of a "soft" interface between an isotropic layer and an anisotropic layer of the layered structure and let $\mathbf{A}_|$ be the base matrix of the matrix $(\mathbf{t}_|)_{\mathrm{STU}}$, calculated from $\mathbf{t}_|$ by formulas (12.6)–(12.8). In common situations, the matrix $\mathbf{A}_|$ at a given α_{inc} is a very slow function of the refractive index of the isotropic medium. We illustrate this property by the following numerical example. In this example, the anisotropic layer has its optic axis parallel to interfaces and principal refractive indices $n_\perp = 1.52$ and $n_\parallel = 1.72$, and we compare values of the matrix $\mathbf{A}_|$ (for $\mathbf{t}_| = \mathbf{t}_{\mathrm{I}\to\mathrm{L}}$) at different values of the refractive index of the isotropic medium, n_{I}. Figure 12.7 shows the degree of discrepancy between the matrices $\mathbf{A}_{|n'}$ ($\mathbf{A}_|$ at $n_{\mathrm{I}} = n'$) and $\mathbf{A}_{|n''}$ ($\mathbf{A}_|$ at $n_{\mathrm{I}} = n''$) at $n' = 1.5$ and $n'' = 1.55$, 1.6, and 1.7; max $\|\mathbf{A}_{|n'}(\alpha_{\mathrm{inc}}) - \mathbf{A}_{|n''}(\alpha_{\mathrm{inc}})\|_{\mathrm{E}}$ is the maximum value of $\|\mathbf{A}_{|n'} - \mathbf{A}_{|n''}\|_{\mathrm{E}}$ as a function of α_{inc} at a given β_{inc}. As is seen from this figure, the matrices $\mathbf{A}_{|n'}$ and $\mathbf{A}_{|n''}$ are very close to each other for all values of β_{inc} from 0° to 70° even at $n'' = 1.7$. Such insensitivity of the matrix $\mathbf{A}_|$ to moderate variations of n_{I} warrants the use of a unified immersion medium in the immersion model and makes the choice of the refractive index of this medium noncritical.

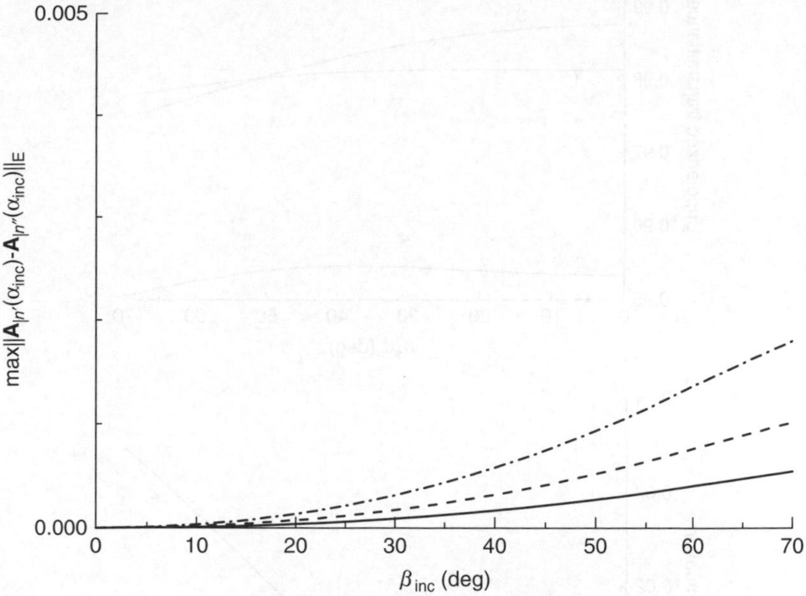

Figure 12.7 Comparison of the polarization transmission matrices for the interface between isotropic and anisotropic media calculated for different values of the refractive index of the isotropic medium; $n' = 1.5$, $n'' = 1.55$ (solid line), 1.6 (dashed line), and 1.7 (dash-dot line)

Polarization Effect of the Electrode–Alignment Layer Systems (EASs)

In the immersion model we neglect the polarization effect of thin isotropic layers that are elements of the real PCS—such layers are not included in this model. Under usual illumination conditions, individual thin layers with "soft" interfaces, elements of real PCSs, as a rule change very little, if at all, the polarization state of the passing light and the neglect of them is fully justified. In the case of the electrode–alignment layer system (EAS), we face a different situation. Because of a relatively high refractive index of ITO (~2), reflection from interfaces of the ITO layer may be significant and the transmission diattenuation at EASs at oblique incidence may be noticeable. Since the net optical effect of EASs is determined by multiple-beam interference, the strength of the transmission diattenuation at an EAS strongly depends on the thicknesses of the layers constituting this system as well as on the refractive indices of these layers. Note that the contribution of the reflection and transmission diattenuation at the alignment layer–LC interface to the transmission diattenuation of the EAS may also be amplified or weakened due to multiple-beam interference. Furthermore, we should note that for a system of conductive and alignment layers, the diattenuation at the alignment layer–LC interface may cause a specific phase-retarding action of the system. Generally speaking, a phase-retarding action is inherent in "thin" layered systems. Even when such a system consists of only isotropic layers and is surrounded by isotropic media, it may exhibit a nonzero retardance, but only at oblique incidence, and behave, in this respect, like a uniaxial layer with optic axis perpendicular to the layer boundaries. An example is given in Figure 12.8. An EAS in an LCD is in contact with the LC layer and, if the surface LC director is not perpendicular to boundaries, may show a nonzero retardance at normal incidence as well, because of the reflection diattenuation at the alignment layer–LC interface. This retardance at normal incidence is usually relatively small (as a rule, less than 1°) and, as a function of the thicknesses of the EAS elements and wavelength, oscillates near zero (see Figure 12.9). In Figures 12.10 and

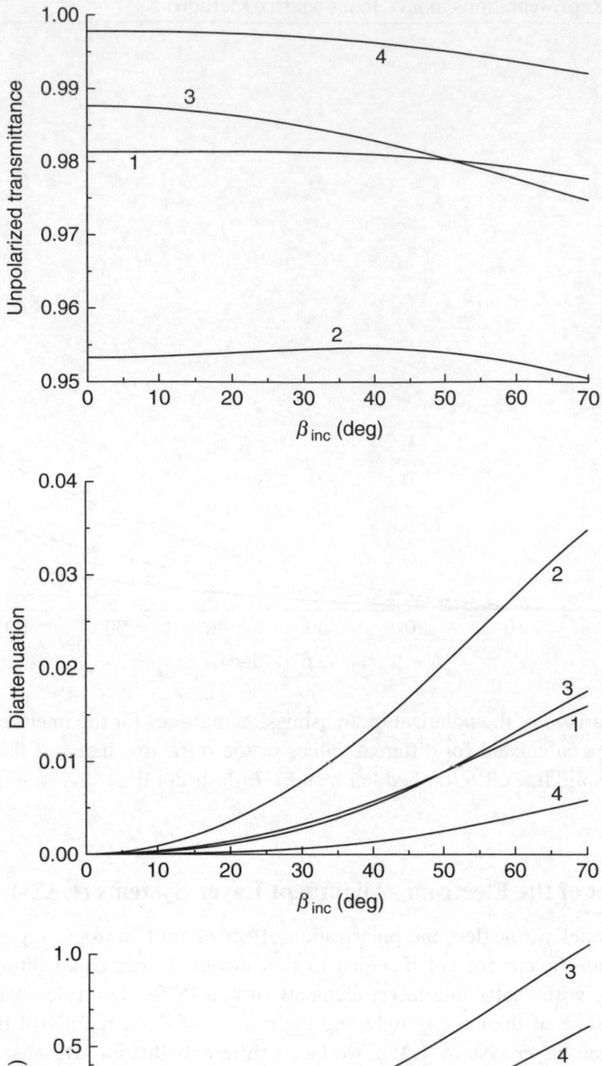

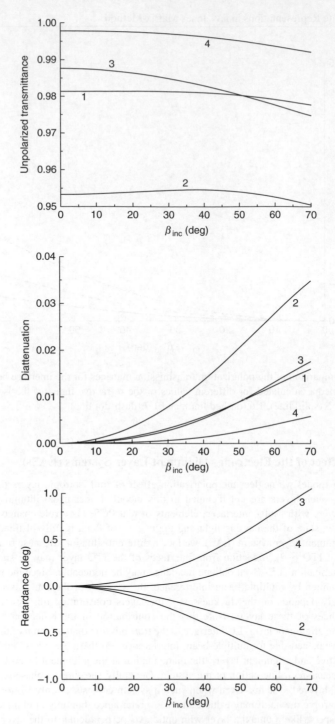

Figure 12.8 Unpolarized transmittance, diattenuation [D, see (12.136)], and retardance of simple (double-layer) EASs with different thicknesses of the electrode (d_E), sandwiched between isotropic media of refractive index 1.52: (1) $d_E = 0.03$ μm, (2) $d_E = 0.06$ μm, (3) $d_E = 0.12$ μm, (4) $d_E = 0.14$ μm. The refractive index of the electrode $n_E = 2$. The refractive index of the alignment layer $n_A = 1.6$, the thickness of the alignment layer $d_A = 0.1$ μm. Wavelength $\lambda = 550$ nm

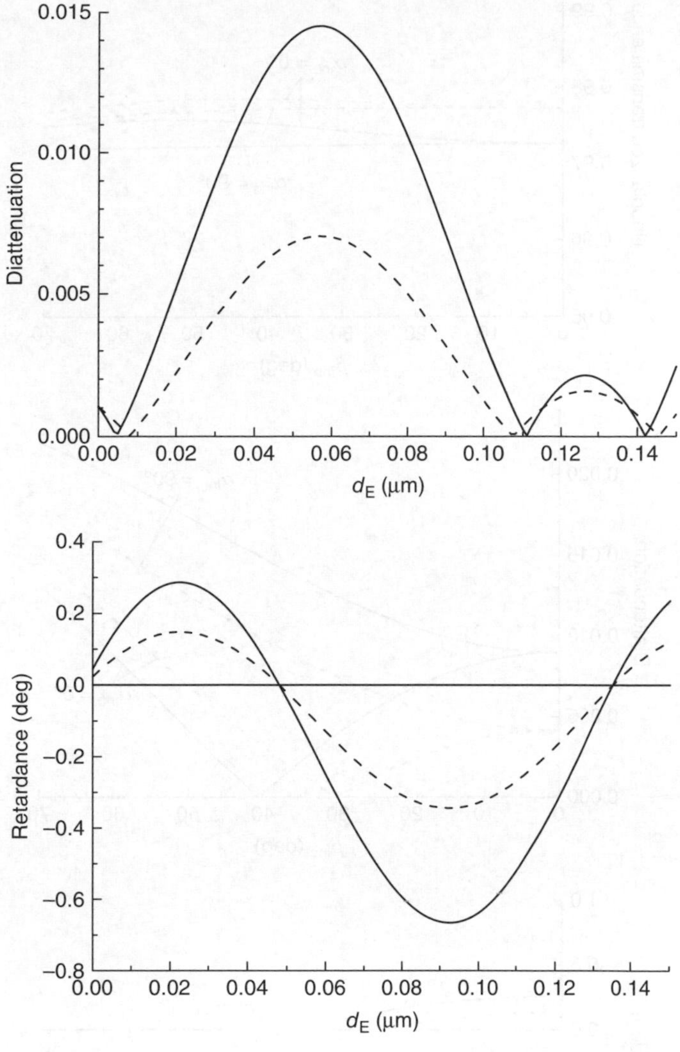

Figure 12.9 Dependence of the transmission diattenuation and retardance of a simple (double-layer) EAS on the electrode thickness d_E at normal incidence. The refractive index of the glass substrate $n_G =$ 1.52. Electrode: $n_E = 2$. Alignment layer: $n_A = 1.6$; $d_A = 0.1$ μm. LC: $n_\perp = 1.52$, $n_\parallel = 1.72$ (solid lines) and $n_\parallel = 1.62$ (dash lines); the LC director is parallel to interfaces ($\theta_1 = 0°$). Wavelength $\lambda = 550$ nm

12.11, we compare data on retardance and diattenuation for the nonantireflected (Figure 12.10) and antireflected (Figure 12.11) EASs appearing in examples 2 and 3 of Section 10.3. As can be seen from Figure 12.11, for antireflected EASs both diattenuation and retardance may be very small for a wide cone of the directions of incidence. The polarization matrix of the antireflected EAS, $\mathbf{A}_{EAS}$, calculated by formulas (12.6)–(12.8) from the true transmission FI–EW Jones matrix of the EAS, is very close to the polarization matrix $\mathbf{A}_|$ of the glass–LC interface that we would obtain taking the thicknesses of all the layers of the EAS equal to zero (see Figure 12.12). Thus, we have good grounds for saying that immersion models accurately represent the properties of real PCSs with antireflected EASs.

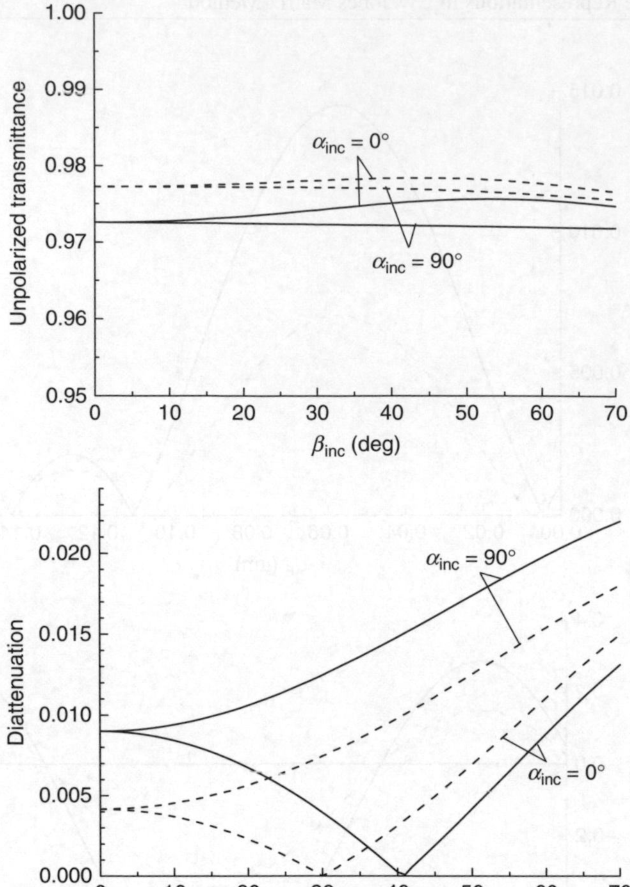

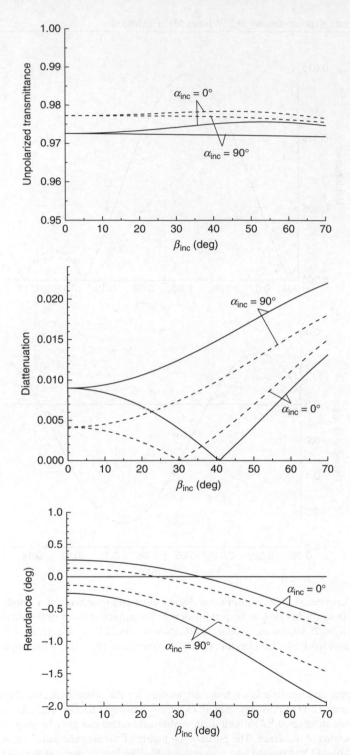

Figure 12.10 Unpolarized transmittance, diattenuation, and retardance for a simple (double-layer) EAS. Glass: $n_G = 1.52$. LC: $n_\perp = 1.52$, $n_\| = 1.72$ (solid lines) and $n_\| = 1.62$ (dashed lines); the LC director is parallel to interfaces ($\theta_1 = 0°$). At $\alpha_{inc} = 0$, the LC director is parallel to the plane of incidence. Wavelength $\lambda = 550$ nm

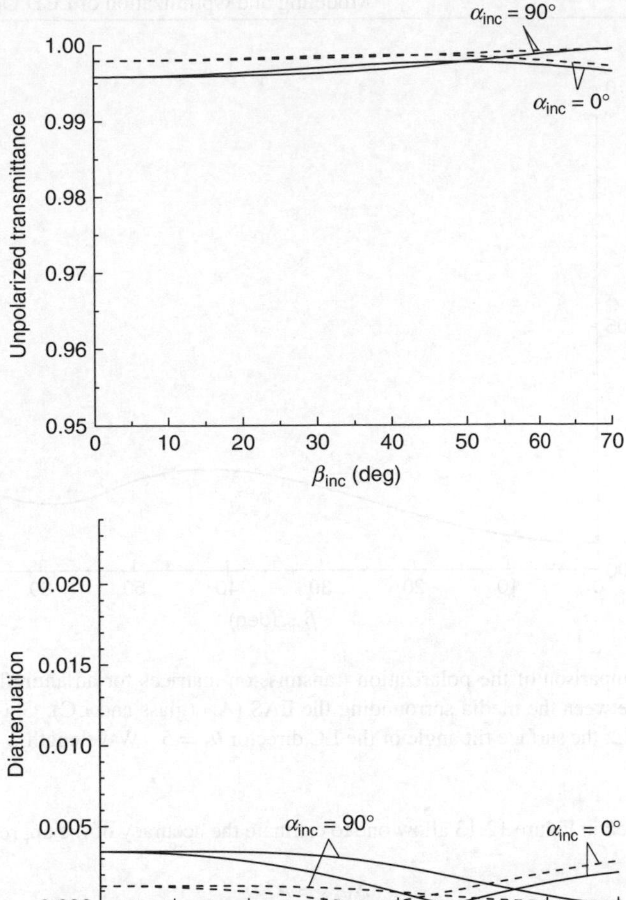

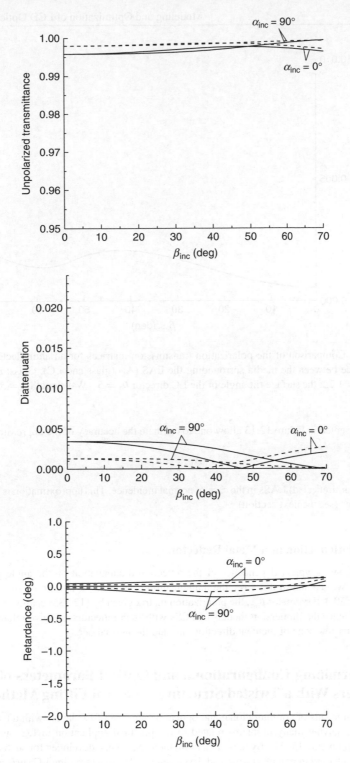

Figure 12.11 Unpolarized transmittance, diattenuation, and retardance for an antireflected EAS. The parameters of the glass substrate and LC are the same as in the previous example (Figure 12.10). Wavelength $\lambda = 550$ nm

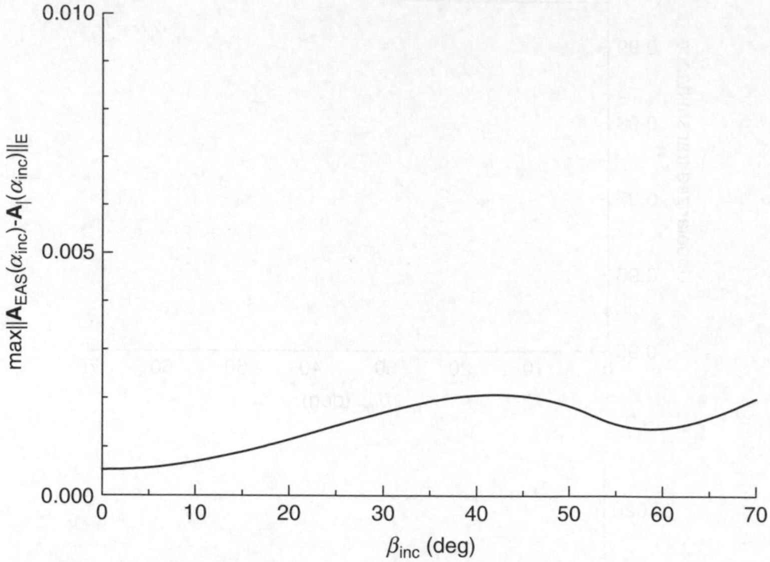

Figure 12.12 Comparison of the polarization transmission matrices for an antireflected EAS ($\mathbf{A}_{EAS}$) and the interface between the media surrounding the EAS ($\mathbf{A}_{|}$) (glass and LC). Glass: $n_G = 1.52$. LC: $n_{\perp} = 1.52$, $n_{\|} = 1.72$; the surface tilt angle of the LC director $\theta_1 = 5°$. Wavelength $\lambda = 550$ nm

The data presented in Figure 12.13 allow one to estimate the accuracy of the approximation

$$\mathbf{A}_{EAS} \approx \mathbf{A}_{|} \tag{12.139}$$

for simple (nonantireflected) EASs in the case of normal incidence. This approximation is used in solving inverse problems (see the next section).

Reflection Diattenuation at a Metal Reflector

Figure 12.14 shows a typical dependence of the reflection diattenuation on the angle β_{inc} for a good metal reflector. We see that in this case, the diattenuation is small ($D < 0.01$) only for relatively small values of β_{inc} (<30°). For greater β_{inc}, the polarization matrix given by (12.135) ceases to be an adequate characteristic. Hence the immersion models of PCSs with such reflectors can give reliable results only for a relatively narrow cone of incident directions around the normal one.

12.5 Determining Configurational and Optical Parameters of LC Layers With a Twisted Structure: Spectral Fitting Method

Accurate measurements of configurational and optical parameters of LC layers with a twisted structure are required for solving many problems related to the practical application of LCs and fundamental studies of their properties [5–21]. By now, a lot of methods have been developed for such measurements. These methods are used in manufacturing and developing LCDs, in examining LC surface alignment, in measurements of surface-anchoring parameters, and so on. Some of the methods are intended only for measurement of the LC layer thickness [5, 6, 18]; they require accurate a priori information on the LC

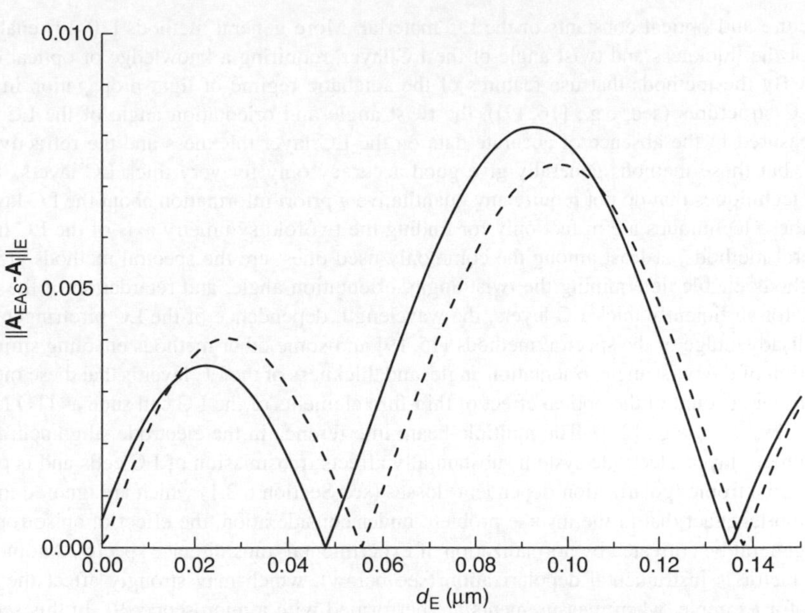

Figure 12.13 Comparison of the polarization transmission matrices for a simple EAS ($\mathbf{A}_{EAS}$) and the interface between the media surrounding the EAS ($\mathbf{A}_{|}$) at normal incidence. Alignment layer: $d_A = 0.1$, $n_A = 1.5$ (dash line) and $n_A = 1.6$ (solid line); electrode: $n_E = 2$, thickness d_E. The other parameters are the same as in the previous example (Figure 12.12)

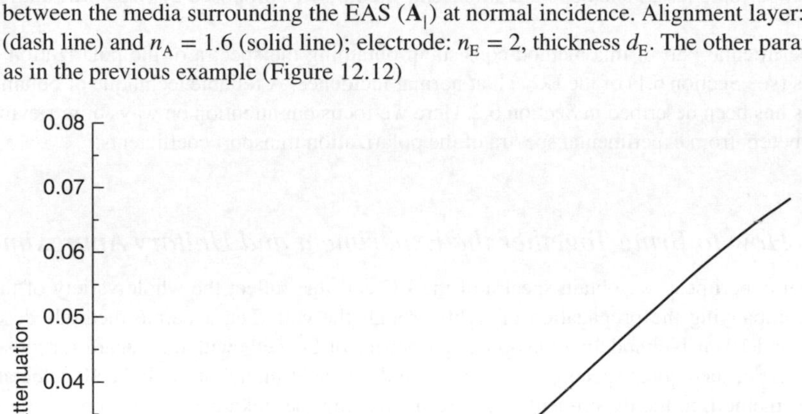

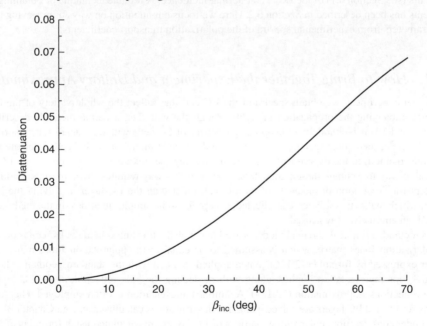

Figure 12.14 Diattenuation at a metal reflector. The data specified are for the interface between a nonabsorbing medium with refractive index 1.6 and a medium with complex refractive index 1.44+i5.23 (aluminum)

layer structure and optical constants of the LC material. More general methods [10, 14] enable determination of the thickness and twist angle of the LC layer, requiring a knowledge of optical constants of the LC. By the methods that use features of the adiabatic regime of light propagation in inhomogeneous LC structures (see, e.g., [16, 17]), the twist angle and orientation angle of the LC structure can be measured in the absence of accurate data on the LC layer thickness and the refractive indices of the LC, but these methods generally give good accuracy only for very thick LC layers. There are symmetry techniques that do not require any quantitative a priori information about the LC layer [8, 9]; however, these techniques are in fact only for finding the twofold symmetry axis of the LC layer. The most general methods, at least among the commonly used ones, are the spectral methods [15, 19, 21]. These methods enable determining the twist angle, orientation angle, and retardation of the LC layer as well as, for sufficiently thick LC layers, the wavelength dependence of the LC birefringence Δn. A common disadvantage of the spectral methods [15, 19] and some other methods enabling simultaneous determination of the twist angle, orientation angle, and thickness of the LC layer is that these methods do not take proper account of the optical effect of thin-film elements of the LC cell such as ITO layers and alignment films (see, e.g., [22]). The multiple-beam interference in the electrode–alignment layer–LC layer–alignment layer–electrode system substantially affects transmission of LC cells and is one of the sources of anisotropic (polarization-dependent) losses (see Section 6.2.1) which are ignored in [15, 19]. It is an important fact that in the inverse problem under consideration, the effect of anisotropic losses cannot in general be corrected by normalization of experimental transmittance spectra. Another usually neglected factor is instrumental depolarization (see below), which may strongly affect the accuracy of results, for example, when measurements are performed with a microscope [9]. In this section, we describe a method, mentioned above in the list of spectral methods, that is tolerant to the presence of moderate anisotropic losses and depolarization. This method was proposed by two of the authors of this book in [21].

The experimental part of this method consists in obtaining the spectra of the polarization transport coefficients (see Section 6.1) of the LC cell at normal incidence. A reliable technique of obtaining these coefficients has been described in Section 6.2. Here we focus our attention on ways of retrieving the LC layer parameters from experimental spectra of the polarization transport coefficients.

12.5.1 How to Bring Together the Experiment and Unitary Approximation

So, in a real experiment, we obtain spectra of the LC cell that reflect the whole variety of the optical effects accompanying the propagation of light through the cell. The accurate methods described in Chapters 8 and 10 enable modeling the optical properties of LC cells with due regard for most of these effects. Fitting of theoretical spectra for a realistic multielement model of the LC cell, calculated using an accurate method, to the experimental spectra by varying the unknown parameters of the LC layer is a possible way to estimate these parameters. However, this way requires involving many additional parameters for description of elements of the LC cell other than the LC layer in solving the inverse problem, which makes this way too complicated, except for some simple situations. In the method being presented, an alternative way is used.

In this method, a simple theoretical model is used in which the transmission of the LC cell is described by a polarization Jones matrix which is assumed to be equal to the polarization Jones matrix of the LC layer expressed by formula (12.126). When applied to a cell with the standard geometry shown in Figure 6.3, this representation implies, along with neglecting the bulk reflection in the LC layer and multiple reflections, approximation (12.139). A similar representation is often employed when inverse problems for twisted LC layers are solved using the classical Jones calculus (see, e.g., Chapter 3). In the method under consideration, polarization matrices of LC layers are calculated using formulas of NBRA (see Section 11.4.1). Recall that the formulas of the classical JC for twisted layers can be derived from those of the NBR approximation on the assumption that the birefringence of the medium is very small (small-birefringence approximation, SBA).

In this consideration, the axes X and Y of the reference system (X,Y,Z) used in optical calculations are assumed to be parallel to the axes x_I and y_I, respectively, of the laboratory reference system (x_I, y_I, z_I) (Figure 6.1) rather than rigidly bound to the LC cell. Therefore, say, the angle φ_1, which specifies the azimuthal orientation of the LC director at the frontal boundary of the LC layer, can be defined as in Figure 6.1.

In the method being presented, from the polarization Jones matrix of the LC layer $\mathbf{t}^{(p)}_{\text{layer}}$, of the form (12.129), the unitary polarization transport coefficients B_{Uj} $(j = 1,2,3,4)$ are calculated by formulas (6.31). The desired parameters of the LC layer are determined by fitting the spectra of B_{Uj} to the experimental spectra of the quantities

$$A_j = B_j/\bar{B} \quad (j = 1,2,3,4), \tag{12.140}$$

where B_j are the polarization transport coefficients of the cell and

$$\bar{B} = \sqrt{B_1^2 + B_3^2} + \sqrt{B_2^2 + B_4^2}. \tag{12.141}$$

This approach is based on the fact that the spectra of the coefficients A_j calculated for realistic models of LC cells, like that shown in Figure 6.3, taking the multiple reflections and other relevant factors into account are usually very well approximated, on average, by the spectra of the coefficients B_{Uj} calculated for these cells as described above. Numerical experiments showed that this approximation remains good even in the presence of significant anisotropic losses in the cells. That the parameters A_j are relatively insensitive to anisotropic losses may be illustrated by the following numerical examples. Figures 12.15 and 12.16 present results of numerical modeling for four hypothetic LC cells. The first cell, called cell A, has a structure as shown in Figure 6.3 and a quasi-planar twisted LC layer with twist angle $\Phi = 240°$, pretilt angles $\theta_1 = \theta_2 = 4°$, and $\varphi_1 = 0°$; the corresponding profiles of θ and φ are labeled by 1 in Figure 11.2. The LC material is absolutely transparent in the spectral region under consideration. The optical constants and thicknesses of layers of this LC cell are given in Section 12.5.3. Each of the other three cells, called cells B, C, and D, differs from cell A in one detail only. Cell B has no electrodes. Cell C has a dichroic LC material. In cell D, one of the alignment films (rear) is dichroic. For the LC in cell C, the perpendicular absorption coefficient is equal to zero, so that the transmittance of the bulk of the LC layer for the ordinary component, the LC structure be nontwisted, would be equal to 1. The parallel absorption coefficient of the LC is such that at the actual LC layer thickness, the bulk transmittance of the LC layer for the extraordinary component, the LC structure be nontwisted and planar, would be equal to 0.7 (throughout the spectral region under consideration).

Similarly, the principal absorption coefficients of the dichroic alignment film in cell D are such that the bulk transmittances of this film for the ordinary and the extraordinary components are equal to 1 and 0.7, respectively. For each of these cells, using the 8×8 transfer matrix method described in Section 10.2, we calculated exact spectra of 16 polarized transmittances used in the technique of determining B_j described in Section 6.2. These polarized transmittances were calculated from Mueller matrices characterizing the overall transmission of the LC cell (see Section 12.5.3 for details). From these rigorously calculated spectra of polarized transmittances, spectra of B_j and A_j were calculated. Figure 12.15 compares the spectra of the polarized transmittances t_{PA} (0°,45°) and t_{PA} (90°,−45°) for these four cells. Recall that in the absence of diattenuation, the spectra of t_{PA} (0°,45°) and t_{PA} (90°,−45°) coincide [see (6.15)]. Figure 12.16 compares the spectra of A_j for the LC cells. As is seen from Figure 12.15, in the case of cell A, the anisotropic losses manifest themselves almost as much as in the experiment illustrated by Figure 6.2 within the transparency region of the material of the alignment layers (SD-1). For cell B the effect of anisotropic losses is far smaller than for cell A, because reflection from the boundaries of the alignment films in cell B is very weak. The effect of diattenuation in cells C and D on the polarized transmittances is very significant. However, as seen from Figure 12.16, in spite of large difference in data on polarized transmittances for these four cells, the spectra of A_j calculated from these data

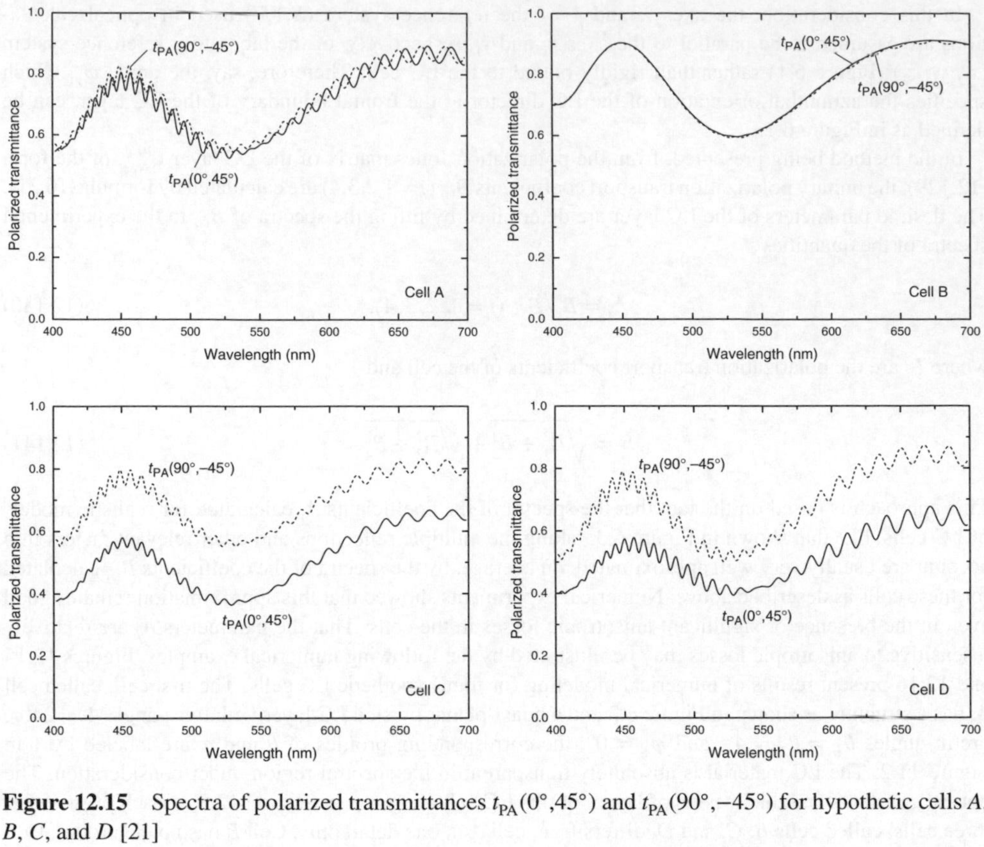

Figure 12.15 Spectra of polarized transmittances $t_{PA}(0°,45°)$ and $t_{PA}(90°,-45°)$ for hypothetic cells *A*, *B*, *C*, and *D* [21]

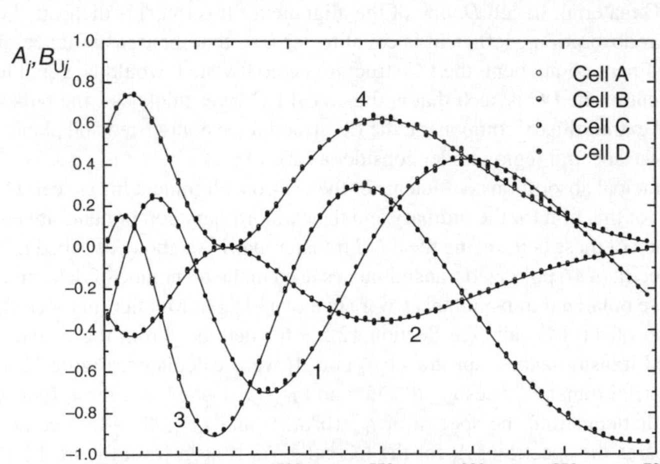

Figure 12.16 Spectra of parameters A_j and B_{Uj} for cells *A*, *B*, *C*, and *D* [21]. The symbols show the rigorously calculated values of A_j. Spectra of B_{Uj} are shown by lines. The labels are the values of the index *j*

Table 12.5 Dependence of the accuracy of the SBA and NBRA representations on the LC layer thickness. $\Phi = 240°$. The cells are without electrodes. n_A is the refractive index of the alignment films

	SD			
	SBA		NBRA	
Thickness, μm	$n_A = 1.5$	$n_A = 1.6$	$n_A = 1.5$	$n_A = 1.6$
2	0.0040	0.0044	0.0008	0.0010
4	0.0026	0.0028	0.0009	0.0010
6	0.0020	0.0024	0.0011	0.0014
8	0.0019	0.0023	0.0012	0.0015

(in Figure 12.16, the values of A_j are shown by circles) turned out to be almost the same and very close to the spectra of the coefficients B_{Uj} (solid lines in Figure 12.16) computed by formulas (6.31), (12.130), (12.131), (11.46), and (11.47) (NBRA + PITL approximation). The root-mean-square deviation of A_j from B_{Uj} is 0.0073 for cell A, 0.0018 for cell B, 0.0072 for cell C, and 0.0103 for cell D.

In Tables 12.5 and 12.6, we compare the degree of closeness of the spectra of B_{Uj} calculated using the NBRA representation and those calculated using the SBA representation (or, what is here the same, the classical JC) (see Section 12.3) to the rigorously calculated spectra of A_j for cells with ideal twisted LC layers in a situation when anisotropic losses are very small. Here we consider cells with the glass substrate–alignment film–LC layer–alignment film–glass substrate structure (as in cell B of the previous example). The refractive index of the glass substrates, thickness of the alignment films, and principal refractive indices of the LC are the same as in the previous example. The tables show the values of the root-mean-square deviation of the spectra of B_{Uj} from the spectra of A_j (SD). Table 12.5 shows the dependences of SD on the LC layer thickness for a cell with an ideal planar ($\theta_c = 0°$) twisted layer with $\Phi = 240°$ at two values of the refractive index of the alignment films (n_A): $n_A = 1.5$ and 1.6. Table 12.6 presents the dependence of SD on the twist angle for such a cell. It is seen from these tables that in all cases, the accuracy of the NBRA representation is very high and varies only slightly with changing Φ, d, and n_A. At small twist angles and large thicknesses of the LC layer, the accuracy of the SBA representation is on the same level as in the case of NBRA. However, in contrast to the NBRA representation, the error of the SBA representation significantly increases with increasing the twist angle and with decreasing the thickness of the LC layer, as might be expected from the comparison of the basic formulas of these representations (Section 12.3).

Table 12.6 Dependence of the accuracy of the SBA and NBRA representations on the twist angle. $d = 4$ μm, $n_A = 1.6$. The cells are without electrodes

	SD	
Φ, °	SBA	NBRA
0	0.0014	0.0014
90	0.0013	0.0011
180	0.0021	0.0009
270	0.0036	0.0011
360	0.0058	0.0012

Thus, we see that the true spectra of A_j are very well approximated by the spectra of B_{Uj} even in the presence of significant anisotropic losses. This leads one to expect that fitting of the spectra of B_{Uj} to the experimental spectra of A_j by varying the unknown parameters of the LC layer can provide sufficiently accurate estimates of these parameters at a typical, relatively low, level of anisotropic losses. The results presented in the next section confirm this expectation. Before proceeding to a detailed description of a fitting technique, let us mention another advantage of this approach to solving the inverse problem.

In Section 6.2.1, we noted that LC cells may partially depolarize quasimonochromatic light due to multiple reflections. In real experiments, the effect of many factors, such as the local spectral averaging (see Section 7.1), a wide angular spectrum of the probe beam (e.g., when the measurements are carried out with a microscope at high magnification), and stray light, on the spectral measurement results is similar to that produced by a partially depolarizing sample. It is this that is called here *instrumental depolarization*. The parameters A_j contain information only on the polarized part of the light transmitted by the cell and are unaffected by the presence of the apparent depolarized component due to normalization involved in (12.140). Owing to this, the effect of any kind of depolarization on the final results in typical situations is insignificant.

12.5.2 Parameterization and Solving the Inverse Problem

The fitting of spectra of coefficients B_{Uj} to experimental spectra of parameters A_j can be carried out by minimizing the following objective function:

$$F(\mathbf{X}) = \sqrt{\frac{1}{4M}\sum_{k=1}^{M}\sum_{j=1}^{4}\left(A_j^{\text{EXP}}(\lambda_k) - B_{Uj}(\lambda_k;\mathbf{X})\right)^2}, \qquad (12.142)$$

where A_j^{EXP} are experimental values of A_j, $\{\lambda_k\}$ $(k = 1,2,\ldots,M)$ is the set of experimental values of λ, and $\mathbf{X}$ is the vector of the LC layer parameters to be determined. This minimization problem is easily solved with the aid of numerical optimization methods. In our data-treatment program, search of local minima of $F(\mathbf{X})$ is performed using the Nelder–Mead simplex method [23]. The global search is carried out using the multistart algorithm with the search for "good" starting points by the Monte Carlo method. This technique has proved to be highly reliable and efficient in this problem, even when very large search intervals to up to five unknowns are assigned (see below). In all tests described below, we used a set $\{\lambda_k\}$ with $M = 62$, covering the range 400–700 nm with intervals of about 5 nm.

In the simplest case, when sufficiently accurate data on the spectral dependence of the principal refractive indices of the LC material are at hand, one can set $\mathbf{X} = \{\varphi_1, L_\theta, \Phi\}$. However, often the required data on spectral dependence of the refractive indices are not available. In this case, solving the inverse problem involves variation of the LC optical constants.

Solving the Inverse Problem With Estimation of the Spectral Dependence of Δn

Considering $n_\perp$ and Δn, rather than $n_\perp$ and $n_\parallel$, as independent variables in (11.46) and (12.133), one can see that the coefficients B_{Uj} are almost unaffected by variations in $n_\perp$ within the range of typical values of this index. This allows the fitting to be performed at a given, not necessarily very accurately, dependence of $n_\perp$ on λ, because even a large inaccuracy in $n_\perp$ cannot affect the final results significantly. For most practical LC materials, the dependence of Δn on λ is accurately represented by the three-coefficient Cauchy equation

$$\Delta n(\lambda) = a_{1\Delta} + \frac{a_{2\Delta}}{\lambda^2} + \frac{a_{3\Delta}}{\lambda^4} \qquad (12.143)$$

(see, e.g., [24]). Using this representation, we specify the wavelength dependence of Δn by the following three parameters:

$$D_{\lambda 1} = \frac{\Delta n_{450}}{\Delta n_{550}},\ D_{\lambda 2} = \frac{\Delta n_{550}}{\Delta n_{450}}\left(1 - \frac{\Delta n_{650}}{\Delta n_{550}}\right), \tag{12.144}$$

and Δn_{550}, where Δn_{450}, Δn_{550}, and Δn_{650} are the values of Δn at wavelengths of 450, 550, and 650 nm, respectively. The values of $D_{\lambda 1}$ and $D_{\lambda 2}$ for practical LC materials lie within narrow ranges[1], which is very convenient in solving the inverse problem. In our data-treatment program, the evaluation of the unknown parameters in this case is performed in the following way. The objective function $F(\mathbf{X})$ is minimized at a fixed value of Δn_{550}, we denote it by $(\Delta n_{550})_{\mathrm{R}}$, for a set of unknowns including $D_{\lambda 1}$ and $D_{\lambda 2}$; thus, in the most general case, the vector of unknowns is $\mathbf{X} = \{\varphi_1, L_\theta, \Phi, D_{\lambda 1}, D_{\lambda 2}\}$. In the minimization process, after each change in $D_{\lambda 1}$ and $D_{\lambda 2}$, new values of Δn_{450} and Δn_{650} are calculated by the formulas $\Delta n_{450} = D_{\lambda 1}\Delta n_{550}$ and $\Delta n_{650} = (1 - D_{\lambda 2}D_{\lambda 1})\Delta n_{550}$; then, from $(\Delta n_{550})_{\mathrm{R}}$ and the new values of Δn_{450} and Δn_{650} the coefficients of (12.143), $a_{1\Delta}$, $a_{2\Delta}$, and $a_{3\Delta}$, are computed which are then used to calculate $B_{\mathrm{U}j}$. We will refer to the values of L_θ, Δn_{450}, and Δn_{650} obtained by fitting at $\Delta n_{550} = (\Delta n_{550})_{\mathrm{R}}$ as the reference values and denote them by $(L_\theta)_{\mathrm{R}}$, $(\Delta n_{450})_{\mathrm{R}}$, and $(\Delta n_{650})_{\mathrm{R}}$. As can be seen from (12.130)–(12.132), (11.46), and (6.31), the parameters $B_{\mathrm{U}j}$ change only slightly under simultaneous variations of Δn and L_θ that keep the product $\Delta n L_\theta$ fixed. Due to this fact, even under relatively large deviations of $(\Delta n_{550})_{\mathrm{R}}$ from the true value of Δn_{550} for the LC material, the value of the product $(\Delta n_{550})_{\mathrm{R}}\cdot(L_\theta)_{\mathrm{R}}$ is very close to the true value of the product $\Delta n_{550}L_\theta$ for the examined LC layer. Therefore, with a knowledge of Δn for the LC material at some wavelength λ_0 within the spectral range under consideration, the parameters Δn_{450}, Δn_{550}, Δn_{650}, and L_θ can be estimated from the reference values as

$$\begin{gathered}\Delta n_{450} \approx \kappa(\Delta n_{450})_{\mathrm{R}},\quad \Delta n_{550} \approx \kappa(\Delta n_{550})_{\mathrm{R}},\quad \Delta n_{650} \approx \kappa(\Delta n_{650})_{\mathrm{R}},\\ L_\theta \approx (L_\theta)_{\mathrm{R}}/\kappa,\\ \kappa = (\Delta n(\lambda_0))_{\mathrm{R}}/\Delta n(\lambda_0),\end{gathered} \tag{12.145}$$

where $(\Delta n(\lambda_0))_{\mathrm{R}}$ is the value of $\Delta n(\lambda_0)$ of the reference spectrum of Δn with $\Delta n(450\ \text{nm}) = (\Delta n_{450})_{\mathrm{R}}$, $\Delta n(550\ \text{nm}) = (\Delta n_{550})_{\mathrm{R}}$, and $\Delta n(650\ \text{nm}) = (\Delta n_{650})_{\mathrm{R}}$. In the absence of sufficiently accurate data on Δn or L_θ, the wavelength dependence of the retardation $(\Delta n L_\theta)$, which can easily be calculated from the retrieved values of $\Delta n_{550}L_\theta$, $D_{\lambda 1}$, and $D_{\lambda 2}$, may be considered as a final result.

Let us give some numerical modeling results confirming the reliability of this approach.

Numerical Tests

In this subsection, we present the results of numerical experiments that we carried out to estimate the accuracy of determining the LC layer parameters by this method in typical situations. In all examples, the rigorously calculated spectra of the polarized transmittances were used as input data (see Sections 12.5.1 and 12.5.3) and all five parameters, φ_1, L_θ, Φ, $D_{\lambda 1}$, and $D_{\lambda 2}$, were retrieved. The search intervals for these parameters were: $-10° < \varphi_1 \leq 80°$, $0.01\ \mu\text{m} \leq L_\theta \leq 20\ \mu\text{m}$, $-270° \leq \Phi \leq 270°$, $1 < D_{\lambda 1} < 1.3$, and $0 < D_{\lambda 2} < 0.5$. Such a large search domain was set in order to estimate the efficiency and reliability of the minimization algorithm used; in practical situations, the search intervals may be, of course, significantly narrowed.

[1] For example, the LC materials considered here have the following $D_{\lambda 1}$ and $D_{\lambda 2}$. HR-8596: $D_{\lambda 1} = 1.093$, $D_{\lambda 2} = 0.043$. ZLI-4792: $D_{\lambda 1} = 1.052$, $D_{\lambda 2} = 0.032$. ZLI-5700–000: $D_{\lambda 1} = 1.085$, $D_{\lambda 2} = 0.038$. MLC-6080: $D_{\lambda 1} = 1.131$, $D_{\lambda 2} = 0.048$. E7: $D_{\lambda 1} = 1.147$, $D_{\lambda 2} = 0.053$.

Table 12.7 Retrieved parameters for the model cells *A*, *B*, *C*, and *D*

Cell	$F_{\min}$	Φ, °	φ_1, °	L_θ, μm	Δn_{450}	Δn_{650}
A	0.0069	240.15	−0.07	8.00	0.1601	0.1398
B	0.0012	240.10	−0.05	7.99	0.1603	0.1397
C	0.0068	240.09	−0.04	7.99	0.1601	0.1398
D	0.0094	240.31	−0.11	8	0.1601	0.1397

Table 12.7 shows the values of the "unknown" parameters of LC layers and the minimum values of the objective function $F(\mathbf{X})$ ($F_{\min}$) that were found by fitting for the model cells *A*, *B*, *C*, and *D* of Section 12.5.1. In this case, the fitting was performed with true values of $n_\perp$ and Δn_{550}. As is seen from the table, the retrieved values of φ_1, L_θ, Φ, Δn_{450}, and Δn_{650} are very close to the true ones (for the LC in this test, $\Delta n_{450} = 0.1602$ and $\Delta n_{650} = 0.1397$), even under such extreme conditions as in the cases of cells *C* and *D* (recall that these cells have dichroic layers). In the next examples, we consider LC cells without dichroic layers.

Table 12.8 shows the absolute values of deviations of retrieved values from true ones ($\delta[\Phi]$, $\delta[\varphi_1]$, etc.) for LC cells with different thicknesses of LC layers. The calculations were performed for two LC materials with significantly different birefringence. One of them has $\Delta n_{550} \sim 0.1$; in this case we used data for the mixture ZLI-4792 (Merck). For the other material, $\Delta n_{550} \sim 0.2$ (E7, BDH). The column "Ref. LC" (Reference LC) of Table 12.8 indicates the LC material whose parameters ($n_\perp(\lambda)$ and Δn_{550}) were taken as the reference ones. One of the reference materials was HR-8596 (Hoffmann-La Roche) with $\Delta n_{550} \sim 0.15$. The wavelength dependences of refractive indices of ZLI-4792, E7, and HR-8596 are shown below in Figures 12.20 and 12.21. In calculations by formulas (12.145), we used data on Δn for $\lambda_0 = 589.3$ nm. The calculations for Table 12.8 were performed for the free-of-electrodes cell configuration (as in cell *B*), that is, under conditions most favorable to obtaining accurate estimates. The presented results support the assertion made above: even when deviations of the reference values of $n_\perp$ and Δn_{550} from the true ones are large, the accuracy of estimates remains relatively good. In all

Table 12.8 Absolute error in the retrieved parameters for LC cells without electrodes. $\delta[\Delta n] = \max\{\delta[\Delta n_{450}], \delta[\Delta n_{650}]\}$

LC	Φ, °	d, μm	Ref. LC	$\delta[\Phi]$, °	$\delta[\varphi_1]$, °	$\delta[L_\theta]$, μm	$\delta[\Delta n]$
E7	240	6	HR-8596	0.05	0.02	≤ 0.01	≤ 0.0001
		4		0.09	0.05		
		3		0.07	0.03		
ZLI-4792	240	9	HR-8596	0.07	0.04		
		7		0.03	0.01		
		6	HR-8596	0.20	0.11		
			ZLI-4792	0.15	0.08		
		4	HR-8596	0.52	0.26		
			ZLI-4792	0.32	0.16		
		3	HR-8596	1.25	0.57		
			ZLI-4792	0.71	0.35		
	120	3	HR-8596	0.10	0.06		
	60			< 0.01	< 0.01		

Table 12.9 Absolute error in the retrieved parameters for LC cells with different surface tilt angles of the LC director. $\Phi = 240°$, $\varphi_1 = 0°$, and $d = 8$ μm. In the column "Electrodes", symbol "o" denotes the absence of electrodes, and symbol "+" the presence of electrodes in the LC cell

Cell parameters			Fitting results				
$\theta_{1,2}$, °	L_θ, μm	Electrodes	$F_{\min}$	$\delta[\Phi]$, °	$\delta[\varphi_1]$, °	$\delta[L_\theta]$, μm	$\delta[\Delta n]$
0	8.00	o	0.0011	0.07	0.04	0.01	0.001
		+	0.0069	0.13	0.07		
4	7.99	o	0.0012	0.10	0.05		
		+	0.0069	0.15	0.07		
15	7.89	o	0.0046	0.57	0.28		
		+	0.0080	0.60	0.30		
30	7.54	o	0.0081	0.38	0.19	0.06	0.002
		+	0.0103	0.35	0.17		

situations presented in Table 12.8, the errors in the retrieved L_θ, Δn_{450}, Δn_{550}, and Δn_{650} turned out to be very small. Numerical tests revealed a tendency for the error in the retrieved Φ and φ_1 to increase with decreasing the ratio $\Delta n_{550}L_\theta/\Phi$, regardless of the choice of the reference parameters. The data for the cells with ZLI-4792 presented in Table 12.8 reflect this tendency. The numerical experiments show that the method can give accuracies of the order of 0.25° and 0.15° for Φ and φ_1, respectively, for LC layers with $\Delta n_{550}L_\theta/\Phi > 0.0025$ μm/deg.

Table 12.9 shows to what extent departures of the LC layer structure from an ideal twisted one affects the accuracy of determining LC layer parameters and allows one to compare results for cells with electrodes and without them. These calculations were made for four configurations of the LC director field: an ideal twisted structure with $\theta_1 = \theta_2 = 0°$ and three equilibrium configurations with $\theta_1 = \theta_2 =$ 4°, 15°, and 30° shown in Figure 11.2. It is seen from Table 12.9 that at large deviations of the LC layer structure from an ideal twisted one, the error in Φ may be a half a degree and larger. The presence of electrodes makes the accuracy worse but not to a large extent.

Experimental Tests

In this subsection, we give some results of experimental testing of the method described.

Table 12.10 and Figures 12.17–12.20 show results for six series of measurements. The experiments were performed on four LC cells. The first cell, *cell E1*, was described in Section 6.2. For this cell, two series of measurements were made. The spectra of the polarized transmittances and B_j obtained in the first series are presented in Figures 6.2 and 6.4a. The second series of measurements was made after rotation of the cell by 45° with respect to its position in the first series (Figure 6.4b). The experimental spectra of A_j and the best fit curves $B_{Uj}(\lambda)$ for these series are presented in Figure 12.17 (in Figures 12.17–12.19, experimental points are shown by symbols, and best fit curves by solid lines). As can be seen from Table 12.10, the retrieved values of L_θ, Φ, Δn_{450}, Δn_{550}, and Δn_{650} for these two series are almost identical. In calculations by formulas (12.145), we used the value of Δn at $\lambda_0 = 589.3$ nm provided by the manufacturer of the LC material (Merck). The retrieved values of φ_1 for these two series differ from each other by 44.8°, which is consistent with the nominal angle of the cell rotation in this experiment to within the accuracy of positioning. It should be noted that the method, using the unitary approximation, because of (6.15), gives solutions for φ_1 in the form $\varphi_1 = \varphi_{1F} + m{\cdot}90°$, where φ_{1F} is the value provided by the method and m is an unknown integer. In real situations, m is usually known a priori.

Table 12.10 Retrieved parameters for the experimental LC cells

N. exp.	Cell	LC	F_{min}	Φ, °	φ_1, °	L_θ, μm	Δn_{450}	Δn_{550}	Δn_{650}
1	*E1*	ZLI-5700–000	0.0064	87.49	−0.06	5.28	0.1362	0.1253	0.1198
2	*E1*		0.0052	87.37	44.74	5.27	0.1362	0.1253	0.1199
3	*E2* (zone 1)		0.0103	−6.60	−3.14	9.17	0.1356	0.1251	0.1201
4	*E2* (zone 2)		0.0095	−63.82	−2.41	8.96	0.1358	0.1251	0.1201
5	*E3*	MLC-6080	0.0097	0.41	1.55	3.55	0.2350	0.2078	0.1969
6	*E4*		0.0072	−90.44	−49.27	3.71	0.2353	0.2081	0.1963

The liquid crystal in the second cell (cell *E2*), as in the first one, is ZLI-5700–000. The LC layer has a thickness of about 9 μm. The alignment layers in this cell are a unidirectionally rubbed polyimide (PI) film ($\theta_1 \approx 2$–4°) and a photoaligned film ($\theta_2 \approx 0$°). On the latter film, areas with various azimuthal orientation of the easy axis were created by photoalignment. Zones with different twist angles were thereby produced in the LC layer. For this cell, two series of measurements were made: one for a zone with $\Phi \approx -7°$ (zone 1) and the other for a zone with $\Phi \approx -64°$ (zone 2). The experimental and best fit curves for these zones are shown in Figure 12.18. The angle φ_1 in this experiment characterizes the LC director orientation at the PI film. The small difference (~0.8°) in the retrieved values of φ_1 for zones 1 and 2 may be caused by finite azimuthal anchoring of the LC at the PI film surface or by nonuniformity of the easy axis orientation on this surface. The ~0.2 μm discrepancy between the retrieved values of L_θ for these zones is explained by nonuniformity of the LC layer thickness. The retrieved values of Δn_{450}, Δn_{550}, and Δn_{650} for zones 1 and 2 are almost identical and very close to those obtained for cell *E1*. Figure 12.20 presents all four dispersion curves obtained in the four series described. As can be seen, these curves are very close to each other. Also shown in the same plot are Merck data on Δn for ZLI-5700–000. It is seen that the Merck data are in very good agreement with our results.

The fifth and sixth series of measurements were performed for cells filled with nematic MLC-6080 (Merck). This LC material has significantly stronger birefringence than ZLI-5700–000: at $\lambda = 589.3$ nm and a temperature of 20°C, $\Delta n = 0.1228$ for ZLI-5700–000 and $\Delta n = 0.2024$ for MLC-6080. In the first cell (cell *E3*), the director configuration is almost uniform ($\Phi \approx 0°$); in the second cell (cell *E4*) $\Phi \approx -90°$. The LC layer thicknesses in these cells are about 3.6 μm. The alignment films in both cells are rubbed polyimide layers ($\theta_1,\theta_2 \approx 2$–4°). The experimental spectra of B_j for cell *E3* are given in Figure 6.9a. The experimental and best fit spectra for cells *E3* and *E4* are shown in Figure 12.19. The retrieved wavelength dependences of Δn for the nontwisted layer and the twisted one are very close to each other (see Table 12.10 and Figure 12.20).

As is seen from Figures 12.17–12.19, in all the cases the experimental spectra of A_j are almost perfectly fitted by the spectra of B_{Uj}; the deviation F_{min} is of the order of 10^{-2} or smaller (see Table 12.10).

Ways to Ease Measurements and Data Treatment

We note two things that allow one to make the measurement procedure and data-treatment procedure easier.

First, the values of the polarized transmittances that are required for calculating A_j for a wavelength can be determined up to a multiplier common for these values of transmittances, because such multipliers cancel in (12.140). This allows using, in place of the polarized transmittances, the corresponding intensities (nonnormalized to the intensity of the incident light) as input data. Thus, the normalization to the source spectrum is not necessary in this method. This is convenient, especially when the measurements are carried out with a microscope.

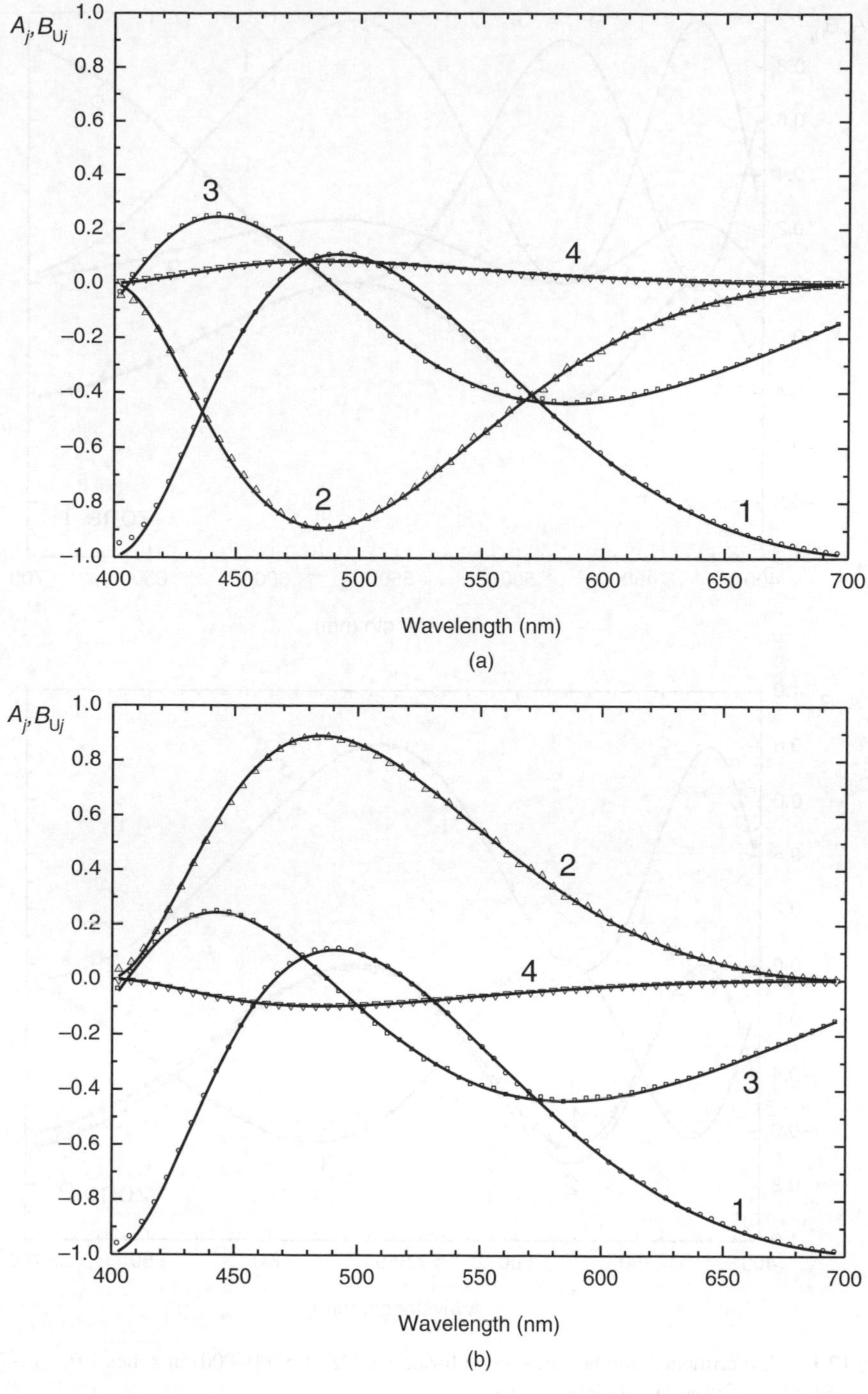

Figure 12.17 Experimental and best fit spectra for cell *E1* in the initial position (a) and after rotation by 45° (b) [21]. The symbols show the experimental points for A_j. The best fit spectra of B_{Uj} are shown by lines. The curves are labeled by values of the index j

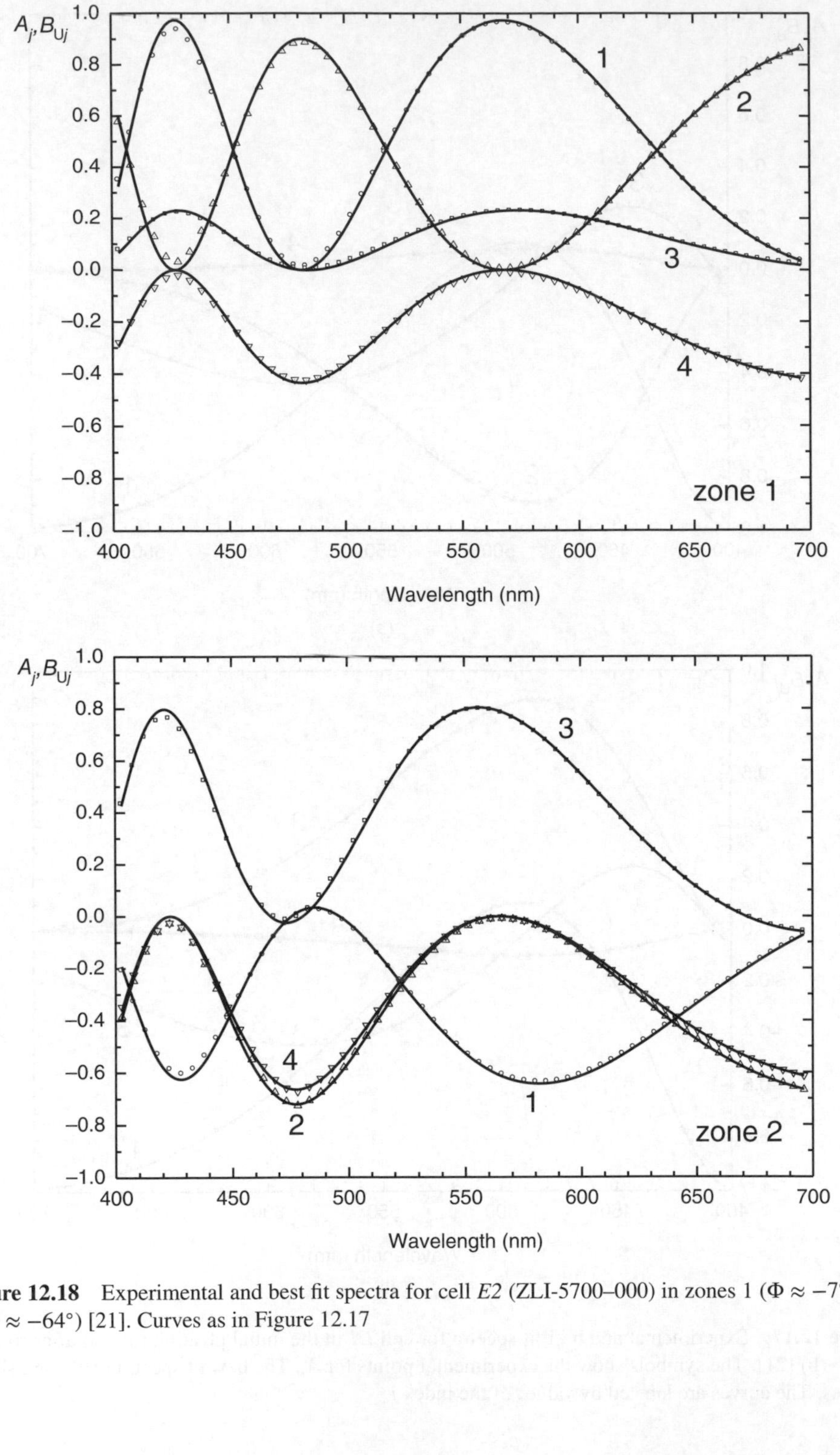

Figure 12.18 Experimental and best fit spectra for cell *E2* (ZLI-5700–000) in zones 1 ($\Phi \approx -7°$) and 2 ($\Phi \approx -64°$) [21]. Curves as in Figure 12.17

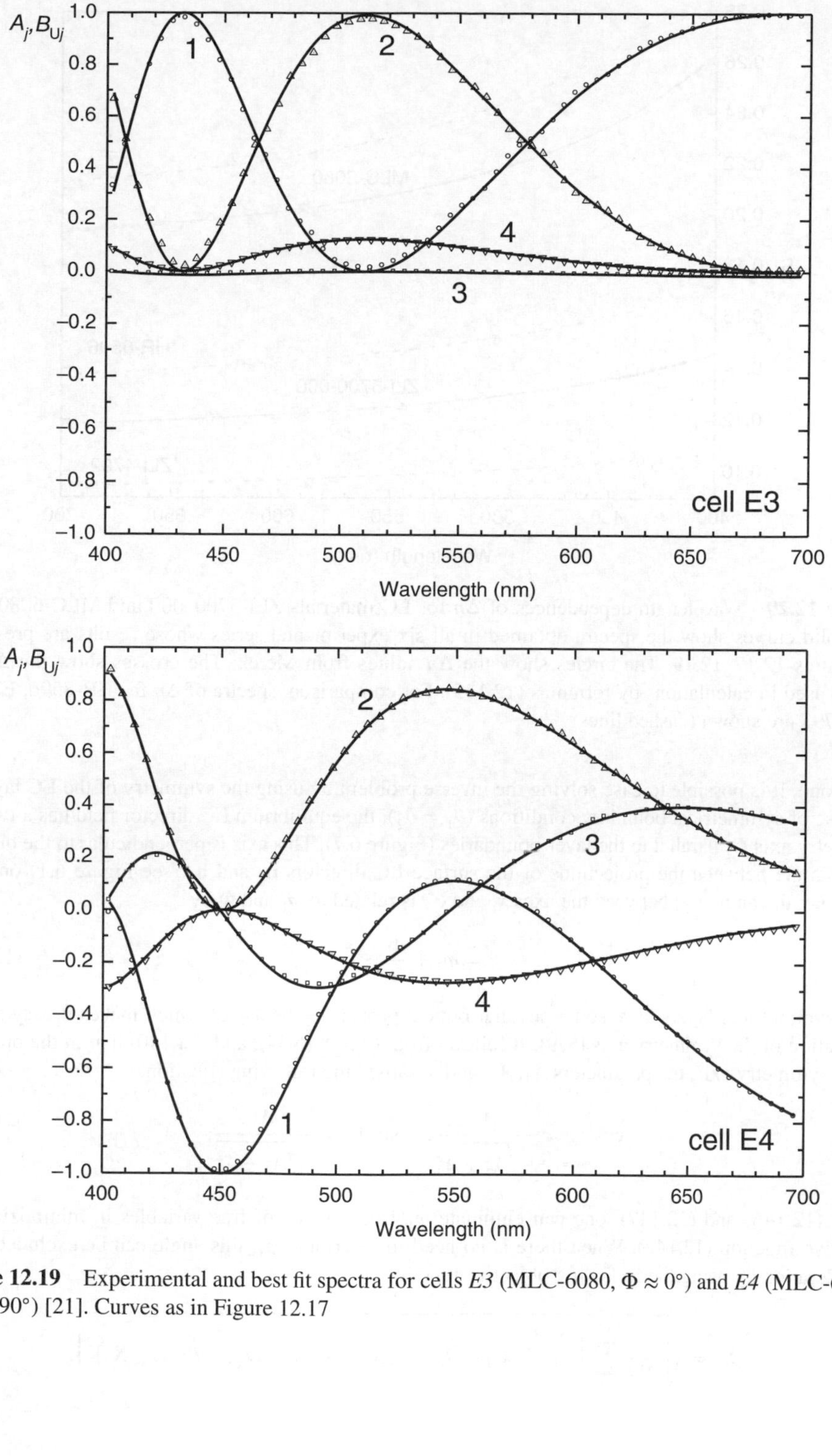

Figure 12.19 Experimental and best fit spectra for cells *E3* (MLC-6080, $\Phi \approx 0°$) and *E4* (MLC-6080, $\Phi \approx -90°$) [21]. Curves as in Figure 12.17

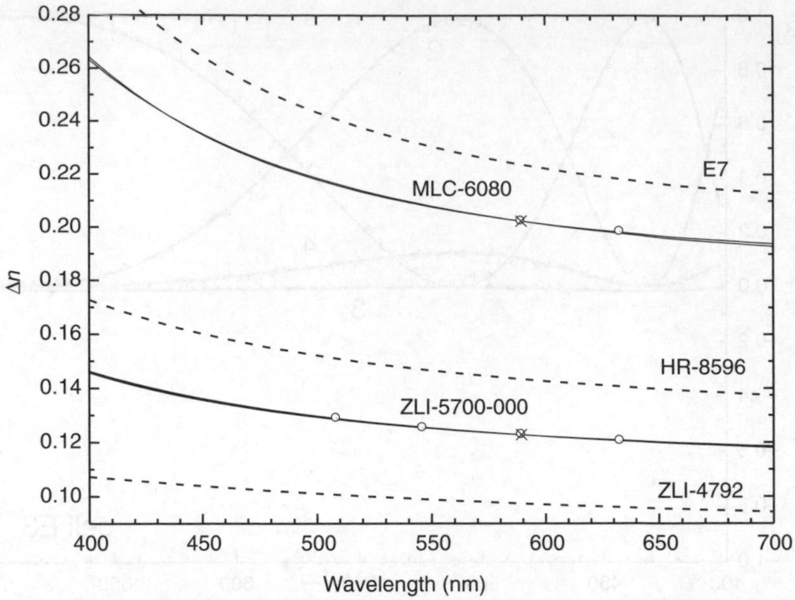

Figure 12.20 Wavelength dependences of Δn for LC materials ZLI-5700–000 and MLC-6080 [21]. The solid curves show the spectra obtained in all six experimental series whose results are presented in Figures 12.17–12.19. The circles show the Δn values from Merck. The crosses show the $\Delta n(\lambda_0)$ values used in calculations by formulas (12.145). For comparison, spectra of Δn for HR-8596, E7, and ZLI-4792 are shown (dashed lines)

Second, it is possible to ease solving the inverse problem by using the symmetry of the LC layer. In the case of symmetrical boundary conditions ($\theta_1 = \theta_2$), the equilibrium LC director field has a twofold symmetry axis C_2 parallel to the layer boundaries (Figure 6.7). This axis is perpendicular to the bisector of the angle between the projections of the surface LC directors $\mathbf{n}_1$ and $\mathbf{n}_2$ (see Figure 6.1) onto the XY-plane; the angle (χ) between the axes x_{I} and C_2 is related to φ_1 and Φ by

$$\chi = \varphi_1 + \frac{\Phi}{2} \pm \frac{\pi}{2}. \tag{12.146}$$

Such symmetrical layers have some peculiar optical properties the use of which makes it easy to find orientation of the symmetry axis [8, 9]. It follows from (6.39), (6.34), and (12.140) that in the presence of the symmetry axis, the parameters A_2, A_4, and χ satisfy the following equations:

$$\cos 4\chi = \frac{A_2}{\sqrt{A_2^2 + A_4^2}}, \quad \sin 4\chi = \frac{A_4}{\sqrt{A_2^2 + A_4^2}}. \tag{12.147}$$

Using (12.146) and (12.147), one can eliminate φ_1 from the set of free variables in minimizing the objective function (12.142). When there is no need to determine φ_1, this angle can be excluded from consideration by using the following objective function:

$$F(\mathbf{X}) = \sqrt{\frac{1}{2M}\sum_{k=1}^{M}\left[\left(A_1^{\mathrm{EXP}}(\lambda_k) - B_{\mathrm{U}1}(\lambda_k;\mathbf{X})\right)^2 + \left(A_3^{\mathrm{EXP}}(\lambda_k) - B_{\mathrm{U}3}(\lambda_k;\mathbf{X})\right)^2\right]}.$$

In contrast to (12.142), this function involves only parameters A_1, A_3, B_{U1}, and B_{U3}. All these parameters are invariant under rotations of the cell about the z_I-axis and hence do not depend on φ_I [see (6.34) and (6.36), Figures 6.4 and 12.17].

12.5.3 Appendix to Section 12.5

Specification of the Model LC Cells

In this section, we specify the components of the model LC cells used in the numerical examples of Sections 12.5.1 and 12.5.2 (subsection *Numerical tests*) that were not specified in those sections.

LC layer. In the numerical examples, we used data on refractive indices for three LC materials: HR-8596 (Hoffmann-La Roche), E7 (BDH), and ZLI-4792 (Merck). The wavelength dependences of the principal refractive indices for these LC materials, which we used in the calculations, are shown in Figure 12.21. The corresponding curves of $\Delta n(\lambda)$ are presented in Figure 12.20. In all examples of Section 12.5.1, we used the data for HR-8596. The thickness of the LC layer in cells *A*, *B*, *C*, and *D* is 8 μm.

Glass substrates. The refractive index of the glass substrates was taken to be 1.52 throughout the spectral region under consideration.

Alignment films. Except for the second alignment film in cell *D*, the alignment films were taken to be isotropic and nonabsorbing. Unless otherwise specified, the refractive index of alignment layers is 1.6 (it is a typical value of the refractive index of PI) for all wavelengths. The parallel and perpendicular real principal refractive indices of the dichroic alignment film in cell *D* are 1.501 and 1.5, respectively, at all wavelengths. In all the model cells, the thickness of alignment films is 0.1 μm.

Electrodes. Electrode films were treated as homogeneous absorbing isotropic layers. The spectra of the real refractive index n_{ITO} and extinction coefficient k_{ITO} of electrodes that we used in the calculations are shown in Figure 12.22. The thickness of the electrode films was taken to be 0.03 μm.

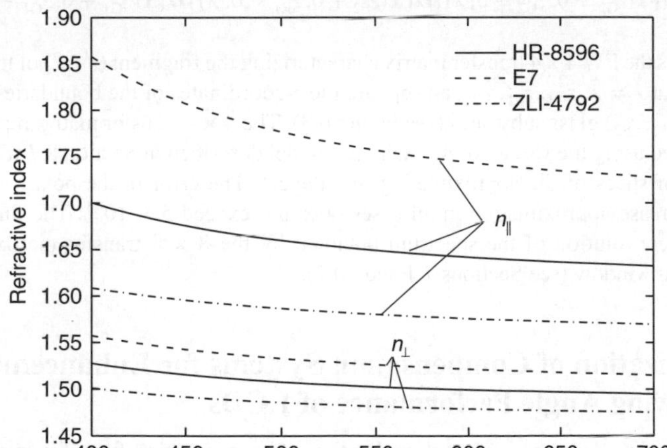

Figure 12.21 The wavelength dependences of the principal refractive indices of LC materials used in the numerical experiments

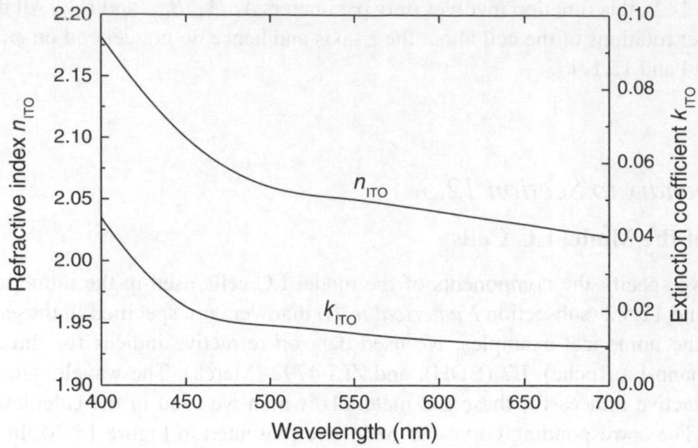

Figure 12.22 The wavelength dependences of the refractive index and the extinction coefficient of the electrode material (ITO) used in the numerical experiments

Rigorous Modeling of the Polarized Transmittance Spectra of LC Cells

In this case, our purpose was to model as accurately as possible the spectral dependences of polarized transmittances of the LC cells for incident quasimonochromatic light with a bandwidth of about 4 nm (it is a typical magnitude of the spectral resolution in examining LC cells), taking into account all significant factors, including multiple reflections and multiple-beam interference. For the calculations, we used the 8 × 8 transfer matrix method (see Section 10.2). The glass substrates were treated as "thick" layers, and the LC layer together with the thin-film systems surrounding it as a "thin" layered system. The 8 × 8 transfer matrix of the cell, $\mathbf{D}_{\text{cell}}$, was calculated as follows:

$$\mathbf{D}_{\text{cell}} = \widehat{\mathbf{D}}\{\mathbf{T}(z_{\text{G2}}+0, z_{\text{G2}}-0; \tilde{\lambda})\}\widehat{\mathbf{D}}\{\mathbf{T}(z_{\text{E2}}+0, z_{\text{E1}}-0; \tilde{\lambda})\}\widehat{\mathbf{D}}\{\mathbf{T}(z_{\text{G1}}+0, z_{\text{G1}}-0; \tilde{\lambda})\},$$

where $\mathbf{T}(z'', z'; \tilde{\lambda})$ is the EW 4 × 4 transfer matrix characterizing the fragment (z', z'') of the approximating stratified medium at $\lambda = \tilde{\lambda}$; z_{G1}, z_{E1}, z_{G2}, and z_{E2} are the z-coordinates of the boundaries of the first (z_{G1}, z_{E1}) and second (z_{G2}, z_{E2}) glass substrates (see Figure 6.3). The 4 × 4 transfer matrix representing the LC layer was calculated using the variant of the staircase model described in Section 8.7 with $N = 801$ (here N is the number of slices of the approximating multilayer). The error in the polarized transmittances caused by the staircase approximation in all cases does not exceed 5×10^{-5}. The final spectrum was calculated as the convolution of the spectrum obtained by the 8 × 8 transfer matrix method with a 4-nm-wide spectral window (see Sections 7.1 and 10.2).

12.6 Optimization of Compensation Systems for Enhancement of Viewing Angle Performance of LCDs

In Section 6.5, we considered examples of application of compensation films for improvement of the LCD performance for the normal viewing direction and introduced a special parameter—the modulation efficiency factor—whose examination enables one to evaluate the ultimate characteristics of the LCD that can be achieved by fitting the compensation system in advance, before optimization calculations. The main purpose of compensation systems in modern LCDs is to enhance the viewing angle performance.

Some basic principles of film compensation for different kinds of LCDs and various types of compensation films are considered in previous books of this series (see, e.g., [25–27]). A thorough analysis of limitations to the viewing angle characteristics of film-compensated LCDs determined by features of electro-optical properties of LC layers was performed by Stallinga et al. [28]. The material presented in [28] is a good addition to the material given in Section 6.5 and this section. Consideration of the mentioned limitations leads to the concept of modulation efficiency. As was shown in Section 6.5, the parameter of modulation efficiency introduced in [28] (it was denoted by Q_Δ in Section 6.5) is closely related to the modulation efficiency factor Q introduced in Section 6.5:

$$Q_\Delta = \sqrt{Q}. \tag{12.148}$$

The physical meaning of both these parameters is explained in Section 6.5.

Angular Dependence of Modulation Efficiency Factor

Within the accuracy of the immersion model—in which the optical action of all PCS elements is described by unitary (polarization) matrices, just as in the analysis that led us to the notion of the modulation efficiency factor in Section 6.5—the notion and definition of the modulation efficiency factor can be extended to the case of oblique incidence. At arbitrary incidence, for a transmissive LC device, the factor Q can be calculated by formulas (6.122) and (6.130) or (6.131) with $\boldsymbol{t}_{\text{LC–B}}$ and $\boldsymbol{t}_{\text{LC–D}}$ being the polarization matrices of the LC layer in the bright and dark states, respectively, calculated by formula (12.134). Since the matrices $\boldsymbol{t}_{\text{LC–B}}$ and $\boldsymbol{t}_{\text{LC–D}}$ in this case have determinant 1, calculation of Q can be performed by the simpler formula (6.130). The parameter Q_Δ can be calculated by (12.148). We note that this method of calculating Q_Δ is more accurate than that proposed in [28], because the latter uses the small-birefringence approximation.

Recall that the modulation efficiency factor Q is equal to the maximum reduced transmittance of the LCD panel in the bright state subject to the condition that the polarizers and compensators maintain a zero transmittance in the dark state. Examining the angular and spectral dependences of Q, one can ascertain whether the desired viewing angle performance can be attained or not with the given LC layer at the given working voltages.

Let us consider several examples. In these examples, the transmission of an LCD panel for a given direction of incidence is characterized by the following parameters:

(i) Average transmittances of the panel in the red (t_R), green (t_G), and blue (t_B) regions that are defined as

$$t_j = \frac{\int t(\lambda) f_j(\lambda) \mathrm{d}\lambda}{\int f_j(\lambda) \mathrm{d}\lambda}, \quad j = \text{R, G, B}, \tag{12.149}$$

where t is the unpolarized transmittance of the LCD panel, and $f_\text{R}(\lambda)$, $f_\text{G}(\lambda)$, and $f_\text{B}(\lambda)$ are the weighting functions shown in Figure 12.23. The values of t_R, t_G, and t_B for a bright state will be denoted by $t_{\text{R–b}}$, $t_{\text{G–b}}$, and $t_{\text{B–b}}$, respectively, and those for a dark state by $t_{\text{R–d}}$, $t_{\text{G–d}}$, and $t_{\text{B–d}}$.

(ii) Contrast ratios in the red, green, and blue regions:

$$C_\text{R} = \frac{t_{\text{R–b}}}{t_{\text{R–d}}}, \quad C_\text{G} = \frac{t_{\text{G–b}}}{t_{\text{G–d}}}, \quad C_\text{B} = \frac{t_{\text{B–b}}}{t_{\text{B–d}}}.$$

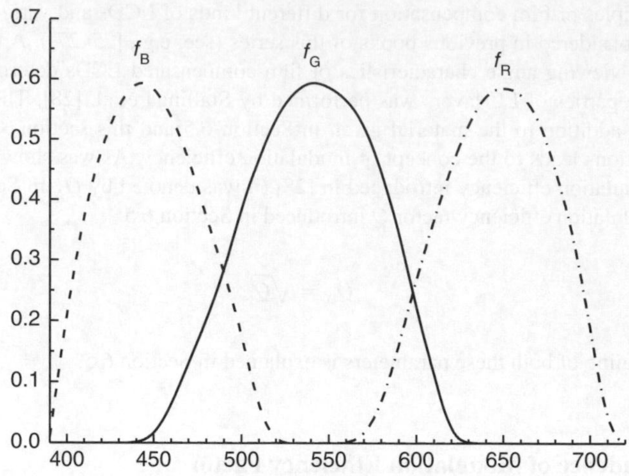

Figure 12.23 Weighting functions $f_R(\lambda)$, $f_G(\lambda)$, and $f_B(\lambda)$ used in calculations of averages in the red, green, and blue regions

(iii) Average modulation efficiency factors in the red (Q_R), green (Q_G), and blue (Q_B) regions:

$$Q_j = \frac{\int Q(\lambda) f_j(\lambda) d\lambda}{\int f_j(\lambda) d\lambda}, \quad j = \mathrm{R, G, B},$$

where Q is the modulation efficiency factor of the LC layer.

The listed quantities will be considered as functions of the angles β_{inc} (the polar angle of incidence) and α_{inc} (the azimuthal angle) shown in Figure 9.1. In the polar diagrams in Figures 12.25–12.34, the radial coordinate represents β_{inc}, and the azimuthal one α_{inc}.

The first example is for a 90° TN LCD. In this example, we consider an LC cell with the following parameters of the LC layer: $K_{11} = 1.3 \times 10^{-6}$ dyn, $K_{22} = 7.1 \times 10^{-7}$ dyn, $K_{33} = 1.95 \times 10^{-6}$ dyn, $\varepsilon_{\parallel} = 15.1$, $\varepsilon_{\perp} = 3.8$, $d/p_0 = 0.35$, where d is the thickness of the LC layer and p_0 is the natural helix pitch of the LC. The principal refractive indices of the LC are the following:

Wavelength (nm)	$n_{\perp}$	$n_{\parallel}$
400	1.5277	1.7005
550	1.5027	1.6493
700	1.4949	1.6324

The easy axis tilt angle is equal to 4° for both boundaries. The X-axis of the reference system XYZ (Figure 9.1), attached to the LCD panel, is oriented so that the easy axis on the frontal boundary is parallel to the XZ-plane and makes an angle of 4° with the positive direction of the X-axis, that is, for this easy axis, $\varphi = 0°$ and $\theta = 4°$ (see Figure 9.3; the orientation of specific axes of other elements is also specified as in Figure 9.3). For the easy axis on the second LC layer boundary, $\varphi = 90°$ and $\theta = 4°$.

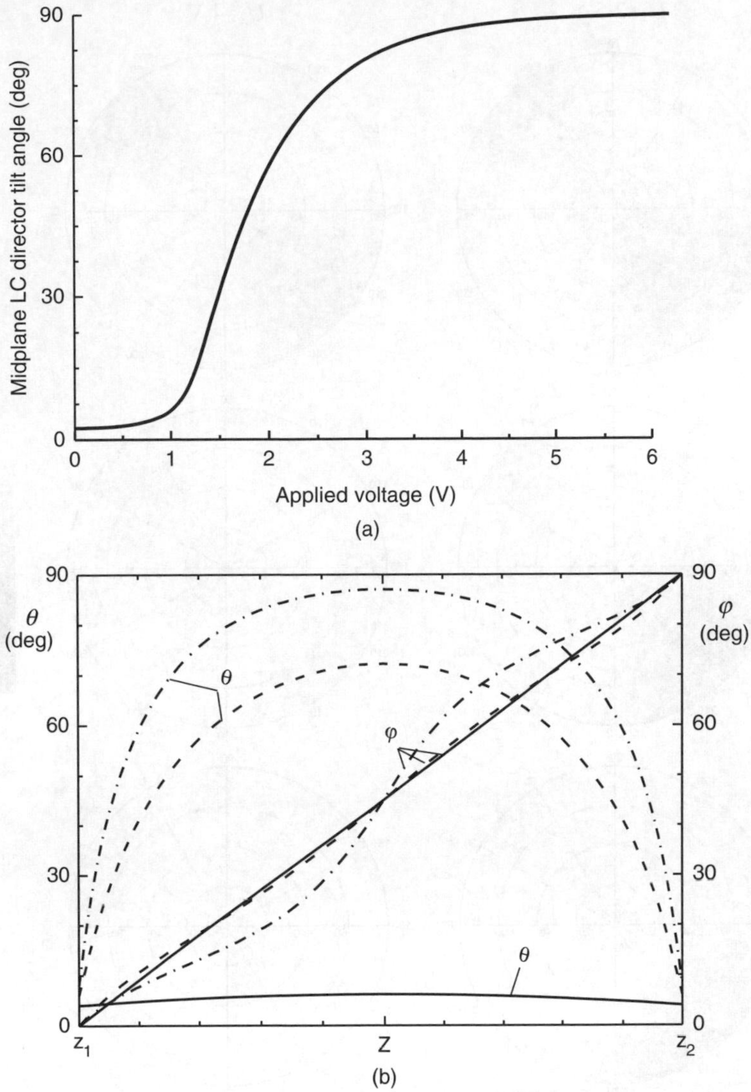

Figure 12.24 Midplane LC director tilt angle versus applied voltage U (a) and profiles of the LC director at $U = 0.89$, 2.5, and 4 V (b) for the model TN LC layer at $d = 3.07$ μm

For both boundaries, the polar anchoring strength is 1 erg/cm^2 and the azimuthal anchoring is infinitely strong.

Figure 12.24 shows the dependence of the midplane LC director tilt angle on the voltage U applied to the LC layer and the LC director field configurations at $U = 0.89$, 2.5, and 4 V for $d = 3.07$ μm. Figure 12.25 presents the angular dependences of the factors Q_R, Q_G, and Q_B for the layer with $d = 3.07$ μm for two cases. In the first case, state 1 of the LC layer corresponds to $U = 0.89$ V, and state 2 to $U = 2.5$ V, that is, $U_1 = 0.89$ V, $U_2 = 2.5$ V (see Section 6.5). In the second case, the working voltages are $U_1 = 0.89$ V and $U_2 = 4$ V. Comparing diagrams in Figure 12.25, we see that with $U_2 = 4$ V we

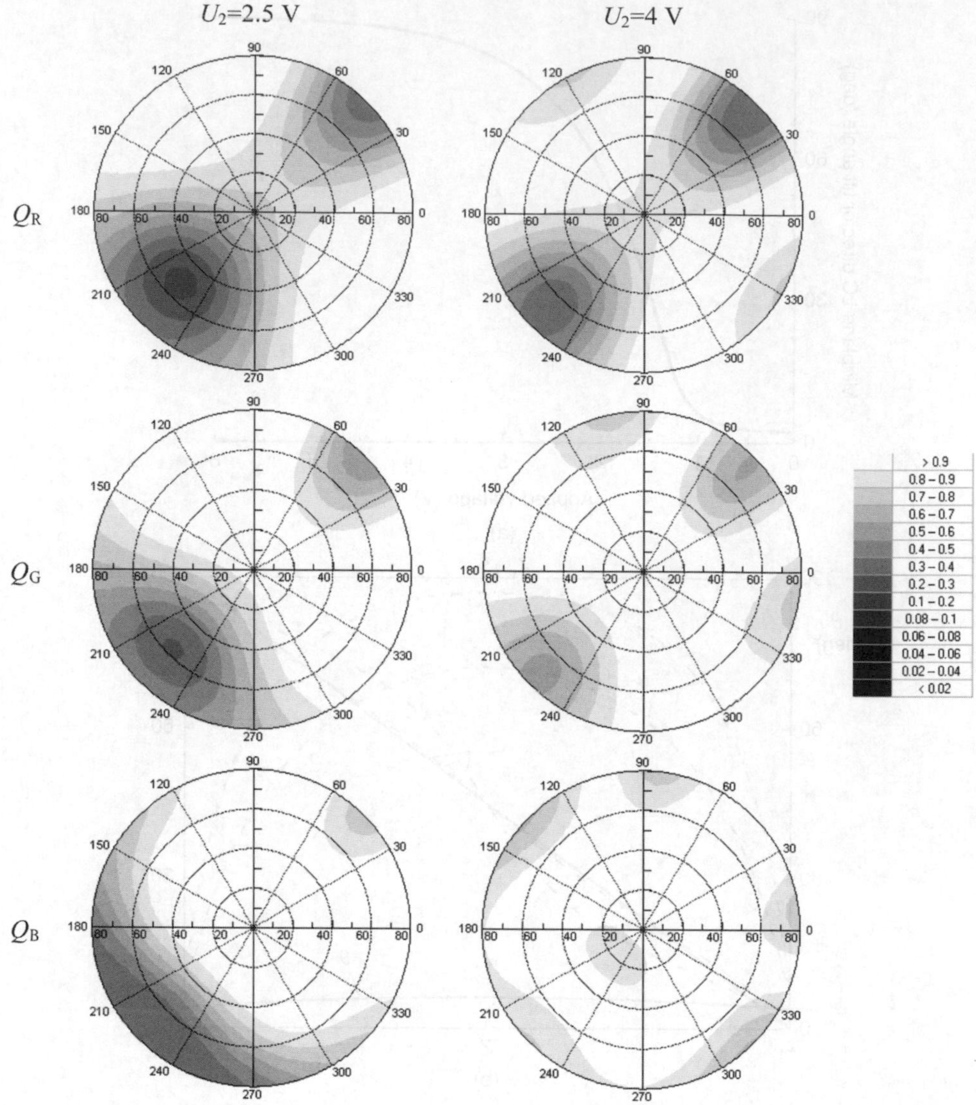

Figure 12.25 Viewing angle dependences of the average modulation efficiency factors Q_R, Q_G, and Q_B of the model TN LC layer with $d = 3.07$ μm at $U_1 = 0.89$ V and $U_2 = 2.5$ and 4 V

can obtain much better black–white switching in a wide viewing cone than with $U_2 = 2.5$ V. We see also that with $U_2 = 2.5$ V it is impossible to obtain simultaneously high contrast and a high bright-state transmittance in the sector $180° < \alpha_{inc} < 270°$. Note that at $U_2 = 4$ V, the values of Q_B are close to 1 for all incident directions in the range under consideration, that is, in the blue region it is possible, in principle, to obtain practically perfect dark–bright switching for all viewing directions with $\beta_{inc} < 80°$.

In Figure 12.26, we compare the angular dependences of Q_R, Q_G, and Q_B at $d = 3.07$, 3.3, and 3.55 μm for $U_1 = 0.89$ V and $U_2 = 4$ V. As can be seen from this figure, the increase of the LC layer thickness

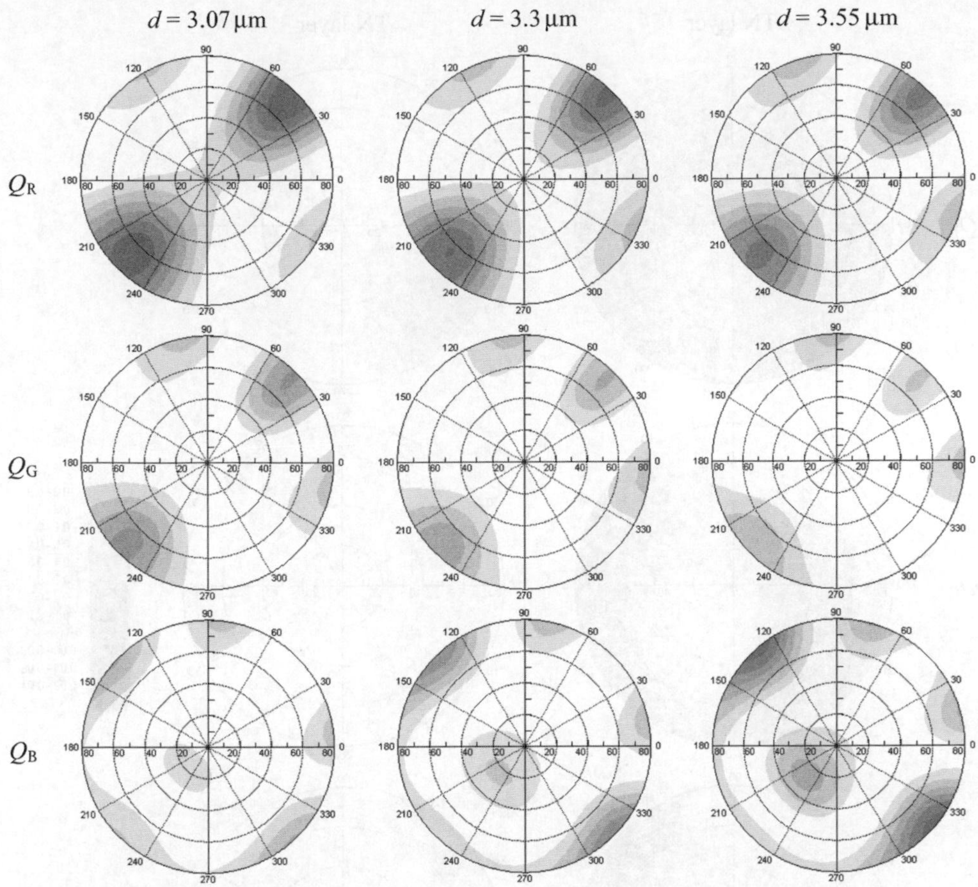

Figure 12.26 Viewing angle dependences of the factors Q_R, Q_G, and Q_B of the model TN LC layer at $d = 3.07$, 3.3 and 3.55 μm; $U_1 = 0.89$ V, $U_2 = 4$ V

in this case somewhat improves the angular dependences of Q_R and Q_G, but worsens that of Q_B. The variant with $d = 3.3$ μm seems to be the best compromise.

Figure 12.27 shows the angular dependences of Q_R, Q_G, and Q_B for the STN layer that we dealt with in the examples of Section 6.5 (Figures 6.16–6.19) at working voltages $U_1 = 2.35$ V and $U_2 = 2.47$ V. These dependences are compared with those for the above TN layer at $d = 3.3$ μm, $U_1 = 0.89$ V, and $U_2 =$ 4 V. We see that in the case of the STN layer, the viewing cone for which good black–white switching can be achieved is relatively narrow (with a half-angle of about 25°) and has an inclined central axis.

Now we present examples of computer optimization of compensation systems for enhancement of the viewing angle performance for TN and STN LCDs. Solutions presented below were found with the help of the optimization engines of MOUSE-LCD.

TN LCD

Figure 12.28 shows the angular dependences of the bright-state transmittance t_{G-b} and contrast ratio C_G for uncompensated normal-white LCDs with the described TN layer at $d = 3.3$ μm, bright-state voltage

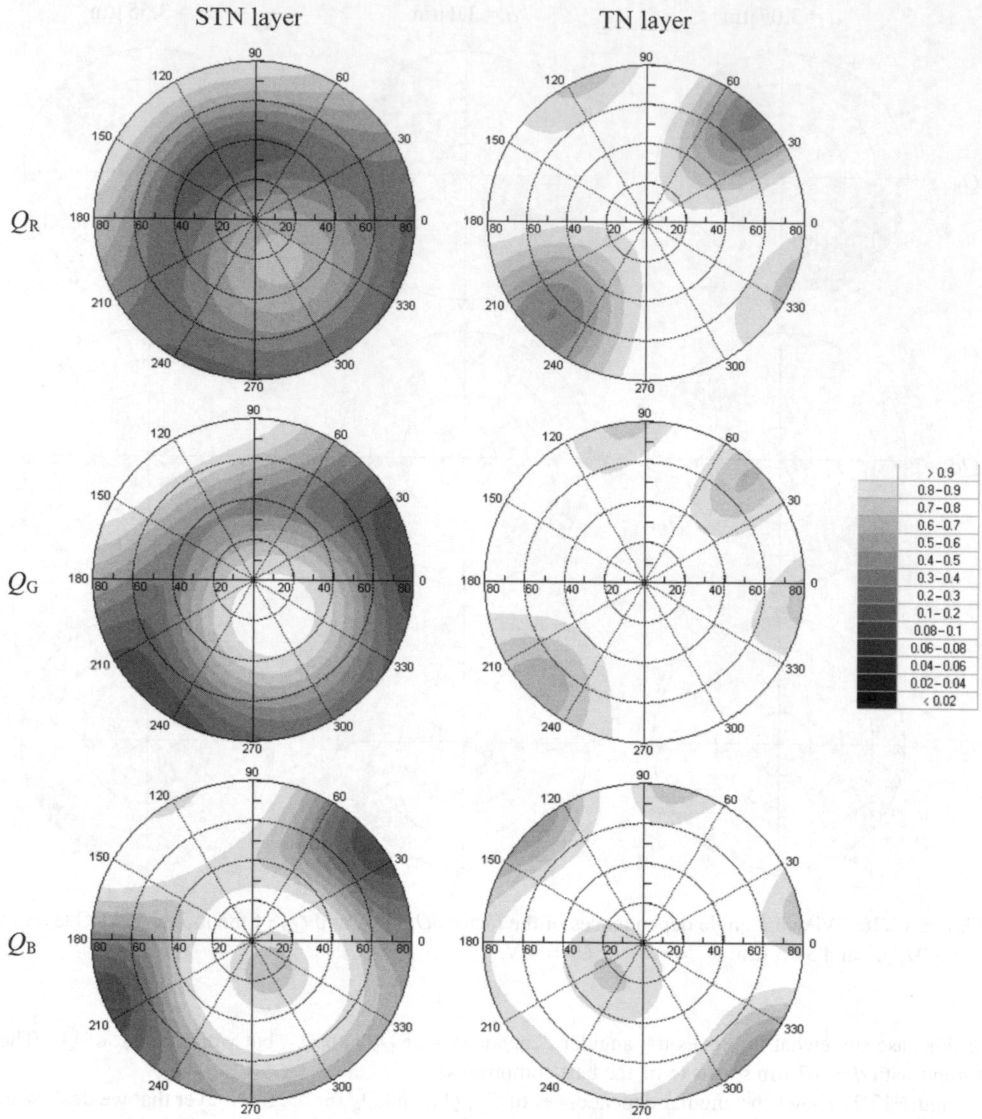

Figure 12.27 Viewing angle dependences of the average modulation efficiency factors Q_R, Q_G, and Q_B of the model TN and STN layers

$U_B = 0.89$ V, and dark-state voltage $U_D = 4$ V. The data are presented for two modes, *e*-mode and *o*-mode, different from each other in polarizer orientation. In the case of *e*-mode, the transmission axis of the frontal polarizer is parallel to the *X*-axis and that of the rear polarizer is parallel to the *Y*-axis. In the case of *o*-mode, the polarizers are rotated by 90° with respect to their position in *e*-mode. Here and in the following examples, we use quasi-ideal polarizers of *o*-type with complete extinction of the extraordinary components, $\mathrm{Re}(n_\parallel) = 1.5001$, and $\mathrm{Re}(n_\perp) = 1.500$. For both polarizers, the spectral dependence of the imaginary part of the complex refractive index $n_\perp$ is such that the transmittance of the polarizer, when it is surrounded by air, for normally incident linearly polarized light with polarization direction parallel

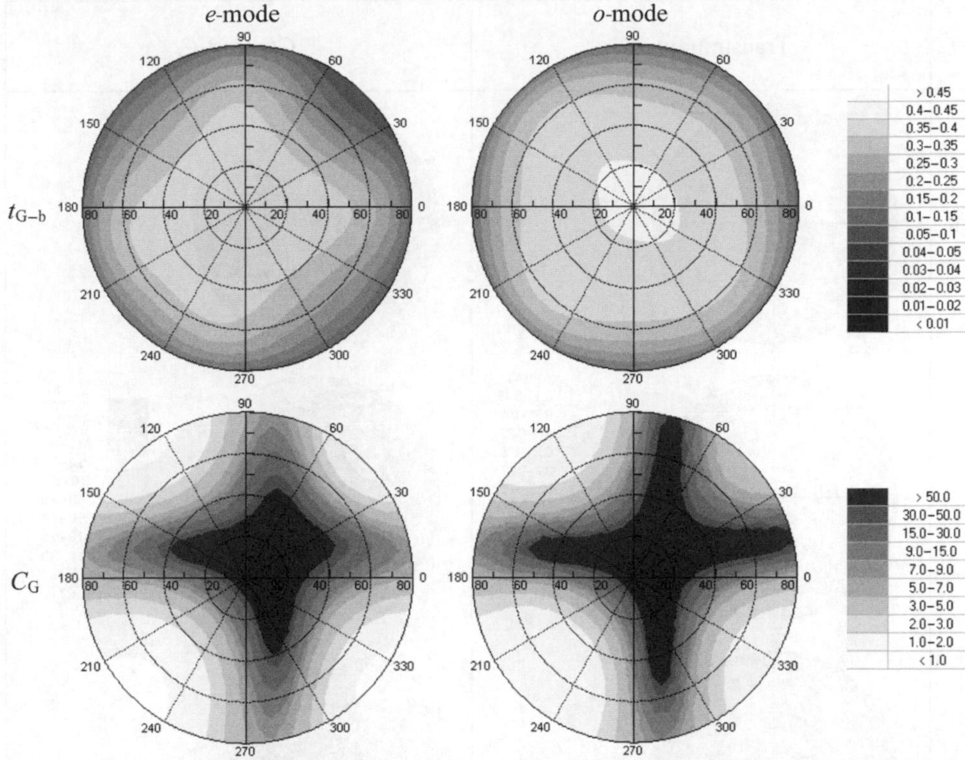

Figure 12.28 Viewing angle dependences of the bright-state transmittance t_{G-b} and contrast ratio C_G for uncompensated normal-white TN LCDs with the model TN layer, operating in *e*-mode and *o*-mode. Normal direction: for *e*-mode $t_{G-b} = 0.391$ and $C_G = 169.4$, for *o*-mode $t_{G-b} = 0.402$ and $C_G = 169.9$

to the transmission axis of the polarizer (i.e., perpendicular to the optic axis of the polarizer) is equal to 0.9 throughout the visible region. The fitting of $\mathrm{Im}(n_\perp)$ was performed using expression (7.43). In the model LCDs, polarizers are in contact with adjacent layers of the LCD panel (glass substrates, as in this example, or compensation films). The model LC cells in this example and the following examples of this section have a structure as shown in Figure 6.3. The thickness of the alignment layers is 0.1 μm, the refractive index of these layers is 1.6. The parameters of the conductive layers are the same as in the computational tests of the previous section (see Section 12.5.3). Transmittance t [see (12.149)] was taken to be the transmittance of the "useful" channel of the LCD panel, the alignment layer–conductive layer systems being treated as OTR units (see Section 7.1). This transmittance was calculated using exact EW Jones matrices.

Figures 12.29 and 12.30 show the angular dependences of the transmittances t_{R-b}, t_{G-b}, and t_{B-b} and contrast ratios C_R, C_G, and C_B for two compensated LCDs with the same LC cell at the same working voltages ($U_B = 0.89$ V, $U_D = 4$ V). In either case, the compensation system consists of four homogeneous uniaxial films. Two of them are situated before the LC layer, and the other two after the LC layer. In both cases, all compensation films have the same principal refractive indices. In the first case, the films are positive ($n_\parallel > n_\perp$). They have the same principal refractive indices as the compensation films in the examples of Section 6.5 but tilted optic axes. Parameters of the films and the polarizer orientation for this case are shown in Table 12.11; the thicknesses, optic axis tilt angles, and azimuthal orientation of compensation films and orientation of polarizers were found by MOUSE-LCD. In Table 12.11 and next

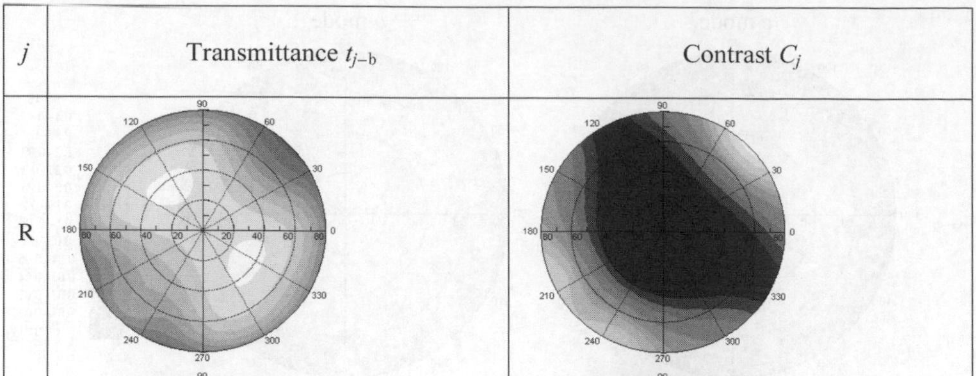

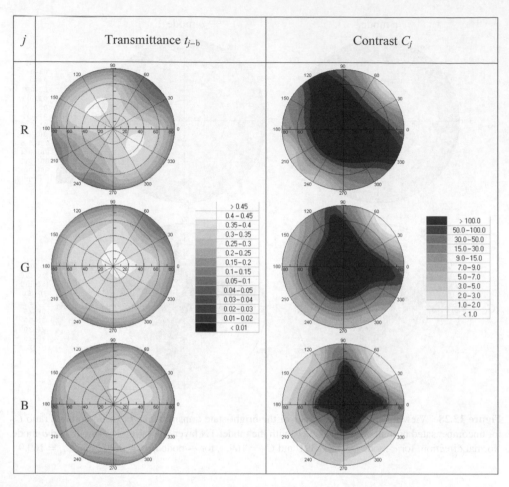

Figure 12.29 Viewing angle dependences of the bright-state transmittances t_{R-b}, t_{G-b}, and t_{B-b} and contrast ratios C_R, C_G, and C_B for the TN LCD with an optimized compensation system consisting of positive uniaxial films (Table 12.11). For the normal direction, $t_{R-b} = 0.379$, $t_{G-b} = 0.402$, $t_{B-b} = 0.346$, $C_R = 631$, $C_G = 586$, and $C_B = 357$

tables, we use the same notation as in Table 6.1. The tilt angle of the optic axis (θ in Figure 9.3) of a compensation film is denoted by θ_L. The results for the first case are shown in Figure 12.29. This is one of the best variants that we found for this compensation scheme with positive compensation films. Much better results were obtained with negative compensation films ($n_{\parallel} < n_{\perp}$) (Figure 12.30). The spectral dependence of the principal refractive indices of the films that we used in this case is:

Wavelength (nm)	$n_{\perp}$	$n_{\parallel}$
400	1.5578	1.5483
550	1.5461	1.5368
700	1.5407	1.5317

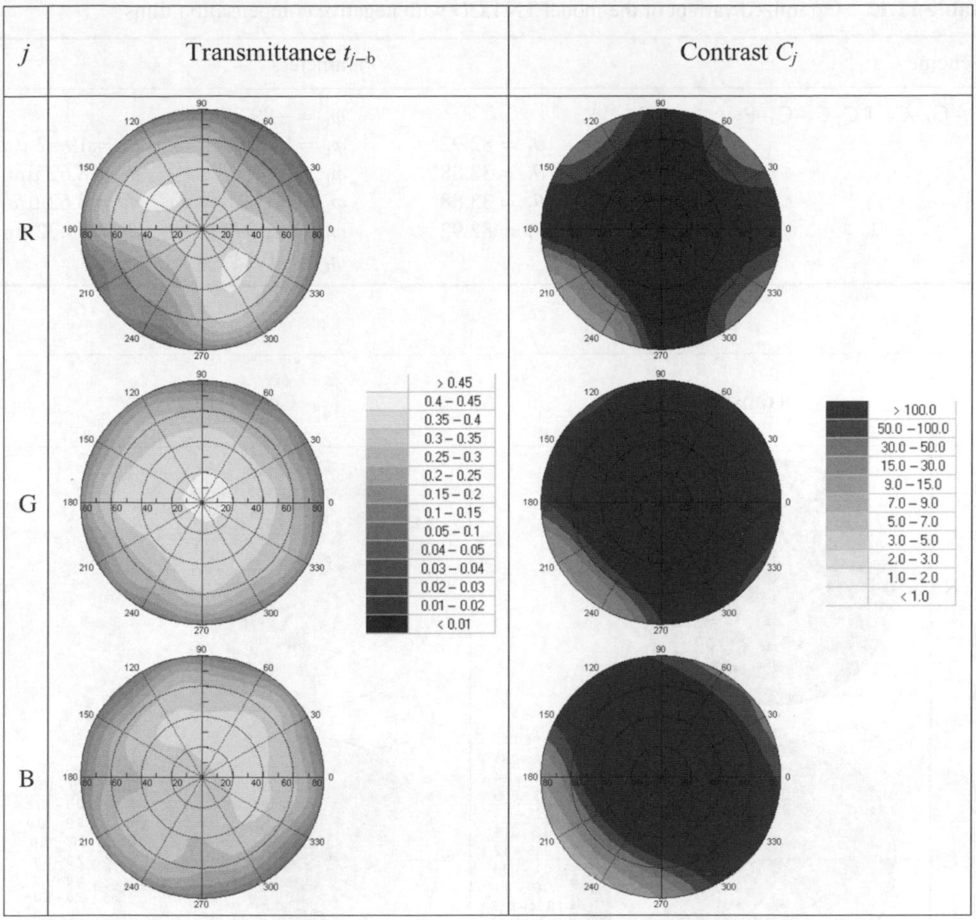

Figure 12.30 Viewing angle dependences of the bright-state transmittances t_{R-b}, t_{G-b}, and t_{B-b} and contrast ratios C_R, C_G, and C_B for the TN LCD with an optimized compensation system composed of negative uniaxial films (Table 12.12). For the normal direction, $t_{R-b} = 0.384$, $t_{G-b} = 0.402$, $t_{B-b} = 0.338$, and C_R, C_G, $C_B > 1000$

Table 12.11 Optimized variant of the model TN LCD with positive compensation films

Scheme		Parameters		
P_1–C_1–C_2–LC–C_3–C_4–P_2	P_1		$\varphi_L = 90.87°$	
	C_1	$\theta_L = 16.55°$	$\varphi_L = -88.59°$	$d_L = 25.95$ μm
	C_2	$\theta_L = 18.77°$	$\varphi_L = 179.79°$	$d_L = 5.59$ μm
	C_3	$\theta_L = 18.77°$	$\varphi_L = -89.79°$	$d_L = 5.59$ μm
	C_4	$\theta_L = 16.55°$	$\varphi_L = 178.59°$	$d_L = 25.95$ μm
	P_2		$\varphi_L = -0.87°$	

Table 12.12 Optimized variant of the model TN LCD with negative compensation films

Scheme		Parameters		
P_1–C_1–C_2–LC–C_3–C_4–P_2	P_1		$\varphi_L = -89.87°$	
	C_1	$\theta_L = 82.92°$	$\varphi_L = 91.97°$	$d_L = 19.32$ μm
	C_2	$\theta_L = 33.88°$	$\varphi_L = 3.38°$	$d_L = 8.62$ μm
	C_3	$\theta_L = 33.88°$	$\varphi_L = 86.62°$	$d_L = 8.62$ μm
	C_4	$\theta_L = 82.92°$	$\varphi_L = -1.97°$	$d_L = 19.32$ μm
	P_2		$\varphi_L = -0.13°$	

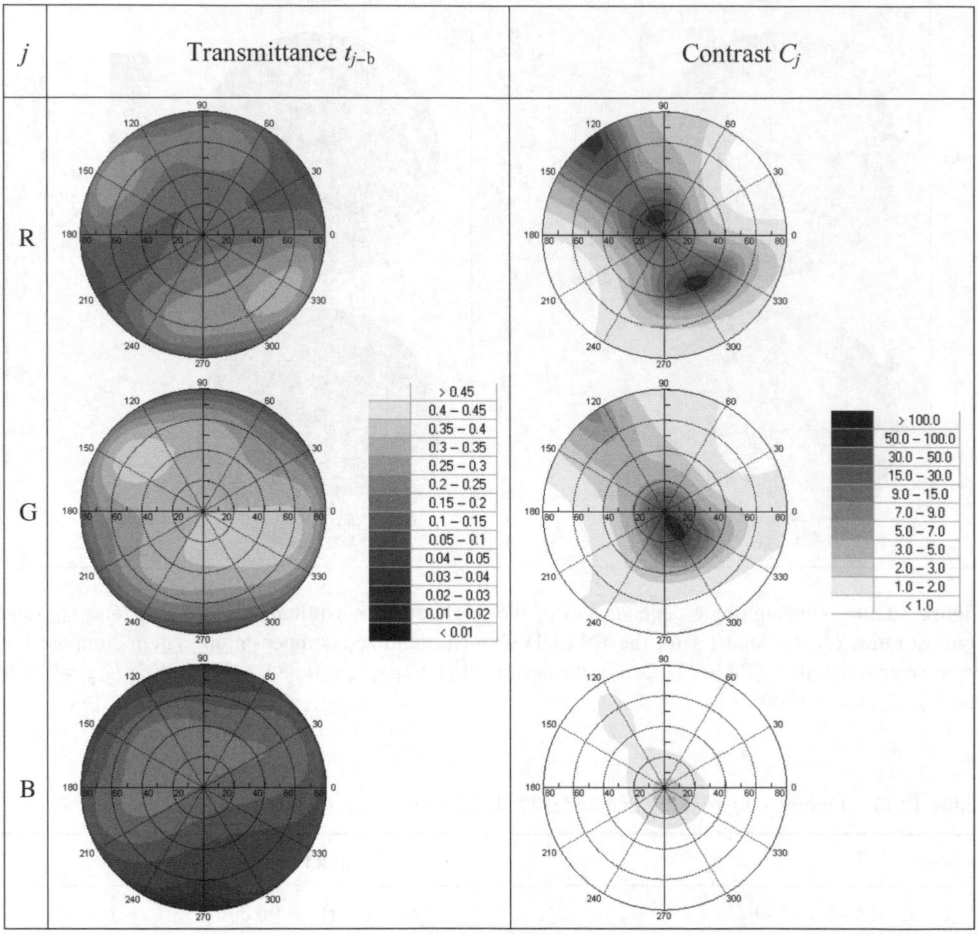

Figure 12.31 Viewing angle dependences of the bright-state transmittances t_{R-b}, t_{G-b}, and t_{B-b} and contrast ratios C_R, C_G, and C_B for the uncompensated variant (variant 1) of the model STN LCD (Table 6.1). For the normal direction, $t_{R-b} = 0.17$, $t_{G-b} = 0.346$, $t_{B-b} = 0.195$, $C_R = 18.5$, $C_G = 36.9$, and $C_B = 2.53$

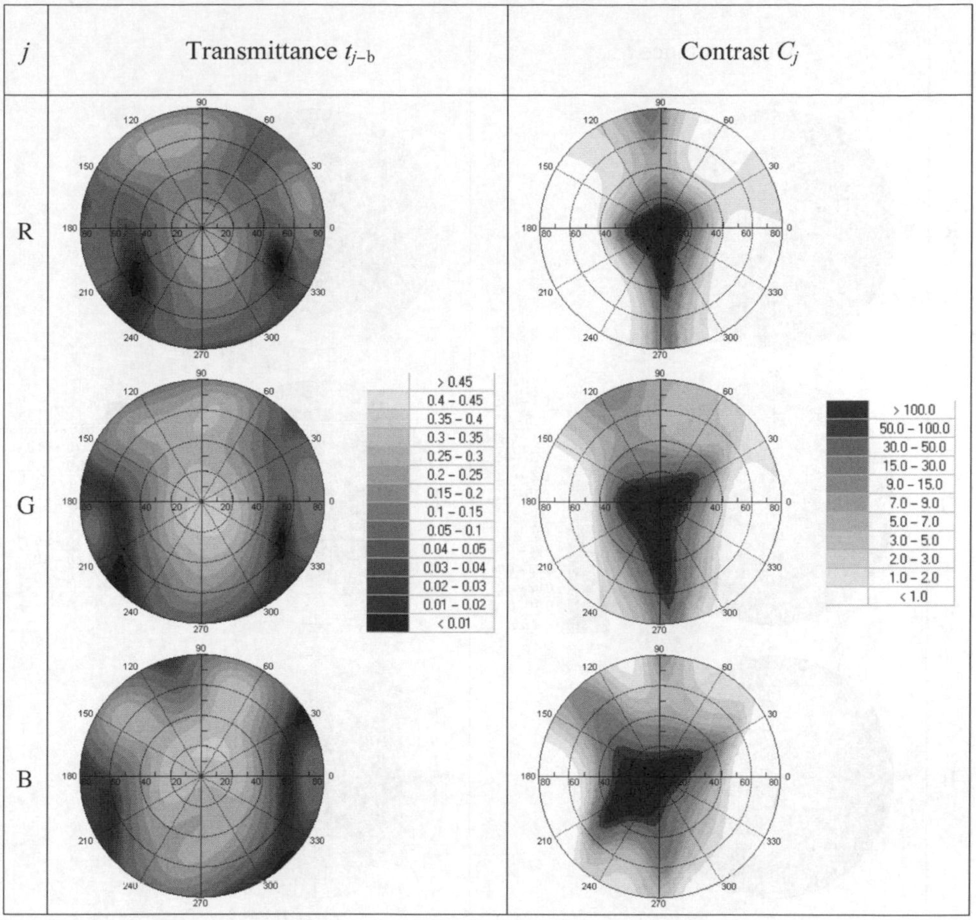

Figure 12.32 Viewing angle dependences of the bright-state transmittances $t_{\text{R-b}}$, $t_{\text{G-b}}$, and $t_{\text{B-b}}$ and contrast ratios C_{R}, C_{G}, and C_{B} for variant 4 of the model STN LCD (Table 6.1, positive compensation films). For the normal direction, $t_{\text{R-b}} = 0.251$, $t_{\text{G-b}} = 0.363$, $t_{\text{B-b}} = 0.321$, and $C_{\text{R}}, C_{\text{G}}, C_{\text{B}} > 1000$

The other parameters of the compensation films and the orientation of the polarizers for this case are shown in Table 12.12. As is seen from Figure 12.30, this compensation system provides high contrast with rather good bright-state transmission for all viewing directions with $\beta_{\text{inc}} < 80°$.

STN LCD

In this example, we consider the same LC layer as in the examples of Section 6.5 (Figures 6.16–6.19) at the same working voltages: $U_1 = 2.35$ V and $U_2 = 2.47$ V. The angular dependence of the modulation efficiency factors of this layer for these working voltages is shown in Figure 12.27. Figures 12.31 and 12.32 demonstrate the angular dependences of the transmittances $t_{\text{R-b}}$, $t_{\text{G-b}}$, and $t_{\text{B-b}}$ and contrast ratios C_{R}, C_{G}, and C_{B} for the uncompensated variant of the device (variant 1 in Table 6.1) and for variant 4 with

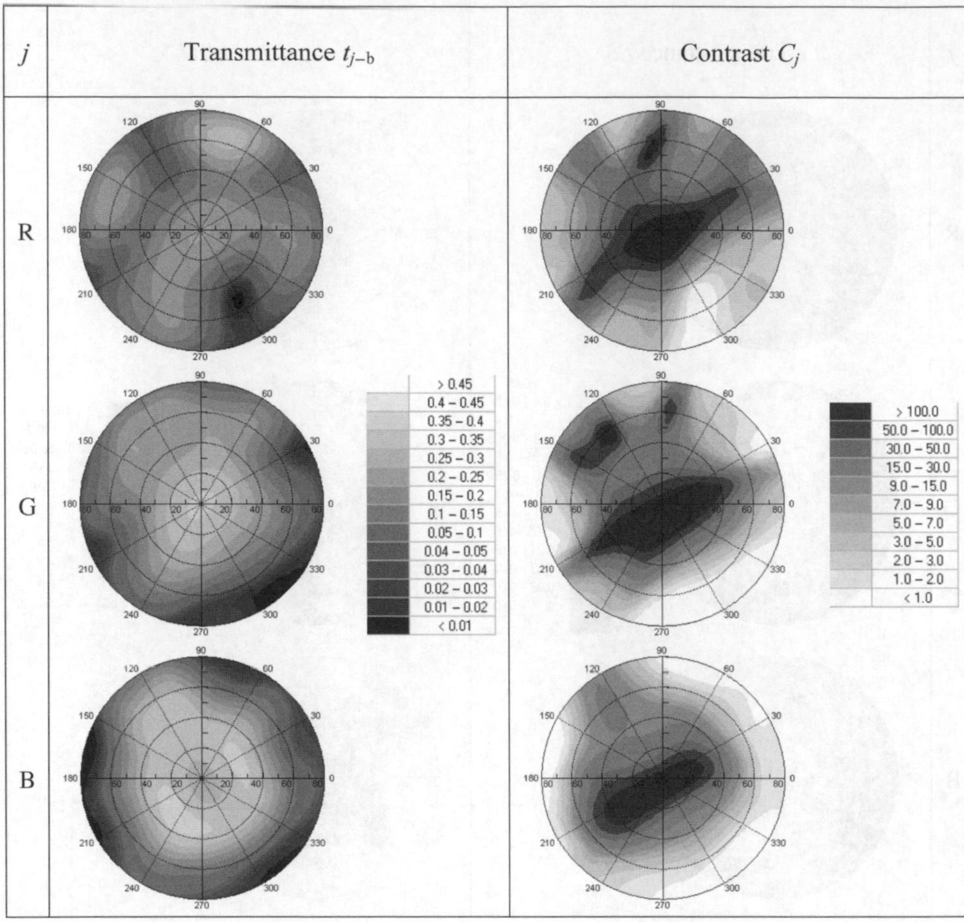

Figure 12.33 Viewing angle dependences of the bright-state transmittances t_{R-b}, t_{G-b}, and t_{B-b} and contrast ratios C_R, C_G, and C_B for variant 5 of the model STN LCD (Table 12.13, positive compensation films). For the normal direction, $t_{R-b} = 0.253$, $t_{G-b} = 0.357$, $t_{B-b} = 0.277$, $C_R = 260$, $C_G = 213$, and $C_B = 275$

compensation (Table 6.1). Recall that the compensation system of variant 4 was designed considering the device transmission at normal incidence only. Figures 12.33 and 12.34 show results for two variants, variant 5 and variant 6, of the device with enhanced viewing angle characteristics. In both these variants, as in variant 4, the normal-black mode is realized, so that $U_D = 2.35$ V and $U_B = 2.47$ V. The compensation scheme in these variants is the same as in variant 4 and the compensated TN devices considered above. In variant 5 (Table 12.13, Figure 12.33), we use positive compensation films with nontilted optic axis ($\theta_L = 0°$). In variant 6 (Table 12.14, Figure 12.34), negative compensation films with tilted optic axis, $\theta_L = 5.43°$, are used. The principal refractive indices of the positive and negative compensation films are the same as in the previous cases. As can be seen from Figures 12.33 and 12.34, better results in this case, as in the above example for a TN LCD, are achieved with negative compensation films.

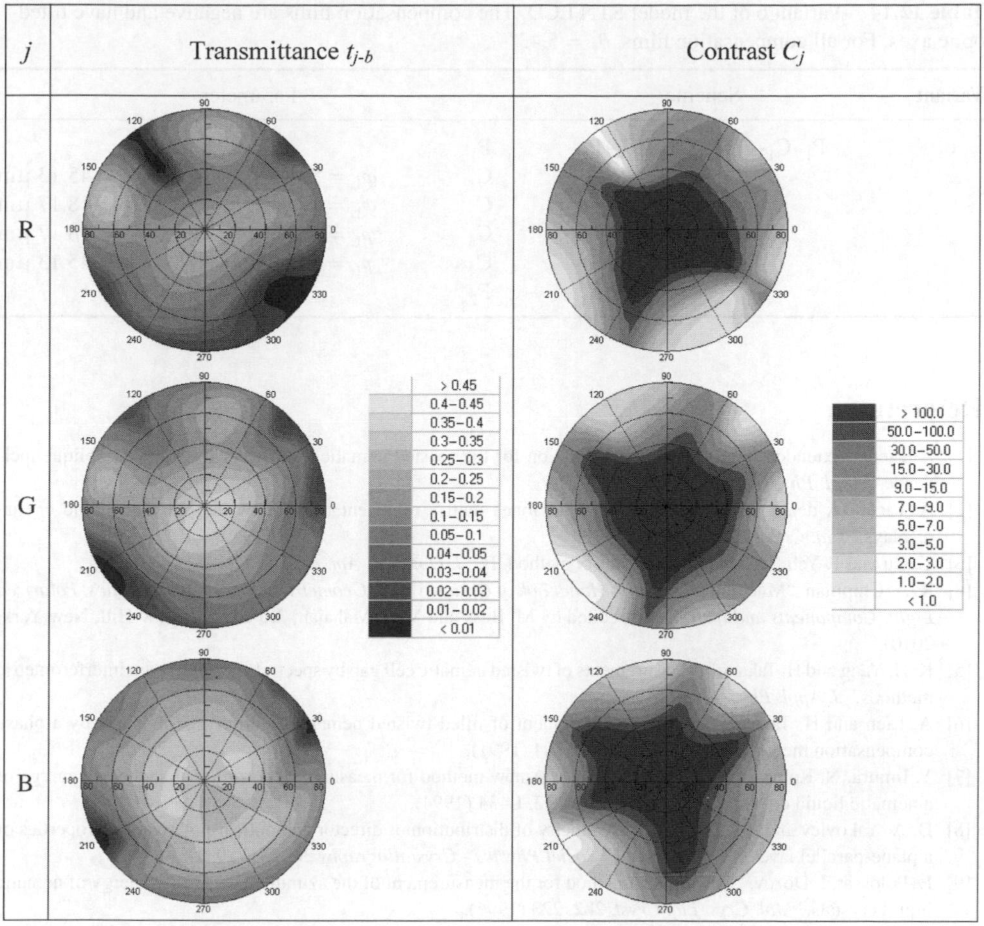

Figure 12.34 Viewing angle dependences of the bright-state transmittances t_{R-b}, t_{G-b}, and t_{B-b} and contrast ratios C_R, C_G, and C_B for variant 6 of the model STN LCD (Table 12.14, negative compensation films). Normal direction: $t_{R-b} = 0.246$, $t_{G-b} = 0.365$, $t_{B-b} = 0.310$, C_R, $C_G > 1000$, and $C_B = 473$

Table 12.13 Variant 5 of the model STN LCD. The compensation films are positive and have nontilted optic axes ($\theta_L = 0°$)

Variant	Scheme	Parameters		
5	P_1–C_1–C_2–LC–C_3–C_4–P_2	P_1	$\varphi_L = 51.17°$	
		C_1	$\varphi_L = -8.47°$	$d_L = 41.85$ μm
		C_2	$\varphi_L = 45.89°$	$d_L = 26.72$ μm
		C_3	$\varphi_L = 14.38°$	$d_L = 26.72$ μm
		C_4	$\varphi_L = 68.76°$	$d_L = 41.85$ μm
		P_2	$\varphi_L = 99.14°$	

Table 12.14 Variant 6 of the model STN LCD. The compensation films are negative and have tilted optic axes. For all compensation films, $\theta_L = 5.43°$

Variant	Scheme	Parameters		
6	P_1–C_1–C_2–LC–C_3–C_4–P_2	P_1	$\varphi_L = 11.70°$	
		C_1	$\varphi_L = 80.75°$	$d_L = 45.13\ \mu m$
		C_2	$\varphi_L = 27.9°$	$d_L = 18.17\ \mu m$
		C_3	$\varphi_L = -146.07°$	$d_L = 18.17\ \mu m$
		C_4	$\varphi_L = 165.26°$	$d_L = 45.13\ \mu m$
		P_2	$\varphi_L = -38.37°$	

References

[1] A. Lien, "Extended Jones matrix representation for the twisted nematic liquid crystal display at oblique incidence," *Appl. Phys. Lett.* **57**, 2767 (1990).

[2] A. Lien, "A detailed derivation of extended Jones matrix representation for twisted nematic liquid crystal displays," *Liq. Cryst.* **22**, 171 (1997).

[3] C. Gu and P. Yeh, "Extended Jones matrix method. II," *J. Opt. Soc. Am. A* **10**, 966 (1993).

[4] R. A. Chipman, "Mueller matrices," in *Handbook of Optics, Vol. 1. Geometrical and Physical Optics, Polarized Light, Components and Instruments*, edited by M. Bass and V. N. Mahajan, 3rd ed. (McGraw-Hill, New York, 2010).

[5] K. H. Yang and H. Takano, "Measurements of twisted nematic cell gap by spectral and split-beam interferometric methods," *J. Appl. Phys.* **67**, 5 (1990).

[6] A. Lien and H. Takano, "Cell gap measurement of filled twisted nematic liquid crystal displays by a phase compensation method," *J. Appl. Phys.* **69**, 1304 (1991).

[7] Y. Iimura, N. Kobayashi, and S. Kobayashi, "A new method for measuring the azimuthal anchoring energy of a nematic liquid crystal," *Jpn. J. Appl. Phys.* **33**, L434 (1994).

[8] D. A. Yakovlev and I. S. Lin'kova, "Symmetry of distribution of director and optical-polarization properties of a plane-parallel layer of liquid crystal," *Soviet Physics - Crystallography* **36**, 551 (1991).

[9] E. Polossat, I. Dozov, "New optical method for the measurement of the azimuthal anchoring energy of nematic liquid crystals," *Mol. Crys. Liq. Cryst.* **282**, 223 (1996).

[10] T. Akahane, H. Kaneko, and M. Kimura, "Novel method of measuring surface torsional anchoring strength of nematic liquid crystals," *Jpn. J. Appl. Phys.* **35**, 4434 (1996).

[11] E. L. Wood, G. W. Bradberry, P. S. Cann, and J. R. Sambles, "Determination of azimuthal anchoring energy in grating-aligned twisted nematic liquid-crystal layers," *J. Appl. Phys.* **82**, 2483 (1997).

[12] V. P. Vorflusev, H.-S. Kitzerow, and V. G. Chigrinov, "Azimuthal anchoring of liquid crystals at the surface of photo-induced anisotropic films," *Appl. Phys. A* **64**, 615 (1997).

[13] V. P. Vorflusev, H.-S. Kitzerow, and V. G. Chigrinov, "Azimuthal surface gliding of a nematic liquid crystal," *Appl. Phys. Lett.* **70**, 3359 (1997).

[14] Y. Zhou, Z. He, and S. Sato, "An improved Stokes parameter method for determination of the cell thickness and twist angle of twisted nematic liquid crystal cells," *Jpn. J. Appl. Phys.* **37**, 2567 (1998).

[15] S. T. Tang and H. S. Kwok, "Transmissive liquid crystal cell parameters measurement by spectroscopic ellipsometry," *J. Appl. Phys.* **89**, 80 (2001).

[16] J. G. Fonseca and Y. Galerne, "Simple method for measuring the azimuthal anchoring strength of nematic liquid crystals," *Appl. Phys. Lett.* **79**, 2910 (2001).

[17] S. Faetti, G. C. Mutinati, "Light transmission from a twisted nematic liquid crystal: accurate methods to measure the azimuthal anchoring energy," *Phys. Rev. E* **68**, 026601 (2003).

[18] J. S. Gwag, K.-H. Park, G.-D. Lee, T.-H. Yoon, and J. C. Kim, "Simple cell gap measurement method for twisted-nematic liquid crystal cells," *Jpn. J. Appl. Phys.* **43**, L30 (2004).

[19] J. S. Chae and S. G. Moon, "Cell parameter measurement of a twisted-nematic liquid crystal cell by the spectroscopic method," *J. Appl. Phys.* **95**, 3250 (2004).

[20] S.V. Pasechnik, V. G. Chigrinov, D. V. Shmeliova, V. A. Tsvetkov, V. N. Kremenetsky, Liu Zhijian, and A. V. Dubtsov, "Slow relaxation processes in nematic liquid crystals at weak surface anchoring," *Liq. Cryst.* **33**, 175 (2006).

[21] D. A. Yakovlev and V. G. Chigrinov, "A robust polarization-spectral method for determination of twisted liquid crystal layer parameters," *J. Appl. Phys.* **102**, 023510 (2007).

[22] T. Akahane, "The influence of multiple-beam interference in a liquid crystal cell on the determination of the optical retardation and the twist angle," *Jpn. J. Appl. Phys.* **37**, 3428 (1998).

[23] J. A. Nelder and R. Mead, "A simplex method for function minimization," *Comput. J.* **7**, 308 (1965).

[24] J. Li, S.-T. Wu, S. Brugioni, R. Meucci, and S. Faetti, "Infrared refractive indices of liquid crystals," *J. Appl. Phys.* **97**, 073501 (2005).

[25] D.-K. Yang and S.-T. Wu, *Fundamentals of Liquid Crystal Devices* (John Wiley & Sons, Ltd, Chichester, 2006).

[26] M. G. Robinson, J. Chen, and G. D. Sharp, *Polarization Engineering for LCD Projection* (John Wiley & Sons, Ltd, Chichester, 2005).

[27] *Mobile Displays: Technology and Applications*, edited by A. Bhowmik, Z. Li, and P. Bos (John Wiley & Sons, Ltd, Chichester, 2008).

[28] S. Stallinga, J. M. A. van den Eerenbeemd, and J. A. M. M. van Haaren, "Fundamental limitations to the viewing angle of liquid crystal displays with birefringent compensators," *Jpn. J. Appl. Phys.* **37**, 560 (1998).

13

A Few Words About Modeling of Fine-Structure LCDs and the Direct Ray Approximation

The modeling of LC displays operating in the in-plane switching (IPS) mode, fringe-field switching (FFS) mode, patterned vertical alignment (PVA) mode, multi-domain vertical alignment (MVA) mode, and other modes that use patterned intra-pixel electrode structures or/and intra-pixel multi-domain surface alignment necessarily involves calculation of transmission characteristics for a pixel within which the LC director field is inhomogeneous in two or three dimensions and has significant lateral variations at distances of the order of optical wavelengths. The latter circumstance makes diffraction on the transverse inhomogeneities of the LC layer an important factor and greatly complicates the rigorous optical simulation. There are several electromagnetic methods that, in certain situations, can provide reliable simulation of light propagation in media with such kind of inhomogeneity. The best known of them are the grating method (GM) [1,2], the finite-difference time-domain method [3–7], the wide-angle beam propagation method [8–10], the reduced-order grating method [10–12], and the finite difference in frequency domain (FDFD) method [12] (the given references are to publications where the mentioned methods are applied to LC structures). Discussion of these methods is outside the scope of this book. Their potentialities in LCD modeling are reviewed in Reference 12.

In many cases, the listed methods are capable of giving sufficiently accurate solutions of corresponding electromagnetic problems. However, even the most efficient of these methods, the reduced-order grating method, in standard situations is computationally very expensive. Because of this, in solving most optimization problems for fine-structure LCDs fast approximate methods are used which are based on the optical methods developed for 1D-inhomogeneous media. These "one-dimensional" methods are adapted for the modeling of 2D- and 3D-inhomogeneous structures with invoking a set of heuristic assumptions referred to in this book as the direct ray approximation (DRA) (see Section 7.1). Because of the heuristic nature of DRA, the question of applicability and accuracy of this approximation is very acute and should be considered individually for each specific case. Rigorous modeling is an appropriate means for finding the right answer to this question. In this chapter, we will illustrate this statement by some examples.

In addition to our basic modeling program MOUSE-LCD (see Chapter 4) we developed a program MOUSE-LCD2D [13] which enables modeling the electro-optical performance of fine-structure LCDs with due regard for the diffraction effects. Moreover, this program allows simulation of various optical

Modeling and Optimization of LCD Optical Performance, First Edition.
Dmitry A. Yakovlev, Vladimir G. Chigrinov and Hoi-Sing Kwok.

Website Companion: www.wiley.com/go/yakovlev/modelinglcd

experiments with fine-structure LC cells. In particular, simulation of measurements carried out with a polarizing microscope can be performed.

In MOUSE-LCD2D, we employ the flexible adding technique (see Section 7.2.2) within the grating method framework [1, 2] (see also References 14 and 15) for rigorous optical calculations. The laterally inhomogeneous domain of the LC panel under consideration, the domain including the LC layer and adjacent thin films (ITO, alignment layers), is considered as a medium whose local optical parameters are independent of one of the transverse coordinates (*Y*) and are periodic functions of the other transverse coordinate (*X*). A large-dimension transmission matrix characterizing this 2D-inhomogeneous domain is calculated which relates the scalar complex amplitudes of plane-wave components of the incident and transmitted fields. From this matrix, output optical characteristics are calculated using methods of Fourier optics and the theory of coherency [16–18]. Some tools of MOUSE-LCD2D offer the comparison of results obtained by the grating method with those obtained by using DRA. For the DRA calculations, the EW Jones matrix method, described in Chapters 8 and 11, is used. The vector $\mathbf{I}_a$ (see the description of DRA in Section 7.1) is calculated by formula (7.61).

The LC director field configuration is calculated in MOUSE-LCD2D by using the successive relaxation method with usual division of each iteration step into two stages. The first stage is the calculation of the electric potential distribution at a given LC director configuration. The second stage is the calculation of partial relaxation of the LC director field in the electric field calculated in the first stage. The "elastic" part of the problem is solved with the help of a finite-difference method. The electric field distributions are calculated using a finite-element method. Finite surface anchoring can be taken into account; both symmetric and asymmetric boundary conditions can be considered.

13.1 Virtual Microscope

One of the tools included in MOUSE-LCD2D is Virtual Microscope (VM). The microscopic image of an inhomogeneous LC layer with a fine transverse structure strongly depends on observation conditions, in particular, on the aperture of the condenser of the illuminating system and the aperture of the microscope objective (Figure 13.1). Virtual Microscope simulates microscopic images of LC layers that could be observed with a polarizing microscope under given observation conditions. The observation conditions can be changed by the user.

Wishing to evaluate quantitatively some parameters of an inhomogeneous LC layer by using microphotometry or microspectrometry, as a rule, we must first estimate the applicability of DRA in the situation at hand because it is DRA that gives relatively simple formulas for optical characteristics of inhomogeneous LC layers in terms of their parameters, formulas suitable for solving inverse problems. Virtual

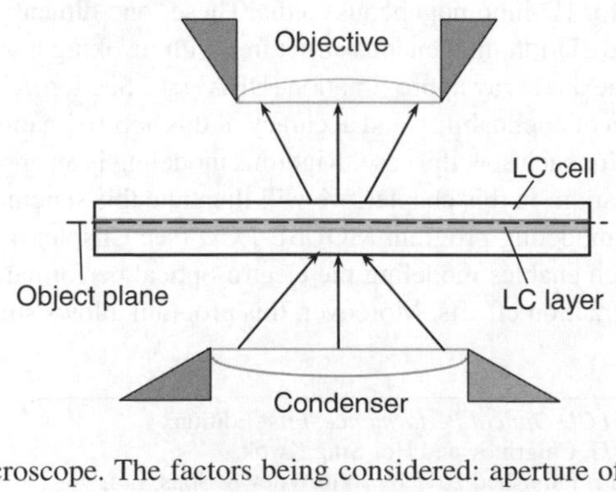

Figure 13.1 Virtual Microscope. The factors being considered: aperture of the condenser of the illuminating system of the microscope, aperture of the objective, object plane position

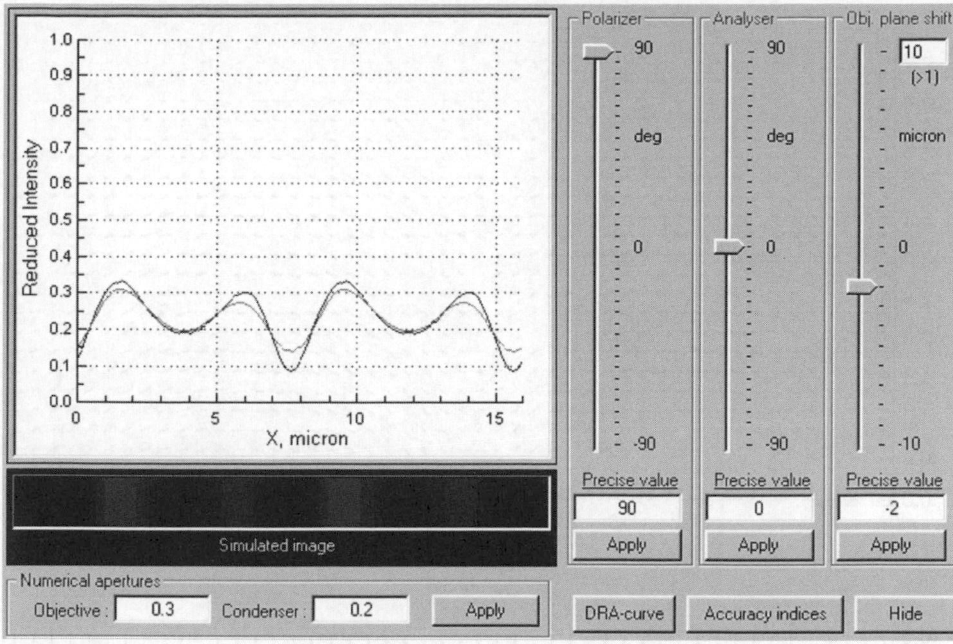

Figure 13.2 Dialog box **Virtual Microscope**. The plot shows the calculated dependences of the reduced intensity of the image and the normal DRA transmittance of the sample on the lateral coordinate *X*. One can rotate the polarizer and analyzer as well as move the object plane, thereby scanning the sample in depth

Microscope helps to ascertain whether DRA is applicable or not and find the experimental conditions under which DRA can be used.

Figure 13.2 shows a dialog box **Virtual Microscope** of the program. The plot on this box shows dependences of a reduced intensity of the image of the sample and its normal DRA transmittance on the lateral coordinate (*X*). The reduced intensity is defined here as the ratio of the intensity of the sample image to the intensity of the empty field image (when the sample, polarizer, and analyzer are removed) at a given point of the image plane; this quantity can easily be measured in a real experiment. The normal DRA transmittance, calculated by the DRA method, is an idealized, unaffected by diffraction, local transmittance of the polarizer–sample–analyzer system for normally incident unpolarized light.

Let us give some examples demonstrating capabilities of the virtual microscope. In these examples, we deal with two model LC cells, we call them IPSS cell and IPSVA cell, each having a 6.1-μm-thick LC layer. The electrode pattern period in both cells is equal to 8 μm; the electrode width is equal to 3 μm. IPSS cell in the field-off state has a homogeneous LC director field with 88° azimuthal angle (with respect to the *X*-axis) and 2° tilt angle of the LC director. IPSVA cell in the field-off state has a homeotropic alignment of the LC. The following parameters were used in the calculations. IPSS cell: for the liquid crystal (a pure nematic), elastic constants $K_{11} = 1.32 \times 10^{-6}$ dyn, $K_{22} = 6.5 \times 10^{-7}$ dyn, and $K_{33} = 1.38 \times 10^{-6}$ dyn, low-frequency principal permittivities $\varepsilon_{\parallel} = 8.3$ and $\varepsilon_{\perp} = 3.1$; voltage between neighboring electrodes in the field-on state $U_{\mathrm{ON}} = 6$ V. IPSVA cell: $K_{11} = 1.32 \times 10^{-6}$ dyn, $K_{22} = 7.1 \times 10^{-7}$ dyn, $K_{33} = 1.95 \times 10^{-6}$ dyn, $\varepsilon_{\parallel} = 5.1$, $\varepsilon_{\perp} = 3.8$; $U_{\mathrm{ON}} = 8$ V. The calculated director profiles for the field-on states of the cells are shown in Figure 13.3. Optical calculations were performed for quasimonochromatic light with mean wavelength 550 nm. Refractive indices: liquid crystals—$n_{\parallel} = 1.581$, $n_{\perp} = 1.482$; ITO layers—$n_{\mathrm{ITO}} = 2.05$; alignment layers—$n_{\mathrm{AL}} = 1.6$; glass substrates—$n_{\mathrm{G}} = 1.52$. The electrode thickness

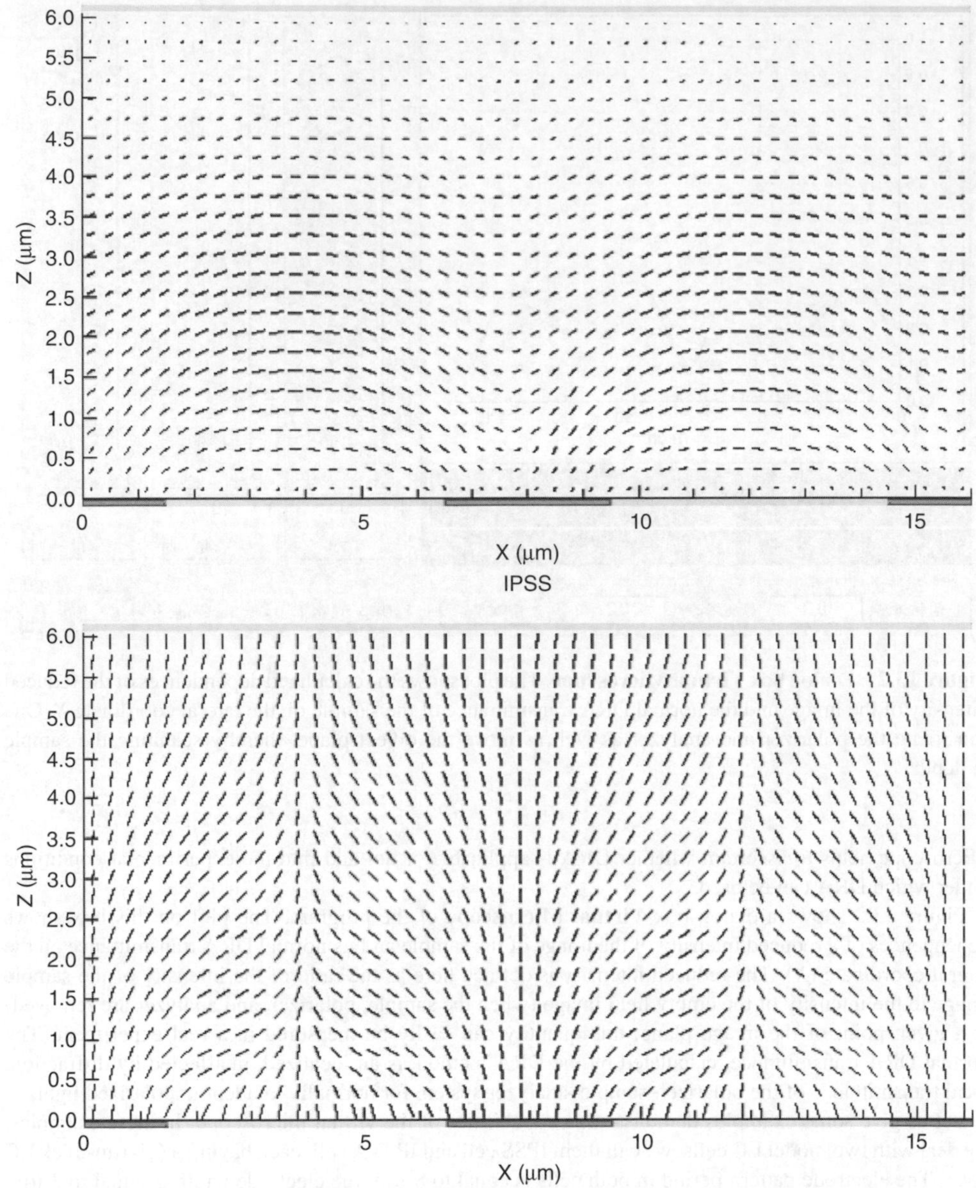

Figure 13.3 Calculated director profiles (the field-on states) and electrode configurations in IPSS and IPSVA cells

was taken to be 0.03 μm. The thickness of alignment layers is equal to 0.1 μm over electrodes and to 0.13 μm between electrodes.

In Figure 13.4 we show results for the simplest situation when the LC director is aligned uniformly throughout the layer and the electrode pattern is the only factor that makes the LC cell transversally inhomogeneous. The calculations were carried out for IPSS cell in the field-off state; the polarizer and

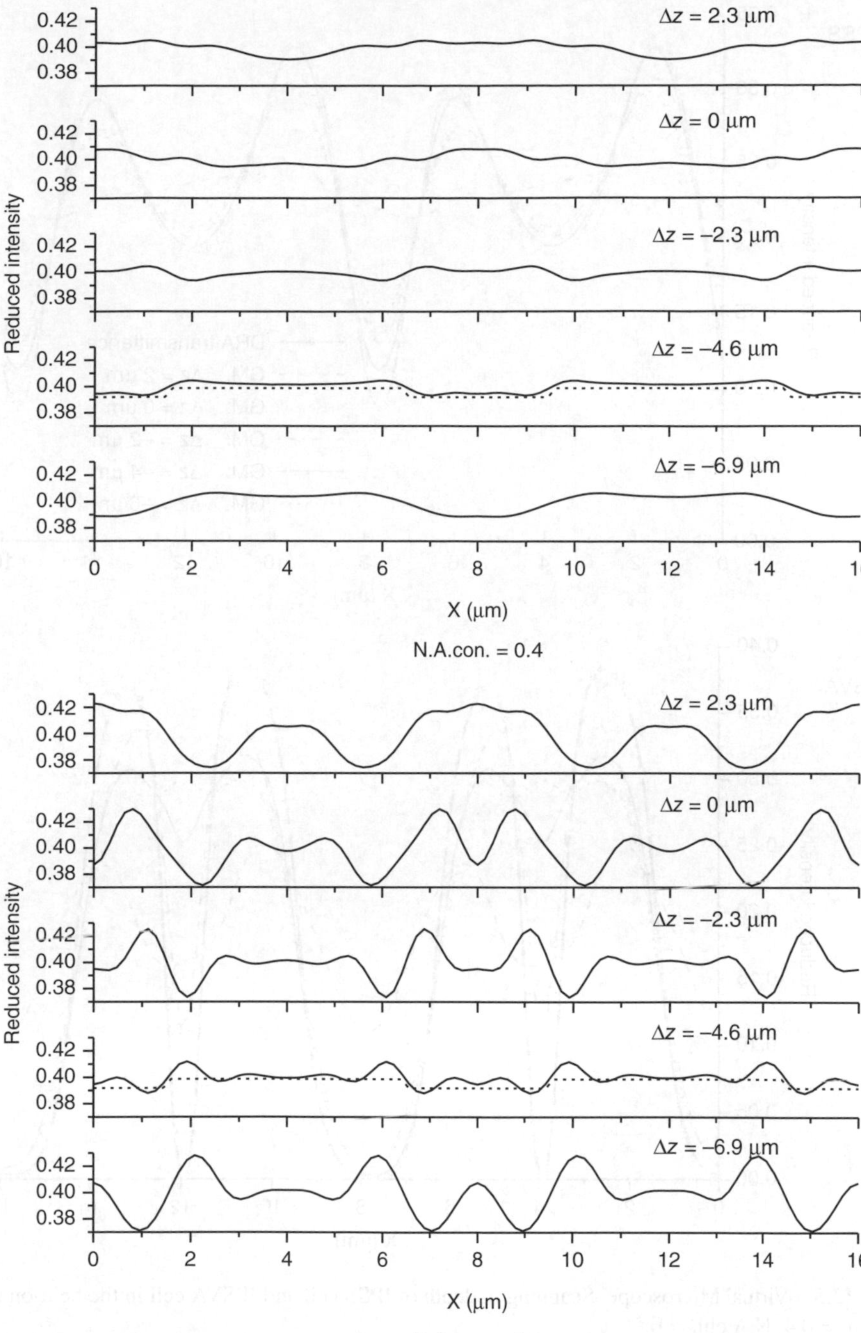

Figure 13.4 Virtual Microscope. Scanning in depth of IPSS cell in the field-off state at different numerical apertures of the condenser, N.A.con. The numerical aperture of the objective, N.A.obj., is equal to 0.5. With N.A.con. = 0.4, if the object plane is positioned so that $\Delta z = -4.6$ μm, we can see a relatively sharp image of the electrodes. With these settings, the curve of the reduced intensity is very close to the DRA transmittance curve (dotted line). Decreasing N.A.con. blurs the image

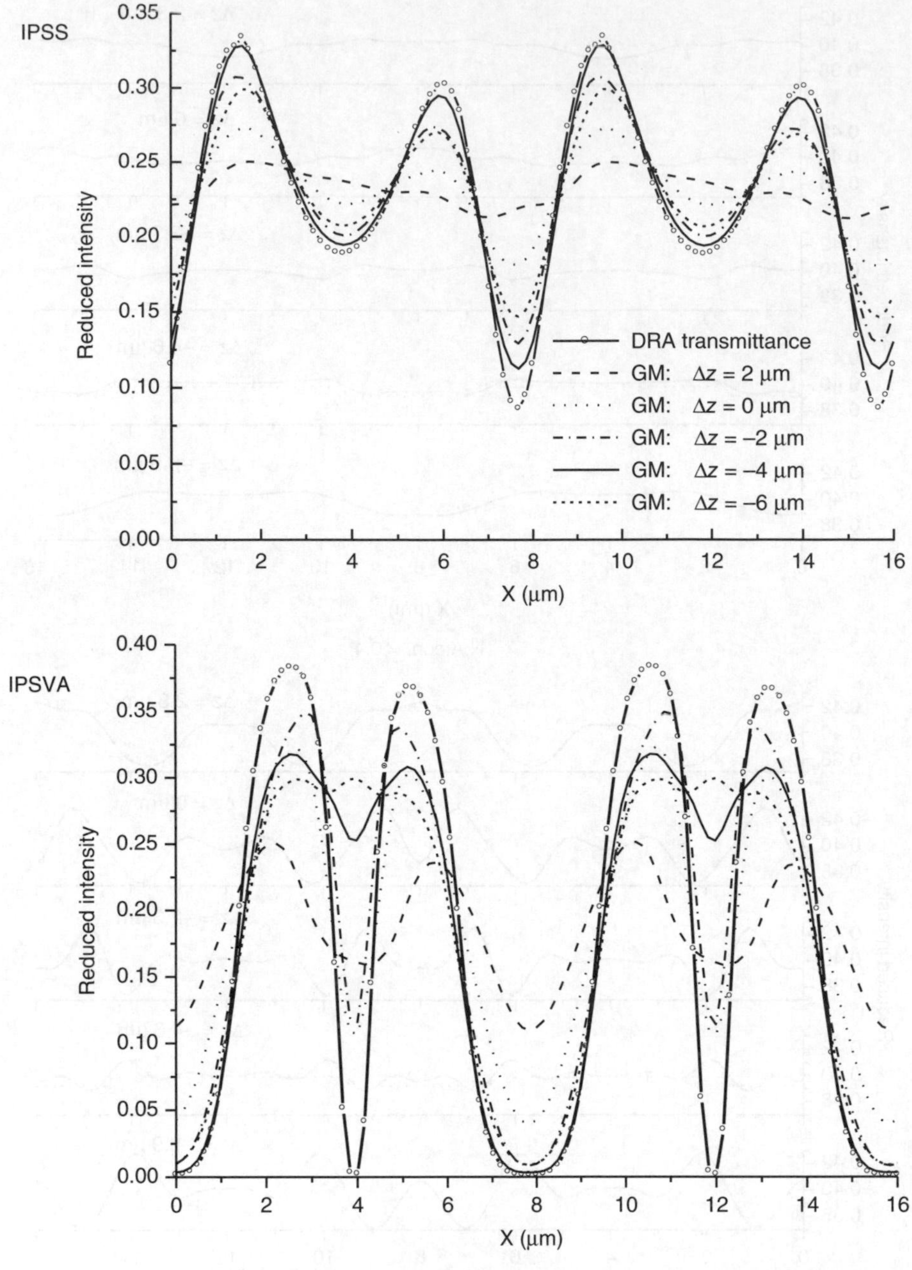

Figure 13.5 Virtual Microscope. Scanning in depth of IPSS cell and IPSVA cell in the field-on states; N.A.con. = 0.4, N.A.obj. = 0.5

analyzer transmission axes are parallel to the *X*-axis. Figure 13.4 demonstrates that the virtual microscope indeed operates like a real one. Moving the object plane we can bring the electrodes, which are situated on the bottom substrate (Figure 13.3), into focus. The parameter Δz in Figure 13.4 is the difference of the current value of the working distance (D_C)—here the working distance is the distance between the LC cell and the microscope objective—and the value of the working distance at which the microscope is focused at the lower boundary of the upper glass substrate of the LC cell (D_0); that is, $\Delta z = D_C - D_0$. As can be seen from Figure 13.4, a sharp image can be obtained only if the numerical apertures of the condenser and objective are sufficiently large, just as for a real microscope.

Figure 13.5 presents the results for IPSS and IPSVA cells in the field-on states (Figure 13.3), when the LC layers have strong lateral variation of the LC director field. In both cases, the polarizer and analyzer are crossed. In the case of IPSS cell, the analyzer is parallel to the *X*-axis. For IPSVA cell, the analyzer transmission axis is at 45° to the *X*-axis. The figure illustrates a search for the position of the object plane for which the curve of the reduced intensity (recall that it is a measurable quantity) most closely approaches the DRA transmittance curve.

13.2 Directional Illumination and Diffuse Illumination

A transversally inhomogeneous LC layer scatters light. When such an LC layer is illuminated by a collimated light beam, it redirects a part of the transmitted light in directions different from the propagation direction of the incident light (Figure 13.6a). On the other hand, being diffusely illuminated

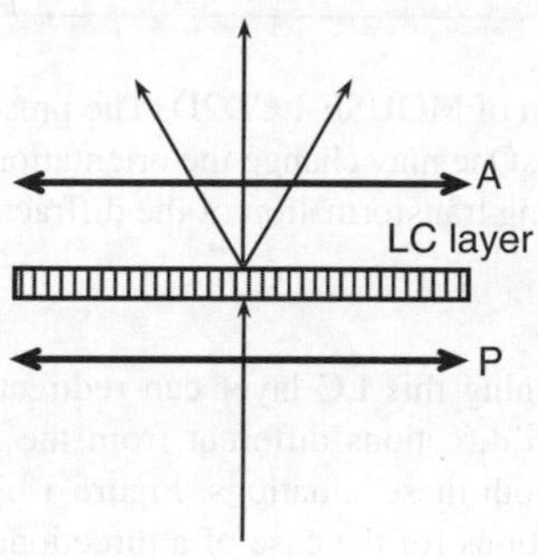

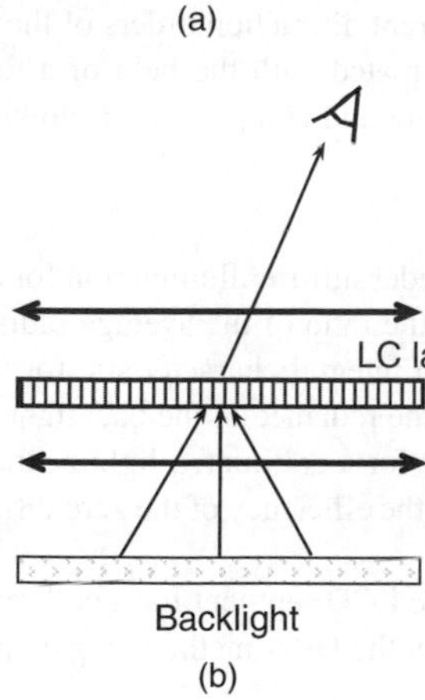

Figure 13.6 Manifestations of light scattering on an LC layer under directional (a) and diffuse (b) illumination. Simulated experiments. P and A are polarizers

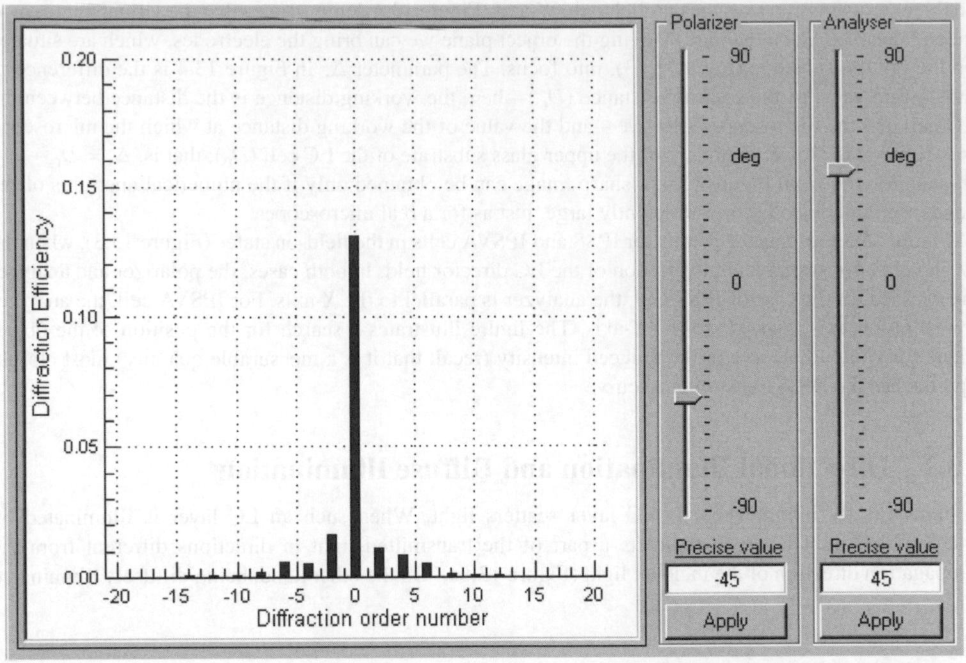

Figure 13.7 Dialog box **Diffraction** of MOUSE-LCD2D. The presented results are for IPSVA cell in the field-on state at normal incidence. One may change the orientation of the polarizer and analyzer and immediately observe the corresponding transformation of the diffraction pattern

by a backlight, the LC panel containing this LC layer can redirect toward the viewer some portion of the light incident on the panel in directions different from the viewing direction (Figure 13.6b). MOUSE-LCD2D helps to analyze both these situations. Figure 13.7 shows a dialog box **Diffraction** which presents the results of calculations for the case of a directional (collimated-beam) illumination. This box shows the efficiencies of different diffraction orders of the transmitted light.

The case of diffuse illumination is treated with the help of a tool called IME (Illumination Mode Effect). By using IME, one can calculate and compare the following characteristics of a transmissive LCD:

(i) Reduced brightness of the LCD under diffuse illumination for a given viewing direction. The term "reduced brightness" means here the ratio of the average radiance of the light emerging from the LCD in a given direction—it is an integral characteristic for the whole 2D-inhomogeneous LCD segment under consideration—to the radiance of the backlight for this direction.
(ii) Regular transmittance of the LCD for a collimated light beam incident on it in a given direction. This transmittance is calculated as the efficiency of the zero diffraction order of the light transmitted by the LCD.
(iii) Area-averaged transmittance of the LCD segment for a collimated beam incident on the LCD in a given direction that is calculated by the DRA method, neglecting diffraction.

The results of application of IME to IPSS and IPSVA LCDs are shown in Figure 13.8. In the case of the IPSVA LCD, the polarizers are oriented as in the above example illustrated by Figure 13.5. In the

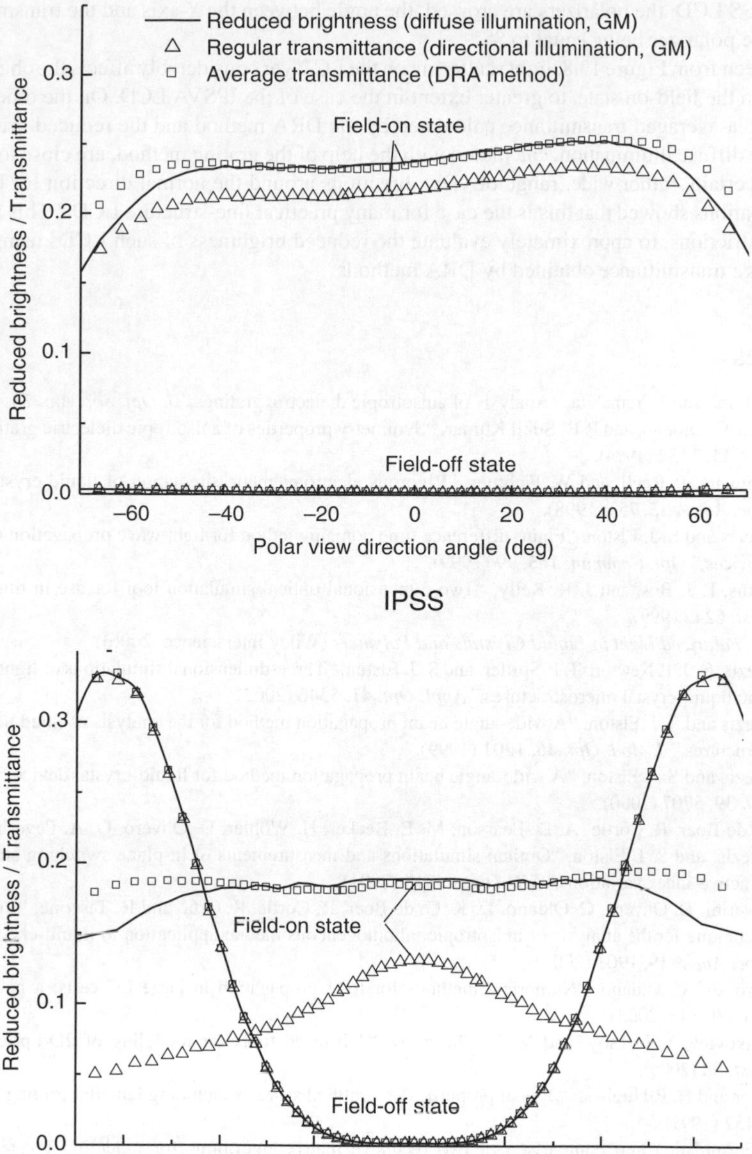

Figure 13.8 Viewing characteristics of the IPSS and IPSVA LCDs for directional and diffuse illumination, simulated using IME. The viewing plane is perpendicular to the *Y*-axis

case of the IPSS LCD, the polarizers are crossed, the angle between the X-axis and the transmission axis of the entrance polarizer being equal to 88°.

As can be seen from Figure 13.8, light scattering on the LC layer considerably affects the characteristics of the LCDs in the field-on state, to greater extent in the case of the IPSVA LCD. On the other hand, we see that the area-averaged transmittance calculated by the DRA method and the reduced brightness for the case of the diffuse illumination, calculated with the help of the grating method, are close to each other in value in a certain, rather wide, range of view directions around the normal direction for both model LCDs. Calculations showed that this is the case for many practical fine-structure LCDs. This allows one, with some restrictions, to approximately evaluate the reduced brightness of such LCDs using estimates of their average transmittance obtained by DRA methods.

References

[1] K. Rokushima and J. Yamakita, "Analysis of anisotropic dielectric gratings," *J. Opt. Soc. Am.* **73**, 901 (1983).

[2] P. Galatola, C. Oldano, and P. B. Sunil Kumar, "Symmetry properties of anisotropic dielectric gratings," *J. Opt. Soc. Am. A* **11**, 1332 (1994).

[3] B. Witzigmann, P. Regli, and W. Fichtner, "Rigorous electromagnetic simulation of liquid crystal displays," *J. Opt. Soc. Am. A* **15**, 753 (1998).

[4] E. E. Kriezis and S. J. Elston, "Finite-difference time domain method for light wave propagation within liquid crystal devices," *Opt. Commun.* **165**, 99 (1999).

[5] C. M. Titus, P. J. Bos, and J. R. Kelly, "Two-dimensional optical simulation tool for use in microdisplays," *SID Digest*, 624 (1999).

[6] T. Scharf, *Polarized Light in Liquid Crystals and Polymers* (Wiley Interscience, 2006).

[7] E. E. Kriezis, C. J. P. Newton, T. P. Spiller, and S. J. Elston, "Three-dimensional simulations of light propagation in periodic liquid-crystal microstructures," *Appl. Opt.* **41**, 5346 (2002).

[8] E. E. Kriezis and S. J. Elston, "A wide angle beam propagation method for the analysis of tilted nematic liquid crystal structures," *J. Mod. Opt.* **46**, 1201 (1999).

[9] E. E. Kriezis and S. J. Elston, "A wide-angle beam propagation method for liquid-crystal device calculations," *Appl. Opt.* **39**, 5707 (2000).

[10] D. K. G. de Boer, R. Cortie, A. D. Pearson, M. E. Becker, H. Wöhler, D. Olivero, O. A. Peverini, K. Neyts, E. E. Kriezis, and S. J. Elston, "Optical simulations and measurements of in-plane switching structures with rapid refractive-index variations," *SID Digest*, 818 (2001).

[11] O. A. Peverini, D. Olivero, C. Oldano, D. K. G. de Boer, R. Cortie, R. Orta, and R. Tascone, "Reduced-order model technique for the analysis of anisotropic inhomogeneous media: application to liquid-crystal displays," *J. Opt. Soc. Am. A* **19**, 1901 (2002).

[12] D. Olivero and C. Oldano, "Numerical methods for light propagation in large LC cells: a new approach," *Liq. Cryst.* **30**, 345 (2003).

[13] D. A. Yakovlev, V. I. Tsoy, and V. G. Chigrinov, "Advanced tools for modeling of 2D-optics of LCDs," *SID Digest*, 58 (2005).

[14] C. D. Ager and H. P. Hughes, "Optical properties of stratified systems including lamellar gratings," *Phys. Rev. B* **44**, 13452 (1991).

[15] L. Li, "Formulation and comparison of two recursive matrix algorithms for modeling layered diffraction gratings," *J. Opt. Soc. Am. A* **13**, 1024 (1996).

[16] J. W. Goodman, *Introduction to Fourier Optics*, 2nd ed. (McGraw-Hill, New York, 1996).

[17] J. W. Goodman, *Statistical Optics* (Wiley, New York, 2000).

[18] G. Gallatin, J. C. Webster, E. C. Kintner, and C. J. Morgan, "Scattering matrices for imaging layered media," *J. Opt. Soc. Am. A* **5**, 220 (1988).

Appendix A

LCD Modeling Software MOUSE-LCD Used for the HKUST Students Final Year Projects (FYP) from 2003 to 2011

A.1 Introductory Remarks

This Appendix provides several examples of solving research and optimization problems for LCDs and serves as an illustration of the potential of MOUSE-LCD modeling software. The calculations were performed by students of the Hong Kong University of Science and Technology (HKUST) for their final-year projects from 2003 to 2011.

A.2 Fast LCD

A.2.1 TN Cell

First, the LCD parameters are fixed as in Table A.1 and the results are provided in Figs. 1-7 and Tables A.3–A.9. [1].

Input polarizer = 0°, output polarizer = 90°, $V_{on} = 0$ V, $V_{off} = 5$ V, $d = 4.8$ μm and hence $d\Delta n = 0.475$ μm. $d\Delta n = 0.475$ μm is chosen because this is the first maximum in green light for TN cell. This can be derived by the following equation:

$$\frac{d\Delta n}{\lambda} = \frac{\sqrt{3}}{2}, \quad \text{where } \lambda = 550 \text{ nm}$$

The following three equations are used to explain the results:

$$\tau_{on} = \frac{\gamma_1^*}{\Delta\varepsilon \frac{E^2}{4\pi} - \tilde{K}\frac{\pi^2}{d^2}} \qquad \tau_{off} = \frac{\gamma_1^* d^2}{\tilde{K}\pi^2} \qquad \tau_{total} = \tau_{on} + \tau_{off}, \tag{A.1}$$

Modeling and Optimization of LCD Optical Performance, First Edition.
Dmitry A. Yakovlev, Vladimir G. Chigrinov and Hoi-Sing Kwok.

Website Companion: www.wiley.com/go/yakovlev/modelinglcd

Table A.1 Parameters of TN LC cell [1]

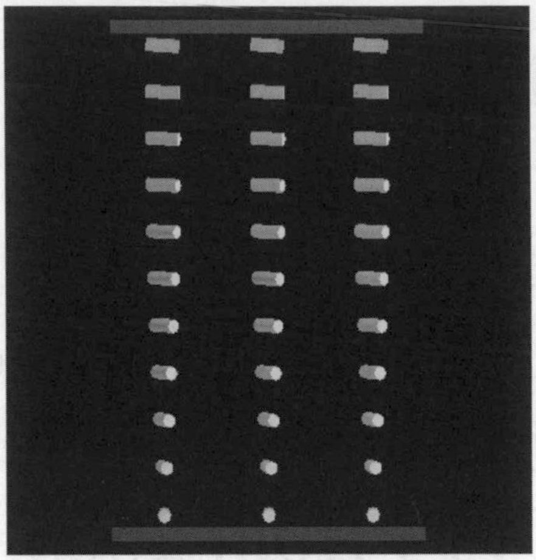

Wavelength, nm	$n_{\|\|}$	$n_{\perp}$	Δn
436	1.5987	1.4939	0.1048
550	1.5809	1.4819	0.099
633	1.5734	1.4774	0.096

	Parameter	Value	Unit
1	K_{11}	1.28E-06	dyne
2	K_{22}	0.0000007	dyne
3	K_{33}	1.95E-06	dyne
4	$\varepsilon_{\|\|}$	10.1	
5	$\varepsilon_{\perp}$	3.8	
6	γ_1	0.15	CGS unit

	Parameter	Value	Unit
1	Pretilt angle 1 (1st boundary)	4	degree
2	Pretilt angle 2 (2nd boundary)	4	degree
3	Twist angle	90	degree
4	Natural helix pitch P	20.86087	mcm
5	L/P ratio	0.23	
6	Anchoring strength 1 (1st boundary)	1	Erg/cm^2
7	Anchoring strength 2 (2nd boundary)	1	Erg/cm^2

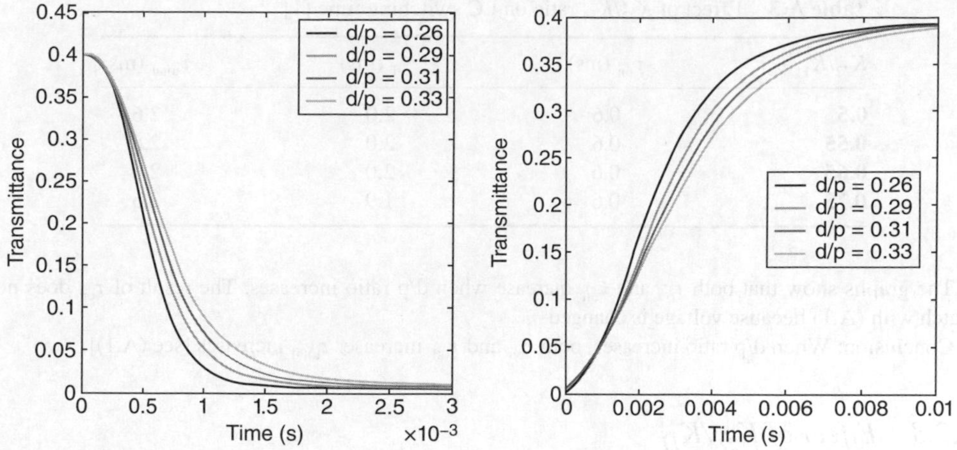

Figure A.1 Left: switch-on times for different d/p ratios; right: switch-off times for different d/p ratios [1]

where τ_{on} is the response time in on-state, τ_{off} is the response time in off-state, τ_{total} is the total response time, γ_1^* is the effective rotational viscosity, d is the thickness, $\Delta\varepsilon$ is the dielectric anisotropy, E is the energy, $\tilde{K}$ is function of K_{11}, K_{22} and K_{33}, K_{11} is splay distortion elastic constant, K_{22} is twist distortion elastic constant and K_{33} is bend distortion elastic constant.

$\tilde{K}$ is function of K_{11}, K_{22}, and K_{33} because it involves splay, twist, and bend effect.

The following parameters are considered for simulation:

1. d/p ratio (d, LC cell thickness; p, equilibrium helix pitch)
2. K_{22}/K_{11}
3. K_{33}/K_{11}
4. $\Delta\varepsilon$
5. γ_1
6. anchoring strength W

A.2.2 Effect of d/p Ratio

All values of LC parameters are fixed and changed d/p ratio only to see the effect on response time (Table A.2 [1]). For 90° TN, the d/p ratio is between 0 and 0.5. V_{on} is changed and V_{off} is fixed at 5 V.

Table A.2 Effect of d/p ratio on LC switching time [1]

d/p ratio	V_{on}	τ_{on} (ms)	τ_{off} (ms)	τ_{total} (ms)
0.26	1.76	0.69	4.62	5.31
0.29	1.91	0.77	5.08	5.85
0.31	2.0	0.92	5.08	6.0
0.33	2.09	1.0	6.15	7.15

Table A.3 Effect of K_{22}/K_{11} ratio on LC switching time [1]

K_{22}/K_{11}	τ_{on} (ms)	τ_{off} (ms)	τ_{total} (ms)
0.5	0.6	2.0	2.6
0.55	0.6	2.0	2.6
0.65	0.6	2.0	2.6
0.75	0.6	1.9	2.5

The graphs show that both τ_{on} and τ_{off} increase when d/p ratio increases. The result of τ_{on} does not match with (A.1) because voltage is changed.

Conclusion: When d/p ratio increases, both τ_{on} and τ_{off} increase. τ_{total} increases [see (A.1)].

A.2.3 Effect of K_{22}/K_{11}

All values of LC parameters are fixed and K_{22}/K_{11} ratio is changed only to see the effect on response time (Table A.3) [1].

The graphs show that K_{22}/K_{11} have not much effect on τ_{on}. This is because when voltage applied is much larger so that $\frac{|\Delta\varepsilon|E^2}{4\pi} \gg \frac{\tilde{K}\pi^2}{d^2}$ [refer to (A.1)], effect of K_{22}/K_{11} is not dominant. When K_{22}/K_{11} increases, τ_{off} decreases. The result is matched with (A.1).

Conclusion: As K_{22}/K_{11} ratio increases, τ_{off} decreases. So, τ_{total} decreases.

A.2.4 Effect of K_{33}/K_{11}

All values of LC parameters are fixed and K_{33}/K_{11} ratio is changed only to see the effect on response time (Table A.4) [1].

The graphs show that when K_{33}/K_{11} ratio increases, τ_{on} increases and τ_{off} decreases. This is because when K_{33} is larger, more energy would be consumed in order to bend LC molecules inside TN cell.

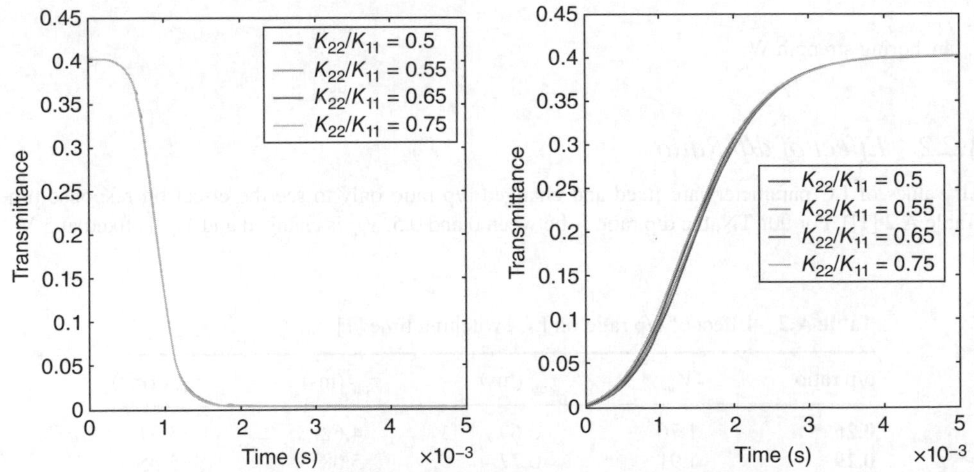

Figure A.2 Left: switch-on times for different K_{22}/K_{11} ratios; right: switch-off times for different K_{22}/K_{11} ratios [1]

Table A.4 Effect of K_{33}/K_{11} ratio on LC switching time [1]

K_{33}/K_{11}	τ_{on} (ms)	τ_{off} (ms)	τ_{total} (ms)
0.7	0.46	3.00	3.46
1.3	0.46	2.08	2.54
1.9	0.54	1.69	2.23
2.6	0.54	1.38	1.92

Therefore, τ_{on} increases. However, there is no much energy needed to make LC molecule return to initial position. Therefore, τ_{off} decreases.

Conclusion: As K_{33}/K_{11} ratio increases, τ_{on} increases and τ_{off} decreases.τ_{total} decreases.

A.2.5 Effect of $\Delta\varepsilon$

All values of LC parameters are fixed and $\Delta\varepsilon$ is changed only to see the effect on response time (Table A.3) [1].

The Table A.5 [1] shows that τ_{on} decreases when $\Delta\varepsilon$ increases. τ_{off} does not depend on $\Delta\varepsilon$, so there is no much effect on it. These results match with (A.1).

Conclusion: $\Delta\varepsilon$ increases, τ_{on} decreases. So τ_{total} decreases.

A.2.6 Effect of γ_1

As $\gamma_1^* = \gamma_1 = \alpha_3 - \alpha_2$ in this case, γ_1 is tested. All values of LC parameters are fixed and γ_1 is changed only to see the effect on response time (Table A.6) [1].

The graphs show that both τ_{on} and τ_{off} increase when γ_1 increases. This is because γ_1 is proportional to τ_{on} and τ_{off}.

Conclusion: When γ_1 increases, both τ_{on} and τ_{off} increase. So, τ_{total} increases.

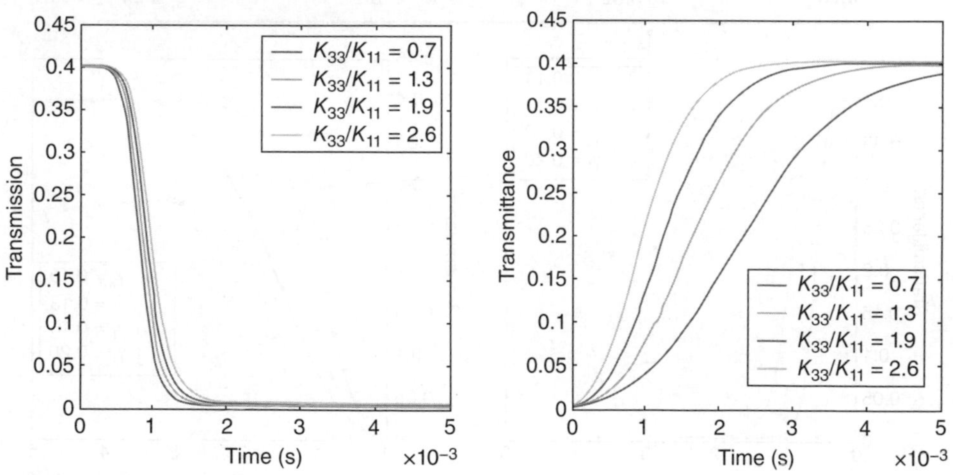

Figure A.3 Left: switch-on times for different K_{33}/K_{11} ratios; right: switch-off times for different K_{33}/K_{11} ratios [1]

Table A.5 Effect of $\Delta\varepsilon$ on LC switching time [1]

$\Delta\varepsilon$	τ_{on} (ms)	τ_{off} (ms)	τ_{total} (ms)
5.3	0.62	2.0	2.62
7.3	0.38	2.0	2.38
9.3	0.35	2.0	2.35
11.3	0.35	2.0	2.35

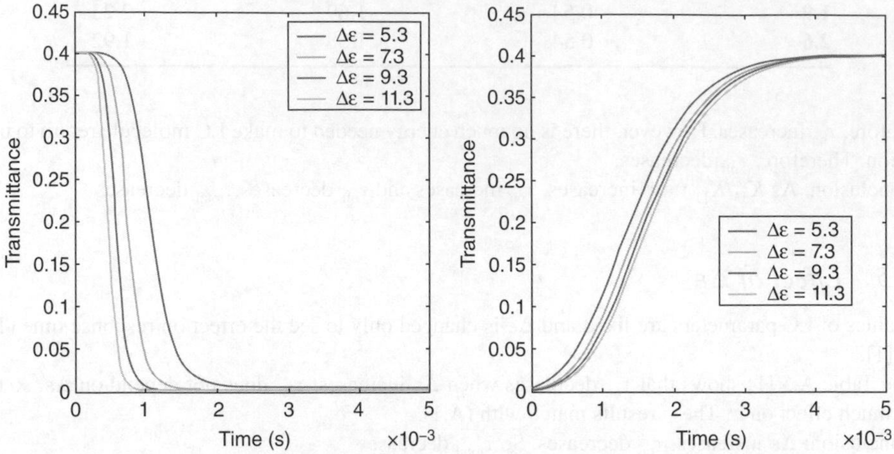

Figure A.4 Left: switch-on times for different $\Delta\varepsilon$ values; right: switch-off times for different $\Delta\varepsilon$ values [1]

Table A.6 Effect of γ_1 on LC switching time [1]

γ_1	τ_{on} (ms)	τ_{off} (ms)	τ_{total} (ms)
0.10	0.35	1.23	1.58
0.13	0.38	1.62	2.00
0.16	0.54	2.00	2.54
0.20	0.62	2.62	3.24

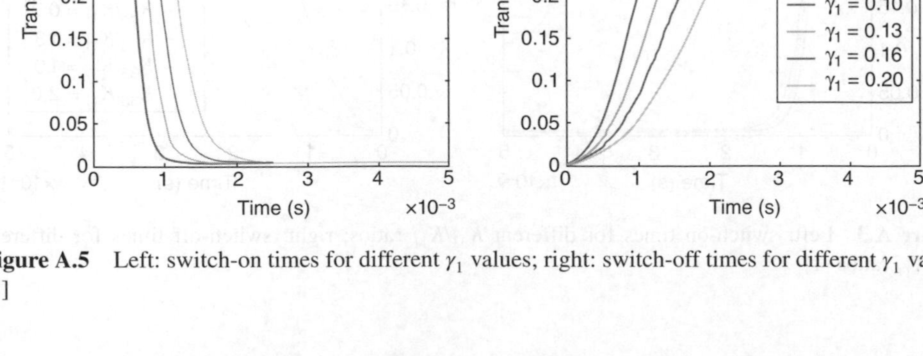

Figure A.5 Left: switch-on times for different γ_1 values; right: switch-off times for different γ_1 values [1]

Table A.7 Effect of W on LC switching time [1]

W (Erg/cm^2)	τ_{on} (ms)	τ_{off} (ms)	τ_{total} (ms)
0.03	0.31	2.54	2.85
0.05	0.35	2.23	2.58
0.08	0.38	2.15	2.53
0.1	0.38	2.00	2.38
0.3	0.38	2.00	2.38

A.2.7 Effect of Anchoring Strength W

All values of LC parameters are fixed and anchoring energy is changed only to see the effect on response time (Table A.7) [1].

When $\frac{Wd}{K} \ll 1$ where W is the anchoring strength,

$$\tau_{on} = \frac{\gamma_1}{\Delta\varepsilon\frac{E^2}{4\pi} - \frac{2W}{d}} \quad \text{and} \quad \tau_{off} = \frac{\gamma_1 d}{2W}. \tag{A.2}$$

From the equations, when W increases, τ_{on} increases and τ_{off} decreases.

Conclusion: When anchoring strength increases, τ_{on} increases and τ_{off} decreases. τ_{total} decreases as a result.

A.2.8 Optimized TN Cell With Fast Response Time

From the above result, TN cell is optimized (Table A.8) [1].

Conclusion: When K_{22}, K_{33} and $\Delta\varepsilon$ are increased, the total response time is 1.76 ms. For TN cell, the effect of d/p ratio, γ_1, and K_{33} are the highest.

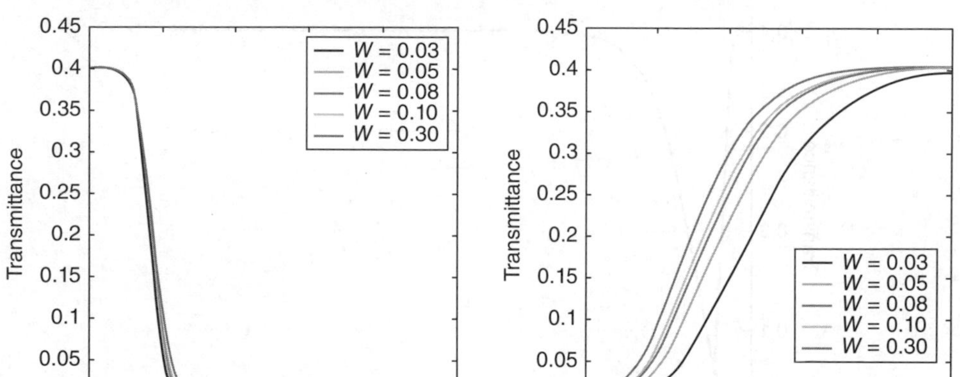

Figure A.6 Left: switch-on times for different W values; right: switch-off times for different W values [1]

Table A.8 Parameters of optimized TN cell with fast switching time [1]

Parameter	Value	Unit
K_{11}	1.28E-06	dyne
K_{22}	9.6E-07	dyne
K_{33}	3.328E-06	dyne
$\varepsilon_{\parallel}$	15.1	
$\varepsilon_{\perp}$	3.8	
γ_1	0.15	CGS unit

Parameter	Value	Unit
Pretilt angle 1 (1st boundary)	4	degree
Pretilt angle 2 (2nd boundary)	4	degree
Twist angle	90	degree
Natural helix pitch P	20.86087	mcm
L/P ratio	0.23	
Anchoring strength 1 (1st boundary)	1	Erg/cm^2
Anchoring strength 2 (2nd boundary)	1	Erg/cm^2

A.2.9 Other LC Modes

Finally based on the results of the calculations of various LC electro-optical modes, the following final results were obtained. The key parameters, which affected LC response time, were found to be γ_1, W, and K_{33}/K_{11}. The best results are summarized in Table A.9.

A.3 Color LCD [2]

A.3.1 The Super-Twisted Nematic Cell

The typical parameters of super-twisted nematic (STN) LC cell are given in Table A.10 [2, 3].

Chromatic to chromatic switching of the six electro-optic LC modes:

(a) TN (Twisted Nematic);
(b) STN (Super-Twisted Nematic);
(c) π cell (Bend Cell);
(d) ECB (Electrically Controlled Birefringence);
(e) VAN (Vertically Aligned Nematic); and
(f) HAN (Hybrid Aligned Nematic)

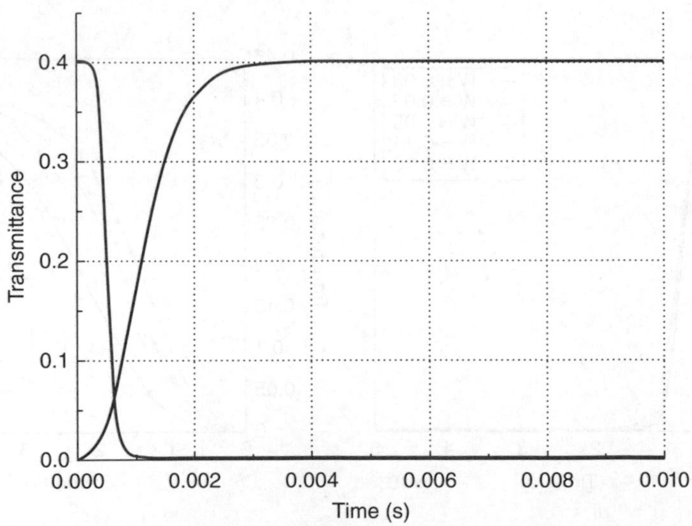

Figure A.7 Switching time of optimized TN cell: $\tau_{on} = 0.38$ ms, $\tau_{off} = 1.38$ ms, $\tau_{tot} = 1.76$ ms [1]

Table A.9 The optimization of LC response time, using MOUSE-LCD software [1]

LC mode	τ_{on} (ms)	τ_{off} (ms)	τ_{total} (ms)	Contrast ratio	Transmittance, %
TN cell	0.38	1.38	1.76	267.65	40.14
STN cell[a]	9.23	9.23	18.46	81.87	32.06
HAN cell	0.3	3	3.3	27.774	39.36
Half-π cell	0.3	2.2	2.5	20.493	40.2
ECB cell	0.1	1	1.1	194.27	40.36
VAN cell	0.6	0.9	1.5	444.18	40.65

[a]The above table shows that the response time of STN cell is much longer than the others. This is because the multiplexing ratio is important in STN cell.

Two uniaxial compensators were added between the output polarizer and the LC layer. The other configuration is that one of the uniaxial compensators added on the top and bottom of the LC layer. In the third configuration, one of the uniaxial compensators added on the top and bottom of the LC layer. A biaxial compensator was also added between the output polarizer and the LC layer with the similar result as the two uniaxial compensators. Some typical constructions are shown in Figure A.8.

The combinations of the color include

- Blue-yellow switching in transmissive mode
- Blue-yellow switching in reflective mode
- Green-magenta switching in transmissive mode
- Green-magenta switching in reflective mode
- Red-cyan switching in transmissive mode
- Red-cyan switching in reflective mode

Some birefringent color based on STN-LCD are shown in Figures A.9–A.12 [2].

In conclusion, optimization of the six cells on blue-yellow switching, green-magenta switching and red-cyan switching, and FLC cells has been successfully accomplished. The color filters, which considerably decrease LCD resolution, its light efficiency, with a high additional cost are not required in this type of LCD.

A.3.2 STN Birefringent Colors in Transmissive and Reflective Modes

The birefringent color configurations and the corresponding color triangle in transmissive and reflective LC electro-optical modes are shown in Figures A.12 and A.13 [3].

A.4 Transflective LCD

A.4.1 Vertical Aligned Nematic Cell

The typical parameters of vertical aligned nematic (VAN) LC cell are given in Table A.11 [4].

The modeled transflective VAN structure is shown in Figure A.14 [4].

The optimized transflective LCD uses the same LC layer for both transmissive and reflective modes. That means that the optimized structure uses the same LC properties in both modes (except for the LC thickness, which can be vary by using alignment glass of different thickness), but different layer configurations such as retardation layers. The structure can be manufactured with pixelated polarizer technology (Figure A.14) [4].

Table A.10 The typical parameters of STN LC cell [2]

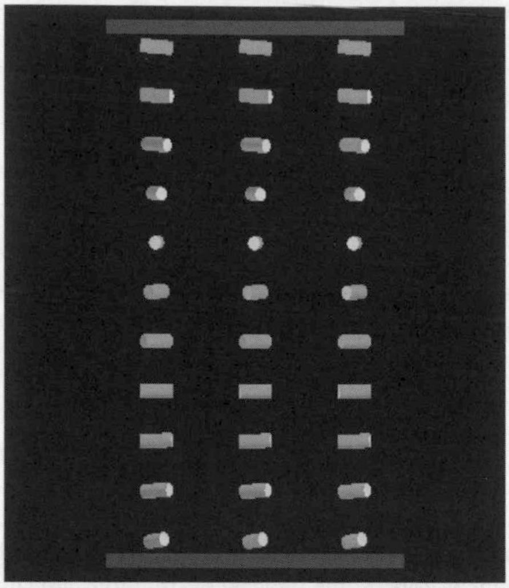

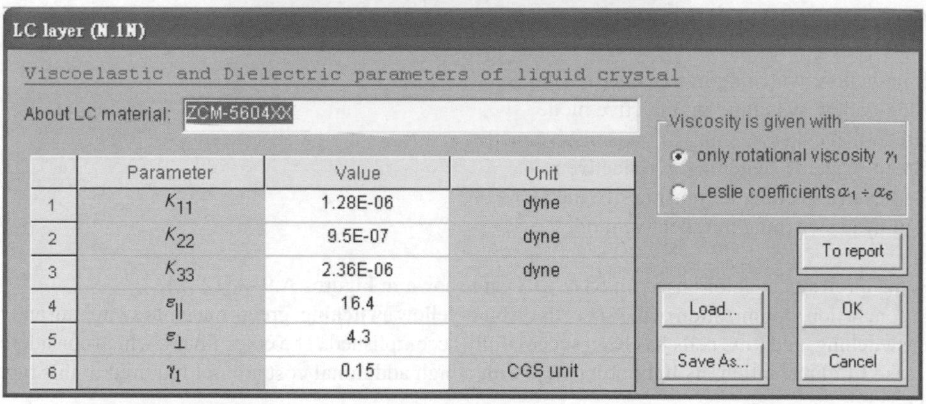

LC layer (N.1N)

Viscoelastic and Dielectric parameters of liquid crystal

About LC material: ZCM-5604XX

Viscosity is given with
only rotational viscosity γ_1
Leslie coefficients $\alpha_1 \div \alpha_6$

	Parameter	Value	Unit
1	K_{11}	1.28E-06	dyne
2	K_{22}	9.5E-07	dyne
3	K_{33}	2.36E-06	dyne
4	$\varepsilon_{\|\|}$	16.4	
5	$\varepsilon_{\perp}$	4.3	
6	γ_1	0.15	CGS unit

To report
Load...
OK
Save As...
Cancel

	Parameter	Value	Unit
1	Pretilt angle 1 (boundary 1)	4	degree
2	Pretilt angle 2 (boundary 2)	4	degree
3	Twist angle	240	degree
4	Natural helix pitch P	12.6	mcm
5	L/P ratio	0.5	
6	Pol.anchoring strength 1(boundary 1)	10	Erg/cm^2
7	Pol.anchoring strength 2(boundary 2)	10	Erg/cm^2
8	Az.anchoring strength 1(boundary 1)	10	Erg/cm^2
9	Az.anchoring strength 2(boundary 2)	10	Erg/cm^2

Element	N.	Data	Commentary
AR layer	1	Def.	
AR layer	2	Def.	
AR layer	3	Def.	
Polarizer	1		
Uniaxial compensator	1		
Uniaxial compensator	2	Def.	
Glass plate	1	Def.	
Electrode	1		
Aligning layer	1		
LC layer (nonabsorbing)	1		
Aligning layer	2		
Electrode	2		
Glass plate	2	Def.	
Polarizer	2		

Element	N.	Data	Commentary
Polarizer	1		
Glass plate	1	Def.	
Biaxial compensator	1	Def.	
Electrode	1	Def.	
Alignment layer	1	Def.	
LC layer (nonabsorbing)	1		
Alignment layer	2	Def.	
Electrode	2	Def.	
Glass plate	2	Def.	
Polarizer	2		

Figure A.8 Typical configurations of transmissive LCD for birefringent color optimizations. MOUSE-LCD software was used to choose the proper angles of the polarizers, and the parameters of the phase retardation plates [2]

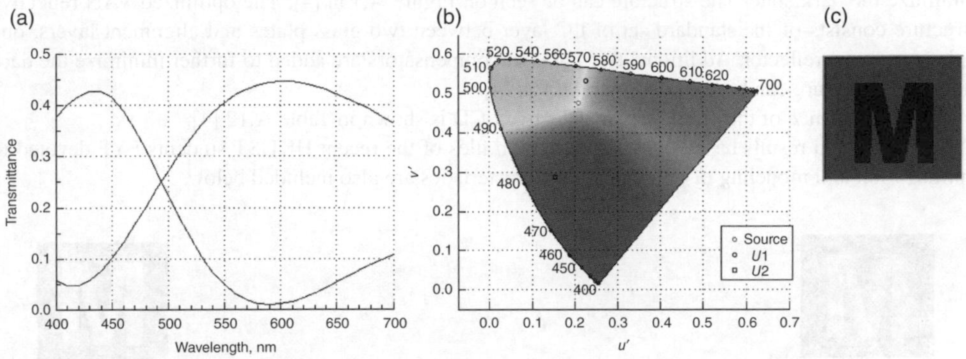

Figure A.9 Blue-yellow switching in STN transmissive mode (positive contrast). (a) Transmittance, (b) chromaticity diagram, (c) simulated result [2]

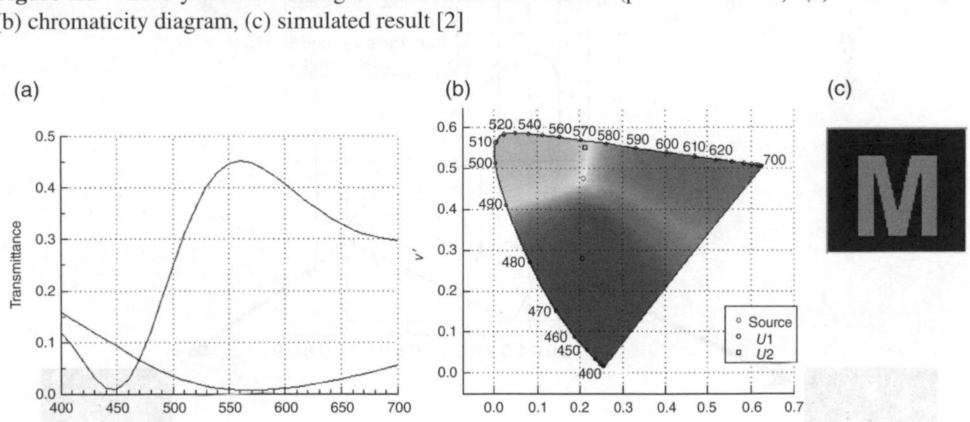

Figure A.10 Blue-yellow switching in STN transmissive mode (negative contrast). (a) Transmittance, (b) chromaticity diagram, (c) simulated result [2]

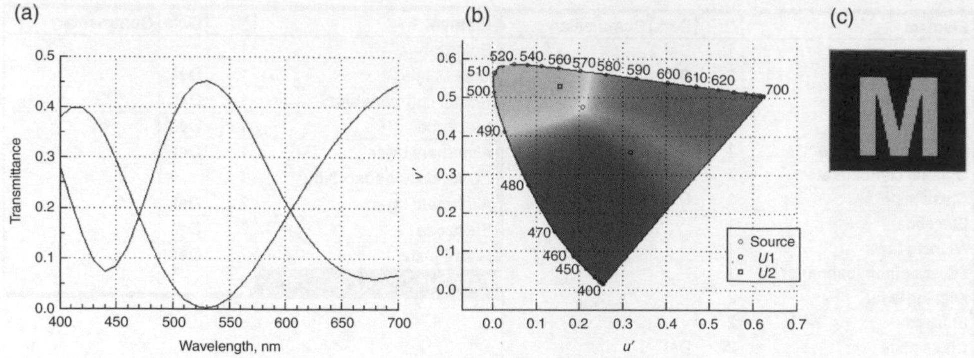

Figure A.11 Green-magenta switching in STN transmissive mode (negative contrast). (a) Transmittance, (b) chromaticity diagram, (c) simulated result [2]

The optimized VAN transmissive structure consists of the standard set of LC layer between two glass plates and alignment layers, two polarizers and an additional two uniaxial compensators to further minimize the dark state. The structure can be seen on Figure A.15a [4]. The optimized VAN reflective structure consists of the standard set of LC layer between two glass plates and alignment layers, one polarizer, and a reflector. Additional two uniaxial compensators are added to further minimize the dark state. The structure can be seen on Figure A.15b [4].

The performance of the optimized transflective LCD is shown in Table A.12 [4].

Some selected results based on the presentation files of the recent HKUST students FYP devoted to optimization and modeling of optimal LCD characteristics are also included below.

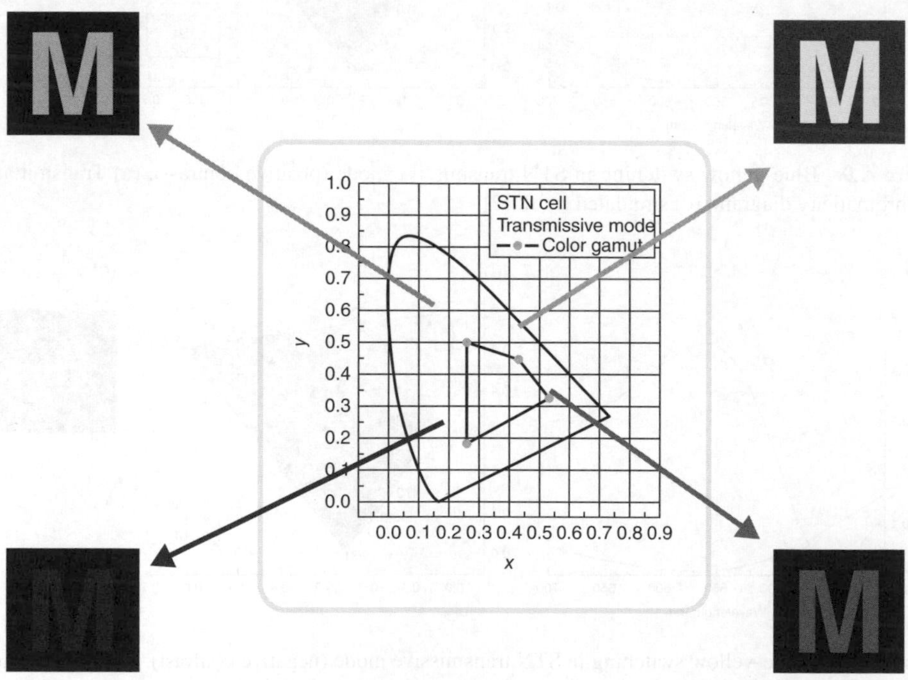

Figure A.12 The birefringent color triangle in transmissive STN mode [3]

Table A.11 The typical parameters of VAN LC cell [4]

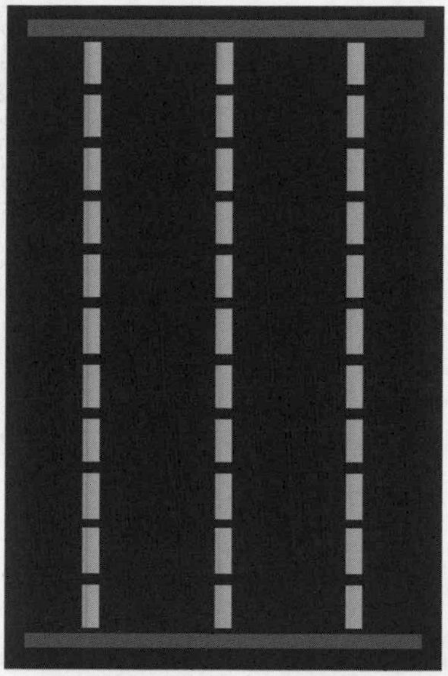

Wavelength, nm	$n_{\|\|}$	$n_{\perp}$	Δn
400 - 700	1.703	1.503	0.2

$d = 2$ mcm

	Parameter	Value	Unit
1	K_{11}	0.000002	dyne
2	K_{22}	0.0000011	dyne
3	K_{33}	0.000002	dyne
4	$\varepsilon_{\|\|}$	14	
5	$\varepsilon_{\perp}$	20.2	
6	γ_1	0.15	CGS unit

	Parameter	Value	Unit
1	Pretilt angle 1 (1st boundary)	89.5	degree
2	Pretilt angle 2 (2nd boundary)	89.5	degree
3	Twist angle	0	degree
4	Natural helix pitch P	INFINITY	mcm
5	L/P ratio	0	
6	Anchoring strength 1 (1st boundary)	1	Erg/cm^2
7	Anchoring strength 2 (2nd boundary)	1	Erg/cm^2

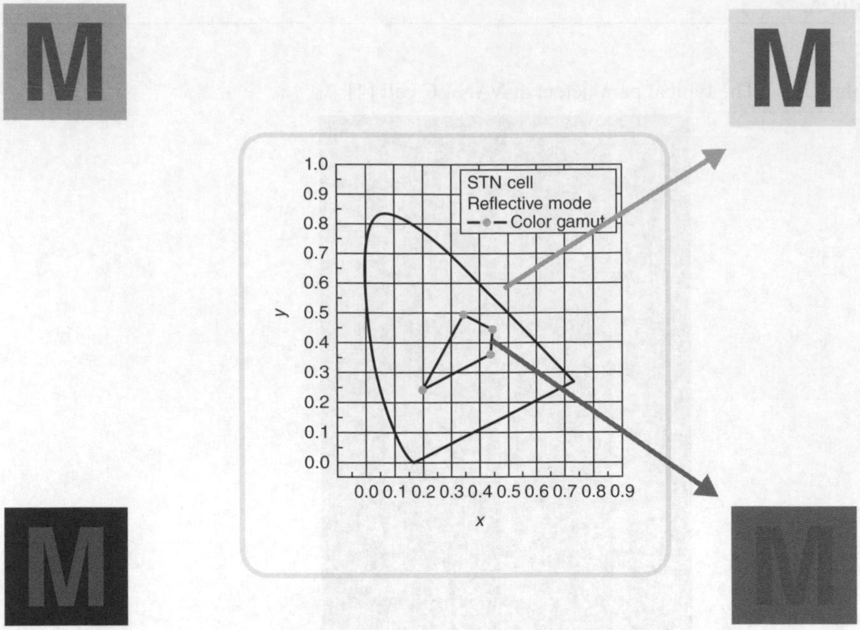

Figure A.13 The birefringent color triangle in reflective STN mode [3]

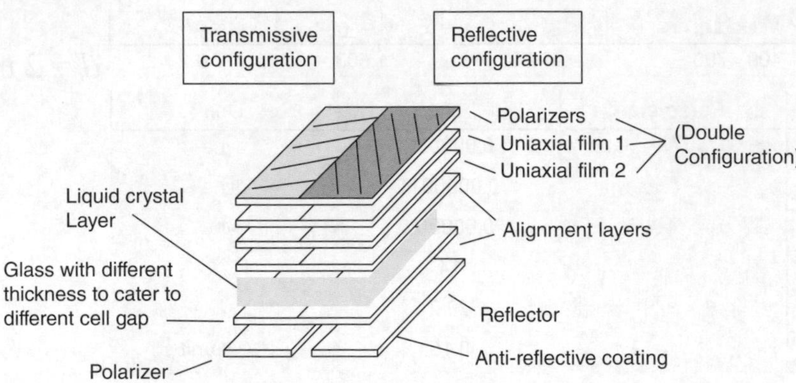

Figure A.14 The modeled transflective VAN structure [4].

Element	N.	Data
Polarizer	1	
Uniaxial compensator	1	
Uniaxial compensator	2	
Glass plate	1	Def.
Electrode	1	Def.
Alignment layer	1	Def.
LC layer (nonabsorbing)	1	
Alignment layer	2	Def.
Electrode	2	Def.
Glass plate	2	Def.
Polarizer	2	

(a)

Element	N.	Data
AR layer	1	
Polarizer	1	
Uniaxial compensator	1	
Uniaxial compensator	2	
Glass plate	1	
AR layer	2	
Electrode	1	Def.
AR layer	3	
Alignment layer	1	Def.
LC layer (nonabsorbing)	1	
Alignment layer	2	Def.
AR layer	4	
Electrode	2	Def.
AR layer	5	
Glass plate	2	Def.
Reflector	1	Def.

(b)

Figure A.15 The optimal transmissive (a) and reflective (b) configuration of the trasnsflective VAN LCD [4].

Table A.12 The performance of the optimized transflective LCD [4]

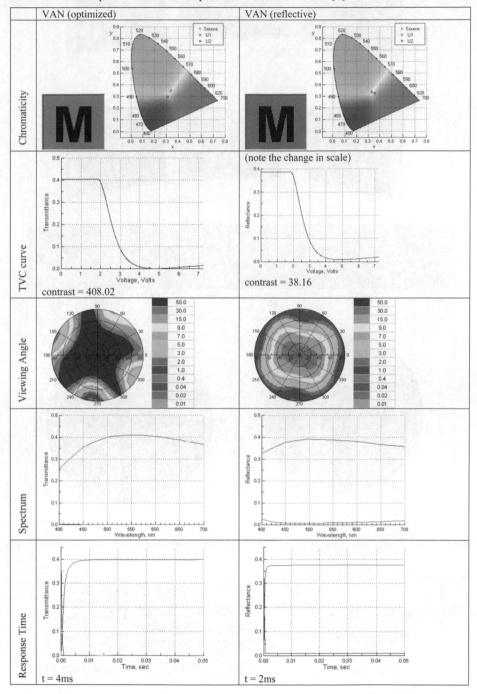

	VAN (optimized)	VAN (reflective)
Chromaticity	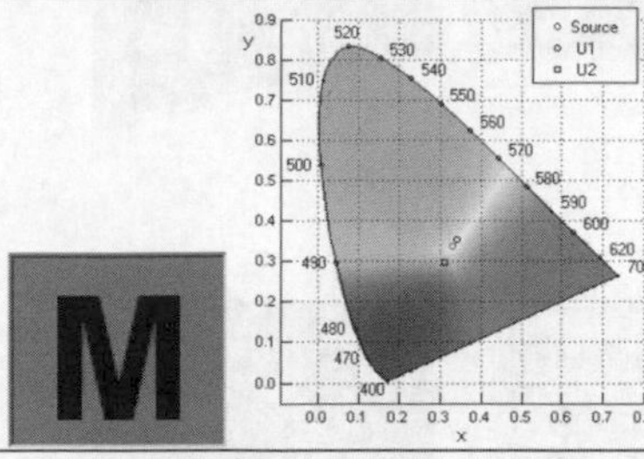	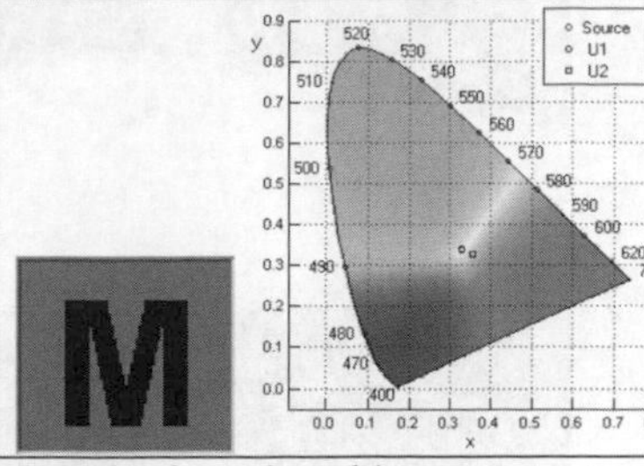
TVC curve	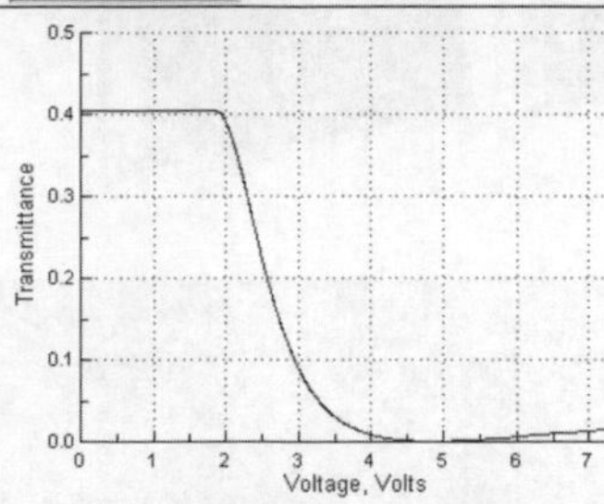 contrast = 408.02	(note the change in scale) 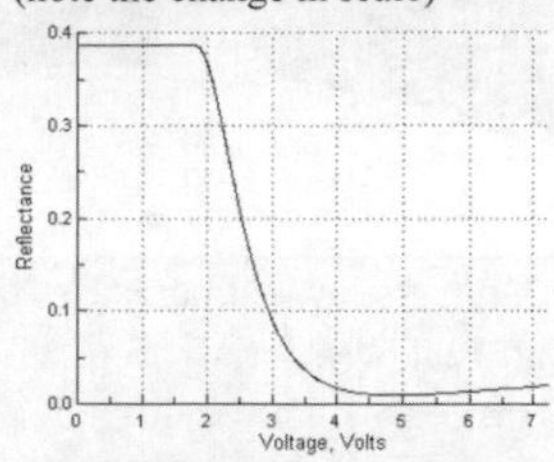 contrast = 38.16
Viewing Angle	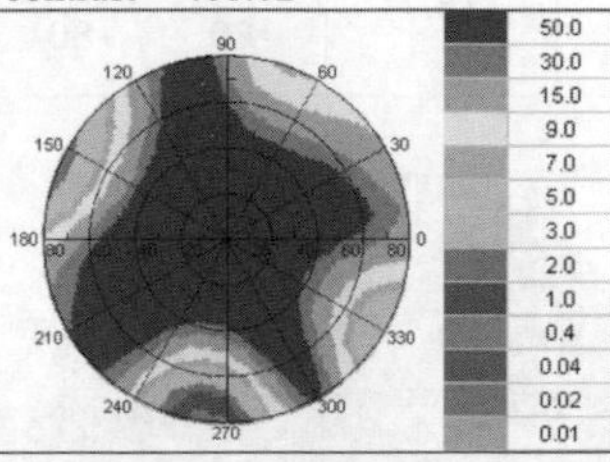	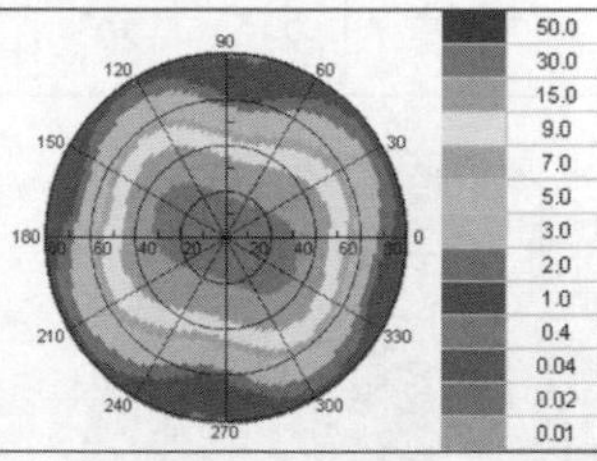
Spectrum	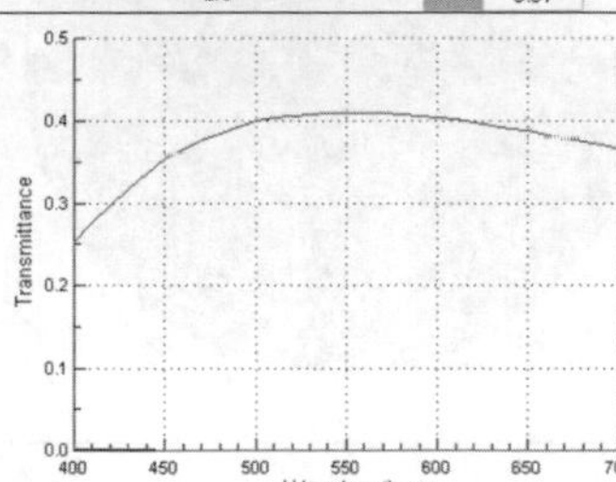	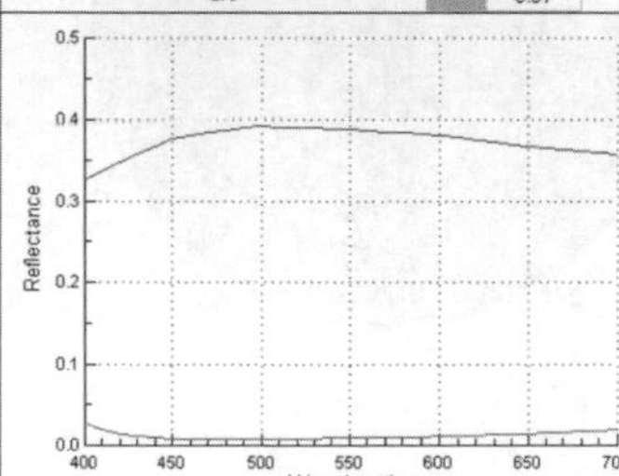
Response Time	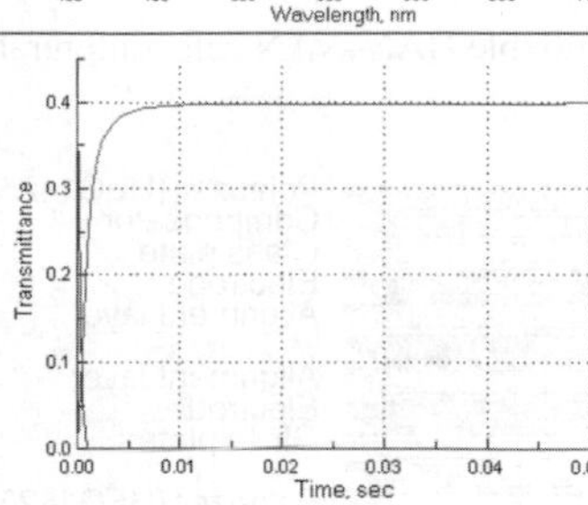 t = 4ms	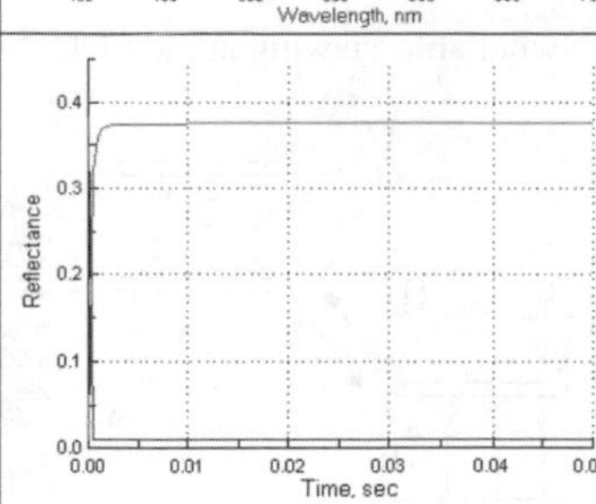 t = 2ms

Max. transmittance (narrow)	Max. contrast ratio (narrow)	Viewing angle (narrow)	Max. transmittance (wide)	Max. contrast ratio (wide)	Viewing angle (wide)
0.3042	319.25	-21° -- +20°	0.3042	318.71	-80° -- +80°

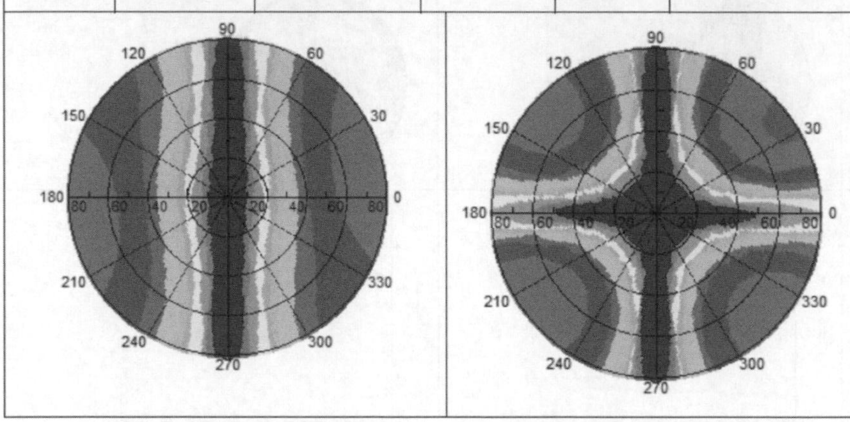

Figure A.16 Switchable viewing angle LCD, based on double HAN–VAN cell configuration [5]

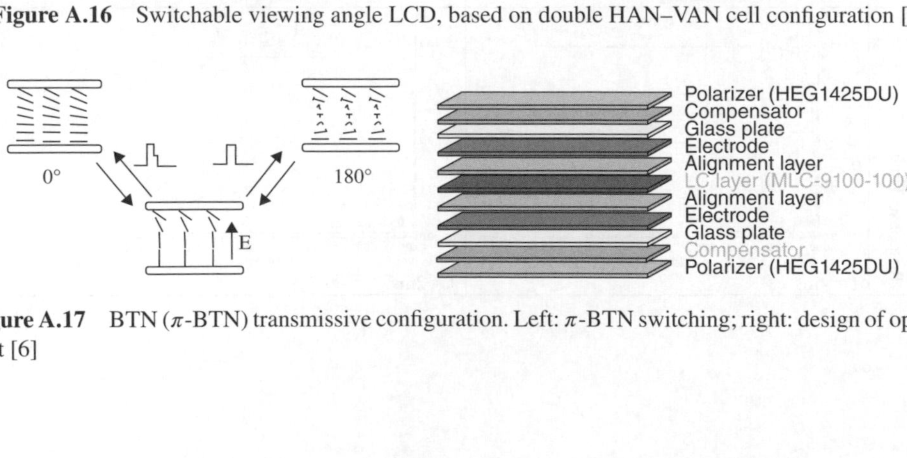

Figure A.17 BTN (π-BTN) transmissive configuration. Left: π-BTN switching; right: design of optical part [6]

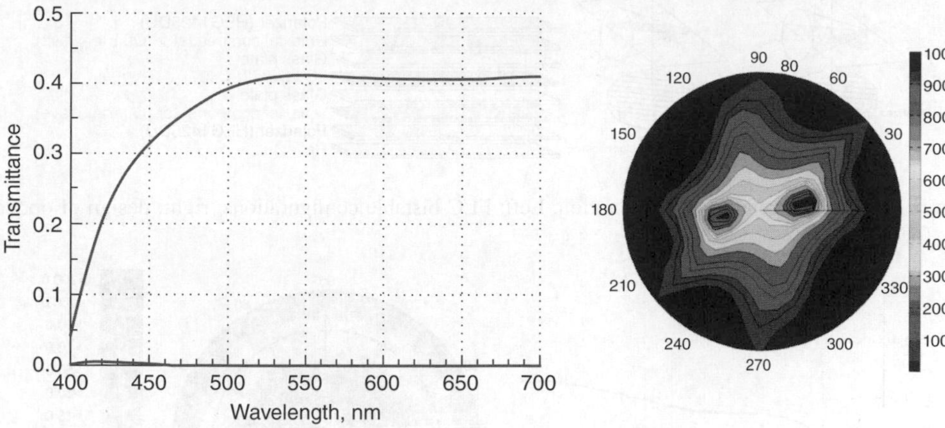

Figure A.18 Optimization results of BTN (π-BTN) transmissive configuration. Left: π-BTN transmittance in dark and bright bistable states; right: viewing angles [6]

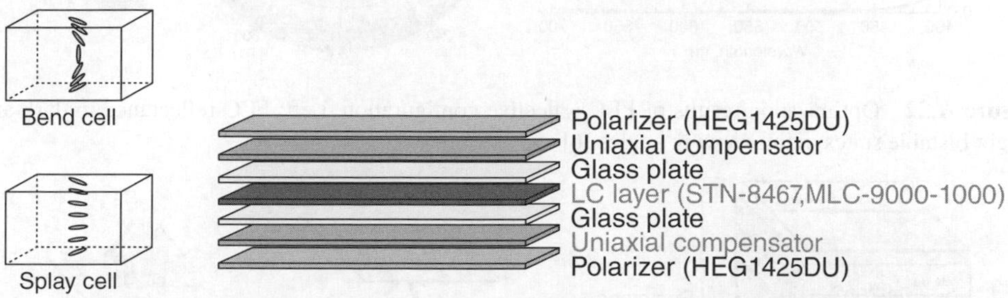

Figure A.19 BBS transmissive configuration. Left: BBS bistable configurations; right: design of optical part [6]

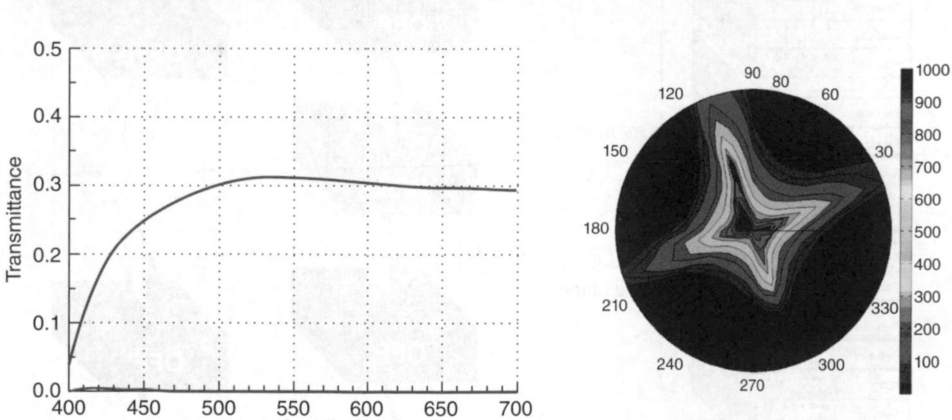

Figure A.20 Optimization results of BBS transmissive configuration. Left: BBS transmittance in dark and bright bistable states; right: viewing angles [6]

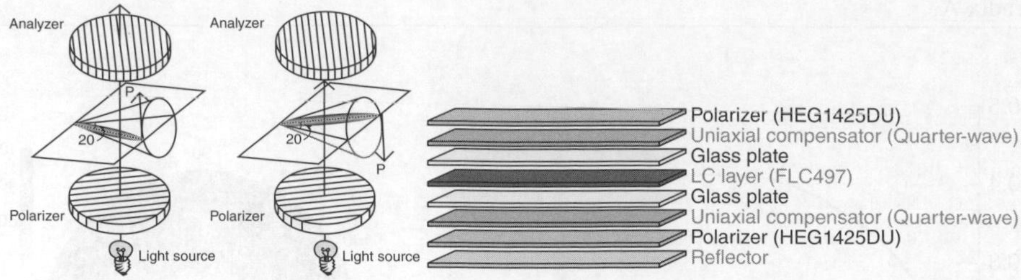

Figure A.21 FLC reflective configuration. Left: FLC bistable configurations; right: design of optical part [6]

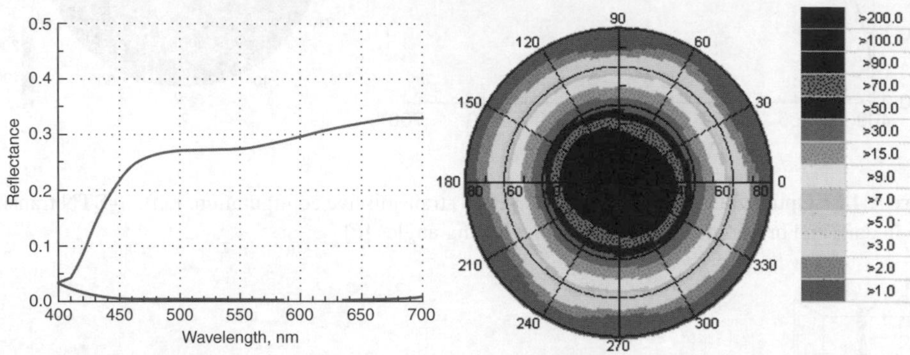

Figure A.22 Optimization results of FLC reflective configuration. Left: FLC reflectance in dark and bright bistable states; right: viewing angles [6]

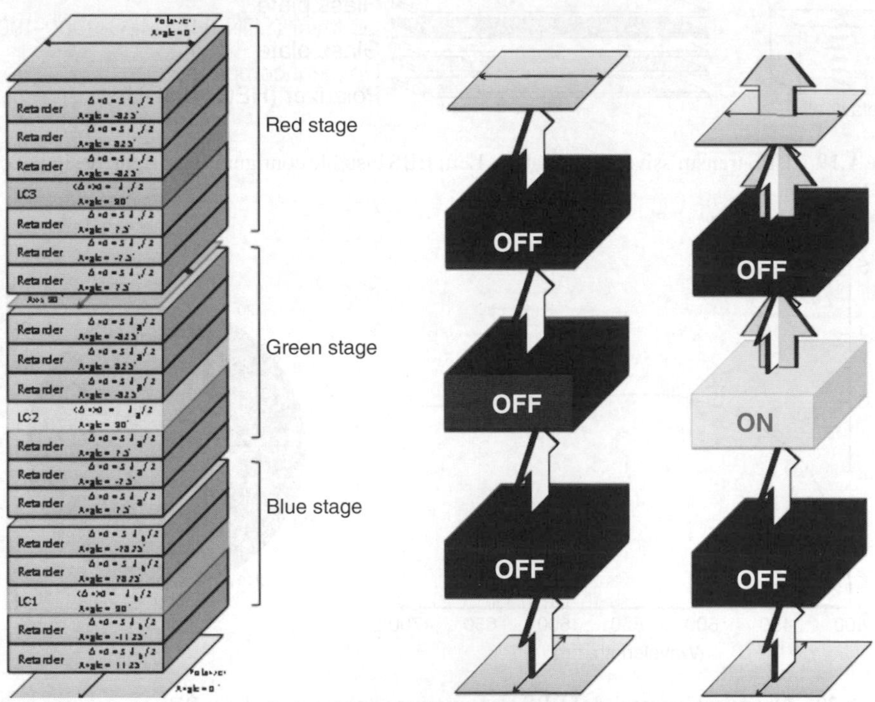

Figure A.23 Solc filter configuration principle. Left: Solc filter configuration; right: principle of green light switching [7]

A.5 Switchable Viewing Angle LCD

The aim of the work was (i) to discover factors affecting viewing angle performance of LCD; (ii) simulate narrow and wide viewing angle performance with different configurations; and (iii) optimize and model an LCD that can be switchable between narrow and wide viewing mode (Figure A.16) [5].

A.6 Optimal e-paper Configurations

The aim of the work was (i) to find out factors and conditions affecting the performance of three popular types of e-paper, including *Bistable twisted nematic (BTN), Bistable bend splay (BBS), Ferroelectric*

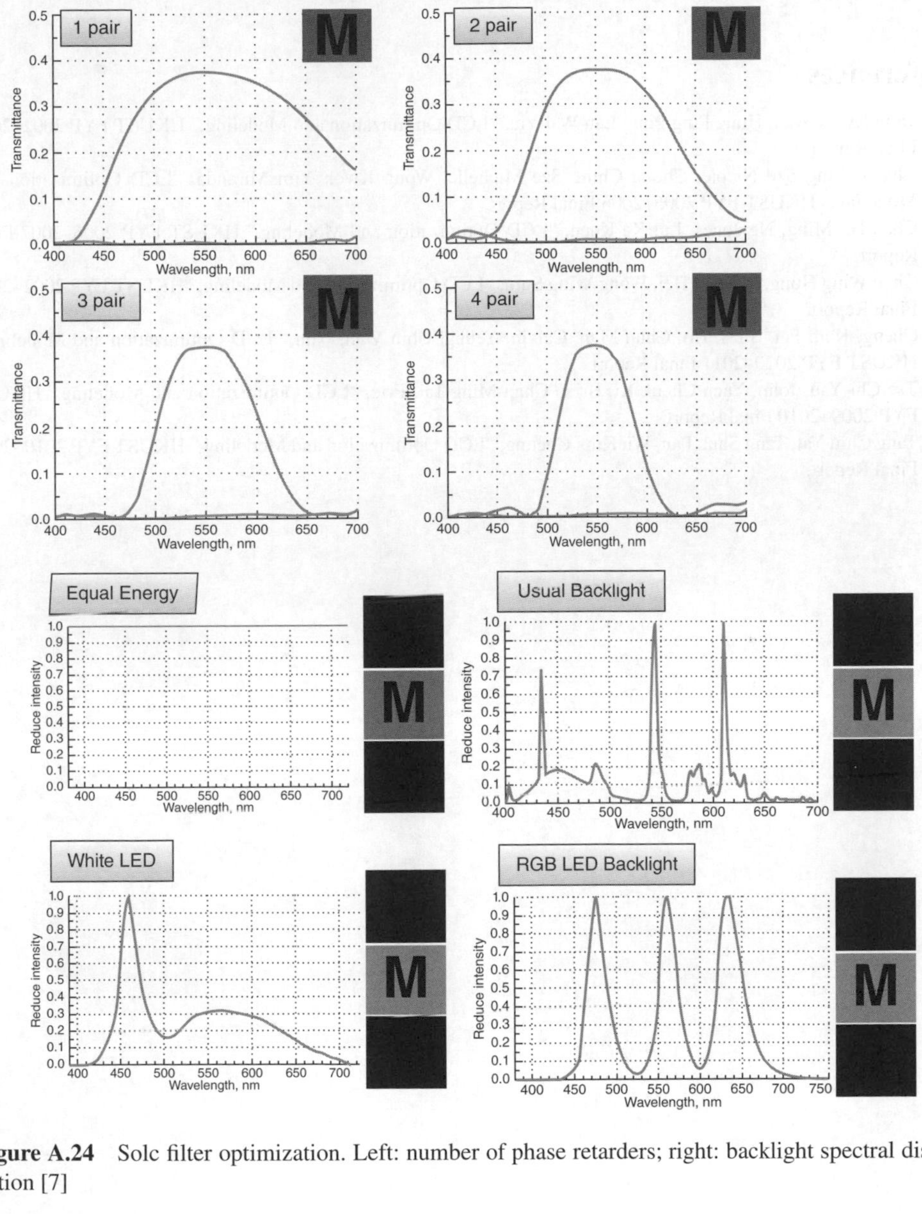

Figure A.24 Solc filter optimization. Left: number of phase retarders; right: backlight spectral distribution [7]

liquid crystal (FLC); (ii) to find out a better bistable LCD configuration for e-paper; (iii) to optimize the display performance of the three types of e-paper in terms of contrast ratios and viewing angles [6]. The optimization procedure in each configuration includes (i) choosing an optimal LC material; (ii) setting the polarizer and analyzer angle; (iii) finding an optimal LC layer thickness and pretilt angle. The optimization results are shown in Figures A.17–A.22 [6].

A.7 Color Filter Optimization

The aim of the work was (i) to investigate new LC configuration with good color production ability; (ii) improve color production ability using multiple LC layer structure [7]. In particular, the results of the optimization of Solc Filter are given in Figures A.23 and A.24 [7].

References

[1] Tang Kwok Wah, Hung Ping Ping, Lau Wai Yin, "LCD Optimization and Modeling," HKUST FYP 2002–2003 Final Report.

[2] Chang Wing Sze Nicole, Chong Ching Sze Michelle, Wong Kwok Yim Miranda, "LCD Optimization and Modeling," HKUST FYP 2005–2006 Final Report.

[3] Chan Ho Ming, Ng Jones, Lui Ka Kuen, "LCD Optimization and Modeling," HKUST FYP 2006–2007 Final Report.

[4] Chau Wing Hong, Li Sung Hei, Wong Wing Sum, "LCD Optimization and Modeling," HKUST FYP 2002–2003 Final Report.

[5] Cheng, Kam Fai, Silas, Ho, Chun Man, Calvin, Yeung, Chin Wan, Ivan, "LCD Optimization and Modeling," HKUST FYP 2010–2011 Final Report.

[6] Tse Chi Yau, John, Yuen Cheuk Ki, Brian Chan Ming Tai, Eric, "LCD Optimization and Modeling," HKUST FYP 2009–2010 Final Report.

[7] Tang Chiu Yat, Tang Shuk Fan, Wu King Cheung, "LCD Optimization and Modeling," HKUST FYP 2010–2011 Final Report.

Appexdix B
Some Derivations and Examples

B.1 Conservation Law for Energy Flux

In this section, we show that equation (8.17) for the energy flux is consistent with conditions (8.10) for the material tensors.

Using the vector identity

$$\nabla(\mathbf{a} \times \mathbf{b}) = \mathbf{b} \cdot (\nabla \times \mathbf{a}) - \mathbf{a} \cdot (\nabla \times \mathbf{b}), \tag{B.1}$$

we can write

$$\nabla \cdot (\mathbf{E} \times \mathbf{H}^*) = \mathbf{H}^* \cdot (\nabla \times \mathbf{E}) - \mathbf{E} \cdot (\nabla \times \mathbf{H}^*). \tag{B.2}$$

Multiplying the complex conjugate of equation (8.1) scalarly by **E**, and (8.2) by $\mathbf{H}^*$, we obtain

$$\mathbf{E} \cdot (\nabla \times \mathbf{H}^*) = ik_0 \mathbf{E} \cdot \mathbf{D}^*, \tag{B.3}$$

$$\mathbf{H}^* \cdot (\nabla \times \mathbf{E}) = ik_0 \mathbf{H}^* \cdot \mathbf{B}. \tag{B.4}$$

Substitution of (B.3) and (B.4) into (B.2) gives

$$\nabla \cdot (\mathbf{E} \times \mathbf{H}^*) = ik_0(\mathbf{H}^* \cdot \mathbf{B} - \mathbf{E} \cdot \mathbf{D}^*). \tag{B.5}$$

For the real part of $\nabla \cdot (\mathbf{E} \times \mathbf{H}^*)$ we have

$$\mathrm{Re}[\nabla \cdot (\mathbf{E} \times \mathbf{H}^*)] = \nabla \cdot \mathrm{Re}(\mathbf{E} \times \mathbf{H}^*), \tag{B.6}$$

$$2\mathrm{Re}[\nabla \cdot (\mathbf{E} \times \mathbf{H}^*)] = \nabla \cdot (\mathbf{E} \times \mathbf{H}^*) + \nabla \cdot (\mathbf{E}^* \times \mathbf{H}). \tag{B.7}$$

From (B.5)–(B.7) we obtain

$$2\nabla \cdot \mathrm{Re}(\mathbf{E} \times \mathbf{H}^*) = ik_0(\mathbf{H}^* \cdot \mathbf{B} - \mathbf{E} \cdot \mathbf{D}^* - \mathbf{H} \cdot \mathbf{B}^* + \mathbf{E}^* \cdot \mathbf{D}). \tag{B.8}$$

Modeling and Optimization of LCD Optical Performance, First Edition.
Dmitry A. Yakovlev, Vladimir G. Chigrinov and Hoi-Sing Kwok.

Website Companion: www.wiley.com/go/yakovlev/modelinglcd

After substitution of the expressions for **D** and **B** from the constitutive relations (8.5) into (B.8), it is easy to ascertain, invoking the tensor-vector identity

$$\mathbf{a} \cdot \alpha\mathbf{b} = \mathbf{b} \cdot \alpha^{\mathrm{T}}\mathbf{a}, \tag{B.9}$$

that the bracketed quantity on the right-hand side of (B.8) is equal to zero identically (at any **E** and **H**) if the material tensors meet (8.10). In view of (8.16), this means that in media whose material tensors satisfy conditions (8.10) the conservation law (8.17) holds.

B.2 Lorentz's Lemma

Let us write the Maxwell equations for the fields $\{\mathbf{E}_1, \mathbf{H}_1\}$ and $\{\mathbf{E}_2, \mathbf{H}_2\}$ entering into (8.19):

$$\begin{aligned}\nabla \times \mathbf{H}_1 &= -ik_0\mathbf{D}_1,\\ \nabla \times \mathbf{E}_1 &= ik_0\mathbf{B}_1,\\ \nabla \times \mathbf{H}_2 &= -ik_0\mathbf{D}_2,\\ \nabla \times \mathbf{E}_2 &= ik_0\mathbf{B}_2.\end{aligned}$$

Multiplying scalarly these equations by $\mathbf{E}_2$, $\mathbf{H}_2$, $-\mathbf{E}_1$, and $-\mathbf{H}_1$, respectively, and adding all four together, we obtain

$$\begin{aligned}&\mathbf{E}_2 \cdot (\nabla \times \mathbf{H}_1) + \mathbf{H}_2 \cdot (\nabla \times \mathbf{E}_1) - \mathbf{E}_1 \cdot (\nabla \times \mathbf{H}_2) - \mathbf{H}_1 \cdot (\nabla \times \mathbf{E}_2)\\ &= ik_0(-\mathbf{E}_2 \cdot \mathbf{D}_1 + \mathbf{H}_2 \cdot \mathbf{B}_1 + \mathbf{E}_1 \cdot \mathbf{D}_2 - \mathbf{H}_1 \cdot \mathbf{B}_2).\end{aligned} \tag{B.10}$$

According to (B.1), the left-hand side of (B.10) is identically equal to $\nabla \cdot (\mathbf{E}_1 \times \mathbf{H}_2 - \mathbf{E}_2 \times \mathbf{H}_1)$. Using the constitutive relations (8.5) and the identity (B.9), the quantity in brackets on the right-hand side of (B.10) can be transformed as follows:

$$\begin{aligned}&-\mathbf{E}_2 \cdot \mathbf{D}_1 + \mathbf{H}_2 \cdot \mathbf{B}_1 + \mathbf{E}_1 \cdot \mathbf{D}_2 - \mathbf{H}_1 \cdot \mathbf{B}_2\\ &\quad = \mathbf{E}_1 \cdot (\varepsilon - \varepsilon^{\mathrm{T}})\mathbf{E}_2 - \mathbf{H}_1 \cdot (\rho^{\mathrm{T}} + \rho')\mathbf{E}_2 + \mathbf{H}_2 \cdot (\rho^{\mathrm{T}} + \rho')\mathbf{E}_1 + \mathbf{H}_2 \cdot (\mu - \mu^{\mathrm{T}})\mathbf{H}_1.\end{aligned}$$

Therefore, from (B.10), we have

$$\begin{aligned}&\nabla \cdot (\mathbf{E}_1 \times \mathbf{H}_2 - \mathbf{E}_2 \times \mathbf{H}_1)\\ &= ik_0[\mathbf{E}_1 \cdot (\varepsilon - \varepsilon^{\mathrm{T}})\mathbf{E}_2 - \mathbf{H}_1 \cdot (\rho^{\mathrm{T}} + \rho')\mathbf{E}_2 + \mathbf{H}_2 \cdot (\rho^{\mathrm{T}} + \rho')\mathbf{E}_1 + \mathbf{H}_2 \cdot (\mu - \mu^{\mathrm{T}})\mathbf{H}_1].\end{aligned} \tag{B.11}$$

As can be seen from (B.11), equation (8.20) is satisfied identically only if the material tensors meet the conditions (8.8).

B.3 Nonexponential Waves

In certain situations, the modal basis for a homogeneous layer cannot be formed only of waves with a pure exponential spatial dependence ($e^{i\mathbf{kr}}$) and must involve nonexponential waves. Here is a simple example of such a situation.

Let a plane wave be incident on an isotropic nonabsorbing medium A of refractive index n from an optically denser isotropic nonabsorbing medium at the critical angle of total reflection. In this case, $\zeta = n$ (see Section 8.1.3) and equation (8.73) has four identical roots

$$\sigma_1 = \sigma_2 = \sigma_3 = \sigma_4 = 0$$

[see (9.18)]. In this case, it is impossible to choose the vibration vectors for four solutions of the form (8.21) so that these solutions are linearly independent. The fundamental system of solutions for (8.58) in this situation can be composed of the following linearly independent solutions:

$$\begin{pmatrix} \mathbf{E}_1(\mathbf{r}) \\ \mathbf{H}_1(\mathbf{r}) \end{pmatrix} = A_1 \begin{pmatrix} \mathbf{z} \\ -\zeta\mathbf{y} \end{pmatrix} \exp\left(ik_0\mathbf{br}\right), \tag{B.12}$$

$$\begin{pmatrix} \mathbf{E}_2(\mathbf{r}) \\ \mathbf{H}_2(\mathbf{r}) \end{pmatrix} = A_2 \begin{pmatrix} \mathbf{y} \\ \zeta\mathbf{z} \end{pmatrix} \exp\left(ik_0\mathbf{br}\right), \tag{B.13}$$

$$\begin{pmatrix} \mathbf{E}_3(\mathbf{r}) \\ \mathbf{H}_3(\mathbf{r}) \end{pmatrix} = A_3 \left[\begin{pmatrix} -\zeta^{-1}\mathbf{x} \\ \widehat{\mathbf{0}} \end{pmatrix} + ik_0\left(z - z_0\right)\begin{pmatrix} \mathbf{z} \\ -\zeta\mathbf{y} \end{pmatrix}\right] \exp\left(ik_0\mathbf{br}\right), \tag{B.14}$$

$$\begin{pmatrix} \mathbf{E}_4(\mathbf{r}) \\ \mathbf{H}_4(\mathbf{r}) \end{pmatrix} = A_4 \left[\begin{pmatrix} \widehat{\mathbf{0}} \\ -\mathbf{x} \end{pmatrix} + ik_0\left(z - z_0\right)\begin{pmatrix} \mathbf{y} \\ \zeta\mathbf{z} \end{pmatrix}\right] \exp\left(ik_0\mathbf{br}\right), \tag{B.15}$$

where z_0 is the z-coordinate of the interface between the media; $\mathbf{x}$, $\mathbf{y}$, and $\mathbf{z}$ are the unit vectors along the axes of the coordinate system (x, y, z); $\widehat{\mathbf{0}}$ denotes the zero vector. The first and second waves are usual surface waves. The fields of the third and fourth waves have a *linear* dependence on z. Making use of this fundamental system of solutions, we obtain the following expression for the Berreman transfer matrix of the layer $(z_0, z_0 + d)$ of the medium A:

$$\mathbf{P}(z_0, z_0 + d) = \begin{pmatrix} 1 & 0 & 0 & 0 \\ ik_0n^2d & 1 & 0 & 0 \\ 0 & 0 & 1 & ik_0d \\ 0 & 0 & 0 & 1 \end{pmatrix}_{xyz}. \tag{B.16}$$

We can verify by a straightforward calculation that Berreman's formalism by itself gives the same result. The matrix $\boldsymbol{\Delta}$ of the medium A in the situation under consideration ($\zeta = n$) has the following form:

$$\boldsymbol{\Delta} = \begin{pmatrix} 0 & 0 & 0 & 0 \\ n^2 & 0 & 0 & 0 \\ 0 & 0 & 0 & 1 \\ 0 & 0 & 0 & 0 \end{pmatrix}_{xyz}.$$

It is easily seen that the matrix $\boldsymbol{\Delta}^2$ is zero. Therefore, from (8.131), we have

$$\mathbf{P}(z_0, z) = \mathbf{U} + ik_0z\boldsymbol{\Delta},$$

which leads to (B.16).

One may notice that it is impossible to select among the solutions (B.12)–(B.15) two forward propagating and two backward propagating waves. Therefore, the methods requiring such a selection fail in this case. When using such methods, the simplest way to cope with this situation is to escape from it. For this, it suffices to slightly change ζ or optical constants of the medium (say, by 10^{-5}).

B.4 To the Power Series Method (Section 11.3.3)

Let us consider the operator equation

$$\frac{d\mathbf{t}(0,\eta)}{d\eta} = i\kappa\tilde{\mathbf{N}}(\eta)\mathbf{t}(0,\eta) \quad [\mathbf{t}(0,0) = \mathbf{U}], \tag{B.17}$$

where $\tilde{\mathbf{N}}(\eta)$ is a real 2×2 matrix of the form

$$\tilde{\mathbf{N}}(\eta) = \begin{pmatrix} F'(\eta) & F''(\eta) \\ F''(\eta) & -F'(\eta) \end{pmatrix} \tag{B.18}$$

at any η from the range (0,1), κ is a scalar real parameter, and $\mathbf{U}$ is the unit matrix. Assume that we need to calculate the dependence of the matrix $\tilde{\mathbf{t}} = \mathbf{t}(0, 1)$ on κ. Because of the special form of the operator $\tilde{\mathbf{N}}(\eta)$, the matrix $\tilde{\mathbf{t}}$ at any κ is unitary and has the form

$$\tilde{\mathbf{t}} = \begin{pmatrix} c_1 & c_2 \\ -c_2^* & c_1^* \end{pmatrix}, \tag{B.19}$$

therefore to determine the matrix $\tilde{\mathbf{t}}$ it suffices to calculate the complex parameters c_1 and c_2. According to (B.17), the matrix $\tilde{\mathbf{t}}$ can be expressed as follows:

$$\tilde{\mathbf{t}} = \mathbf{U} + \sum_{j=1}^{\infty} (i\kappa)^j \tilde{\mathbf{N}}_j(1), \tag{B.20}$$

$$\tilde{\mathbf{N}}_1(\eta) = \int_0^{\eta} \tilde{\mathbf{N}}(\bar{\eta})d\bar{\eta}, \tag{B.21a}$$

$$\tilde{\mathbf{N}}_j(\eta) = \int_0^{\eta} \tilde{\mathbf{N}}(\bar{\eta})\tilde{\mathbf{N}}_{j-1}(\bar{\eta})d\bar{\eta} \quad j = 2, 3, 4, \ldots \tag{B.21b}$$

Due to the symmetry properties of the matrix $\tilde{\mathbf{N}}(\eta)$ (B.18), the matrices $\tilde{\mathbf{N}}_j(\eta)$ at any η have the form

$$\tilde{\mathbf{N}}_j(\eta) = \begin{pmatrix} \mathrm{f}'_j(\eta) & \mathrm{f}''_j(\eta) \\ (-1)^{j-1}\mathrm{f}''_j(\eta) & (-1)^j \mathrm{f}'_j(\eta) \end{pmatrix} \tag{B.22}$$

[which can be verified by substituting (B.18) into (B.21)], where $\mathrm{f}'_j(\eta)$ and $\mathrm{f}''_j(\eta)$ are real. This allows one to specify the function $\tilde{\mathbf{N}}_j(\eta)$ with the aid of the following scalar complex-valued function

$$\mathrm{f}_j(\eta) = \mathrm{f}'_j(\eta) + i\mathrm{f}''_j(\eta).$$

According to (B.18) and (B.21a), the function $\mathrm{f}_1(\eta)$ can be expressed as

$$\mathrm{f}_1(\eta) = \int_0^{\eta} F(\bar{\eta})d\bar{\eta}, \tag{B.23}$$

where

$$F(\eta) = F'(\eta) + iF''(\eta). \tag{B.24}$$

The functions $\mathrm{f}_j(\eta)$ with $j > 1$ can also be expressed in terms of $F(\eta)$. Using (B.18), (B.22), and (B.21b), one can find that

$$\mathrm{f}_j(\eta) = \begin{cases} \int_0^\eta F(\bar{\eta})^* \mathrm{f}_{j-1}(\bar{\eta}) \mathrm{d}\bar{\eta} & j = 2, 4, 6, \ldots \\ \int_0^\eta F(\bar{\eta}) \mathrm{f}_{j-1}(\bar{\eta}) \mathrm{d}\bar{\eta} & j = 3, 5, 7, \ldots \end{cases} \tag{B.25}$$

It follows from (B.19), (B.20), and (B.22) that the parameters c_1 and c_2 specifying the matrix $\tilde{\mathbf{t}}$ can be calculated by the formulas

$$c_1 = 1 + \sum_{j=1}^{\infty} (i\kappa)^j \mathrm{Re}\left(\mathrm{f}_j(1)\right), \quad c_2 = \sum_{j=1}^{\infty} (i\kappa)^j \mathrm{Im}\left(\mathrm{f}_j(1)\right). \tag{B.26}$$

This representation underlies the power series method (PSM) described in Section 11.3.3.

B.5 One of the Ways to Obtain the Explicit Expressions for Transmission Jones Matrices of an Ideal Twisted LC Layer

Both the classical differential Jones calculus and the NBR approximation enable one to obtain simple analytical expressions for transmission Jones matrices (respectively the classical-Jones-calculus Jones matrices and EW Jones matrices) of ideal twisted LC layers (see Sections 2.1, 11.1.1, and 11.4.1). In either case, such an expression can be derived by solving an operator equation of the form

$$\frac{\mathrm{d}\mathbf{t}(z', z)}{\mathrm{d}z} = \mathbf{N}\mathbf{t}(z', z) \quad [\mathbf{t}(z', z') = \mathbf{U}], \tag{B.27}$$

where $\mathbf{N}$ is a 2×2 matrix independent of z. The solution of (B.27) can be written in the form of the matrix exponential

$$\mathbf{t}(z', z) = \exp\left[\mathbf{N}\left(z - z'\right)\right] \equiv \mathbf{U} + \sum_{j=1}^{\infty} \frac{(z - z')^j}{j!} \mathbf{N}^j.$$

The operator $\mathbf{t}(z', z'')$ for a given interval (z', z'') is then expressed as

$$\mathbf{t}(z', z'') = \exp\left(\mathbf{N}d\right) = \mathbf{U} + \sum_{j=1}^{\infty} \frac{d^j}{j!} \mathbf{N}^j, \tag{B.28}$$

where $d = z'' - z'$. Suppose that the matrix $\mathbf{N}$ has two different eigenvalues τ_1 and τ_2. Then this matrix can be expressed as

$$\mathbf{N} = \Psi_\tau \mathbf{N}_\tau \Psi_\tau^{-1}, \tag{B.29}$$

where

$$\mathbf{N}_\tau = \begin{pmatrix} \tau_1 & 0 \\ 0 & \tau_2 \end{pmatrix}, \tag{B.30}$$

and Ψ_τ is a 2×2 matrix composed of eigenvectors of the matrix **N**:

$$\Psi_\tau = \begin{pmatrix} \mathbf{j}_1 & \mathbf{j}_2 \end{pmatrix}, \tag{B.31}$$

where $\mathbf{j}_k$ is an eigenvector corresponding to the eigenvalue τ_k ($k = 1, 2$). The representation (B.29) follows from the definition of eigenvalues and eigenvectors. Substitution of (B.29) into (B.28) gives the following expression for the matrix $\mathbf{t}(z', z'')$

$$\mathbf{t}(z', z'') = \Psi_\tau \begin{pmatrix} \exp(\tau_1 d) & 0 \\ 0 & \exp(\tau_2 d) \end{pmatrix} \Psi_\tau^{-1}. \tag{B.32}$$

The representation (B.32) was used by R.C. Jones in his classical work [1], where he presented his differential calculus. In (8.132), a representation of this kind is used for the Berreman transfer matrix of a homogeneous layer.

As an example, we use expression (B.32) to find the Jones matrix $\mathbf{t}_{Ux'-y'}(z', z'')$ from (11.35). In the case of an ideal twisted layer, the matrix $\mathbf{N}_{Ux'-y'}$ entering into this equation can be expressed as

$$\mathbf{N}_{Ux'-y'} = \begin{pmatrix} ik & q \\ -q & -ik \end{pmatrix}, \tag{B.33}$$

where

$$k = G/d, \quad q = \varphi_z = \Phi/d \tag{B.34}$$

with d and Φ being respectively the thickness and twist angle of the layer. The parameter G is defined under (11.42). The eigenvalues of $\mathbf{N}_{Ux'-y'}$ are

$$\tau_1 = ip, \quad \tau_2 = -ip, \tag{B.35}$$

where

$$p = \sqrt{k^2 + q^2}. \tag{B.36}$$

The corresponding eigenvectors can be chosen as follows:

$$\mathbf{j}_1 = \begin{pmatrix} k - i\tau_1 \\ iq \end{pmatrix} = \begin{pmatrix} a \\ iq \end{pmatrix}, \quad \mathbf{j}_2 = \begin{pmatrix} iq \\ k + i\tau_2 \end{pmatrix} = \begin{pmatrix} iq \\ a \end{pmatrix}, \tag{B.37}$$

where

$$a = k + \sqrt{k^2 + q^2}. \tag{B.38}$$

In this case,

$$\Psi_\tau = \begin{pmatrix} a & iq \\ iq & a \end{pmatrix}, \quad \Psi_\tau^{-1} = \frac{1}{\Delta} \begin{pmatrix} a & -iq \\ -iq & a \end{pmatrix}, \tag{B.39}$$

where

$$\Delta = a^2 + q^2 = 2\sqrt{k^2+q^2}\left(k+\sqrt{k^2+q^2}\right) = 2ap.$$

From (B.32), on carrying out the matrix multiplications, we obtain

$$\mathbf{t}_{\mathrm{U}x'-y'}(z',z'') = \frac{1}{\Delta}\begin{pmatrix} a^2e^{iQ}+q^2e^{-iQ} & -iaq\left(e^{iQ}-e^{-iQ}\right) \\ iaq\left(e^{iQ}-e^{-iQ}\right) & a^2e^{-iQ}+q^2e^{iQ} \end{pmatrix},$$

where $Q = pd = \sqrt{G^2+\Phi^2}$, and then, on substituting $e^{\pm iQ} = \cos Q \pm i\sin Q$,

$$\mathbf{t}_{\mathrm{U}x'-y'}(z',z'') = \begin{pmatrix} \cos Q + i\dfrac{a^2-q^2}{\Delta}\sin Q & \dfrac{2aq}{\Delta}\sin Q \\ -\dfrac{2aq}{\Delta}\sin Q & \cos Q - i\dfrac{a^2-q^2}{\Delta}\sin Q \end{pmatrix}.$$

Having noted that

$$a^2 - q^2 = 2\left(k^2 + k\sqrt{k^2+q^2}\right) = 2k\left(k+\sqrt{k^2+q^2}\right) = 2ak,$$

we can rewrite the latter expression as

$$\mathbf{t}_{\mathrm{U}x'-y'}(z',z'') = \begin{pmatrix} \cos Q + i\dfrac{k}{p}\sin Q & \dfrac{q}{p}\sin Q \\ -\dfrac{q}{p}\sin Q & \cos Q - i\dfrac{k}{p}\sin Q \end{pmatrix}. \tag{B.40}$$

Upon substituting (B.34) into (B.40), we arrive at the well-known expression

$$\mathbf{t}_{\mathrm{U}x'-y'}(z',z'') = \begin{pmatrix} \cos Q + i\dfrac{G}{Q}\sin Q & \dfrac{\Phi}{Q}\sin Q \\ -\dfrac{\Phi}{Q}\sin Q & \cos Q - i\dfrac{G}{Q}\sin Q \end{pmatrix}. \tag{B.41}$$

Substitution of (11.21) and (B.41) into (11.34) gives expression (11.41) for the "full" Jones matrix of the layer.

Reference

[1] R. C. Jones, "A new calculus for the treatment of optical systems. VII. Properties of the N-matrices," *J. Opt. Soc. Am.* **38**, 671 (1948).

Index

Modeling and Optimization of LCD Optical Performance, First Edition.
Dmitry A. Yakovlev, Vladimir G. Chigrinov and Hoi-Sing Kwok.

Website Companion: www.wiley.com/go/yakovlev/modelinglcd